Biochemistry of Insects

Contributors

MOISES AGOSIN

FRANCISCO J. AYALA

MURRAY S. BLUM

P. S. CHEN

G. MICHAEL CHIPPENDALE

W. C. DAUTERMAN

ROGER G. H. DOWNER

ERNEST HODGSON

A. E. NEEDHAM

R. D. O'BRIEN

A. GLENN RICHARDS

LYNN M. RIDDIFORD

WENDELL L. ROELOFS

JAMES W. TRUMAN

NEVIN WEAVER

Biochemistry of Insects

EDITED BY

Morris Rockstein

Department of Physiology and Biophysics
School of Medicine
University of Miami
Coral Gables, Florida

ACADEMIC PRESS New York San Francisco London 1978

A Subsidiary of Harcourt Brace Jovanovich, Publishers

To **SUZI**
and
MADY

ACADEMIC PRESS, INC.
111 Fifth Avenue, New York, New York 10003

United Kingdom Edition published by
ACADEMIC PRESS, INC. (LONDON) LTD.
24/28 Oval Road, London NW1 7DX

Library of Congress Cataloging in Publication Data

Main entry under title:

Biochemistry of insects.

Includes bibliographies and index.
1. Insects--Physiology. I. Rockstein, Morris.
[DNLM: 1. Insects--Metabolism. 2. Insects--Physiology. QX500.3 B615]
QL495.B56 595.7'01'92 77-11221
ISBN 0-12-591640-X

PRINTED IN THE UNITED STATES OF AMERICA

Contents

11 BIOCHEMICAL DEFENSES OF INSECTS

MURRAY S. BLUM

12 THE BIOCHEMISTRY OF TOXIC ACTION OF INSECTICIDES

R. D. O'BRIEN

List of Contributors

Numbers in parentheses indicate the pages on which the authors' contributions begin.

MOISES AGOSIN (93), Department of Zoology, University of Georgia, Athens, Georgia

FRANCISCO J. AYALA (579), Department of Genetics, Agricultural Experiment Station, College of Agricultural and Environmental Sciences, University of California, Davis, California

MURRAY S. BLUM (465), Department of Entomology, University of Georgia, Athens, Georgia

P. S. CHEN (145), Institute of Zoology and Comparative Anatomy, University of Zurich, Zurich, Switzerland

G. MICHAEL CHIPPENDALE (1), Department of Entomology, College of Agriculture, University of Missouri–Columbia, Columbia, Missouri

W. C. DAUTERMAN (541), Department of Entomology, North Carolina State University, Raleigh, North Carolina

ROGER G. H. DOWNER (57), Department of Biology, Faculty of Science, University of Waterloo, Waterloo, Ontario, Canada

ERNEST HODGSON (541), Department of Entomology, North Carolina State University, Raleigh, North Carolina

A. E. NEEDHAM (233), Department of Zoology, Oxford University, Oxford, England

R. D. O'BRIEN (515), Section of Neurobiology and Behavior, Langmuir Laboratory, Cornell University, Ithaca, New York

A. GLENN RICHARDS (205), Department of Entomology, Fisheries, and Wildlife, University of Minnesota Twin Cities, St. Paul, Minnesota

LYNN M. RIDDIFORD (307), Department of Zoology, University of Washington, Seattle, Washington

WENDELL L. ROELOFS (419), Department of Entomology, Entomology–Plant Pathology Laboratory, New York State Agricultural Experiment Station, Cornell University, Geneva, New York

JAMES W. TRUMAN (307), Department of Zoology, University of Washington, Seattle, Washington

NEVIN WEAVER (359, 391), Department of Biology, University of Massachusetts–Boston, Harbor Campus, Boston, Massachusetts

Preface

The science of biochemistry may be viewed as microanatomy studied at its ultimate level. Indeed, knowledge of an organism's intimate molecular makeup, and of its component parts as well, adds an important dimension to our understanding of its life processes.

Insect biochemistry as we know it today evolved from earlier efforts of such *biologists* as Keilin, Wigglesworth, and Fraenkel. Later studies by Chadwick, Dethier, Williams, and Richards (and by their students) broadened the range and extended the depth of our knowledge of insectan biochemistry, especially over the past two decades. More recently, increasing numbers of *biochemists* have exploited the wide versatility of this largest single taxonomic group for basic biochemical studies *per se*. Taken as a whole, their combined published researches in insectan biochemistry have dictated the need for a summary volume such as this edited work, which was intended as an in-depth updating of the only prior substantive book, by Darcy Gilmour, published in 1961. The extensive coverage provided by this book reflects the progress made since that time in insect biochemistry as a generic science.

Designed to serve as a basic textbook in field, this volume should be equally useful as an auxiliary text for most relevant courses in insect biology, particularly insect physiology, insect ecology, insect control, and economic entomology. To this end, the authors have attempted to present the material at the level of a teaching aid. But in a few chapters of highly technical subject matter, has this not been completely feasible. In any case, a background in general biochemistry is a prerequisite. The book should also serve as an important reference source for the advanced student, the research scientist, and the professional entomologist seeking authoritative details of relevant areas of subject matter. Most chapters include two lists of references. One is a list of general references intended particularly for students, and includes books, chapters in books, and review articles. The second is a detailed list of specific, original publications intended for advanced students and research scientists. In some cases, where the literature is more limited, only one list of references, mostly general in nature, has been included.

As in any general work in which the names of insects are cited, the problem of synonymy arises. The knowledgeable entomologist will experience some dismay at the occasional use of different names for the same insect. These differences naturally are due to the fact that, when the need arises, taxonomists must correct previously employed scientific names. The average reader will not view this as a problem. To the purist, the insect taxonomist in particular, such encounters will prove understandably painful.

The Editor acknowledges with thanks the advice of his former mentor and present colleague, Dr. Glenn Richards, for his advice in the early stages of planning the scope and contents of this book. To all of the contributors goes my sincere appreciation for their cooperation in helping to maintain the originally planned schedule for the publication of this compendious work.

Morris Rockstein

Biochemistry
of Insects

1

The Functions of Carbohydrates in Insect Life Processes

G. MICHAEL CHIPPENDALE

I. INTRODUCTION*

Carbohydrates, along with proteins and lipids, form the principal classes of organic compounds that are found in insects and other organisms. Carbohydrates contribute to the structure and functions of all insect tissues, and can be found in the nuclei, cytoplasm, and membranes of cells, as well as in the extracellular hemolymph and supporting tissues. The term "carbohydrate" is applied to polyhydroxy aldehydes and ketones and their derivatives. The classification of carbohydrates as monosaccharides, disaccharides, oligosaccharides, and polysaccharides according to the number of sugar residues found in the molecule has been well established. Insect carbohydrates are present both in free form and combined with other molecules, including purines, pyrimidines, proteins, and lipids. Carbohydrates are involved at all levels of cellular organization.

In general, carbohydrate metabolism in insects is similar to that found in vertebrates. However, insects, unlike vertebrates, possess an exoskeleton that is rich in the aminopolysaccharide, chitin. They also contain a disaccharide, trehalose, which acts as a storage form of glucose. Special features of insect carbohydrate metabolism, therefore, center around the synthesis and hydrolysis of these compounds.

The function of carbohydrates in regulating and maintaining the life processes of insects is examined in this chapter. Included are accounts of dietary carbohydrate requirements, the digestion and absorption of carbohydrates, the involvement of carbohydrates in excretory processes, the nature of structural carbohydrates and carbohydrate pigments, and the metabolic interconversions and breakdown of carbohydrates. In addition, the origin, nature, and fate of carbohydrates in specific tissues during

*The following standard abbreviations are used in the text: ADP, adenosine 5′-diphosphate; ATP, adenosine 5′-triphosphate; CoA, coenzyme A; cyclic AMP, adenosine 3′:5′-monophosphate; DNA, deoxyribonucleic acid; FAD, flavin-adenine dinucleotide; FMN, flavin mononucleotide; NAD^+, $NADP^+$, nicotinamide-adenine dinucleotide (phosphate); RNA, ribonucleic acid; UDP, uridine 5′-diphosphate; UTP, uridine 5′-triphosphate.

growth, metamorphosis, flight, reproduction, embryonic development, diapause, and coldhardiness are examined. The importance of carbohydrates in regulating feeding behavior and in the synthesis of defensive secretions and pheromones is also stressed. For additional information several well-documented review articles dealing with insect carbohydrates are available. These include reviews about the biochemistry of carbohydrates (Wyatt, 1967), the general metabolism of carbohydrates (Friedman, 1970), the intermediary metabolism of carbohydrates (Chefurka, 1965), carbohydrate fuel supply to flight muscles (Bailey, 1975), and the regulation of carbohydrate metabolism in flight muscles (Sacktor, 1970, 1975). In addition, comparative aspects of carbohydrate metabolism are documented in reviews about trehalose metabolism (Elbein, 1974), and fuel utilization and metabolism by muscles (Crabtree and Newsholme, 1975). Our understanding of insect carbohydrates continues to expand rapidly with the development and application of increasingly sophisticated analytical techniques.

II. CLASSIFICATION OF INSECT CARBOHYDRATES

Insect carbohydrates can be classed as (1) "pure" polyhydroxy aldehydes and ketones present as monomers (monosaccharides) or polymers (di-, oligo-, and polysaccharides); (2) derived carbohydrates, in which the monomer has been to various degrees modified chemically; and (3) conjugated carbohydrates, in which pure and derived carbohydrates are combined with other noncarbohydrate molecules.

The "pure" carbohydrates include the simple sugars or monosaccharides and their polymers which contain residues of identical or different monosaccharides. *Monosaccharides*, such as xylose and ribose (pentoses), and glucose and fructose (hexoses), have an empirical formula of $C_nH_{2n}O_n$ and cannot be hydrolyzed into simpler molecules. Although most monosaccharides have at least one asymmetric carbon atom and therefore can exist as stereoisomers in the D or L form, the naturally occurring sugars are usually D isomers. In solution, monosaccharides that have five or more carbon atoms form stable rings. Five-membered rings are called furanoses and six-membered rings are called pyranoses. They exist in α and β anomeric forms, which differ in their configuration about the hemiacetal carbon atom. An internal hemiacetal group containing the anomeric or glycosidic hydroxyl group forms during closure of the furanose and pyranose rings. The most abundant monosaccharide is D-glucose, which is the parent monosaccharide from which all other sugars can be derived. It is present in solution as an equilibrium mixture of about one-third α-D-glucopyranose and two-thirds β-D-glucopyranose.

Disaccharides contain two residues of monosaccharides, which can be either identical or different. If the glycosidic bond is formed through the anomeric carbon atom of each residue, the disaccharide is nonreducing. Examples include the nonreducing disaccharides trehalose (two glucose residues) and sucrose (glucose and fructose residues), and the reducing disaccharide maltose (two glucose residues). The most common disaccharide in insects is trehalose.

Oligosaccharides are normally defined as carbohydrates containing from two to ten monosaccharide residues. However, the most common oligosaccharides are the di-, tri-, and tetrasaccharides. In insects, it is uncommon to find oligosaccharides that contain more than two sugar residues. Short-chain oligosaccharides, such as maltosyl fructoside, may be formed in the intestinal lumen by transglucosylation reactions. These reactions involve the transfer of glucose units to a parent oligosaccharide, and seem to be responsible for the production of at least some of the oligosaccharides found in insect honeydew.

Polysaccharides, or glycans, are high molecular weight polymers containing many residues of monosaccharides or monosaccharide derivatives. Insect polysaccharides include glycogen, a branched-chain polymer of glucose residues; chitin, a straight-chain polymer of 2-acetamido-2-deoxy-D-glucose residues; and glycosaminoglycans or mucopolysaccharides, straight-chain polymers containing amino sugar and uronic acid residues found in connective tissues.

Derived carbohydrates are found in large numbers in insect cells because monosaccharides readily undergo enzymatic reactions, including oxidation, reduction, esterification, and substitution. These derivatives include deoxy sugars (e.g., 2-deoxy-D-ribose), phosphoric acid esters (e.g., D-glucose 1-phosphate), amino sugars (e.g., 2-amino-2-deoxy-D-glucose), sugar acids (e.g., glucuronic acid, ascorbic acid), alditols (e.g., glycerol, threitol, ribitol, sorbitol, mannitol), and cyclitols (isomers of inositol, especially *myo*-inositol).

Insect carbohydrates and their derivatives are also found conjugated to other molecules in nucleic acids (DNA, RNA), nucleosides (e.g., adenosine, uridine), nucleotides (e.g., ATP, UTP), coenzymes (e.g., CoA, FAD, FMN, NAD^+, $NADP^+$), phospholipids (cerebrosides, phosphatidylinositol), glycoproteins, and glycosides. For the most part, "trivial" rather than systematic names will be used to describe carbohydrates in this text. An account of the widely accepted rules for carbohydrate nomenclature is available in the report of the IUPAC-IUB Commission (1972a).

III. CARBOHYDRATES IN INSECT NUTRITION

To date, nutritional studies have focused primarily on determining which carbohydrates are necessary to support optimum growth, development, reproductive activity, and survival of individual species. For most species, glucose, fructose, and sucrose are nutritionally adequate sugars, but much interspecific variation exists in the utilization of other dietary carbohydrates. Nutritional studies have also shown that quantitative requirements for carbohydrates often vary according to age, sex, and metamorphic stage.

Some carbohydrates have been shown to be nutritionally ineffective because they act as feeding deterrents or because they are incompletely hydrolyzed or absorbed. Still other carbohydrates are readily absorbed, but are not metabolized. At high concentrations they may even inhibit enzymatic reactions, including those involved in glycolysis and gluconeogenesis. For example, dietary pentoses, including arabinose, ribose, and xylose, have been shown to inhibit larval growth of the southwestern corn borer, *Diatraea grandiosella*, and the yellow mealworm, *Tenebrio molitor*. These larvae are typical of many insects which are unable to use these five-carbon sugars to sustain normal oxidative metabolism. Similarly, the dietary hexose, galactose, is poorly utilized by larvae and adults of the boll weevil, *Anthonomus grandis,* even though it acts as a feeding stimulant and is readily absorbed. Nettles and his associates have shown that fasted adults of *A. grandis* metabolize only small amounts of absorbed galactose to trehalose and glycogen because galactose is converted to galactitol. This sugar alcohol is not further metabolized and accumulates in the hemolymph. In addition, the hexose, sorbose, which is present in nature only as a fermentation product of sorbitol in various ripe berries, is commonly found to be nonutilizable, presumably because of its low rate of conversion to glucose and trehalose.

In many cases, the reason for some carbohydrates not supporting growth, development, and survival still remains unclear. Those nutritional studies that have been supplemented with feeding behavioral analyses to determine the insect's response to the carbohydrate and with biochemical analyses to determine the metabolic effects of the carbohydrate, have to date yielded the most far-reaching and satisfactory results. It is therefore now evident that an insect's ability to utilize any specific carbohydrate depends upon a complex interrelationship including gustation, digestive hydrolysis, absorption, and metabolism.

A. Larval Carbohydrate Requirements

Insect larvae differ in the amounts of dietary carbohydrates they require for normal growth and development. The larvae of some species do not require carbohydrates, since they are able to substitute dietary protein or lipid for carbohydrate and meet their energy needs for growth and development from amino acid and fatty acid oxidation. Larvae that can grow satisfactorily without ingesting any carbohydrate include the coleopterans,

Dermestes maculatus (the hide beetle) and *Tribolium castaneum* (the red flour beetle), and the dipterans, *Aedes aegypti* (the yellow fever mosquito), *Musca domestica* (the house fly), and *Phormia regina* (the blow fly).

Other species require only moderate amounts of carbohydrate in their larval diet. For most of these species, glucose, fructose, or sucrose, comprising 15–30% of the dry weight of the diet, serves as an adequate carbohydrate source. While pentoses and sorbose have generally proved to be of no nutritional value, much species variation exists in the utilization of galactose, mannose, oligosaccharides, and polysaccharides. The limiting factors in carbohydrate utilization appear to be the rate of hydrolysis of polysaccharides and oligosaccharides in the intestinal lumen, and the rate of conversion of other monosaccharides to glucose or fructose in the tissues.

For most species, the optimum concentration of dietary carbohydrate falls within fairly narrow limits. For example, Fig. 1 illustrates how three different concentrations of dietary D-glucose affect the larval growth of *D. grandiosella*. When the dietary sugar was absent from its diet, the borer's growth rate was retarded significantly, whereas it reached an optimum growth rate with the addition of 3.3% glucose (20% dry weight basis). Yet, when a higher concentration of glucose (5%) was tested, it did not further accelerate the growth rate. Similar quantitative requirements have been demonstrated for other mandibulate plant-feeding insects, including the locusts, *Locusta migratoria* and *Schistocerca gregaria,* the European corn borer, *Ostrinia*

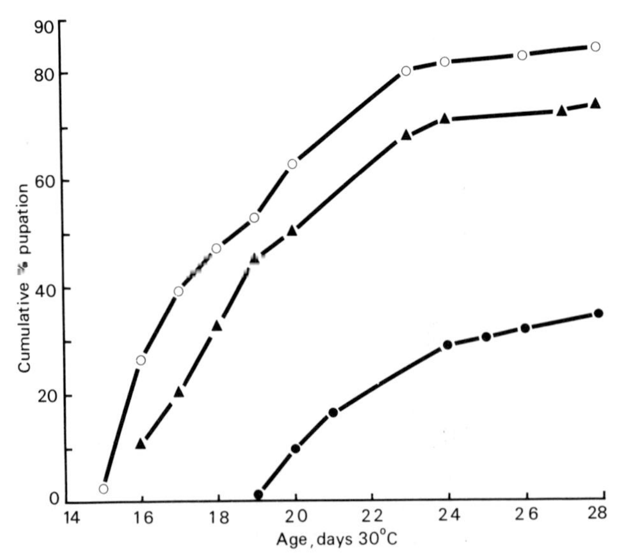

Fig. 1. Effect of three glucose concentrations in an agar-wheat germ diet on larval growth and development of the corn borer, *Diatraea grandiosella*. ●, 0% glucose; ▲, 1.65% glucose; ○, 3.30% glucose. (Redrawn from Chippendale and Reddy, 1974.)

nubilalis, the silkworm, *Bombyx mori,* and the spruce budworm, *Choristoneura fumiferana.*

The requirement for a high concentration of dietary carbohydrate appears to be restricted to those larvae which normally feed on starch-containing stored products. These larvae require a dietary concentration that can reach 70% carbohydrate on a dry weight basis. Such a requirement is especially well demonstrated in larvae of the Angoumois grain moth, *Sitotroga cerealella,* and the grain weevils, *Sitophilus* spp., which are obligatory in-kernel feeders. A high dietary starch concentration is necessary to meet the chemical and physical feeding requirements of these larvae. In addition, it has been shown that larvae of grain-feeding insects utilize the branched-chain, amylopectin, portion of starch more efficiently than they utilize the straight-chain, amylose, portion. Their digestive amylase appears to generate absorbable glucose at a higher rate from amylopectin than it does from amylose (see Section IV,A,1).

The recent development of liquid media for artifically culturing aphids, planthoppers, and leafhoppers was facilitated by the finding that larvae and adults ingest sugary fluids through a Parafilm membrane. Chemically defined diets have now been developed which maintain several species of these haustellate plant feeding insects through many generations. Sucrose has been shown to be the most suitable sugar for maintaining these insects in culture. For example, satisfactory larval growth of the aster leafhopper, *Macrosteles fascifrons,* and the brown planthopper, *Laodelphax striatellus,* has been obtained with liquid diets containing only 5% sucrose (w/v). In contrast, aphids, such as *Aphis fabae, Myzus persicae,* and *Neomyzus circumflexus* require dietary sucrose concentrations from 15 to 35% (w/v) for optimum larval growth and development. Besides meeting the aphid's carbohydrate requirement, the high sucrose concentration controls feeding stimulation, osmotic pressure, and microbial growth in the liquid media.

B. Adult Carbohydrate Requirements

Insects which feed as adults commonly consume large quantities of carbohydrates. These ingested carbohydrates, together with nutrient reserves carried over from the larval stage, are necessary to meet the insect's energy demands for flight, reproduction, and longevity. Adults most commonly obtain their dietary carbohydrates from plant nectars and grain kernels and therefore consume mainly sucrose and starch. Most female Diptera, including *P. regina, Sarcophaga bullata,* and *M. domestica,* must consume carbohydrate and protein before they are able to lay viable eggs.

Table I summarizes the effect of dietary carbohydrates on the longevity of adult females of three species of mosquitoes. Female mosquitoes feed on vertebrate blood, nectar, and honeydew and rely largely on the last two

TABLE I

Effect of Dietary Carbohydrates on the Life Span of Females of Three Species of Mosquitoes [a,b]

Dietary carbohydrate	Mean survival time (days, 27°C)		
	Aedes taeniorhynchus	*Aedes aegypti*	*Culiseta inornata*
Water (control)	4	4	4
Arabinose	5	4	4
Ribose	8	5	3
Xylose	9	4	9
Fructose	37	25	28
Galactose	33	24	5
Glucose	39	35	24
Mannose	26	30	15
Sorbose	3	4	3
Lactose	4	4	4
Maltose	35	35	30
Sucrose	39	37	30
Raffinose	28	28	9
Glycerol	11	6	3
Sorbitol	25	31	16

[a]From Nayar and Sauerman (1971).
[b]Newly emerged females were starved for 2 days and then fed a 10% carbohydrate solution.

substrates for dietary carbohydrates. Table I shows that all three species are able to utilize the sugars which are major constituents of nectar and honeydew, but that differences exist in their ability to utilize sugars which they do not normally encounter in nature. It is clear that while all the species utilize fructose, glucose, sucrose, and maltose, differences exist in their capacity to utilize galactose, mannose, raffinose, and sorbitol, and that arabinose, ribose, xylose, sorbose, lactose, and glycerol are essentially not utilized at all. The nectar-feeding adult of the cabbage root fly, *Erioischia brassicae,* shows a similar spectrum of utilizable dietary carbohydrates that includes fructose, glucose, sucrose, maltose, melezitose, mannitol, and sorbitol.

Adult fecundity and longevity of several grain-feeding Coleoptera have also been shown to depend upon adequate starch consumption. For example, long-term survival of the grain weevils, *Sitophilus* spp., is possible only if they consume amylopectin-rich starches which are readily hydrolyzed by their digestive amylase. Rice, wheat, and maize kernels, therefore, are satisfactory feeding substrates (see Section IV,A,1).

C. Carbohydrate Growth Factors

Besides having a dietary requirement for sugars and starches, insects have also been found to need a dietary source of some conjugated and derived carbohydrates in the form of vitamins and related compounds. All insects are known to require vitamin B_2 (riboflavin). This water-soluble vitamin contains a D-ribitol residue and is a constituent of flavoprotein coenzymes (FAD, FMN) which mediate hydrogen transfer reactions, notably in the respiratory chain. While insects may also prove to have a general requirement for vitamin B_{12} (cyanocobalamin), the required amounts are so small that an exact determination is difficult to obtain. Cyanocobalamin contains a D-ribose residue in a complex molecule which also contains cobalt, a corrin ring, phosphate, and an imidazole unit. The coenzyme of vitamin B_{12}, 5-deoxyadenosylcobalamin, regulates several metabolic systems including transmethylations and gluconeogenesis (see Section VIII,A,4).

Some insects, especially plant feeders, also require dietary L-ascorbic acid and *myo*-inositol for optimum growth and development. The indispensability of carbohydrate growth factors will be illustrated by describing the special requirements some species have for ascorbic acid and *myo*-inositol.

1. L-Ascorbic Acid

L-Ascorbic acid (vitamin C) is the γ-lactone of a hexonic acid and has an enediol structure at carbon atoms two and three. The enediol gives the molecule its reducing and acidic properties. The molecule is highly unstable in solution and is rapidly oxidized to dehydroascorbic acid. Ascorbic acid is widely distributed among insect tissues, and most insects are able to synthesize adequate amounts of the substance from glucose. Although the biosynthesis of ascorbic acid in insects has not yet been studied thoroughly, the pathway appears to be similar to that found in vertebrates. For example, ascorbic acid synthesis in rat liver occurs as a side reaction of the glucuronic acid cycle resulting in the oxidation of L-gulonic acid via L-gulonolactone to 2-keto-L-gulonolactone, which spontaneously isomerizes to L-ascorbic acid. Those vertebrates (primates, guinea pigs, and bats) which have a dietary requirement for ascorbic acid lack the liver enzyme, L-gulonolactone oxidase,* which catalyzes the oxidation of L-gulonolactone to 2-keto-L-gulonolactone. Of special interest is the finding that the intermediates from the pathway of ascorbic acid synthesis in vertebrates (D-glucuronic acid, D-glucuronolactone, and L-gulonolactone) do not substitute for ascorbic acid in those insects which require the vitamin. Therefore, these insects also may be deficient in L-gulonolactone oxidase.

A dietary requirement for ascorbic acid in insects was first demonstrated by Dadd in 1957. He showed that dietary ascorbic acid was necessary for the normal larval growth of *S. gregaria*. Subsequently, Chippendale and Beck in 1964 showed that ascorbic acid was an essential nutrient for larvae of *O.*

*The names applied to enzymes throughout this chapter are those recommended in the reports of the IUPAC-IUB Commission (1972b, 1976).

nubilalis, and that it provided a suitable substitute for a dietary supplement of freeze-dried plant tissue. Dietary ascorbic acid is now known to be required for maintaining normal growth, molting, and fertility of several other plant-feeding insects, including *B. mori, D. grandiosella, A. grandis,* and the aphid, *M. persicae* (see Chippendale, 1975). Figure 2 illustrates the

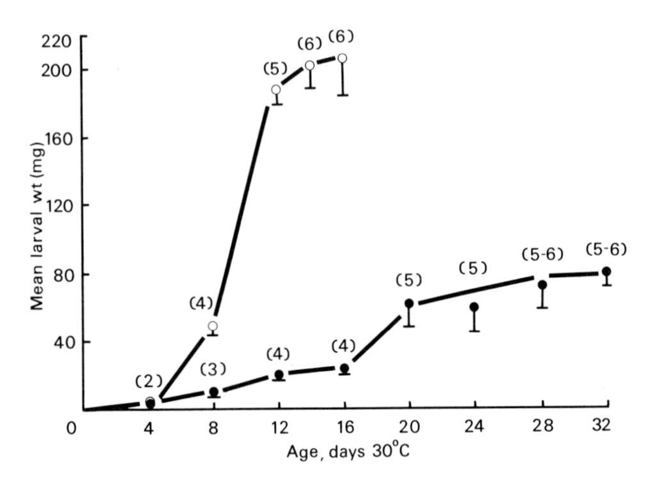

Fig. 2. Effect of L-ascorbic acid incorporated into an agar-wheat germ diet on larval growth of the corn borer, *Diatraea grandiosella*. Each vertical line represents a single standard error of the mean (n= 10). Parenthetic numbers refer to larval instars. ●, 0% ascorbic acid; ○, 0.5% ascorbic acid. (Redrawn from Chippendale, 1975.)

effect of omitting ascorbic acid from the larval diet of *D. grandiosella*. The omission of dietary ascorbic acid resulted in a low larval growth rate and few fertile adults were produced. Although ascorbic acid has been found to be an essential nutrient for larvae of several species of plant-feeding insects, others including the pink bollworm, *Pectinophora gossypiella,* and the redbanded leafroller, *Argyrotaenia velutinana,* can develop normally without a dietary source of the substance.

Although the precise functions of ascorbic acid in insects are not known, it probably has specific metabolic functions and also serves as a nonspecific antioxidant. Specific functions are implicated by the finding that the reducing agents, L-cysteine and glutathione, are not effective substitutes for ascorbic acid in those species which have a dietary requirement for the vitamin. In vertebrates, metabolic functions of ascorbic acid include the regulation of enzymatic hydroxylations, adipose tissue lipase, and protein catabolism. Perhaps one function of ascorbic acid in insects is to promote enzymatic hydroxylations such as in the formation of the tanning agent, *N*-acetyldopamine, from tyrosine, and the molting hormone, ecdysone, from cholesterol.

2. Myo-inositol

Although inositol (hexahydroxycylohexane) can exist in several isomeric forms, only four of them occur naturally (i.e., *myo-*, *scyllo-*, *neo-*, and *chiro*-inositol). In insects and other organisms, the *myo* isomer is the most important because it is the only isomer which is incorporated into phosphatidylinositol, an essential structural component of cell and subcellular membranes. Phosphatidylinositol, therefore, is widely distributed among insect tissues, but it is present at a much lower concentration than the two principal insect phospholipids, phosphatidylcholine and phosphatidylethanolamine.

Most insects appear to synthesize adequate amounts of the lipogenic *myo*-inositol from glucose. For example, the fat body of the adult American cockroach, *Periplaneta americana,* contains enzyme systems capable of synthesizing and interconverting the naturally occurring isomers of inositol. However, other insects, especially some plant feeders, have become dependent upon a dietary source of the substance. Species which have been found to be dependent upon dietary *myo*-inositol include *L. migratoria, S. gregaria, M. persicae, A. grandis,* and the bollworm, *Heliothis zea.* The hemolymph of *L. migratoria* has also been shown to contain *scyllo*-inositol in amounts up to 1.7 m*M*. The *scyllo* isomer is derived from the *myo* isomer in the fat body, but it is not incorporated into phosphatidylinositol, and currently its function remains unknown.

IV. DIGESTION OF CARBOHYDRATES

A. Common Digestive Carbohydrases

Insect digestive carbohydrases hydrolyze poly-, oligo-, and disaccharides to their constituent monosaccharides in preparation for absorption. These enzymes are secreted mainly by the salivary glands and the epithelium of the midgut. Although digestive carbohydrases have been examined in many species, much remains to be learned about the biochemical characteristics of the individual enzymes. Recent research, however, notably by Droste and Zebe (1974) and Morgan (1976) has placed needed emphasis on determining the physicochemical properties of insect carbohydrases. Two categories of carbohydrases are commonly recognized: α- and β-amylases, which hydrolyze the glucosidic bonds of polysaccharides, and glycosidases, which hydrolyze the glycosidic bonds of oligo- and disaccharides.

1. Digestive Amylases

Insects usually contain a digestive amylase which hydrolyzes dietary starch (amylose, amylopectin) and glycogen. Current evidence suggests that the enzyme is an α-amylase or endoamylase (1,4-α-D-glucan glucanohydrolase) which attacks the interior glucosidic linkages of starch or glycogen, thereby generating a mixture of dextrins of varying chain lengths. The rate of hydrolysis of starch by α-amylase is controlled by such factors as molecular weight and degree of branching of the substrate. The branched-chain amylopectin molecule, which makes up about 75% of starch, is usually hydrolyzed at a higher rate than the straight-chain amylose, which makes up the remaining 25% of most starches. A salivary or intestinal α-amylase, together with an α-glucosidase and an oligo-1,6-glucosidase (isomaltase), completely hydrolyzes dietary starch or glycogen to absorbable glucose. Recently Buonocore and his colleagues (1976) have isolated and characterized a typical calcium-containing α-amylase (endoamylase) from whole larvae of *T. molitor*. The enzyme is made up of a single polypeptide chain and has a molecular weight of 68,000 and an isoelectric point of 4.0. Under their assay conditions the amylase had a pH optimum of 5.8, a temperature optimum of 37°C, and was activated by chloride ions and inhibited by fluoride ions.

The salivary amylases of plant-sucking insects have received special attention because of their importance in extraintestinal hydrolysis of plant starch. For example, Hori has shown that the saliva of the polyphagous hemipteran, *Lygus disponsi,* contains an α-amylase which hydrolyzes plant starch, thereby making it available to the insect. The amylolytic activity of the salivary glands was found to be much higher than that of the midgut, suggesting that the ventricular amylase originates from the salivary glands. Figure 3 shows that the amylase is activated by chloride and nitrate ions, which are commonly present in green plants. Since these activators extend the amylase activity over a wide pH range they may contribute towards stabilizing the enzyme against pH changes in the intestinal lumen.

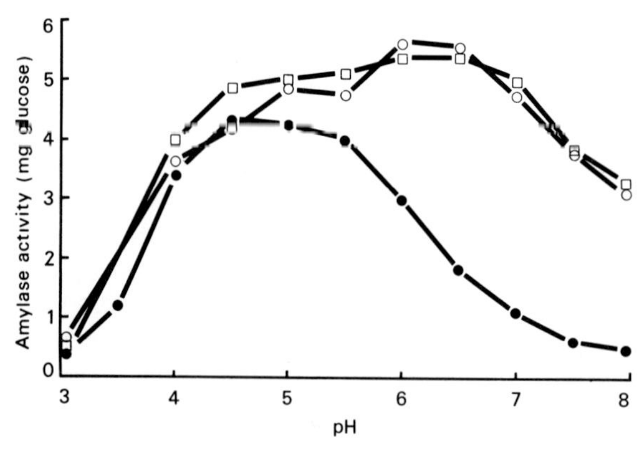

Fig. 3. Effect of pH and chloride and nitrate ions on the activity of salivary amylase at 37°C from adults of the hemipteran, *Lygus disponsi*. ●, control medium; ○, medium containing 0.01 *M* sodium chloride; □, medium containing 0.01 *M* potassium nitrate. (Redrawn from Hori, 1969.)

2. Digestive Glycosidases

Although digestive glycosidases have been found in many insect species, quantitative data about their biochemical characteristics remain scarce. Based on the nature of the glycosidic bond and the form of linkage of the substrate, five digestive glycosidases are known to be capable of hydrolyzing all oligosaccharides and glycosides containing the common monosaccharides, glucose, galactose, and fructose, which most insects are likely to ingest. These enzymes are *α-glucosidase* (hydrolyzes maltose, sucrose, trehalose, melezitose, raffinose, stachyose), *β-glucosidase* (cellobiose, gentiobiose, methyl β-glycoside), *α-galactosidase* (melibiose, raffinose, stachyose), *β-galactosidase* (lactose), and *β-fructofuranosidase* (sucrose, raffinose). Although one or more of these glycosidases may be absent from the digestive fluid of any given species, all five enzymes have been demonstrated to exist within the class Insecta. In addition to a broad-spectrum α-glucosidase, some species also contain a specific digestive α,α-trehalase. Table II summarizes one of the few available studies in which the

TABLE II

Characteristics of Intestinal Carbohydrases Isolated from Adults of *Locusta migratoria*[a]

Enzyme	Substrate tested	Molecular weight	pH optimum	K_m
α-Glucosidase	Sucrose	115,000	6.0	$3.3 \times 10^{-2} M$
α-Glucosidase	Maltose	115,000	6.6	$5.8 \times 10^{-3} M$
β-Glucosidase	Cellobiose	202,000	5.6	$5.5 \times 10^{-3} M$
α-Galactosidase	Raffinose	240,000	5.3	$1.6 \times 10^{-3} M$
α-Galactosidase	Melibiose	240,000	5.3	$2.4 \times 10^{-3} M$
β-Galactosidase	Lactose	202,000	6.0	$8.3 \times 10^{-2} M$
α,α-Trehalase (soluble)	Trehalose	133,000	5.6	$3.4 \times 10^{-3} M$
α,α-Trehalase (particulate)	Trehalose	–	6.5	$1.8 \times 10^{-3} M$
α-Amylase	Starch	68,000	6.0	–
Oligo-1,6-glucosidase	Isomaltose	185,000	6.0	$2.3 \times 10^{-2} M$

[a]From Droste and Zebe (1974).

physicochemical characteristics of the digestive carbohydrases of an individual species have been determined.

B. Special Digestive Carbohydrases

Insects which are adapted to consume carbohydrate-rich foods such as plant nectar and cellulose have evolved specialized carbohydrate digestive capabilities. Foremost among these insects are honeybees and termites.

1. Alpha-Glucosidases of the Honey Bee

The worker adult of the honey bee, *Apis mellifera,* is dependent upon dietary nectar, containing mainly glucose, fructose, and sucrose, as a carbohydrate source. Huber and Mathison (1976) have isolated and characterized highly active sucrases from *A. mellifera* that hydrolyze dietary sucrose to meet the insect's needs for tissue sugars and for honey production. Since the sucrases show specificity for the α-glucosidic bond, they may be classed as α-glucosidases. They have been shown to hydrolyze both sucrose and maltose, but not trehalose. The products of sucrose hydrolysis, glucose (ca. 180 m*M*) and fructose (ca. 90 m*M*) are the principal carbohydrates present in the hemolymph of the worker honey bee. These sugars are more abundant than trehalose, which seldom exceeds a concentration of 30 m*M* in the hemolymph. Whole-body analyses have shown that the α-glucosidases are ten times more active than trehalase. Two types of α-glucosidase that can be distinguished by their solubility properties in ammonium sulfate have been isolated. A fairly insoluble α-glucosidase appears to be restricted to the pharyngeal glands and to the honey stomach where it may function mainly in honey production. This enzyme has properties similar to the α-glucosidase present in honey. In addition, a highly soluble α-glucosidase has been found in the abdomen of all three honey bee castes. This enzyme appears to be a true digestive α-glucosidase and has the sucrose of nectar as its principal substrate.

2. Cellulases of Termites

Termites contain a digestive cellulase which hydrolyzes ingested wood and lignified tissues. The digestive cellulase may be synthesized by intestinal microorganisms or by the insect's own ventricular cells. The cellulase, aided by a β-glucosidase, hydrolyzes cellulose via cellodextrins, to cellotriose, cellobiose, and glucose. In lower termites the gut contains specially modified compartments for cellulase-secreting protozoa, which may comprise from 16 to 36% of the live weight of the termite. For example, the hindgut of the Formosan subterranean termite, *Coptotermes formosanus,* contains a protozoan species which hydrolyzes wood anaerobically via glucose to acetic acid which is then available to the termite for energy production and for lipid synthesis. Mauldin and his colleagues (1972) have shown that the protozoan enzyme is necessary for normal cellulose catabolism, lipid synthesis, and survival of the worker termite. In higher termites, cellulose-digesting protozoa are absent and cellulase is produced by the termite's own ventricular cells. For example, the midgut cells of the harvester termite, *Trinervitermes trinervoides,* produce a cellulase which is active both in the cell and in the lumen. This purified cellulase has been shown to have a random mode of action on cellulose. The main hydrolytic product is cellotriose, which is further hydrolyzed by a β-glucosidase to absorbable glucose.

Digestive cellulases are also secreted by the salivary glands and in the intestinal lumen of omnivorous and specific wood-consuming cockroaches. For example, Wharton and his colleagues (1965) have shown that the cellulases of *P. americana* originate from the cells of the salivary gland and from microorganisms present in the intestinal lumen. Comparative studies dealing with the contribution these two enzymes make to cellulose digestion in cockroaches have yet to be undertaken.

V. ABSORPTION OF CARBOHYDRATES

The physiology of absorption of nutrients from the lumen of the intestine to the hemolymph of insects has not yet been studied in great detail. Those investigations which deal with sugar absorption have largely been limited to *in vivo* and *in vitro* studies of the intestinal tissues of *P. americana, S. gregaria,* and *P. regina.* To date, two important limiting factors which control the rate of sugar absorption have been uncovered: the rate of release of fluid from the crop into the midgut, and the rate of conversion of absorbed glucose into trehalose. The first factor regulates the availability of glucose to the midgut epithelium, while the second regulates the rate of diffusion of glucose through the midgut epithelium by controlling the concentration gradient of glucose between the midgut lumen and the hemolymph.

The control of fluid release from the crop into the midgut protects the midgut epithelium from being saturated by an excess of glucose. Neural or hormonal, or both pathways are involved in controlling the release of fluids from the crop. For example, crop emptying in *P. americana* is known to be under nervous control through a pathway involving pharyngeal osmoreceptors, the stomodeal nervous system, and the proventriculus. Gelperin (1966), in a study of sugar absorption in the adult *P. regina,* showed the importance of the rate of crop emptying in controlling sugar absorption. He showed that even at the highest concentration of glucose tested (3.0 *M*), release of glucose into the midgut occurred at no higher rate than 0.015 μmoles/min despite findings from other *in vitro* experiments which showed that the midgut was capable of absorbing 0.15 μmoles/min of glucose.

The synthesis and accumulation of hemolymph trehalose appear to play a central role in carbohydrate absorption by maintaining a steep concentration gradient of glucose between the intestinal lumen and the hemolymph, thereby facilitating glucose uptake by the physical process of diffusion. After monosaccharides, most commonly glucose and fructose, have been absorbed by the epithelial cells and transferred to the hemolymph, they are then transported to the fat body for trehalose synthesis (Section VIII, A, 2). Trehalose is then released into the hemolymph. This results in the maintenance of a high trehalose and low glucose concentration in the hemolymph that, in turn, facilitates glucose absorption by maintaining a steep concentration gradient between the intestinal lumen and the hemolymph. In addition, the presence of an active α,α-trehalase in the midgut cells appears to limit the back-diffusion and loss of trehalose into the gut lumen, and to aid in maintaining a favorable concentration gradient for glucose absorption. Glucose is more rapidly absorbed because the fat body converts it into trehalose at a higher rate than it does other monosaccharides.

These findings imply that sugar absorption in insects does not depend upon an energy-requiring active transport of glucose across the midgut epithelium. The mechanism of carbohydrate absorption therefore differs fundamentally from that in vertebrates which have been shown to absorb sugars by active transport. However, an active transporting system may be present in some insects which have a high concentration of glucose and low concentration of trehalose in their hemolymph. Larvae of the dipterans, *Agria affinis* and *P. regina,* fit into this category. Sugars may be absorbed by active transport in these species because diffusion of glucose across the midgut epithelium is not facilitated by the presence of a high trehalose concentration in their hemolymph.

VI. EXCRETION AND DETOXICATION

A. Regulation of Carbohydrate Excretion

The Malpighian tubules and hindgut, especially the rectum, are the most important excretory organs of insects. It has long been thought that the Malpighian tubules produce an ultrafiltrate of the hemolymph solutes that passes into the hindgut, where selective resorption of useful small molecules such as glucose, trehalose, and amino acids takes place. However, recent research examining insect excretory mechanisms in greater detail has begun to provide evidence which suggests the need for reappraising this simple secretion–reabsorption mechanism. Increasing evidence has emerged suggesting that the Malpighian tubules themselves regulate the excretion of sugar. It is possible that an enzymatically controlled selective permeability of the Malpighian tubules is involved in the conservation of hemolymph sugars. For example, the isolated Malpighian tubules from adult blow fly, *Calliphora vomitoria,* have been shown to restrict their output of trehalose and glucose. Trehalose may be hydrolyzed in the tubule wall, and glucose may be reabsorbed by the tubule itself (see Table III). Alternatively, the

Malpighian tubules *in vivo* may be impermeable to disaccharides because, as in the case of the adult *C. vomitoria,* only a small proportion of [U-^{14}C]glucose or [U-^{14}C]trehalose injected into the hemocoel is excreted. Much additional research is necessary to determine precisely how the excretory system prevents the loss of the principal hemolymph sugars, trehalose and glucose.

The excretion of a carbohydrate-rich honeydew by homopterous insects illustrates a special adaptation which has evolved to accommodate a high carbohydrate and low nitrogen diet. Aphids and coccids have an intestinal filter chamber which concentrates ingested nitrogen compounds at the expense of sugars, many of which pass through the alimentary canal without being absorbed. Honeydew has been shown to contain monosaccharides, including fructose and glucose, disaccharides, including sucrose and maltose, and several oligosaccharides, including maltosylfructose, melezitose, raffinose, and stachyose. Ribitol has also been found in the honeydew of the scale insect, *Ceroplastes* spp. Selective absorption of some of the ingested sugars, by an as yet unknown mechanism, may also take place. For example, the scale insect, *Ceroplastes pseudoceriferus,* ingests fructose and sucrose from the tissues of its host tea plant, but these sugars do not appear in the honeydew, which contains mainly glucose, maltose, raffinose, stachyose, and ribitol. The oligosaccharides present in honeydew may originate from the tissues of the host plant or may be formed in the intestinal lumen by transglycosylation reactions (i.e., glycosyl transfer to sugar acceptors). For example, the major sugars present in the honeydew of the lime aphid, *Eucallipterus tiliae,* are (as % of total sugars) melezitose (40), sucrose (39), and fructose (14). Since sucrose is the principal sugar present in the leaves of the lime tree, the trisaccharide melezitose appears to be formed from the ingested sucrose exclusively by transglucosylation in the intestinal lumen of the aphid.

Some insect species have evolved special mechanisms for using unabsorbed or excreted carbohydrates and eliminated digestive carbohydrases in ways which have provided them with adaptive value. For example, some Homoptera are able to use eliminated carbohydrates to produce spittle or protective chambers. The Malpighian tubules of cecropid larvae secrete a glycosaminoglycan which has been found to improve the spreading properties of spittle. This glycosaminoglycan is made up of glucuronic acid, glucosamine, glucose, and rhamnose. In addition, tube-dwelling cecropids of the family Machaerotidae build protective tubes composed mainly of an acid glycosaminoglycan, present as microscopic fibrils. Upon hydrolysis, the fibrils yield glucuronic acid, glucosamine, and glucose. Another use for eliminated carbohydrates is illustrated by larvae of certain Australian genera of psyllid homopterans which construct an elaborate protective cover or "lerp." The lerps of *Cardiaspina albitextura* and *Lasiopsylla rotundipennis* consist mainly of starch (87%) and water (11%) in association with small amounts of protein and lipid. The starch is mainly amylose which originates from the soluble starch present in the host plant, eucalyptus. The starch-rich feces are molded into a protective cover as they solidify on exposure to air (White, 1972).

Insect feces may contain enzymes, including carbohydrases, particularly at the end of the larval feeding period. These fecal enzymes may represent only excess digestive enzymes which are eliminated with other waste materials, or they may also have some adaptive value for the insect. For example, attine ants such as the leaf-cutting ant, *Atta colombica tonsipes,* which culture "fungus gardens," produce fecal enzymes which hydrolyze plant tissue and mycelial debris. The fecal enzymes, including α-amylase, chitinase, and proteinase, liberate soluble sugars and amino acids from the debris, thereby permitting rapid growth of the fungal mycelia. The ants feed on the fluid contained within the mycelia of the cultivated fungus. These examples illustrate how economically insects are able to use available carbohydrate substrates and digestive enzymes.

B. Glycosides in Detoxication

Detoxication reactions which transform foreign organic compounds into less toxic metabolites often involve the formation of glycosides. A glycoside is formed when the toxicant or its metabolite (the aglycone) combines with a monosaccharide residue (usually glucose) through an acetal linkage at carbon atom one. This reaction usually takes place in the fat body. Glycosides are important in detoxication mechanisms because they are more polar and less toxic than the unconjugated aglycone and are, therefore, more rapidly excreted. Detoxication involving the formation of a β-glucoside typically takes place as follows:

UTP + D-glucose 1-phosphate = pyrophosphate + UDPglucose (glucose-1-phosphate uridylyltransferase) (1)

UDPglucose + a phenol = UDP + aryl β-D-glucoside (UDPglucosyltransferase) (2)

These reactions have been shown to occur in the fat body of many insects including *P. americana, L. migratoria, S. gregaria,* and *M. domestica.* For example, a recent comparative study showed that five species conjugated 1-naphthol to naphthylglucoside via the glucosyltransferase reaction. It should be noted that the sugar nucleotide, UDPglucose, which takes part in β-glucoside detoxication reactions, serves as the glucose source for normal glycogen and trehalose synthesis in insect tissues (see Section VIII, A, 1, 2). Additional information about the involvement of β-glucosides in insect detoxication is provided in Chapter 13.

VII. CARBOHYDRATES IN STRUCTURAL COMPONENTS AND PIGMENTS

A. Chitin

The principal structural component of the insect body is chitin, consisting mainly of unbranched chains of β-(1,4)-2-acetamido-2-deoxy-D-glucose (*N*-acetyl-D-glucosamine). Chitin forms an integral structural component of the insect's endo- and exocuticle, tracheae, intestine (including the peritrophic membrane), and reproductive tract, making this aminopolysaccharide the major carbohydrate present in insect tissues.

Chitin is synthesized most rapidly during the pharate phase of each instar and immediately after ecdysis has taken place. During the pharate phase of the instar the epidermal cells secrete new cuticle under the remnants of the old cuticle which is shed at ecdysis. Integumental chitin is synthesized in the epidermal cells from UDP-2-acetamido-2-deoxy-D-glucose. The immediate precursors of this sugar nucleotide (2-acetamido-2-deoxy-D-glucose 1(6)-phosphate and 2-amino-2-deoxy-D-glucose 6-phosphate) may also be synthesized in the epidermis or in the fat body. The following series of reactions is thought to be involved in the synthesis of chitin from glucose in insects:

ATP + D-glucose = ADP + D-glucose 6-phosphate (3)
(hexokinase)

D-Glucose 6-phosphate = D-fructose 6-phosphate (4)
(glucosephosphate isomerase)

D-Fructose 6-phosphate + L-glutamine = 2-amino-2-deoxy-D-glucose 6-phosphate + L-glutamate (5)
(glucosaminephosphate isomerase)

Acetyl-CoA + 2-amino-2-deoxy-D-glucose 6-phosphate = CoA + 2-acetamido-2-deoxy-D-glucose 6-phosphate (6)
(glucosaminephosphate acetyltransferase)

UTP + 2-acetamido-2-deoxy-D-glucose 1-phosphate = pyrophosphate + UDP-2-acetamido-2-deoxy-D-glucose (UDPacetylglucosamine pyrophosphorylase) (7)

UDP-2-acetamido-2-deoxy-D-glucose + [1,4-(2-acetamido-2-deoxy-β-D-glucosyl)]$_n$ = UDP + [1,4-(2-acetamido-2-deoxy-β-D-glucosyl)]$_{n+1}$ (8)
(chitin synthase)

The present status of our knowledge suggests that the mechanism of chitin biosynthesis in insects is generally similar to that operating in crustaceans and fungi. However, the sequence of reactions has been studied in only a few insects including *S. gregaria, L. migratoria,* and the southern armyworm, *Prodenia eridania*. Problems in unravelling the precise mechanism of chitin synthesis include demonstrating (1) the exact precursors for the formation of 2-amino-2-deoxy-D-glucose, (2) the conversion of 2 acetamido-2-deoxy-D-glucose 6-phosphate to the 1-phosphate (an unidentified acetylglucosamine phosphomutase is probably involved), and (3) the mode of action and subcellular location of chitin synthase. It has proved especially difficult to isolate and characterize chitin synthase in insects. The enzyme may be bound to subcellular membranes and be readily deactivated during tissue preparation. Additional information about the synthesis, organization, and degradation of chitin is presented in Chapter 5.

B. Glycosides and Sclerotization

Insect cuticle hardens and darkens when a protein–quinone complex (sclerotin) is formed from proteins and oxidized polyhydric phenols present in the outer layers of the cuticle. This process of cuticular sclerotization or tanning, which is widespread among insects, has been most thoroughly studied using fly puparia and cockroach egg cases or ootheca. Glycosides are not known to be involved in regulating the formation of integumental sclerotin such as that present in the fly puparium. However, they prevent premature sclerotization of the precursors of the cockroach ootheca by masking the functional group of the tanning agent, a dihydric phenol. This glycoside is stored in one of the paired colleterial glands of the female reproductive system.

Although the eggs of all cockroaches are enclosed within oothecae composed of sclerotin, the process of oothecal formation has been most thoroughly studied in *P. americana* and *Blattella germanica*. A β-glucosidase controls the synthesis and formation of the sclerotin in the genital vestibulum as follows: the left colleterial gland secretes soluble proteins, 3,4 dihydroxybenzoic acid or alcohol as the 4-*O*-β-D-glucoside, and a phenol oxidase. The right colleterial gland secretes a β-glucosidase which hydrolyzes the glucoside when the secretions mix in the genital vestibulum. The liberated tanning agent is then oxidized to a *o*-quinone which combines with the N-terminal groups of the secreted proteins to form the hard and dark sclerotin of the ootheca. Evidence has been presented that the bulk of the phenolic glucoside which accumulates in the left colleterial gland of *P. americana* is synthesized in the fat body and then transported to the gland. The discovery of the sequence of reactions which result in the formation of the cockroach ootheca has led to a clearer understanding of the process of insect integumental sclerotization in general.

C. Glycosidic Pigments and Cardiac Glycosides

The color of insects largely depends upon the presence of noncarbohydrate pigments including sclerotin, melanin, ommochromes, pterins, carotenoids, and bile pigments present in the integument or internal tissues, especially the hemolymph and fat body (see Chapter 6). However, glycosidic pigments contribute to the protective coloration of insects in the suborder Homoptera. Based on pigment composition, this suborder can be divided into the arylglycoside-containing Aphidoidea and Coccoidea, and the pterin-containing remainder. Aphids, scale insects, and mealybugs commonly contain pigments which are glycosides derived by the coupling to naphthalenic and anthracenic phenols, and structurally related quinones. These pigments consist of β-D-glucosides or 6-acetyl-β-D-glucosides. Aphid glucosidic pigments are all related compounds and have been classed as aphins and aphinins. Aphins give a dark body color. For example, the hemolymph of the black bean aphid, *Aphis fabae,* contains a brownish yellow protoaphin, ($C_{36}H_{38}O_{16}$), which is purple in alkaline solution. After death this protoaphin is converted to a fluorescent yellow aglycone, xanthoaphin, ($C_{30}H_{26}O_{10}$), except when tissue β-glucosidase is inactivated. Aphinin, a blue-green glucoside with an identical empirical formula, is also readily converted to an aglycone. Aphinin may occur alone in the hemolymph, or in combination with aphin, frequently protoaphin, and its titer varies according to the species, climatic conditions, and season.

Several brightly colored insect species have been shown to store glycosides from their plant hosts that prove toxic to potential predators, especially birds. Other species have evolved superficial resemblances to these nonpalatable species and through this capacity for mimicry are assumed to derive a high degree of protection from predation. Warningly-colored (aposematic) insects have perhaps been most thoroughly studied among species which feed on milkweed plants (Asclepiadaceae). For example, the large milkweed bug, *Oncopeltus fasciatus,* is brightly colored because it contains several pterin pigments, including xanthopterin, erythropterin, and biopterin. The species is, however, protected from predation because it contains cardiac glycosides (cardenolides) sequestered from its host, milkweed. The cardenolides are potent heart drugs and cause severe vomiting if they are consumed by predators. Both larvae and adults store cardenolides and in the adult most of the deposits are located in the dorsolateral spaces of the thorax and abdomen. If the larvae feed on sunflower seeds they do not contain cardenolides and are not emetic to bird predators, thereby demonstrating the plant origin of the toxins. Similarly, larvae of the monarch butterfly, *Danaus plexippus,* store cardenolides from milkweed plants. If sufficient cardenolides are consumed during the larval stage, then larval, pupal, and adult tissues are all emetic and bird predators vomit immediately after their ingestion. The cardenolides ingested from the milkweed plant have been shown to differ in emetic potency, and are not distributed uniformly within the body or between sexes. Abdominal cardenolides have a higher emetic potency than those present in the remainder of the body, and females usually are more emetic than males.

D. Glycosaminoglycans in Connective Tissues and Cement

There are relatively few connective tissues found in insects because the integument provides the principal support for the body organs. Ashhurst and Costin (1976) have demonstrated that thin connective tissues containing acid and neutral glycosaminoglycans (mucopolysaccharides), frequently covalently linked to protein (proteoglycans) and associated with collagen, surround insect organs and support them in the hemocoel. Basement membranes form a continuous lining of the hemocoel and sometimes additional membranous strands run between organs. While extensive analyses of the chemical nature of the glycosaminoglycans have not yet been undertaken, hyaluronic acid has been found to be present in basement and peritrophic membranes of the wax moth, *Galleria mellonella,* and glycosaminoglycans are widely distributed among larvae of *M. domestica* and *P. regina*. In addition, chondroitin and dermatan sulfates, and possibly keratan sulfate and neutral proteoglycans, are present in the collagenous connective tissue which surrounds the ejaculatory duct of the male *L. migratoria*. Whole body analyses of the adult blow fly, *Calliphora erythrocephala,* have shown that hyaluronic acid, chondroitin, and chondroitin sulfate are present, but dermatan sulfate is absent. The existence of collagenous connective tissues in insects is now well established, but much remains to be learned about the nature and distribution of insect glycosaminoglycans and proteoglycans.

The histochemistry of the fibrous connective tissue surrounding the central nervous system of Orthoptera has been studied in more detail than of other orders, and has been shown to contain several glycosaminoglycans. For example, the neural lamella of the central nervous system of adults of *L. migratoria, P. americana* and the stick insect, *Carausius morosus,* appears to contain a mixture of chondroitin, dermatan, and keratan sulfates. However, only hyaluronic acid has been detected in the glial lacunar system of the central nervous system of these species. The glial lacunar system forms extracellular channels around the neuropile of the ganglion, and the hyaluronic acid in the system may bind cations, especially sodium ions, and thereby assist in the maintenance of a favorable ionic environment in the ganglion. In contrast, the sulfated glycosaminoglycans present in the neural lamella may act as a molecular sieve, by retarding the passage of large molecules with anionic charges, and thereby contribute to the selective permeability of the neural sheath.

Insect glycosaminoglycans and proteoglycans also function as a cement which is used internally to bind cells together, or externally to glue down puparia. For example, these polymers cement hemocytes together to encapsulate parasites and foreign bodies in the hemocoel. Once the foreign body is encapsulated no further defensive reactions occur, leading Salt (1970) to suggest that glycosaminoglycans are involved in conferring some immunity to insects. Since immunoglobulins have not been found in insects, Salt believes that glycosaminoglycans could very well control cellular defense reactions in insects. However, experimental evidence to show such an involvement is still lacking. Glycosaminoglycan production has also been well established in the larval salivary glands of *Drosophila* spp. The glycosaminoglycans which are secreted shortly before pupariation function as a glue which fixes the puparium to its substrate. The secretion of this glue appears to be stimulated by the presence of β-ecdysone in the hemolymph immediately prior to pupariation. To date, most of our limited information about insect glycosaminoglycans has been obtained using histochemical techniques. Additional biochemical studies are needed to confirm the chemical identity of these compounds.

VIII. CARBOHYDRATES IN INSECT METABOLISM

A. Synthesis, Nature, and Storage of Carbohydrate Reserves

In most insects carbohydrate reserves are present as glycogen and trehalose which can be readily converted into glucose. In addition, various amounts of glycoproteins may be present, especially in the hemolymph. The processes of gluconeogenesis (i.e., glucose synthesis from noncarbohydrate substrates) and fatty acid synthesis from glucose will also be discussed in this section.

1. Glycogen Synthesis

Glycogen, a branched chain polysaccharide composed of α-D-glucopyranose residues, represents the major reserve polysaccharide of insects. The molecular weight of glycogen ranges from about 10^6 to 10^8. The straight-chain portion of the glycogen molecule is made up of α-D-glucopyranose residues linked through 1,4-glucosidic bonds with branching occurring in the molecule as a result of 1,6-glucosidic bond formation. Since only twelve to eighteen glucose residues are present between branching points, glycogen is a highly branched molecule.

The reactions involved in the synthesis of glycogen from glucose in insects are similar to those found in vertebrates. Glycogen is synthesized when glucose units are donated from UDPglucose to a 1,4-α-glucan chain primer in a reaction catalyzed by glycogen synthase. Glucose units are added through 1,4-glucosidic bonds to the nonreducing end of the primer. In conjunction with this chain elongation a "branching" enzyme catalyzes the transfer of a 1,4-α-glycan chain to a similar chain through the formation of 1,6-glucosidic bonds, thereby producing the branched structure of glycogen. The series of reactions involved in glycogen synthesis from glucose is summarized as follows:

$$\text{ATP} + \text{D-glucose} = \text{ADP} + \text{D-glucose 6-phosphate} \quad \text{(hexokinase)} \tag{9}$$

$$\text{D-Glucose 6-phosphate} = \text{D-glucose 1-phosphate} \quad \text{(phosphoglucomutase)} \tag{10}$$

$$\text{UTP} + \text{D-glucose 1-phosphate} = \text{pyrophosphate} + \text{UDPglucose} \quad \text{(glucose-1-phosphate uridylyltransferase)} \tag{11}$$

$$\text{UDPglucose} + (1,4\text{-}\alpha\text{-D-glucosyl})_n = \text{UDP} + (1,4\text{-}\alpha\text{-D-glucosyl})_{n+1} \quad \text{(glycogen synthase)} \tag{12}$$

Glycogen reserves are most abundant in insect fat body, flight muscles, and intestinal tissues (see Table V). To date, the enzymes involved in glycogen synthesis have been studied most thoroughly in the insect fat body. For example, studies conducted on glycogen synthesis in the larval fat body of the silkmoth, *Hyalophora cecropia,* and *A. mellifera* have shown that glycogen synthase is bound to glycogen particles in the cytosol.

Electron microscopic analyses have shown that insect glycogen can occur either as single β-particles or in aggregations (α-particles, also called glycogen rosettes) depending upon individual species. For example, while glycogen has been found to be present as rosettes in the flight muscles of *P. regina,* most of the glycogen present in the flight muscles of the black fly, *Simulium vittatum,* is present as β-particles. Insect glycogen may also be covalently linked to proteins or polypeptides. For example, the inner portion of glycogen molecules extracted from whole pupae of *C. erythrocephala* was found to contain polypeptides which appear to retard glycogen hydrolysis by α-amylase.

Although the insect fat body, flight muscles, and intestine provide the most important glycogen deposits, substantial amounts can also be found in other tissues with the notable exception of the hemolymph. Glycogen deposits appear to provide an important energy source for the central nervous system. For example, glycogen has been detected in the perineurial cells, interaxonal glial membranes, and neuropile of the nerves of *P. americana.* The extent of glycogen deposits varies between species and depends further on the developmental stage and activity level of the insect. To illustrate, the flight muscles of Diptera typically contain substantial deposits of glycogen, but the flight muscles of the tsetse fly, *Glossina morsitans,* represent an important exception. In this species, flight muscles oxidize the amino acid, proline, for energy and contain only 1% glycogen on a fresh weight basis. In contrast, the intestinal tissues of *G. morsitans* oxidize glycogen to provide energy for digestion and absorption and contain fourteen times more glycogen than the flight muscles.

The mobilization of glucose and trehalose from glycogen reserves is under metabolic and hormonal regulation. The phosphorolytic and hydrolytic enzymes involved and the extent of the controls are described in Section VIII, B, D.

2. Trehalose Synthesis

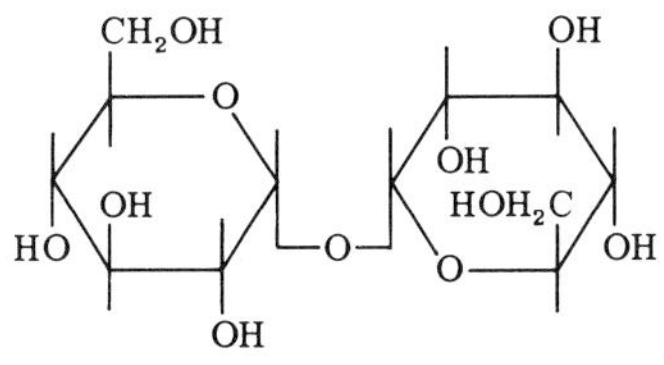

CH_2OH OH O OH OH HOH_2C HO O O OH OH

Structure of α, α-trehalose
(α-D-glucopyranosyl-α-
D-glucopyranoside)

The naturally occurring isomer of trehalose, α,α-trehalose, was first identified as an important reserve carbohydrate of insects in 1956 when Wyatt and Kalf isolated and identified the trehalose present in the hemolymph of the silkmoth, *Telea polyphemus*, and Howden and Kilby made a similar determination using *S. gregaria*. Since then, this nonreducing disaccharide of glucose has been shown to occur in many other insects as well (see Wyatt, 1967).

Although trehalose is the predominant sugar in most insect species which have been studied to date, in others it is present only at a low concentration. For example, the trehalose concentration in the hemolymph of the dermapterans, *Anisolabis littorea,* and *Forficula auricularia* is $\leq$ 2 m*M*. Trehalose has also been found in other organisms where it may be a free reserve sugar or a component of various glycolipids. Other trehalose-containing organisms besides insects are bacteria, algae, yeasts, fungi, nematodes, annelids, and crustaceans. A few ferns and seed plants have also been shown to contain trehalose (Elbein, 1974).

As in the case of glycogen synthesis, trehalose synthesis is an energy-requiring process involving the incorporation of glucose donated from UDPglucose. D-glucose is transferred from UDPglucose to glucose 6-phosphate to form the 6-phosphate of α,α-trehalose and UDP in a reaction catalyzed by α,α-trehalose-phosphate synthase. Free trehalose is liberated from trehalose 6-phosphate by the action of a phosphatase. Trehalose is therefore synthesized from glucose or fructose as follows:

ATP + D-glucose = ADP + D-glucose 6-phosphate (hexokinase) (13)

ATP + D-fructose = ADP + D-fructose 6-phosphate (hexokinase) (14)

D-Fructose 6-phosphate = D-glucose 6-phosphate (glucose-phosphate isomerase) (15)

UDPglucose + D-glucose 6-phosphate = UDP + α,α-trehalose 6-phosphate (α,α-trehalose-phosphate synthase) (16)

$$\text{Trehalose 6-phosphate} + H_2O = \text{trehalose} + \text{orthophosphate} \quad (17)$$
$$\text{(trehalose-phosphatase)}$$

Candy and Kilby in 1959 first demonstrated the presence of this biosynthetic pathway in the fat body of *S. gregaria*. Since then this pathway has been shown to exist in several other species, including *L. migratoria, H. cecropia,* and *P. regina*. The synthesis has been shown to take place in the cytosol of the fat body cells. Although by far the highest rate of trehalose synthesis occurs in the fat body, in some species muscle and intestinal tissues may synthesize lesser amounts of trehalose (see Bailey, 1975).

Trehalose is an important reserve disaccharide because it is readily hydrolyzed to glucose, which is in turn oxidized to provide energy, especially for insect flight. Although trehalose is widely distributed among tissues, its highest concentration is normally found in insect hemolymph. The concentration of trehalose in the hemolymph usually ranges from 23 to 175 m*M* (8 to 60 mg/ml), depending upon the species, developmental stage, and sex (Wyatt, 1967). The high concentrations of trehalose in the hemolymph provide several advantages to the insect which appear to more than compensate for the high energy cost of its synthesis. A high concentration of trehalose provides adaptive advantages such as (1) few nonspecific glucosylations because the aldehyde group of glucose is masked, (2) lower osmotic effects than from equivalent concentrations of glucose, and (3) a promotion of glucose absorption by facilitated diffusion. A high concentration of trehalose further ensures that local fluctuations in its concentration do not limit the rate of flight metabolism. Trehalose concentrations in the hemolymph are regulated homeostatically by metabolic and hormonal controls operating on its synthesis and hydrolysis (Section VIII, D).

Some investigators have examined hemolymph trehalose concentrations for cyclical changes which might indicate an underlying circadian rhythmicity in the synthesis or metabolism of trehalose. Endogenous rhythms that have a period of about 24 hr and that persist even in the absence of external synchronizers are classed as circadian rhythms. Nowosielski and Patton (1964) first demonstrated an apparent circadian rhythm in the hemolymph trehalose concentration of the house cricket, *Acheta domesticus*. A similar rhythm may also be present in *P. americana*. Evidence has been presented which shows a maximum concentration of hemolymph trehalose in the second half of the scotophase when adults were held under short day conditions (12 hr L : 12 hr D). This trehalose concentration declines rapidly just before the onset of the photophase and does not peak again until the succeeding scotophase. Whether these fluctuations represent a true circadian rhythm under endogenous control has not been finally determined. However, the maximum concentration of trehalose does not coincide with feeding or activity cycles. Although a circadian fluctuation in the titer of some cellular components, such as nucleic acids, may be associated with the mechanism of the biological clock, any circadian changes in trehalose concentration are more likely to be a by-product of the primary driving mechanism.

3. Glycoproteins

In addition to the proteoglycans present in their connective tissues, insects contain various amounts of glycoproteins in other tissues, especially in their hemolymph. The presence of hemolymph glycoproteins has been demonstrated in several species. For example, the larval hemolymph of *D. grandiosella* and the larval and adult hemolymph of the blow fly, *Calliphora stygia,* have been shown to contain two glycoproteins, which are separable by electrophoresis. In addition, the major protein in the hemolymph of larvae of the Japanese beetle, *Popillia japonica,* has been shown to be a lipoglycoprotein which contains mannose and glucose as its main carbohydrate components. Two glycoprotein fractions have also been isolated from the hemolymph of larvae of *P. americana*. Since these glycoproteins contain mannose and glucosamine, one of their functions may be to serve as an amino sugar reserve for chitin synthesis. More recently, the stonefly, *Pteronarcys californica,* has also been shown to contain a glycoprotein which binds the heavy metal, cadmium. Since stonefly species which are insensitive to cadmium contain significantly more glycoprotein than those which are sensitive, the conjugate may function in detoxication.

While insect glycoproteins have not yet been studied extensively, it is likely that they have important functions in insect metabolism. Glycoproteins may function as storage molecules, enzymes, or to protect their constituent monosaccharides from enzymatic attack. A transport function is also possible, but the carbohydrates which are commonly present in insect hemolymph are freely soluble in water. Since glycoproteins present in the blood of antarctic fish have been shown to lower the freezing temperature and thereby protect the fish from death, it would be of special interest to determine whether hemolymph glycoproteins present in diapausing insects exposed to subzero temperatures have similar properties.

4. Gluconeogenesis

In insects, gluconeogenetic reactions are those which convert metabolites such as lactate, glycerol, and glucogenic amino acids including serine, glycine, alanine, and glutamic acid to glucose. Gluconeogenesis is important because it meets the needs for glucose when the dietary carbohydrate supply is limiting, and it clears the hemolymph of metabolites such as lactate and glycerol which are not as readily oxidized as glucose. Although gluconeogenetic reactions in insects require much further study, they have been demonstrated in several species including the mosquito, *Aedes taeniorhynchus*, and *S. gregaria* (Bailey, 1975). The enzymes involved have been found in fat body and muscle. A simple reversal of glycolysis is not possible because the reactions catalyzed by the enzymes phosphofructokinase and pyruvate kinase are essentially irreversible. During gluconeogenesis phosphoenolpyruvate resynthesis from pyruvate is catalyzed by the enzymes pyruvate carboxylase and phosphoenolpyruvate carboxykinase, and fructose 6-phosphate resynthesis is catalyzed by the enzyme, hexosediphosphatase. Pyruvate kinase also functions to regulate the relative rates of glycolysis and gluconeogenesis. In the muscle and fat body of adults of the cricket, *A. domesticus*, pyruvate kinase has been shown to be activated by fructose 1,6-diphosphate and magnesium ions, but to be inhibited by alanine and ATP.

In contrast with microorganisms and plants, insects do not carry out gluconeogenesis from lipid substrates because the glyoxylate cycle is either totally or partially inoperative. In insects,the key enzymes in the glyoxylate cycle, namely, isocitrate lyase and malate synthase, appear either to be absent or to be present in very low concentrations.

5. Lipid Synthesis from Glucose

The insect fat body contains the enzymes systems which are necessary to convert glucose into fatty acids and glycerol which are then in turn incorporated into neutral and phospholipids, especially triglycerides, phosphatidylcholine, and phosphatidylethanolamine. The extent of this conversion of carbohydrate to lipid depends upon the chemical nature of the nutrient reserves, the developmental stage, the nutritional state, and the activity level of the insect. For example, experiments have shown that larvae of *B. mori*, which store large amounts of triglycerides, incorporate as much as 8.4% of labeled glucose into palmitic, palmitoleic, stearic, and oleic acids, but pharate adults of *H. cecropia* incorporate only 2.5% of labeled glucose into glycerides, mostly into the glycerol moiety of triglycerides. Lipid synthesis from glucose has been compared in resting and active adults of the blow fly, *Phaenicia sericata*, and shown to be controlled by the insect's locomotory demands for glucose. Mathur and Yurkiewicz (1969) showed that 1% of the label of [U-^{14}C]glucose was recovered in lipids following 3 hr of rest after the injection. In contrast, only 0.5% of the radiolabel was recovered in lipids following 3 hr of continuous flight after the injection. Flight activity, therefore, decreased the rate of conversion of glucose to lipid in *P. sericata*.

Several workers have studied the enzyme systems involved in the conversion of glucose to fatty acids. For example, Walker and Bailey have shown that lipogenesis from glucose is tightly linked to citrate metabolism in the cytosol of the fat body cell (see Bailey, 1975). Glucose is oxidized to acetyl-CoA and citric acid via enzymes of the glycolytic and tricarboxylic acid cycles. Mitochondrial citrate passes into the cytosol and is hydrolyzed by the citrate-cleaving enzyme, ATP citrate lyase, to yield oxaloacetic acid and acetyl-CoA. Fatty acids are synthesized from acetyl-CoA in the cytosol. Reducing equivalents (NADPH) for fatty acid biosynthesis are provided by the pentose phosphate pathway and $NADP^+$-dependent isocitrate dehydrogenase. An insect's ability to convert glucose into triglyceride reserves provides value because the oxidation of fatty acids generates more calories and metabolic water than does the oxidation of equivalent amounts of glucose or glycogen.

B. Mobilization of Carbohydrate Reserves

Glycogen and trehalose reserves are mobilized by enzymatically controlled reactions which generate glucose for the support of all life processes.

1. Glycogenolysis

Glycogenolysis refers to the phosphorolytic and hydrolytic reactions which degrade glycogen to its constituent glucose units. Phosphorolytic cleavage which generates glucose 1-phosphate from glycogen deposits is the most important reaction. The enzyme involved, glycogen phosphorylase, has been most thoroughly studied in insect fat body and muscle, but it is present in all tissues which contain glycogen deposits. The enzyme catalyzes the phosphorolytic cleavage of the α-(1,4)-glucosidic bonds of the reducing terminals of glycogen yielding glucose 1-phosphate as follows:

$$\underset{\text{orthophosphate}}{(1{,}4\text{-}\alpha\text{-D-glucosyl})_n + } = \underset{\substack{\text{glucose 1-phosphate} \\ \text{(glycogen phosphorylase)}}}{(1{,}4\text{-}\alpha\text{-D-glucosyl})_{n-1} + \alpha\text{-D-}} \quad (18)$$

Glycogen phosphorylase is an allosteric enzyme which exists in an active (*a*) and inactive form (*b*). The forms of the enzyme are interconverted as follows:

$$\text{Phosphorylase } a + 4\,H_2O = \underset{\text{(phosphorylase phosphatase)}}{2 \text{ phosphorylase } b + 4 \text{ orthophosphate}} \quad (19)$$

$$2 \text{ Phosphorylase } b + 4\,\text{ATP} = \underset{\text{(phosphorylase kinase)}}{\text{phosphorylase } a + 4\,\text{ADP}} \quad (20)$$

The rate of glycogenolysis is therefore controlled by the interconversion of phosphorylase *a* and *b* which is catalyzed by the enzymes phosphorylase phosphatase and phosphorylase kinase. The activity of phosphorylase kinase is controlled by the availability of ATP, cyclic AMP, magnesium ions, and a hyperglycemic hormone (Sacktor, 1975). The rate of phosphorolytic cleavage of glycogen is closely regulated through metabolic and hormonal controls to maintain trehalose homeostasis in the hemolymph (Section VIII, D).

Besides the well documented enzymatic phosphorolysis of glycogen, an α-amylase may also hydrolyze tissue glycogen. Doane and her colleagues (1975) have carried out a detailed study of the genetic control of amylase isoenzymes in the tissues of the pomace flies, *Drosophila melanogaster* and *D. hydei*. They have shown that although the larval midgut represents the major site of amylase synthesis, amylase isoenzymes are also present in the hemolymph, fat body, muscle, and reproductive system. The molecular characterization of these isoenzymes is providing valuable new information about the genetic control of insect enzymes at the translational level. In addition, an active extraintestinal amylase has been found in larvae of *A. aegypti*. This extraintestinal amylase accounts for up to two-thirds of the total tissue amylase activity. The precise function of extraintestinal α-amylase activity as of now remains obscure. It is possible that this enzyme, in conjunction with an α-glucosidase, hydrolyzes glycogen released from tissues during periods of histolysis.

2. Trehalose Hydrolysis

α,α-Trehalose is hydrolyzed to its constituent glucose units by the enzyme α,α-trehalase (α,α-trehalose glucohydrolase) which has been purified and characterized from several insect tissues. The reaction takes place as follows:

$$\alpha,\alpha\text{-Trehalose} + H_2O = 2\ \text{D-glucose}\ (\alpha,\alpha\text{-trehalase}) \quad (21)$$

While the enzyme has been found to be present in most tissues, it appears to be especially active in insect fat body, muscle, intestine, and salivary glands.

Table III shows that trehalase is widely distributed between larval, pupal, and adult tissues. This enzyme has been found to exist in two distinct forms in insect tissues: free in the cytosol (soluble trehalase), and bound to membranes of cell organelles (particulate trehalase). The soluble enzyme is found primarily in the intestine, whereas the particulate one is associated mainly with muscle mitochondria. Although the subcellular locations of the particulate trehalase need further study, it is becoming increasingly evident that the mitochondrial membrane is an important location. For example, cytochemical techniques have shown that trehalase is associated with the outer surface of the inner mitochondrial membrane of flight muscles removed from adults of the dipterans, *P. regina* and *S. bullata*. Since the outer membrane is permeable to sugars, both trehalose and glucose exchange readily with the cytosol and the space between the inner and outer mitochondrial membranes.

Trehalase has the important function of liberating glucose to provide energy for flight, especially among adult Diptera. Endogenous muscle tre-

TABLE III

Soluble Trehalase Activity in Various Tissues of Larvae, Pupae, and Adults of the Tobacco Hornworm, *Manduca sexta*[a]

	Mean α,α-trehalase activity (units/mg protein)[b]				
Tissue	Mid 5th larva	Mature larva	20-Day pupa	New ♀ adult	4-Day ♀ adult
Salivary glands	0.56	0.44	–	2.07	1.30
Midgut	0.74	0.26	–	0.17	0.09
Hindgut	1.57	1.41	–	–	–
Malpighian tubules	2.76	0.79	0.03	0.24	0.28
Hemolymph	0.20	0.06	0.10	–	–
Fat body	0.52	0.05	0.11	0.10	0.04
Muscle, abdominal	0.15	0.43	0.04	–	–
Muscle, thoracic	–	–	0.04	0.17	0.29
Ganglia, abdominal	1.26	0.99	–	0.35	1.14

[a]From Dahlman (1970).
[b]A unit was defined as the amount of enzyme which liberated 1 μmole of glucose from trehalose in 15 min. at 32°C. Data based on 2–4 replicates each from at least 6 insects.

halose, like that originating from the fat body and hemolymph, is hydrolyzed by soluble and particulate muscle trehalase. For example, the soluble trehalase in the muscles of the fleshfly, *Sarcophaga barbata,* has been shown to be located in the hemolymph, trapped between muscles, and within the extracellular spaces.

Although the presence of a soluble trehalase has been detected in the hemolymph of several insects, its precise function remains obscure. Hemolymph trehalase may be inhibited during intermolt periods and therefore be unavailable to hydrolyze the substantial trehalose reserves in the hemolymph, and disturb the carbohydrate homeostasis. Evidence exists that the hemolymph trehalase of the adult *P. regina* originates from the midgut and is inhibited in the hemolymph by a metalloprotein. In immature insects, hemolymph trehalase may be activated during molting cycles, liberating glucose for uptake by tissues deficient in trehalose.

Since trehalose utilization is dependent on the presence of trehalase, the enzyme is present in all tissues which rely upon glucose generated from trehalose reserves to meet their energy needs. In addition, trehalase may function to regulate sugar absorption and excretion (Sections V and VI,A).

C. Glycolysis and the Pentose Phosphate Pathway

In insects the process of glycolysis, whereby sugars are metabolized through their phosphates to pyruvic acid, is similar to that of other organisms (Crabtree and Newsholme, 1975). Pyruvic acid is then converted to acetyl-CoA, which is oxidized to carbon dioxide and water through the tricarboxylic acid cycle and terminal oxidation. The conversion of 1 mole glucose to pyruvic acid yields a net 2 moles of ATP. The enzymology of glycolysis has been well studied in insect fat body and muscle where phosphofructokinase and pyruvate kinase have been shown to be key regulatory enzymes. The glycerol 3-phosphate shuttle is an important side reaction of glycolysis and permits a high rate of oxidative metabolism in insect flight muscles. The operation of this shuttle is described in Section X.

Although glycolysis provides the main pathway for the anaerobic conversion of glucose to pyruvate, the pentose phosphate pathway or hexose monophosphate shunt also operates in insects (Chefurka, 1965; Friedman, 1970). This pathway is not important for energy generation in insects, but it can result in the complete oxidation of glucose 6-phosphate to carbon dioxide and water without the involvement of glycolysis and the tricarboxylic acid cycle. In insects the main functions of the pentose phosphate pathway are to provide NADPH for fatty acid synthesis, pentose phosphate for nucleic acid synthesis, and triose phosphate for glycogen synthesis. It has been estimated that the pathway accounts for 35% of the total glucose catabolism in larvae of *B. mori,* 39% in adults of the two-striped grasshopper, *Melanoplus bivittatus,* 15% in adults of *O. fasciatus,* and 17% in adult males of *P. americana.* The relatively active pentose phosphate pathway found in *B. mori* and *M. bivittatus* may be especially important in promoting fatty acid synthesis in these species.

D. Regulation of Carbohydrate Metabolism

Enzymatic regulation via metabolites, ions, and hormones controls the rate of interconversion of glycogen, trehalose and glucose in insects. These controls have been most thoroughly studied in fat body and muscle preparations and are especially important in maintaining homeostasis of hemolymph sugars. Several enzyme activators and inhibitors have been identified which regulate the flow of substrates through glycogen and trehalose synthesis and degradation. Figure 4 summarizes some of the metabolic controls and illustrates the central role of glucose 1-phosphate and glucose 6-phosphate in controlling carbohydrate metabolism. Beyond these metabolic controls,a hyperglycemic (glucagon-like) hormone, and perhaps a hypoglycemic (insulin-like) hormone also regulate carbohydrate metabolism.

The presence of a hyperglycemic hormone originating from the neuroendocrine system of insects was first demonstrated by Steele in 1961 (see Steele, 1976). He showed that a saline extract of the corpora cardiaca of *P. americana* caused an increase in the titer of hemolymph trehalose and a concomitant decrease in fat body glycogen after it was injected into adult

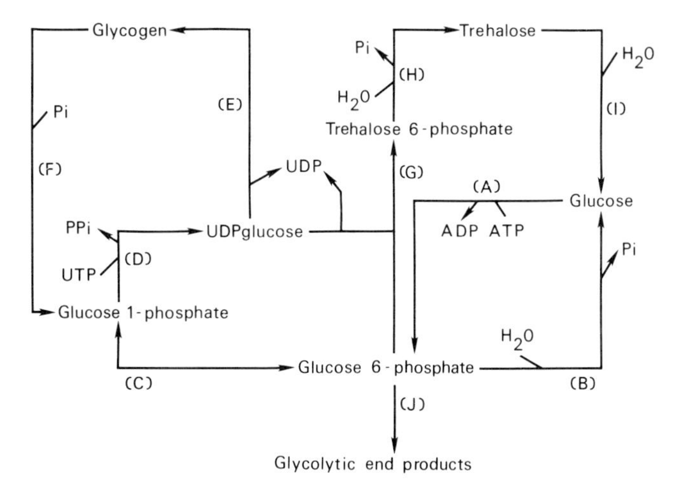

Fig. 4. A simplified diagram of the interactions of glucose, trehalose, and glycogen in the regulation of carbohydrate metabolism. Parenthetic letters refer to enzymes as follows: (A) hexokinase (inhibited by glucose 6-phosphate); (B) glucose-6-phosphatase (activated by trehalose); (C) phosphoglucomutase; (D) glucose-1-phosphate uridylyltransferase; (E) glycogen synthase (activated by glucose 6-phosphate); (F) glycogen phosphorylase (inhibited by ATP); (G) α,α-trehalose-phosphate synthase (activated by Mg^{2+}, inhibited by trehalose); (H) trehalose phosphatase; (I) α,α-trehalase (control mechanism as yet unknown); (J) glycolytic enzymes (control at phosphofructokinase and pyruvate kinase). (Redrawn from Sacktor, 1970.)

males. Since then, hyperglycemic activity has been demonstrated to exist in several other species including *L. migratoria, C. erthyrocephala,* and *A. mellifera*. The hormone has been partially characterized and shown to be a polypeptide secreted by the corpora cardiaca. It is believed to be synthesized in the intrinsic neurosecretory cells of the corpora cardiaca or in the cerebral neurosecretory cells. The hormone appears to act through cyclic AMP to activate glycogen phosphorylase (Section VIII, B, 1). Table IV illustrates the effect of cardiacectomy and allatectomy on the hemolymph trehalose concentration of adults of *C. erythrocephala*. The results show that corpora cardiaca removal or denervation causes a significant decrease in the trehalose concentration in the hemolymph. The main target of this hyperglycemic hormone is the fat body, where it promotes glycogen degradation leading to an increased rate of trehalose synthesis, thereby maintaining an optimum trehalose concentration in the hemolymph. The hormone may also act on nerve cord glycogen, but appears to have little effect on glycogen reserves in muscle and intestine. The generation of glucose 1-phosphate from glycogen reserves present in the nerve cord of *P. americana* has been shown to be controlled by an endogenous phenolic amine (octopamine) and the hyperglycemic hormone. Octopamine, like the hyperglycemic hormone, appears to activate glycogen phosphorylase by

TABLE IV

Effect of Cardiacectomy and Allatectomy on the Hemolymph Trehalose Concentration of Adults of the Blow Fly, *Calliphora erythrocephala*[a]

Treatment[b]	Number of adults	Mean hemolymph trehalose concentration (mg/ml ± SE)
Control, untreated	10	21.6 ± 1.3
Control, sham-operated	10	21.2 ± 1.1
Allatectomized	9	22.2 ± 2.1
Cardiacectomized	8	3.1 ± 0.3
Corpora cardiaca denervated	10	3.7 ± 1.0

[a]From Vejbjerg and Normann (1974).
[b]Operations performed 60 min before onset of flight. Hemolymph samples drawn after 45 min of flight.

increasing the rate of synthesis of cyclic AMP, and to be the primary glycogenolytic agent in the nerve cord. The glucose 1-phosphate liberated from glycogen remains in the nerve cord to provide an energy source for neural metabolism.

Evidence is beginning to accumulate that insects secrete a hypoglycemic hormone from their corpora cardiaca. For example, a radioimmunoassay has detected an insulin-like material in larvae of *D. melanogaster* and in adults of the tobacco hornworm, *Manduca sexta*. The hormone appears to cause a decrease in hemolymph trehalose concentration. As in vertebrates, this hypoglycemic hormone may function to facilitate the transport of glucose across cell membranes and to promote glycogen synthesis. The extent of hyper- and hypoglycemic hormones and their precise involvement in carbohydrate metabolism in the different orders of insects remain to be determined.

IX. CARBOHYDRATES IN METAMORPHOSIS

Metamorphic changes in both endopterygote and exopterygote insects are usually accompanied by substantial depletions of their carbohydrate reserves. During this period glycogen and trehalose supply glucose which provides an energy source and a substrate for the synthesis of pupal and adult tissues, especially the cuticle. In preparation for metamorphosis, insects usually attain their maximum carbohydrate content as mature nonfeeding larvae. At this time their carbohydrate reserves are present mainly as glycogen in the fat body and trehalose in the hemolymph. Currently, most is known about the utilization of nutrient reserves during the metamorphosis of Lepidoptera and Diptera, although even in these orders relatively few species have been examined in detail. Representatives which have been most extensively studied include the moths, *G. mellonella, H. cecropia,* and *B. mori,* and the blow flies, *Lucilia cuprina* and *P. regina.*

Table V illustrates the tissue distribution of those pure soluble carbohydrates which are present at the beginning of metamorphosis in larvae of the

TABLE V

Distribution of Carbohydrates between the Tissues of Fully Grown Larvae of the Wax Moth, *Galleria mellonella*[a]

Tissue	Number of observations	Mean concentration (mg/g wet wt. ± SD)		
		Glucose	Trehalose	Glycogen
Intestine	9	Trace	1.32 ± 0.39	0.56 ± 0.06
Hemolymph	9	0.08	15.10 ± 0.33	0.36 ± 0.09
Fat body	9	0.09	2.36 ± 0.30	5.38 ± 0.60
Carcass	9	Trace	1.50 ± 0.24	1.05 ± 0.18

[a]From Lenartowicz *et al.* (1967).

lepidopteran, *G. mellonella.* Since this fully grown larva contains only about 1% carbohydrate on a fresh weight basis, it must rely mainly upon lipids as an energy source during metamorphosis. These limited carbohydrate reserves are present mainly in the fat body as glycogen, and hemolymph as trehalose. During metamorphosis these reserves appear to be used as an immediate energy source and for the synthesis of the chitin present in the pupal and adult cuticle. These carbohydrate reserves are supplemented by the amino sugar (2-amino-2-deoxy-D-glucose) derived from the chitin of larval and pupal cuticle during the pupal and adult molts. Studies with another lepidopteran, *H. cecropia,* also suggest that the amino sugar derived from cuticular chitin contributes to the general carbohydrate pool during metamorphosis. In this species Bade and Wyatt determined that there was a decrease in the glycogen and trehalose content during the spinning period of the last larval instar, followed by an apparent increase at the time of the pupal ecdysis. The increase appeared to occur because soluble carbohydrates were synthesized from the amino sugar derived from chitin. An examination of the distribution of the label from [^{14}C]glucose injected into larvae of *H. cecropia* during the penultimate instar showed that most of the label was present in chitin at the end of the final larval instar. However, a significant amount of the label was transferred to glycogen, trehalose, and lipid during the pharate pupal stage (see Wyatt, 1967).

At the onset of metamorphosis dipterous larvae usually contain larger carbohydrate reserves than lepidopterous larvae. However, even among Diptera, little comparative information is available about the utilization of glycogen, trehalose, and the products of cuticular hydrolysis during metamorphosis. The few data which are available suggest that there are

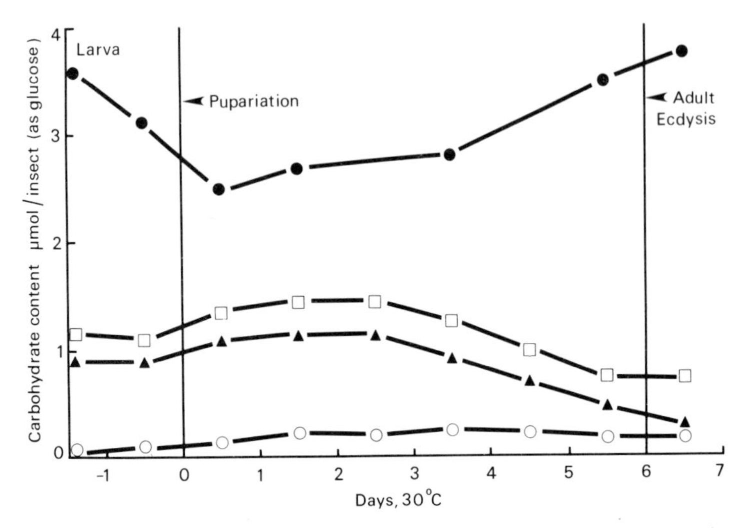

Fig. 5. Variations in the amounts of whole body carbohydrates during metamorphosis of the blowfly, *Lucilia cuprina*. Data are based on analyses of extracts of 20–30 insects at each stage. ●, chitin; □, total soluble carbohydrates; ▲, glycogen; ○, trehalose. (Redrawn from Crompton and Birt, 1967.)

considerable differences among the species. The data remain insufficient to allow for any broad generalizations at this time. For example, Fig. 5 illustrates the relationship which exists between structural and soluble carbohydrates during the metamorphosis of *L. cuprina*. In this insect, lipids and proteins rather than carbohydrates provide most of the energy requirements for metamorphosis. The carbohydrate content decreases by about 20% during pupariation and then remains relatively constant during the pharate adult stage. The synthesis of integumental chitin during the pharate adult stage takes place at the expense of glycogen. Together, these polysaccharides account for 94% of the total tissue carbohydrates. The small amount of trehalose present showed little change during pupariation or pharate adult development. Similar studies conducted by Wimer and his colleagues on metamorphosing individuals of the related *P. regina* also showed a decline in glycogen content during pharate adult development, but in this insect there was a sharp increase in trehalose content just before adult eclosion. During the pharate adult development of *P. regina*, glycogen appears to provide glucose units for chitin synthesis and for the synthesis of trehalose as a flight fuel in the newly eclosed adult. In another experiment the fate of uniformly labeled [^{14}C]glucose injected into mature larvae of *P. regina* was examined throughout metamorphosis. During this period the ^{14}C-label was metabolized as follows: respired carbon dioxide (41.5%), amino acids (31.7%), protein (8.9%), chitin (8.3%), soluble carbohydrate, mainly glycogen (7.6%), and lipid (3%). These results also suggest a substantial involvement of the carbohydrate reserves in energy production, chitin synthesis and other biosynthetic processes during metamorphosis (see Tate and Wimer, 1974).

Beyond the dependency of insects upon carbohydrate reserves to meet energy demands and to supply substrates for biosynthetic processes during metamorphosis, carbohydrates may also be secreted with silk in insect cocoons. During preparation for metamorphosis varying amounts of carbohydrates, including chitin, have been found mixed with the silk of insect cocoons. For example, larvae of the sciarid fly, *Rynchosciara americana*, secrete a communal cocoon which contains about 13% carbohydrate. Acid hydrolysis of this carbohydrate showed that mainly glucose, galactose, and galactosamine were present. In contrast, the cocoon of *B. mori* contains only about 1.5% carbohydrate, mainly as amino sugar, mannose, and galactose. Although much species variation appears to exist in the relative utilization of glycogen and trehalose reserves during metamorphosis, it is clear that metamorphosing insects rely on carbohydrates sequestered during the larval period as an energy source and as substrates for the synthesis of chitin and other cellular components.

X. CARBOHYDRATES IN FLIGHT MUSCLES

The initiation and maintenance of insect fight require a highly efficient oxidative metabolism which is fueled by carbohydrates and lipids. In the past, unnecessary emphasis has been placed on differentiating whether carbohydrates or lipids serve as the flight fuels in the various orders. Table VI compares hexokinase and carnitine palmitoyltransferase activities in flight muscles from several insect species. These enzyme activities provide an estimate of the relative rates of oxidation of glucose and fatty acids in the flight muscles. Table VI shows that many species of Hymenoptera, Diptera, and Orthoptera utilize carbohydrate predominantly as a flight fuel, whereas many species of Hemiptera, and other Orthoptera use a more balanced combination of carbohydrate and lipid. Until recently, the flight muscles of Lepidoptera were believed to rely exclusively upon lipid as an energy source, but it has now been shown that the flight muscles of some moths have a hexokinase and a phosphofructokinase which are as active as those in the flight muscles of Diptera and Hymenoptera.

Insects may use different energy sources according to the stage of flight. For example, although the oxidation of fatty acids provides the energy for sustained flight in locusts, carbohydrate is used as a readily mobilized substrate to initiate flight. Recently Jutsum and Goldsworthy (1976) have shown that adult males of *L. migratoria* rely upon energy derived from the oxidation of carbohydrates during the first 30 min of flight, when hemolymph trehalose is oxidized at a rate of 120 μg/min. After 30 min of flight, energy

TABLE VI

Comparison of the Activities of Key Enzymes in Glycolysis and Fatty Acid Oxidation Present in the Flight Muscles of Insects from Five Orders[a]

Order	Genus and species	Enzyme activity (μmole/min/g muscle at 25°C)[b] Hexokinase	Carnitine palmitoyltransferase	Ratio
Orthoptera	*Locusta migratoria*	8.0	6.3	1.3
Orthoptera	*Periplaneta americana*	18.0	0.18	100.0
Orthoptera	*Schistocerca gregaria*	11.5	4.9	2.4
Lepidoptera	*Noctua pronuba*	32.0	2.0	16.0
Lepidoptera	*Plusia gamma*	50.0	2.0	25.0
Lepidoptera	*Vanessa urticae*	4.8	1.5	3.2
Hemiptera	*Lethocerus cordofanus*	4.0	6.2	0.6
Hymenoptera	*Apis mellifera*	29.0	0.35	82.9
Hymenoptera	*Bombus hortorum*	114.0	1.3	87.7
Diptera	*Calliphora erythrocephala*	35.0	0.2	175.0
Diptera	*Protophormia terranova*	14.0	0.3	46.7
Diptera	*Sarcophaga barbata*	17.0	0.4	42.5

[a]From Crabtree and Newsholme (1975).

[b]The activity of hexokinase is assumed to represent the maximum capacity for the use of glucose and that of carnitine palmitoyltransferase for the use of fatty acids. For comparison under conditions of equivalent rates of ATP production, the activity of carnitine palmitoyltransferase has been multiplied by 3.5/2. The factor 3.5 arises since the oxidation of 1 mole of fatty acid produces approximately 3.5 times more ATP than does the oxidation of 1 mole of glucose. The factor 0.5 arises since the enzyme participates at two points in the overall pathway.

was shown to be provided mainly by fatty acid oxidation, as carbohydrate oxidation fell to 9% of its initial rate.

Most adult Diptera rely upon carbohydrate reserves to provide energy for the maintenance of flight. A hundredfold increase in glycolytic flux has been shown to occur at the onset of flight of *P. regina*. However, obligatory blood-feeding Diptera use the amino acid, proline, as the principal substrate for oxidative metabolism to maintain flight. Since vertebrate blood contains a high protein and extremely low carbohydrate concentration the oxidation of proline represents a metabolic adaptation of the flight muscle to a substrate derived from consumed blood. Bursell (1975) has compared the capacity of the flight muscles of several species of Diptera to oxidize pyruvate and proline. As a result of this study the species were divided into three categories: those adults which do not feed on blood, or in which females only are blood feeders, oxidize pyruvate; those which are facultative blood feeders oxidize both pyruvate and proline; and those species of *Glossina* which are obligatory blood feeders, oxidize proline to provide energy for flight. Other biochemical analyses have shown that the mitochondria isolated from the flight muscles of the obligatory blood-feeding *Glossina* spp. oxidize proline at a sufficiently high rate to meet the energy requirements for flight.

An active glycerol 3-phosphate* shuttle operates in the flight muscles of those insects which rely upon carbohydrate substrates to meet their energy demands for sustained flight. The key reactions in the shuttle are illustrated in Fig. 6. This shuttle, which was first demonstrated by Zebe and McShan in

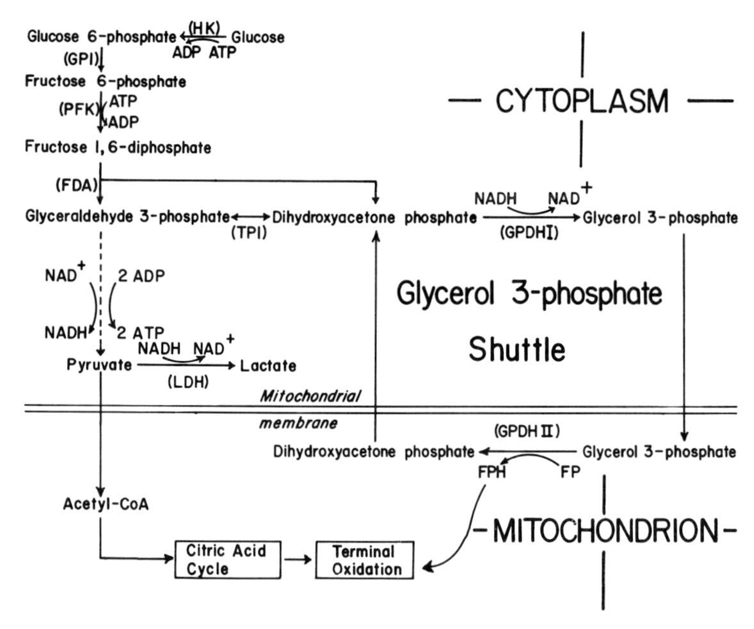

Fig. 6. A simplified diagram of the involvement of the glycerol 3-phosphate shuttle in carbohydrate metabolism in insect flight muscles. The arrows represent the usual direction of flux in the system. Parenthetic letters refer to enzymes as follows: FDA, fructose-diphosphate aldolase; GPDH I, glycerol-3-phosphate dehydrogenase (NAD^+ coenzyme); GPDH II, glycerol-3-phosphate dehydrogenase (flavoprotein (FP) coenzyme); GPI, glucosephosphate isomerase; HK, hexokinase; LDH, lactate dehydrogenase; PFK, phosphofructokinase; TPI, triosephosphate isomerase.

1957, functions primarily to keep NAD^+ available for glycolysis (see Sacktor, 1975). Since glycolysis takes place in the cytoplasm and terminal oxidation takes place in the mitochondria, and the pyridine nucleotides, e.g.,

*Glycerol 3-phosphate is the name applied to the naturally occurring stereoisomer of glycerolphosphate, L-α-glycerolphosphate, using the stereospecific numbering system described in the report of the IUPAC-IUB Commission (1967). Accordingly, the more specific term, glycerol 3-phosphate shuttle, is substituted for α-glycerophosphate cycle.

NAD^+, NADH) do not penetrate the inner mitochondrial membrane, an indirect mechanism (i.e., the glycerol 3-phosphate shuttle) is used to transfer reducing equivalents into the mitochondrion. The shuttle, therefore, permits a high rate of reoxidation of NADH, and the complete oxidation of glucose to carbon dioxide and water without any accompanying accumulation of lactate. Although the reduction of pyruvate to lactate reoxidizes NADH, the lactate so formed is not reoxidized in the mitochondria, and glycolysis rapidly becomes "uncoupled" from terminal oxidation. In contrast, glycerol 3-phosphate is reoxidized in the mitochondria through the glycerol 3-phosphate shuttle which links glycolysis with terminal oxidation. Flight muscles with an active glycerol 3-phosphate shuttle are characterized by the presence of well developed mitochondria (sarcosomes) and a complex network of oxygen-supplying tracheoles.

Once an insect has attained a steady state during flight, an active glycerol 3-phosphate shuttle enables muscle mitochondria to oxidize pyruvate and glycerol 3-phosphate as rapidly as they are formed. Most or all of the NADH produced during glycolysis is reoxidized through the shuttle. The NADH formed in the oxidation of glyceraldehyde 3-phosphate to 3-phospho-D-glyceroyl phosphate, which is catalyzed by glyceraldehyde-phosphate dehydrogenase, is reoxidized by serving as the cofactor for cytosolic glycerol 3-phosphate dehydrogenase (GPDH) which catalyzes the reduction of dihydroxyacetone phosphate to glycerol 3-phosphate. This reaction is linked to the mitochondrial respiratory chain because the glycerol 3-phosphate penetrates the outer mitochondrial membrane and is then reoxidized. The glycerol 3-phosphate is reoxidized by a separate flavoprotein-linked GPDH which is bound to the outer surface of the inner mitochondrial membrane. The reduced flavoprotein, after serving as the coenzyme for this mitochondrial GPDH, is then transported across the inner membrane to the respiratory chain. The dihydroxyacetone phosphate re-enters the cytosol and becomes available as a substrate for further oxidation of NADH, thereby completing the shuttle. The difference in redox potential between the NAD^+ coenzyme of the cystosolic GPDH and the flavoprotein coenzyme of the mitochondrial GPDH indicates that the main function of the shuttle is to transport reducing equivalents into the mitochondrion. The cycle therefore regenerates NAD^+ through a system tightly linked to glycolysis and the electron transport chain.

The absence or low titer of lactate dehydrogenase (LDH) and the presence of a high titer of GPDH in the flight muscles of species from several orders have been well established. Table VII illustrates the high GPDH to LDH ratio found in the flight muscles of young adults from three orders. It has been found that the specific activity of flight muscle GPDH rises sharply following adult eclosion. For example, Fig. 7 compares the specific activity of GPDH in flight and leg muscles of the grasshopper, *Schistocerca vaga,* during the

TABLE VII

Comparison of the Activities of Glycerol-3-Phosphate Dehydrogenase and Lactate Dehydrogenase Present in the Flight Muscles of Three Insect Species[a]

Order	Genus and species	Days after adult ecdysis	μmol/min/mg protein[b] GPDH	LDH	Ratio
Orthoptera	*Locusta migratoria*	4	0.49	0.01	49
Lepidoptera	*Philosamia cynthia*	3	0.19	0.02	10
Diptera	*Calliphora erythrocephala*	1	0.81	~0	–

[a]From Beenakkers *et al.* (1975).
[b]Mean enzyme activities calculated from at least five determinations at 25°C.

last larval and adult stages. The specific activity of GPDH in the powerful flight muscles (the longitudinal dorsal muscles and most of the lateral pterothoracic muscles) rises six- to eightfold and reaches its maximum level in the 5-day-old adult. In contrast, only a twofold increase in the specific activity of GPDH was found in the flexor and extensor muscles of the metathoracic leg during the same period. Unlike the flight muscles, these leg muscles are incapable of sustaining forceful contractions for long time intervals, and have a less active glycerol 3-phosphate shuttle.

The biochemical and genetic characteristics of both cytosolic and mitochondrial GPDH have been examined. Cytosolic GPDH has been isolated in crystalline form from the thorax of adult worker bees, *A. mellifera*. How-

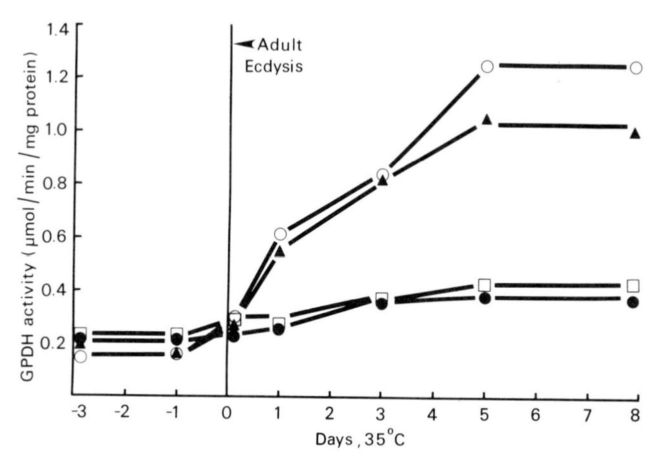

Fig. 7. Activity of glycerol-3-phosphate dehydrogenase in flight and leg muscles of the grasshopper, *Schistocerca vaga*, during the last larval and adult stages. Enzyme activity is expressed as μmoles substrate reduced per min per mg muscle protein at 37°C. Values are means of determinations from two insects, range <±10% mean. ○, dorsal longitudinal flight muscle; ▲, lateral pterothoracic flight muscle; □, metathoracic flexor leg muscle; ●, metathoracic extensor leg muscle. (Redrawn from Brosemer, 1965.)

ever, to date, most information about the characteristics of GPDH is available for *Drosophila* spp. (Bewley *et al.*, 1974; Collier *et al.*, 1976). The production of mutants of *D. melanogaster* which are deficient in cytosolic GPDH and incapable of flight has shown the importance of the glycerol 3-phosphate shuttle in flight metabolism. The enzyme has a homodimeric structure being made up of two subunits each with a molecular weight of about 32,000. The cytosolic GPDH of *D. melanogaster* has been separated into several isoenzymes which, although derived from the same structural gene, are present in different proportions in larval and adult tissues. The isoenzymes of GPDH differ in certain physicochemical properties. For example, they are separable by electrophoresis and have distinct pH optima. The cytosolic GPDH of the adult is present mainly in the thoracic flight muscles, whereas 50% of that present in the larva is found in the fat body. During the larval stages the enzyme may function in lipid metabolism by providing glycerol 3-phosphate for phosphatidic acid, phospholipid, and triglyceride synthesis. The physiological significance of the different isoenzymatic forms of GPDH still needs further clarification.

XI. CARBOHYDRATES IN REPRODUCTION AND EMBRYONIC DEVELOPMENT

Carbohydrates are necessary for the normal functioning of the male and female reproductive systems, as well as for the development of the embryo. In males, sugars form an important constituent of the testes and the seminal plasm. In females, the accumulation of carbohydrate yolk in the oocyte is an essential preparatory step for the successful development of the embryo.

A. Male and Female Reproductive Systems

As yet, little comparative information is available about the involvement of carbohydrates in the male reproductive system of insects because of the difficulty in obtaining uncontaminated tissue samples and an adequate amount of semen for analysis. Most is presently known about the nature and distribution of the carbohydrates in the male reproductive system of the drone bee, *A. mellifera*. Blum and his colleagues (1962) found that most of the carbohydrates of the reproductive system were present in the testes. Fructose and glucose were most abundant and made up 83% of the testicular carbohydrates. Within the semen most of the carbohydrates, which include fructose, glucose, and trehalose, were present in the plasm rather than the spermatozoa. Fructose was shown to be rapidly metabolized by the spermatozoa, and is therefore unlikely to be involved in their long-term storage in the spermatheca. Glucose and trehalose, together with amino acids, probably serve as an energy source for sperm maintenance in both the seminal vesicle and spermatheca. Since the sugars in the reproductive system were shown to be depleted during flight, they are in equilibrium with the hemolymph sugars and do not appear to be synthesized by the tissues of the reproductive system.

In the female system, carbohydrates are necessary for vitellogenesis and for the formation of the glycosaminoglycans present in the vitelline membrane and the chorion. Vitellogenesis involves the accumulation within the oocyte of carbohydrate, lipid, and protein yolk to meet the structural and metabolic needs of the developing embryo. A substantial portion of the yolk is derived from nutrient reserves of the fat body and hemolymph. Glycogen, which serves as the principal carbohydrate yolk, is usually synthesized in the ovary from glucose and trehalose derived from the fat body and hemolymph. In those insects where vitellogenesis occurs during the pharate adult stage, the carbohydrate yolk is laid down prior to eclosion at the expense of fat body glycogen and hemolymph trehalose. For example, Fig. 8 illustrates the fate of [^{14}C]glucose injected into the hemocoel of pharate adults of *B. mori*. The insect rapidly incorporated radioactive glucose into hemolymph trehalose and fat body and ovarial glycogen. A relatively stable distribution of the radioactivity was reached after 4 hr and ovarial glycogen was found to contain about eight times as much label as fat body glycogen. These findings imply that glycogen stored in the fat body during the larval

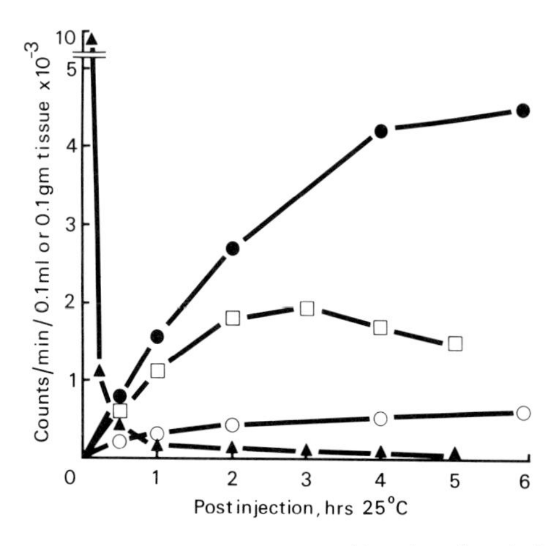

Fig. 8. Rate of incorporation of [^{14}C]glucose into hemolymph trehalose, and fat body and ovary glycogen of pharate adults of the silkworm, *Bombyx mori*. Injections of [^{14}C]glucose (0.1 μCi/insect) were made 5 days after the pupal–adult apolysis. Each point represents the mean activity of two replicates of pooled samples from five insects. ▲, hemolymph glucose (cpm/0.1 ml); □, hemolymph trehalose (cpm/0.1 ml); ○, fat body glycogen (cpm/0.1 gm fresh tissue); ●, ovary glycogen (cpm/0.1 gm fresh tissue). (Redrawn from Yamashita and Hasegawa, 1974.)

instars of *B. mori* is released into the hemolymph as trehalose and taken up by the developing oocytes in the pharate adult.

Glycogen is synthesized in insect ovaries during the terminal phase of vitellogenesis. For example, measurements of [^{3}H]glucose uptake by ovaries of the queen bee, *A. mellifera,* and *M. domestica* have shown that glycogen is synthesized in the ooplasm during the last stages of vitellogenesis. At the same time, little glycogen synthesis was detected in the trophocytes and follicle cells. Glycogen synthase was shown to be present in an inactive form during the early and mid phases of vitellogenesis, but could be activated by a 1–3-day exposure to 4°C. Since the enzyme was activated by exposure to a low-temperature, precocious glycogen synthesis may be inhibited by hormonal and metabolic controls until the terminal phase of vitellogenesis.

In addition to carbohydrate yolk, insect oocytes may also store carbohydrates as prosthetic groups of vitellogenic proteins. For example, the vitellogenin found in the hemolymph and oocytes of adult females of the cricket, *A. domesticus,* is a glycoprotein, and the purified vitellogenin of *B. germanica* has been shown to contain 8% carbohydrate and 7.6% lipid. Yamasaki (1974) has also shown that a glycoprotein is the principal carbohydrate and protein reserve in eggs of *L. migratoria.* In this species the glycoprotein makes up 77% of the dry weight of the eggs and contains 14% carbohydrate as mannose and glucosamine in a molar ratio of 1 to 7. Beyond this, vitellogenins sequestered by developing oocytes may be associated with glycosaminoglycans. Some evidence also exists that glycosaminoglycans released from the follicle cells of the Colorado potato beetle, *Leptinotarsa decemlineata,* react with hemolymph vitellogenins. The resulting complex may form a precipitate and then condense on the oocyte surface, where it may be treated as a foreign particle and be absorbed by pinocytosis. Although considerable information is available about the synthesis and composition of glycogen and glycoprotein yolk, the origin and nature of ovarial glycosaminoglycans have not yet been examined systematically.

B. Embryonic Development

During embryonic development, carbohydrate and lipid yolk provide the main substrate for energy production. Measurements of the respiratory quotient (i.e., the ratio of the volume of carbon dioxide released to the volume of oxygen taken up per unit time) of whole eggs have shown a general tendency for the quotient to decrease from about 1.0 to 0.7 during embryonic development, suggesting that carbohydrate yolk is utilized before lipid yolk. However, virtually all the yolk is utilized by the time hatching occurs, and first stage larvae can survive for only short periods without food.

In the developing egg, glycogen is mobilized as glucose and trehalose which are present together with small amounts of other sugars. For example, trehalose makes up about 90% of the total free sugars present in the early embryo of the grasshopper, *Aulocara ellioti*. Mannose, glucose, fructose, mannitol, and glycerol make up the remainder of the free sugars. Trehalose has also been detected in the embryos of *A. aegypti*, *C. morosus*, the grasshopper, *Melanoplus differentialis*, the oak silkmoth, *Antheraea pernyi*, and *B. mori*. Carbohydrate metabolism has also been examined in the embryos of the ovoviviparous Madeira cockroach, *Leucophaea maderae*. The embryos appear to synthesize glycogen during the early and mid stages of their development and mobilize it just before parturition. The substrates for this glycogen synthesis are derived from maternal tissues.

Besides serving as an important energy source, glycogen yolk also provides glucose for the synthesis of the chitin which is laid down in the cuticle of the first stage larva. Glycogen yolk also provides precursors for the synthesis of polyols, especially glycerol and sorbitol, in those species which undergo an embryonic diapause (see Section XII, B). In general, however, our information about carbohydrate metabolism during embryonic development is rather sketchy due mainly to difficulties in obtaining sufficiently large and homogeneous samples of developing eggs for biochemical analyses.

XII. CARBOHYDRATES IN DIAPAUSE AND COLDHARDINESS

Diapause is a state of developmental arrest which enables an insect to survive under adverse environmental conditions and synchronizes the active stages of its life cycle with available food sources. Depending on the species, diapause can intervene in the egg, larval, pupal, or adult stage. In those species which exhibit a facultative diapause, the state is instituted following the exposure of sensitive prediapausing stages to inductive environmental conditions, primarily short days and low temperatures in temperate zone species. Diapause is, therefore, a genetically controlled life phase for which preparatory biochemical adjustments, including the accumulation of lipid and carbohydrate reserves, and behavioral adaptations often including the location of a protective site, occur in advance. Once diapause intervenes, the respiratory rate falls dramatically and may be as low as 2% of the rate in prediapausing individuals.

A. Diapause Maintenance

Most diapausing insects rely upon reserves of triglycerides, glycogen, and trehalose, and to a lesser extent upon proteins and amino acids to support their metabolism. In diapausing larvae, pupae, and adults these reserves are accumulated mainly in the fat body and hemolymph during the active feeding period which precedes the onset of diapause. Glycogen and trehalose reserves appear to be most valuable during the early stages of diapause when they are utilized preferentially. They supply glucose to meet the limited energy demands, and serve as precursors for the synthesis of polyhydric alcohols. Polyols accumulate in the tissues of many diapausing insects and function by contributing to an increased coldhardiness. The long-term survival of diapausing insects usually depends upon an adequate lipid reserve. Large amounts of triglycerides are commonly available and their oxidation generates more calories per gram (9) than the oxidation of either carbohydrate (4), or protein (4).

The nature and quantity of the carbohydrate and lipid reserves present during diapause have been examined in several species. For example, newly diapaused larvae of *D. grandiosella* are known to contain about 52% lipid and 5% glycogen on a dry weight basis. Most of these reserves are present in the fat body, which contains about 11 mg lipid and 0.8 mg glycogen per larva. Similarly, newly diapaused adults of the coccinellid beetle, *Semiadalia undecimnotata,* have been shown to contain about 40% lipid and 2% glycogen on a dry weight basis. On the other hand, diapausing pupae of the cabbage butterfly, *Pieris brassicae,* accumulate substantial amounts of trehalose which presumably functions as a nutrient reserve and cryoprotectant. A maximum concentration of 12% trehalose and 4% glycogen on a dry weight basis is present in diapausing pupae, compared with respective values of 6% and 10% in nondiapausing pupae. The metabolism of diapausing insects is such that sufficient reserves remain available at diapause termination to provide energy for a postdiapause molt and other life processes until feeding resumes.

B. Polyols in Coldhardiness

Diapausing insects in temperate and cold regions are able to withstand low temperatures for relatively long periods through a process of coldhardening which usually takes place at the onset of the resting period. Although the precise mechanism of coldhardening is still being investigated, it appears that the increased production of polyols is the principal metabolic adjustment (see Baust and Morrissey, 1976). Polyols appear to protect by enhancing supercooling in those species where freezing would prove fatal (freezing-susceptible insects), and by minimizing freezing damage in freezing-tolerant insects. Both freezing-susceptible and freezing-tolerant insects are capable of supercooling. Insects supercool when their temperature falls below the crystallization equilibrium temperature of their tissues without accompanying ice formation. Insects can supercool about 30 to 35°C, and remain in this state for long periods if ice crystal nuclei do not form in their tissues. For insects to survive air temperatures of about -30°C they must have a low freezing point and supercool maximally, or be freezing-

diapause for one month at 4°C. Ziegler and Wyatt (1975) have demonstrated that exposure to low temperatures activates glycerol synthesis from fat body glycogen. The key regulatory enzyme appears to be glycogen phosphorylase, which is activated at low temperatures, but which is sensitive to feedback inhibition by high glycerol concentrations.

Over the last 20 years our understanding of the carbohydrate metabolism associated with the embryonic diapause of *B. mori* has been advanced by several Japanese investigators, most notably Chino, Hasegawa, Yamashita, Takahashi, and Kageyama. The polyols, sorbitol and glycerol, present in diapausing eggs have been shown to be formed from glycogen which is stored in the oocytes during the pharate adult stage. This storage of glycogen is controlled by a diapause hormone secreted by the neurosecretory cells of the subesophageal ganglion. The diapause hormone promotes glycogen synthesis in the ooplasm of the oocytes of the pharate adult by stimulating the enzyme, trehalase, thereby providing the necessary glucose substrate for glycogen synthesis (Yamashita and Hasegawa, 1976). The substrate for the ovarial trehalase originates from glycogen stored in the fat body. Once embryonic diapause intervenes, an increase in the activity of glycogen phosphorylase *a* results in glycogenolysis and subsequent polyol formation.

At the onset of the embryonic diapause of *B. mori* a substantial decrease in oxygen consumption takes place and about 60% of the stored glycogen is converted into sorbitol, glycerol, and lactate. It has been calculated by Kageyama (1976) that the relative rates of glycolysis and the pentose phosphate pathway in diapausing eggs are 62% and 38%, respectively. The operation of these two anaerobic pathways, combined with a low rate of terminal oxidation imposed by the low rate of oxygen consumption, appears to cause the accumulation of precursors for sorbitol and glycerol synthesis, including glucose 6-phosphate, fructose 6-phosphate, glyceraldehyde 3-phosphate, and NADPH. In addition, two types of $NADP^+$-specific polyol dehydrogenase that reduce sugars or sugar phosphates have been isolated from the eggs of *B. mori.* It appears, therefore, that polyol synthesis can be accounted for without assuming a deficiency in any specific enzymes in the electron transport chain. Whether this mechanism of polyol formation is found in other species undergoing an embryonic diapause remains to be shown. It does, however, illustrate that the diapause state is closely associated with changes in the relative rates of carbohydrate metabolism.

XIII. CARBOHYDRATES AND INSECT BEHAVIOR

A. Carbohydrates in Feeding Behavior

Carbohydrates present in food sources not only provide insects with necessary nutrients, but also may have a significant effect on feeding behavior (see Section III). Carbohydrates, as well as other chemical and physical factors, may be involved at all levels of the insect's feeding behavioral responses, including orientation as attractants or arrestants, biting and piercing as incitants, and the maintenance of feeding as stimulants. Many secondary plant substances occur naturally as glycosides, which have been shown to regulate insect feeding activity and therefore control host plant selection. For example, Verschaffelt in 1910 first reported that mustard oil glucosides were attractants for the cabbage butterflies, *P. brassicae* and *P. rapae* (see Schoonhoven, 1973). Since then mustard oil glucosides such as sinigrin and glucoheirolin have been shown to control the host plant selection of several other cruciferous feeding insects. Representatives can be found among several orders including the diamond-back moth, *Plutella xylostella,* the mustard weevil, *Phaedon cochleariae,* the flea beetles, *Phyllotreta* spp., and the aphid, *Brevicoryne brassicae.* Similarly, the oxasteroid glycosides present in milkweed tissues have been shown to serve as potent attractants for *O. fasciatus.*

Glucose, fructose, and sucrose have been shown to be highly active feeding incitants and stimulants for many insects, including larvae of *O. nubilalis, D. grandiosella,* the cabbage looper, *Trichoplusia ni,* and the cotton leafworm, *Spodoptera littoralis.* The relatively stimulatory activities and threshold concentrations of these three sugars vary across species. Other sugars may also function as feeding stimulants. For example, larvae of *S. littoralis* also respond positively to maltose, melibiose, and raffinose. In addition, a synergistic effect on feeding behavior between amino acids or fatty acids and sugars has been detected. For example, the feeding stimulatory effect of amino acids has been shown to be synergized by the presence of sugars in *O. nubilalis,* and the clearwinged grasshopper, *Camnula pellucida.* Similarly, linoleic, oleic, and palmitic acids act as feeding stimulants for larvae of the Indian meal moth, *Plodia interpunctella,* only in combination with sucrose.

The electrophysiology of insect reception of glycosides, sugars, and other regulatory chemicals has been studied in plant feeding lepidopterous larvae. Sensilla styloconica on the maxilla have been shown to serve as gustatory receptors. Table VIII compares the spectrum of taste reception of the four cells of each maxillary sensillum styloconicum of five species of Lepidoptera. The receptors respond to sugars (glucose, sucrose, sorbitol, inositol), salts, acids, and various secondary plant chemicals. Related compounds other than those listed also stimulate the receptors, but usually they must be present at high concentrations. The presence of both fructose and sucrose receptors probably enables larvae to obtain more detailed information about sugar mixtures than would be possible with only one type of sugar receptor.

Sugar solutions control the initiation and maintenance of feeding of adult Diptera through their feeding stimulatory and osmotic properties. For exam-

TABLE VIII

Taste Reception of the Four Cells of Each Sensillum Styloconicum of the Maxilla in Larvae of Five Species of Plant-Feeding Lepidoptera[a]

		Cells of maxillary sensilla styloconica[b]							
		Medial				Lateral			
Family	Genus and species	1	2	3	4	1	2	3	4
Lymantriidae	*Lymantria dispar*	SALT	–	DET	SORB	SALT	SU/GL	–	–
Pieridae	*Pieris brassicae*	SALT	SALT	DET	MOG I	AMA	SU/GL	MOG II	ANTH
Saturniidae	*Philosamia cynthia*	SALT	CON	GL	INOS	AMA	SU/GL	GL	INOS
Sphingidae	*Manduca sexta*	SALT/A	–	ALK	INOS	SALT/A	SU/GL	SA	INOS
Tortricidae	*Adoxophyes reticulana*	SALT	SALT	–	–	AMA	SU	PHL	INOS

[a]From Schoonhoven (1973).

[b]Abbreviations of stimulatory chemicals: A, acid; ALK, alkaloids; AMA, amino acids; ANTH, anthocyanins; CON, conessine; DET, deterrents; GL, glucose; INOS, inositol; MOG I, II, mustard oil glucosides; PHL, phloricin; SA, salicin; SORB, sorbitol; SU, sucrose.

ple, the crop volume in males of *P. regina* is regulated at a constant level at a given dietary sugar concentration. Long-term sugar consumption appears to depend on a negative feedback between the osmotic pressure of the hemolymph and the rate of food passage through the intestine. The sugar solutions stimulate chemoreceptory sensilla trichodea on the labella and tarsi of adult flies. The hairs are usually supplied with five sensory cells (one mechanoreceptor and four chemoreceptors). The dendrites of the chemoreceptors contact external solutions through a pore at the hair tip. Four dendrites have been found within a single hair that correspond to sugar, water, and two salt receptors. The primary process of sugar reception may involve the formation of a sugar–glucosidase complex. Recent evidence suggests the presence of a pyranose and furanose site on the receptor. For example, the outside of the labella and tarsi of the flies, *P. regina* and *Boettcherisca peregrina,* shows α-glucosidase activity. The enzyme appears to be located on the membrane of the sensory dendrite at the tip of the chemosensory hair. The receptor has been shown to respond to some pentoses, hexoses, α-glucosides, di- and trisaccharides (α bonds), sugar alcohols, and some amino acids.

B. Glycosides in Exocrine Glands

Insect exocrine glands have been shown to contain the precursors of some defensive secretions and pheromones in the form of glycosides. The glycoside masks the reactive group of the secretion, prevents autotoxicity, and maintains solubility. For example, benzoquinones, which are widely distributed arthropod defensive secretions, may be generated from phenolic glycosides which are hydrolyzed and oxidized at the time of secretion. The defensive glands of the tenebrionid beetles, *Eleodes longicollis* and *Tribolium castaneum,* have been shown to secrete benzoquinones derived from *p*-diphenols which are present in the gland as glucosides. The action of a β-glucosidase and phenoloxidase generates the free benzoquinone from the glucoside at the time of secretion. Similarly, the male sex pheromones of the noctuid moths, *Pseudaletia separata* and *Mamestra configurata,* are stored as glycosides of benzyl alcohol and phenyl ethanol, respectively. Within 24 hr of eclosion, free pheromone is present on the abdominal hair pencils. The cap cells at the base of the hair pencil produce a β-glycosidase which liberates the free pheromone after the glycoside is released from the gland.

XIV. CONCLUSION

In 1928 B. P. Uvarov wrote in the first comprehensive review of insect nutrition and metabolism: "The literature on carbohydrates in the insect body is remarkable for its scantiness. Only scattered data exist on the few main substances of this group." This chapter has provided testimony that substantial progress in our understanding about insect carbohydrates has been made in the last 50 years. Papers dealing with insect carbohydrates and their functional involvement in the life processes of insects continue to be published at a high rate. Currently, hundreds of papers dealing with insect carbohydrates are available, and they form a significant part of modern insect biochemistry as a whole.

Besides a continuing interest in the biochemistry of insect cuticle, recent literature on insect carbohydrates has focused on the involvement of glycogen, trehalose, glycerol 3-phosphate and polyols in growth, development, and flight. Analyses of individual tissues, especially the fat body, hemolymph, muscle, intestine, nerve cord, and ovary have provided much new data on the distribution and interrelationships of insect carbohydrates. As a result, a clearer understanding is beginning to emerge about the hormonal and metabolic integration of carbohydrate utilization. Beyond this, substantial progress has recently been made in increasing our understanding about the mechanisms of sensory perception of sugars and related compounds. However, on a more cautionary note, it is obvious that carbohydrates have, as yet, been studied in only a few of the million or so described insect species which among them show a greater diversity of form and habitat than can be found in any other group of organisms. To date, biochemical studies have focused on accessible species among the five major orders: Coleoptera, Diptera, Hymenoptera, Lepidoptera, and Orthoptera. Few comparative studies involving representatives from several orders have as yet been undertaken. Until additional comparative information is available, all generalizations about the functional role of carbohydrates must continue to be made with caution.

ACKNOWLEDGMENTS

This chapter is a contribution from the Missouri Agricultural Experiment Station, Columbia. I thank Dr. D. L. Dahlman for his helpful review of the manuscript.

GENERAL REFERENCES

Bailey, E. (1975). *In* "Insect Biochemistry and Function" (D. J. Candy and B. A. Kilby, eds.), pp. 89–176. Chapman & Hall, London.

Chefurka, W. (1965). *In* "The Physiology of Insecta" (M. Rockstein, ed.), 1st Ed., Vol.2, pp. 581–667. Academic Press, New York.

Crabtree, B., and Newsholme, E. A. (1975). *In* "Insect Muscle" (P. N. R. Usherwood, ed.), pp. 405–500. Academic Press, New York.

Elbein, A. D. (1974). *Adv. Carbohydr. Chem. Biochem.* **30,** 227–256.
Friedman, S. (1970). *In* "Chemical Zoology" (M. Florkin and B. T. Scheer, eds.), Vol. 5. Part A, pp. 167–197. Academic Press, New York.
Sacktor, B. (1970). *Adv. Insect Physiol. 7,* 267–347.
Sacktor, B. (1975). *In* "Insect Biochemistry and Function" (D. J. Candy and B. A. Kilby, eds.), pp. 1–88. Chapman & Hall, London.
Wyatt, G. R. (1967). *Adv. Insect Physiol.* **4,** 287–360.

REFERENCES FOR ADVANCED STUDENTS AND RESEARCH SCIENTISTS*

Ashhurst, D. E., and Costin, N. M. (1976). *J. Cell Sci.* **20,** 377–403.
Baust, J. G., and Morrissey, R. E. (1976). *Proc. 15th Int. Congr. Entomol. Washington* pp. 173–184.
Beenakkers, A. M. T., van den Broek, A. T. M., and de Ronde, T. J. A. (1975). *J. Insect Physiol.* **21,** 849–859.
Bewley, G. C., Rawls, J. M. Jr., and Lucchesi, J. C. (1974). *J. Insect Physiol.* **20,** 153–165.
Blum, M. S., Glowska, Z., and Taber III, S. (1962). *Ann. Entomol. Soc. Am.* **55,** 135–139.
Brosemer, R. W. (1965). *Biochim. Biophys. Acta* **96,** 61–65.
Buonocore, V., Poerio, E., Silano, V., and Tomasi, M. (1976). *Biochem. J.* **153,** 621–625.
Bursell, E. (1975). *Comp. Biochem. Physiol.* **52(B),** 235–238.
Chippendale, G. M. (1975). *J. Nutr.* **105,** 499–507.
Chippendale, G. M., and Reddy, G. P. V. (1974). *J. Insect Physiol.* **20,** 751–759.
Collier, G. E., Sullivan, D. T., and MacIntyre, R. J. (1976). *Biochim. Biophys. Acta* **429,** 316–323.
Crompton, M., and Birt, L. M. (1967). *J. Insect Physiol.* **13,** 1575–1592.
Dahlman, D. L. (1970). *Ann. Entomol. Soc. Am.* **63,** 1563–1565.
Doane, W. W., Abraham, I., Kolar, M. M., Martenson, R. E., and Deibler, G. E. (1975). *In* "Isozymes 4-Genetics and Evolution"(C. L. Markert, ed.), pp. 585–608. Academic Press, New York.
Droste, H. J., and Zebe, E. (1974). *J. Insect Physiol.* **20,** 1639–1657.
Friedman, S. (1978). *Ann. Rev. Entomol.* **23,** 389–407.
Gelperin, A. (1966). *J. Insect Physiol.* **12,** 331–345.
Hori, K. (1969). *J. Insect Physiol.* **15,** 2305–2317.
Huber, R. E., and Mathison, R. D. (1976). *Can. J. Biochem.* **54,** 153–164.
IUPAC-IUB Commission (1967). Report on the nomenclature of lipids. *Biochem. J.* **105,** 897–902.
IUPAC-IUB Commission (1972a). Report on the tentative rules for carbohydrate nomenclature. *J. Biol. Chem.* **247,** 613–635.
IUPAC-IUB Commission (1972b). "Enzyme Nomenclature," 443 pp. Elsevier, Amsterdam.
IUPAC-IUB Commission (1976). Supplement 1 on enzyme nomenclature. *Biochim. Biophys. Acta* **429,** 1–45.
Jutsum, A. R., and Goldsworthy, G. J. (1976). *J. Insect Physiol.* **22,** 243–249.
Kageyama, T. (1976). *Insect Biochem.* **6,** 507–511.
Lenartowicz, E., Zaluska, H., and Niemierko, S. (1967). *Acta Biochim. Polon.* **14,** 267–275.
Mathur, C. F., and Yurkiewicz, W. J. (1969). *J. Insect Physiol.* **15,** 1567–1571.
Mauldin, J. K., Smythe, R. V., and Baxter, C. C. (1972). *Insect Biochem.* **2,** 209–217.

*Together with Crabtree and Newsholme (1975) and Sacktor (1970) already cited under general references, this list includes the sources for the tables and figures used in this chapter.

Morgan, M. R. J. (1976). *Acrida* **5,** 45–58.
Nayar, J. K., and Sauerman, D. M., Jr. (1971). *J. Insect Physiol.* **17,** 2221–2233.
Nowosielski, J. W., and Patton, R. L. (1964). *Science, Washington* **144,** 180–181.
Salt, G. (1970). "The Cellular Defence Reactions of Insects." 118 pp. University Press, Cambridge.
Schoonhoven, L. M. (1973). *Symp. R. Entomol. Soc. London* **6,** 87–99.
Sømme, L. (1969). *Nor. Entomol. Tidsskr.* **16,** 107–111.
Steele, J. E. (1976). *Adv. Insect Physiol.* **12,** 239–323.
Tate, L. G., and Wimer, L. T. (1974). *Insect Biochem.* **4,** 85–98.
Uvarov, B. P. (1928). *Trans. R. Entomol. Soc. London* **76,** 255–343.
Vejbjerg, K., and Normann, T. C. (1974). *J. Insect Physiol.* **20,** 1189–1192.
Wharton, D. R. A., Wharton, M. L., and Lola, J. E. (1965). *J. Insect Physiol.* **11,** 947–959.
White, T. C. R. (1972). *J. Insect Physiol.* **18,** 2359–2367.
Yamasaki, K. (1974). *Insect Biochem.* **4,** 411–422.
Yamashita, O., and Hasegawa, K. (1974). *J. Insect Physiol.* **20,** 1749–1760.
Yamashita, O., and Hasegawa, K. (1976). *J. Insect Physiol.* **22,** 409–414.
Ziegler, R., and Wyatt, G. R. (1975). *Nature, London* **254,** 622.

2

Functional Role of Lipids in Insects

ROGER G. H. DOWNER

I. INTRODUCTION*

Lipids have assumed considerable functional significance during the evolutionary history of the class Insecta. They are essential structural components of the cell membrane and cuticle, they provide a rich source of metabolic energy for periods of sustained energy demand, they facilitate water conservation both by the formation of an impermeable cuticular barrier and by yielding metabolic water upon oxidation, and they include important hormones and pheromones. The purpose of the present chapter is to assimilate the diverse literature which has been published on the subject of insect lipids into a form that will readily identify our current understanding of the topic and reveal those areas where knowledge remains fragmentary or is nonexistent.

At the outset it is useful to define the classes of compound that may be considered to be lipids. Most frequently the term is applied to a heterogeneous group of compounds which share two properties: (1) relative insolubility in water; (2) high solubility in nonpolar solvents such as chloroform and acetone. Thus the classification of lipids has a practical, rather than structural basis, and that used in the present chapter is taken from White *et al.* (1973).

Lipid biochemistry of insects has been the subject of several reviews (Scoggin and Tauber, 1950; Strong, 1963; Fast, 1964, 1970; Tietz, 1965; Gilby, 1965; Gilmour, 1965; Gilbert, 1967; Gilbert and O'Connor, 1970; Turunen, 1974; Gilbert and Chino, 1974; Svoboda *et al.*, 1975; Downer and Matthews, 1976; Jackson and Blomquist, 1976) and these, along with more specialized reviews that are cited in context, have provided a valuable resource for this chapter.

II. LIPID REQUIREMENTS

The dietary requirements of an organism reflect biochemical emphases and deficiencies; thus it is appropriate that a discussion of functional aspects of lipid biochemistry should consider the organism's lipid requirements. The need for lipid in the diet of insects has been detailed in several reviews (Lipke and Fraenkel, 1956; Dadd, 1963, 1970; Altman and Dittmer, 1968; Rodriguez, 1972; House, 1973). These accounts demonstrate that polyun-

*The following abbreviations are used in this chapter: C10:0, *n*-decanoic acid, capric acid; C12:0, *n*-dodecanoic acid, lauric acid; C14:0, *n*-tetradecanoic acid, myristic acid; C16:0, *n*-hexadecanoic acid, palmitic acid; C16:1, *cis*-9-hexadecanoic acid, palmitoleic acid; C18:0, *n*-octadecanoic acid, stearic acid; C18:1, *cis*-9-octadecanoic acid, oleic acid; C18:2, *cis,cis*-9,12-octadecadienoic acid, linoleic acid; C18:3, *all-cis*-6,9,12-octadecatrienoic acid, linolenic acid; C20:0, *n*-eicosanoic acid, arachidic acid.

saturated fatty acids and sterols comprise the two major lipid classes for which insects display an absolute requirement, although the fat-soluble vitamins, carotene and α-tocopherol, are also essential dietary constituents.

A. Polyunsaturated Fatty Acids

The inclusion of saturated and monounsaturated fatty acids in the diet has a beneficial effect upon the growth of most species, although the ability of insects to synthesize these compounds from nonlipid precursors may indicate nonessentiality. However it is important to distinguish between the ability to synthesize a compound and the capacity to synthesize that compound at the required rate. Thus the demonstration of biosynthetic potential does not eliminate the possibility of a dietary requirement, particularly during periods of high metabolic demand.

Polyunsaturated fatty acids, especially those of the C18 series are definite requirements in the diet of most species. Indeed only certain Diptera appear capable of normal growth in the absence of C18:2 and C18:3. It is interesting to note that these insects contain a high proportion of C16:1, the fatty acid that increases in content in certain species, when raised on C18:3-deficient diets. The free fatty acid content of tissues is relatively low and those of dietary origin are rapidly esterified into complex lipids, with the various biosynthetic pathways exhibiting specificity with regard to fatty acid "selection." The possible sparing effect of C16:1 on polyunsaturated fatty acid requirement may reside in the nature of the specificity of one or more biosynthetic pathways.

The nutritional requirement for polyunsaturated fatty acids has been demonstrated by diet-deletion studies in which physiological manifestations of fatty acid deficiency have been observed in deprived insects, and by metabolic studies of biosynthetic capacity. The two experimental approaches are discussed separately below.

1. Nutritional Studies

Data obtained from classical diet-deletion studies require cautious interpretation. It is difficult to completely eliminate trace amounts of the substances under investigation from prepared diets, and some earlier claims for the nonessential nature of polyunsaturated fatty acids may require reassessment because the absence of trace contaminants was not adequately confirmed. Furthermore, the observation that certain fatty acids are passed from mother to offspring requires that studies be extended through several generations before any dietary component may be considered nonessential. Finally, the presence of intestinal and/or intracellular symbionts, which possess biosynthetic capacities that are not present in the host, introduces a further complication. However, the demonstration of morphological and physiological symptoms associated with dietary insufficiency provides an indication that the missing component is an essential part of the diet. The effects of polyunsaturated fatty acid deficiency have been demonstrated in many species and from these data it is evident that the deficiency may be manifest in one or both of the following symptoms: (1) impaired growth and development, the most extreme expression of which involves unsuccessful emergence, failure of wings to expand and loss of wing- and body-scales; (2) reduced reproductive capacity. It is difficult to assess if similar physiological mechanisms are contributing to both effects, or even if prolonged exposure to deficient diets would result in the appearance of both effects. C18:2 and C18:3 accumulate in membrane phosphoglycerides and particularly at the 2-position of the molecule, and their absence will undoubtedly result in impaired membrane function. However the precise physiological consequences of the membrane impairment await elucidation.

Arachidonic acid (C20:4) is required for normal growth and development in the beetle *Trogoderma granarium,* but has not been investigated as a possible dietary requirement in many species. This fatty acid is a precursor of prostaglandins in vertebrates, and the recent demonstration of prostaglandins in the accessory glands of crickets leads to speculation on a role for this class of compounds in insects. If prostaglandins serve a similar function in insects as in vertebrates, C20:4 may prove to be a very important dietary component.

2. Biochemical Studies

The inability of insects to synthesize polyunsaturated fatty acids has been confirmed in many species by studies which have traced the incorporation of radiolabeled precursors, such as acetate or glucose, into saturated and monounsaturated fatty acids but have detected no incorporation of the label into the polyunsaturated acids. In three species, however, incorporation of [^{14}C]acetate into C18:2 and/or C18:3 has been reported. Limited incorporation was obtained in the mosquito, *Aedes sollicitans*, and in the aphid *Myzus persicae*, 3.8% of the recovered label was found to be associated with the C18:3 fraction. In *Periplaneta americana* 9.9% (male) and 11.5% (female) of label was associated with C18:2 and 4.3% (male) and 4.6% (female) with C18:3. The possibility of a residual biosynthetic capacity in these species cannot be discounted, although other explanations may be advanced to account for the observations. Lambremont (1971) recognized the technical difficulties associated with isolating C18:2 from C18:1 and C18:3 from C20:0, and cautioned that additional identification is required following separation of these acids by gas–liquid chromatographic procedures. It is also possible that intracellular bacteria, contained within mycetocytes, may be responsible for synthesizing polyunsaturated fatty acids from the acetate precursor.

B. Sterols

A dietary need for sterol is believed to be a general characteristic of the class Insecta. The requirement has been demonstrated by nutritional studies and by biochemical investigation of cholesterogenic capacity. Various aspects of insect sterol biochemistry are discussed in several reviews (Clayton, 1964, 1971; Gilbert, 1967; Svoboda *et al.,* 1975). From these accounts it is evident that sterols serve a variety of functions in the insect including (1) essential component of subcellular membranes; (2) precursor of the molting and vitellogenic hormone, ecdysone; (3) constituent of surface wax of insect cuticles; and (4) constituent of lipoprotein carrier molecules. This multiplicity of steroid function is evidenced by the observation that, if subminimal concentrations of cholesterol are provided in the diet of some species, along with another sterol, normal growth will occur. The cholesterol-sparing property of certain sterols indicates that cholesterol is not required for all steroid functions and that some steroidal tasks may be performed by other sterols.

1. Nutritional Studies

Among phytophagous insects, phytosterols are often an adequate and, in some cases, better substitute for cholesterol in the diet. However cholesterol remains the major sterol in the body tissues of these species, thus indicating a capacity for conversion of phytosterols to cholesterols. Some of the metabolic pathways involved in these conversions have been elucidated and are included in the review of Svoboda *et al.* (1975). The $\Delta^{5,7,24}$-cholestatrien-3β-ols are intermediates in several interconversions. Other intermediates which have been identified include fucosterol and fucosterol-24,28-epoxide in the conversion of sitosterol to cholesterol, Δ^{24}-methylene cholesterol in the conversion of campesterol to cholesterol, and $\Delta^{5,7,24}$ cholestatrien-3β-ol in the conversion of stigmasterol to cholesterol. Tentative pathways for these conversions have been proposed and are illustrated in Fig. 1 (numbers in parentheses refer to those species in which the intermediates involved have been reported; see legend).

It is evident that the effectiveness of a sterol in sparing the dietary requirement for cholesterol depends upon the ability of the insect to effect the metabolic transformations necessary to convert the sterol into a usable form. Synthesis of ecdysone requires the presence of a Δ^7-bond in the steroid nucleus, and insects must either possess the biochemical capacity for this step or be dependent upon a dietary source of Δ^7-sterols. This requirement has been demonstrated for two species, each of which depends upon a plant for supply of Δ^7-sterols. *Drosphila pachea* feeds on the senita cactus, *Lophocereus schotti*, which contains high concentrations of stigmasten-3β-ol, whereas the beetle *Xyleborus ferrugineus* symbiotically hosts the

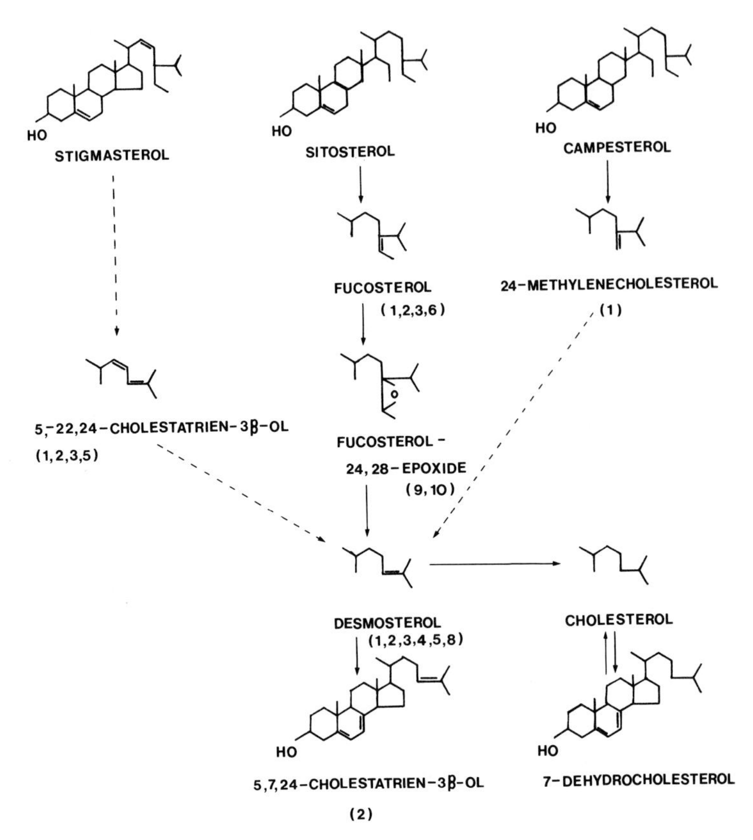

Fig. 1. Proposed pathways of phytosterol metabolism in insects. Numbers in parentheses identify those species in which the intermediate compound has been reported, according to the following key: (1) *Manduca sexta,* (2) *Tribolium confusum,* (3) *Blattella germanica,* (4) *Periplaneta americana,* (5) *Heliothis zea,* (6) *Eurycotis floridana,* (7) *Thermobia domestica,* (8) *Spodoptera frugiperda,* (9) *Bombyx mori,* (10) *Locusta migratoria.* (Adapted, with permission, from Annual Review of Entomology, Volume 20. Copyright © 1975 by Annual Reviews, Inc. All rights reserved.)

ergosterol-rich fungus *Fusarium solani* in its gut. Many insects rely on gut symbionts for steroids, and it is probable that further studies on species which have adapted to highly specialized ecological niches will reveal additional idiosyncrasies of sterol interconversion and metabolism.

2. Cholesterogenic Studies

The dietary requirement for sterols implies the absence, in insects, of the cholesterogenic pathway that has been proposed for vertebrates and microorganisms (Bloch, 1965). An important intermediate in this pathway is mevalonic acid, and the biosynthesis of this substrate from acetate has been demonstrated in both axenic and nonaxenic blow flies, *Sarcophaga bullata,* by R. D. Goodfellow and his associates. The cofactor requirements were found to be the same in the insect as in rat liver, and it appears that a similar biosynthetic pathway exists in insects and vertebrates. The next stage of vertebrate cholesterogenesis involves condensation of six 5C derivatives of mevalonic acid to yield the sterol precursor squalene. The sequence of reactions leading to squalene formation is illustrated in Fig. 2. Several studies have reported the failure of acetate and/or mevalonate to be incorporated into squalene in insects. The enzyme mevalonate kinase has been described in *S. bullata* and additional studies have demonstrated the incorporation of label from [2-^{14}C]mevalonic acid into the isoprenoid compounds, geraniol, farnesol, nerolidol, and geranylgeraniol. These results, and the importance of such intermediates as geranyl pyrophosphate and farnesyl pyrophosphate in other anabolic processes in insects, suggest that one metabolic block to cholesterol synthesis occurs at

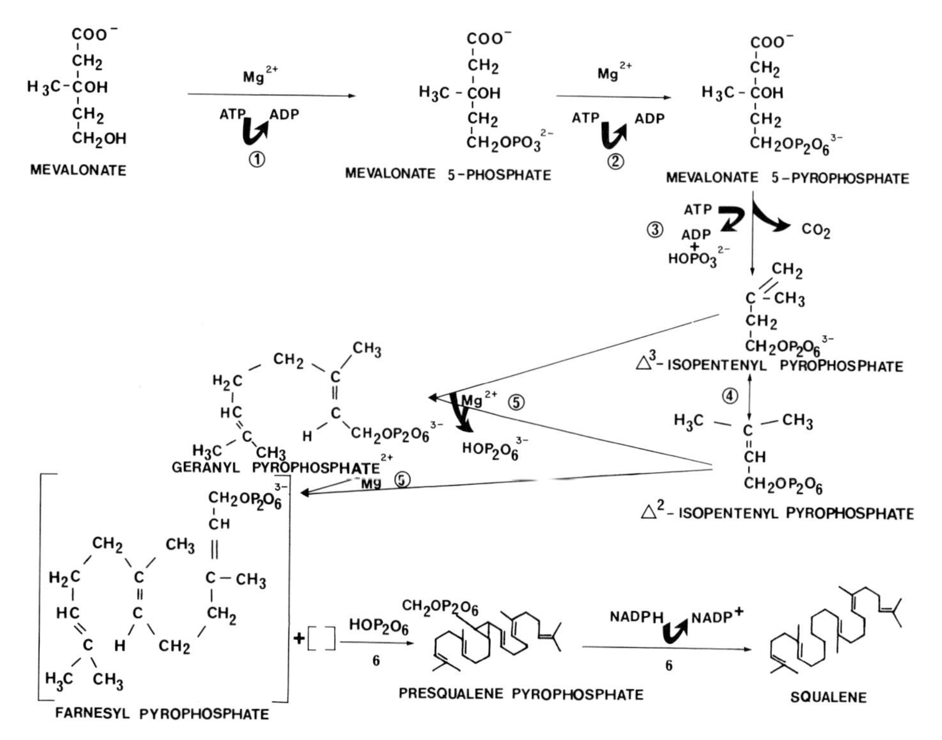

Fig. 2. Pathway of biosynthesis of squalene from mevalonate. Numbers in circles refer to enzymes according to the following key: (1) mevalonate kinase, (2) phosphomevalonate kinase, (3) pyrophosphomevalonate decarboxylase, (4) isopentenyl pyrophosphate isomerase, (5) geranyl transferase, (6) squalene synthetase.

the level of reductive dimerization of farnesyl pyrophosphate. This compound is an important intermediate in the production of juvenile hormone, and it is interesting to speculate on the possibility that the evolution of this unique hormone was dependent upon a block occurring in the cholesterogenic pathway at the level of farnesyl pyrophosphate. In this regard it has been proposed that the enzyme squalene synthetase is lacking or nonoperative in insects, and another cholesterogenic block has been shown to occur at the level of squalene cyclization. In this study it was found that although radioactivity from [^{14}C]squalene was incorporated into a squalene-2,3-oxide-like compound in *S. bullata*, the intermediate was not cyclized to yield lanosterol as occurs in vertebrates. Lanosterol can, however, substitute for cholesterol in the diet of some species. Absence of the cyclization step may also have evolutionary significance in the development of juvenile hormone by protecting the hormone molecule, which possesses a 2,3-epoxide linkage, from possible cyclization.

III. DIGESTION AND ABSORPTION

With the exception of a few highly specialized species, insects feed on a diet of macromolecules, many of which do not penetrate the gut wall. In order that these molecules may be used by the tissues they must be reduced to an absorbable form by the process of digestion, and these digestive products then cross the gut wall and are distributed to the tissues. The associated processes of digestion and absorption have been reviewed by Day and Waterhouse (1953), Waterhouse and Day (1953), Treherne (1967), Dadd (1970), and House (1973). These reviews necessarily concentrate on nonlipid aspects of digestion and absorption; this serves to emphasize the paucity of information that is available on the fate of dietary lipids in the insect alimentary tract.

A. Digestion

Studies with dietary sterols indicate that no digestion or modification of sterols is required in order for these molecules to be absorbed. The principal lipid class that is digested in the gut is triacylglycerol, the digestion of which involves hydrolysis of long-chain fatty acylglycerol esters. Lipolysis is effected in the presence of lipases, a class of hydrolytic enzymes that has been defined in terms of specificity for long-chain fatty acylglycerol esters (Gilbert, 1967) and capacity to hydrolyze the esters of emulsified glycerides at an oil–water interface (Jensen, 1971). Neither definition should be regarded as absolute (Hipps and Nelson, 1974), however, they serve to distinguish lipases from the more general group of hydrolytic enzymes, the esterases. Esterases include all enzymes that catalyze the hydrolysis of ester linkages irrespective of chain length or solubility. Many studies of insect esterases have been reported (Wagner and Selander, 1974), but in only a few cases has the demonstration of esterase activity been related to digestive function. Frequently the activity has been demonstrated against phenolic esters because of the simplicity and sensitivity of the associated colorimetric assay, but it is difficult to base physiological conclusions on such data. Thus there are a limited number of studies in which true lipase activity has been demonstrated.

Triacylglycerol hydrolysis has been investigated in only a few species, and as digestive function often reflects the diet and habits of the species, it is not possible to generalize for the class from such a restricted data base. Cockroaches have been the most popular group of insects for studies of lipid digestion, and a number of early studies indicating lipolytic activity in the fore- and midgut of cockroaches are discussed by Eisner (1955) and Treherne (1958). Eisner (1955) confirmed early reports of hydrolysis of triolein in both the fore- and midgut, and further demonstrated that the lipase originates in the epithelial cells of the midgut and caecae. Thus the presence of lipolytic activity in the foregut results from regurgitation of midgut contents into the foregut. An indication of the rate of lipolysis was provided by Treherne (1958) who measured the radioactivity associated with the various lipid fractions at varying times after feeding tripalmitin ([1-^{14}C]palmityl) glycerol. He found that approximately 77% of glyceryltripalmitin was unhydrolyzed after 20 hrs with most of the remaining activity associated with the free palmitic acid fraction. Both workers suggest that the incomplete hydrolysis observed at 20 hrs may be due to displacement of the enzyme by hydrolytic end products accumulating at the oil–water interface on which the enzyme acts. The midgut lipase of *P. americana* displays dual pH optima with activity peaks corresponding to the pH found in fore- and midgut. This enzyme also shows enhanced activity in the presence of calcium ions, a property which is analogous to mammalian pancreatic lipase. Bollade *et al.* (1970) showed additional similarities between insect gut lipase and mammalian pancreatic lipase by demonstrating preferential cleavage, by the lipase, of fatty acids located at the 1 and 3 positions of the triacylglycerol molecule. These workers used doubly labeled 1,3-dipalmitoyl, 2-oleoyl glycerol ([^{3}H]glycerol and [^{14}C]oleic acid) to demonstrate this positional specificity, and thus did not accommodate the possibility of fatty acyl specificity influencing the rate of hydrolysis. However, stereospecific analysis of the lipolytic products of tri-[1-^{14}C]oleoyl glycerol hydrolysis in gut homogenates has recently confirmed the conclusions of the French group.

Few attempts have been made to purify or characterize the midgut lipase(s), and consequently no information is presently available concerning their absolute chemical properties. Cook *et al.* (1969) used electrophoretic techniques to isolate an enzyme, from the gastric caecum and midgut, that readily hydrolyzes glyceryl tripalmitin to free fatty acid, diacylglycerol, and monoacylglycerol. Enzyme activity was found to increase in the presence of calcium and manganese ions. However, the enzyme also hydrolyzes naphthyl acetate and thus differs from mammalian pancreatic lipase. Hipps and Nelson (1974) obtained a more complete purification of seven esterases from the midgut and caecae of the cockroach. By studying the inhibition kinetics of the enzymes they concluded that in at least some cases the esterases exist as molecular aggregates or in some form having multiple active centers. This observation corresponds with reports that mammalian lipase operates optimally when oversaturation of the substrate solution induces some kind of molecular aggregation.

Triacylglycerol hydrolysis has also been studied in the migratory locust *Locusta migratoria* (Weintraub and Tietz, 1973). In this species glyceryl tripalmitin is an unsatisfactory substrate for hydrolysis and is excreted in unhydrolyzed form. By contrast glyceryl triolein is rapidly hydrolyzed at a rate which exceeds that found in cockroaches. The specificity of the enzyme for C18:1 is of particular interest in view of the report that the mammalian lipase preferentially cleaves unsaturated fatty acids from the 1 and 3 positions (Morley *et al.*, 1974). Indeed the difference in the hydrolytic rates reported for cockroaches and locusts may merely reflect the choice of artificial substrate employed in the investigation. However, the interspecific difference can be rationalized by consideration of the habits of the two species. Locusts feed prior to and during migration in order to provide the prodigious amounts of metabolic substrate that are required for extended periods of flight. Thus rapid digestion and absorption of foodstuff are desirable. The cockroach, by contrast, does not require rapid assimilation and utilization of dietary materials and these processes may be effected at a slower rate.

Fifth instar larvae of *Pieris brassicae* hydrolyze triacylglycerol in the lumen of the gut at a rate similar to that reported for the cockroach (Turunen, 1975). Fifteen hours after feeding the test meal, 60% of the label remained associated with the triacyglycerol fraction.

Although it is tempting to dismiss discrepancies in reported data in terms of interspecific differences, variations in experimental technique should not be neglected as a possible explanation for some of the apparent anomalies. Desnuelle and Savary (1963) have drawn attention to the importance of emulsification in obtaining optimal enzyme activity, and therefore variation in the manner in which the substrate is presented to the enzyme may result in significant differences in observed activity.

Few data are available on the nature of emulsifying agents in insects, although the bile salts which perform this function in vertebrates do not appear to be present in arthropods. Indeed, in the absence of sterol biosynthesis it would be difficult to provide adequate dietary sterol to satisfy such a digestive requirement. In Crustacea the emulsifying agents are derivatives of fatty acyltaurine and fatty acyl dipeptides and a similar structure has been proposed for the diving beetle *Dytiscus* (Vonk, 1969). Fatty acylamino acid complexes have also been shown to effect emulsification in the cricket, *Gryllus bimaculatus,* with up to fourteen amino acids and fatty acids of chain length C_8 to C_{20} contributing to the effect (Collatz and Mommsen, 1974). It is apparent that future studies of gut lipase require that substrates be prepared using some of these natural emulsifying agents.

Lipolytic activity has also been reported in the salivary secretion of some insects (Dadd, 1970; House, 1973), and it is possible that this activity may contribute to digestive lipolysis, especially in those species that practice extraintestinal digestion. However no data are presently available to substantiate this possibility. The contribution of intestinal symbionts to the process of lipid digestion is similarly unknown.

The wax moth, *Galleria mellonella*, feeds entirely upon the wax of honeybee combs, and the specialized digestion and absorption of this substrate are discussed in the reviews of Gilbert (1967) and Dadd (1970).

B. Absorption

1. Lipolytic Products

The process of lipid absorption in insects is poorly understood and the few relevant studies that have been reported have discussed experimental findings in light of the situation that prevails in vertebrates. Indeed the similar properties of mammalian pancreatic lipase and insect gut lipase (Section III,A) encourage speculation on the possibility that the products of triacylglycerol hydrolysis are absorbed by a process in insects similar to that which has been proposed for mammals. The mammalian mechanism has been described by Johnston (1970) and requires that dietary triacylglycerol is hydrolyzed to 2-acylglycerol and free fatty acid in the presence of pancreatic lipase. These products, in combination with bile salts, form negatively charged polymolecular aggregates (micelles) that penetrate the mucosal cell wall. Most of the free fatty acid and 2-acylglycerol is then resynthesized to triacylglycerol by the monoglyceride pathway described in Fig. 3. The remaining free fatty acid is converted to phosphoacylglycerol or triacylglycerol through the α-glycerophosphate pathway which is also shown in Fig. 3. The reconstituted triacylglycerol and phosphoacylglycerols, together with cholesterol, cholesterol esters, and protein, are incorporated into a distinct particle, the chylomicron, which is discharged into the extracellular space surrounding the intestinal cells.

The high residual levels of diacylglycerol that are normally present in

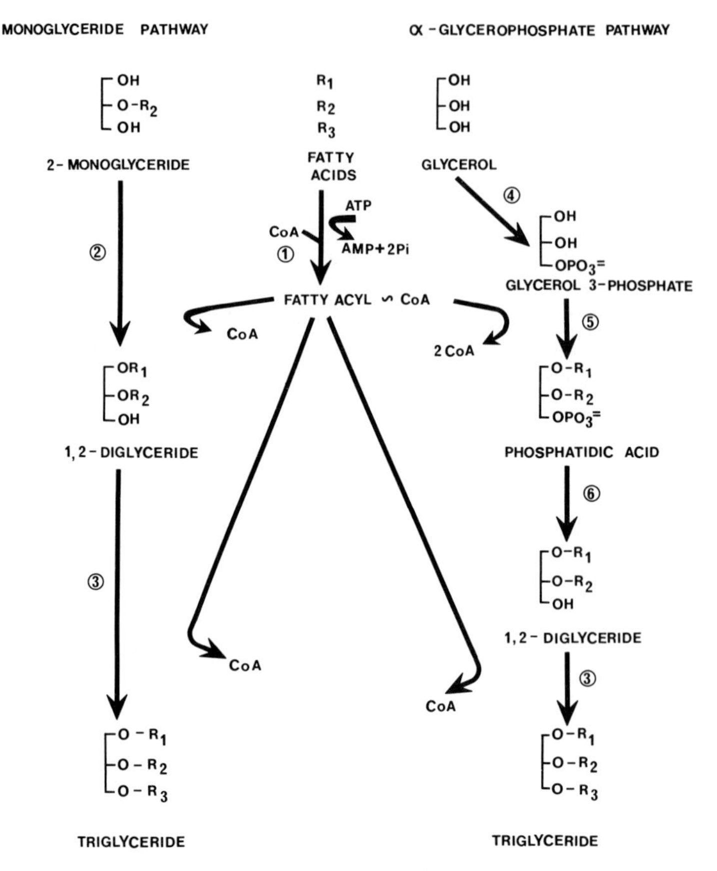

Fig. 3. Proposed pathways for glycerolipid biosynthesis in the intestinal mucosa of mammals (after Johnston, 1970). Numbers in circles refer to enzymes according to the following key: (1) acyl-CoA synthetase, (2) monoglyceride acyltransferase, (3) diglyceride acyltransferase, (4) α-glycerophosphatase, (5) glycerolphosphate acyltransferase, (6) phosphatidate phosphohydrolase.

insect hemolymph (Section IV,A) introduce a major deviation from the situation that exists in vertebrates, and suggest that the insect may use a different absorptive mechanism from that found in vertebrates. The few studies which have been conducted to date indicate that this is the case. In insects the products of glyceride digestion do not appear to be resynthesized to triacyglycerol in the gut, and chylomicrons are not released into the hemolymph. Instead the lipolytic products may enter the hemolymph in the form of diacylglycerol which is bound to a protein carrier. Weintraub and Tietz (1973) followed the fate of ingested tri-[1-^{14}C]oleoyl [2-^{3}H]-glycerol in the locust and found that glycerol, formed by lipolytic activity in the gut lumen, is rapidly taken up by the gut and passed unchanged into the hemolymph. Free fatty acid was synthesized by the gut into phosphoacylglycerol and acylglycerol fractions, but only diacylglycerol appeared in appreciable amounts in the hemolymph. These observations are strongly indicative of diacylglycerol synthesis being effected through the α-glycerophosphate pathway. The synthesis of phosphoacylglycerols from dietary glyceride has also been reported in *G. mellonella* and *P. brassicae*, and has prompted the speculation that phosphoacylglycerol synthesis is an integral part of the fatty acid absorption process with phospholipolytic activity preceding release of diacylglycerol into hemolymph. However, Weintraub and Tietz (1973) have suggested that the observed biosynthesis is more probably related to cell wall synthesis of the active gut cells. In this regard it is unfortunate that in most of the feeding studies that have been conducted so far, only trace amounts of radiolabeled triacylglycerol have been administered. Incorporation of label into various lipid classes under these conditions does not necessarily reflect the rates of synthesis of the same classes when a larger amount of lipid is ingested. These studies serve only to demonstrate the existence of particular metabolic pathways, but do not indicate the relative importance of the pathways under normal feeding conditions.

Preliminary studies from this laboratory indicate that monoacylglycerol acyltransferase activity is also present in the crop and midgut of the cockroach, indicating that the monoglyceride pathway may also be involved in lipid absorption in insects. This pathway requires the incomplete hydrolysis of triacylglycerol to 2-monoacylglycerol and free fatty acids, and the previously indicated slow rate of digestion suggests that this condition is fulfilled.

The major site of absorption of dietary glyceride is the anterior region of the midgut and the associated caecae (Treherne, 1958), although some fatty acid absorption has been shown to occur across the crop in the cockroach (Hoffman and Downer, 1976).

2. Sterols

In the mammal, free cholesterol is absorbed from the lumen by intestinal cells, with the molecule undergoing intracellular esterification prior to being incorporated into the chylomicron and discharged into the extracellular space. In the insect, the principal site of cholesterol absorption is the crop, but as sterol esters are absorbed more readily than free cholesterol, esterification does not appear to be a prerequisite for sterol absorption. Indeed the low cholesterol ester:free cholesterol ratios observed in insect hemolymph make it unlikely that large amounts of esterified cholesterol are released from the alimentary tract. Chino and Gilbert (1971) confirmed this conclusion in the moth *Philosamia cynthia* by showing that a lipoprotein is required for cholesterol release and that the cholesterol associated with the lipoprotein is present in the free, rather than the esterified form. In contrast to the situation that has been described for mammals, lipoprotein synthesis does not occur concomitantly with cholesterol release in the insect.

IV. TRANSPORT

The lipid profile of insect hemolymph varies with the physiological state of the animal and reflects the role of this tissue as a medium for the transportation of materials from sites of absorption or synthesis to sites of utilization or storage. An understanding of lipid transport requires knowledge of the form in which lipids are present in hemolymph and an appreciation of the mechanisms by which these molecules are released from and taken into cells. In the account that follows, these topics are treated separately although the entire process is obviously integrated. Additional accounts are presented in the reviews of Gilbert and Chino (1974) and Downer and Matthews (1976), and the reader should consult these works for original references.

A. Hemolymph Lipids

The major lipid classes that may be isolated from insect hemolymph are insoluble in water and consequently are not present in hemolymph as free molecules, but rather as components of macromolecular complexes, the lipoproteins. A variety of lipoprotein fractions have been identified in insects and two major classes have received particular attention. These lipoproteins have been named diglyceride-carrying lipoproteins I and II (DGLP-I and DGLP-II) and have been isolated from the moths *P. cynthia* and *Hyalophora cecropia*. Detailed chemical analyses of the molecules reveal many similarities in molecular structure and composition between the two species and also serve to identify differences in the lipid composition of insect hemolymph and vertebrate blood. Of particular interest in this regard is the predominance of diacylglycerol as the major lipid component of hemolymph in all but a few species of insects, and the presence of nonesterified cholesterol in the insect lipoprotein. Studies of the physiological role of DGLP-I indicate that it accepts diacylglycerol from fat body and free sterol from midgut, and functions to carry these substrates between tissues. The phospholipids, phosphatidylcholine, phosphatidylethanolamine and sphingomyelin are also associated with DGLP-I, and the lipoprotein may also serve as a vehicle of phospholipid transport. Studies with delipidized, high-density lipoprotein from human serum indicate that phosphoacylglycerol must be taken up by the carrier protein before neutral lipids are accepted, and phosphoacylglycerol may play a similar role in effecting diacylglycerol uptake by insect DGLP-I. The rate of phosphoacylglycerol uptake by DGLP-I is not as rapid as that of diacylglycerol and sterols.

A lipoprotein with structural and physiological similarities to lepidopteran DGLP-I has been isolated from *L. migratoria*. The molecule is required in order for diacylglycerol to be released from fat body, and it appears that this lipoprotein serves a similar function to that described for the lepidopteran DGLP-I.

The second of the two major lipoproteins isolated from moth hemolymph, DGLP-II, does not accept diacylglycerol from the fat body. The lipoprotein is found only in female pupae and increases in concentration during ovarian development. These observations suggest that DGLP-II may function as a vitellogenin; this proposal is further supported by the structural similarities of the lipoprotein and the vitellogenins of *H. cecropia* and *Leucophaeae maderae*. Recently, appreciable quantities of DGLP-II have been isolated from the eggs of *P. cynthia,* thus supporting the proposed role of the molecule as a vitellogenin (Chino *et al.,* 1976). The lipid content of DGLP-II is not, however, sufficient to account for the large amounts of lipid present in eggs, and it is probable that the major vehicle of lipid transport to the egg is DGLP-I (Chino *et al.,* 1977).

Thus two insect hemolymph lipoproteins have been characterized and their physiological functions determined. Other hemolymph lipoproteins are involved in hormone transport but no structural or functional data are available for the remainder. Nor is information available concerning the form in which the other lipid classes of hemolymph are transported. Hydrocarbon has been reported to be associated with DGLP-I and DGLP-II in *H. cecropia* and with a third, low-density lipoprotein isolated from this species. These observations suggest that the answer to questions regarding transportation of other lipid classes may result from analyses of as yet undetermined hemolymph lipoproteins.

B. Release and Uptake of Hemolymph Lipids

1. Release

The primary source of hemolymph diacylglycerol is the triacylglycerol store contained within fat body, although the absorption of dietary lipid contributes some diacylglycerol to the hemolymph at certain times (see Section III,B). The release of diacylglycerol from fat body requires hydrolysis of the stored triacylglycerol and lipolytic activity has been reported in the fat body of several species. The fatty acid composition of hemolymph diacylglycerol differs from that of fat body tricylglycerol in *H. cecropia* and *L. migratoria*. This observation precludes simple monoacyl cleavage as the sole progenitive event in diacylglycerol formation and has resulted in the proposal that diacylglycerols with specific fatty acid compositions are synthesized in an active compartment of the fat body using the products of triacyglycerol hydrolysis. This interesting hypothesis is supported by double-labeling experiments and analogous systems in vertebrate adipose tissue, but in the absence of definitive studies on the specificity and intracellular localization of fat body lipase, and the diacylglycerol biosynthetic capacity of this tissue, no positive conclusions can be reached. Indeed, Spencer and Candy (1976) have suggested that partial hydrolysis of triacylglycerol to diacylglycerol and fatty acid may be the dominant process, with reesterification of fatty acid assuming a lesser, albeit important, role.

The release of diacylglycerol from fat body requires the presence of a hemolymph protein acceptor (Section IV,A), and it was of obvious interest to determine if the lipoprotein complex was synthesized in the fat body prior to being released into hemolymph. This possibility was investigated by A. Tietz and her associates, who studied the twin processes of protein synthesis and diglyceride release in the locust fat body. They demonstrated that the addition of the protein synthesis inhibitor, cycloheximide, to incubating fat body inhibited the incorporation of amino acids into protein but did not affect the incorporation of palmitic acid into glycerides. Furthermore they were able to show that the release of glycerides from fat body was not impaired by treatment with cycloheximide and may be presumed to proceed independently of protein synthesis. It should be noted that the vitellogenic protein, DGLP-II, does not accept diacylglycerol from the fat body and the loading of this fraction with lipid must either occur prior to its release from fat body or from some other source.

Although the release of glyceride from fat body in the form of diacylglycerol has been clearly demonstrated in at least two species, it is possible that in some insects, under certain physiological conditions, the lipid reserve may be released in the form of triacylglycerol and/or free fatty acid.

2. Uptake

Lipids comprise an essential structural and metabolic component of all cells and as such are taken up from hemolymph by all tissues. However, the mechanisms of uptake are poorly understood and only fragmentary pieces of evidence are available to elucidate the process. The triacylglycerol high-density and very-low-density lipoproteins in vertebrate serum are hydrolyzed by a specific group of lipoprotein lipases that are found in various tissues, including adipose tissue. The enzyme releases free fatty acids from the lipoprotein complex and these bind to serum albumins and enter the cell. The enzyme is activated by the mucopolysaccharide, heparin, and this substance is considered a necessary component of the enzyme complex.

Recently a heparin-activated lipase has been reported in homogenized fat bodies of the reduviid bug *Triatoma maculata,* with a 66% decrease in enzyme activity observed in the absence of albumin. In light of this observation it is interesting to note the enhanced diglyceridase activity reported for moth flight muscle when the diacylglycerol substrate is associated with a lipoprotein carrier. If uptake of neutral lipids is preceded by extracellular hydrolysis in the immediate vicinity of the tissue, increased rates of hydrolysis would be expected during periods of increased energy demand. Thus the increase in circulating levels of free fatty acid that has been observed during flight in the locust, lends support to this proposal although studies of fatty acid turnover under such conditions are required before definitive conclusions can be drawn. The lipolytic activity of moth flight muscle has been investigated under *in vitro* conditions, but no studies are available on the mechanisms of activation of this (these) enzyme(s), nor is the site of hydrolysis (extra- or intracellular) known.

Uptake of the lipid-containing vitellogenin by the ovary has been studied in several species and appears to proceed through selective sequestration of the specific hemolymph lipoprotein. However as indicated above (Section IV,A) an additional source of lipid is required to explain the large amounts of triacylglycerol found in eggs, and this is probably derived from DGLP-I, for which the mechanism of uptake is not known.

Again little information is available concerning the uptake of other lipid classes although it seems that the cuticular lipids (paraffin hydrocarbons and waxes), which are synthesized near the surface, are transported to the cuticle through pore canals.

V. UTILIZATION OF LIPIDS

The evolutionary success of the class Insecta has been facilitated by the animals' ability to store large quantities of fat and use this substrate as a source of energy for sustained muscle activity and as a source of metabolic water to supplement strategies of water conservation. Fat is stored and transported principally in the form of tri- and diacylglycerol, and ester hydrolysis is a necessary prelude to oxidation of the fatty acid.

A. Extradigestive Lipases

Lipolytic activity has been reported in a number of tissues including fat body and Malpighian tubules (Wills, 1965). However, ambiguity concerning the definition of a lipase (see Section III,A) coupled with questionable experimental procedures prevent absolute endorsement of all reports. For example, the use of emulsified tributyrin as a substrate for lipolysis fails to satisfy the criterion of chain-length specificity and, as commercial preparations of this substrate are notoriously contaminated with the diacyl and monoacyl derivatives, the solubility requirement also may not be fulfilled. Physical differences in the nature of the emulsified substrates cause variation in observed rates of lipolysis, and the multifarious nature of emulsification procedures employed to study extradigestive lipolysis in insects makes a comparative assessment of published data unproductive at this time.

Few data are available concerning the glyceride specificity of extradigestive lipases and those that have been published suggest that differences exist between individual species. Stevenson (1972) found homogeneity in glyceride specificity of fat body lipase(s) in the southern armyworm moth, but reported preferential hydrolysis of monoacylglycerol in the flight muscle. The monoglyceridase activity of moth flight muscle concurs with the findings of Crabtree and Newsholme (1972) who observed that in the locust, water bug, fly, and bee the enzyme specificity for diacylglycerol is also high. Increased hydrolysis of diacylglycerol by moth flight muscle is observed when the substrate is presented to the tissue in association with the natural lipoprotein carrier (Gilbert and Chino, 1974). Four electrophoretically distinguishable lipases occur in cockroach muscle and it is possible that these differ in their glyceride specificity. However until the different molecular species are characterized with regard to specificity and mechanisms of control of muscle lipases are elucidated, extrapolations from *in vitro* observations to *in vivo* rates of hydrolysis are highly speculative.

Other aspects of lipase specificity remain virgin research areas and it is unfortunate that no information is available concerning positional specificity, fatty acid specificity, and stereospecificity of extradigestive lipases. These data are required in order for mechanisms of metabolic supply to be understood.

Some indication of a capacity for lipolytic regulation is provided by the observation that fat body lipase activity is higher in lysed tissue than intact tissue (Stevenson, 1972), and from demonstrated changes of lipolytic activity during various treatments of incubated eggs (Krysan and Guss, 1973). However no information is available concerning the nature of the regulatory mechanism.

B. Oxidation of Fatty Acids

Fatty acid oxidation has been demonstrated to occur in the fat body of many insect species, and a substantial volume of evidence indicates that fatty acid oxidation proceeds by sequential shortening of the fatty acid chain according to the sequence described in Fig. 4. [See the reviews of Gilbert (1967), Sacktor (1970, 1975), and Bailey (1975) for this evidence.]

Carnitine (1,3-hydroxy-4-trimethylammonium butyrate) is an obligatory

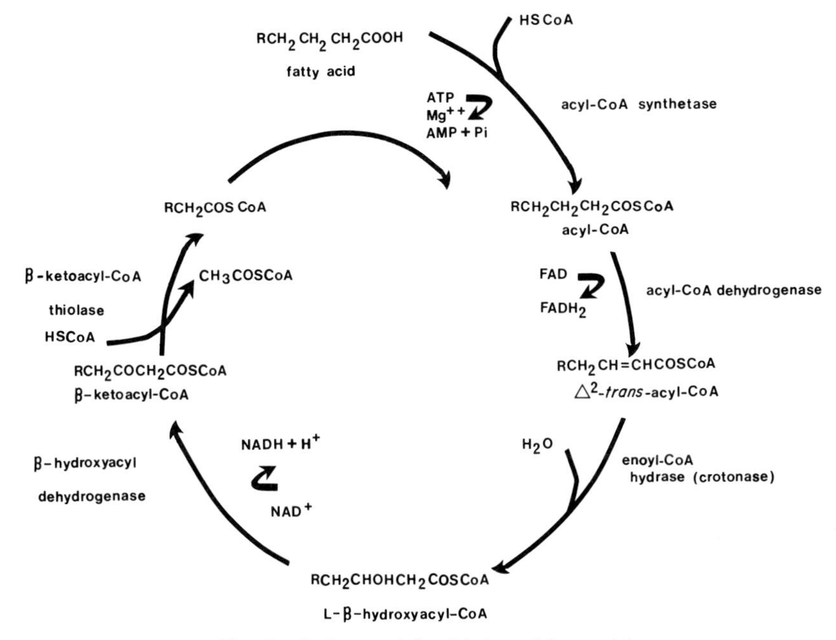

Fig. 4. Pathway of β-oxidation of fatty acids.

cofactor in some species, serving to facilitate the transport of acyl groups into and out of mitochondria. This facilitation is achieved through the action of an enzyme, carnitine acyltransferase, which catalyzes the formation of an acylcarnitine compound according to the reaction:

$$\text{Acyl-S-CoA} + \text{carnitine} \rightarrow \text{acylcarnitine} + \text{CoA}$$

The reaction is reversible and acyl-S-CoA is reconstituted on the other side of the mitochondrial membrane. Two carnitine acyltransferases have been described in rat liver, one of them catalyzing the acylation of carnitine with short-chain fatty acids whereas the other catalyzes the reaction with long-chain fatty acids. Circumstantial evidence suggests that fatty acid specificity may influence the reaction in insect flight muscle. Preferential acylation of carnitine with long-chain fatty acids occurs in the locust, and a carnitine-induced stimulation of palmitate oxidation, but not of butyrate oxidation, has been reported in thoracic homogenates of *Lucilia cuprina*.

Carnitine acyltransferase is believed to be contained within the mitochondria. Evidence in support of this conclusion is provided by the demonstration that the partitioning of the enzyme among fractions of muscle homogenate parallels that of intramitochondrial marker enzymes, and has led to the proposal that the transferase may occur in two intramitochondrial compartments separated by a crista membrane that is impermeable to acetyl-CoA.

Although carnitine is an obligatory cofactor for fatty acid oxidation in some species, it is not required universally among insects. Mitochondria from moth flight muscle are capable of rapid oxidation of palmitate in the presence of malate, and the rate of oxidation is unaffected by the addition of carnitine. Furthermore, no carnitine palmitoyltransferase activity has been detected in this species, an enzyme deficiency that has also been reported in the bee. Mitochondria from *Phormia* flight muscle fail to oxidize palmitate or hexanoate in the presence or absence of carnitine even though this tissue is a rich source of carnitine and possesses acetyl carnitinetransferase activity. These studies suggested that the acetylcarnitine ester is formed during the initial phases of flight in the blow fly, with the carnitine serving as an acceptor for acetyl groups from acetyl-CoA. Thus, in this species, carnitine may lower the acetyl-CoA:CoA ratio and permit the continuous formation of acetyl-CoA from pyruvate by relieving the inhibitory effect of acetyl-CoA upon pyruvate decarboxylase.

Therefore, it appears that in insects which do not use lipid as a primary energy source (e.g., bees and flies), carnitine plays little or no role in fatty acid oxidation. In insects that are major users of lipid, a dichotomy of carnitine dependence exists, with moths being independent of this requirement whereas locusts are unable to oxidize fatty acids at adequate rates in the absence of the compound. Locusts that are maintained at 40°C are able to oxidize long-chain fatty acids more readily than insects maintained at 30°C. The permeability of membranes for long-chain fatty acids increases with temperature and it is possible that the carnitine requirement of locusts is more pronounced at lower temperatures and during the initial stages of flight before body temperatures rise. Moths "warm-up" for several minutes before undertaking prolonged flights and this preflight activity raises the thoracic temperature by several degrees. One advantage of the "warm-up" may be to facilitate transport of long-chain fatty acid derivatives across the mitochondrial membrane in the absence of carnitine acetyltransferase. Measurements of rates of fatty acid oxidation by isolated mitochondria at various temperatures would provide useful information in this regard.

C. Fate of Glycerol

In addition to fatty acids, glycerol is also produced by the action of lipolytic enzymes on neutral fats. Glycerol kinase is present in the muscles of several insect species but in the locust the activity of the enzyme is not sufficient to account for all the glycerol produced by muscle hydrolysis of diacylglycerol (Candy *et al.*, 1976). These workers propose that during flight, glycerol is transported to the fat body where it provides a source of glycerol 3-phosphate for reesterification of the additional fatty acid moiety resulting from the conversion of stored triacylglycerol to diacylglycerol.

D. Ketone Body Formation

Although fatty acid oxidation normally proceeds to completion without the accumulation of intermediates, under certain conditions the ketone bodies, acetoacetate and D-3-hydroxybutyrate, may be formed. Synthesis occurs in the fat body, and the biosynthetic pathway is probably the hydroxymethyl–glutaryl-CoA pathway which has been described in the mammal (Bailey, 1975).

VI. BIOSYNTHESIS OF LIPIDS

A. Fatty Acids

Insects resemble most other life forms in their capacity to form long-chain saturated and monounsaturated fatty acids from acetyl-CoA. Thus any compound that can yield acetyl-CoA may be considered a potential precursor of fatty acid. In vertebrates and microorganisms the major fatty acid synthesizing system has been elucidated and synthesis shown to proceed according to the following reaction:

$$\text{Acetyl-CoA} + 7\text{malonyl-CoA} + 14\text{NADPH} + 14\text{H}^+$$
$$\rightarrow 1\text{palmitic acid} + 7\text{CO}_2 + 8\text{CoA} + 14\text{NADP}^+ + 6\text{H}_2\text{O}$$

The biosynthetic pathway for this scheme is detailed in Fig. 5, and the evidence in support of a similar pathway existing in insects is documented in the reviews of Gilbert (1967) and Bailey (1975). Consistent with this pathway is the dominance of even-numbered fatty acids and the reviews of Fast (1964, 1970) confirm that this condition is satisfied in insects. It is known that propionyl-CoA can replace acetyl-CoA to some degree in the initial condensation reaction and it is assumed that this mechanism accounts for the presence of small amounts of odd-numbered fatty acids in many species.

In vertebrates and microorganisms the enzymes of fatty acid synthesis are bound to protein in a cytoplasmic enzyme complex, and a similar fatty acid–synthetase complex has recently been isolated from whole insect homogenates of the blow fly *Lucilia sericata* (Thompson *et al.*, 1975). The sole pyridine nucleotide requirement for the insect synthetase is NADPH, a finding that concurs with earlier observations of fatty acid synthesis in the fat body of a moth and a locust.

The cytoplasmic fatty acidsynthetase system is responsible for *de novo* synthesis of saturated fatty acids, but additional systems function to elongate and desaturate the products of the cytoplasmic system and fatty acids of dietary origin. A mitochondrial system results in elongation of preformed fatty acids by successive additions of acetyl-CoA along a pathway that is

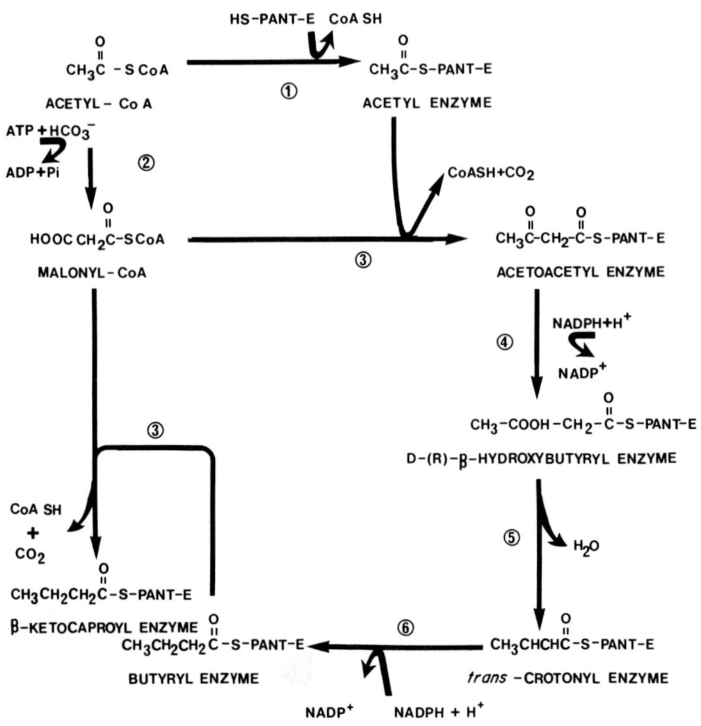

Fig. 5. Proposed pathway of cytoplasmic fatty acid synthesis. (After White *et al.*, 1973.) Notes: (1) Purified acetyl-CoA carboxylase contains covalently bound biotin which is essential for the reaction. (2) Acyl intermediates are bound as thiol esters to a heat-stable acyl carrier protein in the same fashion as to CoA, which also contains 4^1-phosphopantetheine as its acyl-binding site. Numbers in circles refer to enzymes according to the following key: (1) acetyl transacylase, (2) acetyl-CoA carboxylase, (3) condensing enzyme, (4) acetoacetyl enzyme reductase, (5) enoyl enzyme hydrase, (6) enoyl enzyme reductase.

essentially a reversal of that described for β-oxidation of fatty acids (Section V,B). Under some conditions this pathway may achieve fatty acid synthesis from acetyl-CoA, and Goldin and Keith (1968) demonstrated this capacity in isolated mitochondria of *Drosophila melanogaster*. These workers also reported a capacity within mitochondrial preparations for desaturation of stearate to oleate, a finding which is at variance with that of Tietz and Stern (1969) for isolated mitochondria of locust fat body. Monounsaturated fatty acids are normally formed in a microsomal system with the saturated fatty acid being directly desaturated to the monoenoic equivalent. Tietz and Stern (1969) incubated fat body microsomes aerobically in the presence of reduced pyridine nucleotides and found ready desaturation of stearate and palmitate to the equivalent Δ^9-monoenoic acids, an observation that has been confirmed in several other species. An alternative route of monounsaturated fatty acid synthesis has been described in microorganisms and vertebrates, and involves β-α dehydration of medium-chain β-hydroxy acids with subsequent chain elongation of the resulting Δ^3-enoates by the fatty acid–synthetase complex. Keith (1967) demonstrated the existence of an alternative pathway for oleate production in axenic cultures of *D. melanogaster* by blocking the direct desaturation of stearate. A subsequent study in which stearate desaturase activity was similarly blocked tested the ability of all even-chained fatty acids (C_2–C_{18}) to serve as precursors of C16:1 and C18:1. The results indicated that C10:0 and C12:0 were favored precursors for C16:1 and C18:1 respectively, and supported the proposal that formation of the Δ^9-monoenoic acid may proceed according to the following sequences:

1. 10:0 → Δ^3-10:1 → Δ^5-12:1 → Δ^7-14:1 → Δ^9-16:1
2. 12:0 → Δ^3-12:1 → Δ^5-14:1 → Δ^7-16:1 → Δ^9-18:1

Further evidence in support of this pathway occurring in *Drosophila* was obtained by isolating *cis*-5-tetradecenoic (Δ^5-14:1) and *cis*-7-tetradecenoic (Δ^7-14:1) acids from the insect.

Thus it is apparent that several mechanisms for fatty acid synthesis occur in insects and this permits considerable variation in the fatty acid composition between and within species. The factors that influence fatty acid synthesis in the fruit fly *Ceratitis capitata* have been extensively investigated in whole insect homogenates by Professor A. M. Municio and his associates. These studies indicate marked differences between developmental stages in biosynthetic capacity, cofactor requirements, and in the nature of the fatty acids that are synthesized. The data suggest that different fatty acid–synthetase systems operate in larval and adult insects. The differing habits of larvae, pharate adults, and adults predispose these stages to varying fatty acid requirements, and the observations on *C. capitata* may be presumed to be manifestations of these differences. It must be recognized, however, that observations on whole insect homogenates do not necessarily reflect the situation in individual tissues. Insects host a variety of symbiotic microorganisms and it is possible that some developmental differences attributed to the insect system may be due to changes in symbiont populations.

Dietary influences on fatty acid synthesis have been reported in several species and it is probable that other environmental perturbations influence the biosynthetic pathway to alter the final fatty acid composition of the insect.

B. Acylglycerols

The free fatty acid content of insect tissues is relatively low, with most fatty acids present in esterified form as di- or triacylglycerols or phosphoglycerides. Developmental changes in the fatty acid complement of triacylglycerols of the fruit fly *C. capitata* have been demonstrated and this variation attributed to substrate differences in acyltransferase activity. Two pathways for the formation of acyl esters of glycerol in the gut have been described (Section III,B: Fig. 3) and these pathways appear to exist also in other insect tissues as evidenced by studies on the locust (Tietz *et al.*, 1975). These workers have demonstrated that a major site of di- and triacylglycerol synthesis is the fat body, the microsomal fraction of which contains the enzymes of the α-glycerophosphate pathway and the transacylase pathway. These studies suggest that the transacylase system serves to synthesize diacylglycerols for release into hemolymph and subsequent utilization in energy-requiring processes, whereas the glycerophosphate pathway functions in lipid deposition. It has been suggested that the availability of glycerol 3-phosphate to serve as an acyl acceptor may determine which of the two pathways predominates. This suggestion is based upon the reasoning that the substrate will be present in low concentrations during periods of high energy demand, at which time fatty acyl-CoA will enter the transacylase pathway for synthesis to energy-supplying diacylglycerols; however, during periods of feeding, when glucose levels are high, glycerol 3-phosphate will be present in appreciable quantities to participate in the formation of storage triacylglycerols. This hypothesis is not supported by recent observations in the cockroach, which indicate elevated levels of glucose, and presumably glycerol 3-phosphate, during exercise. Furthermore the hydrolysis of triacylglycerol to 2-monoacylglycerol and subsequent acylation of this compound to diacylglycerol would result in elevated free fatty acid levels within the cell if glycerol 3-phosphate were unavailable to accept the free acyl group. No accumulation of fatty acid has been demonstrated and thus it is reasonable to suggest that the additional free fatty acid moiety is synthesized to tri- or diacylglycerol by the α-glycerophosphate pathway. It is possible that the glycerophosphate pathway operates at a rate that is dependent on the availability of free fatty acid and glycerol 3-phosphate, whereas the rate of the transacylase pathway is related to the availability of 2-monoacylglycerol, which in turn is dependent upon the lipolytic activity of the tissue.

C. Phosphoglycerides

The complement of phosphoglycerides found in insects is similar to that of other animal groups although quantitative differences occur in certain species. Dictyoptera, Lepidoptera, and Hymenoptera resemble vertebrates in having phosphatidylcholine as the major phospholipid type, but in Diptera a large proportion of the phospholipid is present as phosphatidylethanolamine.

The major site of phosphoacylglycerol synthesis appears to be the fat body and specifically the microsomal fraction of this tissue, although the cytosol must also be present. Williams *et al.* (1974) studied the incorporation of [^{32}P]orthophosphate into the major phosphoglycerides of the microsomes and mitochondria and concluded that most of the mitochondrial phosphoglycerides are synthesized in the endoplasmic reticulum from which they are subsequently transferred to the mitochondria. The enhanced phosphoacylglycerol biosynthesis observed when mitochondria are added to the incubation is thought to result from mitochondrial phosphatidic acid biosynthesis.

Substantial evidence exists to support the proposal that phosphatidylcholine and phosphatidylethanolamine are synthesized by the analogous triple enzyme pathways illustrated in Fig. 6. By this sequence, free choline, derived from a dietary source, is phosphorylated to form phosphorylcholine, which subsequently reacts with cytidine triphosphate to yield cytidinediphosphate choline. This compound is then converted to phosphatidylcholine by reaction with 1,2-diacylglycerol, formed as previously described from phosphatidic acid. Studies supporting the existence of this pathway have been performed under *in vivo* conditions and using tissue

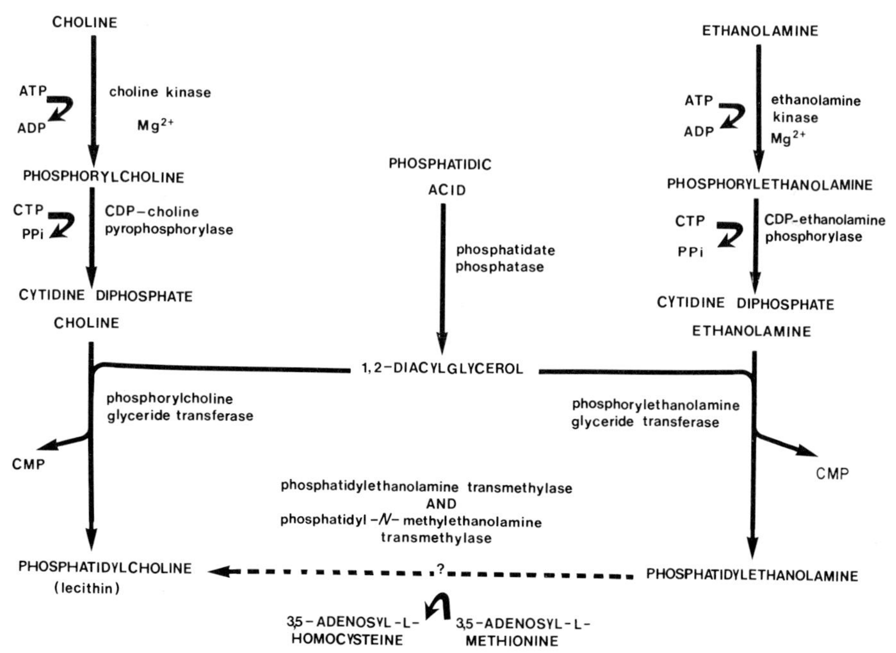

Fig. 6. Proposed pathway of synthesis of phosphatidylcholine and phosphatidylethanolamine in insects.

homogenates. The choline kinases of cockroach gut and blow fly larval fat body have been partially purified and characterized in the laboratories of E. Hodgson and R. W. Newburgh. These studies indicate that the enzyme(s) has (have) requirements for magnesium and adenosine triphosphates similar to those of the vertebrate enzyme. The choline kinase of blow fly was further shown to have affinity for ethanolamine and, on the basis of observations of enzyme stability, pH curves, and mutual inhibition experiments, it was concluded that two separate kinase enzymes are present, both of which can phosphorylate choline, but only one of which can phosphorylate ethanolamine.

An additional pathway for *de novo* synthesis of phosphatidylcholine from phosphatidylethanolamine has been described in vertebrates. This pathway involves the stepwise methylation of ethanolamine with *S*-adenosylmethionine (Fig. 6) and the presence of a transmethylation system capable of effecting this transformation has been reported in the dried fruit moth. More recently the participation of adenosylmethionine methyltransferase in the biosynthesis of phosphatidylcholine has been described in larvae and pharate adults of the fly *C. capitata*. These findings are at variance with a substantial volume of evidence that suggests that in many species the transmethylation pathway does not operate. In the light of such conflicting evidence it would be imprudent to accept or deny existence of the transmethylation pathway. It must be stated, however, that substantiation of the pathway will be achieved only when the possible involvement of intestinal symbionts has been satisfactorily excluded.

Williams *et al.* (1974) studied the incorporation of [^{32}P]phosphate and [^{14}C]ethanolamine into flight muscle phospholipids of the blow fly in the presence and absence of the protein synthesis inhibitor, cycloheximide. They found that the inhibitor prevented incorporation of inorganic phosphate, but not ethanolamine, into phosphatidylethanolamine, and concluded that different biosynthetic pathways are involved. They suggested that ethanolamine enters the phosphatidylethanolamine preferentially by base-exchange reaction whereas inorganic phosphate enters by a pathway of *de novo* synthesis.

Lambremont (1972a,b) investigated the synthesis of acyl and ether bonds in the phosphoglycerides of *Heliothis virescens,* and showed that fatty acids and fatty alcohols are readily incorporated into both acyl and ether bonds in insect phospholipids, with acyl-bond formation predominating over synthesis of ether bonds. The results of these studies suggest that insects synthesize glyceryl ethers by a metabolic pathway similar to that found in other organisms. Thus glyceryl ether is formed by reaction of long-chain alcohols (formed by reduction of fatty acids) with dihydroxyacetone phosphate, followed by reduction of the keto group, acylation by a fatty acyl-CoA, and finally liberation of phosphate.

The metabolism of the lesser phospholipids, ceramide phosphorylethanolamine, phosphatidylinositol, phosphatidylserine, and phosphatidylglycerol was investigated in housefly larvae (Hildenbrandt *et al.*, 1971). It was found that microsomal preparations (40,000–90,000 *g* sediment) have the capacity to convert these compounds to the respective glycerophosphoryl derivatives. In this study it was also shown that the metabolism of ceramide phosphorylethanolamine by these insects proceeds along the following pathways in an analogous manner to that described for mammalian systems:

$$\text{Ceramide phosphatidylethanolamine} + \text{HOH} \rightarrow \text{ceramide} + \text{phosphatidylethanolamine}$$
$$\text{Ceramide} + \text{HOH} \rightarrow \text{fatty acids} + \text{sphingosine}$$

Apart from this study and isolated reports of an exchange reaction between phosphatidylserine and phosphatidylethanolamine the biosynthesis of lesser phospholipids has been neglected by insect biochemists.

D. Cuticular Lipids

The lipid content of insect cuticle has been described by previous authors (Locke, 1973; Hackman, 1973; Jackson and Blomquist, 1976). Of the several lipid classes that have been extracted from insect cuticle, hydrocarbons comprise the major proportion. The hydrocarbon content includes alkanes (*n*-alkanes, 3-methylalkanes, internally branched monomethylalkanes and internally branched dimethylalkanes), alkenes, and alkadienes.

In spite of the plethora of information available on cuticular lipid composition, few definitive studies have elucidated the biosynthetic pathways; accordingly, we must depend upon circumstantial evidence to conclude that alkane synthesis in insects proceeds according to the elongation–decarboxylation pathway proposed for plants (Kolattukudy, 1968). This pathway involves the elongation of a preformed fatty acid by 2C units on an enzyme complex until the chain length of the acid reaches C_{30} or C_{32}. It then undergoes decarboxylation with subsequent release of the alkane. During the elongation process some long-chain fatty acids are released prematurely from the complex. These may be reduced to fatty alcohols or esterified, and may ultimately serve as precursors for wax ester formation. The elongation–decarboxylation pathway could also account for the presence of unsaturated paraffins through the elongation of an unsaturated fatty acid. Thus the dominant 6,9-heptacosadiene of cockroaches may arise by elongation and subsequent decarboxylation of 18:1. Similarly, branched alkanes may be formed by the substitution of methylmalonyl-CoA, or other precursor, for malonyl-CoA during the elongation pathway. The incorporation of labeled acetate and propionate into cuticular hydrocarbon of the cockroach has indicated that the major pathway for synthesis of 3-methylalkanes involves the incorporation of propionate during the penultimate step of the elongation decarboxylation pathway.

A dietary source of *n*-alkanes has been demonstrated in the grasshopper and the tobacco hornworm, although in both cases the insect is capable of synthesizing the paraffins independently. Myriapods, by contrast, are entirely dependent on the diet for their supply of *n*-alkanes. In addition to the influence of diet on hydrocarbon synthesis, the hydrocarbon content of insects has been shown to be affected by age, sex and circadian fluctuations.

Few data are available on the biosynthesis of other cuticular lipid fractions. Secondary alcohol formation involves the incorporation of one atom of molecular oxygen through the action of a mixed-function oxygenase-type enzyme. Reduction of palmitic acid and stearic acid to the corresponding fatty alcohol has been demonstrated, and it is proposed that a similar reduction of longer chain fatty acids results in the formation of cuticular primary alcohols.

VII. ENDOCRINE REGULATION OF LIPID METABOLISM

The lipid content of insects fluctuates during the animal's life history. The changes reflect variations in the metabolic balance between lipid synthesis (including intake) and lipid utilization, and result from direct or indirect action of hormones and/or neurohormones. Thus such processes as development, migration, diapause, reproduction, and flight, all of which are under endocrine control, influence lipid metabolism. Gross fluctuations have been documented in previous reviews (Fast, 1964, 1970; Gilbert, 1967; Chen, 1973; Agrell and Lundquist, 1973; Downer and Matthews, 1976) and only demonstrated instances of endocrine involvement will be described below. Various aspects of endocrine regulation of lipid metabolism have been discussed in the recent reviews by Gilbert and Chino, 1974; Goldsworthy and Mordue, 1974; Bailey, 1975; Steele, 1976).

A. Adipokinetic Factor(s)

Aqueous extracts of corpora cardiaca influence lipid levels in the fat body and hemolymph of several species. In locusts and the mealworm this influence is manifested in elevated hemolymph diacylglycerol levels at the expense of fat body reserves of triacylglycerol, whereas in the cockroach the gland extracts cause a hypolipemic response with hemolymph triacylglycerol and diacylglycerol levels being depressed and fat body triacylglycerol elevated. The cockroach response is most pronounced when the donor glands are derived from female insects (Steele, 1976), a finding that is particularly interesting in light of the sexual differences in response to corpus allatum hormone that have been reported for fat body of *D. melanogaster*. The interspecific differences in response to corpora cardiaca extracts do not appear to result from evolutionary changes in the chemical nature of the adipokinetic factor(s). Cockroach extracts injected into locusts elicit the typical locust hyperlipemic response, whereas locust extracts produce hypolipemia when injected into cockroaches; thus the evolutionary changes have occurred most probably at the site of action of the hormone.

Extensive studies of the release and action of adipokinetic factor(s) in the locust have been conducted by G. J. Goldsworthy and his associates, and these have provided useful information on the probable physiological significance of the adipokinetic effect in this insect (see review in Goldsworthy and Mordue, 1974). The adipokinetic factor(s) is (are) stored in and released from the glandular lobes of the locust corpus cardiacum and the amount present varies with the developmental stage of the insect. Hemolymph from flown locusts causes elevated diacylglycerol levels when injected into resting insects, and additional evidence supporting the release of adipokinetic factor(s) from corpora cardiaca during flight is provided by the demonstration that diacylglycerols of similar fatty acid composition are released by fat body in response to aqueous extracts of corpora cardiaca and during flight. The precise physiological role of the factor(s) remains undetermined, although it appears to involve more than simple homeostatic regulation of hemolymph lipid levels; rather the factor(s) may serve to provide appropriate substrates for metabolism by working flight muscle. It has long been recognized that locust flight muscle catabolizes carbohydrate during the early stages of flight with lipid assuming increasing dominance as flight continues. Removal of the glandular lobes of the corpora cardiaca results in impaired flight function, and it has been shown that the factor(s) contained within these lobes effect not only the mobilization of fat body triacylglycerol, but also stimulate the utilization of lipid by muscle with concomitant suppression of carbohydrate metabolism. Thus the metabolic switch from carbohydrate to lipid utilization that is observed in the locust may be regulated by the secretions of the corpora cardiaca. The apparent duplicity of action of "adipokinetic hormone" may reflect the cumulative effects of two or more factors. The corpus cardiacum is known to contain and release several neurohormones (Goldsworthy and Mordue, 1974; Steele, 1976) and until these factors have been isolated, specific physiological roles cannot be ascribed to individual molecules (see Stone *et al.*, 1976, for characterization of hormone).

Lipid is a better substrate than carbohydrate for long migratory flights because of its higher caloric content per unit weight and the greater yield of metabolic water following oxidation. Thus the switch to lipid metabolism during flight is metabolically appropriate. However the rationale for utilization of carbohydrate during the initial stages of flight is not understood. Flight speed is greater when carbohydrate is being used and it is possible that use of this substrate may enable the thoracic temperature to achieve optimal levels more readily (see Section V,B). It is also appropriate to mention that hemolymph diacylglycerol levels are low in the resting insect whereas muscle glycogen and hemolymph trehalose levels are high; thus carbohydrate constitutes a more readily available reservoir of metabolic energy during the early stages of flight.

The observation that hemolymph trehalose levels drop markedly during the early stages of flight suggested that the release of adipokinetic factor(s) might be influenced by hemolymph trehalose levels. However, the injection of flying locusts with trehalose failed to produce any differences in hemolymph lipid concentrations from those observed in insects receiving saline injections.

B. Corpus Allatum Hormone

The corpus allatum synthesizes and releases hormones that play important roles in developmental and reproductive regulation (Gilbert and King, 1973). Changes in lipid metabolism effected through the corpus allatum reflect the functional role of this gland in controlling developmental and reproductive processes. The active molecule has been isolated and characterized in several insect species, and these studies reveal that often more than one molecular form of the basic homosesquiterpenoid structure is represented in an individual species. However no physiological significance has yet been ascribed to the different molecules.

It has long been recognized that allatectomy results in the accumulation of fat body triacylglycerol, and the hypertrophy can be reversed by implantation of active corpora allata into the allatectomized individuals: however the *modus operandi* of this effect awaits elucidation. Corpus allatum hormone induces vitellogenin synthesis in several species. This action appears to involve the induction of specific protein synthesis and has led to the suggestion that lipid accumulation in allatectomized insects results from the lack of an appropriate protein acceptor to effect glyceride release from fat body. However, the proposal does not accommodate the observation that lipid release from fat body occurs independently of vitellogenin concentrations. Indeed the major vitellogenic lipoprotein fails to take up [^{14}C]label from fat body that has been prelabeled with [^{14}C]fatty acid, whereas DGLP-I efficiently accumulates label. In addition it has been demonstrated that diglyceride release from locust fat body into hemolymph can proceed even when protein synthesis is blocked (Section IV, B). Thus it is necessary to involve some additional mechanism to explain the allatectomy-induced accumulation of lipids in fat body. Gilbert (1967) reported that inclusion of corpora allata with incubated fat body tissue suppresses the incorporation of preformed fatty acids into triglycerides, thus suggesting a direct effect of the corpus allatum hormone upon lipid biosynthesis. However, these observations might also be explained in terms of altered rates of oxidative metabolism (Steele, 1976), possibly resulting from the contamination of corpus allatum tissue with the closely associated corpus cardiacum tissue.

Fat body hypertrophy following allatectomy does not occur in all species. The operation has no apparent effect on lipid deposition or utilization in the mosquito, *Aedes taeniorhynchus* and implantation of female corpora allata into male *Drosophila* causes hypertrophy of fat cells in the male hosts. A more general anomaly occurs in the larval stages of most species. The fat body of these immature stages accumulates lipid in spite of a high titer of juvenile hormone. The intraspecific differences in hormonal response are difficult to reconcile, although it is interesting to reflect on the observation that several different molecular forms of the hormone may be present in the corpus allatum of an individual species. If these are not physiological homologues, but distinct hormones with separate physiological roles, the anomalous observations could be explained by postulating sexual and developmental differences in the relative amounts of the different molecules.

A final consideration in assessing the contribution of the corpus allatum to lipid metabolism in insects concerns the interrelationships of the neurosecretory cells and the corpus allatum. Allatectomy is a drastic operation and among the side effects associated with removal of the glands is reduced release of neurosecretory material from the cerebral neurosecretory cells. Similarly when active corpora allata are implanted into starved locusts, increased synthesis of neurosecretory products and increased axonal transport are observed. Among the neurosecretions that may be affected in this way are the adipokinetic hormone(s) and the factors that influence hemolymph volume and consequently hemolymph lipid concentrations. These considerations combined with the lack of reliable *in vitro* systems and purified neurohormones prevent an absolute understanding of the precise action of corpus allatum hormone on lipid metabolism.

C. Ecdysone

The role of the insect molting hormone, ecdysone, in insect metamorphosis has been established for many years (see review in Gilbert and King, 1973). The hormone is released as α-ecdysone from the prothoracic glands in response to a trophic brain hormone. Although α-ecdysone is capable of eliciting a specific independent response from target tissues, this hormone is believed to function primarily as a precursor, being converted to the more active β-ecdysone. The most obvious effect of β-ecdysone is to initiate the apolytic processes whereby epidermal tissue is retracted from the old cuticle prior to the molt. In addition it has been suggested that β-ecdysone may favor cuticle deposition (Sedlak and Gilbert, 1975), and Armold and Regnier (1975a,b) have shown that the hormone stimulates hydrocarbon biosynthesis in the blow fly *S. bullata* at the time of pupariation. This effect was most pronounced when isolated integuments were investigated for their hydrocarbon-synthesizing capacity. Inclusion of β-ecdysone in the incubation medium enhanced hydrocarbon synthesis by almost tenfold.

The influence of ecdysone is not limited to the cuticle, and a vitellogenic property has been attributed to this hormone in the mosquito. In *A. aegypti* the stimulus of a blood meal brings about release of an egg-development neurosecretory hormone from the brain (Hagedorn, 1974). This hormone causes the resting ovary to release α-ecdysone, which induces vitellogenin synthesis in the fat body. The effect can be reproduced with β-ecdysone and it seems probable that the α-ecdysone is converted to the β-form by the fat body. This observation may help to explain the enhanced lipid synthesis reported in male *Drosophila* when a single ovary is transplanted into a male host although, as has been pointed out previously (Section IV,A), the lipid content of vitellogenin in Lepidoptera is not sufficient to account for all the lipid deposited in the egg.

VIII. CONCLUSIONS

Several metabolic pathways that occur in vertebrates have been described in insects; however it is also important to recognize those aspects of lipid metabolism in which the two animal classes differ. The hemolymph of most insect species contains high concentrations of diacylglycerol and low levels of free fatty acid, and this reflects different mechanisms of lipid release and uptake by insect tissues compared with vertebrate tissues. The lack of steroidogenesis in insects provides another major difference from the vertebrate system, and may have contributed to the development of the corpus allatum hormones. Finally, the important role of lipids in protecting insects from desiccation has resulted in certain lipoidal compounds and pathways assuming greater prominence in insects than in vertebrates. These differences, combined with the unique physiology of insects serve to identify the distinctive nature of lipid biochemistry in insects and caution against direct extrapolation from the vertebrate system.

The development of reliable microanalytical techniques during the last decade has helped to advance our understanding of the subject and has permitted investigations to be conducted at the tissue and organ level. It remains for these studies to be extended and, in an animal class that shows enormous diversity of habit and function, for the many exceptions to each generalized rule to be identified.

ACKNOWLEDGMENTS

This chapter is dedicated to the memory of my friend and colleague Dr. Jonathan R. Matthews whose untimely death prevented this contribution from being enriched by his perspicacity and scholarship. Dr. L. L. Jackson and Dr. J. E. Steele kindly provided me with manuscripts of unpublished review articles. I thank A. G. D. Hoffman for several useful discussions and Drs. H. Chino, M. Hoshi and S. M. Smith for critical reading of the manuscript. Studies from my laboratory were financed by an operating grant from the National Research Council of Canada.

GENERAL REFERENCES

Agrell, I. P. S., and Lundquist, A. M. (1973). *In* "The Physiology of Insecta" (M. Rockstein, ed.), 2nd Ed., Vol. 1, pp. 159–247. Academic Press, New York.
Altman, P. L., and Dittmer, D. S. (1968). "Metabolism," Committee on Handbooks pp. 148–167. Fed. Am. Soc. Exp. Biol., Washington, D.C.
Armold, M. T., and Regnier, F. E. (1975a). *J. Insect Physiol.* **21,** 1581.
Armold, M. T., and Regnier, F. E. (1975b). *J. Insect Physiol.* **21,** 1827.
Bailey, E. (1975). *In* "Insect Biochemistry and Function" (D. J. Candy and B. A. Kilby, eds.), pp. 89–176. Chapman & Hall, London.
Bloch, K. (1965). *Science* **150,** 19.
Bollade, D., Paris, R., and Moulins, M. (1970). *J. Insect Physiol.* **16,** 45.
Candy, D. J., Hall, L. J., and Spencer, I. M. (1976). *J. Insect Physiol.* **22,** 583.
Chen, P. S. (1973). "Biochemical Aspects of Insect Development." Karger, Basel.
Chino, H., Downer, R. G. H., and Takahashi, K. (1977). *Biochim. Biophys. Acta* **487,** 508.
Chino, H., and Gilbert, L. I. (1971). *Insect Biochem.* **1,** 337.
Chino, H., Yamagata, M., and Takahashi, K. (1976). *Biochim. Biophys. Acta* **441,** 349.
Clayton, R. B. (1964). *J. Lipid Res.* **5,** 3.
Clayton, R. B. (1971). *In* "Aspects of Terpenoid Chemistry and Biochemistry" (T. W. Goodwin, ed.), pp. 1–27. Academic Press, New York.
Collatz, K. G. and Mommsen, T. (1974). *J. Comp. Physiol.* **94,** 339.
Cook, B. J., Nelson, D. R., and Hipps, P. (1969). *J. Insect Physiol.* **15,** 581.
Crabtree, B., and Newsholme, E. R. (1972). *Biochem. J.* **130,** 697.
Dadd, R. H. (1963). *Adv. Insect Physiol.* **1,** 47.
Dadd, R. H. (1970). *In* "Chemical Zoology" (M. Florkin and B. T. Scheer, eds.), Vol. 5, pp. 117–145. Academic Press, New York.
Day, M. F., and Waterhouse, D. F. (1953). *In* "Insect Physiology" (K. D. Roeder, ed.), pp. 273–330. Wiley, New York.
Desnuelle, P., and Savary, P. (1963). *J. Lipid Res.* **4,** 369.
Downer, R. G. H., and Matthews, J. R. (1976). *Am. Zool.* **16,** 733.
Eisner, T. (1955). *J. Exp. Zool.* **130,** 159.
Fast, P. G. (1964). *Mem. Entomol. Soc. Can.* **37,** 1.
Fast, P. G. (1970). *Prog. Chem. Fats Other Lipids* **2,** 181.
Gilbert, L. I. (1967). *Adv. Insect Physiol.* **4,** 69.
Gilbert, L. I., and Chino, H. (1974). *J. Lipid Res.* **15,** 439.
Gilbert, L. I., and King, D. S. (1973). *In* "The Physiology of Insecta" (M. Rockstein, ed.), 2nd Ed., Vol. 1, pp. 249–370. Academic Press, New York.
Gilbert, L. I., and O'Connor, J. D. (1970). *In* "Chemical Zoology" (M. Florkin and B. T. Scheer, eds.), Vol. 5, pp. 229–254. Academic Press New York.

Gilby, A. R. (1965). *Annu. Rev. Entomol.* **10,** 141.
Gilmour, D. (1965). "The Metabolism of Insects." Freeman, San Francisco, California.
Goldin, H. H., and Keith, A. D. (1968). *J. Insect Physiol.* **14,** 887.
Goldsworthy, G. J., and Mordue, W. (1974). *J. Endocrinol.* **60,** 529.
Hackman, R. H. (1973). *In* "The Physiology of Insecta" (M. Rockstein, ed.), 2nd Ed., Vol. 6, pp. 216–270. Academic Press, New York.
Hagedorn, H. H. (1974). *Am. Zool.* **14,** 1207.
Hildenbrandt, G. R., Abraham, T., and Bieber, L. L. (1971). *Lipids* **6,** 508.
Hipps, P. P., and Nelson, D. R. (1974). *Biochim. Biophys. Acta* **341,** 421.
Hoffman, A. G. D., and Downer, R. G. H. (1976). *Can. J. Zool.* **54,** 1165.
House, H. L. (1973). *In* "The Physiology of Insecta" (M. Rockstein, ed.), 2nd Ed., Vol. 5, pp. 63–117. Academic Press, New York.
Jackson, L. L., and Blomquist, G. J. (1976). *In* "Chemistry and Biochemistry of Natural Waxes" (P. E. Kolattukudy, ed.). Elsevier, Amsterdam.
Jensen, R. G. (1971). *Prog. Chem. Fats Other Lipids* **11,** Part 3, 347–394.
Johnston, J. M. (1970). *In* "Comprehensive Biochemistry" (M. Florkin and E. H. Stotz, eds.), Vol. 18, pp. 1–18. Elsevier, Amsterdam.
Keith, A. D. (1967). *Comp. Biochem. Physiol.* **21,** 587.
Kolattukudy, P. E. (1968). *Science* **159,** 498.
Krysan, J. L., and Guss, P. L. (1973). *Biochim. Biophys. Acta* **296,** 466.
Lambremont, E. N. (1971). *Insect Biochem.* **1,** 14.
Lambremont, E. N. (1972a). *J. Insect Physiol.* **18,** 581.
Lambremont, E. N. (1972b). *Lipids* **7,** 528.
Lipke, H., and Fraenkel, G. (1956). *Annu. Rev. Entomol.* **1,** 17.
Locke, M. (1973). *In* "The Physiology of Insecta" (M. Rockstein, ed.), 2nd Ed., Vol. 6, pp. 123–213. Academic Press, New York.
Morley, N. N., Kuksis, A., and Buchnea, D. (1974). *Lipids* **9,** 481.
Rodriguez, J. G., ed. (1972). "Insect and Mite Nutrition," North-Holland Publ., Amsterdam.
Sacktor, B. (1970). *Adv. Insect Physiol.* **7,** 267.
Sacktor, B. (1975). *In* "Insect Biochemistry and Function" (D. J. Candy and B. A. Kilby, eds.), pp. 1–88. Chapman & Hall, London.
Scoggin, J. K., and Tauber, O. E. (1950). *Iowa State Coll. J. Sci.* **25,** 99.
Sedlak, B. J., and Gilbert, L. I. (1975). *Trans. Am. Microsc. Soc.* **94,** 480.
Spencer, I. M., and Candy, D. J. (1976). *Insect Biochem.* **6,** 289.
Steele, J. E. (1976). *Adv. Insect Physiol.* **12,** 239.
Stevenson, E. (1972). *J. Insect Physiol.* **18,** 1751.
Stone, J. V., Mordue, W., Batley, K. E., and Morris, H. R. (1976). *Nature (Lond.)* **263,** 207.
Strong, F. E. (1963). *Hilgardia* **34,** 43.
Svoboda, J. A., Kaplanis, J. N., Robbins, W. E., and Thompson, M. J. (1975). *Annu. Rev. Entomol.* **20,** 205.
Thompson, S. N., Barlow, J. S., and Douglas, V. M. (1975). *Insect Biochem.* **5,** 571.
Tietz, A. (1965). *In* "Handbook of Physiology" (A. E. Renald and G. F. Cahill, Jr., eds.), Sect. 5, pp. 45–54. Williams & Wilkins, Baltimore, Maryland.
Tietz, A., and Stern, N. (1969). *FEBS Lett.* **2,** 286.
Tietz, A., Weintraub, H., and Peled, Y. (1975). *Biochim. Biophys. Acta* **388,** 165.
Treherne, J. E. (1958). *J. Exp. Biol.* **35,** 862.
Treherne, J. E. (1967). *Annu. Rev. Entomol.* **12,** 43.
Turunen, S. (1974). *Ann. Zool. Fenn.* **11,** 300.
Turunen, S. (1975). *J. Insect Physiol.* **21,** 1521.
Vonk, H. J. (1969). *Comp. Biochem. Physiol.* **29,** 361.
Wagner, R. P., and Selander, R. K. (1974). *Annu. Rev. Entomol.* **19,** 117.

Waterhouse, D. F., and Day, M. F. (1953). *In* "Insect Physiology" (K. D. Roeder, ed.), pp. 331–349. Wiley, New York.
Weintraub, H., and Tietz, A. (1973). *Biochim. Biophys. Acta* **306,** 31.
White, A., Handler, P., and Smith, E. L. (1973). "Principles of Biochemistry," 5th Ed. McGraw-Hill, New York.
Williams, M. L., Bygrave, F. L., and Birt, L. M. (1974). *Insect Biochem.* **4,** 161.
Wills, E. D. (1965). *Adv. Lipid Res.* **3,** 197.

REFERENCES FOR ADVANCED STUDENTS AND RESEARCH SCIENTISTS

Barnes, F. J., and Goodfellow, R. D. (1971). *J. Insect Physiol.* **17,** 1415.
Beenakkers, A. M. T. (1965). *J. Insect Physiol.* **11,** 879.
Beenakkers, A. M. T. (1973). *Insect Biochem.* **3,** 303.
Beenakkers, A. M. T., and Gilbert, L. I. (1968). *J. Insect Physiol.* **14,** 481.
Beenakkers, A. M. T., and Henderson, P. T. (1967). *Eur. J. Biochem.* **1,** 187.
Beenakkers, A. M. T., and Klingenberg, M. (1964). *Biochim. Biophys. Acta* **84,** 205.
Bieber, L. L. (1968). *Biochim. Biophys. Acta* **152,** 778.
Bieber, L. L., and Newburgh, R. W. (1963). *J. Lipid Res.* **4,** 397.
Blomquist, G. J. and Jackson, L. L. (1973). *J. Insect Physiol.* **19,** 1639.
Bridges, R. G. (1972). *Adv. Insect Physiol.* **9,** 51.
Bridges, R. G., and Watts, S. G. (1975). *J. Insect Physiol.* **21,** 861.
Butterworth, F. M., and Bodenstein, D. (1968). *J. Exp. Zool.* **167,** 207.
Butterworth, F. M., and Bodenstein, D. (1969). *Gen. Comp. Endocrinol.* **13,** 68.
Castillon, M. P., Catalan, R. E., Municio, A. M., and Suarez, A. (1971). *Insect Biochem.* **1,** 237.
Castillon, M. P., Catalan, R. E., Jimenez, C., Madariaga, M. A., Municio, A. M., and Suarez, A. (1974). *J. Insect Physiol.* **20,** 507.
Childress, C. C., Sacktor, B., and Traynor, D. R. (1967). *J. Biol. Chem.* **242,** 754.
Crompton, M., and Birt, L. M. (1967). *J. Insect Physiol.* **13,** 1575.
Crone, H. D., Newburgh, R. W., and Mezei, C. (1966). *J. Insect Physiol.* **12,** 619.
D'Costa, M. A., and Birt, L. M. (1966). *J. Insect Physiol.* **12,** 1377.
D'Costa, M. A., and Birt, L. M. (1969). *J. Insect Physiol.* **15,** 1629.
Downer, R. G. H., and Steele, J. E. (1972). *Gen. Comp. Endocrinol.* **19,** 259.
Downer, R. G. H., and Steele, J. E. (1973). *J. Insect Physiol.* **19,** 523.
Fernandez-Sousa, J. M., Municio, A. M., and Ribera, A. (1971). *Biochim. Biophys. Acta* **231,** 527.
Gilbert, L. I., Chino, H., and Domroese, K. A. (1965). *J. Insect Physiol.* **11,** 1057.
Goldsworthy, G. J., and Coupland, A. J. (1974). *J. Comp. Physiol.* **89,** 359.
Goldsworthy, G. J., Coupland, A. J., and Mordue, W. (1973). *J. Comp. Physiol.* **82,** 339.
Goldsworthy, G. J., Johnson, R. A., and Mordue, W. (1972). *J. Comp. Physiol.* **79,** 85.
Goldsworthy, G. J., Mordue, W., and Guthkelch, J. (1972). *Gen. Comp. Endocrinol.* **18,** 545.
Goodfellow, R. D., and Barnes, F. J. (1971). *Insect Biochem.* **1,** 271.
Goodfellow, R. D., Huang, Y. S., and Radtke, H. E. (1972). *Insect Biochem.* **2,** 467.
Goodfellow, R. D., and Liu, G. C. K. (1972). *J. Insect Physiol.* **18,** 95.
Goodfellow, R. D., Liu, G. C. K., Stein, J. P., and Harker, K. (1973). *Insect Biochem.* **3,** 113.
Goodfellow, R. D., Radtke, H. E., Huang, Y. S., and Liu, G. C. K. (1973). *Insect Biochem.* **3,** 61.
Habibulla, A. M., and Newburgh, R. W. (1969). *J. Insect Physiol.* **15,** 2245.
Jackson, L. L. (1970). *Lipids* **5,** 38.

Jackson, L. L., Armold, M. T., and Regnier, F. E. (1974). *J. Insect Physiol.* **4,** 369.
Madariaga, M. A., Mata, F., Municio, A. M., and Ribera, A. (1974). *Insect Biochem.* **4,** 151.
Mayer, R. J., and Candy, D. J., (1969). *J. Insect Physiol.* **15,** 611.
Municio, A. M., Garcia, R., and Perez-Albarsanz, M. A. (1975). *Eur. J. Biochem.* **60,** 117.
Municio, A. M., Odriozola, J. M., Perez-Albarsanz, M. A., and Ramos, J. A. (1974). *Biochim. Biophys. Acta* **360,** 289.
Municio, A. M., Odriozola, J. M., Perez-Albarsanz, M. A., and Ramos, J. A. (1974). *Insect Biochem.* **4,** 401.
Municio, A. M., Odriozola, J. M., Pineiro, A., and Ribera, A. (1972). *Biochim. Biophys. Acta* **280,** 248.
Peled, Y., and Tietz, A. (1973). *Biochem. Biophys. Acta* **296,** 499.
Peled, Y., and Tietz, A. (1974). *FEBS Letts.* **41,** 65.
Peled, Y., and Tietz, A. (1975). *Insect Biochem.* **5,** 61.
Ribeiro, L. P., and Coutardo da Fonseca, C. L. (1975). *Comp. Biochem. Physiol.* **52B,** 523.
Shelley, R. M., and Hodgson, E. (1970). *J. Insect Physiol.* **16,** 131.
Shelley, R. M., and Hodgson, E. (1971). *Insect Biochem.* **1,** 149.
Shelley, R. M., and Hodgson, E. (1971). *J. Insect Physiol.* **17,** 545.
Tietz, A. (1961). *J. Lipid Res.* **2,** 182.
Tietz, A. (1967). *Eur. J. Biochem.* **2,** 236.
Tietz, A. (1969). *Isr. J. Med. Sci.* **5,** 1007.

3

Functional Role of Proteins

MOISES AGOSIN

I. INTRODUCTION*

Of all known chemical compounds, proteins are the most complex and at the same time the most characteristic of living matter. They are present in all viable cells; they are the compounds which, as nucleoproteins, are essential to the process of cell division and, as enzymes and hormones, control many chemical reactions in the metabolism of cells. As the only constituent of viruses, nucleoproteins are synonyms of the most elementary form of life. They are found either free in the cell cytoplasm or associated with other molecules in the various cellular organelles. The recent spectacular progress in our understanding of the biosynthetic process for protein synthesis has been due to the reconstitution of most of its key features in cell-free systems. The sequences of genetic replication are performed by the method of negative–positive image formation through hydrogen bonding between two pairs of purine and pyrimidine bases. There are two nucleic acid transcriptions: DNA to DNA and DNA to RNA (messenger RNA). As a result, polynucleotide sequences are produced which are then translated into amino acid sequences through hydrogen binding between triplet nucleotide codons of messenger RNA and anticodon triplets of aminoacyl-tRNA. This translation is the expression of the genetic code and yields the proteins which characterize an organism.

The traditional classification of proteins is based on solubility, ionic properties, and function. As dynamic constituents of all cells, proteins may be classified on the basis of structural association (Table I). Nonassociated proteins correspond essentially to the globular proteins in the classic classification, which includes the albumins and the globulins. Examples of these proteins are the many enzymes, antibodies, and protein hormones. Other proteins, including enzymes and contractile proteins, are associated with proteins and also correspond to globulins. Nucleoproteins correspond to proteins associated with either nuclear nucleic acids (histones and nonhistones) or with ribosomal nucleic acids. The latter correspond to the protamines of the classic classification. Proteins associated with cellular membranes include lipoproteins, transport proteins, and enzymes. They correspond to the lipo-, glyco- and mucoproteins of the classic classification. Finally, certain specialized proteins, with essentially no enzyme activity and serving as skeleton structures, include the collagens, elastins, keratins and fibroins. They were previously known as insoluble or fibrous proteins. The fibrous proteins may be considered as "passive" proteins consisting of quite

*This broad area of subject matter has been presented as completely as possible within the limitations of space and available information. For additional information on specific subjects related to this present chapter, the reader is referred to Chapters 1, 2, 4, and 5 of this book.

TABLE I
Functional Classification of Proteins

Type	Example
Single proteins	Enzymes Antibodies Hormones Hemoglobins
Associated with proteins	Myosin Actin Tropomyosin Troponin
Associated with membranes	Glycoproteins Lipoproteins Emphores
Structure proteins	Collagen Keratins Elastins Fibroins
Associated with nucleic acids	Ribosomal proteins Histones Nonhistones

rigid molecules containing a high degree of inter- and intrapolypeptide-chain cross-links. In contrast to these fibrous proteins, we may consider the so-called active proteins, which are capable of recognizing and binding other molecules. They include the enzymes, which bind their corresponding substrates; antibodies, which bind their corresponding antigens; transport proteins, which reversibly bind and transport *ligands,* as is the case of hemoglobin; regulatory proteins, which are involved in the binding of ligand leading to activity changes; and the contractile proteins, which by binding a ligand lead to a process resulting in mechanical work. ''Active'' proteins, which are not enzymes but which specifically bind ligands, have been called *emphores.* Since contractile proteins have an important structural purpose, they are usually included in the structural proteins. The same may be said of the histones and nonhistones although we now know they appear to participate in the regulation of gene expression. It is unfortunate that few insect proteins have been obtained in a pure state. As a result, information on the chemical structure of insect proteins is scarce. The subject of this chapter is a correlation of structure and biological function of the proteins, restricted whenever possible to those proteins which have been chemically characterized to a reasonable extent.

II. STRUCTURE PROTEINS

Conveniently grouped under structure proteins are the fibrous or insoluble proteins and some globular proteins that serve a structural role but without substantial enzyme activity. Studies on fibrous proteins are often complicated by the wide variety of cross-links that may join subunits either to one another or to another tissue component. Such cross-links, as in elastin, make it difficult to recognize and isolate the precursor subunit, which makes the examination of primary structure very difficult. In terms of amino acid sequence, it is now clear that structural proteins are no exception to the general rule that primary structure governs the three-dimensional structure. For example, every third residue in the collagen chain must be glycine if the desired triple helix is to form.

A. Contractile Proteins

One of the characteristics of animals is motility, which is associated with the presence of contractile proteins. In insects, as well as other differentiated organisms, these contractile proteins are present in a single tissue, muscle. The major protein components of muscles are actin and myosin, but several other protein components involved in muscle contraction have also been described.

1. Physicochemical Properties

The molecular weight of myosin was determined over 20 years ago and now it appears that a measure of agreement has finally been reached. The high-speed equilibrium method yields values of 4.58×10^2 and 4.68×10^5. Much higher values obtained previously by ultracentrifugation methods are possibly due to the strong tendency of myosin to form a dimer. In denaturing solvents, the myosin molecule dissociates into several polypeptide chains. Reduction and denaturation in guanidine hydrochloride dissociate myosin into heavy and light chains. The heavy chains have a molecular weight which ranges from 1.89×10^5 to 2.12×10^5 depending on the method used. As there is general agreement that the light chains make up less than 15–18% of the total weight, myosin must contain two heavy chains. Under the assumption that the molecular weights of myosin and of the heavy chain are correct, the contribution of the light chains should be $4–7 \times 10^4$ or 9–15% of the total mass. The number of light chains appear to range from two to three according to the source of muscle used. Proteolytic digestion of myosin yields two pieces, light meromyosin (LMM) and heavy meromyosin (HMM). LMM is a fibrous protein, while HMM is a globular one. HMM is also an enzyme (ATPase) catalyzing the hydrolysis of adenosine triphosphate (ATP). The ATPase activity is stimulated by Ca^{2+} and inhibited by Mg^{2+} and is very sensitive to pH. After proteolysis, only HMM retains ATPase activity. Aggregation of myosin leads to the formation of a uniform complex 6000–7000 Å long, with a diameter of 100–200 Å and a particle weight of 6.3×10^6 daltons.

Actin exists in two forms, the globular monomer G-actin and the fibrous polymer with a very high molecular weight, F-actin. Until a few years ago the molecular weight of mammalian muscle G-actin was considered to be $5.5–6.0 \times 10^4$; however, recent determinations support a lower figure, from 4.3 to 4.7×10^4. G-actin contains 374 residues which indicates a molecular weight of 4.1 to 4.2×10^5 daltons. This is similar to the values reported for a variety of animals from trout to invertebrates such as ameba G-actin. The G-actin contains one mole each of ATP and Ca^{2+} (or Mg^{2+}) per mole of actin. Removal of ATP or Ca^{2+} with chelating agents such as EDTA leads to inactivation and dimerization but no further aggregation occurs. If G-actin is exposed to 0.1 *M* salts containing Mg^{2+}, it polymerizes very rapidly to F-actin; in the process, ATP is hydrolyzed to adenosine diphosphate (ADP) and P_i. Since the ADP in F-actin does not readily exchange with ATP, actin cannot be considered as an ATPase. With sonication or mechanical disruption in the presence of ATP, the reaction can be reversed as indicated in Eq. (1).

$$\text{G-Actin-ADP} + \text{ATP} \rightarrow \text{G-actin-ATP} + \text{ADP} \tag{1}$$

Under these artificial conditions, F-actin may be considered an ATPase. It is not clear that hydrolysis of ATP is obligatory for polymerization of G-actin to F-actin; with artificially produced G-actin-ADP, polymerization still takes place. The structure of F-actin appears to correspond to a double-helical array of G-actin molecules, with thirteen to fifteen subunits per turn of the helix. The function of Mg^{2+} and Ca^{2+} in altering the rate of nucleotide exchange by G-actin may be explained as the presence of a single high affinity site for Mg^{2+} or Ca^{2+} in G-actin; occupancy of the site appears to be necessary to stabilize the actin–nucleotide complex. The dissociation rate of ATP is slow when Ca^{2+} is bound and even slower when Mg^{2+} is bound. At low concentration, nucleotide exchange fits a mechanism in which Ca^{2+} dissociates first [Eq. (2) and (3)].

$$\text{Ca}^{2+}\text{-G-actin-ATP} \leftrightarrows \text{Ca}^{2+} + \text{G-actin-ATP} \tag{2}$$

$$\text{G-actin-ATP} \rightarrow \text{G-actin} + \text{ATP} \tag{3}$$

Actomyosin is the molecular complex of actin and myosin. It is obtained by extraction from muscle, first with water to remove glycolytic enzymes and then with 0.6 *M* KCl. After extensive extraction with the second solvent, a highly viscous solution of fairly pure actomyosin is obtained. When this solution is squirted through a narrow hole into a dilute salt solution, the actomyosin precipitates out in the form of long threads. These threads, when exposed to ATP in the presence of K^+ and Mg^{2+}, contract in the most dramatic fashion while ATP is hydrolyzed to ADP and P_i. Insect actomyosin behaves exactly as vertebrate actomyosin although at different ionic strengths; at pH 7.0 and 0°C, it is completely soluble in the presence of KCl concentrations above 0.35 *M* (Gilmour, 1961); it precipitates between 28 and 32% ammonium sulfate in the presence of 0.5 *M* KCl as rabbit actomyosin does. In the presence of ATP, contraction of actomyosin into a plug in the test tube is called "superprecipitation." Insect actomyosin in 0.03–0.11 *M* KCl at pH 7.0 and 25°C shows typical superprecipitation. As in rabbit actomyosin, Mg^{2+} also is required for superprecipitation.

Another muscle protein is tropomyosin A or paramyosin. It corresponds to a highly helical coiled-coil of molecular weight 2.2×10^5 and a length about 130–140 nm. In contrast, tropomyosin B is a completely helical two-stranded coiled-coil of molecular weight $6.3–6.8 \times 10^4$ and a length of 40 nm. The molecule separates into two identical units under reducing conditions.

α-Actinin increases actomyosin ATPase activity under certain conditions, increases the rate of superprecipitation and enhances tension of an actomyosin thread. The protein forms a complex with F-actin, whose formation is inhibited by tropomyosin. The molecular weight is 1.8×10^5 with a sedimentation coefficient of 6.3 S. Apparently, 1 molecule of α-actinin is bound to 10 molecules of G-actin and tropomyosin competes for the α-actinin binding site(s). α-Actinin is apparently associated with the Z line.

Numerous studies have shown that activation of actomyosin ATPase and superprecipitation of actomyosin require a low concentration of Ca^{2+}. The Ca^{2+} requirement is conferred on the system by a combination of tropomyosin B and the protein called troponin. The complete system is known as natural actomyosin (NAM) and the crude tropomyosin B-troponin complex is referred to as natural tropomyosin (NT). Repeated extraction of NAM at low ionic strength removes the factors responsible for the Ca^{2+} requirement, and the resulting system is known as "desensitized" actomyosin (DAM) since ATPase activity is no longer sensitive to low concentrations of Ca^{2+}.

Troponin is not a single protein and it can be separated into three components: troponin I with a molecular weight of 23,000 which inhibits actomyosin ATPase; troponin T which has a strong affinity for tropomyosin, a molecular weight of 37,000, and reverses the inhibition caused by troponin I whether Ca^{2+} is bound or not; and troponin C of molecular weight 18,000, which binds two molecules of Ca^{2+}. When troponin C combines with troponins I and T, it restores the control of actomyosin ATPase by Ca^{2+}. The three troponins appear to be in a ratio of 1:1:1 and one tropomyosin and one troponin per 7 actin monomers.

Electron microscopy studies of insect and vertebrate muscle have shown that many cytological features are common to all fibers: a plasma membrane similar to that limiting other types of cells, variously arranged cylindrical irregular lamillar myofibrils, and a supply of mitochondria ranging from abundant to scarce. Myoglobin, which is present in vertebrate muscle, is absent in insect muscle and is replaced by an extensive complement of tracheoles. The basic elements in all cases are thick and thin filaments. The thick filaments of the myofibril correspond essentially to myosin. Since the thick filament has a bipolar structure, the myosin molecule must be packed in an antiparallel manner in the central region and parallel throughout the rest of the filament. The thin filaments are composed essentially of actin, tropomyosin, and troponin (Taylor, 1972). Antibodies prepared against myosin become associated with the A band while antibodies against actin become associated with the A and I bands. Antibodies against HMM concentrate in the center of the A band and antibodies against LMM concentrate in the distant portions of the A band. When the muscle contracts, the two sets of filaments, thick and thin, slide past one another under the influence of cyclically acting cross-bridges. The latter correspond to projections of the HMM. When the muscle is relaxed, the attachment of HMM to actin is inhibited. Upon stimulation, Ca^{2+} ions are released from the sarcoplasmic reticulum and remove the inhibition of the thin filaments. The myosin can then form cross-bridges to the actin in the thin filament. This attachment is followed by a change in conformation of the myosin that generates tension in the muscle, followed in turn by the release of the cross-bridges from the actin. This repeated cycle of events is associated with the splitting of ATP which provides a source of energy.

2. Insect Contractile Proteins

Insect muscle corresponds to the vertebrate striated muscle. The contractile proteins are in many respects similar to those of vertebrates. The most studied (by Jewell and Rüegg, 1966) is the fibrillar flight muscle which has oscillatory mechanic activity that is not due to activation of the sarcotubular system but is a property of the contractile proteins. Glycerinated fibers of fibrillar muscle require Ca^{2+} for activity as do fibers of nonfibrillar muscle, but they can be further activated by a small amount of stretching. Studies on proteins isolated from fibrillar muscle have shown that the major proteins of the myofibrils are similar to those of vertebrate muscle. Insect actin has been prepared from locust muscle, honey bee thoracic muscle (Gilmour, 1961), and muscle of the blow fly, and of the dung beetle. The amino acid composition of *Phormia regina* G-actin is very close to rabbit muscle actin, although it contains slightly more glutamic acid; however, increases in lysine and amide nitrogen make the net charge of insect invertebrate actin similar. The honey bee myosin is also very similar to rabbit myosin with lengths that vary between 1700 and 2000 Å. The ATPase activity of myosin from honey bee shows essentially the same characteristic of rabbit myosin and a similar amino acid composition. The myosin content in dorsal skeletal muscle of

Lethocerus is 55% of the total myofibril protein. The electrophoretic pattern of the muscle proteins of *Lethocerus cordifanus*, *Lethocerus maximus* and the dung beetle, *Heliocopris japetus* is similar to that of fly myofibrils, except for a greater paramyosin content (Bullard and Reedy, 1973). Extensive extraction of myosin and actin leaves only Z bands. A 95,000 molecular weight component apparently corresponds to α-actinin. The myosin of *L. cordifanus*, *L. maximus*, and *H. japetus* has electrophoretic patterns similar to *P. regina* myosin with a heavy subunit of molecular weight 200,000 and two low molecular weight subunits. An additional protein in *Lethocerus* myosin has a subunit molecular weight of 108,000 and apparently corresponds to paramyosin. There are also light subunits of molecular weight of 17,000 and 30,000 which differ from the molecular weight of the light subunits present in rabbit muscle myosin.

Tropomyosin has been isolated from larval and adult *P. regina*. The sedimentation coefficient of the adult protein is 2.53 S while the sedimentation coefficient for the larval protein is 2.62 S. The molecular weight varies from 65,000 to 84,000 for larval tropomyosin. Apparently, larval tropomyosin is in a more polymerized state than adult tropomyosin; however, the amino acid composition of larval and adult tropomyosin, with the exception of proline, histidine, and phenylalanine, is almost identical. Fifty peptides are essentially the same, but the basic peptides present in the tryptic digest of larval tropomyosin are not found in the adult.

Electron microscopy of actin from *H. japetus* shows filaments with protein bands at intervals of 40 nm. The troponin of vertebrate muscle binds to actin filaments containing tropomyosin at about the same periodicity. The flight muscle from *Sarcophaga bullata* shows cross-bridges projecting from the myosin filaments in groupings which repeat every 14.5 nm along the filament shaft. Each cross-bridge represents one myosin molecule. The groupings are called "crowns" and it has been estimated that there is sufficient myosin for six molecules per crown in *Sarcophaga* and *Lethocerus* (Reedy *et al.*, 1973). Unfractionated extracts of regulatory proteins from flight muscles (tropomyosin, troponins) have an inhibitory effect on the ATPase of desensitized or synthetic actomyosin from rabbit. Fractionation with 40–80% ammonium sulfate yields a fraction containing tropomyosin and proteins of molecular weight of 18,000, 27,000, and 30,000 that correspond to the troponins. These fractions produce Ca^{2+}-sensitive inhibition and only proteins of molecular weight of 18,000 lead to Ca^{2+}-sensitizing activity.

Bullard and Reedy (1973) have extracted and purified paramyosin from the flight muscle of *L. cordifanus*, *L. maximus*, *H. japetus*, and *Pachnoda ephippiata*. The subunit molecular weight estimated by polyacrylamide gel electrophoresis in the presence of sodium dodecyl sulfate is 107,000 and the sedimentation constant, 3.17 S. The molecule as judged by circular dichroism is a two-chain rod. The amino acid composition of insect paramyosin resembles that of molluscan and annelid paramyosins, except that the glutamic/aspartic acids ratio is higher. About 6.3% of the myofibrils is formed by paramyosin in *L. cordifanus* and 9.5% in *Pachnoda*. The ratio of myosin to paramyosin in *L. cordifanus* is 8.2. The amino acid composition of insect muscle proteins is shown in Table II.

Earlier observations on the contractile properties of insect actomyosin have been summarized by Gilmour (1961). In the presence of 0.15 *M* KCl and 1 m*M* $MgCl_2$ a complete clearing of honeybee actomyosin is observed. The drop in viscosity with the addition of ATP and the subsequent recovery as the added ATP is hydrolyzed was first observed in locust actomyosin. The intrinsic viscosity of insect actomyosin is about 2.7 without added ATP and 1.9 with ATP at pH 7.0 and 20°C (½γ0.62) in a mean velocity gradient of 100 sec^{-1}. Ultracentrifugation clearly indicates that insect actomyosin is dissociated into actin and myosin; in addition to the actomyosin (30 S), it reveals a slower component (4–5 S) which corresponds to myosin. Upon addition of ATP in the presence of Mg^{2+}, the myosin peak becomes very sharp and the faster peak, either actomyosin or F-actin, sediments very rapidly.

The ATPase activity of myosin or actomyosin has been extensively studied in insects. The reaction shown in Eq. (4) corresponds to a typical

$$ATP + H_2O \rightarrow ADP + P_i \tag{4}$$

adenosinetriphosphatase. Since most actomyosins from insects are not completely pure, the ATPase shows an apyrase activity rather than an ATPase one, due to the presence of adenylate kinase (Eq. 5). The AMP

$$ATP \rightarrow AMP + PP_i \tag{5}$$

formed is not transformed into IMP since adenylate deaminase is not present in insect actomyosin. Actomyosin ATPase from several insects behaves as the vertebrate enzyme (Chesky, 1975). High concentrations of KCl are inhibitory. The optimal temperature of housefly ATPase is about 45°C, similar to that of other species. One of the characteristics of the Ca^{2+}-activated ATPase is its pH dependency. The activity exhibits a biphasic curve with one peak at pH 9 and another at about pH 6; however, in houseflies, according to Chesky (1975), who has made an extensive study of its kinetics, and related properties, in *Musca domestica,* maximal activity is observed only at alkaline pH. Apparently, the pH dependency may be due to intrinsic characteristics of the light or heavy subunits of the myosin molecules (Maruyama, 1974).

The phenomenon of substrate inhibition of actomyosin ATPase has been well documented in insects, such as in the water bug *L. cordifanus* (Chaplain, 1967). Inhibition of ATPase may be obtained at concentrations of ATP

TABLE II

Amino Acid Composition of Contractile Proteins from Various Insects[a,b]

	Phormic regina			*Heliocopris japetus*		*Pachnoda ephippiata*	
Amino acid	Actin	Myosin	Tropomyosin	Tropomyosin	Paramyosin	Tropomyosin	Paramyosin
Asp	79	88	93	114	72	104	74
Thr	47	30	31	28	44	35	41
Ser	49	36	37	35	43	39	45
Glu	115	158	229	213	204	204	193
Pr	44	19	2.8	3.0	4.8	7.0	7.7
Gly	66	44	16	21	24	39	26
Ala	80	81	102	87	70	94	77
Val	41	30	36	41	53	44	49
Met	24	15	10.6	18	8.2	18	6.8
Ile	40	32	16	23	54	33	57
Le	69	88	98	101	101	99	105
Tyr	28	17	5.8	8.1	13	8.8	16
Phe	27	24	8.5	13	11	19	12
Lys	58	72	87	88	70	84	74
His	16	13	5.8	3.5	16	6.9	17
Arg	46	46	106	56	75	49	71

[a]Adapted from Kominz *et al.* (1962) and Bullard and Reedy (1973).
[b]Results expressed as nmoles/10^5 g of protein.

as low as 1.6 mM in housefly thoracic muscle in the presence of concentrations of Mg^{2+} as low as 5 mM. The K_m values for ATP are about 0.2 mM. ADP inhibits some insect muscle ATPases (Chesky, 1975). Mg^{2+} ions are necessary to activate the ATPase, as shown in *L. maximus*. Maximal activities are obtained at 1 mM Mg^{2+} in the presence of 2×10^{-5} M Ca^{2+}. With a constant concentration of Mg^{2+}, there is a maximal activity at 10^{-6} M Ca^{2+}, but inhibition is observed at 1 mM Ca^{2+}. The activity is inhibited by chelating agents, such as ethylene glycol-bis-(β-aminoethyl ether)-N, N'-tetraacetic acid. From these observations it would appear that the difference between insect and vertebrate actomyosin ATPase is the lower levels of Ca^{2+} required to activate the former. Levels of Ca^{2+} lower than 10^{-7} M only activate insect muscle, and fibrillar muscle actomyosin can be activated by Ca^{2+} as low as 10^{-9} M. The modulation of ATPase activity with necessary Ca^{2+} concentrations in insect muscle involves the troponin system which is essential for oscillatory activity (Lehman *et al.*, 1973). Evidence has been presented for the presence of a dual system of Ca^{2+} regulation in *Lethocerus* muscle associated with the thin filaments and with myosin. It is possible that the low concentrations of Ca^{2+} required for activation of insect actomyosin ATPase may be physiologically related to the reduced size of the sarcoplasmic reticulum (Smith, 1965). In this respect, it has been shown that differences exist between the fibrillar fly muscle and the nonfibrillar leg muscle of the water bug *L. maximus*. Similar observations indicating that smaller and more gradual control of ATPase by Ca^{2+} concentrations is found in other types of fibrillar muscles, those of the honeybee and beetle, while the nonfibrillar muscle of the beetle has a higher and steeper Ca^{2+} sensitivity.

Tryptic degradation of fibrillar muscles from *L. maximus* leads to a loss of sensitivity to Ca^{2+} of actomyosin ATPase. Addition of rabbit tropomyosin partially restores the sensitivity in insect contractile systems; this further suggests that the tropomyosin–troponin regulatory system also is operative in insect muscle. Apparently, there is less tropomyosin in flight muscle than in leg muscle.

The inhibition of insect actomyosin ATPase by ADP appears to be cooperative and this has been interpreted as to correspond to an allosteric phenomenon (Chaplain, 1966). *L. maximus* and *L. cordifanus* have been used to demonstrate this phenomenon. The kinetic for ADP inhibition does not follow the normal competitive form in a Lineweaver and Burk plot; furthermore, the ADP inhibition can be antagonized by the presence of an ATP analogue (β-ribofuranosylpurine 5'-triphosphate) as well as by the substrate in a true competitive way. These observations suggest the presence of at least two stereospecifically distant receptor sites on the actomyosin ATPase: one would bind exclusively the substrate (ATP); the other, the allosteric effectors (ADP). Actomyosin ATPase may exist in two stages, one with low affinity for the substrate and the other with high affinity, the latter being catalytically inactive. Actin changes the equilibrium between the active and inactive states of the ATPase toward the active state; and in the presence of low Ca^{2+} concentration, the inactive state becomes less readily accessible. The type of inhibition produced by ADP also suggests the presence of a third state of the ATPase which has high affinity for ADP. Although the allosteric nature of insect actomyosin ATPase has been challenged, studies with glycerinated fibers show that Ca^{2+} acts in a cooperative manner as measured by ATPase activity and tension development. A characteristic feature is that the activating effect of low Ca^{2+} concentrations on tension development lags behind that on the ATPase activity. Both curves for ATPase activity and tension development are biphasic, but reach a plateau at about 10^{-7} *M* Ca^{2+}. ATPase remains high at 10^{-4} *M* Ca^{2+} but the fibers relax completely. Differences in shape of the tension and ATPase activity curves suggest that they do not involve the same steps in the reaction sequence. This supports the idea that tension in insect fibrillar muscle is generated mainly by the binding of ADP. There is considerable evidence that changes in the sarcoplasmic reticulum Ca^{2+} concentration (from 0.01 to 1.0 μM) during the contraction cycle control the activities of not only myofibrillar ATPase but also of some key enzymes including phosphorylase and outer membrane mitochondrial enzymes (Ebashi and Endo, 1968). Similar changes in Ca^{2+} concentration modify the activity of intramitochondrial enzymes, such as pyruvate dehydrogenase. In addition, mitochondria may play a complementary role to the sarcoplasmic reticulum in the control of changes of intracellular Ca^{2+} concentrations. Thus, changes in intramitochondrial Ca^{2+} in *L. cordifanus* and *Schistocerca gregaria* regulate NAD^+-dependent isocitrate dehydrogenase in order to control energy formation for the contractile process (Zammit and Newsholme, 1976).

The isolation of myosin, actin, and troponin from insect fibrillar muscle (*L. cordifanus, L. maximus, H. japetus*) has revealed some differences between the insect proteins and those of vertebrate muscles. Although the myosins from both kinds of muscle have about the same molecular weight, the light subunit composition differs. Reconstitution of actomyosin, using homologous and heterologous myosins, indicates a low ATPase activity of actomyosins containing insect myosin. The activation of ATPase by tropomyosin, as discussed above, under these conditions may be due to some properties of the insect myosin that prevent its interaction with actin in the absence of tropomyosin. The activation by tropomyosin appears to be the most striking difference between vertebrate and insect muscle.

In summary, it can be stated that only minor differences exist between the contractile proteins of insect and vertebrate striated tissue and that their mechanism of action is essentially similar. The oversimplified model described here leaves many questions unanswered, but it does give some indication as to how a highly sophisticated complex involving an enzyme

(ATPase), two sets of regulatory proteins (tropomyosins and troponins), and an interacting pair of fibrous proteins (actin and myosin) can translate the energy released by the hydrolysis of ATP into mechanical work in insect muscle (Heilmeyer *et al.,* 1976).

B. Fibroins

Fibroins are the protein substances excreted by various species of Arthropoda in special glands and stored there in solution prior to extrusion to form filaments. Silk plays an important part in the lives of many insects; of course the importance of various kinds of silk in the lives of spiders needs no emphasis (Rudall and Kenchington, 1971).

1. Structure

Silk fibroin is most varied in primary and in secondary structures. Apart from the classic extended, parallel β-fibroin types, conformations corresponding to cross-β-types, α-helical types, and even the collagen type have been reported. The silks of hymenopteran aculeata larvae (bees, wasps, ants) are all of the α-helical type. In bee larvae, the silk is stored as microscopic tactoids in the lumen of tubular glands. Both at the optical level and in the electron microscope, these tactoids seem to be cross-striated as if composed of filaments of defined length and with the filament ends all at one level across the tactoids. In bumblebees, some tactoids are present, attached to the interior wall of the gland, but the usual arrangement is that of well-defined laminae or rods. All the fibroins from the Saturniidae have high alanine contents, a somewhat smaller amount of glycine, and serine is the third most abundant component. *Bombyx mori* silk has a somewhat different amino acid composition. The silk of the silkworm has been the most studied. Under natural conditions, it consists of a double filament of fibroin embedded in a globular protein, sericin (Table III). The *B. mori* fibroin is a cocoon protein synthesized in and secreted from the posterior end of the silk glands. As in Bombycidae, the fibroins from the *Anaphe* and *Hypsoides* genera are remarkable in that alanine and glycine form such a high proportion of their mass. Some 94% of the fibroin of *Anaphe moloneyi* consists of these two amino acids and the material is essentially a polymer of these two amino acids, and as such one of the simplest proteins yet investigated (Lucas *et al.,* 1960). The fibroin of the egg stalk of *Chrysopa flava* contains about 40% serine.

The X-ray diffraction patterns of the various insect silks indicate that they differ most markedly in their meridian series of reflection. In the honeybee, bumblebee, colonial wasp, and other related species, measurements indicate an axial period of at least 280 Å. Filaments of *Bombyx* silk appear to be of uniform width, 40 Å, with some evidence that each filament consists of two

TABLE III

Amino Acid Composition of *Bombyx mori* Fibroin and Sericin (Mole %)[a]

Amino acid	Fibroin	Sericin
Gly	44.45	8.60
Ala	29.15	4.00
Val	2.13	1.85
Le	0.49	0.88
Ile	0.64	0.80
Ser	10.42	30.1
Thr	0.73	5.18
Asp	1.49	16.8
Glu	1.01	10.7
Phe	0.64	0.65
Tyr	4.66	3.80
Lys	0.30	3.84
His	0.17	1.39
Arg	0.42	4.09

[a]Modified from Sasaki and Noda (1973), Lucas *et al.* (1958), and Fukuda *et al.* (1955).

rods about 20 Å in diameter. The X-ray photographs of several fibroins studied show patterns of spots and arcs that indicate that parts of the polypeptide chains are mutually arranged in an ordered way so as to form crystals. From measurement of these arcs the spacings of the planes of atoms in the crystal have been calculated, and on this basis, the fibroins may be divided into five groups, the members of each group having the same spacing. Two of these groups, however, may be subdivided on the basis of visual differences in intensity of the two principal equatorial spots. The dimensions of the unit cells for the five main groups are shown in Table IV. The dimensions in two directions are the same for all fibroins. These directions are that along the fiber axis (b) where the repeat distance of 6.95 Å represents two residues of an almost fully extended peptide chain and that between chains,which is the direction of the hydrogen bonds (c). The third dimension (a) is in the direction of the projecting side chains that distinguish the different amino acids, and this dimension varies for the five groups from 9.3 to 15.7 Å. All the fibroins from the Saturniidae belong to the X-ray group 3 (Lucas *et al.*, 1960). They may be divided into groups 3a and 3b and this apparently reflects differences in alanine contents. As it was stated before, the fibroin of *B. mori* (Table III) is unusual in its amino acid composition. A similar type of fibroin appears to be produced in the agrotid, *Bena prasinana*. Group 2a is found among the Thaumetopoeae, in the genera *Anaphe* and *Hypsoides,* where the alanine content is over 50%. The group 2b is found in *Clania sp.*, a member of the Psychidae. The spider silks from *Nephila madagascariaensis* and *Nephila senegalensis* are different in amino

TABLE IV

Unit Cell Dimensions of Silk Fibroins and Synthetic Polypeptides[a]

Polypeptide[b]	Unit cell dimension (Å) (fiber axis)			Species
	a	b	c	
Silk fibroin (1)	9.3	6.95	9.44	*Bombyx mori*
Silk fibroin (2a)	10.0	6.95	9.44	*Anaphe moloneyi*
Silk fibroin (2b)	10.0	6.95	9.44	*Clania* sp.
Silk fibroin (3a)	10.6	6.95	9.44	*Antheraea mylitta*
Silk fibroin (3b)	10.6	6.95	9.44	*Dictyoploeae japonica*
Silk fibroin (4)	15.0	6.95	9.44	*Thaumetopoca pityocampa*
Silk fibroin (5)	15.7	6.95	9.44	*Nephila senegalensis*
$(\text{Ala-Gly})_n$	9.42	6.95	8.87	
$(\text{Ala-Gly-Ala-Gly-Ser-Gly})_n$	9.39	6.85	9.05	
CTP fraction	9.38	6.87	9.13	

[a]Modified from Fraser *et al.* (1966) and Lucas *et al.* (1960).
[b]The figures in parentheses correspond to the various groups of fibroins.

acid composition and in X-ray grouping. Whether this is due to the different functional forms of the fibroins arising from different glands remains to be established. The fibroin from *N. madagascariaensis* is reeled from the spider, while the fibroin from adult *N. senegalensis* is the egg-cocoon fiber.

Earlier work on *Bombyx* silk fibroin indicated that the molecule is not homogeneous but consists of a crystalline moiety and a second amorphous phase that does not contribute to the X-ray pattern. An antiparallel chain β-structure was proposed for the crystalline regions of *Bombyx* fibroin. This structure was based on the hypothesis, suggested by chemical evidence that every second residue in the polypeptide chains of silk is glycine. Lucas and co-workers (Lucas *et al.*, 1958, 1960) isolated a chymotrypsin-resistant (CTP) fraction from solubilized fibroin that is believed to be derived from the crystalline regions. They proposed for this fraction the sequential formula shown in Eq. (6),

$$\text{Gly-Ala-Gly-Ala-Gly[Ser-Gly-(Ala-Gly}_m)]_8\text{Ser-Gly-Ala-Ala-Gly-Tyr} \quad (6)$$

where m varies with an average value of $m = 2$. Polymers having similar sequences have been prepared and studied by X-ray diffraction. Schnabel and Zahn (1958) synthesized the hexapeptide Ser-Gly-Ala-Gly-Ala-Gly and suggested, on the basis of certain X-ray diffraction spacings, that the chain forms are extended β-conformation and that the hexapeptide probably has a crystal structure similar to that suggested for silk fibroin. Fraser and co-workers (Fraser *et al.*, 1966) have synthesized the polymers $(\text{Ala-Gly})_n$ and $(\text{Ala-Gly-Ala-Gly-Ser-Gly})_n$ and have studied the X-ray diffraction patterns of these polymers (Table IV). The X-ray diffraction of the CTP fraction and the polypeptide (Ala-Gly-Ala-Gly-Ser-Gly)$_n$ is very similar, thus supporting the sequence proposed for the CTP fraction. The presence of an additional oxygen atom on every six residues in (Ala-Gly-Ala-Gly-Ser-Gly)$_n$ leads to a small increase in intersheet spacings compared with (Ala-Gly)$_n$. The values obtained are very similar to those reported for fibroins from different insect species.

One of the problems in studying the molecular structure of silk fibroin is the difficulty in solubilizing the fibroin. Most solubilization procedures have involved boiling or treatment with alkali. These drastic procedures may degrade the fibroin molecule, and the molecular weights of 3.3×10^4 and 10^6 earlier reported suggest this possibility. Recently, soluble fibroin has been extracted directly from the silk gland, and a molecular weight of 3.6×10^5 has been obtained. When the fibroin was washed with water to remove the sericin which covers the fibroin, the fibroin was dissociated in the presence of 6–8 *M* guanidine hydrochloride or urea (Sasaki and Noda, 1973). When analyzed by sedimentation equilibrium, two bands were obtained, one with a sedimentation coefficient of 4.5 S which corresponds to sericin and the other with a sedimentation coefficient of 10 S which corresponds to fibroin. A molecular weight of 3.65×10^5 was obtained for fibroin and it did not change either in physiological salt concentrations, 6 *M* guanidine hydrochloride, or 8 *M* urea, which suggests that the fibroin has no subunits bound by noncovalent bonds. Polyacrylamide gel electrophoresis, sedimentation equilibrium, Sephadex G-200 column chromatography, and viscosity measurements indicate that the fibroin molecule is formed by three small components of molecular weight 2.6×10^4 and one large component of molecular weight 2.8×10^5 connected by disulfide bonds. The amino acid composition of the large component and of the component with molecular weight 2.6×10^4 is shown in Table V. Both the sericin and the large polypeptide from fibroin differ in amino acid composition from the small polypeptide from fibroin (Tables III and V). The number of cystine residues is 8–9 in the fibroin molecule. If all are cystine residues, then four disulfide bonds should exist. The presence of one large and three small components requires at least three disulfide bonds; therefore, there are sufficient cystine residues for this purpose.

A feature of the analysis of amino acid composition of cocoon fibroins is the variation in amino acid composition in various insects. The fibroins that provide the structural basis of the cocoons of different insect species perform essentially the same function for each. This function is the production of filaments from which a rigid structure can be made in which the pupae are protected from predators, extremes of temperature, physical shock, microbiological attack, and the general rigors of climate. It is remarkable that the wide variation in amino acid composition that has been found is perfectly compatible with this function.

TABLE V
Amino Acid Composition of Fibroin Subunits[a]

Amino acid	Subunits	
	MW 2.9×10^5	MW 2.6×10^4
Gly	46.6	11.7
Ala	31.4	11.8
Val	2.13	7.2
Le	0.27	7.8
Ile	0.33	5.6
Ser	10.3	6.7
Thr	0.80	1.7
Asp	0.99	11.6
Glu	0.86	11.4
Phe	0.60	2.8
Tyr	4.98	2.4
Lys	0.27	4.9
His	0.13	1.9
Arg	0.33	5.4

[a]Modified from Sasaki and Noda (1973).

2. Biosynthesis

Large-scale synthesis of a particular protein is tantamount to a large-scale expression of a particular gene. Gene expression can be controlled at many levels of "information flow." But whether one is interested in control at the template level (amplification or redundancy), in transcription, in post-transcriptional steps (messenger RNA selection, transport, and stabilization), in translation or posttranslation events (protein cleavage, assembly, turnover), a system involving large-scale expression of a particular gene may allow the identification and isolation of messenger RNA, as it has been shown for hemoglobin, myosin, and other proteins in eukaryotes. Attempts have been made to obtain messenger RNA from insects which direct primarily the synthesis of fibroin. The posterior and middle silk gland of silkworms were fractionated on an ecteola cellulose column and the ability of the RNA to direct amino acid incorporation into protein was determined using an *Escherichia coli* protein synthesis system. It was observed that a fraction of RNA obtained from the middle silk gland directed the incorporation of glycine, alanine, and serine; but the percentages of these amino acids incorporated resembled their percentages of sericin rather than fibroin. Silk protein from *B. mori* contains 45% of glycine residues alternating predominantly with alanine and serine residues, as shown by Sasaki and Noda (1973). A calculation for the assignment of the codes for these amino acids predicts that the fibroin messenger RNA should be at least 57% G + C, a value which is higher than that of DNA or ribosomal RNA of *B. mori*. This higher G + C content should consist of an unusually high G content of about 40% of the base residues. In addition, the repetitive amino acid sequences of the fibroin protein predict a simple repeating amino acid sequence for its messenger RNA. When fifth instar larvae are labeled for 24 hr with ^{32}P-orthophosphate and the RNA sedimenting between 45 S and 65 S is isolated, it is found that the RNA contains 59% of G + C residues and 40% G residues. The RNA is synthesized very slowly and is very stable. About 1.4% of the total RNA of the posterior gland corresponds to the fibroin messenger RNA. The patterns of oligonucleotides liberated from the RNA by digestion with various RNases may be compared with the patterns which can be predicted by writing out a base sequence for a fibroin messenger knowing its amino acid sequence and using accepted codon assignments. The product of digestion closely matches those predicted; as a result, there is no doubt that the isolated RNA is highly pure fibroin messenger RNA, in which glycine is coded by GGU and GGA, alanine by GCU, and serine by UCA. Buoyant density fractionation coupled with molecular hybridization has been used to obtain information on the length of fibroin DNA (Lizardi and Brown, 1974). The fibroin messenger RNA is extracted and its corresponding DNA is synthesized by the use of a reverse transcriptase. The length of the DNA appears to be 6×10^6 daltons of double-stranded DNA. This compares well with the size of fibroin messenger RNA, about $5.5–6 \times 10^6$ daltons. This means that a single fibroin structural gene should consist of $11–12 \times 10^6$ daltons of double-stranded DNA.

A problem closely related to cocoon fibroin is the mechanism utilized by the silkworm and other species to escape from the cocoon. For this purpose specialized cells develop to produce a large amount of a proteolytic hatching enzyme, cocoonase (Kafatos, 1972). The enzyme extrudes to the surface late in adult development, and just before hatching, it can be found as a dry semicrystalline material. In moths (*Antheraea*) 0.1 mg or more can be collected with forceps from each insect. The material consists virtually of pure enzyme. At the time of hatching, the enzyme is redissolved near its pH optimum of 8.3 in isotonic "bicarbonate," which is secreted by another gland; it is brought into contact with the cocoon and is allowed to digest the sericin binding the silk fibers together. The structure of the cocoon is thus loosened and the insect can find its way to the outside. The glia cells of *Antheraea polyphemus* produce cocoonase as an inactive precursor. Differentiation of these cells is under the hormonal control of ecdysone and juvenile hormone. Ecdysone triggers their differentiation; but when juvenile hormone is also present, differentiation is inhibited and the cells remain in the state characteristic of the pupae.

The synthesis of cocoonase has been extensively studied by Kafatos (1972). It has been shown that messenger RNA for noncocoonase protein has a half-life of 1.5–2 hrs; on the other hand, the apparent half-life of the cocoonase message is at least 7 days. The synthesis of cocoonase starts suddenly, probably utilizing a preexisting message. Cocoonase shows a remarkable chemical similarity to mammalian trypsin and other serine proteases. The molecular weight is approximately 24,000, although in *B. mori,* the molecular weight has been estimated to be approximately 20,000 (Hruska and Law, 1970). Its amino acid composition is very similar to that of bovine trypsin. The enzyme has immunological cross-reactivity with trypsin and chymotrypsin. Cocoonase is different from the proteolytic enzyme of the molting fluid which is secreted by epidermal cells throughout the body of the insect for the purpose of digesting the old cuticle during the molt. The differences between cocoonase and molting fluid proteases indicate that both possess specific messenger RNA's.

C. Proteins of the Integument

The integrity of a structural tissue is maintained by the concerted influence of physical and covalent forces between polypeptides and polysaccharides. As it occurs with the skin of vertebrates, the exoskeleton of insects derives stability from interchain linkages between specific amino acid residues from the ion- or water-binding properties of polysaccharides and from the directionality afforded by carbohydrates associated with the fibrous protein elements. Structure proteins and enzymes of the cuticle participate in the tanning process known as sclerotization. Sclerotization involves the hydroxylation of tyrosine to dihydroxyphenylalanine (dopa) which is decarboxylated to dopamine by dopa-decarboxylase. Dopamine is in turn *N*-acetylated to form *N*-acetyldopamine. Through the action of the phenolase system, *N*-acetyldopamine is oxidized to an *o*-quinone which reacts with the amino group in the cuticle proteins. A diphenol oxidase and its activator have also been purified and shown to oxidize *N*-acetyldopamine to the corresponding diquinone. The phenol oxidase occurs as an inactive precursor or pro(phenol oxidase) which has to be activated. It has been suggested that the pro(phenol oxidase) occurs in the larval hemolymph in *Calliphora* and the activator in the cuticle. There is evidence that *o*-diphenol oxidase, often referred to as tyrosinase, phenolase, phenol oxidase, polyphenol oxidase and catechol oxidase, corresponds to a cuticular protein. The cuticle of larval *Drosophila melanogaster* also contains several dehydrogenases, phosphatases and esterases, and aminopeptidase; however, the possibility that the latter enzyme derived from tissues contaminating the cuticle during extraction has not been excluded. The cuticular phenol oxidase in the pupae of *B. mori* can be classified as a laccase; it appears to be a copper protein because of its sensitivity to cyanide diethyldithiocarbamate; it oxidizes *p*-phenylenediamine; it does not oxidize monophenols and it is insensitive to carbon monoxide. The properties of the enzyme are similar to those of the enzyme obtained from puparia of *Drosophila virilis* and *Papillo xuthus*. The *B. mori* enzyme has three isozymes with molecular weights ranging from 62,000 to 70,000. The syntheses of the phenol oxidase activator and dopa-decarboxylase are under the control of ecdysone. Bursicon, a protein hormone of molecular weight of about 40,000, also appears to control the sclerotization process of newly hatched adult flies and freshly molted cockroaches (see Section VII, B). It has also been suggested that bursicon may control the hydroxylation of tyrosine in *Sarcophaga*. The pro(phenol oxidase) of *B. mori* is formed by three subunits of molecular weights 35,000, 47,000 and 58,000, respectively.

Cuticular protein is heterogeneous, and neither the proteins involved in sclerotization nor the structures of the tanned proteins have been completely elucidated.

The heterogeneity of cuticle proteins has been shown by extraction of protein fractions with differing solubilities and amino acid composition, by the electrophoretic mobility of extracted proteins, and by immunological procedures. In addition, proteins extracted from different insect species also differ. In *Periplaneta americana* the number of electrophoretic bands resolved from a cuticle protein extract varies from 7 to 14 depending on the stage of development. The globulin-type protein fraction from larval cuticle of *Sarcophaga crassipalpis* behaves as a single component by sedimentation analysis, with a molecular weight of 7000–8000, although the protein is heterogeneous. The globulin fraction from cuticles of *Acheta domesticus* have been resolved into five components, all with molecular weights which correspond to multiples of 7000. The soluble proteins of *Schistocerca gregaria* larvae comprise 6% of the cuticle dry weight, while the residual protein ranges from 44 to 49% of the dry weight. In other insect species, the soluble protein varies from 13 to 23% of the dry weight; but the insoluble proteins are the same as in *S. gregaria*. In the larval cuticle of *Agrianome spinicollis* (Coleoptera) there are a great many different proteins. The water-soluble fraction contains 13 proteins; a KCl-soluble fraction, 11 proteins; and an urea-soluble fraction, 20 proteins; in addition, 11 more proteins are found in a NaOH-soluble fraction. As expected, proteins in the urea and NaOH fractions form aggregates when dissolved in dilute buffer at pH 7.0. Further characterization is required to assign a function to these proteins as well as to elucidate their biosynthesis.

In contrast to fibroins, cuticular proteins have an abundance of amino acids with bulking side chains. Soluble proteins apparently do not represent the untanned proteins since no tanning process may be obtained *in vitro*. The amino acids reported in cuticle proteins are aspartic and glutamic acids, serine, glycine, threonine, alanine, tyrosine, valine, phenylalanine, leucine, isoleucine, proline, hydroxyproline, lysine, and tryptophan. Cysteine or methionine or both are probably also present, but in low concentrations, in all insect cuticles.

Amino acid analysis of cuticle proteins indicates certain similarities in the composition of insect cuticle; thus, *Agrianome* (Coleoptera) and *Bombyx* (Lepidoptera) have cuticles with proteins of similar amino acid composition, but that differs from that of larval cuticles of *Lucilia* (Diptera). The *Drosophila* species have similar amino acids, but differ from the protein component of the puparia of *Calliphora* and *Lucilia*. The proteins of the cuticle at various developmental stages of *Galleria mellonella* show a high percentage of glycine and alanine. β-Alanine is present in all pupal cuticles.

A special type of cuticle protein is rather like an insoluble protein called resilin, which may be isolated with chitin. The amino acid composition of resilin is different from other structural proteins. No structure has been observed in the electron microscope. Resilin is located in regions where springlike movements occur. The hinges of larval *S. gregaria* contain 86% resilin.

It is possible that part of the cuticle proteins are conjugated with chitin and glycoproteins. The nature of the bonding between chitin and protein has not been elucidated.

D. Collagen

Collagen is the most important insoluble fiber of connective tissues. Its physiological function is to act as an inert, inextensible material, to transmit the tensions exerted by muscles such as in tendons, or as in bone to form a constituent of the skeletal framework. Collagen is markedly inert to proteolytic enzymes, especially in its native state, and is converted into gelatin upon prolonged heating with water or alkali. Collagen occurs in many phyla, including insects. Collagen has a very unusual amino acid composition: about one-third of the residues are glycine and another one-third are either proline or hydroxyproline. The collagen or tropocollagen (soluble or extracted collagen) molecule consists of three chains that are identical or nonidentical according to the tissue and species from which it originates. The chains have the same polarity and associate in an overlapping way to form a collagen fibril so that suitable preparations yield electron micrographs with a characteristic pattern of bands and antibands. The polypeptide chains of the collagen molecule are known as α1, α2, and α3. The different types of collagen are known as I, II, and III. Thus, collagens have been described whose molecules consist of (a) three apparently identical chains, such as $[(\alpha 1)I]_3$, $[(\alpha 1)II]_3$, $[(\alpha 1)III]_3$; (b) three different chains, such as (α1), (α2), (α3); or (c) two identical and one different chain, such as $[(\alpha 1)I]_2\alpha 2$. The several kinds of chains differ in contents of certain amino acids but are, most likely, of closely similar length and molecular weight.

Insect organs are enveloped and, in some cases, partitioned by layers of connective tissue which separate parenchyma from the circulating hemolymph. This stromal element has been particularly studied in the nervous system because its permeability properties bear on certain neurophysiological considerations. Electron microscopy birefringence and X-ray diffraction suggest that the noncellular component of the nerve sheath, called the neural lamella, consists of a matrix in part of which are embedded collagen-like fibrils. A collagen-like material also forms part of the stroma of insect nonnervous system. The occurrence of collagen in the cockroach *L. maderae* and *Blaberus craniifer* has been demonstrated by a variety of techniques. Electron microscopy of the corpora allata and cardiaca and the prothoracic glands, muscle and fat body indicate that the connective tissue fibrils have a banding characteristic of collagen. Apparently, the connective tissue fibrils in insects show considerable variation in ultrastructure. This applies to fibril diameter and axial periodicity. Values as low as about 170 Å have been determined in certain insects for axial periodicity, such as *G. mellonella*. In cockroaches, periodicities of 500–600 Å are characteristic. Amino acid analysis of fractions of the tissue complex consisting of corpora cardiaca and corpora allata show that a water-insoluble residue contains a significant amount of hydroxylysine and a smaller amount of hydroxyproline. These two amino acids are markers for the presence of collagen. Proteolysis can be achieved by the use of collagenase which further substantiates the presence of collagen. The amino acid compositions of sharkfin collagen and of that obtained from cockroach carcasses are shown in Table VI. The composition of cockroach carcass collagen is similar to that reported from other sources. The amount of collagen extracted from carcasses is less than 0.1% of the total protein extracted with trichloroacetic acid, and this suggests that the collagen may be restricted in location.

E. Chromosomal Proteins

Chromosomal proteins, especially histones, have customarily been included among the structural proteins; however, we now know that they also play a role in gene regulation.

1. Histones

Histones are basic proteins associated with DNA. They are major general structural proteins of chromatin, but they also can act as repressors of template activity. Histones are highly conserved proteins; there has been little variation in the amino acid sequences of histones from widely differing sources. They are acetylated, methylated, or phosphorylated; but the consequences of these modifications are not understood at present. Histones are classified according to their chemical characteristics (Elgin *et al.*, 1971) as shown in Table VII.

TABLE VI

Amino Acid Composition of Sharkfin Collagen and Collagen from Cockroach Carcasses[a,b]

Amino acid	Sharkfin	Cockroach
Gly	25.53	23.52
Ala	11.39	10.47
Ser	4.43	6.25
Thr	2.87	4.01
Pr	14.11	8.81
HyPro	9.68	—
Val	2.02	4.4
Ile	2.46	2.53
Le	2.50	3.53
Phe	2.45	2.76
Tyr	1.72	3.17
Trp	0	—
Cys/2	0.22	7.9
Met	2.05	0.38
Asp	6.13	6.64
Glu	12.07	8.65
Arg	9.02	3.45
His	0.70	1.82
Lys	3.59	2.07
HyLys	1.17	6.76

[a]Modified from Tristram and Smith (1963) and Harper *et al.* (1967).
[b]Data expressed as g amino acid per 100 g of protein.

The primary structure of histones from various sources has been determined and detailed information on amino acid composition and sequence of histones obtained. However, there have been few structural studies of the histones of insects, with the exception of *Drosophila*. Fractionation of histones extracted from polytene nuclei of *Drosophila melanogaster* reveals

TABLE VII

Classification of Histones

Class	Fraction	Lysine/arginine ratio	Total residues	MW	N-terminal	C-terminal
Very rich in lysine	H1	2.2	215	21,500	Ac-Ser	Lys
Lysine-rich	H2a	1.17	129	14,004	Ac-Ser	Lys
	H2b	2.5	125	13,774	Pro	Lys
Arginine-rich	H3	0.72	135	15,324	Ala	Ala
	H4	0.79	102	11,282	Ac-Ser	Gly

that three of the five major histone fractions, H1, H2b, and H3 differ in their relative electrophoretic mobility from homologous fractions of mammalian histones. Amino acid analysis of the H1 fraction of *Drosophila* demonstrates that it contains less lysine, proline, and alanine but significantly more histidine, asparagine, serine, valine, and isoleucine than mammalian H1. Differences have been found in the electrophoretic mobility of histones from *Planococcus citri* and *A. domesticus* as compared to histones from calf thymocytes. An arginine-rich histone found exclusively in the late spermatids and spermatozoa of *A. domesticus* appears to have unusual characteristics. Refined studies have been made on the basic nucleoproteins of *A. domesticus, P. citri,* and *Oncopeltus fasciatus*. Histones have been isolated from *D. melanogaster* by a high-salt concentration method (Dick and Johns, 1969). Recently, the five major histone fractions have been purified from *Drosophila,* and again the most important difference appears in H1. This fraction has 23.6% of basic residues instead of 28.6 in calf, as well as a 5% increase in molecular weight. Histones also have been isolated from *Ceratitis capitata* by the procedures of Dick and Johns (1969) and fractionated by using acid or organic selective extractions. *Ceratitis capitata* H1 has less electrophoretic mobility when compared with the corresponding mammalian fraction; however, H4 has the same mobility as mammalian H4, although there are some differences in amino acid composition. Cysteine does not occur in fly H2a, but H3 aggregates readily under oxidizing conditions. The histones of *Drosophila* embryo correspond to about 60% of the chromosome DNA. As the embryo advances in age, the amount increases to about 79% at 6–18 hrs. The *Drosophila* blastula chromatin is deficient in histone H1, but otherwise the histone pattern is constant through the development transition. The mass ratio of DNA, histones, and nonhistones in adult houseflies is 1:16.1:476 (Tsang and Agosin, 1976).

2. Nonhistones

The role of histones in genetic regulation has been deemphasized in recent years. This role has now been assigned to the nonhistone chromosomal proteins. These are the proteins that isolate together with DNA as chromatin (the interphase form) or chromosomes (the metaphase form). It is thought that the nonhistone protein functions include enzymes of chromosomal metabolism, activation/repression molecules related to a specific gene expression, and proteins contributing to the structure of different chromatin forms. Already, many enzyme activities have been found in isolated chromatins, including RNA polymerase, DNA polymerase, nucleases, histone acetylases, histone proteases. As structure proteins, that is, those nonhistone proteins contributing to structure and form of chromatin, these proteins may modify the accessibility of DNA; this has brought implications in control mechanisms. Differences have been found in nonhistone proteins from young *Drosophila* embryos as compared to older embryos. The nonhistone proteins in *Drosophila* embryos chromatin of 0–2 hrs are about 128% of the DNA content and does not change significantly through development. The differences in composition of *Drosophila* embryo chromatin from 0–2 hrs as compared to 16–18 hrs may explain the much higher template activity exhibited by the blastula chromatin.

Chromatin has been prepared from highly purified nuclei from the salivary glands of third instar larvae of *Drosophila hydei*. Such chromatin differs from that of diploid nuclei mainly by differences in certain nonhistone proteins, designated as ψ, λ, and χ_3. On the basis of this result, it has been suggested that these proteins may be important components of constitutive heterochromatin, which is severely underrepresented in polytene chromosomes. Similar patterns are observed in *D. melanogaster*. Nonhistone proteins from *O. fasciatus* embryos also are markedly heterogeneous, with molecular weights ranging from 32,000 to 120,000. Moreover, changes in their concentration with development occur as well (Teng, 1974).

III. INTERSTITIAL PROTEINS

The hemolymph proteins of insects have been studied by a large number of workers for taxonomic purposes as well as for determining the origin and function of several protein components. Studies of hemolymph proteins during metamorphosis have yielded information on metabolic activity associated with differentiation, because the hemolymph is in direct contact with body tissues. In holometabolic insects a high proportion of proteins is synthesized during the larval development and used later for the synthesis of adult organs. The plasma proteins from these insects show a rapid rise during the midlarval stages, then fall down during pupation and the early pharate adult stages. A peak content of 6–8% is common in the Lepidoptera. In Diptera, a maximum level is reached in late larval life at a point determined in relation to the end of the active feeding period. Species such as those of *Calliphora* show a maximum content of about 20%.

The analysis and identification of hemolymph proteins have been facilitated by the use of electrophoretic procedures. Fourteen distinct bands are found in *Phormia rufa* and *P. polyctena* larvae while seventeen bands appear in *P. pratensis*. Polyacrylamide gel electrophoresis of seven species of Lepidoptera (*Hyalophora cecropia, H. gloveri, Callosamia promethea, Philosamia cynthia, Antheraea polyphormus, A. mylitta, M. sexta*) reflects the close relationship between the same genera and the difference between the saturniids and *Sphingia*. Twenty-two bands have been found in insects including *A. domesticus*. Of these twenty-two bands, eight correspond to conjugated proteins. The slowest moving of these conjugated proteins has a molecular weight of 45,000 and an isoelectric point of 6.6–7.1. Three other conjugated proteins with isoelectric points ranging from 4.5–6.85 have molecular weights of 74,000–170,000 while three other proteins have isoelectric points from 4.31–4.91 and molecular weights ranging from 35,100–74,000. It is possible that the protein with a molecular weight of 45,000 may correspond to the growth hormone elaborated by the pars intercerebralis, although no conclusive evidence has been presented to substantiate this claim.

A. Carrier Proteins

1. Juvenile Hormone Binding Protein

Juvenile hormone is synthesized by the corpus allatum and secreted into the hemocoel. To reach the target cells, it must be transported by the hemolymph. The presence of lipoproteins or proteins in the hemolymph which are capable of binding juvenile hormone and its analogues in the concentrations of 10^{-6} to 10^{-5} *M* has been reported in *H. cecropia* pupae and in *Tenebrio molitor*. Gel filtration of hemolymph of the tobacco hornworm *M. sexta* incubated with an aqueous solution of juvenile hormone shows the hormone as a macromolecular complex of protein nature. The protein can be fractionated from the hemolymph by ammonium sulfate precipitation between 20 to 60% saturation, and it has been purified to homogeneity. Binding is specific; no binding occurs with the epoxy acid or the diol; while under similar conditions, 65% of juvenile hormone at a concentration of 2.6×10^{-7} *M* is associated with the binding protein. The molecular weight of the binding protein is 34,000 as shown by gel filtration in Sephadex G-100 columns calibrated with standard proteins of known molecular weights. The binding of juvenile hormone with the protein corresponds to a simple thermodynamic equilibrium, with a dissociation constant K of 2.99×10^{-7} *M*. The molarity of the binding protein in the hemolymph is 7.7×10^{-6} *M*. The binding protein appears to either bind juvenile hormone as soon as it is secreted by the corpus allatum, or to bind juvenile hormone which is stored in the fat body. The concentration of the binding protein in the hemolymph is sufficient to maintain 96% of the hormone in the bound state; thus, the physiological concentration of the binding protein appears to be optimal for almost quantitative complexation of the hormone. Recent studies suggest that the *M. sexta* juvenile hormone-binding protein may protect the hormone from degradation by hemolymph enzymes. Similar low molecular weight carrier proteins have been detected in larvae of *Plodia interpunctella* and *H. cecropia*.

The *M. sexta* larvae hemolymph contains, in addition to the hemolymph-binding protein, a large molecular weight lipoprotein that is also capable of binding the C_{18}-juvenile hormone but with less affinity than the binding protein. Immunoelectrophoretic and gel electrophoretic analysis shows that the fat body contains binding protein; and under appropriate conditions, the fat body is able to incorporate labeled amino acids into the binding protein which is then released. This observation suggests that the binding protein is synthesized in the fat body, as it occurs with many other hemolymph proteins. Indeed, the binding lipoprotein is the major protein in the hemolymph of *M. sexta*.

2. Lipid-Binding Proteins

This type of protein has been extensively studied. In *H. cecropia*, diglycerides released from the fat body become associated with a specific hemolymph lipoprotein. Such specific lipoproteins have been purified from several species of silkmoths. The hemolymph of these insects contain two major classes of lipoproteins which are called diglyceride-carrying lipoproteins I and II, or high density lipoproteins and very high density lipoproteins, respectively. Diglyceride-carrying lipoprotein I accepts diglycerides from the fat body while lipoprotein II, although it has diglyceride associated with it, apparently does not accept released diglycerides from the fat body. Both proteins have been purified to essentially a homogeneous state, as judged by ultracentrifugation, electrophoresis, and electron microscopy. They correspond to globular proteins. The diameter of the molecule is 13.5 nm for lipoprotein I and 10 nm for lipoprotein II. The molecular weight, estimated by sedimentation equilibrium, is 700,000 for lipoprotein I and 500,000 for lipoprotein II. Lipid constitutes 44% of the weight of lipoprotein I and approximately 10% of lipoprotein II. Diglyceride is the major (and possibly only) glyceride present in both lipoproteins. A significant amount of cholesterol is present in both lipoproteins, but no cholesteryl ester or free fatty acids are detected. Phospholipids are high in both proteins, phosphatidylcholine, phosphatidylethanolamine, and sphingomyelin being the major components. The chemical constitution of hemolymph lipoproteins is shown in Table VIII. The amino acid composition of lipoproteins I and II (Table IX) indicates that lipoprotein II is more acidic than lipoprotein I, due to the presence of more glutamic acid in lipoprotein II and less glycine in lipoprotein I.

Lipoprotein II appears to be a lipoglycoprotein, since it contains a carbohydrate moiety. Lipoprotein II is in high concentration in the hemolymph of *H. gloveri* female pupae but in barely detectable concentrations in male pupal hemolymph. During ovarian maturation, the levels of lipoprotein II decrease in the hemolymph. Ovariectomy reverses this effect. These findings suggest that lipoprotein II is a vitellogenin that moves from the hemolymph to the maturing ovary during development and is involved in the transport of protein, carbohydrate, and lipid to the oocytes as prospective

TABLE VIII

Chemical Constitution of Hemolymph Lipoproteins from *Philosamia cynthia*[a]

Component	% Weight	
	Lipoprotein I	Lipoprotein II
Protein	56.0	90.3
Total lipids	44.0	9.7
Triglycerides	< 1.2	—
Diglycerides	56.3	34.5
Monoglyceride	Not detected	—
Cholesterol	13.2	12.7
Cholesteryl esters	Not detected	—
Phospholipids	25.8	49.5

[a]Modified from Chino *et al.* (1969).

TABLE IX

Amino Acid Composition of *Philosamia cynthia* Lipoproteins I and II[a]

Amino acid	mmoles/mole amino acid recovered	
	Lipoprotein I	Lipoprotein II
Asp	126	118
Thr	49	59
Ser	69	78
Glu	104	149
Pr	47	50
Gly	67	48
Ala	63	75
Val	74	66
Met	5	4
Ile	58	48
Leu	90	66
Tyr	28	44
Phe	48	33
His	28	32
Lys	107	90
Arg	37	40

[a]Data from Chino *et al.* (1969).

components of the yolk. Yolk lipoprotein and lipoprotein II from hemolymph appear to be identical as judged by analytical centrifugation.

Hemolymph lipoproteins from the pupae of *H. cecropia* and *H. gloveri* are not identical with those of *P. cynthia*. Lipoprotein I appears to have a lower molecular weight and the lipid content is different, 48% for lipoprotein I and 6% for lipoprotein II. Diglycerides constitute 70% of the total lipid in lipoprotein I and 45% in lipoprotein II. In addition, *Hyalophora* lipoproteins contain significant amounts of sterol ester, as well as carbohydrates. In addition to lipoproteins I and II, other lipoproteins are present in the hemolymph of these insects; but their function is yet to be determined. A lipoprotein responsible for diglyceride transport in *Locusta migratoria* has a molecular weight of 350,000, and its chemical composition and amino acid analysis are similar to those of lipoprotein I.

The hemolymph proteins of the brown blow fly, *Calliphora stygia,* vary from 29 to 200 mg/ml according to the developmental stage. Four major proteins are found in the hemolymph, A, B, C, and D. Protein A has a molecular weight over 400,000 and is formed by two subunits of molecular weight 240,000. This protein is a lipoprotein which is distinct from the other major plasma species. Protein A does not show reversible pH-dependent dissociation, which suggests that its oligomeric form is maintained by hydrophobic bonds rather than electrostatic interactions. Its role corresponds to storage and transport of lipids synthesized in the fat body. Protein B is a glycoprotein of molecular weight 240,000 and is formed by three subunits of molecular weight 81,000. The subunits of protein B appear to be held together essentially by electrostatic forces. It contains calcium at a concentration of 12 matom/mmole. No disulfide bonds are present in the molecule. Protein C appears to be similar to protein II described in *C. erythrocephala,* although its amino acid composition is different in some respects, having a higher histidine and a lower phenylalanine content, and cysteine is absent. Protein B comprises 8% of the total plasma proteins in late-feeding larvae. Subsequent to emergence protein B disappears. Its role is not known at present, but in addition to transport of carbohydrates, it may have a function in the maturation of the adult. Protein D is immunologically and structurally a distinct protein appearing in the hemolymph of adult flies about the time of emergence. It is synthesized at this stage; and as it shares no antigenic determinants with other larval proteins, no direct relationship appears to exist between protein D and larval proteins. Protein D gives rise to a single subunit of molecular weight 79,000 on SDS-polyacrylamide gel electrophoresis.

3. Xenobiotic-Binding Proteins

Lipoproteins that bind xenobiotics are present in the hemolymph of *P. americana*. Insecticides such as DDT and dieldrin are bound by two prominent fractions of the hemolymph, as shown by Sephadex G-200 column chromatography. One of the fractions appears to be more specific for DDT than dieldrin. The fraction that usually binds DDT is composed of two lipoprotein species, the heavier one having a greater affinity for DDT than the lighter one. The fraction that binds dieldrin preferentially consists of only one lipoprotein species and it binds DDT equally well. Its molecular weight is estimated to be 520,000.

A low molecular weight protein that binds phenobarbital has been found in the cytosolic fraction of *M. domestica* (Tsang and Agosin, 1976). This protein may act as a carrier for the drug to interact with carrier proteins at the nuclear level and thus modify the transcription properties of chromatin.

4. Hemoglobins

Hemoglobins are hemoproteins in which the iron of the protoporphyrin is in the reduced divalent state. They bind oxygen reversibly and the iron remains in the divalent form. Oxidation converts the ferroheme into the ferriheme and the resulting methemoglobin can no longer bind oxygen. The heme group is bound to the protein mainly by noncovalent bonds and can be removed by mild extraction procedures. At neutral pH, the globin and the heme recombine to yield hemoglobin. The oxygenation of hemoglobin is thought to correspond to a direct interaction between oxygen and the sixth coordination Fe^{2+}. The first four coordinations bind to the four pyrrole nitrogens and the fifth to a protein-imidazole group. Mammalian hemoglobin has a molecular weight of 65,000 formed by four subunits, each one of molecular weight of about 16,000. Each subunit contains one heme group. The major monomers of hemoglobin are called α, β, and γ; and normal adult hemoglobin is made up of two α and β subunits ($\alpha_2 \beta_2$); normal fetal subunits correspond to $\alpha_2 \gamma_2$. There are important differences in the ligand-binding properties of the α and β chains and available evidence suggests that they are not equivalent as previously thought. The tridimensional structure of these subunits is held together by noncovalent bonds, including salt bridges, hydrogen bonds, and nonpolar interactions. A plot of oxygenation against percent saturation of hemoglobin produces a sigmoidal curve, which is suggestive of cooperative binding. In a Hill plot, the slope which gives a value for *c*, the cooperativity is +2.8. The oxygen affinity for hemoglobin changes with pH, the so-called Bohr effect. The Bohr effect demonstrates a significant difference in the dissociation of at least one ionizable group in hemoglobin and hemoglobin-oxygen; a proton is released from hemoglobin-hydrogen when oxygen is bound which indicates competition between oxygen binding and the binding of a proton. The pK_a values involved are 8.2 for hemoglobin and 6.95 for hemoglobin-oxygen at 20° and ionic strength of 0.2. This pK_a value is characteristic of histidine. In addition to oxygen and other ligands, hemoglobin can bind carbon monoxide. The binding sites of oxygen and carbon monoxide are the same and the kinetic binding with carbon monoxide shows the same characteristics as with oxygen.

The few insects possessing hemoglobin include the larva in the dipteran family Tendipendidae; the larva of *Gastrophilus intestinalis;* and three Hemiptera, *Buocnoa margaritacea, Anisops producta* and *Macrocorixa geoffrey*. Occurrence of hemoglobin in freshwater insects notonectids is well known. It is located in the cytoplasm of the tracheal cells. The hemoglobin of

Anisops nevins shows similar properties as described for other hemoglobins. In the chironomid (Tendipendidae) larvae, hemoglobin is found in the hemolymph. Some years ago it was found that the hemoglobin of individual chironomid larvae actually corresponds to several hemoglobins in many of the chironomid species and that the highly hyperbolic curve of oxygenation of the hemoglobin actually corresponds to the net effect in oxygen binding of a family of hemoglobins differentiated with respect to their individual properties. The heme group of these hemoglobins is apparently similar to the vertebrate hemoglobin but the apoprotein is very distinct in chain lengths and sequence. The chironomid hemoglobins appear to be related in structure and in *Chironomus tentans* they consist entirely of ten to twelve monomers with a molecular weight of approximately 15,900. They are apparently encoded by an equivalent number of structural genes. The total hemoglobins in larval *C. tentans* correspond to about 40% of the total hemolymph proteins. Differences between species usually involve relative quantity of the various hemoglobins, rather than numbers and kinds of different components. Structural relationships among the ten to twelve larval hemoglobins of *Chironomus thummi* and *C. tentans* have been studied in some detail. One of the subunits of the hemoglobin (CTT-III) contains 135–136 amino acids. The sequence is not uniform and contains two neutral exchanges and two insertions (isoleucine and alanine). The fifth position of coordination of the heme group is histidine, while the sixth position is isoleucine. The various chain types show a wide range of similarity and difference but are obviously evolved from a common ancestral form.

Populations from different localities, however, may differ in the number of hemoglobins as well as in their electrophoretic mobility. Hybrids between races have an essentially codominant pattern. Loci encoding two very similar monomeric hemoglobins among the ten to twelve larval hemoglobins of *C. tentans* show opposite, apparently compensatory, changes in activity in the presence of a spontaneous regulatory mutant. The regulatory mutant maps near the structural loci for both hemoglobins in the right tip of chromosome 3. The locus showing regulatory increase in rate of synthesis does show in both *cis* and *trans* alleles. The locus showing decreased activity, however, undergoes nearly complete inactivation in *cis*. It has been suggested that the regulatory mutant may represent a lesion of a promoter serving the latter. Juvenile hormone blocks production of the two forms of hemoglobin when administered to third instar larvae, which suggests that synthesis of important proteins may be coupled with juvenile hormone action and dependent of physiological change.

Chironomid hemoglobins exhibit a normal Bohr effect for pH values about 7.4 to 7.5. Similarly, the oxygen equilibrium curves are indicative of cooperative binding.

The function of these hemoglobins appears to be mainly as an oxygen carrier; however, since they occur in such large amounts in the hemolymph, they also may act as buffers. It has also been suggested that they may facilitate diffusion of oxygen. No good correlation has been found between resistance to lack of oxygen and levels of hemoglobin, and it is generally thought that the hemolymph hemoglobins may play a role as oxygen carriers only under very low oxygen tensions. The hemoglobins are synthesized in the fat body and are degraded during imaginal development; although some proteins remain conjugated with a green pigment. This fact, together with the similarities of amino acid composition with calliphorin (i.e., low cysteine and phenylalanine), suggests that chironomid hemoglobins may also serve as storage proteins, in place of calliphorin (see below), which is not present in chironomids (see Section III, B, 1.).

B. Storage Proteins

1. Calliphorin

About 60% of the total soluble protein present in the late larval stage of *C. erythrocephala* is composed of the protein calliphorin. Calliphorin and other related proteins such as chironomid hemoglobin may correspond to specialized storage proteins which function as a source of the nutrients required for synthesis of adult proteins. Calliphorin is synthesized mainly in the fat body and released into the hemolymph. After the fat body cells complete their period of synthetic activity, hemolymph proteins are in part resorbed, apparently contributing to the formation of the proteinaceous spheres characteristic of the mature larval fat body. During development of the pharate adult, the breakdown of fat body cells releases the storage proteins for degradation to the individual amino acids for the *de novo* synthesis of adult proteins. Protein C (Section III, A, 2) of *C. stygia* apparently corresponds to calliphorin as shown by immunological procedures. The calliphorin from *C. erythrocephala* larvae and pupae has a sedimentation coefficient value of 19.9 S and a molecular weight of 540,000. It is rich in tyrosine and it denatures reversibly with changes in pH. The calliphorin from *C. stygia* has a molecular weight of 250,000 with subunits of about 85,000,suggesting that it corresponds to a trimer, while the calliphorin from *C. erythrocephala* apparently occurs as an hexamer. The arrangement of subunits in calliphorin may correspond to a noncyclic hexameric structure in which the six subunits are arranged as a trigonal prism. It is quite possible that protein C from *C. stygia* may also correspond to an hexamer which has been denatured into trimers during isolation. Disulfide cross-linkages are not involved in maintaining the structure of calliphorin. Calliphorin from *C. stygia* has a high content of aspartic and glutamic acids and their side-chain carbonyl groups are apparently involved in the formation of hydrogen bonds. Preparations of trimeric calliphorin from *C. stygia* dissociate to at least six electrophoretically separate subunits, but each of these subunits may be of greater heterogeneity with respect to the calliphorin subunit patterns. The six bands obtained by electrophoresis of calliphorin from *C. stygia* have the same molecular weight, have similar general properties and possess identical antigenic determinants. Calliphorin has a small amount of carbohydrate and lipid, in addition to tightly bound calcium. The amino acid composition of calliphorin from *C. stygia* and from *C. erythrocephala* are essentially the same (Table X).

TABLE X

Amino Acid Composition of Calliphorin from *Calliphora stygia*[a]

Amino acid	mmoles/g protein
Ala	0.24
Arg	0.23
Asp	0.97
Glu	0.86
Gly	0.44
His	0.25
Ile	0.33
Le	0.55
Lys	0.64
Met	0.31
Phe	0.88
Pr	0.33
Ser	0.34
Thr	0.38
Trp	0.11
Tyr	0.96
Val	0.43

[a]Modified from Munn *et al.* (1971).

Sarcophaga barbata and *Lucilia cuprina* contain an antigen which gives a cross-reaction with calliphorin antibodies obtained from *C. erythrocephala,* which suggests complete identity with calliphorin. Partial cross-reactions are obtained with antigens from extracts of *Gastrophilus intestinalis* and *D. melanogaster.* The *Gastrophilus* antigen appears to be more closely related to calliphorin than that of *Drosophila.* Extracts of larvae of *Chrysopilus cristatus, Rhagio scolpacea* and *Rhanphomya sulcata* (suborder, Brachycera) all give reactions of partial identity with calliphorin. No cross-reactions are obtained with larvae and pupae of *Tenebrio molitor* and larvae of *L. migratoria.*

2. Lipovitellin

The hemolymph of female pupal stages of several insect species contains a major lipoprotein which is present in only small amounts in males. This lipoprotein corresponds to a lipovitellin. The term lipovitellin is used to designate a family of lipoproteins, representatives of which are found as the principal protein in the eggs of all animals that have been studied. The lipovitellins are interesting not only as proteins but as part of a controlled developmental process, vitellogenesis. In insects, this process is under endocrine control by juvenile hormone. Juvenile hormone control of lipovitellins has been shown in several species of insects, such as *Leucophaea maderae, Sarcophaga bullata, Leptinotarsa decemlineata, Byrsotria fumigata, Schistocerca vaga, Divians plexippus,* and *Aedes aegypti.* In addition, female specific proteins have been identified immunologically in *Rhodnius prolixus, Schistocerca gregaria*, and *Pieris brassicae*. Most of the yolk protein in the oocyte of several species examined, especially *L. maderae,* is represented by a single species which is formed during deposition by the aggregation of a subunit. The subunit is synthesized outside of the ovary in the fat body. The lipovitellin from *L. maderae* contains two components with sedimentation constants of 28 S and 14 S. The lipid moiety consists of phospholipids and represents about 6.9% of the total weight. The material contains carbohydrate made up of mannose and hexosamine, corresponding to 6.4–7.0% and about 1.6%, respectively, of the total weight. The hexosamine is primarily glucosamine but another hexosamine is present in small amounts. The amino acid composition of each subunit is the same (Table XI). The molecular weight determined by

TABLE XI

Amino Acid Composition of the 28 S and 14 S Components of *Leucophaea maderae* Lipovitellin[a]

	Residues per arginine residue	
Amino acid	28 S	14 S
Lys	0.82	0.91
His	0.51	0.48
Arg	1.00	1.00
Asp	2.40	2.32
Thr	0.76	0.81
Ser	1.03	1.09
Glu	1.82	1.74
Pro	0.75	0.74
Gly	0.54	0.52
Ala	0.88	0.83
Cys/2	—	—
Val	1.17	1.01
Met	0.31	0.31
Ile	0.81	0.79
Le	1.45	1.40
Tyr	0.58	0.60
Phe	0.67	0.64

[a]Modified from Dejmal and Brookes (1972).

sedimentation equilibrium is 5.59×10^5 for the 14 S component and 1.59×10^6 for the 28 S component. The 14 S component is made up of subunits as shown by gel electrophoresis. It consists of four discrete peptides in the ratio $A_1B_1C_2D_2$ with molecular weights of 118,000, 87,000, 57,500, and 96,000, respectively. The 28 S component appears as a trimer containing the peptides A, B, and C in the ratio of $A_1B_3C_2$. The lipovitellin from *L. maderae* is closer in constitution to that of vertebrates but its carbohydrate content is lower. As in other insects, the protein is transported to the egg by the hemolymph and taken up by the egg by pinocytosis. The 14 S fraction is only found in young oocytes, and maturation to the 28 S fraction probably involves proteolytic conversion of peptide D to B^D, a different peptide of the same size as B, suggesting that the true composition of the 28 S fraction is $A_1B_1B_2{}^DC_2$. The vitellogenin in *P. cynthia* apparently is identical with lipoprotein II from the hemolymph (see Section III, A, 2).

Perhaps it would be appropriate to mention here that proteins associated with molting have been described in several insect species; however, these proteins have not been characterized and their role is uncertain.

C. Enzymes of the Hemolymph

A number of oxidoreductases and hydrolases have been detected in the insect hemolymph. In some instances, there is controversy on whether these enzymes are the result of leakage from tissues or are constituents of the hemolymph proper. In general, the enzymes are found in the form of multiple isozymes (see Section IV, A, 1). A list of enzymes found in the hemolymph of insects has been compiled by Jeuniaux (1971). Of particular interest are trehalase, which appears to be a permanent constituent of the protein pool of the hemolymph and juvenile hormone-specific esterases. These and other enzymes of interest will be briefly discussed.

1. Trehalase

Trehalase hydrolyzes trehalose (1-*O*-α-D-glucopyranosyl-α-D-glucopyranoside) to glucose. In *P. regina*, it appears in the hemolymph at about the same time as does its substrate trehalose. Apparently, there are two distinct species of the enzyme in the fly. Isozyme A is restricted to the midgut and hemolymph and has a pH optimum of 4.5, a K_m for trehalose of 1.5–1.9 m*M* and a molecular weight of 115,000–117,000. Isozyme B is restricted to the head, muscle and rectal papillae with a pH optimum of 5–5.5, a K_m for trehalose of 3.1–3.5 m*M* and a molecular weight of 78,000–90,000. It has been suggested that the blood trehalase activity derives from that of the midgut (Friedman, 1975). The presence of trehalase in the hemolymph and in the midgut makes sense. If glucose is transported across the midgut wall by simple diffusion, any back flow produced by trehalose hydrolysis in the hemolymph would be compensated by the action of the corresponding enzyme in the midgut. The soluble form found in *P. regina* muscle, however, is electrophoretically distinct from the hemolymph enzyme.

2. Juvenile Hormone Esterases

Recently, specific esterases present in the hemolymph that hydrolyze juvenile hormone bound to its carrier protein have been detected. Hemolymph from fifth instar larvae of *M. sexta* contains two families of esterases which can be distinguished by their reactivity with diisopropylphosphorofluoridate. One group consists of general esterases which are capable of hydrolyzing free juvenile hormone but not when complexed to the binding protein and are completely inhibited by 10^{-4} *M* diisopropylphosphorofluoridate. The other group is relatively insensitive to the inhibitor, has little general esterase activity, as measured with 1-naphthylacetate, but hydrolyzes both free- and protein-bound juvenile hormone. The major specific juvenile hormone esterase has a sedimentation coefficient of 4.98 S and a diffusion coefficient of 6.4×10^{-7} cm^2sec^{-1}, with a molecular weight of 6.7×10^4. The general esterases are present throughout the larval stage, but the juvenile hormone-specific esterases appear in significant concentrations only on the fourth day of the fifth instar larvae. Juvenile hormone esterases apparently play an important regulatory function. Prior to the fifth instar larvae, the general esterases cannot attack juvenile hormone because it is protected through binding to its specific binding protein; later on, however, juvenile hormone titers decrease through the action of the specific juvenile hormone esterases.

3. Pro-Phenol Oxidases and Phenol Oxidases

Tyrosine metabolism in insects has received much attention because of its special importance in the tanning of the cuticle at each molt. In the third instar larvae of the blowfly *C. erythrocephala,* phenol oxidases in the blood are responsible for most of the phenol oxidase activity. In *Calliphora* as in other insects, the phenol oxidases of the hemolymph are at this stage in the form of inactive proenzymes, which can be activated by an enzyme that can be obtained from cuticle extracts. Three pro-phenol oxidases have been distinguished in the hemolymph of *Drosophila*. Two have similar electrophoretic mobility and both oxidize L-dopa (3,4-dihydroxyphenylalanine) but not tyrosine; they are dopa oxidases. The third, with a different electrophoretic mobility, oxidizes both tyrosine and dopa; it corresponds to a tyrosinase. Similar enzymes have been distinguished in the hemolymph of *Calliphora*. In *C. erythrocephala* the pro-phenol oxidases are found in the blood of late third instar larvae. One, after activation, catalyzes the oxidation of L-dopa and L-tyrosine; the second pro-phenol oxidase catalyzes the oxidation of L-dopa but not of L-tyrosine. The first pro-phenol oxidase has been purified extensively; it appears to be involved in the conversion of tyrosine into diphenols that are used for cross-linking the cuticle at the molt. The presence of the second pro-phenol oxidase suggests that the different phenolases of insect hemolymph may have clearly distinct roles which have yet to be determined.

An *o*-diphenol oxidase has been isolated and purified from third instar larvae of *C. erythrocephala*. It has been shown that the active *o*-diphenol oxidase in acetone-dried powders is on a protein basis, a minor component of the extract with an apparent molecular weight of 35,000, which suggests that it corresponds to a dimer. It was thought that the molecular weight of this enzyme was 530,000; however, it seems that the protein isolated corresponded mainly to calliphorin contaminated with the diphenol oxidase. The amino acid composition of the *o*-diphenol oxidase from *C. erythrocephala* is similar to that of pro(phenol oxidases) of silkworms.

D. Immunoproteins

It has been considered that immune factors in insects are not proteins. Precipitins have not been demonstrated in several attempts. A spontaneous precipitation of hemolymph proteins of *G. mellonella* and *P. americana* occurs when diffused in agar phosphate. Insect plasma proteins of these two species diffused against each other under conditions where a spontaneous precipitation does not occur, develop precipitated band proteins similar to those observed in mammalian antigen–antibody reactions. Further studies are required to establish whether this phenomenon is linked to immunity.

IV. ENZYMES

The enzyme systems involved in the metabolism of insects have been the subject of numerous studies. It is not our purpose to review the enzymes involved in metabolic pathways nor to describe all the enzymes recorded in the literature but to discuss those enzyme proteins that have been purified and characterized extensively and/or have been the subject of structural studies. Unfortunately, very few insect proteins have been characterized in terms of sequence and structure.

A. Oxidoreductases

1. Ecdysone Oxidase

An interesting enzyme (Koolman and Karlson, 1975). ecdysone oxidase, which catalyzes dehydrogenation of the secondary alcoholic group located at the 3-position of the ecdysone steroid nucleus, has been described recently. The enzyme, as reported in *C. erythrocephala*, is not a dehydrogenase since it does not require a dissociable coenzyme, but a true oxidase. The molecular weight is 240,000 and the K_m for ecdysone is 41 μM, which is consistent with the fact that the level of molting hormone in insects is rather low. The enzyme has been purified approximately one-hundred four-fold. It is also present in *L. migratoria, Aeschna cyanea*, and *Choristoneura fumifera*.

An ecdysone, 20-hydroxylase, is also present in the midgut of *M. sexta*. (Nigg *et al.* 1976). The enzyme has an absolute requirement for NADPH and molecular oxygen. The activity is present only during the fifth larval and early prepupation stages. Its substrates are ecdysone and 3-dehydroecdysone.

2. Tryptophan-2,3-Dioxygenase

This enzyme is an allosteric protein which has been purified about one-hundred-fold from *Phormia regina-novae*. Its molecular weight is 120,000; its optimum pH 8.2, and the $(S)_{0.5}$ is 1.25×10^{-3} *M* for tryptophan and 4.2×10^{-5} *M* for oxygen at 37° and pH 8.2. The substrate-binding curves are sigmoid with Hill coefficients (n) of about 2 and 2.3, respectively. Methyltryptophan is an allosteric effector of the reaction (Schartau and Linzen, 1976).

3. α-Glycerophosphate Dehydrogenase

The enzyme has been purified from *D. melanogaster*. It corresponds to a dimer formed by identical subunits of molecular weight 31,700. A similar molecular weight for the native enzyme (63,000–68,000) has been reported for the enzyme obtained from honey bee and bumblebee. The properties of the enzyme resemble those of the vertebrate.

4. Triosephosphate Dehydrogenases

Triosephosphate dehydrogenase

$$\text{D-Glyceraldehyde 3-phosphate} + \text{phosphate} + \text{DPN}^+ \rightarrow \text{1,3-diphospho-D-glyceric acid} + \text{DPNH}$$

has been purified and characterized from six species of bees and two species of flies. The amino acid composition and tryptic peptides from *Apis mellifica, Sarcophaga bullata, Callitropes hominivorax, Megachile rotundata, Bombus appositus, Bombus occidentalis, Psithyrus suckleyi* have been determined and analogies and differences analyzed. From the evolutionary standpoint, it appears that honey bee triosephosphate dehydrogenase has evolved at a faster rate than the enzyme in the other insects tested. These results are interesting because they indicate that molecular studies may suggest when in evolutionary time a particular behavioral pattern develops in an invertebrate line.

5. DDT-Dehydrochlorinase

This soluble enzyme, responsible for the resistance of a number of insect species to DDT, catalyzes the dehydrochlorination of DDT (1,1,1-trichloro-2,2-bis[*p*-chlorophenyl]ethane) to DDE (1,1-dichloro-2,2-bis[*p*-chlorophenyl]ethylene). It corresponds to a tetramer of molecular weight 120,000 composed of four identical subunits. The enzyme is isolated as monomers which are aggregated in the presence of DDT. The tetramer structure is stabilized in the presence of reduced glutathione but monomers are produced when 2-mercaptoethanol or dithiothreitol are added. The tetramer contains thirty-two cysteinyl residues irregularly buried in hydrophobic regions of the protein matrix. There are no disulfide bonds. None of the sulfhydryl groups is associated with the active site and none is detected with *p*-chloromercuribenzoate after DDT-aggregation. The native enzyme apparently corresponds to a lipoprotein (Perry and Agosin, 1974).

B. Transferases

1. Arginine Kinase

Arginine kinase is the sole phosphagen kinase found in several major invertebrate groups, including arthropods, echinoderms, and molluscs. Arginine kinases from several sources have been found to be dimeric molecules with a molecular weight of approximately 80,000. In contrast, the arginine kinases of arthropods are monomeric with molecular weights of about 40,000. The enzyme has been purified from the housefly *Musca domestica,* and a crystalline preparation has been obtained from the thoraces of the honeybee, *Apis mellifica*. The arginine kinases from *D. melanogaster, D. hydei, D. simulans, D. bifasciata* and others have the same electrophoretic mobility. Only the arginine kinases of the drosophilid fly, *Zaprionus vittiger,* migrate differently. The same enzyme appears to exist in different developmental stages of *D. melanogaster*. The enzyme has been purified from *D. melanogaster* by ammonium sulfate fractionation and DEAE-cellulose and Sephadex G-100 column chromatography (Wallimann and Eppenberger, 1973) about sixty fold. The optimum pH is about 8.6 to 8.9 and the K_m for arginine is 0.15 *M*. Its properties are similar to those of the housefly enzyme, with a molecular weight of about 40,000.

2. RNA Polymerases

DNA-dependent RNA polymerases were first detected in rat liver nuclei and later shown to involve DNA as a primer. They are widely distributed and have been studied in plant, bacterial, mammalian, and insect tissues. The reaction catalyzed by these enzymes is shown in Eq. 7.

$$n\,(ATP + GTP + CTP + UTP) \xrightarrow{DNA} (AMP - GMP - CMP - UMP)_n + 4n\,PP_i \quad (7)$$

In vitro, the enzyme will use either single-stranded or double-stranded DNA template, and in the latter case, both chains are transcribed; *in vivo,* it appears that only double-stranded DNA is used and that only one of the chains is copied.

RNA polymerases consist of a catalytic unit (core enzyme) consisting of β', β and α subunits and a dissociable initiation factor subunit (σ). The enzymes appear to have the composition $\alpha_2\,\beta\beta'\sigma$. The σ factor which is loosely bound to the complex is responsible for the initial binding to the DNA template and, therefore, possibly for the selection of the proper transcribing DNA strand. Once RNA synthesis has started, σ dissociates from the enzyme–template complex and initiates synthesis at a new site. The σ factor has a molecular weight of about 95,000; the core enzymes with or without σ factor over 350,000. In contrast to prokaryotes which have only one RNA polymerase, eukaryotic organisms have usually been found to possess from two to four RNA polymerases, but as many as eight have been reported.

The eukaryotic RNA polymerases can be distinguished from one another by their different response characteristics to varying concentrations of Mg^{2+}, Mn^{2+} (Agosin, 1971); $(NH_4)_2SO_4$; and to transcription inhibiting drugs. RNA polymerase I is located within the nucleolus and RNA polymerases II and III are extranucleolar. RNA polymerase I appears to be responsible for the synthesis of ribosomal RNA, while enzyme II catalyzes the synthesis of the so-called heterogeneous RNA. RNA polymerase III has been shown to be involved in the synthesis of tRNA and 5 S RNA.

RNA polymerases have been extracted from the integument of larvae of the blowfly *C. erythrocephala* and their multiple nature has been demonstrated. One polymerase is resistant to α-ammanitin while a second one is sensitive to the toxin. A third polymerase is also insensitive to α-ammanitin. All three enzymes show maximum activity with Mn^{2+} and are more active with single-stranded DNA than with native DNA.

RNA polymerases have been studied in detail in *D. melanogaster.* RNA polymerase II sensitive to α-ammanitin is present in tissue culture cells or imaginal disks. Two RNA polymerases (I and II) are present in larvae and embryos. Recently, three RNA polymerases have been detected in larvae, one of which corresponds to the appearance of a new RNA polymerase during development. Extraction from 12- to 24-hr-old embryos yields two enzyme species (I_a and II); but when the extraction is carried out from third instar larvae, three enzymes are detected. RNA polymerase II from embryos is identical to RNA polymerase II from larvae in terms of their kinetic behavior in the presence of varying concentrations of cations. The embryo RNA polymerase I_a and larval enzyme $I_{a'}$ have similar Mg^{2+}, $(NH_4)_2SO_4$, and inhibitor sensitivity; but they have different chromatographic behavior. Both appear to correspond to RNA polymerases I of other eukaryotes. It is possible that RNA polymerase I_a is modified during development to become $I_{a'}$. The third RNA polymerase, I_b is very sensitive to rifamycin and thus it differs from other eukaryote polymerases. Polymerases I and II do not appear to change during development in other eukaryotes; therefore, the appearance of I_b in *Drosophila* with development is unique. It also has been suggested that I_b may not correspond to a nuclear polymerase but might be of mitochondrial or bacterial origin.

Recently, the RNA polymerase III from the posterior silk gland of *B. mori* has been purified by chromatography on DEAE-cellulose, DEAE-Sephadex, CM-Sephadex, and phosphocellulose and by sedimentation in sucrose density gradients. The molecular weight of the enzyme is approximately 590,000 to 600,000. The enzyme analyzed under denaturing conditions contains subunits with molecular weights of 155,000, 136,000, 67,000, 62,000, 49,000, 39,000, 36,000, 31,000, 28,000, and 18,000. The subunits are present in molar ratios close to unity, with the exception of the 18,000 molecular weight subunit, which is present in an approximate molar ratio of 2. Under non-denaturing conditions only two components are observed upon electrophoresis. The enzyme appears to differ from the polymerase III of other eukaryotes by its insensitivity to α-ammanitin and by the absence of an 89,000-dalton subunit.

3. DNA Polymerases

These enzymes catalyze the conversion of acid-soluble deoxynucleoside triphosphates into an acid-insoluble product, DNA (Eq. 8).

$$n(\mathrm{dATP} + \mathrm{dGTP} + \mathrm{dCTP} + \mathrm{dTTP}) \xrightarrow[\mathrm{Mg^{2+}}]{\mathrm{DNA}} \mathrm{DNA} + 4n\ \mathrm{PP_i} \qquad (8)$$

The nature of the DNA product is determined by the DNA template and is not affected by altering the ratio of the deoxynucleoside triphosphate reactants. Not many enzymes capable of synthesizing or completing DNA chains in eukaryotes have been purified substantially. Three types of template primers have been found active with crude extracts of *D. melanogaster* embryos: extracted DNA, poly(A)-oligo(dT) and poly(A)-oligo(U). It was thought that since these primers were utilized with varying relative efficiencies at different stages of development that several polymerases could exist; however, a nearly homogeneous preparation obtained from *D. melanogaster* embryos retains the ability of utilizing all three template primers. The polymerase has been purified more than 2300-fold. It forms a single band on gel electrophoresis, has a molecular weight of 87,000 and a pH optimum of 8.5. The activity requires a divalent cation, Mg^{2+} being more efficient than

Mn^{2+}. The enzyme is inactivated by mercurials and polyamines, and certain substrates are also inhibitors. The most efficient primer is native DNA, but homopolymers such as poly(dA)-oligo(dT), poly(A)-oligo(dT) and poly(A)-oligo(U) are also very efficient. The *Drosophila* enzyme as it occurs with other DNA polymerases can replicate single-stranded DNA only under conditions of simultaneous transcriptions by RNA polymerase.

C. Hydrolases

1. Amylases

Only a few insect amylases have been purified and studied extensively. The amylase from *T. molitor* larvae is a single polypeptide with molecular weight 68,000, isoelectric point of 4.0 and a very low content of sulfur-containing amino acids. The enzyme is a Ca^{2+}-protein and behaves as an α-amylase. The enzyme is very similar to that of *Callosobruchus chinensis*.

2. Carboxypeptidases

One of the two carboxypeptidases of the larvae of *Tineola bisselliella* has been purified to some extent. Its molecular weight is 72,000; its pH optimum for the hydrolysis of *N*-benzyloxycarbonylglycylleucine is 7.5–7.7. It is strongly inhibited by diisopropylfluorophosphate, sulfhydryl reagents, and 1,10-phenanthroline but not by EDTA. The enzyme has a strong affinity for neutral amino acid residues and does not hydrolyze *C*-terminal proline, arginine, or lysine. It is a true carboxypeptidase requiring an L-amino acid in the *C*-terminal position with a free COOH group and hydrolyzes peptides sequentially from the *C*-terminal end.

D. Isozymes

Isozymes are enzymes that have similar or identical catalytic functions, but are by one means or another identifiably different with respect to structure. Isozymes are usually detected by polyacrylamide gel electrophoresis. Chironomid hemoglobins are an example of insect isozymes.

The malate dehydrogenase of *D. virilis* exists in two forms, mitochondrial and cytoplasmic. They differ in electrophoretic mobility, thermolability, susceptibility to inhibitors, reactivity to coenzyme analogues, and immunological specificity. Another example of compartmentalization of isozymes is α-glycerophosphate dehydrogenase. Again in *D. melanogaster* one species is cytosolic and the other is mitochondrial. Both enzymes are involved in the so-called Bücher cycle. The cytosolic enzyme oxidizes NADH to NAD by the reduction of dihydroxyacetone phosphate to α-glycerophosphate; α-glycerophosphate in the mitochondria is oxidized to dihydroxyacetone phosphate by the mitochondrial enzyme with the production of NADH. The cytosolic enzyme exists as three electrophoretically separable enzymes. Only form three is detected in larvae, but one and three are present in adults.

Alcohol dehydrogenase in some homozygous strains of *D. melanogaster* consists of three forms separable by electrophoresis or chromatography: ADH 5, 3, 1. All three have similar if not identical primary structures as shown by amino acid analysis and peptide fingerprinting. Treatment of the flies with labeled nicotinamide results in radioactivity associated to ADH 3 and ADH 1 but not with ADH 5. In addition, a substance can be released from mixtures containing ADH 3 and 1 that is capable of converting ADH 5 into ADH 3. These results suggest that ADH 3 and ADH 1 have the same peptide composition as 5 but differ in charge, activity, and stability because of the nicotinamide-containing molecule which is noncovalently bound.

Multiple forms of DDT-dehydrochlorinase have been reported in *M. domestica*. It is not known whether these forms are tetrameric or monomeric.

Changes in gene expression during development of *Anophelis albimanus* have been analyzed by following isozyme esterase systems. An area of great interest is the participation of isozymes in gluconeogenesis in insects. Isozymes provide a valuable tool in insects for the study of gene duplication and regulation as well as for evolutionary genetics and systematics (Wagner and Selander, 1974).

V. Hemoproteins Other than Hemoglobin

An important group of hemoproteins is formed by the cytochromes. Cytochromes are defined as hemoproteins whose principal biological function is electron and/or hydrogen transport by virtue of a reversible valency change of their heme iron. This definition excludes catalases and peroxidases which are also hemoproteins and are active in electron transfer and hemoglobins in which no change of the iron valency occurs during oxygen transport. The earlier literature on insect cytochromes was reviewed by Gilmour (1961), who stated that the cytochrome system, as described extensively in mammals, has also been reported for various stages and tissues of numerous species of insects.

The electron transport chain in the mitochondrion corresponds to Eq. (9).

$$\begin{array}{cccccc} \text{Flavin} \;\; 2\text{H}^{+} & 2\text{ cyt } c\ \text{Fe}^{2+} & 2\text{ cyt } c_1\ \text{Fe}^{3+} & 2\text{ cyt } c\ \text{Fe}^{3+} & 2\text{ cyt } aa_3\ \text{Fe}^{3+} & \text{O}^{2-} \\ \text{Flavin H}_2 & 2\text{ cyt } c\ \text{Fe}^{3+} & 2\text{ cyt } c_1\ \text{Fe}^{2+} & 2\text{ cyt } c\ \text{Fe}^{2+} & 2\text{ cyt } aa_3\ \text{Fe}^{2+} & \frac{1}{2}\text{ O}_2 \end{array}$$

In Lepidoptera (*H. cecropia*) moderately high levels of cytochromes aa_3, b, and c occur in larvae. A microsomal b-type cytochrome (b_5) also occurs. Lower amounts of aa_3 and b_5 are observed in diapause but b_5 and b are not present. Cytochromes a and aa_3, NADH oxidase and succinate-cytochrome c reductase disappear in pupal diapause and reappear on termination. In Lepidoptera (*B. mori*) cytochrome oxidase content is low in ovaries in egg-laying females but increases in developing egg after diapause. No cytochrome c is found in earlier stages, including diapause. A long list of insect species and their variations in cytochrome content has been compiled by Lemberg and Barrett (1973); however, only a few cytochromes from insects have been studied in detail.

A. Cytochrome c

Cytochrome c is a basic protein with a molecular weight near 12,300. It consists of a single polypeptide chain 103–108 amino acid residues long, to which a single heme prosthetic group is covalently attached by thioether bonds formed by the addition of the sulfhydryl group of two cysteinyl residues in positions 14 and 17 across the vinyl side chains of the porphyrin ring. Probably because cytochrome c is a component of the mitochondrial respiratory chain which is obtained in a simple water-soluble form, it has been quite extensively studied in insects. The primary sequences of several types of cytochrome c have been determined. One interesting aspect is the constancy of the heme-binding sites Cys-14 (half-cysteine), Cys-17 and His-18. Residues 19, 20 and 21 also are common to most species with some conservative substitutions in the sense that amino acids with similar chemical properties are present. There also is constancy in the sequence 70–80, with the apparent exception of insect cytochrome c.

The iron content of cytochrome c isolated from tobacco hornworm (*Protoparce sexta*) moths is 0.45%. By assuming one iron atom per molecule of cytochrome c, the molecular weight is about 12,411 which agrees with the value obtained from amino acid analysis. Sequence analysis shows there are four residue differences between cytochrome c from this moth and that isolated from *Samia cynthia,* a moth belonging to the same order. In contrast there are at least twelve differences between the cytochrome c belonging to different orders.

The complete amino acid sequence of cytochrome c from *Ceratitis capitata* (Diptera) is known; similar to other Diptera, three arginine residues are present. Differences with the sequences of cytochrome c from *D. melanogaster* are found in positions 50, 60 and 61, while comparison with *Haematobia irritans* (Coleoptera) shows differences in five positions (9, 76, 50, 60, 61). Vertebrate cytochrome c invariably contains methionyl residues in positions 65 and 80 and in no other. In moth (*S. cynthia*) and in the fly protein, a similar methionyl residue exists in position 80 while phenylalanine occupies position 65, which suggests that position 65 may not be important.

B. Cytochrome b_5

Cytochrome b_5 was first described in an insect, *H. cecropia* (Gilmour, 1961). It is a microsomal cytochrome which may participate in electron transport during hydroxylation of exogenous and endogenous compounds by cytochrome *P*-450-linked reactions (Agosin and Perry, 1974). Cytochrome b_5 has been purified from housefly microsomes. The oxidized form has a maximum absorbance at 414 nm which shifts to 424 nm upon reduction. The reduced hemoprotein shows α and β bands close to the mammalian pigment. The extinction coefficient of cytochrome b_5, determined from the dithionite-reduced versus the ferricyanide–oxidized difference spectrum is 184 $cm^{-1} \cdot mM^{-1}$.

C. Cytochrome *P*-450

The characteristics of insect cytochrome *P*-450 have been reviewed by Agosin and Perry (1974). Cytochrome *P*-450 is the terminal oxidase of a NADPH-linked system that hydroxylates endogenous and exogenous compounds (Agosin, 1976). The system includes, in addition to cytochrome *P*-450, a flavoprotein enzyme, NADPH-cytochrome *P*-450 reductase and possibly phospholipids. Whether cytochrome b_5 and its corresponding reductase, NADH-cytochrome b_5 reductase, participate in insect hydroxylation has not been determined so far. Cytochrome *P*-450 is a *b*-type cytochrome so named because its reduced form binds carbon monoxide resulting in a difference spectrum with a band at 450 nm. It is a microsomal hemoprotein of extreme importance in the detoxication of foreign compounds and as such plays a major role in insecticide resistance.

Several types of cytochrome *P*-450 occur in different house fly strains. A diazinon-resistant strain has a cytochrome *P*-448 type, while the Fc strain contains both cytochrome *P*-450 and *P*-448. Susceptible strains apparently have mainly cytochromes of the *P*-452 type.

It is now apparent that multiple forms of cytochrome *P*-450 exist in a given insect. Two spectrally different forms, cytochromes *P*-450 and *P*-448, have been purified from the Fc house fly strain. Cytochrome *P*-450 has an extinction coefficient of 103 $cm^{-1} \cdot mM^{-1}$, while cytochrome *P*-448 has an extinction coefficient of 86 $cm^{-1} \cdot mM^{-1}$. These values are close to the extinction coefficient of mammalian cytochrome *P*-450, 91 $cm^{-1} \cdot mM^{-1}$. The two forms exist in approximately equimolecular amounts in house fly microsomes, and the contribution of each cytochrome would yield a combined extinction coefficient very close to the mammalian one. The molecular weight of housefly cytochrome *P*-450 is 45,000; cytochrome *P*-450 corresponds to a tetramer with a molecular weight of 160,000, while cytochrome *P*-448 is apparently formed by ten subunits, each of molecular weight 45,000. Two forms of cytochrome *P*-450 of the *P*-452 type have been solubilized from a susceptible housefly strain (Schonbrod and Terriere, 1975). Recently, three cytochrome *P*-450's have been isolated and purified from a susceptible housefly strain by the use of affinity chromatography in ω-amino-*n*-octyl-4B Sepharose columns (Capdevila and Agosin, 1976). One of the species has been obtained close to homogeneity. One species absorbs at 450 nm, the other at 452 nm, and the third one corresponds to a *P*-448 species. The 452 nm species has been obtained with a specific activity of about 14 nmoles per mg protein. Its molecular weight is 43,000. The multiplicity of cytochrome *P*-450 forms in insects makes it difficult to correlate levels of the hemoprotein with insecticide resistance. It is quite possible that only certain hemoprotein species may be involved in detoxication, as it seems to occur with cytochrome *P*-448.

VI. BIOLUMINESCENCE

Certain enzyme proteins have the ability to handle energy and create an electronically excited state. Light is the product of the reaction. The enzymatic mechanisms occurring in bioluminescence, which essentially involves oxidative events, appear to represent variations of oxygenase and peroxidase (Cormier *et al.*, 1975); although the formation of an intermediate adenylate is reminiscent of aminoacyl-tRNA synthetases. Bioluminescence occurs in Collembola, Homoptera, Diptera, and Coleoptera. In Diptera, self-luminescence is seen in *Ceroplatus, Orfelia*, and *Arachnocampa*. Bioluminescence in adult fireflies (*Coleoptera: Lampyridae*) has sexual significance.

Luciferase is the generic term applied to the enzyme that oxidizes the substrate luciferin with the emission of light. The luminescent reaction in firefly extracts has been studied in detail. The reaction involves the conversion of luciferin to luciferyl adenylate in the presence of ATP followed by oxidation to oxyluciferin (Eq. 10).

$$\text{Luciferin } H_2 + \text{luciferase} + \text{ATP} \overset{Mg^{2+}}{\rightleftarrows} \text{luciferase-(luciferin } H_2\text{-AMP)} + PP_i \tag{10}$$

The complex luciferase-(luciferin H_2-AMP) is oxidized in the presence of oxygen by reactions that are not clearly understood at present. ATP is an absolute requirement for the reaction.

Firefly luciferase was crystallized in pure form as early as 1956. Its molecular weight is 100,000 and it is composed of two identical subunits of molecular weight 50,000 which dissociate in the presence of guanidine hydrochloride. It has been shown that luciferase as a 50,000 molecular weight species is fully active rather than the dimeric molecular weight species as previously thought. Amino acid analysis of pure luciferase fails to yield any unusual components. It contains about 10 half-cysteine residues and 5.4 free SH groups. The activity is inhibited reversibly by *p*-mercuribenzoate. Only two SH groups are essential for the reaction.

The *in vivo* firefly flash has a duration of about 0.5 sec and is nerve-controlled; however, the mechanisms regulating the *in vivo* activity are poorly understood.

It may be considered that bioluminescence in insects has a role in the economy of the organism. Numerous kinds of positive functions in this sense are recognized, including mating signals in the firefly; however, the possibility that the luciferase–luciferin reaction results in the formation of activated oxygen intermediates that may be utilized by the organism for biological oxidations should also be considered.

VII. PEPTIDE HORMONES

It has long been apparent in insects that polypeptides and peptides regulate various processes and possess a variety of pharmacological activities. Polypeptides are apparently involved in the regulation of the insect endocrine system. The insect endocrine system comprises both the neuroendocrine and classical endocrine components. The insect brain coordinates the endocrine system with sensory information from the environment. The brain contains two major paired clusters of neurosecretory cells, the median and the lateral neurosecretory cells. A third group, the posterior neurosecretory cells, are found on occasion in the protocerebrum. The major neurosecretory cells send axons that terminate in the paired corpora cardiaca which lies posterior to the brain. It is assumed that polypeptides secreted by the neurosecretory cells stimulate a complement of intrinsic neurosecretory cells present in the corpora cardiaca. The stimulated corpora cardiaca secrete in turn tropic polypeptides which stimulate the prothoracic glands which presumably secrete ecdysone, the hormone that initiates the molting cycle. The direction of the molt, whether it be to another larval stage, to the pupae or the adult is determined by the juvenile hormone, a secretion of the corpora allata, the paired glands attached to the corpora cardiaca. The brain–corpora cardiaca complex of insects is homologous to the hypothalamic–neurohypophyseal complex of vertebrates and the polypeptides secreted by the insect complex are collectively known as "brain hormone."

A. Brain Hormone

The protein nature of the brain hormone was contested for some time, and it was assumed to be identical to cholesterol; however, brain hormone activity is found in aqueous extracts from brains of *Bombyx*. The extract can be dialyzed without losing any activity and can be inactivated by heat under special conditions. It also can be fractionated by ammonium sulfate precipitation, and the activity is destroyed by pronase, the bacterial protease obtained from *Streptomyces griseus*. The brain hormone has been purified about 8000-fold, but the activity still is associated with heterogeneous molecular forms of molecular weight 9000, 12,000, and 31,000. The brain hormone from *Periplaneta* appears to be of molecular weight 20,000–40,000 and is similar to the hormone extracted from *Bombyx* brains. Recently, the brain hormone from *B. mori* has been fractionated by column chromatography and a molecular weight of no less than 5000 obtained.

Two prothoracicotropic polypeptides or factors I and II have been demonstrated as part of the brain hormone in *P. americana*. Factor I stimulates significantly the synthesis of various RNA species in the prothoracic glands.

The brain hormone regulates cuticular tanning, adult blood sugar levels, cardiac activity, water balance, and behavior. In some cases the polypeptide nature of neurosecretions regulating these activities appears to be more or less substantiated, as discussed below.

B. Bursicon

The brain is responsible for the initiation of tanning by a bloodborne factor. The tanning agent appears to originate in the ganglia of the ventral nerve cord, is not species specific, and has received the name of bursicon. The material is precipitated by trichloroacetic acid, ethanol, and acetone with loss of activity; ammonium sulfate fractionation yields an active product. It is inactivated by heat and proteolytic enzymes. Thus, bursicon corresponds to a polypeptide. Its molecular weight is estimated to be 40,000. Bursicon has been detected in *P. americana, P. regina*, and *L. migratoria*. Bursicon coregulates with ecdysone the molting process in insects since it affects cuticular development *in toto* and is not simply a tanning agent.

C. Diuretic Hormone

The peptides or polypeptides implicated in the control of urine formation are products of either the cerebral or medial neurosecretory systems. The phenomenon has been studied extensively in *Rhodnius prolixus*. Antidiuretic factors also have been described. Their nature and origin are obscure, although in *Corethra* the thoracic ganglia appear to be involved. In locusts, diuretic and antidiuretic factors appear to be released from the corpora cardiaca. It has been suggested that adenosine 3′,5′-cyclic monophosphate (cyclic AMP) may act as a "second messenger" in the *R. prolixus*, peptide-regulated diuretic system. It is possible that cyclic AMP may play a role at other hormone-regulated sites in insects. Cyclic AMP mediates the effect of a variety of peptide hormones and other biological peptide compounds. These exert their effect by stimulation or inhibition of adenylate cyclase and cyclic AMP phosphodiesterase. Adenylate cyclase, a membrane-bound enzyme catalyzes the formation of cyclic AMP and pyrophosphate from ATP. The hydrolysis of cyclic AMP to 5′-AMP is catalyzed by the phosphodiesterase.

D. Polypeptides Affecting Cardiac Activity

Two polypeptides (neurohormone C and D) are produced by the corpora cardiaca. Neurohormone D affects both the rate and amplitude of the cardiac beat; whereas neurohormone C increases the rate but decreases the amplitude. Both hormones have been extracted from the brain–corpora cardiaca complex of *Periplaneta americana*.

E. Hyperglycemic, Hypoglycemic, and Adipokinetic Peptides

The level of blood sugar, mainly trehalose, is regulated by a hyperglycemic factor(s) originated essentially in the corpora cardiaca. It has been suggested that several peptides are involved in the regulatory phenomenon, some of which may be of brain origin. The phenomenon has been observed in *P. americana*, in species of Orthoptera, in Diptera and Hymenoptera. Extracts of corpora cardiaca from *Phormia* and *Calliphora* and the silkmoth *Cecropia* seem to be inactive; but the hormone is liberated from the *Calliphora* gland by electrical and mechanical stimulation *in situ*. The hyperglycemic factor corresponds to the vertebrate glucagon; the hormone increases carbohydrate levels in the hemolymph *in vivo*, and stimulates glycogenolysis and the activation of phosphorylase in the fat body *in vitro*. A highly acidic small peptide that behaves like glucagon has been identified from the corpora cardiaca of *Manduca sexta*. Radioimmunoassay also has shown the presence of insulinlike peptides in extracts of the corpus cardiacum–corpus allatum complex.

Peptides have been implicated in the regulation of lipid levels in the hemolymph. The factor(s) or adipokinetic peptide appears to originate in the corpora cardiaca. Both the hyperglycemic and the adipokinetic factor seem to correspond to small peptides, but their structures have not been determined so far.

F. Sex Peptides

A ninhydrin-positive male substance has been detected and termed sex peptide. This substance is localized in the accessory gland (paragonia) secretion. Its function is the stimulation of egg laying, inhibition of female receptivity, and successful sperm transfer. Two sex-specific substances have been found in the paragonia of *Drosophila funebris*. The first one (PS-1) lowers receptivity when injected into virgin females, and it corresponds to an eicosaheptapeptide and its amino acid sequence is known. The second substance (PS-2) is a glycerine derivative which enhances oviposition. Other species of *Drosophila* do not show the presence of similar peptides. PS-1 has a molecular weight of 2700 and its amino acid sequence has been established tentatively, the *N*-terminal amino acid being aspartate. The notable features of PS-1 are the high content of alanine and the absence of aromatic residues, the absence of sulfur-containing amino acids, and the absence of glycine, histidine, and isoleucine, and particularly the alternating sequence of Ala-Asn, occurring four times within the twenty-seven residues. The biological activity of PS-1 is destroyed by hydrolysis with pepsin, chymotrypsin, trypsin, aminopeptidase M, and carboxypeptidase A. Blockage of amino groups, removal of the *N*-terminal aspartyl residue, and limited HCl hydrolysis also reduces the biological activity. This is probably the first characterization of a peptide which controls sexual behavior in insects.

G. Proctolin

A small peptide with potent myotropic activity on the proctodeal muscle of *P. americana* recently has been isolated and characterized. The neurotransmitter is a pentapeptide having the sequence Arg-Tyr-Le-Pro-Thr (Starratt and Brown, 1975). The peptide is not species specific, having been found in *Hemiptera, Coleoptera, Lepidoptera, Diptera* and *Hymenoptera*. The proctodeal muscle of *Periplaneta* reacts to proctolin at threshold level of about 0.5 mg per ml. Thus, with a molecular weight of about 500, proctolin is effective in the 10^{-9} *M* range.

VIII. CONCLUDING REMARKS

Whenever possible we have discussed the insect proteins essentially in terms of their structure and function; however, only in a few cases, structure and sequence studies have been conducted with insect proteins and indeed few have been isolated and purified to homogeneity. The literature on the subject is profuse, but most of it is related to the activity of enzymes in several metabolic pathways. This is regrettable since a proper understanding of the biological function of macromolecules such as proteins is most likely to be obtained when their structures are chemically and physically defined. However, this "state of the art" of our knowledge of insect proteins is, in part, due to the fact that only very small amounts of material are obtainable from insects, which precludes the extensive purification and characterization of many proteins.

GENERAL REFERENCES

Agosin, M. (1971). *Insect Biochem.* **1,** 363.
Agosin, M. (1976). *Mol. Cell. Biochem.* **12,** 33.
Agosin, M., and Perry, A. S. (1974). *In* "The Physiology of Insecta" (M. Rockstein, ed.), 2nd Ed., Vol. 5, pp. 537–596. Academic Press, New York.
Bullard, B., and Reedy, M. K. (1973). *Cold Spring Harbor Symp. Quant. Biol.* **37,** 423.
Capdevila, J., and Agosin, M. (1976). *Proc. Int. Congr. Biochem., 10th, Hamburg* p. 264.
Chaplain, R. A. (1966). *Biochem. Biophys. Res. Commun.* **22,** 248.
Chaplain, R. A. (1967). *Arch. Biochem. Biophys.* **121,** 154.
Chesky, J. A. (1975). *Insect Biochem.* **5,** 509.
Chino, H., Murakami, S., and Harashima, K. (1969). *Biochim. Biophys. Acta* **176,** 1.
Cormier, M. J., Lee, J., and Wampler, J. E. (1975). *Annu. Rev. Biochem.* **44,** 255.
Dejmal, R. K., and Brookes, V. J. (1972). *J. Biol. Chem.* **247,** 869.
Dick, C., and Johns, E. W. (1969). *Comp. Biochem. Physiol.* **31,** 529.
Ebashi, S., and Endo, M. (1968). *Prog. Biophys. Mol. Biol.* **18,** 123.
Elgin, S. C. R., Froehner, S. C., Smart, J. E., and Bonner, J. (1971). *Adv. Cell Mol. Biol.* **1,** 1.
Fraser, R. D. B., MacRae, T. P., and Stewart, F. H. C. (1966). *J. Mol. Biol.* **19,** 580.
Friedman, S. (1975). *Insect Biochem.* **5,** 151.
Fukuda, T., Kirimura, J., Matsuda, M., and Suzuki, T. (1955). *J. Biochem. (Tokyo)* **42,** 341.
Gilmour, D. (1961). "Biochemistry of Insects." Academic Press, New York.
Harper, E., Seifter, S., and Scharrer, B. (1967). *J. Cell Biol.* **33,** 385.
Heilmeyer, L. M. G., Jr., Rüegg, J. C., and Wieland, T. H. (1976). "Molecular Basis of Motility," 222 pp. Springer-Verlag, Berlin and New York.
Hruska, J. F., and Law, J. H. (1970). *In* "Methods in Enzymology," Vol. 19: Proteolytic Enzymes (G. Perlmann and L. Lorand, eds.), p. 221. Academic Press, New York.
Jeuniaux, C. H. (1971). *In* "Chemical Zoology" (M. Florkin and B. T. Scheer, eds.) Vol 6 (Pt. B), 36–118. Academic Press, New York.
Jewell, B. R., and Rüegg, J. C. (1966). *Proc. R. Soc., Ser. B* **164,** 428.
Kafatos, F. C. (1972). *In* "Topics in Developmental Biology" (A. A. Moscana and A. Monroy, eds.), Vol. 7, pp. 125–191. Academic Press, New York.
Kominz, D. R., K. Maruyama, L. Levenbook ,and M. Lewis (1962). *Biochim. Biophys. Acta* **63,** 106.
Koolman, J., and Karlson, P. (1975). *Hoppe-Seyler's Z. Physiol. Chem.* **356,** 1131.
Lehman, W., Kendrick-Jones, J., and Szent-Györgyi, A. G. (1973). *Cold Spring Harbor Symp. Quant. Biol.* **37,** 319.
Lemberg, R., and Barrett, J. (1973). "Cytochromes," 580 pp. Academic Press, New York.
Lizardi, P. M., and Brown, D. D. (1974). *Cold Spring Harbor Symp. Quant. Biol.* **38,** 701.
Lucas, F., Shaw, J. T. B., and Smith, S. G. (1958). *Adv. Protein Chem.* **13,** 107.
Lucas, F., Shaw, J. T. B., and Smith, S. G. (1960). *J. Mol. Biol.* **2,** 339.
Maruyama, K. (1974). *In* "The Physiology of Insecta" (M. Rockstein, ed.), 2nd Ed., Vol. 4, pp. 237–267. Academic Press, New York.

Munn, E. A., Feinstein, A., and Greville, G. D. (1971). *Biochem. J.* **124,** 367.
Nigg, H. N., Svoboda, J. A., Thompson, M. J., Dutky, S. R., Kaplanis, J. N., and Robbins, W. E. (1976). *Experientia* **32,** 438.
Perry, A. S., and Agosin, M. (1974). *In* "The Physiology of Insecta" (M. Rockstein, ed.), 2nd Ed., Vol. 6, pp. 3–121. Academic Press, New York.
Reedy, M. K., Bahr, G. F., and Fischman, D. A. (1973). *Cold Spring Harbor Symp. Quant. Biol.* **37,** 397.
Rudall, K. M., and Kenchington, W. (1971). *Annu. Rev. Entomol.* **16,** 73–96.
Sasaki, T., and Noda, H. (1973). *Biochim. Biophys. Acta* **310,** 91.
Schartau, W., and Linzen, B. (1976). *Hoppe-Seyler's Z. Physiol. Chem.* **357,** 41.
Schnabel, E., and Zahn, H. (1958). *Justus Liebig's Ann. Chem.* **614,** 141.
Schonbrod, R. D., and Terriere, L. C. (1975). *Biochem. Biophys. Res. Commun.* **64,** 829.
Smith, D. (1965). *J. Cell Biol.* **27,** 379.
Starratt, A. N., and Brown, B. E. (1975). *Life Sci.* **17,** 1253.
Taylor, E. W. (1972). *Annu. Rev. Biochem.* **41,** 577.
Teng, C. S. (1974). *Biochim. Biophys. Acta* **366,** 385.
Tristram, G. R., and Smith, R. H. (1963). *In* "The Proteins" (H. Neurath, ed.), Vol. 1, pp. 45–50. Academic Press, New York.
Tsang, V., and Agosin, M. (1976). *Insect Biochem.* **6,** 425.
Wagner, R. P., and Selander, R. K. (1974). *Annu. Rev. Entomol.* **19,** 117.
Wallimann, T., and Eppenberger, H. M. (1973). *Eur. J. Biochem.* **38,** 180.
Zammit, V. A., and Newsholme, E. A. (1976). *Biochem. J.* **154,** 677.

4

Protein Synthesis in Relation to Cellular Activation and Deactivation

P. S. CHEN

I. INTRODUCTION

In insects, as in other animals, one major biochemical event resulting from cellular activation is protein synthesis. Modern molecular biology has provided a clear overall picture of the mechanism by which genetic information encoded in DNA can be expressed as structural and catalytic proteins which play an essential role in growth, differentiation, and function of all living cells. In the past two decades, experiments designed to unravel the many steps involved in protein biosynthesis have mainly been carried out in prokaryotes. Although the general scheme is no doubt applicable to all organisms, recent progress in biochemical genetics of higher animals has revealed various unique features in regard to both the transcriptional units and control mechanisms. Thus, it has been proposed that a large part of DNA of the eukaryotic genome bears no structural information, but exerts only a regulatory function. Estimates of the number of genes, by analysis of complementation groups in lethal mutations or the size range of polypeptide gene products, all indicate that only a small proportion (probably less than 5%) of eukaryotic DNA codes for proteins. From their study on the processing of heterogeneous nuclear RNA (hnRNA), which is mostly transcribed from noncoding DNA, Lengyel and Penman (1975) concluded that in *Drosophila* only 20%, and in *Aedes* even only 3.3% of hnRNA is converted to mRNA. In addition, protein synthesis in eukaryotes can be influenced by specific external signals resulting from changes in cellular conditions and physiological states of the organism (Smith, 1976). This has repeatedly been confirmed from recent work on the biochemistry of *Drosophila* (see review by Mitchell, 1967).

In a discussion of protein synthesis in insects, several specific points should not be overlooked. First, insects belong to a class of animals which exhibit the most diversified forms. The metabolic states and cellular conditions in animals of each species often show remarkable specialization. In general, not only the proteins produced in homologous tissues and organs, but also the activation processes leading to the initiation of their production appear to be highly species-specific. Second, both morphogenesis and reproduction in insects are well known to be under the control of hormones. The chemical nature of several major insect hormones has already been elucidated, and the pertinent information on the molecular basis of action of these compounds is now available (Wyatt, 1972; Burdette, 1974; Novak, 1975). For an understanding of the biochemistry of cellular activation and deactivation in insects, a closer examination of the mechanisms of hormone-mediated protein synthesis is in order. Third, one characteristic feature of holometabolous insects is the transformation from larva to adult at metamorphosis. It is conceivable that at this period there is a selective qualitative control of protein synthesis through a coordinated switching of different sets of genes. Although our present knowledge about gene regulation systems in most insects is still very limited, there is no doubt that tissue- and stage-specific patterns of protein synthesis are correlated with changes in gene activity at different phases of the insect life cycle.

This chapter will focus on some current studies on the synthetic activity in several tissues and organs which show the most pronounced cellular activation and deactivation as a response to a variety of intrinsic and extrinsic factors. Earlier literature on protein metabolism during insect development can be found in two previous reviews (Chen, 1966, 1971). A detailed treatment of the molecular mechanism of protein synthesis in insects, mostly based on studies of the mealworm, *Tenebrio molitor,* has been presented by Ilan and Ilan (1974).

II. PROTEIN SYNTHESIS AND NUCLEOCYTOPLASMIC INTERACTIONS DURING EARLY EMBRYOGENESIS

There is strong evidence indicating that during early development of the insect egg nucleocytoplasmic interactions take place, and that these interactions are of primary importance in the determination of the fate of the dividing nuclei and the formation of the embryonic anlage. According to Seidel (cited in Chen, 1971) nuclear division begins at the cleavage center located at about the anterior one-third of the egg. Some of the cleavage nuclei, which are still totipotent at this stage, move into the posterior polar region. Through certain yet unknown interreactions between the incoming nuclei and the polar plasm, a so-called activation center is formed. This newly formed center is in turn responsible for the initiation of a differentiation center at the location of the future embryo by producing a certain agent which diffuses in the anterior direction. In addition, in many insect species the posterior polar cytoplasm of the developing egg contains RNA-rich granules which probably play an essential role in the formation of the pole cells, the primordial germ cells. As we shall see in Section II, B, various experiments have been carried out to clarify the molecular mechanism of this early activation process as well as to explore the nature and mode of interactions between the different developmental centers.

A. Utilization of Amino Acids

The insect egg can be considered as a closed system, and its supply of amino acids for protein synthesis depends on degradation of the yolk reserve. Analysis of the level and utilization of free amino acids in the course of development may provide valuable insight into the basic events that underlie embryonic differentiation.

In general, all amino acids which are commonly found in protein occur in insect eggs (see references in Chen, 1971). The concentrations of aspartic acid, glutamic acid, glutamine, alanine, and glycine are usually high. Other amino acids of rather rare occurrence such as aminobutyric acid, methionine sulfoxide, ornithine, and citrulline have been identified. In a number of species several phosphate esters containing ethanolamine, serine, and threonine have been demonstrated. Of particular interest is the occurrence of various peptides which are apparently linked to the breakdown of yolk proteins. However, owing to the low concentration and the difficulty in the isolation of these compounds, a closer analysis of their amino acid composition is not easy.

There are significant qualitative and quantitative changes of the free amino acid pool during embryonic development. A rapid increase in the level of all major amino acids at the onset of embryogenesis has been reported in *Bombyx, Drosophila, Sphaerodema, Culex, Schistocerca, Teleogryllus* and *Aulocara*. (see Chen 1966 and 1971, for references). This corresponds to the periods of blastoderm formation and germ band elongation. The total concentration remains high at the beginning of blastokinesis and then declines during later embryonic differentiation. The overall variation represents no doubt the balance between release of amino acids from yolk degradation and their utilization for *de novo* protein synthesis. The fact that some amino acids are missing at certain embryonic stages is probably due to their low levels.

More recently, McGregor and Loughton (1977) performed a detailed analysis on the net utilization rate of individual amino acids during embryogenesis of the locust, *Locusta migratoria migratorioides*. On the basis of the degradation rate of yolk protein and the levels of free amino acids, a daily amino acid balance sheet was calculated. Their results indicate that during initial embryonic growth the free amino acid pool increases rapidly with a concomitant decrease in the amino acids available in the yolk protein (Fig. 1). But beginning from blastokinesis (stage 4) until hatching (stage 9), the increase in yolk breakdown does not yield an elevation of the pool size which, instead, declines continuously, suggesting that the rate of amino acid utilization exceeds that of supply from protein degradation.

The general pattern reported by McGregor and Loughton (1977) is in agreement with that mentioned immediately above for other insect embryos. Serine, glycine, and tyrosine are, however, exceptions. As shown in Fig. 2a, from a comparison of the net metabolism and pool size of serine at stages 3 and 4, it is evident that the release of this amino acid from the yolk protein is insufficient to maintain its level in the free amino acid pool during these periods. Since the level of free glycine (which is a major amino acid of the yolk protein) remains fairly constant at these stages in spite of the release of a large amount of it from the yolk reserve (Fig. 2b), it appears probable that glycine is converted to serine. At stage 7 the level of free tyrosine increases as a result of reduced utilization of this amino acid (Fig. 2c). Its subsequent decline prior to hatching is thought to be related to the sclerotization of the pharate larval cuticle. In a previous study on the *Culex* embryo by Chen and

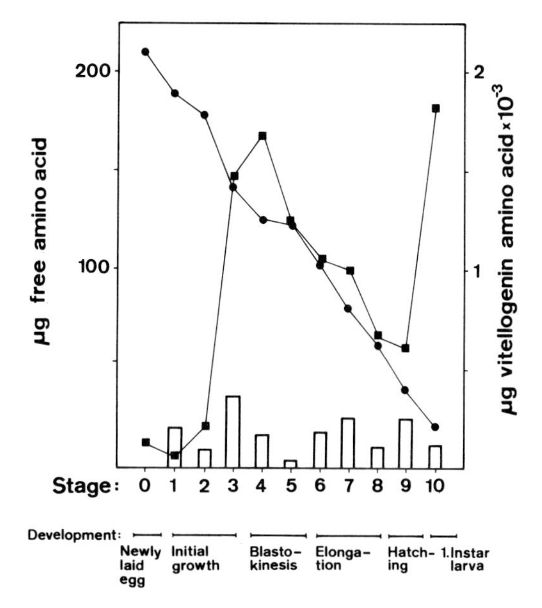

Fig. 1. Changes in total free amino acids (■) and total amino acids remaining in the yolk protein (●) during embryonic development of the locust, *Locusta migratoria migratorioides*. The amount of amino acids released from the yolk during each 24 hr period is indicated by the histogram. [From McGregor and Loughton (1977). Description of the developmental stages from McGregor and Loughton (1974).]

Briegel (1965) a decline in the level of tyrosine due to tanning of the exochorion and differentiation of the cuticular structures has also been noted.

McGregor and Loughton (1976) further reported that in the developing locust egg, labeled leucine and arginine administrated by topical application onto the dechorionated egg surface resulted in labeling of other amino acids and ammonia, whereas no interconversion took place when the same amino acids were incorporated into the yolk protein following injection into the maternal hemocoel. Although the reason for such a difference is unclear, the observation indicates that amino acids supplied exogenously to the egg system may have a different metabolic fate as those originally present in the endogenous pool.

B. Protein Synthesis

The major biochemical process underlying cellular differentiation is protein synthesis. Two types of experiments have been carried out for analyses of protein metabolism during insect embryogenesis: isotopic labeling and electrophoresis. In the first type of experiment labeled amino acids are introduced into the egg system either by injection or by soaking and topical

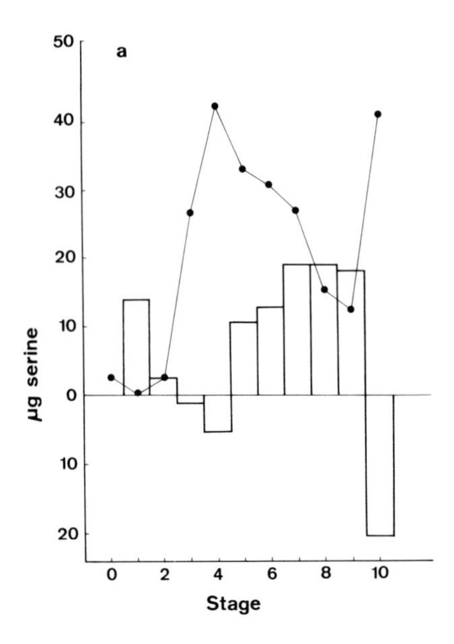

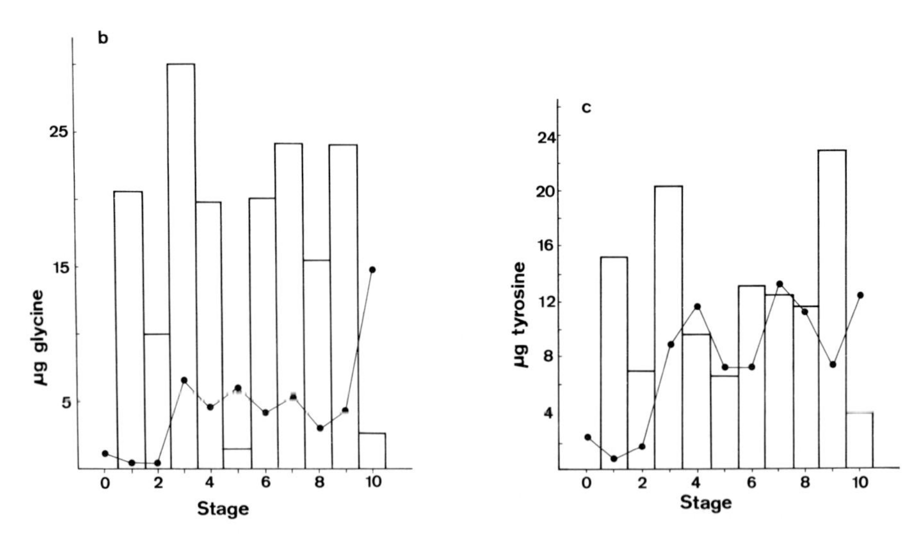

Fig. 2. Changes in the free amino acids serine (a), glycine (b) and tyrosine (c) during embryonic development of the locust, *Locusta migratoria migratorioides.* The net metabolism of each amino acid at each stage indicated by the histogram was calculated by subtracting the amount in the free amino acid pool from the sum of the preexisting amount of free amino acid and the amount released from yolk degradation. (From McGregor and Loughton, 1977.)

application following dechorionation. The main handicap is that the egg membranes are permeable to only low molecular weight and nonpolar compounds. Because of the small size and low consistency of the egg cytoplasm the loss of material by puncturing the egg can hardly be avoided. The dechorionated egg may easily be dehydrated or lose its ionic components by diffusion. Thus, each of these methods has its inherent difficulties and should be used only under carefully controlled conditions.

Electrophoretic procedures in combination with histo- or immunochemical techniques have a high sensitivity and a great resolving power. In view of the limited amount of protein in insect eggs these are no doubt the methods of choice in experiments designed to analyze the protein patterns during embryonic development. However, a clear distinction between protein components derived from yolk degradation and those newly synthesized by embryonic tissues presents certain intractable problems, unless additional techniques such as specific staining or immunological tests are used. As will be discussed in Section III,B the physicochemical properties of several insect vitellogenins have recently been described. These results may serve as guidelines in future studies by the techniques of electrophoretic separation.

We have already mentioned that in the early embryonic development of insects interactions between the migrating cleavage nuclei and the egg cytoplasm take place, and a sequential activation of various morphogenetic centers leads to the determination of the germ anlage. In the egg of the house cricket, *Acheta domestica,* differences in the protein pattern between anterior and posterior parts have been found at the onset of development, suggesting that the activation processes may have their origin in the predetermined biochemical architecture of the egg system (see Chen, 1971, p. 18).

More recently, in an attempt to explore the nature of interactions between the different egg regions, extensive studies of protein synthesis during early embryogenesis of the beetle, *Dermestes frischi,* have been carried out by Küthe (1972, 1973a, b). Incorporation of labeled amino acids injected into eggs aged 1 hr after oviposition showed initially an even distribution of radioactivity in the ooplasm. As nuclear division proceeds, the cortical region gradually became more heavily labeled. In eggs with 256 cleavage nuclei at 4 hr radioactivity was found to be limited in the periplasm. These results indicate that *de novo* protein synthesis begins already at the early cleavage stage, and there is a continuous transfer of the newly synthesized protein to the peripheral region of the egg (Küthe, 1973a). The total incorporation is, however, low at these early stages, but shows a twofold increase following migration of the nuclei into the periplasm at 6 hr (Fig. 3). It reaches a maximum at the time when the blastoderm is formed. Morphologically this period is characterized by the formation of the cell membranes and the appearance of nucleoli in the blastodermal nuclei. In spite of a slight decline

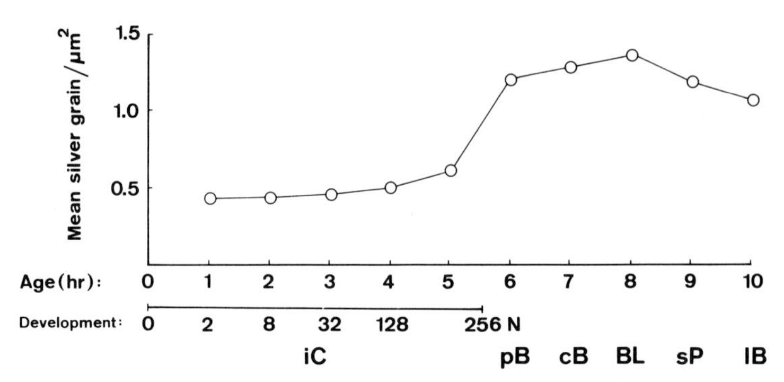

Fig. 3. Incorporation rate of ^{3}H-labeled amino acids into protein during early embryogenesis of the beetle, *Dermestes frischi*. Abbreviations: N, nuclei; iC, intravitelline cleavage; pB, plasmodial preblastoderm; cB, blastoderm following cell membrane formation; Bl, blastoderm; sP, formation of secondary periplasm; lB, late blastoderm. (Redrawn after Küthe, 1973a.)

in the total incorporation during the subsequent 2 hr, an intense labeling of the blastoderm and a newly formed secondary periplasma could be observed (Küthe, 1973a). But up to this stage the incorporation appeared homogeneous, and regional differences became detectable only at the time of germ layer formation which occurs at about 10½ hr after oviposition (Küthe, 1973b),

The formation of a so-called pseudoblastoderm in the *Dermestes* egg can be induced by UV irradiation prior to the migration of the cleavage nuclei into the periplasm at about 4½–5 hr (Küthe, 1972). The total irradiated superficial area remains free of nuclei which form a blastoderm-like layer within the yolk-containing endoplasm (Fig. 4). Nuclear division is inhibited in the pseudoblastoderm which shows no visible differentiation, and the embryo becomes lethal. A detailed electrophoretic analysis at the pseudoblastoderm stage revealed that protein band B_1, which is synthesized during early cleavage and later localized only in the peripheral ectoplasm during normal development, was totally missing (Fig. 5). Bands B4–6 which are normally present in the endoplasm up to the preblastoderm stage could also not be observed. Similarly, bands C4 and C6 which appear at the time of blastoderm formation have not been detected in the irradiated embryo. These results suggest that the cortical part of the developing egg contains certain factors which control the synthesis of various protein components in the yolk-endoplasmic system. The functional condition of the periplasm is apparently essential to a number of biochemical–physiological events underlying early embryogenesis in this insect.

In the egg of another coleopteran, *Dermestes maculata*, and the housefly, *Musca domestica*, protein synthesis also begins as early as the cleavage stage, and increases distinctly during preblastoderm formation. In the house

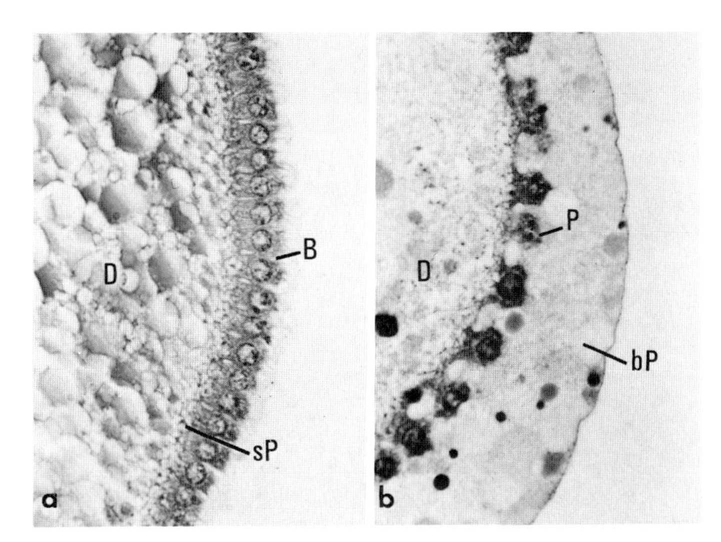

Fig. 4. Cross sections through the peripheral region of the developing egg of the beetle, *Dermestes frischi,* at 9 hr after oviposition. (a) Normal egg at late blastoderm stage with high cylindrical blastodermal cells. (b) Formation of pseudoblastoderm following UV irradiation at intravitelline cleavage stage. Abbreviations: B, blastoderm; D, yolk-endoplasmic system; sP, secondary periplasm; bP, irradiated ectoplasmic region; P, pseudoblastoderm. (From Küthe, 1972.)

cricket, *Acheta domestica,* although protein synthesis is already detectable at the early cleavage stage, a significant increase occurs only at the time when the germ anlage is formed. This is because both coleopteran and dipteran eggs belong to the early differentiation type in which the activation of the differentiation center and the determination of metameric organization in the blastoderm coincide with a maximum rate and high complexity of protein formation (see discussion in Küthe, 1973a).

For an understanding of protein metabolism during early insect embryogenesis, pertinent information on RNA synthesis appears to be essential. In *Dermestes* and *Leptinotarsa* there is no incorporation of labeled uridine into RNA prior to the migration of cleavage nuclei into the periplasm. Similarly, in the milkweed bug, *Oncopeltus fasciatus*, the blowfly, *Phormia regina,* and the cricket, *Acheta domestica, de novo* rRNA becomes detectable only at about the time of blastoderm formation. Thus, it seems that protein synthesis at the cleavage stage utilizes only preexisting stable messenger and ribosomal RNA's which are supplied to the oocyte by the nurse cells (for specific references and discussion, see Chen, 1971, pp. 25–31).

In summary, the activation of morphogenetic centers in the developing insect egg depends probably upon regional differences in the protein struc-

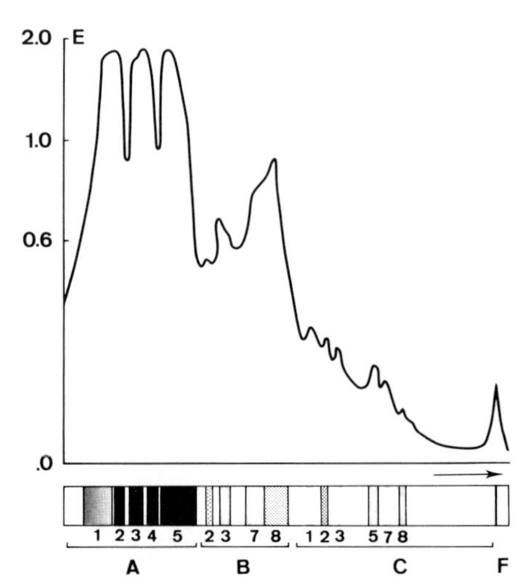

Fig. 5. Densitometric curve (above) and electrophoretic protein pattern (below) of the *Dermestes* egg with pseudoblastoderm formation at 8 hr after oviposition. (From Küthe, 1972.)

ture of the ooplasm. There is evidence that protein synthesis at the cleavage stage is under the control of factors confined to the cortical egg region. The interaction between the periplasm and the yolk-endoplasmic system involves, however, no initiation of gene action, as *de novo* RNA synthesis takes place only at the onset of blastoderm formation.

III. SYNTHESIS AND STORAGE OF PROTEINS IN FAT BODY

The insect fat body is a major organ of multiple metabolic processes and appears to be most conspicuous in the larvae of holometabolous insects. It represents 65% of the total body weight in the mature larva of the honeybee, 50% of that in the late larva of the blow fly and about 40% of that in the *Cecropia* silkmoth pupa. During larval development this organ is responsible for the synthesis of various major hemolymph proteins and serves at the same time as the place of storage of these components, in addition to carbohydrates and lipids. In the female adult it is the site where most yolk proteins or vitellogenins are produced. Hence, the fat body in insects fulfills a variety of functions similar to the hepatopancreas in molluscs and in crustaceans or the liver in mammals. As will be discussed in the following sections, all these activities appear to be stage-specific and are under hormonal control. The protein metabolism of insect fat body has been the subject of two more recent reviews by Price (1973) and Wyatt (1975).

A. Hemolymph Proteins

Conclusive evidence showing that insect hemolymph proteins are synthesized in the fat body was first presented by Shigematsu (1958), who found that incubation of *Bombyx* larval fat body with labeled amino acids resulted in the release of radioactive proteins into the medium. Paper electrophoresis indicated that these were indeed hemolymph proteins. His results have since been confirmed in *Schistocerca, Rhodnius, Calliphora, Drosophila,* and *Tenebrio* (see Chen, 1966). However, as suggested by Chippendale and Kilby (1969), at least in *Pieris* other tissues such as the midgut, pericardial cells, and hemocytes may also take a part in the production of hemolymph proteins.

In general, the rate of protein synthesis in the fat body is high in early growing larvae, and declines rapidly with the advance of larval life. In *Drosophila* (Fig. 6) we observed that fat body incubated *in vitro* with labeled valine decreased distinctly in its incorporation capacity during the period from the second larval instar to the time of puparium formation (Rüegg, 1968). Comparable results have also been reported for the *Calliphora* larva. Electron microscopic studies on the silk moth *Philosamia cynthia,* the flesh fly *Sarcophaga bullata* and the blowfly *Calliphora erythrocephala* demonstrated that fat body cells have an extensive endoplasmic reticulum rich in ribosomes, a well-developed Golgi complex, and numerous mitochondria at early larval stages, but show a drastic reduction of these organelles when the

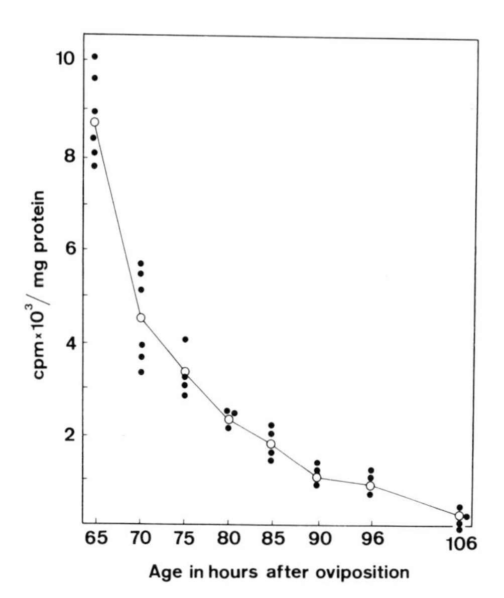

Fig. 6. Changes in the *in vitro* incorporation rate of ^{14}C-valine into fat body protein during larval development of *Drosophila melanogaster*. (From Rüegg, 1968.)

larva becomes mature and approaches pupariation (for references, see Price, 1973). The stage-specificchanges in the capacity of protein synthesis are also reflected in the level of RNA. For example, in *Philosamia cynthia ricini* the RNA content of the fat body is high at the onset of the fifth larval instar, and thereafter declines to a low level by the time of pupariation. Similarly the RNA level in the larval fat body of *Calliphora erythrocephala* shows a two-fold increase from day 3 to day 4, and falls at day 5 (Price, 1965). However, it increases again during the late third larval instar in spite of a distinct reduction in the number of ribosomes. Its RNA/DNA ratio (41.0) has been found to be unusually high compared to that of the rat liver (2.7). Changes in the appearance of ribonucleoprotein bodies included in the nuclei of the larval fat body cells in *Calliphora stygia* are thought to be linked to changes in the rate of protein synthesis (Thomson and Gunson, cited in Price, 1973).

As a consequence of active protein synthesis in the fat body during early larval stages, the total protein concentration in the hemolymph increases rapidly and reaches a maximum in the fully grown larvae of *Bombyx mori, Samia cynthia, Phormia regina, Pieris brassicae,* and *Pyrrhocoris apterus* (see Price, 1973, for references). In general, qualitative changes of the protein pattern are less dramatic compared to the increase in total concentration. Many hemolymph proteins serve as vehicles for the transport of lipids, carbohydrates, and hormones, but they may also function as enzymes. The more recent study by Hürlimann (1974) on the enzymatic activity of hemolymph proteins in the *Phormia* larvae in our laboratory is in accordance with this conclusion.

When the larvae cease to feed and approach puparium formation, the protein concentration in the hemolymph drops abruptly with a concomitant appearance of protein granules in the fat body cells. From their study of the skipper, *Calpodes ethlius,* Locke and Collins (1968) suggested that proteins are sequestered from the hemolymph by the fat body during the last larval instar and are stored as intracellular multivesicular bodies. This suggestion has since been confirmed in a number of insects including *Calliphora erythocephala, Ephestia kühniella, Calliphora stygia, Pieris brassicae, Locusta migratoria migratorioides, Diatraea grandiosella,* and *Galleria mellonella* (references in Price, 1973; Wyatt 1975). The sequestration process appears to be selective, and the proteins stored are apparently utilized during adult differentiation.

More recently, Wyatt and associates (Wyatt, 1975 and personal communication) observed that, in the *Cecropia* silkmoth, protein granules appear in the fat body cells at the beginning of cocoon spinning, are most abundant around the time of larval ecdysis, and remain visible through the pupal stage. Electron microscopy revealed that these granules, which can clearly be distinguished from the urate granules, are crystalline in structure with a dense inner and a less dense outer zone. Electrophoretic analysis of extracts prepared from the protein granules and the whole fat body revealed two major protein bands of similar mobility with a molecular weight of about 250,000 (Fig. 7). Prior to the beginning of cocoon spinning the two protein fractions accumulate in the hemolymph, but decline in concentration following the appearance of granules in the fat body cells. Similar protein granules have also been found in the fat body of *Drosophila* larvae (Thomasson and Mitchell, 1972).

The synthesis and release of specific hemolymph proteins by the fat body of the growing larvae and the removal and storage of these proteins in the same tissue during later development are apparently under hormonal control. In *Calliphora*, β-ecdysone has been shown to activate protein synthesis in the larval fat body (Thomson *et al.*, 1971). In the oakworm, *Antheraea pernyi*, injection of α-ecdysone into the fourth instar larva resulted in a stimulation of fat body protein synthesis and injection of juvenile hormone into fifth instar larvae of the milkweed bug, *Oncopeltus fasciatus*, yielded an increase in the rate of protein synthesis in the fat body (see Price, 1973). As noted previously, there is a rapid decline in the rate of protein synthesis in the insect fat body with the advance of larval age. It appears probable that such a decline is causally related to a decrease in the level of juvenile hormone as the larva approaches metamorphosis.

The formation of protein granules has been shown to be related to the action of ecdysone. In *Calpodes ethlius* the first appearance of protein granules in the fat body coincides with the time at which the prothoracic gland becomes active in its secretory process. By experiments involving ligation and injection of synthetic ecdysone, evidence has been provided that this hormone induces granule formation. Larval fat body of *Drosophila* cultured in the abdomen of male adult flies has been shown to produce protein granules only when the ring gland secreting ecdysone was implanted at the same time. By contrast, for fat body cultured in the abdomen of female flies, no ring gland was needed for granule formation (Butterworth and Bodenstein, cited in Price, 1973). It is thought that the ovary may produce a hormone which takes over the function of ecdysone. As we shall see later, recent work with the mosquito, *Aedes aegypti*, showed that the ovary indeed produces ecdysone (Hagedorn and Fallon, 1973). Thomasson and Mitchell (1972) were able to show that in *Drosophila* ecdysone induces the production of protein granules in the fat body. However, based on the fact that the granules appear several hours earlier than the active secretion of ecdysone during development, they consider that this hormone may not be the normal trigger for granule formation.

In summary, the hitherto studies of various insects have proved beyond doubt that the larval fat body is active in the synthesis and secretion of hemolymph proteins during the period of growth. When the larvae cease to feed, these proteins are again selectively removed from the hemolymph and

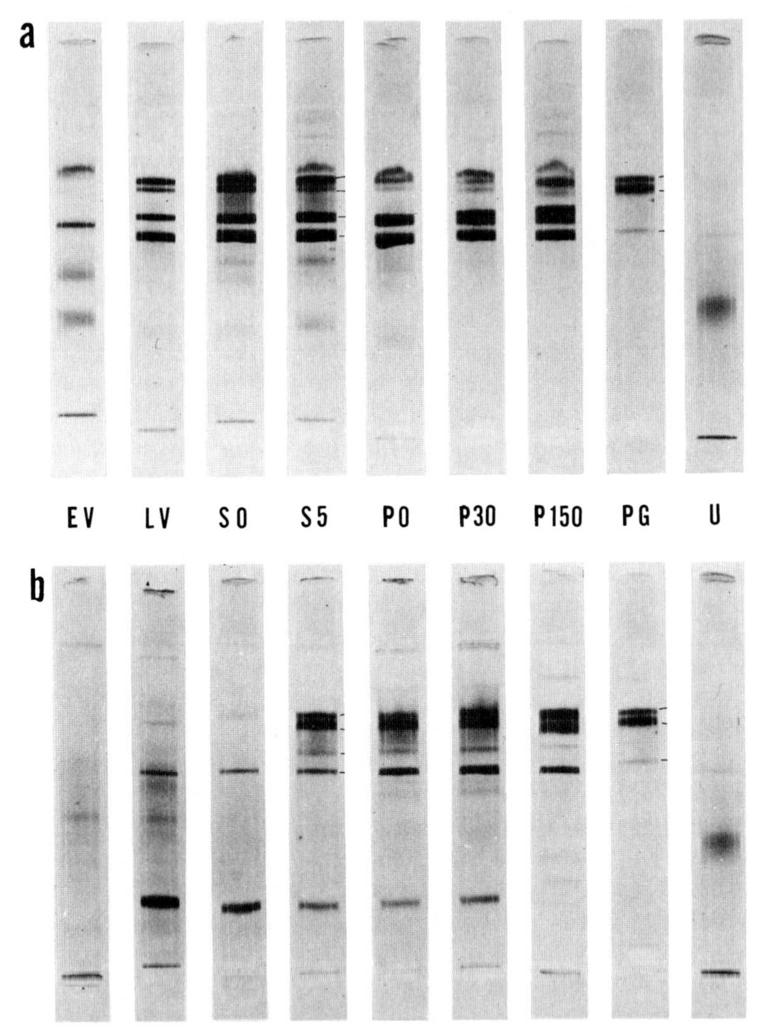

Fig. 7. Separation of *Cecropia* silkmoth proteins by polyacrylamide gel electrophoresis. (a) Hemolymph proteins; (b) fat body proteins. Abbreviations: EV, early fifth instar larva; LV, late fifth instar larva; SO, larva on the day of spinning; S5, 5 days after spinning; PO, pupa on the day of ecdysis; P30, 30-day-old pupa; P150, 150-day-old pupa; PG, protein prepared from purified protein granules; UG, protein prepared from purified urate granules. 1 and 2 are storage granule proteins; 3 and 4 are other hemolymph proteins. (Courtesy Dr. Sumio Tojo and Dr. G. R. Wyatt, personal communication.)

stored as intracellular granules for use at metamorphosis. There is considerable evidence that this cyclic change in the activity of the fat body cells is regulated by juvenile hormone and ecdysone, the two major insect hormones.

B. Yolk Proteins (Vitellogenins)

The major function of fat body in adult female insects is the synthesis of yolk proteins which are released into the hemolymph and are taken up by the growing oocytes. The regulation of these processes by hormones and other mechanisms has been studied in many insects (see reviews by Wyatt, 1972; Doane, 1973; de Wilde and de Loof, 1973). The production of vitellogenins is correlated with the reproductive cycle and indicates remarkable activation and deactivation of the fat body cells in response to changing demands.

Insect vitellogenin was first identified by immunochemical methods in the Cecropia silkmoth by Telfer (1954). This so-called female-specific protein becomes detectable in the pharate pupa and is later deposited in the oocyte with a decrease of its concentration in the hemolymph of the adult female. It is synthesized by the fat body from the female, but not by that from the male. There are two distinct periods of vitellogenin synthesis: first in the late pharate pupal stage and then at the time when protein is sequestered by the developing oocytes (Pan, 1971). So far there is no evidence for a hormonal control of its synthesis which proceeds probably as an ontogenetically programmed process. The more recent analysis of the purified vitellogenin showed that it is a lipophosphoprotein with a molecular weight of 500,000 (Pan and Wallace, 1974).

In contrast to the *Cecropia* silkmoth, in another lepidopteran, the Monarch butterfly *Danaus plexippus,* oogenesis has been found to be clearly under hormonal control (Pan and Wyatt, 1976). Newly emerged females ligated at the neck region and maintained by glucose injection showed no vitellogenin formation which, however, could be induced by injection of juvenile hormone. By using ^{3}H-leucine it was found that about 40% of the synthesized hemolymph protein was vitellogenin (Fig. 8). The fact that up to 92% of its synthesis was inhibited by actinomycin D suggests that induction of the formation of this specific protein in the fat body involves *de novo* RNA synthesis.

In the desert locust, *Schistocerca gregaria,* female-specific protein, which has been purified, is also synthesized in the fat body and removed by the growing oocyte. The rate of synthesis is correlated with the stage of the oocyte development. In the fat body of allatectomized locusts the production of yolk protein ceases, but can be induced by the administration of juvenile hormone. This is apparently also true for the migratory locust,

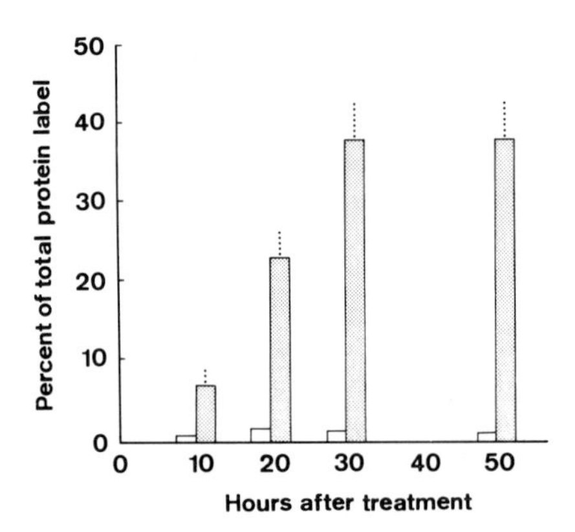

Fig. 8. Juvenile hormone-induced synthesis of vitellogenin in the Monarch butterfly, *Danaus plexippus*. C-18 Juvenile hormone (4 μg) dissolved in olive oil was injected into newly emerged butterflies which were ligated at the neck region and maintained by injection of glucose solution. ^{3}H-Leucine was then injected 4 hr before termination. Based on the radioactivity in total hemolymph protein and that in vitellogenin the percentage of label in vitellogenin was calculated. Shaded columns, hormone-treated animals; open columns, control animals injected with olive oil. (From Pan and Wyatt, cited in Wyatt, 1975.)

Locusta migratoria (Wyatt, 1975). The yolk protein or lipovitellin of this insect has a molecular weight of 571,000 (McGregor and Loughton, 1974).

Essentially the same results have been reported for the cockroaches *Periplaneta americana, Nauphoeta cinerea,* and *Leucophaea maderae* (see Lüscher, 1968 for references). Their synthesis of yolk protein in the fat body during each sexual cycle needs the action of the corpus allatum hormone. In *Leucophaea* the hormone-induced production of the specific protein in allatectomized females has been shown to the inhibited by actinomycin D, suggesting new synthesis of mRNA. In addition, cells from fat body of active protein synthesis are rich in ribosomes as well as polysomes associated with membranes of the endoplasmic reticulum (Engelmann, 1974). According to Dejmal and Brookes (1972), the major yolk protein of this insect is a lipoglycoprotein consisting of a large component with a molecular weight of 1.59 $\times 10^6$ (28 S) and a small component with a molecular weight of 5.9×10^5 (14 S). Their data indicate that the 28 S component is formed by aggregation of the 14 S component. The lipid moiety (6.9%) consists of phospholipid, and the carbohydrate moiety (6.4–7.0%) is primarily made of mannose and glucosamine.

Recent studies of egg maturation in the mosquito, *Aedes aegypti,* disclosed a quite different regulatory mechanism which has so far not been found in other insects. It is known that yolk production in the mosquito is initiated by blood ingestion (Hagedorn *et al.*, 1973). The ingested blood leads to the release of neurosecretion from brain which induces the production of a hormone by the ovary. The induced ovarian hormone in turn acts as a trigger for the synthesis of yolk protein in the fat body (Lea, 1972; Hagedorn and Fallon, 1973). That ecdysone can initiate ovarian development in *Aedes* without the ingestion of a blood meal has been confirmed by the more recent study of Fallon *et al.* (1974) who injected β-ecdysone into unfed adult female mosquitoes and then cultured the fat body from the treated animals in the presence of ^{3}H-phenylalanine. As shown in Fig. 9, a maximum synthesis

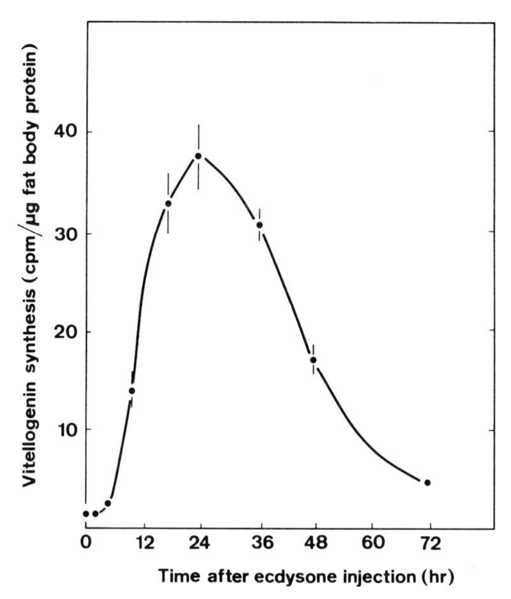

Fig. 9. Ecdysone-induced synthesis of vitellogenin in the adult female mosquito, *Aedes aegypti*. Animals without a blood meal were injected with 5 μg of β-ecdysone. At various time intervals after injection fat body was prepared, incubated with ^{3}H-phenylalanine, and label in isolated vitellogenin determined. (From Fallon *et al.*, 1974.)

of vitellogenin could be observed at 24 hr after injection of the hormone. Evidence is now available indicating that adult females of *Aedes*, after the ingestion of a blood meal, indeed synthesize β-ecdysone which is derived from α-ecdysone produced in the ovary (Hagedorn *et al.*, 1975).

In the Colorado beetle, *Leptinotarsa decemlineata*, the pattern of hemolymph protein in adult females is influenced by day length (de Loof and de Wilde, 1970). Only under long-day condition yolk protein is produced and taken up by the oocyte. It has been suggested that both neurosecretory cells and corpora allata are needed for the synthesis of this protein, while its uptake requires only the action of juvenile hormone. In the mealworm, *T. molitor*, both neurosecretion and corpora allata hormone are thought to be essential for yolk production (Mordue, 1965). Unfortunately, the relationships between these two endocrine systems remain unclear (see Doane, 1973).

In summary, in all insect species which have so far been subjected to a careful analysis the major yolk proteins are synthesized in the fat body, secreted into the hemolymph and taken up by the oocytes. The activation of the synthetic process may involve various hormones, the release of which in turn depends on a variety of external and internal factors such as the state of maturation, food uptake, photoperiod, and mating. Apparently different mechanisms are utilized by insects to regulate the activity of fat body cells for the synthesis of a reproductively important protein.

IV. REGULATION OF PROTEOLYTIC DIGESTIVE ENZYMES

Insect larvae depend on the uptake and utilization of exogenous protein for both growth and development. In adult insects ingestion of protein is a prerequisite for egg maturation. As in other animals, the ingested protein must be first broken down into amino acids before being absorbed. The site for both digestion and absorption is the midgut. Despite numerous studies on insect digestion, many aspects of the digestive process, including the synthesis and secretion of proteases, are still poorly understood (see references in House, 1974).

In an attempt to understand the developmental-genetic control over the synthesis of proteases, we have analyzed in detail the activity of these enzymes during the life cycle of *Drosophila*. The total activity increases rapidly during the first 3 days of intense larval growth, and then falls off during later development. With the onset of pupariation it declines to a very low level which persists throughout the pupal stage. However, a slight increase could be detected shortly before adult emergence. By using synthetic substrates and specific inhibitors, the presence of trypsin, carboxypeptidase A, and leucine aminopeptidase has been demonstrated. We have so far detected no chymotrypsin, carboxypeptidase B, or pepsin at any stage of *Drosophila,* though these enzymes have been identified in a number of other insects (see references in Chen, 1971). The lethal mutant *l(2)me,* which ceases to grow at the early third larval instar, has an exceptionally low protease activity. But this is most likely due to a general impairment of protein synthesis resulting from the mutational effect (Kubli, 1970). Recently, by taking advantage of affinity chromatography we succeeded in isolating trypsin from mature Drosophila larvae (Fig. 10). Our preliminary data on the purified enzyme indicate that it has a molecular weight of about 24,000, similar in size to trypsin from the tobacco hornworm, *Manduca sexta* (Miller *et al.,* 1974). On polyacrylamide gel (10%, pH 4.5) it moves

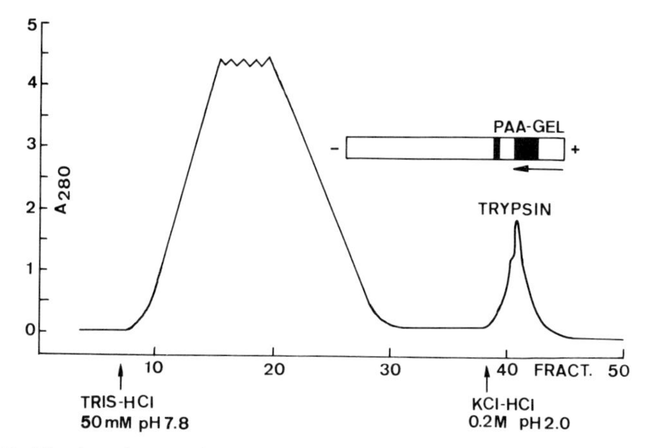

Fig. 10. Purification of trypsin from mature *Drosophila* larvae by affinity chromatography. The larvae were homogenized in Tris buffer (50 m*M*, pH 8.0) and centrifuged at 15,000 rpm for 15 min. The supernatant fraction (15 ml) was applied to a small column of CNBr activated Sepharose-4B coupled with soybean or lima bean trypsin inhibitor. The column was first washed with 50 m*M* Tris buffer, pH 7.8 (3 × 9.5 ml), and the enzyme bound to the inhibitor eluted with 0.2 *M* KCl–HCl, pH 2.0. Electrophoretic separation of the purified trypsin in 10% polyacrylamide (PAA) gel yielded one minor and one major band. (From Gruber, 1977.)

distinctly more slowly in the cathodal direction than bovine trypsin, suggesting the enzymes from the two sources are not identical.

The mosquito, *Culex pipiens,* differs from *Drosophila* in having a fairly high activity of chymotrypsin in the developing larva in addition to the proteases mentioned above (Spiro-Kern, 1974). On the other hand, we found no detectable amount of this enzyme at the adult stage (Fig. 11). Trypsin and chymotrypsin can be distinctly separated on polyacrylamide gel (Fig. 12). We have recently isolated a chymotrypsin inhibitor from this insect by using activated Sepharose coupled with the purified enzyme (Fig. 13). The presence of trypsin inhibitors in the blood of several vertebrates and the interactions of these inhibitors with *Aedes* trypsin under *in vitro* conditions have been demonstrated by Huang (1971). But according to Briegel and Lea (1975) such antitryptic factors have *in vivo* no detectable interference with proteolytic digestion in the 3 genera of the mosquito tested.

From the above studies it may be concluded that the pattern and activity of proteases are stage-specific. This is not surprising since in many insect species the larva and adult have totally different nutritional conditions and feeding habits. As to the regulation of digestion, both food and hormone have been considered as possible causes for the stimulation of enzyme secretion. Thomsen and Møller (1963) first reported that protease production in the gut of adult female *Calliphora* is controlled by the medial neurosecre-

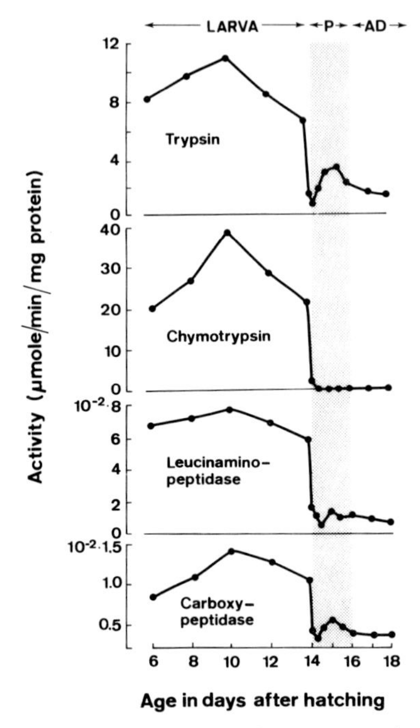

Fig. 11. Changes in the activity of proteases during postembryonic development of the mosquito, *Culex pipiens pipiens*. P, pupa; AD, adult. (From Spiro-Kern, 1974.)

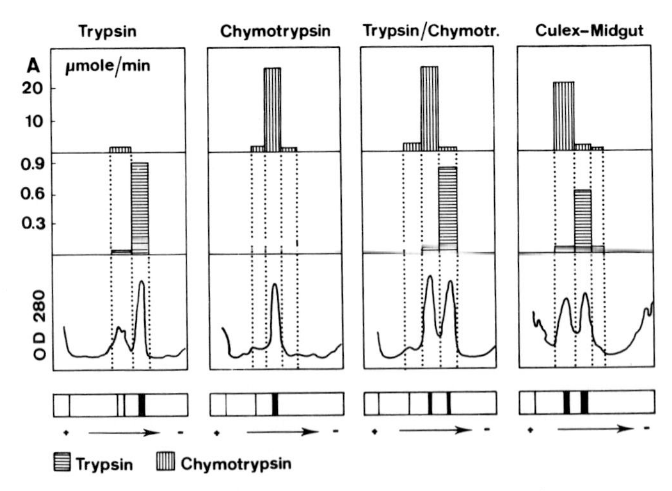

Fig. 12. Polyacrylamide gel electrophoresis of bovine trypsin, chymotrypsin, and midgut homogenate from 8-day-old *Culex p. pipiens* larvae. Upper part: enzyme activities in the corresponding gel segments. Middle part: densitometric tracings of the unstained gels at 280 nm. Lower part: patterns of the enzyme protein bands as revealed by staining the gel with amido black. (From Spiro-Kern, 1974.)

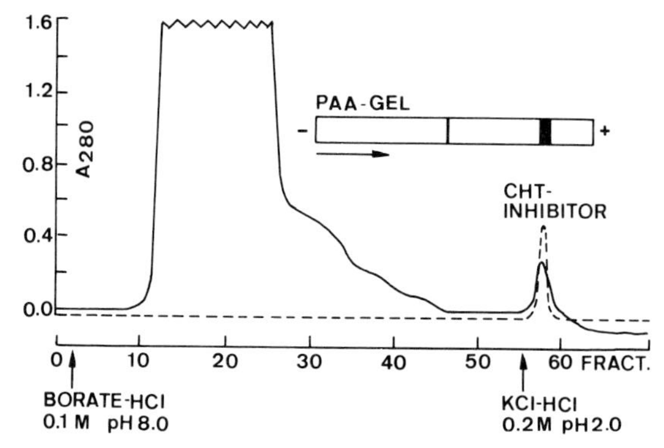

Fig. 13. Purification of a chymotrypsin inhibitor from the adult mosquito, *Culex pipiens fatigans*. A sample (3 ml) prepared from about 3 gm of adult mosquitoes was applied to a small column of Sepharose-4B activated with CNBr and coupled with previously purified *Culex* chymotrypsin. The column was first washed with 0.1 *M* borate buffer, pH 8.0, and the inhibitor bound to the enzyme then removed with 0.2 *M* KCl–HCl, pH 2.0. Following electrophoresis of the purified fraction in 7.5% polyacrylamide (PAA) gel and staining with Coomassie brillant blue, a minor and a major band could be observed. Solid line: absorption at 280 nm. Broken line: inhibitory effect on the activity of *Culex* chymotrypsin (1 unit = reduction in the rate of hydrolysis of *N*-glutaryl-L-phenylalanine-nitroanilide for 1 μmole/min). (From Spiro-Kern and Chen, unpublished observations.)

tory cells in the brain. If food plays a stimulatory role, it may act either directly as a secretagogue stimulus, or indirectly by activating the nervous and/or endocrine system through distension of the gut. Hosbach *et al.* (1972) in my laboratory observed in *Drosophila* that the protease activity in the larval gut is directly correlated with the protein concentration in the food. Similarly, in *Leucophaea* and *Sarcophaga* Engelmann (1969) and Engelmann and Wilkens (1969) noted that the level of proteolytic enzyme activity is proportional to the quantity of ingested proteinaceous food. Since removal of endocrine organs had no influence on the production of gut enzymes, a hormonal control of protease synthesis appears, according to Engelmann (1969), to be ruled out.

Further evidence for a secretagogue stimulation of protease secretion in mosquitoes has recently been provided by Briegel and Lea (1975). Measured volumes of nutrient solutions with known concentrations of blood or plasma protein were administrated to adult female *Aedes aegypti* by anal injection (enema), and the protease activity in posterior midgut after various time intervals was determined. As shown in Fig. 14, in a first experimental series, females were each given a different volume of solution containing equal amount of blood protein. The maximal protease activity and the rate of enzyme secretion were virtually identical in these animals. In a second series the females were each given a constant volume of solution with, however, increasing amounts of blood protein. These animals showed a nearly linear

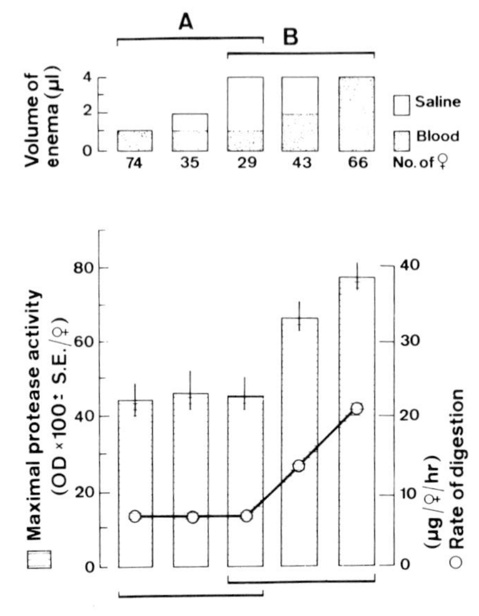

Fig. 14. Dopagogue stimulation of protease secretion in midgut of the mosquito, *Aedes aegypti*. In experiment A, female mosquitoes of each of the three groups were given increasing volumes of solutions by enemas, but all containing the same quantity of protein (1 µl blood). In experiment B, female mosquitoes of each of the three groups received the same volume of solutions by enemas, but with increasing quantities of protein. Data in lower part of the figure demonstrate that in experiment A the maximal protease activity and rate of digestion remain at the same level among the three groups, whereas in experiment B both values rise with increasing amounts of protein. (From Briegel and Lea, 1975.)

increase in both the maximal protease activity and rate of digestion. From these results it is clear that proteolytic activity is not affected by stretching of the midgut or abdomen, but related to the presence of protein. Thus, a neural stimulus with respect to protease secretion following blood ingestion is excluded.

Additional support for protein stimulus has been obtained by injecting female mosquitoes with solutions containing blood, protein hydrolysates, and nonprotein compounds. From the data summarized in Table I it can be seen that nonprotein solutions yielded a very low level of protease activity, whereas all proteins excepting gelatine and histone showed higher activities. It seems that only intact globular proteins which have molecular weights higher than histone exert a stimulatory effect on protease secretion.

It should be pointed out that in spite of the overwhelming evidence in support of a secretagogue stimulation by protein, there is yet no sufficient reason to reject totally a hormonal influence on protease activity. Most recently, Briegel and Lea (personal communication) observed that when adult females of *Aedes aegypti* were given a blood meal by enema following removal of the medial neurosecretory cells or ovaries, their protease activity

TABLE I
Protease Activity in Midguts of Female *Aedes aegypti* after Enemas with Various Protein and Nonprotein Solutions[a]

	Time after enema (hr)	OD × 100/ 0.1 mg protein[b]	n
Protein solutions			
Heparinized blood	6	9.2	5
	12	15.3	16
	18	15.8	12
Lactalbumin (5%)	8	7.3	5
	12	8.7	5
Egg albumin (5%)	4	1.9	5
	8	11.9	4
	12	15.5	5
	16	12.9	4
	20	23.1	4
	24	21.9	5
Blood plasma	24	19.8	8
Blood plasma (50%)	24	18.4	11
Gelatin (5%)	8	0.9	5
	12	1.7	5
	16	0.9	8
	24	1.1	5
Histone (5%)	8	0.5	7
	12	1.1	8
	18	2.3	6
	Time after enema (hr)	**OD × 100/ 0.1 mg substance**	**n**
Nonprotein solutions			
Blood hydrolysate (6.7%)	6	2.0	6
	12	2.3	7
	18	3.0	9
Lactalbumin hydrolysate	6	3.1	5
(7.5%)	12	3.8	7
Peptone (5%)	12	2.3	13
Glutathion (2%)	12	0.7	1
Sucrose (5%)	0–4	1.7	25
NaCl (0.7%)	0–4	0.8	34

[a]From Briegel and Lea (1975).
[b]Activities are expressed as OD × 100 per unit of weight of the substance injected.

was reduced to about 50% of that in the controls. Normal enzyme activity could be restored by reimplantation of one of the extirpated organs or by injection of β-ecdysone, an ovarian hormone in the mosquito. A reasonable interpretation of these and previous results is that although a protein substrate is needed to initiate protease production, a maximal enzyme activity can be achieved only in the presence of this hormone.

Another aspect of protease regulation in *Aedes* is the occurrence of active trypsin in excreta (Briegel, 1975). When the protease secretion reaches a maximum and about 80% of the protein is digested, proteolytic activity in the midgut declines rapidly. This rapid decline is due to both the cessation of enzyme synthesis and excretion. Following depletion of the protease, a new cycle of production is triggered by the next blood meal.

From the studies discussed above, a general conclusion on the regulation of proteolytic enzymes, at least for the adult dipteran insects examined, can be summarized as follows: The predominant enzyme in the midgut is trypsin, whereas other proteases occur only at one stage or another and are usually of rather low level. According to data available, insect trypsin is not identical with bovine trypsin. Although the existing evidence is in favor of a secretagogue stimulation of protease production, a hormonal influence has to be considered. To date the molecular mechanism involved in such a stimulation has been little studied.

V. SYNTHESIS OF SECRETION PROTEINS AND PUFFING OF POLYTENE CHROMOSOMES IN SALIVARY GLANDS

The presence of polytene chromosomes has made the salivary gland of Diptera a favorable material for cytogenetic studies at the morphological, and more recently, at the biochemical–molecular level. Under certain conditions the bands (chromomeres) of these chromosomes undergo the process of puffing which is usually correlated with the production of exportable proteins in the glandular cells. Extensive studies of this phenomenon in various laboratories have demonstrated that puffs show cell- and stage-specific patterns (Beermann, 1972) and represent the sites of active RNA synthesis. It is now generally accepted that puffing reflects gene activation.

In the midge, *Chironomus pallidivitatus,* a total of seven secretory proteins could be separated by polyacrylamide gel electrophoresis. One of these is synthesized only in a group of three or four morphologically distinguishable cells. The protein pattern in *C. tentans* is quite similar. The molecular weights of these proteins have been estimated to range from 1.5×10^5 to 5.69×10^5 daltons (Grossbach, 1973). By a combination of interspecific hybridization experiments and cytogenetic techniques, evidence has been presented for a correlation between genes responsible for the synthesis of the secretion proteins and a special class of puffs, the Balbini rings, which are all localized on chromosome IV. Daneholt and Hosick (1973) have shown that in *C. tentans* a 75 S RNA which amounts to about 1.5% of the total RNA in the cell is transcribed in the Balbini ring 2 region. It is subsequently transmitted to the cytoplasm without apparent reduction in size and appears to be remarkably stable. This RNA could serve as a messenger for the most abundant high molecular weight protein with the highest synthetic rate (see Grossbach, 1973).

In another midge, *Acricotopus lucidus,* cells in the main and lateral lobes of the salivary gland differ from those in the anterior lobe by the presence of a large puff (the Balbini ring 2), and by the synthesis of a specific secretory protein containing hydroxyproline (Baudisch and Hermann, 1972). Administration of gibberellin A_3 resulted in the regression of the large puff with a concurrent inhibition of RNA synthesis at this site. The treated animals ceased to synthesize hydroxyproline and the specific protein, suggesting localization of the gene in the puff region (Baudisch and Panitz, 1968).

Similar experiments designed to correlate chromosome puffs with protein synthesis in the salivary gland have been carried out in *Drosophila melanogaster.* Ultrastructural studies (von Gaudecker, 1972) have shown that secretion granules appear in the salivary gland cells of this insect at about the middle of the third larval instar, and become more numerous as the larva approaches the nonfeeding period. The transport of secretion into the gland lumen occurs about 3 hr before pupariation, and thereafter only a few granules are visible in the cytoplasm. It has been an accepted view that the gland soon begins to degenerate and becomes histolyzed within about 10 hr following puparium formation. However, this does not mean that the salivary gland ceases to function during this period. In fact, according to a recent study by Mitchell *et al.* (1977a) it remains as a highly active, secretory organ throughout and beyond the prepupal stage. The cells do appear to be in a state of disintegration each time following extrusion of their cytoplasmic inclusions into the lumen, but they soon become recovered and appear quite normal (Fig. 15). This cycle repeats several times until histolysis takes place suddenly a few hours after pupation. At the survival period the cells show nuclear blebs which are probably involved in the exchange of material between the nucleus and cytoplasm. The events just described are in agreement with the previous observation by Ashburner (1967) that changes of normal puffing patterns continue in the polytene chromosomes of the prepupal salivary glands. Additional evidence in support of this view has been provided by experiments on the synthesis of proteins related to puffs induced by heat shocks (Tissières *et al.,* 1974; Mitchell and Lipps, 1975). These will be dealt with in more detail in a later section.

Among a total of four secretion proteins detected in the larval salivary gland of *D. melanogaster* two were found to be correlated with puffing activities (Korge, 1975). The structure gene for one protein (fraction 3) could be localized at position 68C on the third chromosome, and that for another protein (fraction 4) at position 3C8-3Dl on the X chromosome. The puffs within these sections are active at least 5 hr before the appearance of secretion proteins and throughout the period of active synthesis. They become inactive prior to the end of the third larval instar when production of

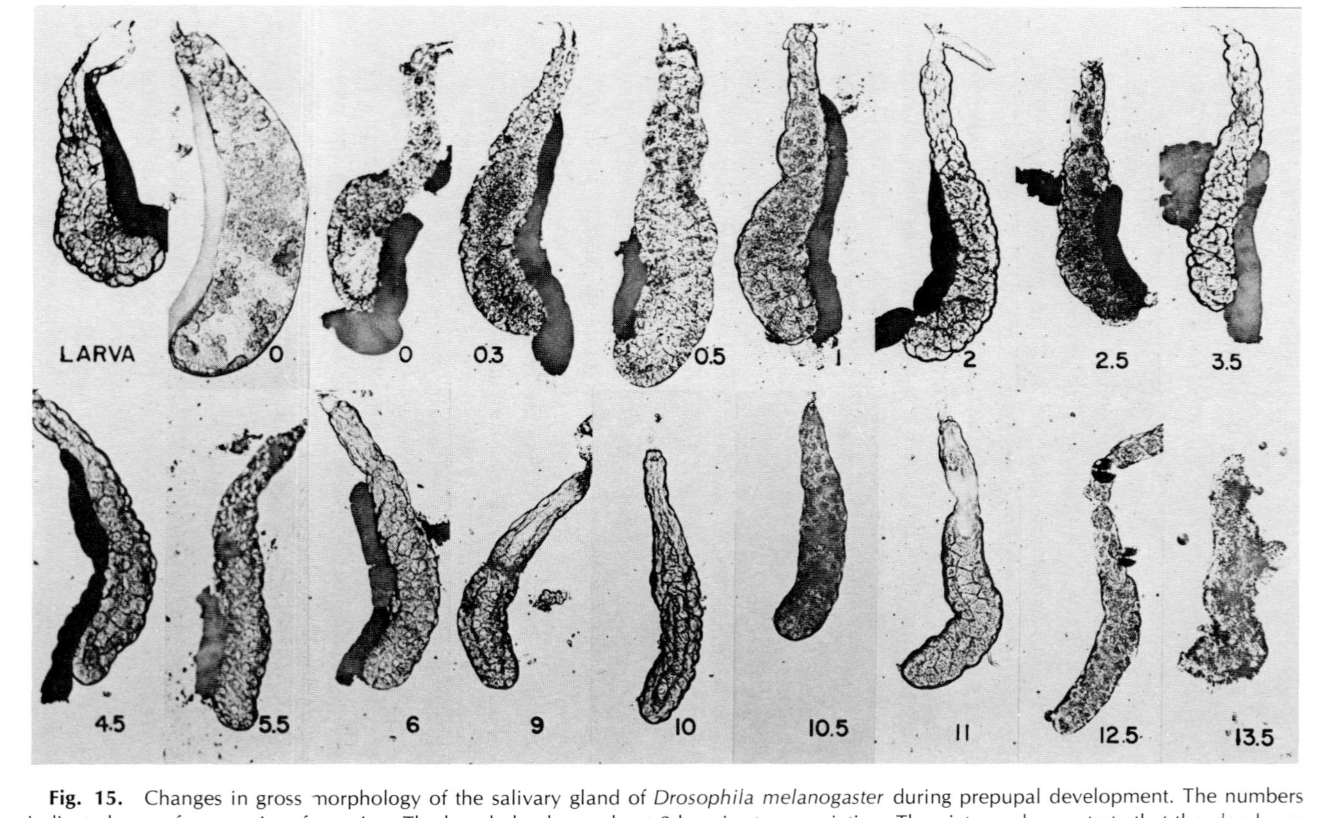

Fig. 15. Changes in gross morphology of the salivary gland of *Drosophila melanogaster* during prepupal development. The numbers indicate hours after purarium formation. The larval gland was about 2 hr prior to pupariation. The pictures demonstrate that the glands are transparent and contain rounded cells with numerous secretion granules in the mature larva and also at 2, 3.5, 4.5, 6, and 11 hr after puparium formation. Each time following release of the secretion material they appear opaque, fragile and have rare normal cell structures, as shown at 0, 1, 2.5, 5.5, 10.5 and 12.5 hr. This cycle of secretion process ceases when the glands become finally histolyzed a few hours after pupation which occurs at 11–12 hr. (From Mitchell *et al.*, 1977a.)

the secretion proteins ceases. Estimation of the amount of fraction 4 produced in duplication and deficiency genotypes indicated that there is dosage compensation of the structure gene (*Sgs*-4).

It is known that modifications of the puffing pattern can be induced by administration of the hormone ecdysone, changes in ion composition, recovery from anaerobiosis, and heat shock treatment. Tissières *et al.* (1974) reported that when *Drosophila* larvae were kept at 37.5°C for 20 min at about the time of puparium formation changes in puffing pattern and accumulation of RNA at the puffing sites were associated with a drastic modification of the pattern of protein synthesis (Fig. 16). Compared to salivary glands during normal development at 23°C, at least six new electrophoretically separable protein bands appeared, and some other bands prior to heat shock diminished. These six induced bands accounted for about 30% of the total incorporation of ^{35}S-methionine (Table II). One of these with a molecular weight of about 70,000 accounted for about 15% of the total label. The same heat shock effect has been observed in several other tissues including brain, Malpighian tubules, and wing imaginal discs (see also Bonner and Pardue, 1976).

By incubating the salivary glands from heat-shocked larvae with ^{3}H-uridine, Tissières *et al.* (1974) detected strong labeling at eight sites on the chromosomes corresponding to positions of the induced puffs. Since RNA synthesis was much more intense at position 87B on the right arm of the third chromosome than at other positions, it seems that the most heavily labeled 70,000 dalton protein induced by heat shock is coded by a gene in this region.

In order to gain further information on the roles of specific proteins in the induction and regression of puffs, Mitchell and Lipps (1975) compared the protein composition between purified nuclei and whole cells of salivary gland with or without heat shock. Their pulse-chase experiments revealed a high turnover rate of nonhistone chromosomal proteins. The patterns of rapidly labeled proteins are different between nuclei and whole cells without heat shock. By contrast, in samples following the heat shock treatment the patterns are remarkably similar. This similarity can be interpreted as due to the induction of new puffs and the regression of preexisting ones. In addition, there is a rapid protein exchange between the nucleus and cytoplasm.

The results of Tissières *et al.* (1974) have been confirmed in both *D. melanogaster* and *D. hydei* by Lewis *et al.* (1975). By using cultured *Drosophila* cells (Schneider's line 2) McKenzie *et al.* (1975) found that heat shock caused the disappearance of nearly all previously existing polysomes and induced the formation of new polysomes (Fig. 17). Poly(A)-containing ^{3}H-RNA prepared from the heat-induced polysomes hybridized *in situ* mainly to position 87C on chromosome 3R. Spradling *et al.* (1975, 1977) also observed a striking alteration in the pattern of gene transcription following heat shock in their *in situ* hybridization experiments with poly(A)-containing

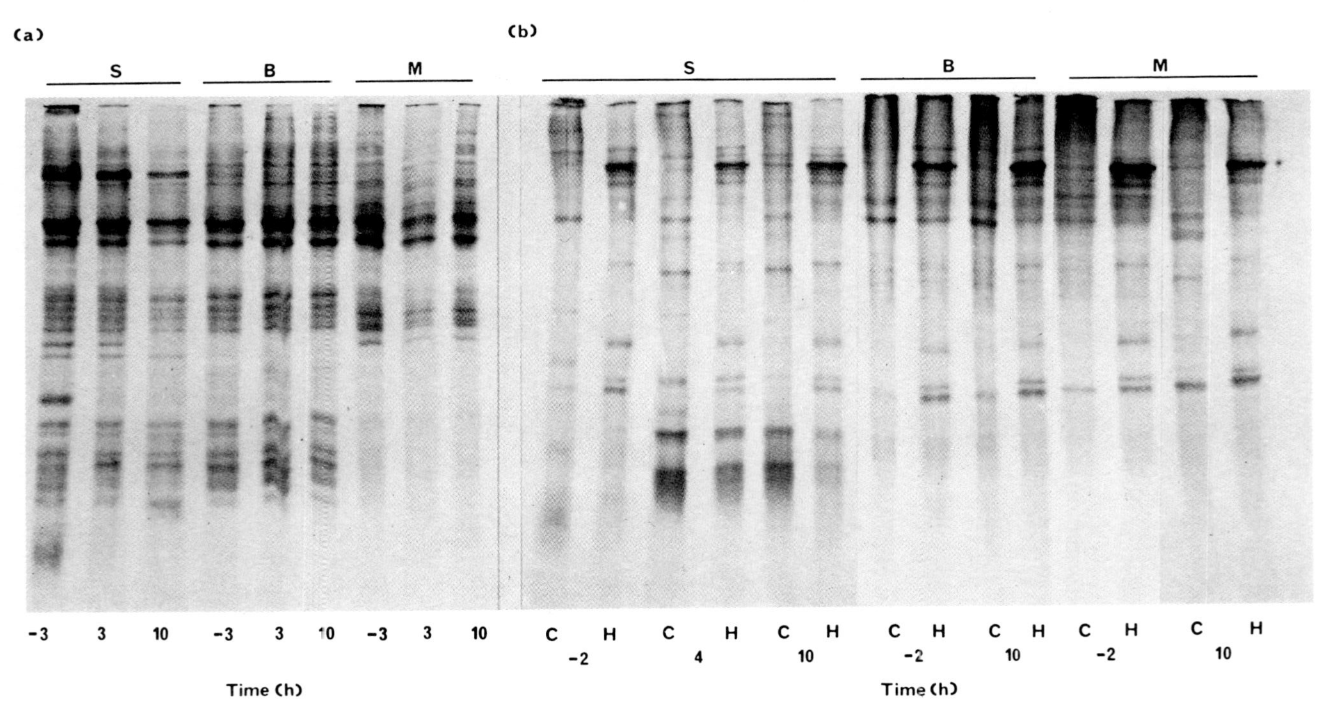

Fig. 16. Polyacrylamide gel electrophoresis of proteins in salivary glands (S), brain (B) and Malpighian tubules (M) and the effect of heat shock on the pattern of labeled proteins in the same tissues during prepupal development of *Drosophila melanogaster*. (a) Protein bands stained with Coomassie blue. (b) Autoradiographs of proteins labeled with ^{35}S-methionine. The numbers given at the bottom are hours before (−) and after puparium formation. C, controls maintained at 25°C; H, animals treated with heat shock at 37.5°C for 20 min. (From Tissières *et al*., 1974.)

TABLE II
Distribution of Label in Several Protein Bands of Salivary Glands, Brain, and Malpighian Tubules of *D. melanogaster* Induced by Heat Shock[a]

Distance of band migration on gel (mm)	Salivary glands −2 hr	Salivary glands +9 hr	Brain −2 hr	Malpighian tubules +9 hr
9.5	19.0	16.0	15.0	17.0
12	2.2	2.4	3.5	2.3
26	1.4	2.2	2.7	0.7
38	1.7	1.7	3.5	2.5
44	1.7	1.7	3.5	2.5
46	4.5	2.7	6.2	6.7

[a]From Tissières *et al.* (1974). Values are given as percentage of total label in each sample. These were obtained by measuring the surface of peaks on densitometer tracings from autoradiographs.

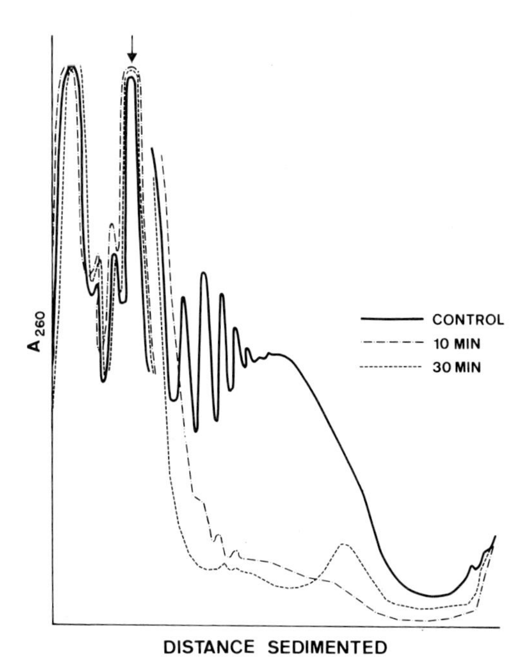

Fig. 17. Sedimentation patterns of polysomes in cultured cells of *Drosophila melanogaster* (Schneider's line 2) at various times following exposure to a temperature of 37°C. Nearly all preexisting polysomes disappeared at 10 min after temperature elevation, and a new peak of polysomes containing about 25 ribosomes per message became visible at 30 min. The arrow indicates a fivefold expansion of the ordinate scale after the monosome peak. (From McKenzie *et al.*, 1975.)

mRNA from cultured *Drosophila* cells. Most of the mRNA synthesis detected in the controls was shut off after heat treatment, and synthesis of new mRNA took place. The newly synthesized mRNA's hybridized to seven sites on the salivary gland chromosomes that did not bind mRNA's from the control cells. Most recently, Tissières and associates (personal communication) demonstrated that each of these mRNA's codes *in vitro* for one of the heat shock proteins.

The fact that nonpolytene cells from both imaginal discs and tissue cultures undergo the same pattern of heat-induced gene expression as do the specialized polytene chromosomes indicates that results from analysis of the puffing phenomenon are of general validity. The susceptibility appears to be locus-specific, since the same protein components are induced in all the tissues and probably at all developmental stages. Furthermore, the elevated temperature is within the range of that in natural environment to which the animals may possibly be exposed. It seems that selection by temperature variations in nature is linked in some way to a differential control of gene activity.

At present the functional roles of the proteins secreted by the larval and prepupal salivary glands during normal development, as well as those produced under specific conditions, must remain a subject for speculation. However, several indications do exist. The mucopolysaccharide-containing secretion produced by the gland shortly before pupariation has been suggested to serve as "glue" for adherence of the pupal case to the substrate (Korge, 1975). According to Mitchell and associates one protein component in this gland may participate in the formation of the phenoloxidase complex (see Chen, 1971). It is secreted into the hemolymph in the late third larval instar and appears to be rate-limiting in the activation of the enzyme essential for the sclerotization of the pupal case. The rapid decrease of the enzyme activity after pupariation is thought to be due to the cessation of secretion of this component. Furthermore, analysis of the pattern and synthesis of proteins in the pupal thoracic hypoderm indicated that an 80K protein component secreted by the late prepupal salivary gland is deposited between the adult and prepupal cuticle to yield a matrix layer which may possibly play an important role in bristle formation (Mitchell *et al.*, 1977b).

VI. ACTIVATION OF PROTEIN SYNTHESIS IN THE MALE ACCESSORY GLAND

The male accessory gland (paragonial gland) of insects has been the subject of numerous investigations in relation to its specific roles in reproduction (see reviews in Chen, 1971; de Wilde and de Loof, 1973). Adult female insects show characteristic alterations in their behavior and physiology following mating. The most significant changes are the increase of oviposition and the reduction of receptivity. There is evidence that various substances secreted by the male accessory glands that are transferred to the female during copulation are responsible for these changes. In addition, the secretion appears to be essential for the transfer, storage, and utilization of the sperm. Following each copulation the secretory products must be replenished. It is thus of particular interest to understand the control over the synthesis of large quantities of exportable compounds in this highly specialized gland.

The developmental and functional aspects of the paragonial gland in *Drosophila* have been discussed in detail in a more recent review by Fowler (1973). Extensive studies carried out in our laboratory demonstrated that the secretion of this gland contains a large number of amino acids, peptides, and proteins. In agreement with the earlier transplantation experiments, a major ninhydrin-positive component containing phosphoethanolamine in *D. melanogaster* and a glycine derivative in *D. funebris* have been found to stimulate egg production (Chen, 1971; Baumann, 1974). Furthermore, a polypeptide with a total of twenty-seven amino acid residues has been shown to inhibit female receptivity in *D. funebris* (Baumann *et al.,* 1975).

As illustrated in Fig. 18, we observed at least twelve electrophoretically separable proteins in the paragonial secretion in *D. melanogaster* (von Wyl and Chen, 1974; von Wyl, 1976). The separation on 10% SDS-polyacrylamide gel yielded even forty-one bands with molecular weights ranging from 12,000 to 120,000 daltons (Fig. 19). A polymorphism of one major protein band, which is autosomal in orgin, could be detected (see arrow in Fig. 18). More recently, in a survey of eleven *Drosophila* species we found that the electrophoretic patterns of the secretion proteins are highly species-specific (Chen, 1976). But there is no correlation between the protein pattern and the taxonomic relationship of the species examined. It has been reported that transplantation of the male accessory gland between different dipteran species resulted in an elevation of oviposition of the female host. In view of the species-specificity of the protein pattern it is difficult to understand the stimulatory effect of such heterologous transplants. It could be that the proteins are more concerned with the transfer and utilization of the sperm rather than with the stimulation of fecundity (see discussion in Fowler, 1973).

When adult males of *Drosophila* are prevented from mating, there is a progressive accumulation of secretion in the lumen of the accessory gland. According to our estimation about three-quarters of the total soluble protein consists of secretory proteins. As can be seen in Fig. 20, under *in vitro* conditions the accessory glands from mated males showed a rapid increase in the incorporation of labeled amino acids into protein (von Wyl, 1974). The incorporation of labeled uridine into RNA was also increased, though to a

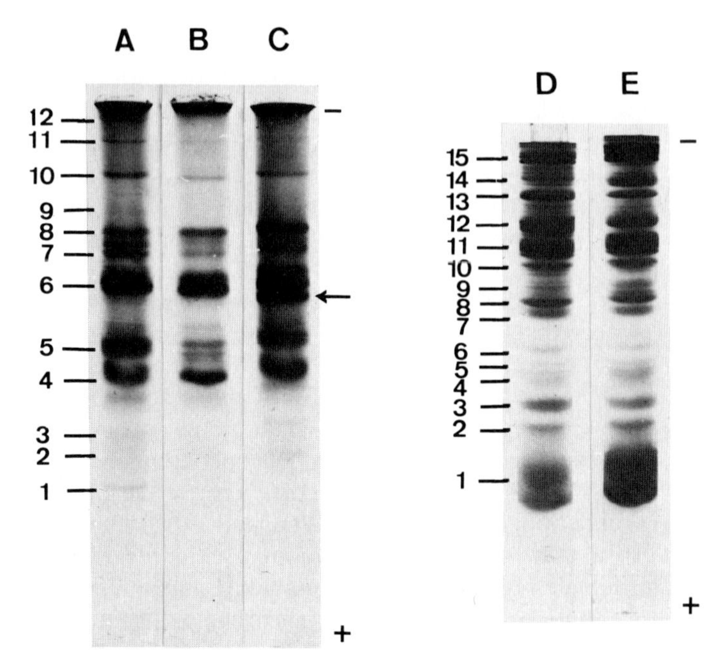

Fig. 18. Electrophoresis of proteins in the male accessory gland of *Drosophila melanogaster* in 7.5% polyacrylamide gel (A-C) and 15% SDS polyacrylamide gel (D, E). (A) and (D) whole gland extract of the wild type (Sevelen). (B) and (E) secretion proteins, wild type. (C) whole gland extract of XO males from cross between the wild-type Sevelen and the stock C(1),y $w^a/\widehat{XY}$,y v f. Arrow indicates polymorphism of band 6 which is autosomal in origin. (From von Wyl and Chen, 1974.)

much lesser extent. Addition of actinomycin D to the incubation medium, which inhibited 90% of RNA synthesis, had no effect on the incorporating rate, suggesting the presence of stable mRNA. Since mechanical depletion of the glands prepared from unmated flies yielded no such increase, the stimulus must be provided by copulation.

Further support of the above conclusion came from parallel studies of the accessory gland in *D. nigromelanica*. Adult male flies of this species are characterized by accumulating a large quantity of glutamic acid in the paragonial secretion which may comprise some 60–70% of the total ninhydrin-reacting components (Chen and Oechslin, 1976; Chen and Baker, 1976). The accumulation is causally linked to a high level of the enzyme L-alanine aminotransferase in the secretory tissue of the accessory gland. Its activity in *D. nigromelanica* has been estimated to be 2–3 times higher than that in *D. melanogaster* which exhibits no accumulation of glutamic acid (Table III). A high level of L-alanine aminotransferase in the sperm of *D. hydei* has been reported by Geer *et al.* (1975). It is suggested that *Drosophila*

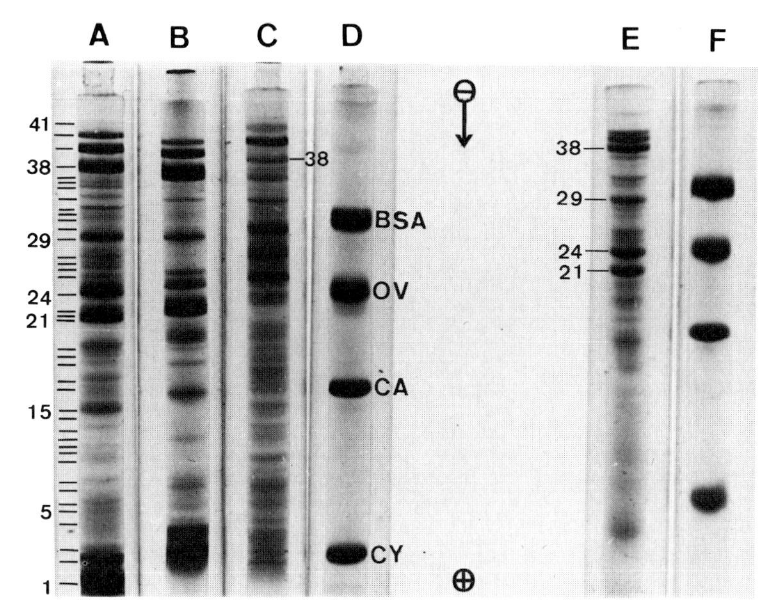

Fig. 19. Electrophoretic separation of accessory gland proteins in adult male *Drosophila melanogaster* according to the system of either Laemmli (A-D) or Weber-Osborn (E, F). A gel containing SDS and 10% acrylamide was used. (A) and (E) whole extract from 10 pairs of glands. (B) secretion from 20 pairs of glands. (C) epithelial extract from 20 pairs of glands. (D) and (F) molecular weight markers, including cytochrome c (CY), carbonic anhydrase (CA), ovalbumin (OV) and bovine serum albumin (BSA). (From von Wyl, 1976.)

sperm may have the capacity to utilize amino acids for energy production. In *D. nigromelanica* we also observed that, following copulation, enzyme activity in the gland was distinctly elevated (Table IV). Since no such elevation could be detected in male flies which received an injection of actinomycin D immediately after mating, the stimulatory effect involves obviously new synthesis of enzyme protein.

The immediate cause of the stimulatory effect just mentioned is not yet fully understood. Our results from reciprocal transplantations of the accessory glands from male flies before and after mating into mated or unmated male hosts indicate that a neurohumoral factor is probably released by the act of copulation and thus triggers the synthesis of secretion proteins (von Wyl, 1974). In *Rhodnius, Schistocerca, Locusta,* and *Periplaneta* the secretory activity of the accessory gland has been shown to be regulated by juvenile hormone (Gillott and Friedel, 1976). There is also evidence indicating that in both *Periplaneta* and *Tenebrio* the activity of aminotransferase, though not specifically localized in the accessory gland, is under the control

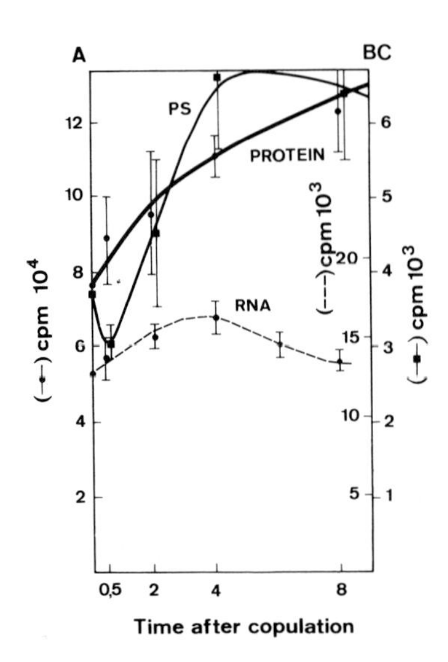

Fig. 20. Stimulation of *in vitro* synthesis of accessory gland components in adult males of *Drosophila melanogaster* by copulation. A, incorporation of ^{14}C-labeled amino acids into protein. B, incorporation of ^{14}C-labeled uridine into RNA. C, incorporation of ^{3}H-labeled ethanolamine into a major ninhydrin-positive paragonial substance (PS). In each experiment 20 pairs of accessory gland from 8-day-old flies at various time intervals following copulation were incubated in 200 μl of a simplified Bobb's medium at 25°C for 1 or 2 hr. (From von Wyl, 1974.)

TABLE III
Activity of L-Alanine Aminotransferase in Accessory Glands of Unmated Adult Male Flies of *D. nigromelanica* and *D. melanogaster*[a]

Species	Age (days)	*n*	Protein (μg/gland)	Total activity (OD/min/gland $\times 10^3$	Specific activity (μmoles /min/mg protein)
D. nigromelanica	7	4	1.24	2.45	0.375
	10	5	2.18	3.70	0.266
	13	6	2.69	3.67	0.229
	15	4	3.17	3.65	0.194
D. melanogaster	6	4	1.97	1.61	0.133
	7	4	2.44	2.04	0.135
	12	4	2.86	1.99	0.111
	14	4	2.40	1.51	0.102
	15	4	2.83	1.59	0.092

[a]Simplified from Chen and Oechslin (1976).

TABLE IV
Activity of L-Alanine Aminotransferase in Accessory Glands of Unmated and Mated Adult Male Flies of *D. nigromelanica*[a]

	Unmated male				Mated male			
Age (days)	n	Protein (μg/ gland)	Total activity (OD/min/ gland $\times 10^3$)	Specific activity (μmoles/ min/mg protein)	n	Protein (μg/ gland)	Total activity (OD/min/ gland $\times 10^3$)	Specific activity (μmoles/ min/mg protein)
12	4	1.87	3.67	0.366	4	1.29	4.24	0.576
13	11	1.85	3.32	0.318	9	1.04	5.04	0.796
14	14	2.31	3.63	0.276	13	1.40	4.37	0.519

[a]Simplified from Chen and Baker (1976).

of this hormone. However, in *Pyrrhocoris* neither the juvenile hormone nor its synthetic analogue has any effect on the activity of this enzyme. More sophisticated experiments are needed to establish the various regulatory possibilities.

Besides *Drosophila,* information on the functional significance of the paragonial secretion is available in two other dipteran insects. In *Aedes aegypti* two protein components, α and β, which have molecular weights of 60,000 and 30,000, respectively, have been isolated. The α component alone can promote oviposition, wheras both α and β components are responsible for female monogamy. In *Musca domestica,* which has no accessory gland per se, the male ejaculatory duct secretes a low molecular weight compound for the stimulation of oviposition and the induction of monocoitic behavior in adult female flies (see Baumann *et al.,* 1975, for references).

The male accessory gland of the house cricket, *Acheta domestica,* consists of a large number of tubules and serves to produce proteins for the formation of spermatophores which are used to transfer sperm to the female during copulation. As several spermatophores may be formed within a period of 24 hr and each of which amounts to about 10% of the total gland weight, the accessory gland in this insect has an unusually large capacity of protein production.

In a more recent study Kaulenas *et al.* (1975) reported that differentiation and growth of the accessory gland in *Acheta* take place in the last nymph instar, but its full capacity of protein synthesis is attained only after imaginal molt. Analysis of the total protein product on SDS acrylamide gels revealed about thirty protein bands with molecular weights ranging from 10,000 to over 100,000 daltons, similar to that found for the paragonial proteins in *Drosophila*. One interesting finding of Kaulenas *et al.* (1975) is that the secretion products in tubules of different lengths do not exhibit the same electrophoretic pattern (Fig. 21). There is apparently a regional differentiation of the gland tubules for the synthesis of specific proteins. This would mean that in different regions of the same gland different sets of genes are activated, obviously as a consequence of cytodifferentiation. In a further experiment actinomycin D was injected into the animals 2 hr before the administration of ^{3}H-uridine or ^{3}H-leucine. Although the antibiotic inhibited most incorporation of uridine into RNA, it did not affect the synthesis of protein. Instead, a superinduction was observed. Thus the presence of long-lived mRNA is indicated. This has been confirmed by analysis of the turnover of labeled poly(A)-containing RNA synthesized *in vivo*.

As mentioned above, the accessory gland in *Acheta* acquired its fully differentiated function of protein synthesis only at the time of imaginal molt. This may suggest that juvenile hormone is responsible for its activation. Fallon and Wyatt (1975) found an extraordinarily high content of guanosine 3′, 5′-monophosphate (cyclic GMP) in the male accessory gland of the house cricket. The parallel increase in its level and that of protein following adult molt indicate that this compound may play a role in the processing or stabilization of the secretion protein. Whether it exerts a regulatory function

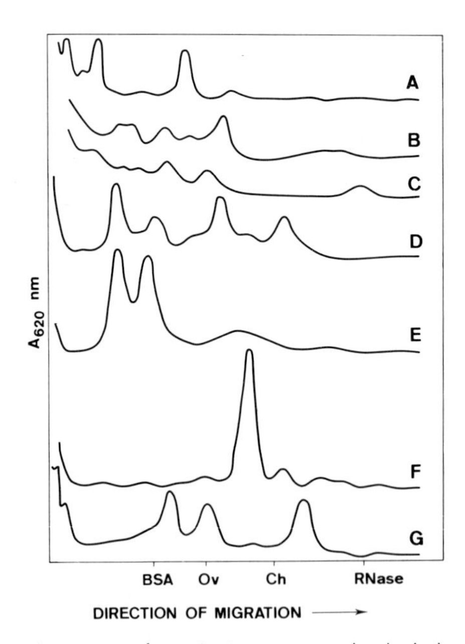

Fig. 21. Electrophoretic patterns of proteins in accessory gland tubules of various lengths of the male house cricket, *Acheta domestica*. (A) (B) small to medium length; (C) (D) (E) medium length; (F) (G) long tubules. Each protein sample was prepared from about 20 tubules and separated on SDS-acrylamide gels. Molecular weight markers were bovine serum albumin (BSA), ovalbumin (Ov), chymotrypsin (Ch) and ribonuclease (RNase). (From Kaulenas *et al.*, 1975.)

as cyclic AMP in hormone actions or plays a part in the maintenance and utilization of sperm as in vertebrates, remains to be established.

The development and metabolism of the male accessory gland in the grasshopper, *Melanopus sanquinipes*, have recently been investigated by Gillott and Friedel (1976). The gland is fully developed within 7 days after adult emergence and shows a sixfold increase in its protein content by day 14. Injection of ^{14}C-glycine revealed a rapid increase of its incorporation into the gland protein during the growth period. This was inhibited following allatectomy, but could be restored by topical application of juvenile hormone. Since both the fat body and hemolymph behaved in a similar manner, it is thought that the secretion proteins are synthesized in fat body and released into the hemolymph before being taken up by the glandular cells. This is analogous to what is known about the yolk proteins (see Section III, B). Furthermore, as in *Drosophila*, copulation caused an increase in the incorporation of labeled glycine into the secretion protein of the accessory gland. The stimulatory effect is ascribed to the induced release of protein from the fat body. As admitted by the authors, the interpretations proposed by them are based so far solely on circumstantial evidence. Further experiments under *in vitro* conditions and straightforward immunochemical identification of individual protein components appear to be desirable.

In a subsequent study Friedel and Gillott (1976) found that two protein fractions in the grasshopper accessory gland stimulate oviposition and are controlled by the corpus allatum hormone. This is consistent with the previous observation that glands from normal males transplanted into virgin females yielded an elevation of egg production, whereas those from allatectomized individuals failed to do so.

In conclusion, one distinguished feature of the male accessory gland in insects is its capacity for synthesizing large quantities of secretion proteins which are transferred to the female during copulation. Several specific functions of these proteins have been established. The immediate control over their synthesis appears to be the corpus allatum hormone which is released by the act of mating. The molecular mechanism involved in this cycle has not yet been fully explored.

VII. PROTEIN SYNTHESIS AS A BASIC PROBLEM OF AGING

Recent progress in studies of aging in insects has been reviewed in detail by Rockstein and Miquel (1973). In spite of the large body of information in this interesting research field, only a very limited number of investigations have been carried out at the molecular level. Several hypotheses have been proposed to explain the molecular-genetic basis of aging. One of these is the error catastrophe hypothesis of Orgel (1963), according to which random damages of cellular components involved in protein synthesis result in a progressive accumulation of altered molecules in the aging organism. Such molecules become eventually catalytic in effect and lead to the loss of vitality and death. In the codon restriction theory of Strehler *et al.* (1971) it is suggested that accumulation of synthetase repressors causes a reduced synthesis of certain proteins containing in their mRNA's those triplets for which the required aminoacyl tRNA's are no more available. If some of the affected proteins are synthetases, this would result in a further decline in the capacity of protein production. According to Sueoka and Kano-Sueoka (1970) an increasing deficiency of specific isoaccepting tRNA's in the process of aging may also set a limit to the translation of certain mRNA's. Although there is yet no conclusive evidence to support either one of the proposed hypotheses, they all imply the protein synthesis is the major event involved in the aging phenomenon. In this respect insects are no exception.

One piece of evidence in support of Orgel's hypothesis has been reported by Harrison and Holliday (1967) who found that when *Drosophila* larvae were fed amino acid analogues, the life span of adult flies developed from such larvae was shortened, probably due to increases in errors of proteins produced. However, the validity of this conclusion has recently been questioned (Baird *et al.,* 1975; but see also Holliday, 1975).

In *Drosophila subobscura* the incorporation rate of labeled leucine into protein has been reported to be two times higher in adult males at day 60 than in flies at day 20. By contrast, the incorporation of labeled glycine, lysine, and alanine in adult male flies of *D. melanogaster* aged 50 days amounted to only 37–42% of that in animals aged 3 days (see Chen, 1971, for references). The difference is apparently due to the fact that in the last-mentioned study the turnover of free amino acids, which is also distinctly reduced in aged flies, has been considered in the evaluation of the incorporation rate. By using a similar approach, essentially the same result has been obtained by Levenbook and Krishna (1971) for the blow fly, *Phormia regina*. In the locust, *Schistocerca gregaria,* there is no age-dependent increase in the incorporation of labeled valine into wing protein. In fact, a 45% reduction in protein synthesis over a period of 14 weeks has been determined. Extensive work on the housefly, *Musca domestica,* by Rockstein and associates demonstrated clearly that the failure of flight ability during aging is accompanied by a decline in activity of various enzyme systems (see references in Rockstein, 1972). Obviously the same is true for *Drosophila* and *Phormia* (Baker, 1975).

Since all studies mentioned above were carried out under *in vivo* conditions, protein synthesis may possibly be complicated by a variety of factors which are not easily controlled. For this reason we have currently used the *in vitro* system for gaining further insight into the molecular mechanism responsible for the regression of synthetic capacity in aging *Drosophila*

(Hosbach *et al.*, 1977; Hosbach, 1977). We first examined whether components involved in aminoacylation undergo specific age-dependent changes. For this purpose tRNA from male flies of *D. melanogaster* aged 5, 22, and 35 days was extracted with phenol and purified by column chromatography. Samples of aminoacyl-tRNA synthetases were also prepared. The tRNA's were then charged *in vitro* with different amino acids. In view of the codon restriction theory of Strehler *et al.* (1971) we examined only those tRNA's which have three or more corresponding codons in genetic code.

Under the conditions which we have used, old flies, at day 35, were found to contain about 35% less extractable tRNA than younger flies at day 5. A comparison of the charging capacity of tRNA's from different ages for nine amino acids with synthetase from 5-day-old animals revealed no essential change up to day 22. However, at day 35 an average decline of 18% could be detected with the exception of $tRNA^{Leu}$, the capacity of which showed as much as a 52% reduction compared to that at day 5 (Table V).

The thermostability of tRNA's from different ages was also analyzed. As indicated by the aminoacylation data on tRNA's exposed briefly to 80°C, there is a slight increase in sensitivity in flies aged 35 days.

There are apparently no distinct qualitative and quantitative changes in the chromatographic profiles of $tRNA^{Ser}$, $tRNA^{Ala}$, and $tRNA^{Met}$ between young and old flies (Fig. 22). We found only a single major isoacceptor of

TABLE V
Maximal Aminoacylation of tRNA from Adult Male Flies of *D. melanogaster* Aged 5 and 35 Days after Emergence[a]

Amino acid	Aminoacylation (pmoles/OD 260 nm) 5 days	35 days	Reduction (%)
Alanine	66	52	21
Arginine	47	41	13
Aspartic acid	61	54	11
Glycine	68	50	26
Histidine	32	28	12
Isoleucine	16	13	19
Leucine	17	8	52
Proline	8	7	11
Serine	82	64	22
Threonine	55	47	15
Tyrosine	18	16	11
Valine	68	59	13
Total	538	439	18

[a]From Hosbach *et al.* (1977).

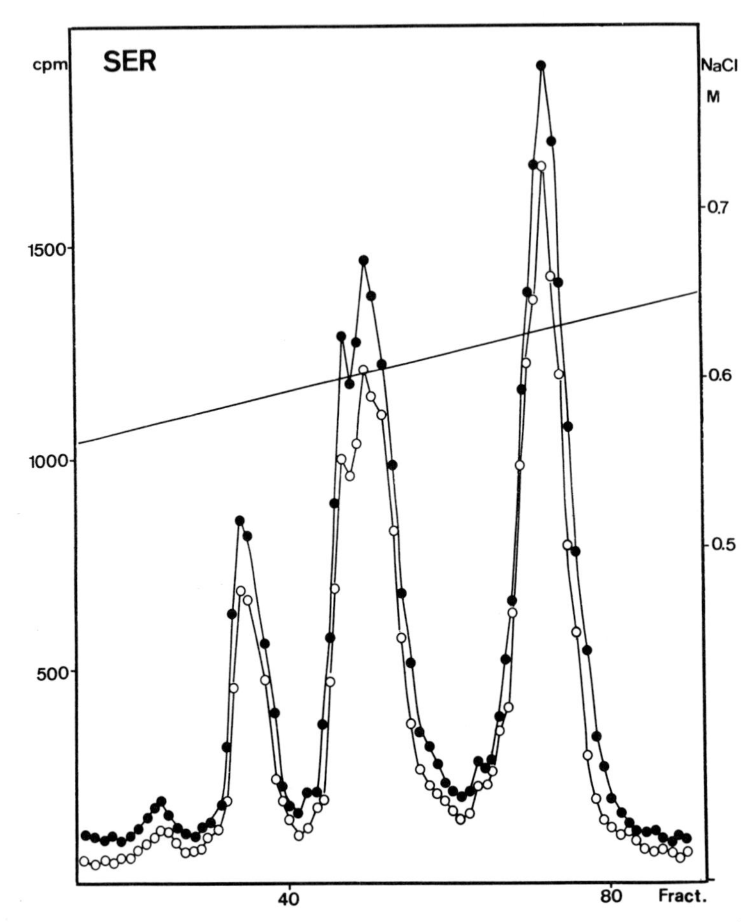

Fig. 22. Reverse phase-chromatographic analysis of the isoacceptor pattern of $tRNA^{Ser}$, $tRNA^{Ala}$, $tRNA^{Met}$ and $tRNA^{Leu}$ in adult male flies of *Drosophila melanogaster* aged 5 (—●—) and 35 (--○--) days after emergence. (From Hosbach, 1977.)

$tRNA^{Met}$, whereas White *et al.* (1973) detected two, one of which is an initiator-tRNA. The reason for this difference is not clear, but could be due to different procedures used. The pattern of $tRNA^{Leu}$ of the three different ages examined by us remain qualitatively the same, but all isoacceptors are greatly reduced in quantity with aging (Fig. 22). Thus, the decline in charging capacity of this tRNA mentioned above is not due to the absence of some specific isoaccepting species.

The chromatographic profiles of $tRNA^{His}$ and $tRNA^{Tyr}$ show significant and similar changes from young to old flies (Fig. 23). The early eluted isoacceptor (σ form of White *et al.*, 1973) increases with age, whereas the late

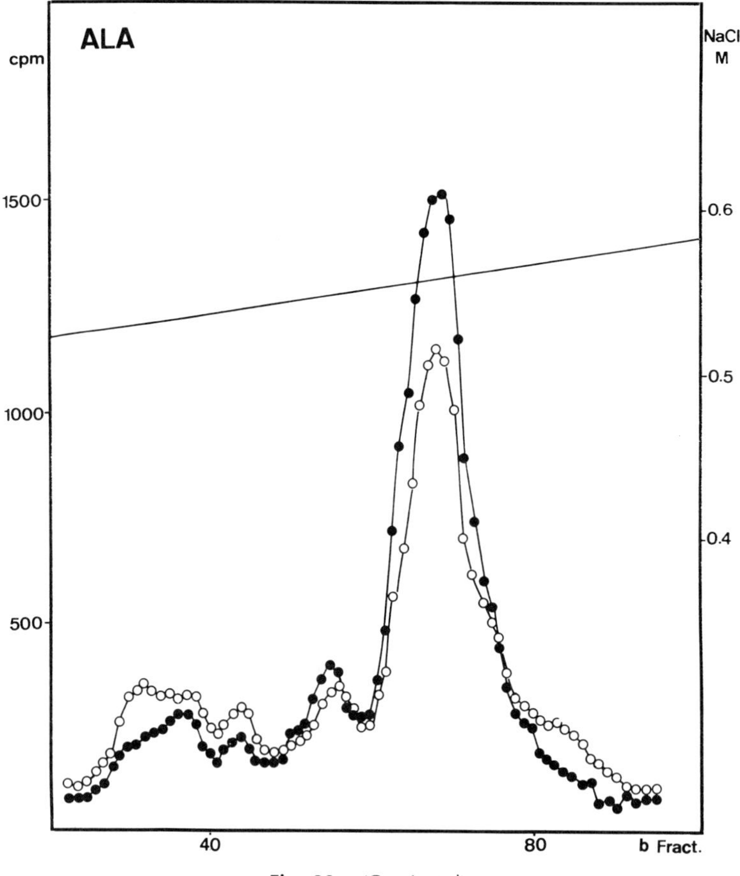

Fig. 22. (Continued)

eluted isoacceptor (γ form) decreases. Since White *et al.* observed the same variation during development from larvae to adults, it seems that aging is at least in part a continuation of the ontogenetic process. Assuming that the enzyme postulated by these authors converts G to Q, it would implicate that its activity increases with age. Conversely, if the enzyme converts Q to G, it would mean that it becomes gradually inactivated with age. A further point of interest is the regulatory significance of the altered proportion of isoaccepting tRNA's at the translational level, since the anticodon QUN of the σ form pairs preferentially with the codon NAU, whereas the anticodon GUN of the γ form with the codon NAC.

With respect to the properties of synthetases from different ages, our *in vitro* experiments have yielded the following results. The enzyme preparations from young and old flies contained about the same amount of protein.

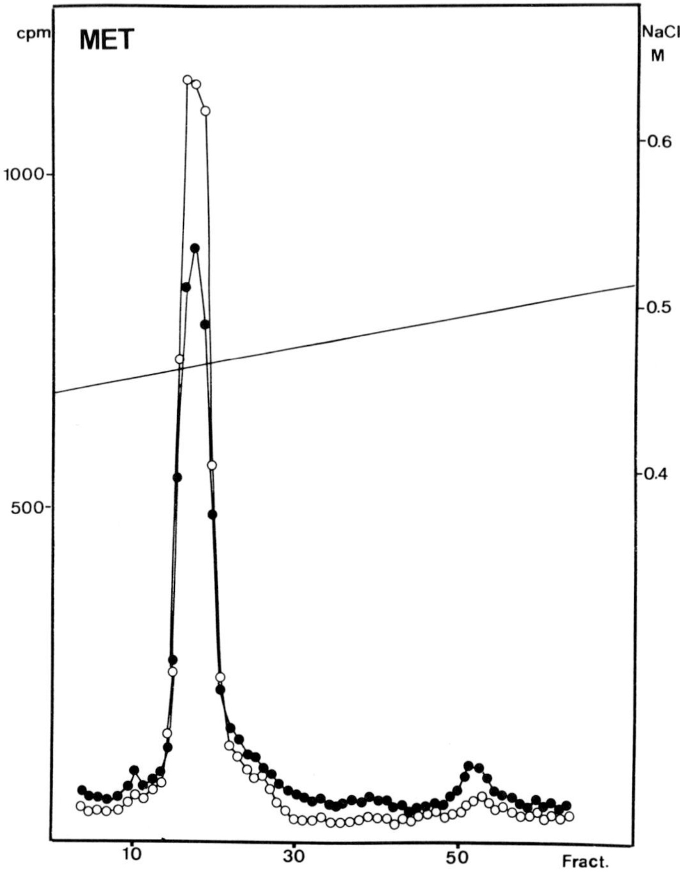

Fig. 22. (Continued)

In spite of this, their abilities to effect aminoacylation are different. The kinetic study showed that a maximal charge of $tRNA^{Asp}$ at day 5 with different concentrations of synthetases from the same age was reached within 20 min, and a charging rate of 58 pmole/min/mg protein was found (Fig. 24). However, with synthetase from old flies the same maximum could never be reached even after an incubation of 60 min, and the charging rate was estimated to be only 13 pmole/min/mg protein. There is further evidence that the thermostability of the aminoacylating enzyme decreases with age.

In a parallel study, age-related structural changes in ribosomes were also examined by Schmidt (1976) in my laboratory. It was found that on a per gram weight basis the total extractable 80 S ribosomal material decreased from 2.46 mg in adult males at day 4 to 1.90 mg in flies at day 30. The ribosome preparations were further treated with different concentrations of

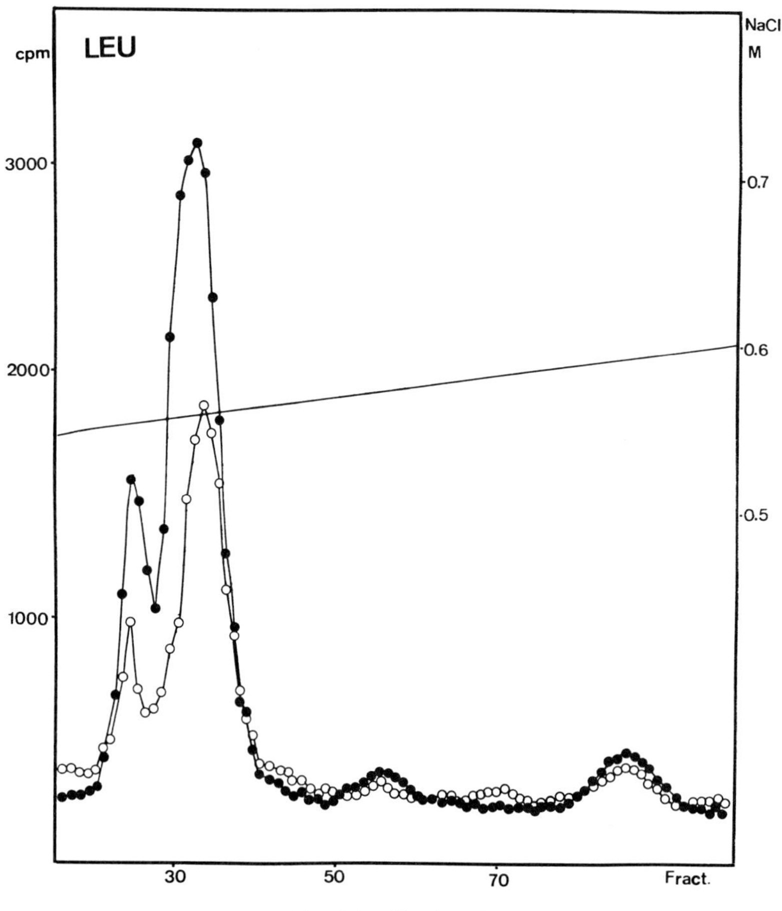

Fig. 22. (Continued)

KCl and the amount of protein dissociated was determined. It is worthy of note that, under the conditions used in our laboratory, about 3 to 5 times more protein could be removed from ribosomes in old flies aged 30 days than from those in young flies aged 4 days (Fig. 25). However, among a total of 67 ribosomal proteins separated by two-dimensional polyacrylamide gel electrophoresis, we observed no qualitative difference in the patterns between these two ages. It could be that the electrophoretic procedure used by us was not sensitive enough to detect minor changes in the protein components. In any case, the elevated amount of extractable ribosomal protein indicates that there must be some alteration in the structural integrity of the rRNA–protein complex with age. Further work is needed to determine whether this is due to a change in the complementary fidelity of the ribosomal protein or rRNA

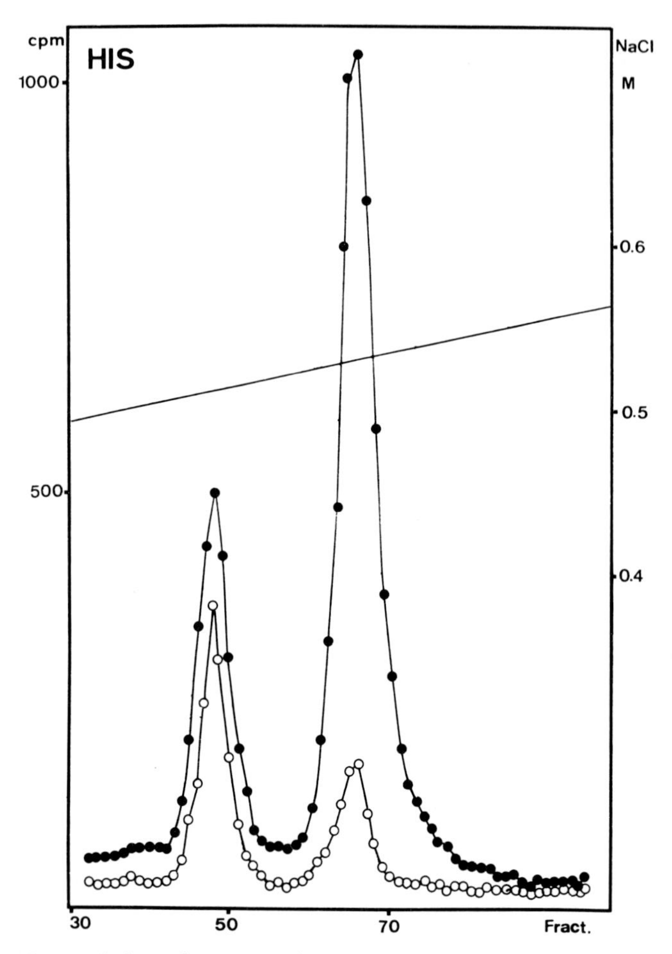

Fig. 23. Reversed phase-chromatographic analysis of the isoacceptor pattern of $tRNA^{His}$ and $tRNA^{Tyr}$ in adult male flies of *Drosophila melanogaster* aged 5 (—●—) and 35 (--○--) days after emergence. (From Hosbach, 1977.)

or both. It would be also of interest to know to what extent the translational capacity of these ribosomes is affected by such a structural alteration.

Clearly, our results are still not extensive enough to allow any definite conclusion on the molecular basis of aging, but have, nevertheless, brought us a step further to understand the mechanism underlying the regression of protein synthesis in aged *Drosophila*. As mentioned above, there is a distinctly reduced charging capacity of tRNA's in old flies. This reduction could be an adaptation to the low level of protein synthesis, or may itself represent a limiting factor in the translation process, as postulated by Sueoka and Kano-Sueoka (1970). We have also observed that both specific activity and thermostability of the aminoacyl-tRNA synthetases become reduced with

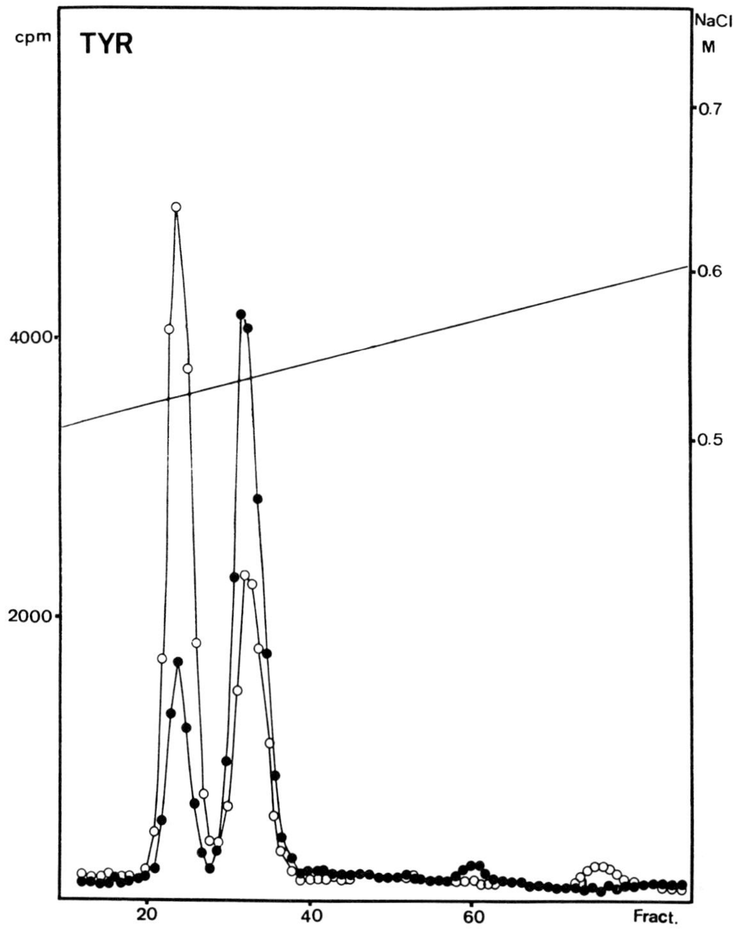

Fig. 23. (Continued)

age, an observation which may yield some support of the error catastrophe hypothesis of Orgel (1963). However, the enzyme preparations so far used by us were not sufficiently pure to allow us to draw the conclusion that these changes are really due to alterations of the molecular structure.

VIII. MORPHOGENETIC CONSEQUENCE RESULTING FROM MUTATIONAL EFFECT ON PROTEIN SYNTHESIS

Biochemical–genetic research has yielded a large volume of information on the nature of gene actions and gene products. Numerous studies on the biochemical effects of mutations have clearly demonstrated that genes are

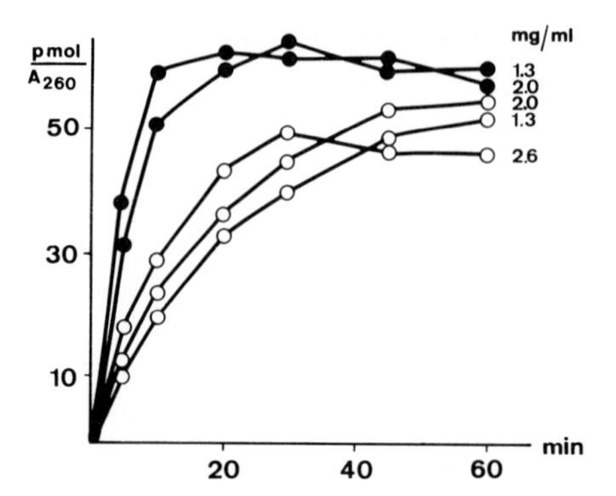

Fig. 24. Kinetics of *in vitro* aminoacylation of $tRNA^{Asp}$ with different concentrations of synthetase from 5-day(—●—) and 35-day(--○--)-old flies of *Drosophila melanogaster*. (From Hosbach *et al.*, 1977.)

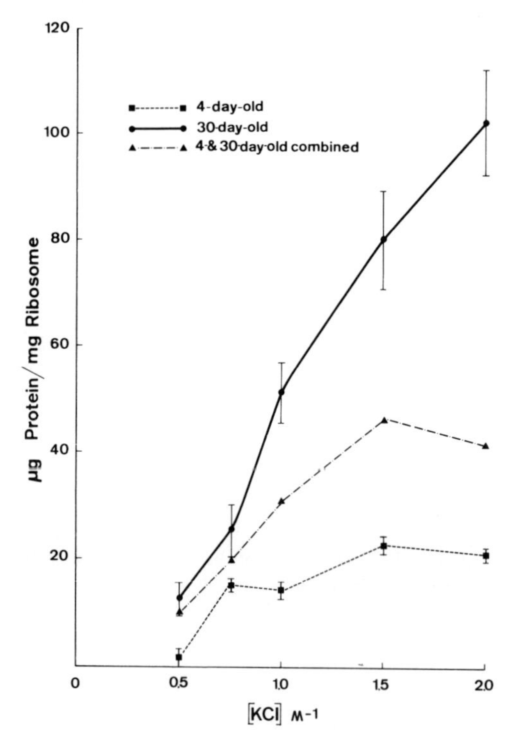

Fig. 25. Dissociation of protein from ribosomes prepared from 4-day, 30-day, as well as 4- and 30-day-old male flies of *Drosophila melanogaster* in the presence of different concentrations of KCl. (From Schmidt, 1976.)

involved in the regulation of specific metabolic pathways, either by acting directly or by determining enzyme specificity. The explanation of gene action can usually be achieved by a systematic analysis of the steps leading from a mutant gene to a mutant phene. Even though most mutations in eukaryotes have rather complex phenotypic effects, in many instances the causes can be traced to an interference with the synthesis of some structural or enzymatic proteins which are closely associated with a particular morphogenetic event. In the following, we shall consider a few examples from recent work in *Drosophila* to illustrate this point.

A. *cryptocephal (crc)*

The mutation *cryptocephal (crc)* of *D. melanogaster* has been studied in detail by Gloor (1945). Animals homozygous for *crc* develop normally up to pupariation, but show no eversion of head at the prepupal stage. The flies thus cannot hatch, though fully differentiated within the pupal case. Additional morphological phenes are short and malformed legs and bristles, as well as abnormal elongation of the abdomen. Gloor showed that the pleiotropic pattern of *crc* could be phenocopied by temperature shock prior to the normal time of head emergence at the prepupal period. *crc* strains differing significantly in penetrance and expressivity could also be selected.

Rizki (1960) produced phenocopy of *crc* by feeding glucosamine (GA) to *Drosophila* larvae. Since GA is a precursor of chitin, the major component of insect cuticle, he suggested that the action of *crc* involves an interference with the formation of the integumental structure. This explanation was later confirmed by Fristrom (1965) who found that the GA content of the cuticle of the *crc* homozygotes was 50–60% higher than that of the wild type, though the protein moiety appears normal. Feeding GA to wild-type larvae resulted in an increase in both GA content and rigidity of their cuticle to levels comparable to the mutant cuticle. In normal development, head eversion requires the creation of an internal pressure by contraction of the abdominal muscles. In the *crc* mutation there is apparently an excessive incorporation of GA into the cuticle, the high rigidity of which prevents the process of head eversion.

More recently, Rapport (1975) reported that *crc* phenocopy could be induced even more efficiently by feeding hydroxyproline (OHP) to wild-type larvae aged 65–70 hr. In the OHP-fed larvae the level of free proline (PRO) decreased greatly (Table VI). Since PRO, in contrast to other amino acids, accumulates during development of *Drosophila* larvae and has a high content in insect cuticle, *crc* effect of OHP has been interpreted as due to inhibition of PRO synthesis, resulting in the production of aberrant cuticular protein. But there is as yet no information on the protein composition of the *crc* and wild-type cuticle. Furthermore, the metabolic interrelationship be-

TABLE VI
Ratio of Free Amino Acid to Free Glycine in Hydroxyproline (OHP)- and Proline (PRO)-Fed Larvae and Pupae of *Drosophila*[a]

	Larvae				Pupae		
Free amino acid	H_2O	OHP	PRO	OHP + PRO	H_2O	OHP	PRO
Hydroxyproline	0.58	60.99	0.0	44.27	0.54	45.17	0.0
Aspartic acid	0.40	0.43	0.18	0.44	0.08	0.09	0.08
Threonine	0.49	0.68	0.65	0.83	0.50	1.0	0.46
Serine	0.52	0.61	0.65	0.61	0.92	0.82	0.92
Asparagine + glutamine	6.0	6.50	6.09	5.91	4.25	1.55	3.08
Sarcosine	0.91	1.04	0.94	1.17	0.0	0.0	0.0
Proline	1.03	0.42	16.6	14.86	4.08	0.18	3.54
Glutamic acid	2.06	1.64	1.53	2.26	1.33	1.09	1.31
Glycine	1.0	1.0	1.0	1.0	1.0	1.0	1.0
Alanine	2.64	2.04	5.21	3.13	5.0	4.27	4.46
Valine	0.19	0.11	0.29	0.22	1.0	1.55	0.62
Methionine	0.0	0.07	0.09	0.09	0.0	0.18	0.08
Isoleucine	0.12	0.11	0.21	0.13	0.33	0.73	0.31
Leucine	0.21	0.32	0.27	0.22	0.50	1.0	0.46
Tyrosine	1.46	1.71	1.41	1.70	0.25	1.27	0.23
Phenylalanine	0.09	0.04	0.06	0.13	0.17	0.36	0.15
β-Alanine	0.52	0.86	0.65	0.91	2.25	0.36	2.62
Lysine	0.67	1.11	0.32	0.74	1.0	2.0	1.0
Histidine	1.88	2.32	1.53	2.35	0.83	2.18	1.31
Arginine	1.55	1.85	1.38	1.78	1.58	2.27	1.08

[a]From Rapport and Yang (1974).

tween GA and OHP is unknown, though both are effective in inducing *crc* phenocopy.

B. *Abnormal abdomen* (A^{53g})

The mutant *Abnormal abdomen* (A^{53g}), which is a sex-linked dominant mutation with phenotypic abnormalities ranging from ragged appearance of the lateral edges of the tergites to complete absence of tergites on the dorsal abdominal surface, has been shown to be related to an interference with adult histoblast differentiation and folding of the abdominal hypodermis (Hillman, 1973). Genetic analysis demonstrated that the mutant phenotype is caused by an interaction of a major gene, A^{53g}, on the X chromosome and a variety of modifier genes localized throughout the genome.

Biochemical studies of this mutation revealed that the mutant flies have consistently a higher concentration of total soluble protein than the wild-type flies (Table VII), and the increase in protein content is already detect-

TABLE VII
Correlation of the Contents of Soluble Proteins and Expressivity of *Abnormal abdomen* (A^{53g}) in Adult Flies Aged 24 to 48 Hr after Emergence[a]

	Mean protein (mg/mg wet weight/fly)			
	+ // + ♀	+/∧ ♂	$A^{53g}//A^{53g}$ ♀	A^{53g}/∧ ♂
Class I normal phenotype	0.039	0.033	—	—
Class II slightly abnormal	—	—	0.046	0.039
Class III moderately abnormal	—	—	0.065	0.049
Class IV extremely abnormal	—	—	0.070	0.054

[a]From Shafer and Hillman (1974).

able at the beginning of adult hypodermis differentiation at 50 hr following pupariation (Shafer and Hillman, 1974). Inhibition of protein synthesis by adding either cycloheximide (Table VIII) or amino acid analogues (Table IX) to the diet is directly correlated to a reduction of expressivity of the mutant phenotype (Hillman *et al.*, 1973). Similarly, interference with RNA synthesis by growing larvae on diet containing 5-bromouridine results in a normalization of the phenotypic effect (Table X). Thus, it seems that an elevated protein synthesis leads to an alteration in the abdominal structure. This conclusion is consistent with the observation that both amino acid incorporation and tRNA aminoacylation proceed at a higher rate in the mutant than in the wild type (Rose and Hillman, 1973).

The sequence of events in the *Abnormal abdomen* mutation can be summarized as follows: At the time when the abdominal hypodermal cells begin to differentiate and to synthesize cuticular protein and chitin needed for the formation of adult integument, an excessive protein production occurs, probably by stimulation of tRNA aminoacylation under the influence of the

TABLE VIII
Effect of Cycloheximide on the Penetrance and Expressivity of *Abnormal abdomen* (A^{53g}) in *Drosophila*[a]

	Number		Penetrance		Expressivity		
Mg/ml	♀	♂	♀	♂	♀	♂	Viability
0.00	152	159	1.00	0.89	2.34	1.97	0.48
0.15	52	45	0.98	0.89	2.08[b]	1.96	0.48
0.20	56	38	0.93	0.89	2.04[b]	1.92	0.52
0.30	75	68	0.79	0.73	1.80[b]	1.71[b]	0.41
0.40	31	33	0.81	0.61	1.84[b]	1.61[b]	0.30

[a]From Hillman *et al.* (1973).
[b]Effect significant at the 5% level.

TABLE IX
Effect of the Amino Acid Analogues, Azetidine-2-Carboxylic Acid and Ethionine, on the Penetrance and Expressivity of *Abnormal abdomen* (A^{53g}) in *Drosophila*[a]

	Number		Penetrance		Expressivity		
Mg/ml	♀	♂	♀	♂	♀	♂	Viability
			Azetidine-2-carboxylic acid				
0.0	194	209	0.98	0.82	2.20	1.86	0.52
1.0	80	83	0.94	0.82	2.01[b]	1.84	0.65
2.0	82	120	0.95	0.76	2.09[b]	1.77	0.52
4.0	43	47	0.91	0.66	1.91[b]	1.66[b]	0.28
			Ethionine				
0.0	218	230	0.99	0.83	2.17	1.86	0.60
1.0	46	47	0.91	0.77	1.98[b]	1.79	0.62
2.0	48	73	0.94	0.56	1.94[b]	1.56[b]	0.48
4.0	32	51	0.72	0.41	1.72[b]	1.41[b]	0.33

[a]From Hillman *et al.* (1973).
[b]Effect significant at the 5% level.

TABLE X
Effect of Inhibition of RNA Synthesis by 5-Bromouridine on the Penetrance and Expressivity of *Abnormal abdomen* (A^{53g}) in *Drosophila*[a]

	Number		Penetrance		Expressivity		
Mg/ml	♀	♂	♀	♂	♀	♂	Viability
0.00	112	141	1.00	0.96	2.42	2.08	0.46
0.15	51	59	0.98	0.93	2.31	2.00	0.55
0.30	58	62	0.98	0.81	2.17[b]	1.86	0.40
0.60	75	60	0.95	0.78	2.03[b]	1.78[b]	0.45
1.20	66	71	0.92	0.76	1.98[b]	1.76	0.55

[a]From Hillman *et al.*, (1973).
[b]Effect of the inhibitor significant at the 5% level.

modifier genes. The protein produced is then modified in its structure or function by the action of the A^{53g} major gene. The modifiers alone have no detectable morphological effect, and the major gene alone gives rise to only a slight abnormality.

C. *rudimentary (r)*

Another interesting case is the recessive sex-linked mutation *rudimentary* (*r*), the phenotype of which includes abnormal wings, female sterility, and reduced viability during development. Nutritional studies demonstrated that the *r* mutant has a requirement for pyrimidines (Nørby, 1970), and fertility of the *r* females can be restored by raising them on pyrimidine-enriched medium. The fact that exogenously supplied carbamyl aspartate is utilized by the mutant suggests that the mutational effect interferes with the synthesis of carbamyl phosphate and/or carbamyl aspartate, the first two steps in thc biosynthctic pathway of pyrimidines.

Interallelic complementation at the *r* locus has been reported for both wing phenotype and female sterility. In addition, complementation for the wing phenotype has been shown to be paralleled by restoration of female sterility (see references in Nørby, 1973). In a further analysis of the complementation phenomenon Nørby (1973) found that the activity of aspartate transcarbamylase (ATCase) in the mutant strain r^{39k} was comparable to that in the wild type, whereas virtually no enzyme activity could be detected in two other strains r^{54c} and r^{C}. Since r^{39k} and r^{54c} exhibit complete complementation with regard to both wing phenotype and pyrimidine requirement and r^{C} complements neither of the two (Table XI), it is thought that r^{39k} is deficient in the enzyme carbamyl phosphate synthetase (CPSase), r^{54c} in ATCase, and r^{C} in both (Table XII). This implication that *r* is the structural locus for the

TABLE XI
Complementation for Wing Phenotypes between the Three Alleles r^{C}, r^{39k} and r^{54c} of the Mutation *rudimentary (r)* in *Drosophila*[a,b]

	r^{54c}	r^{39k}	r^{C}
r^{C}	–	–	–
r^{39k}	++	–	
r^{54c}	–		

[a]From Nørby (1973).
[b]++ , complete complementation; – , no complementation.

TABLE XII
Proposed Phenotypes with Respect to Activities of Carbamylphosphate Synthetase (CPSase) and Aspartate Transcarbamylase (ATCase) in the Three Allelic Strains r^{39k}, r^{54c} and r^{C} of the Mutation *rudimentary (r)* in *Drosophila*[a]

Strain	Phenotype[b]	
r^{39k}	$CPSase^{-}$	$ATCase^{+}$
r^{54c}	$CPSase^{+}$	$ATCase^{-}$
r^{C}	$CPSase^{-}$	$ATCase^{-}$

[a]From Norby (1973).
[b]+ , normal activity; – , deficient activity.

two enzymes has been confirmed by Jarry and Falk (1974). According to Rawls and Fristrom (1975) the activity of dihydroorotase (DHOase) is also under the control of the *r* locus (Fig. 26).

In summary, it appears that the complex *r* locus codes for altogether three enzyme proteins and hence controls the first three steps of pyrimidine

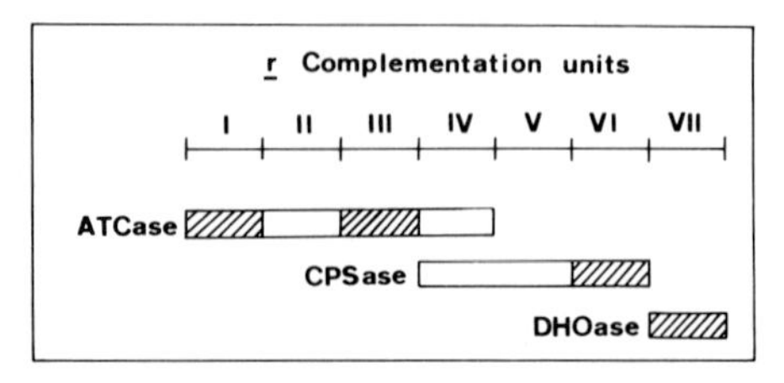

Fig. 26. Regions of the *rudimentary (r)* locus in *Drosophila* which specify the enzymes aspartate transcarbamylase (ATCase), carbamyl phosphate synthetase (CPSase), and dihydroorotase (DHOase) in the biosynthesis of pyrimidines. The cross-hatched segments of the horizontal bars indicate those complementary units for which control of the enzyme levels has been demonstrated. (From Rawls and Fristrom, 1975.)

synthesis in *Drosophila*. The reason that a sufficient supply of pyrimidine is needed for normal wing formation is unknown, but may be inferred from a pertinent observation in the *Cecropia* silkmoth which has a high concentration of the conjugate uridinediphosphate-*N*-acetylglucosamine in the wing hypodermis.

D. *lethal-translucida (l(3)tr)*

Drosophila larvae homozygous for the lethal mutation *l(3)tr* are characterized by an enormous accumulation of hemolymph (see earlier reviews in Hadorn, 1961; Chen, 1971). They pupariate with a delay of about 24 hr and remain usually at the prepupal stage without further development. Some of them can undergo a partial differentiation in the head and thorax, but none emerges. Among the most prominent biochemical phenes are high levels of free amino acids and low concentration of hemolymph proteins. RNA and protein synthesis still proceed normally, though somewhat delayed. There is at the same time a rapid protein degradation caused by a high protease activity.

Recently, we performed a comparative analysis of the isozyme patterns of a total of nine enzyme systems between the mutant and wild-type larvae and pupae (Borner and Chen, 1975). Of special interest is our finding that among a total of eight isozymes of alcohol dehydrogenase (ADH) detected in the wild type, four are consistently missing in the lethals (Fig. 27). The total ADH activity remains at a high level following puparium formation in the wild type, but falls off rapidly in *l(3)tr* (Fig. 28). Since Ursprung and Madhavan (1971) demonstrated that *Drosophila* ADH utilizes the two juvenile hormone analogues, farnesol and farnesal, as substrates, we suspect that the delay of pupariation of the *l(3)tr* mutant could be linked to a defect of this enzyme system. Further information on the levels of juvenile hormone in both genotypes at the phenocritical period is needed to confirm this conclusion.

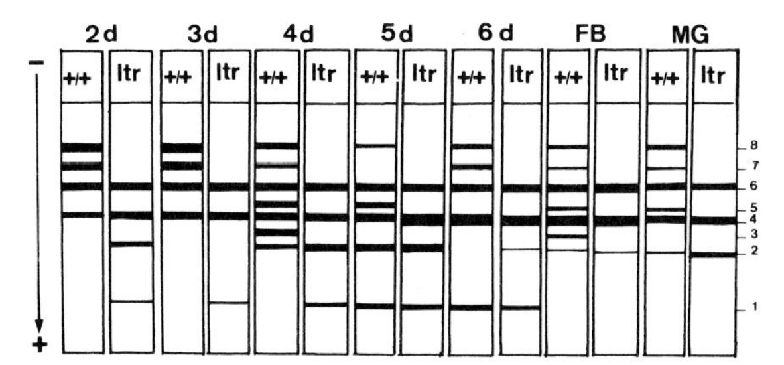

Fig. 27. Polyacrylamide gel electrophoresis of alcohol dehydrogenase isozymes in total homogenates of larvae and pupae aged 2–6 days (d) and in homogenates of fat body (FB) and midgut (MG) of the wild type (+/+) and the mutant *l(3)tr* of *Drosophila melanogaster*. (From Borner and Chen, 1975.)

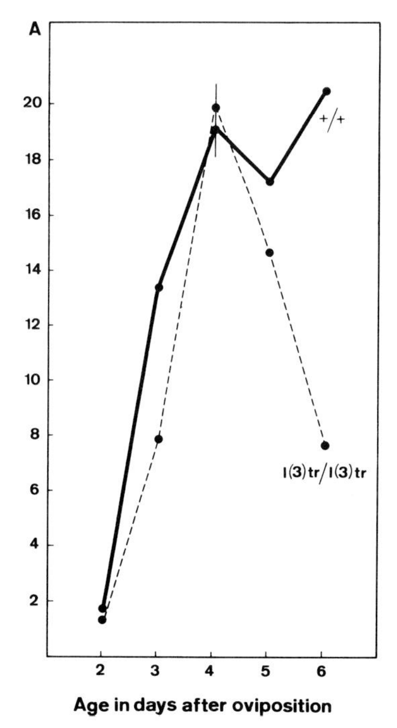

Fig. 28. Changes in total activity of alcohol dehydrogenase during larval and pupal development of the wild type (——) and the mutant *l(3)tr* (----) of *Drosophila melanogaster*. A, activity defined as increase of 0.001 absorption unit per minute. (From Borner and Chen, 1975.)

E. *lethal-meander (l(2)me)*

The lethal factor *l(2)me* differs from *l(3)tr* in having an earlier effect on *Drosophila* larvae. Individuals homozygous for *l(2)me* cease to develop at about the early third larval instar, remain alive for several days, and then die without pupation. The biochemical characteristics of this mutation include a reduced protease activity, a low level of protein synthesis, and a diminished RNA/DNA ratio.

The isozyme patterns of several enzyme systems in the *l(2)me* mutant and the wild type have recently been studied in detail by Kubli and co-workers in the author's laboratory (Züst *et al.*, 1972; Dübendorfer *et al.*, 1974). The pattern of the hexokinase isozyme Hex-3 is of particular interest, because of its significance in relation to the *l(2)me* mutation. As illustrated in Fig. 29,

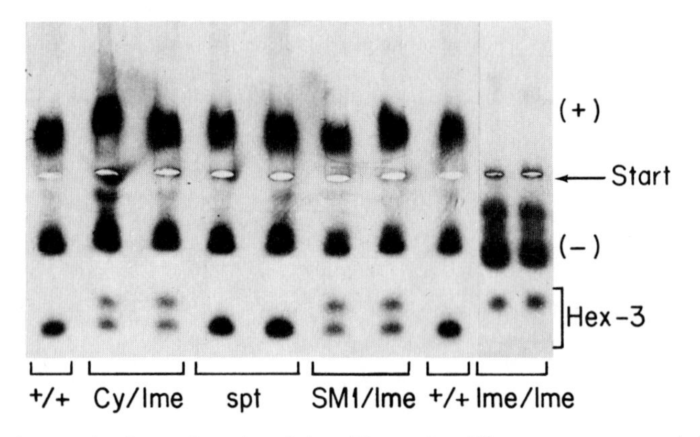

Fig. 29. Agar gel electrophoresis of hexokinase in different genotypes of *Drosophila melanogaster*. For all genotypes homogenates of adult flies were used, with exception of the mutant *l(2)me* for which homogenates were prepared from homozygous lethal larvae aged 96 hr after oviposition. (From Dübendorfer *et al.*, 1974.)

the wild-type flies show only a single form of Hex-3, whereas the heterozygotes *Cy/l(2)me* and SM_1, al^2 *Cy* sp^2*/l(2)me* have two variants, one migrating at the same rate as the wild-type isozyme, and the other at a slightly retarded rate. Homozygotes *l(2)me/l(2)me* larvae exhibit only the slower variant. The gene for Hex-3 has been localized on the second chromosome at position 73.5 and 73±; evidence from *in situ* hybridization has indicated that tRNA cistrons are also localized on the second chromosome at position 72± (see references in Dübendorfer *et al.*, 1974). Since both are very close to the *l(2)me* locus at position 72±, it appears reasonable to postulate that this lethal mutation is a deletion including one part of the tRNA cistrons and extending into the Hex-3 region (Fig. 30). If this is the case, the lethal effect of *l(2)me* could be explained as being due to a deficiency of stage-specific tRNA(s), and the altered electrophoretic mobility of the Hex-3 isozyme as being caused by a terminal deficiency in amino acid component(s) of the enzyme molecule. The proposed hypothesis can be tested by straightforward biochemical analysis of tRNA and Hex-3 isozyme prepared from the *l(2)me* larvae.

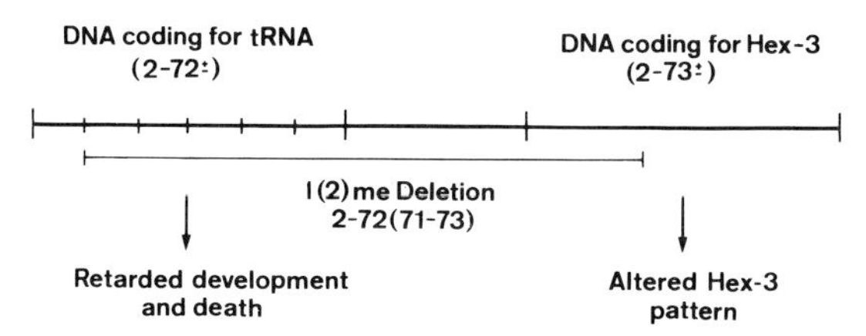

Fig. 30. A hypothetical deletion which would account for the developmental failure and altered hexokinase pattern of the mutation *l(2)me*. (From Züst *et al.*, 1972.)

From the few examples discussed in the previous sections it is evident that metabolic steps leading to the expression of morphological changes and to the arrest of development are highly complex, and there is often a long and obscure way between primary gene action and phenotypic effect. Although the exact nature of the actions of these mutations remains to be elucidated, the experimental results demonstrate clearly that alterations of protein synthesis at a specific stage and in a specific tissue may have profound morphogenetic consequence. This type of information is most valuable for our understanding of the regulation of cellular activity at both the developmental-genetic and molecular level.

IX. CONCLUDING REMARKS

In this chapter we have discussed protein synthesis in several selected tissues and organs of various insect species. Emphasis has been laid on regulatory systems which insects may utilize to meet their changing demands. As already pointed out, cellular activation can be initiated by hormones, feeding, copulation, photoperiod, temperature, as well as other internal and external factors. But it should be kept in mind that all these factors act merely as triggers to which the cells respond according to their genetic makeup. Biochemical–genetic studies, as discussed in the last section, showed conclusively that cell- and stage-specific proteins, structural or enzymatic, are ultimately specified by genetic loci. Additional evidence for regulation at the transcriptional and translational level has been provided by recent work on the synthesis of fibroin in the posterior silk gland of *Bombyx mori* (Suzuki and Suzuki, 1974) and the production of chorion proteins in the follicular epithelium of *Antheraea polyphemus* and *Drosophila melanogaster* (Paul and Kafatos, 1975; Petri *et al.*, 1976).

Similar to other areas in insect biochemistry, work on protein synthesis has been eased by recent progress in the development of new and more sophisticated analytical methods. These include improved conditions for cell and organ cultures, preparative procedures for collecting single tissues and organs, availability of highly active insect hormones as well as affinity chromatography for the isolation and purification of biologically important compounds. Our continuing efforts will no doubt enable us to clarify many of the still poorly understood aspects of this major biochemical event related to cellular activation and deactivation.

ACKNOWLEDGMENTS

The original work done in my laboratory has been supported by grants from the Swiss National Science Foundation and the Georges and Antoine Claraz-Schenkung. I should like to thank Drs. H. K. Mitchell, E. Kubli and H. Briegel for reading various parts of this chapter.

GENERAL REFERENCES

Beermann, W. (1972). *In* "Results and Problems in Cell Differentiation" (W. Beermann, ed.), Vol. 4, p. 1. Springer-Verlag, Berlin and New York.
Burdette, W. J. (1974). "Invertebrate Endocrinology and Hormonal Heterophylly." Springer-Verlag, Berlin and New York.
Chen, P. S. (1966). *Adv. Insect Physiol.* **3,** 53.
Chen, P. S. (1971). "Biochemical Aspects of Insect Development." Karger, Basel.
de Wilde, J., and de Loof, A. (1973). *In* "The Physiology of Insecta" (M. Rockstein, ed.), 2nd. Ed., Vol. 1, pp. 97–157. Academic Press, New York.
Doane, W. W. (1973). *In* "Insects: Developmental Systems" (S. J. Counce and C. H. Waddington, eds.), Vol. 2, pp. 291–497. Academic Press, New York.
Fowler, G. L. (1973). *Adv. Genet.* **17,** 293.
Grossbach, U. (1973). *Cold Spring Harbor Symp. Quant. Biol.* **38,** 619.
Hadorn, E. (1961). "Developmental Genetics and Lethal Factors." Methuen, London.
House, H. L. (1974). *In* "The Physiology of Insecta" (M. Rockstein, ed.), 2nd. Ed. Vol. 5, pp. 63–117. Academic Press, New York.
Ilan, J., and Ilan, J. (1974). *In* "The Physiology of Insecta" (M. Rockstein, ed.), 2nd. Ed., Vol. 4, pp. 355–422. Academic Press, New York.
Mitchell, H. K. (1967). *Annu. Rev. Genet.* **1,** 185.
Novak, V. J. A. (1975). "Insect Hormones." Chapman & Hall, London.
Price, G. M. (1973). *Biol. Rev. Cambridge Philos. Soc.* **48,** 33.
Rockstein, M. (1972). *In* "Molecular Genetic Mechanisms in Development and Aging" (M. Rockstein and G. T. Baker, eds.), pp. 1–10. Academic Press, New York.
Rockstein, M., and Miquel, J. (1973). *In* "The Physiology of Insecta" (M. Rockstein, ed.), 2nd. Ed. Vol. 1, pp. 371–478. Academic Press, New York.
Smith, A. E. (1976). "Protein Synthesis." Chapman and Hall, London.
Wyatt, G. R. (1972). *In* "Biochemical Actions of Hormones" (G. Litwack, ed.), Vol. 2, pp. 385–490. Academic Press, New York.
Wyatt, G. R. (1975). *Verh. Dtsch. Zool. Ges.* 1974, p. 209.

REFERENCES FOR ADVANCED STUDENTS AND RESEARCH SCIENTISTS

Ashburner, M. (1967). *Chromosoma* **21,** 398.

Baird, M. B., Samis, H. V., Massie, H. R. and Zimmerman, J. A. (1975). *Gerontologia* **21,** 57.

Baker, G. T. (1975). *Exp. Gerontol.* **10,** 231.

Baudisch, W. and Heimann, F. (1972). *Acta. Biol. Med. Germ.* **29,** 521.

Baudisch, W., and Panitz, R. (1968). *Exp. Cell Res.* **49,** 470.

Baumann, H. (1974). *J. Insect Physiol.* **20,** 2347.

Baumann, H., Wilson, K., Chen, P. S., and Humber, R. E. (1975). *Eur. J. Biochem.* **52,** 521.

Bonner, J. J. and Pardue, M. L. (1976). *Cell* **8,** 43.

Borner, P., and Chen, P. S. (1975). *Rev. Suisse Zool.* **82,** 667.

Briegel, H. (1975). *J. Insect Physiol.* **21,** 1681.

Briegel, H., and Lea, A. O. (1975). *J. Insect Physiol.* **21,** 1597.

Chen, P. S. (1976). *Experientia* **32,** 549.

Chen, P. S., and Baker, G. T. (1976). *Insect Biochem.* **6,** 441.

Chen, P. S. and Briegel, H. (1965). *Comp. Biochem. Physiol.* **14,** 463.

Chen, P. S., and Oechslin, A. (1976). *J. Insect Physiol.* **22,** 1237.

Chippendale, G. M., and Kilby, B. A. (1969). *J. Insect Physiol.* **15,** 905.

Daneholt, B., and Hosick, H. (1973). *Proc. Natl. Acad. Sci.* U.S. **70,** 442.

Dejmal, R. K., and Brookes, V. J. (1972). *J. Biol. Chem.* **247,** 869.

de Loof, A., and de Wilde, J. (1970). *J. Insect Physiol.* **16,** 1455.

Dübendorfer, K., Nöthiger, R., and Kubli, E. (1974). *Biochem. Genet.* **12,** 203.

Engelmann, F. (1969). *J. Insect Physiol.* **15,** 217.

Engelmann, F. (1974). *Am. Zool.* **14,** 1195.

Engelmann, F., and Wilkens, J. L. (1969). *Nature* (London) **222,** 798.

Fallon, A. M., and Wyatt, G. R. (1975). *Biochim. Biophys. Acta* **411,** 173.

Fallon, A. M., Hagedorn, H. H., Wyatt, G. R., and Laufer, H. (1974). *J. Insect Physiol.* **20,** 1815.

Friedel, T., and Gillott, G. (1976). *J. Insect Physiol.* **22,** 489.

Fristrom, J. W. (1965). *Genetics* **52,** 297.

Geer, B. W., Kelley, K. R., Pohlman, T. H., and Yemm, S. J. (1975). *Comp. Biochem. Physiol.* ***B*50,** 41.

Gillott, C., and Friedel, T. (1976). *J. Insect Physiol.* **22,** 365.

Gloor, H. (1945). *Arch. Julius-Klaus-Stiftung* **20,** 209.

Gruber, C. (1977). Diplom.Thesis, University of Zürich, Zürich.

Hagedorn, H. H., and Fallon, A. M. (1973). *Nature* **244,** 103.

Hagedorn, H. H., Fallon, A. M., and Laufer, H. (1973). *Dev. Biol.* **31,** 285.

Hagedorn, H. H., O'Conner, J. D., Fuchs, M. S., Sage, B., Schlaeger, D. A., and Bohm, M. K. (1975). *Proc. Natl. Acad. Sci.* U.S.A. **72,** 3255.

Harrison, B. J. and Holliday, R. (1967). *Nature* (London) **213,** 990.

Hillman, R. (1973). *Genet. Res.* **22,** 37.

Hillman, R., Shafer, S. J., and Sang, J. H. (1973). *Genet. Res.* **21,** 229.

Holliday, R. (1975). *Gerontologia* **21,** 64.

Hosbach, H. A. (1977). Ph. D. Thesis, University of Zürich, Zürich.

Hosbach, H. A., Egg, A. H. and Kubli, E. (1972). *Rev. Suisse Zool.* **79,** 1049.

Hosbach, H. A., Kubli, E. and Chen, P. S. (1977). *Rev. Suisse Zool.* **83,** 964.

Huang, C. T. (1971). *Insect Biochem.* **1,** 207.

Hürlimann, F. (1974). Ph. D. Thesis, University of Zürich, Zürich.

Jarry, B., and Falk, D. (1974). *Mol. Gen. Genet.* **135,** 113.

Kaulenas, M. S., Yenofsky, R. L., Potswald, H. E., and Burns, A. L. (1975). *J. Exp. Zool.* **193,** 21.
Korge, G. (1975). *Proc. Natl. Acad. Aci. U.S.A.* **72,** 4550.
Kubli, E. (1970). *Z. Vgl. Physiol.* **70,** 175.
Küthe, H. W. (1972). *Wilhelm Roux' Arch.* **170,** 165.
Küthe, H. W. (1973a). *Wilhelm Roux' Arch.* **171,** 301.
Küthe, H. W. (1973b). *Wilhelm Roux' Arch.* **172,** 58.
Lea, A. O. (1972). *Gen. Comp. Endocriol. Suppl.* **3,** 602.
Lengyel, J., and Penman, S. (1975). *Cell* **5,** 281.
Levenbook, L., and Krishna, I. (1971). *J. Insect Physiol.* **17,** 9.
Lewis, M., Helmsing, P. J., and Ashburner, M. (1975). *Proc. Natl. Acad. Sci. U.S.A.* **72,** 3604.
Locke, M., and Collins, J. V. (1968). *J. Cell Biol.* **36,** 453.
Lüscher, M. (1968). *J. Insect Physiol.* **14,** 499.
McGregor, J. A., and Loughton, B. G. (1974). *Can. J. Zool.* **52,** 907.
McGregor, J. A., and Loughton, B. G. (1976). *Wilhelm Roux' Arch.* **179,** 77.
McGregor, J. A., and Loughton, B. G. (1977). *Wilhelm Roux' Arch.* **181,** 113.
McKenzie, S. L., Henikoff, S., and Meselson, M. (1975). *Proc. Natl. Acad. Aci. U.S.A.* **72,** 1117.
Miller, J. W., Kramer, K. J. and Law, J. H. (1974). *Comp. Biochem. Physiol. B* **48,** 117.
Mitchell, H. K., and Lipps, L. S. (1975). *Biochem. Genet.* **13,** 585.
Mitchell, H. K., Tracy, U. M., and Lipps, L. S. (1977a). *Biochem. Genet.* **15,** 563.
Mitchell, H. K., Lipps, L. S., and Tracy, U. M. (1977b). *Biochem. Genet.* **15,** 575.
Mordue, W. (1965). *J. Insect Physiol.* **11,** 617.
Nørby, S. (1973). *Hereditas* **73,** 11.
Orgel, L. E. (1963). *Proc. Natl. Acad. Sci.* U. S. **49,** 517.
Pan, M. L. (1971). *J. Insect Physiol.* **17,** 677.
Pan, M. L. and Wallace, R. A. (1974). *Am. Zool.* **14,** 1239.
Pan, M. L., and Wyatt, G. R. (1976). *Dev. Biol.* **54,** 127.
Paul, M., and Kafatos, F. C. (1975). *Dev. Biol.* **42,** 141.
Petri, W. H., Wyman, A. R., and Kafatos, F. C. (1976). *Dev. Biol.* **49,** 185.
Price, G. M. (1965). *J. Insect Physiol.* **11,** 869.
Rapport, E. W. (1975). *J. Exp. Zool.* **192,** 213.
Rapport, E. W., and Yang, M. K. (1974). *Comp. Biochem. Physiol. B* **49,** 165.
Rawls, J. M., and Fristrom, J. M. (1975). *Nature (London)* **255,** 738.
Rizki, T. M. (1960). *Biol. Bull.* (Woods Hole, Mass.) **118,** 308.
Rose, R., and Hillman, R. (1973). *Genet. Res.* **21,** 239.
Rüegg, M. K. (1968). *Z. Vgl. Physiol.* **60,** 275.
Schmidt, T. (1976). Diplom. Thesis, University of Zürich, Zürich.
Shafer, S. J., and Hillman, R. (1974). *J. Insect Physiol.* **20,** 223.
Shigematsu, H. (1958). *Nature (London)* **182,** 880.
Spiro-Kern, A. (1974). *J. Comp. Physiol.* **90,** 53.
Spradling. A., Pardue, M. L., and Penman, S. (1977). *J. Mol. Biol.* **109,** 559.
Spradling, A., Penman, S., and Pardue, M. L. (1975). *Cell* 4, 395.
Strehler, B., Hirsch, G., Gusseck, D., Johnson, R., and Bick, M. (1971). *J. Theor. Biol.* **33,** 429.
Sueoka, N., and Kano-Sueoka, T. (1970). *Prog. Nucleic Acid Res. Mol. Biol.* **10,** 23.
Suzuki, Y., and Suzuki, E. (1974). *J. Mol. Biol.* **88,** 393.
Telfer, W. H. (1954). *J. Gen. Physiol.* **37,** 539.
Thomasson, W. A. and Mitchell, H. K. (1972). *J. Insect Physiol.* **18,** 1885.
Thomsen, E., and Møller, IB (1963). *J. Exp. Biol.* **40,** 301.
Thomson, J. A., Kinnear, J. F., Martin, M. D., and Horn, D. H. S. (1971). *Life Sci.* **10,** 203.
Tissières, A., Mitchell, H. K., and Tracy, U. M. (1974). *J. Mol. Biol.* **84,** 389.
Ursprung, H. and Madhavan, K. (1971). *Eur. Drosophila Res. Conf. 2nd.,* Zürich.

von Gaudecker, B. (1972). *Z. Zellforsch. Mikrosk. Anat.* **127,** 50.
von Wyl, E. (1976). *Insect Biochem.* **6,** 193.
von Wyl, E. (1974). Ph. D. Thesis, University of Zürich, Zürich.
von Wyl, E., and Chen, P. S. (1974). *Rev. Suisse Zool.* **81,** 655.
White, B. N., Tener, G. M., Holden, J., and Suzuki, D. T. (1973). *J. Mol. Biol.* **74,** 635.
Züst, H., Egg, A. H., Hosbach, H. A., and Kubli, E. (1972). *Verh. Schweiz. Naturforsch. Ges.* p. 172.

5

The Chemistry of Insect Cuticle

A. GLENN RICHARDS

I. INTRODUCTION

The insect cuticle forms the exoskeleton of the insect (and to some extent an internal skeleton), supporting linings for foregut, hindgut, respiratory system, reproductive ducts, and some gland ducts. Important functions of the cuticle are support, including muscle attachment, protection, and permeability barriers.

Insect cuticle is a complex secretion from differentiated epidermal cells of ectodermal origin. On stimulus from hormones, these cells secrete sequentially first one product, then another, then another. Moreover, there is a large amount of difference in what is secreted depending on the species, the stage in development, and the precise area being considered. Many of these cells have no other obvious functions, but some additionally differentiate for essentially unrelated functions (e.g., sensation and pheromone production).

The insect cuticle (Fig. 1) consists of an outermost set of layers lacking chitin; these are collectively termed the epicuticle (layers 1–3). The epicuticle is seldom more than 1 μm in thickness. Underlying the epicuticle is the chitin/protein portion of the cuticle that may be homogeneous in appearance (many membranes and many larvae) or may become differentiated by sclerotization into a hard and unstainable outer portion called exocuticle (which is further subdivisible, as will be treated later), a middle portion called mesocuticle, and an inner portion called endocuticle (layers 4–6). The endocuticle appears histologically similar to the original procuticle. This chitin/protein portion ranges in thickness from about 1 μm to several mm. In membranous regions (between segments or between sclerites) the cuticle may be of similar thickness but it is usually not distinguishable histologically from the endocuticle beneath sclerites. In some cases flexible membranes may have an outer layer of mesocuticle, or very thin exocuticle, or islands of sclerotization (not included in Fig. 1). Beneath the cuticle is the essentially unilaminar epidermis, the thickness of which (squamous, cuboidal, or columnar cells) and the state of the cytoplasmic projections of which may both vary as a function of stage within the molt cycle. In many but not all cuticles, long cytoplasmic projections become embedded in cuticle to give rise to pore canals (shown but not labeled in Fig. 1) which in later stages presumably provide a route from the cells to but not through the epicuticle. Beneath the epidermis is the basal labyrinth and the basement membrane, perhaps secreted by special cells. These separate the epidermal cells from the blood.

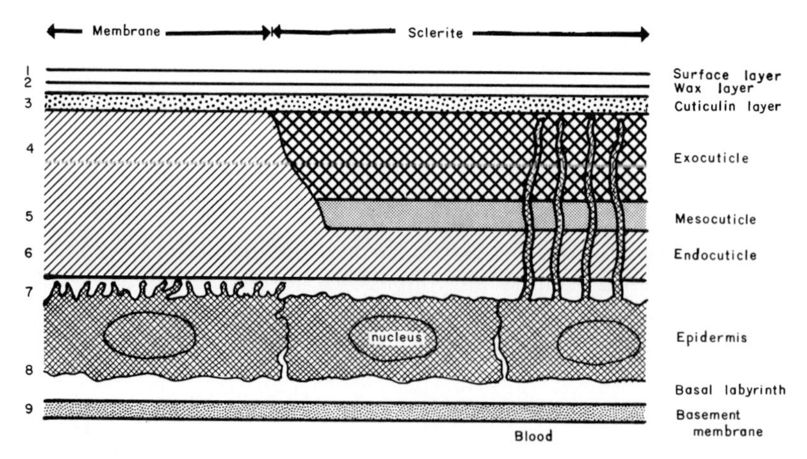

Fig. 1. A diagrammatic section through the integument of an insect showing the main types of cuticle and epidermal cells of various types (or at various stages).

The basement membrane can be penetrated by small tracheae. Muscle fibers also penetrate, extend through the epidermis, and continue as tonofibrillae in the cuticle.

The following treatment will generalize insofar as possible but it will not cover all the recorded components or dwell on the amount of our ignorance. More details can be found in the reviews by Richards (1951) and Hackman (1974). It will, however, mention biological phenomena awaiting chemical explanation. Of necessity, much of the work with structural macromolecules deals with extracts, and these involve uncertainties as to how closely the isolated components portray details in the complex system. Ultrastructure will not be treated but some points will be mentioned because the electron microscope and X-ray diffraction studies provide an intergradation from structure to chemistry. That is so because when structure gets to the molecular level the distinction between anatomy and chemistry disappears.

II. THE COMPONENTS

A. Chitin

Chitin as such is thought not to be a naturally occurring compound; it is a product of the chemical degradation of a naturally occurring glycoprotein. In other words, chitin is the separable prosthetic group from a particular glycoprotein that occurs in a number of animal phyla and also in many fungi. This glycoprotein has sometimes been referred to as "native chitin" in contrast to so-called "pure chitin." But it is not known if all the chitin chains are linked to protein. Chitin usually accounts for 25–40% of the dry weight of insect cuticle but can range from <10% to about 60% [see tables of such values in Jeuniaux (1971) and in Richards (1951)].

Most studies on chitin have been performed with material "purified" with hot dilute alkali (commonly 1 *N* sodium hydroxide at 100° C for several days or to constant weight). Such chitin is colorless, insoluble in water, dilute acids or alkalies, and organic solvents. It is dispersed by concentrated mineral acids with rapid degradation; also by hot concentrated solutions of lithium thiocyanate from which it can be precipitated. Identification of hydrolyzed products show it to be composed of mostly β-polymeric *N*-acetylglucosamine residues where the β-type linkage results in a biose unit (Fig. 2). In modern chemical terminology this is β-(1,4)-linked 2-acetamido-2-deoxy-D-glucose. Actually, enzymatic degradation resulting in obtaining some glucosamine makes it seem probable that some of the units can be present as nonacetylated units, perhaps up to as many as one in six or seven in some cases. Indeed, it has been suggested that the 3.1-nm spacing shown by X-ray diffraction along the chitin chain could be due to every sixth residue being nonacetylated.

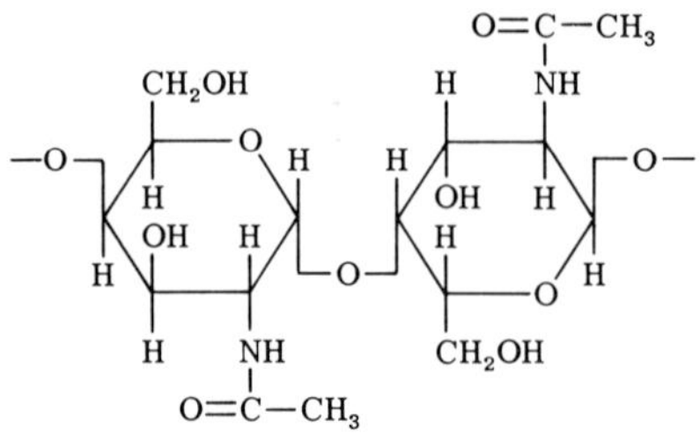

Fig. 2. The repeating chitobiose unit of a chitin chain.

The actual length of chitin chains is quite variable. In the larval cuticle of a blow fly, Strout *et al.* report a continuous range from <50 to >1000 residues (MW 10,000–150,000, 5,000–60,000 daltons) has been recorded (See Hepburn, 1976).

Purified chitin, especially from tendons, is sufficiently crystalline to permit determination of atomic arrangements and spacings by X-ray diffraction methods. Such studies, coupled with studies of hydrogen bonding by absorption of polarized infrared light, led to the three-dimensional atomic diagram proposed by Carlstrøm (1957) (Fig. 3). The location of hydrogen bonds and

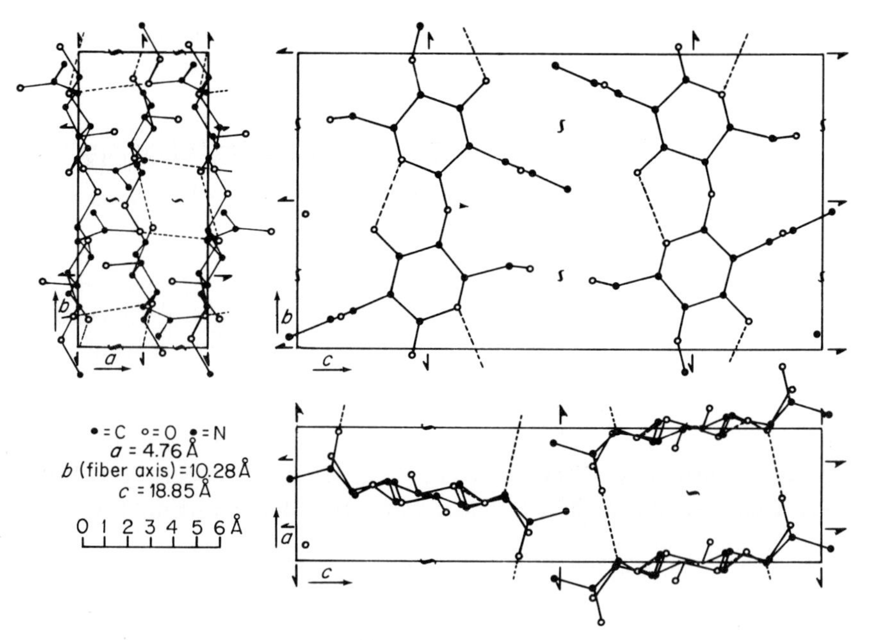

Fig. 3. Three views or projections of the unit cell of purified α-chitin as deduced by Carlstrøm (1957). The three cardinal axes are labeled *a, b,* and *c; b* is the fiber axis. Solid lines represent covalent bonds; dotted lines hydrogen bonds. The angstrom unit, customarily used in crystallography, = 0.1 nm.

covalent bonds implies relatively easy swelling by water molecules in the *c* direction, difficult swelling in the *a* direction, and little or no swelling in the *b* or fiber direction. Note both in this figure and in Fig. 7 the important hydrogen bond between O^{-3} and O^{-5} that stabilizes this polymer. Also note that the pyranose rings are tilted along the fiber axis with the result that two pyranose rings of 0.545 nm length give a repeat distance along the fiber axis of only 1.028 nm. But this is definitely not the same as the chitin lattice in fresh normal cuticle, and unfortunately, such studies on purified chitin cannot show how the other components are inserted in cuticle.

Of more biological interest is the fact that no sample of "purified" chitin has been found to be free of amino acids. Small amounts of any or all of seven to eight amino acids have been reported after hydrolysis by Strout *et al.* (See Hepburn, 1976). These presumably represent the connecting links between chitin and protein chains in cuticle, with the covalent bond involved being so strong that the linking amino acid residues remain with the chitin chains when protein is removed. Actually, determinations are not sufficiently precise to permit one to say that each chitin chain is linked to a protein chain; moreover, it has not yet been possible to dissociate cuticle into a suspension where the chitin is present as a glycoprotein.

In X-ray diffraction patterns (Fig. 4) a perfect crystal shows sharp spots arranged in circles; the same material in powder form shows corresponding concentric circles. When any material is imperfectly crystalline, crystalline in some but not all parts, then there are spots but the spots are diffuse. X-ray diffraction studies of cuticle show a fair degree of regularity to the chitin chains. This is considerably improved by the removal of other components showing both that chitin chains can move somewhat when intervening mate-

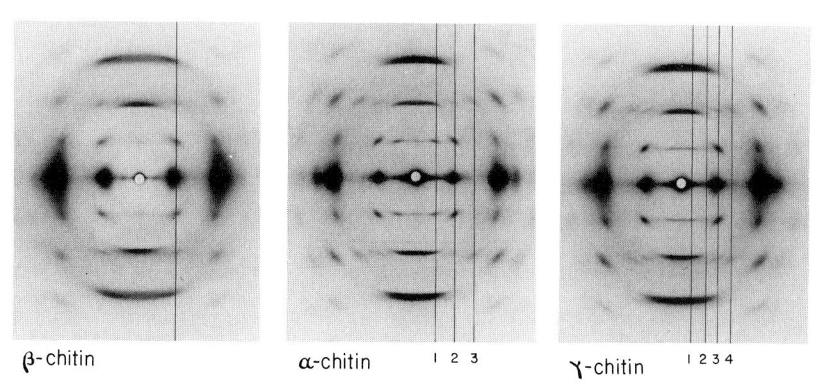

Fig. 4. X-ray diffraction patterns of α, β and γ chitin with the approximate positions of the row lines 1, 2, 3 and 4 indicated (Rudall, 1962).

rial is removed, and that at least some of them then align more closely. Diffuseness of the spots indicates that the degree of crystallinity is seldom high (the oral spines of the chaetognath worm *Sagitta* are an outstanding exception). This semicrystallinity is probably due to the existence of small highly ordered regions (19 or 21 chains) separated by areas of relatively random arrangement. The finest microfibrils of chitin in soft cuticle are about 3 nm in diameter; this is consistent with the suggestion (or led to the suggestion) by Rudall that such microfibrils consist of three by seven chitin chains (see Hepburn). Hard cuticle has microfibrils about twice this diameter (Fig. 5). It is thought that the chitin microfibrils do not react with the heavy metals used as stains for electron microscopy (EM) but that the associated proteins do (it has indeed been demonstrated that purified chitin does not stain with the usual EM stains). In the electron micrographs, therefore, it is

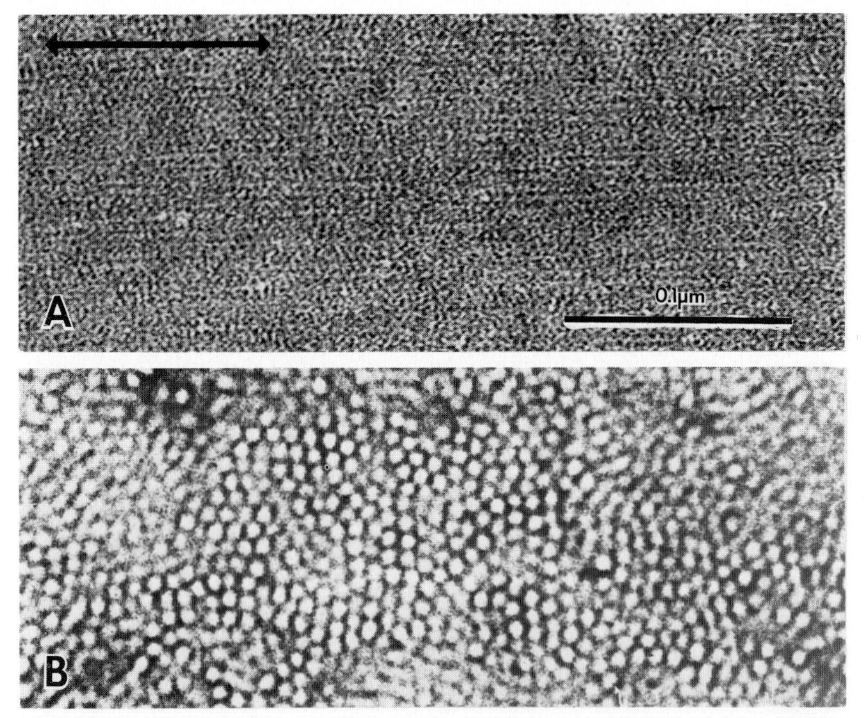

Fig. 5. High-resolution electron micrographs of insect cuticle (Rudall, 1967). (A) Stretched larval cuticle of a blow fly (*Calliphora*) stained with osmic acid vapor at 92% RH for 12 days and sectioned perpendicular to the direction of stretching. Shows microfibrils of about 2.5 nm diameter in random and layered array, the layers having a spacing of 4.5 nm. Layers seen best when viewed at near grazing incidence in direction of arrow. Bar = 100 nm.

(B) Cross section of hard ovipositor of an ichneumonid wasp (*Megarhyssa nortoni*) stained with aqueous uranyl acetate. The large microfibrils are in hexagonal array in some areas but more frequently in irregular rows. Resolution in this picture is to <1 nm.

thought that the clear, seemingly empty areas represent bundles of chitin chains in cross section, and that the intervening dark areas represent protein.

X-ray diffraction studies have also resulted in the recognition of three distinct crystallographic types of purified chitin (Fig. 4). These types all involve the same types of chitin chains; it is only a matter of how the chains aggregate and where water molecules are interpolated. In chronological order of their description, these have been labeled α, β, and γ (Fig. 6), α-chitin being the stable form into which β and γ can be converted by treatment with acid or lithium thiocyanate. α-Chitin is also the form occurring in insect cuticles, whereas β- and γ-chitins in insects are limited to peritrophic membranes of certain species. All three types are normally associated with protein. But it is not yet known whether all chitin chains are bound to protein. The electron micrographs (Fig. 5) could be interpreted as suggesting that only chitin chains on the periphery of a microfibril can be bound to protein. If this is true, then either the association of chitin and protein occurs after the formation of chitin microfibrils or there would be a modification and partial separation of a preceding glycoprotein. The difficulty of solubilizing chitin interferes with attempts to determine when linkages are formed between chitin and protein in soft cuticles.

In connection with the differences in the above three types, and the conversion of β- and γ-chitins to the α form, Rudall and Kenchington (1973) have presented an atomic diagram pointing out the bonds at which rotation is permissible for changes in conformation (Fig. 7).

Jeuniaux (1963, 1971) has elaborately developed the concepts of "free" and "bound" chitin. By this he means that while purified chitinase (usually from a microbial source) will digest purified chitin, it will digest only a portion of the chitin from whole cuticles, that which is attacked by the enzyme is called "free," that which is not is called "bound." Percentages vary greatly from one cuticle to another. It is not known whether free chitin

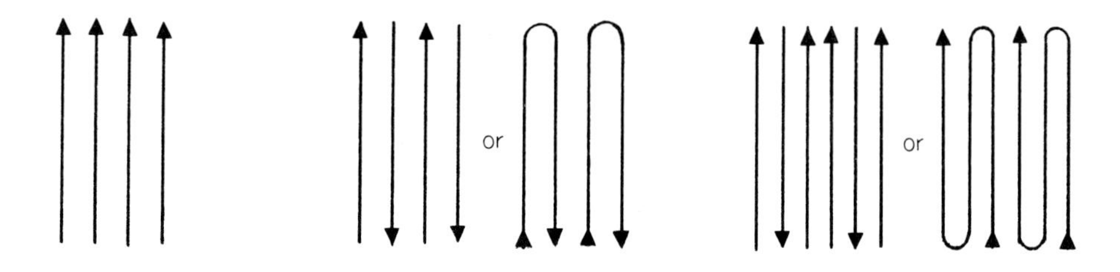

Fig. 6. Rudall's suggestions for the arrangement of chitin chains in the unit cells of the three crystallographic types of chitin.

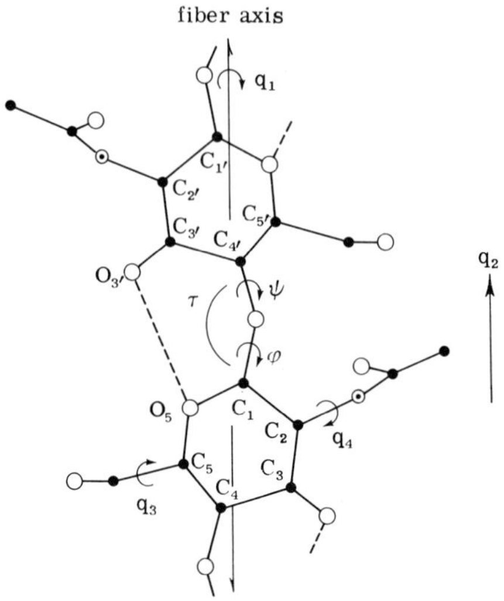

Fig. 7. Diagram of linked *N*-acetylglucosamine residues showing the angles ϕ and ψ, rotation of which leads to the main changes in conformation. ● = carbon, ○ = oxygen, ⊙ = nitrogen. (From Rudall and Kenchington, 1973.)

is truly free or simply available to the enzyme. However, it does not seem possible at the present time to reconcile the electron micrographs and their interpretation (Fig. 5) with the data on free and bound chitins.

In this connection, digestive juices that decompose cuticles in nature all contain at least the hydrolytic enzymes chitinase, chitobiase, and proteinases. The same must be true for molting fluids, but there is a difference between the microbial enzymes and those of molting fluids because the entire cuticle is broken down in nature but only nonsclerotized cuticle by molting fluid. It follows that insect exuviae consist of epicuticle plus the sclerotized portions of the chitin/protein cuticle. [Those conversant with French would do well to consult Jeuniaux's (1963) monograph.]

The pathway for synthesis of chitin has been partly worked out in the past decade. Glucose, usually derived from glycogen via trehalose, is phosphorylated and then aminated (perhaps with glutamine) and acetylated (with acetyl-CoA). This monomer is activated by attachment to uridine diphosphate, and the resulting compound (UDPAG) in the presence of chitin synthetase (chitin-UDP acetylglucosaminyltransferase) adds the monomer to the end of a chitin chain. Presumably the same enzyme can initiate chitin chains too but at a slower rate, and presumably it can also start with *N*-acetylglucosamine from the digestion of endocuticle of the preceding instar. Yet unknown are questions such as: "What is secreted? What changes occur after secretion? How do the chains become glycoprotein?"

The situation is further complicated by recent reports that compounds which interfere with chitin synthetase (polyoxins and the substituted ureas known as TH-6040, DU-19111) may either block cuticle deposition or result in a normal looking cuticle that is said by Muzzarelli to be somewhat deficient in or actually devoid of chitin (see Hepburn, 1976). The situation is certainly more complicated because treated insects may not only have normal-appearing cuticles and peritrophic membranes in sections examined by electron microscopy; they may also be strongly positive to the van Wisselingh test for chitin (see below). However, further studies with these compounds may have far-reaching effects on our understanding of both cuticle ultrastructure and the secretion of cuticular components.

Chitins can be easily deacetylated to give a polymeric glucosamine termed chitosan. This is usually accomplished with hot concentrated KOH solutions. Chitosans are of interest in commerce because the exposed amine groups give them desirable properties not possessed by cellulose. Also chitosans react with iodine to give a characteristic violet color that is the most commonly used qualitative test for the presence of chitin (the van Wisselingh test; see Richards, 1951, for details). The insolubility of chitin and chitosan is such that all except extremely delicate cuticular structures retain their microscopic structure through the purification and conversion process.

Quantitative estimates of chitin are made by gravimetric determinations of insoluble residues assumed to be pure chitin, or by estimates of acetylglucosamine or glucosamine following purification and hydrolysis. Estimates of acetylglucosamine must be preceded by purification because this monomer is found in many compounds (e.g., casein, egg albumen). Use of the supposedly sensitive fluorescent chitinase technique requires highly purified chitinase which is not commercially available (see Richards, 1978). Also, the method has not been adequately evaluated (e.g., is it valid with both free and bound chitin) and has not been used to date except as a qualitative test.

The literature on chitin has been extensively reviewed by Jeuniaux (1963, 1971) and more recently by Muzzarelli (1977).

B. Proteins

Proteins usually account for >50% of the dry weight of insect cuticles and are certainly involved in determining physical properties, but we have no really detailed information such as amino acid sequences or even what percentage of the protein is linked to chitin. In 1941, Trim proposed the term arthropodin for the soluble proteins and the term sclerotin for the insoluble ones. These terms are of dubious value since we now know that there are many different proteins and at least several ways of making them insoluble.

Gel electrophoresis has resulted in separating the proteins into numerous bands, in the extreme case >50 bands from the cuticle of a single beetle larva. It has been suggested that this large number of proteins may represent protein polymorphism.

In keeping with one of the current vogues in biochemistry, Hackman (see Hepburn, 1976) suggests that if one groups the amino acids of soft, nonspecialized cuticles according to the second letter of their genetic codes one finds what he considers to be fair agreement. The range of values tabulated as acceptable by Hackman are uracil 178–248, cytosine 282–362, adenine 285–338, and guanine 123–179. To accept this suggestion one has to assume that the considerable discrepancies are due to differences in preparation of samples and to analytical errors, and that exceptions (values not within the above range values) are all specialized cuticles. With no knowledge of amino acid sequences this is gratuitous guessing.

Structural macromolecules are in general not soluble, at least not without degradation. The proteins of cuticle are exceptional in that they are water-soluble until sclerotized. One can make a series of extractions using what has come to be known as Hackman's (1958) series of solvents. Cuticle from which the cells have been swabbed are extracted successively with (1) cold water at pH 7.0 (48 hr), (2) cold 0.167 *M* Na_2SO_4 at pH 7.0 (48 hr), (3) cold 7 *M* urea at pH 7.0 (48 hr), (4) cold 0.01 *N* NaOH (5 hr), and finally (5) 1 *N* NaOH at 50–60° C (5 hr). It is thought that the first extraction removes soluble unbound compounds, the second removes compounds bound by weak forces such as van der Waals forces, the third removes compounds that are hydrogen bonded, the fourth removes compounds held by electrovalent bonding, and the fifth removes compounds held by strong covalent bonds (leaving the insoluble chitin as a purified residue). For the soft cuticle of a beetle larva (*Agrianome*) Hackman (1974) reports that these fractions represent respectively 14, 2, 25, 3, and 56% of the protein.

The above extractions primarily reflect the nature of the binding forces, but gel electrophoresis shows that the several fractions are qualitatively different. It follows that some of the protein species are more likely to be bound in one particular manner.

The proteins have low isoelectric points (pH 3–6) in keeping with their high percentage of dicarboxylic amino acids. The extractable ones are also of relatively low molecular weight (10,000–80,000 daltons with emphasis on ones <20,000), with the high values in some cases seeming to be and in other cases seeming not to be simple multiples of some basic polypeptide unit. But, of course, the introduction of cross bonds during sclerotization makes such values meaningless. In the extreme case, an entire sclerite could be viewed as one molecule.

Numerous amino acid analyses have been reported on hydrolyzed cuticles or hydrolyzed extracts. Tryptophan may be present in small amount but is seldom reported because it is destroyed by the acid hydrolysis usually used to prepare amino acid digests from protein. A representative few sets of values are presented in Table I. These do show a general similarity but almost random variation. Until more is known about the proteins from which these came, little can be said. Among the things that can be said, the epidermal cells may be programmed to secrete different proteins at different times as is elegantly shown by the alternating layers of chitin/protein and resilin in some structures (and at least in *Limulus* the amino acids of epicuticle have been shown to be distinctly different from those of underlying chitin/protein membrane). It can also be said that there is little if any change in amino acid percentages when a cuticle becomes sclerotized.

A seeming anomaly is the finding of β-alanine in the hydrolysates of some sclerotized cuticles. It seems most reasonable to accept the suggestion that this is a degradation product, likely arising from partial decarboxylation of aspartic acid. Note that β-alanine was recorded from the puparium of *Lucilia* (Table I) but not from the larval cuticle from which the puparium is formed. The most interesting point emerging from the amino acid analyses is the high percentage of units with bulky side chains. Values are from 20 to 60% (average >40%) in contrast to e.g., silk fibroin which has <15%. This must mean that relatively poor packing of chains is possible. Interference with good packing may be worse with proline where the amine is part of a heterocyclic ring than with the phenyl, etc., side chains. Values for tyrosine are notably high, especially in epicuticle, and this is presumably related to sclerotization, but it is still not clear whether sclerotization involves only derivates of free tyrosine or whether tyrosine residues in a protein chain can also participate. Why the high values for proline and dicarboxylic acids is unknown.

One of the most amazing proteins known is resilin—an almost perfect rubber found in elastic ligaments involved in flight and jumping (Andersen, 1971). This protein may be found in reasonably pure form or intercalated with chitin/protein layers. Unlike the protein of soft cuticle it is not readily soluble but is easily hydrolyzed; it does not become sclerotized or otherwise changed after formation; the amino acid values are different from those of cuticular proteins (Table I); and there are differences in amino acid composition for resilins from different insects and even from different parts of a single insect. In electron micrographs resilin shows no internal structure. It can commonly be located by its strong blue fluorescence. The characteristic physical properties of resilin are due to randomly coiled peptide chains cross-linked by a small percentage of dityrosine and trityrosine units (Fig. 8) at approximately every 40–50 residues (spacing corresponding to an intervening molecular weight of 3200). Calculations show that a significantly greater frequency of cross-links would give a rigid structure. Clearly, the study of cuticular proteins is in its infancy.

TABLE I
The Amino Acid Composition of Proteins from Certain Cuticles[a]

	Periplaneta adult[b]	*Schistocerca* adult[c]	*Anabrus* exuviae[d]	*Rhodnius* larva[e]	*Tenebrio* adult[f]	*Bombyx* larva[b]	*Lucilia* larva[b]	*Lucilia* puparium[b]	Resilin[g]
Simple									
Glycine	10.1	8.0	10.2	16.2	25.3	11.2	14.1	12.7	37.9
Alanine	24.3	35.1	15.8	20.6	12.7	12.4	13.7	7.7	10.9
β-alanine	2.5	—	—	—	—	—	—	8.9	—
Serine	5.1	4.7	—	6.5	4.4	7.5	1.4	7.6	7.9
Threonine	3.3	2.4	5.0	8.2	3.8	6.5	5.8	4.0	3.0
Valine	11.8	8.6	8.2	14.3	11.2	8.2	8.3	5.6	2.8
Leucine	4.1	4.6	7.0	6.7	9.0	3.5	4.4	4.0	2.3
Isoleucine	2.0	3.8	4.0	0.4	5.4	4.3	2.3	1.8	1.6
Phenyl									
Tyrosine	7.7	6.3	12.2	7.9	1.2	7.9	4.4	3.7	2.7[h]
Phenylalanine	1.6	0.9	3.2	0.8	0.9	2.4	4.3	3.3	2.5
Heterocyclic									
Histidine	3.3	2.2	2.6	2.9	1.9	1.3	3.2	4.3	0.9
Tryptophan	—	—	0.5	—	—	—	—	—	0
Proline	9.8	11.1	12.8	8.0	8.8	9.9	9.4	8.8	7.6
Dicarboxylic									
Aspartic acid	5.1	4.0	7.6	3.0	3.9	3.9	9.4	8.2	10.1
Glutamic acid	5.1	3.2	6.8	2.0	3.9	9.4	12.4	9.5	4.6
Diamino									
Arginine	1.9	3.7	4.4	1.9	2.9	2.4	1.7	5.0	3.5
Lysine	1.9	1.7	4.0	0.7	4.5	3.5	4.4	3.6	0.5
Sulfur-containing									
Methionine	—	0	0.6	—	—	0.2	0.3	0.2	0
Cystine	0.6	0	4.6	—	—	0.6	0.5	1.3	0

[a]Values are in % of total amino acids.
[b]Hackman and Goldberg (1971).
[c]Andersen (1973).
[d]Johnson *et al.* (1952).
[e]Hackman (1975).
[f]Andersen *et al.* (1973).
[g]Andersen (1966).
[h]Plus 0.8 dityrosine and 0.3 trityrosine.

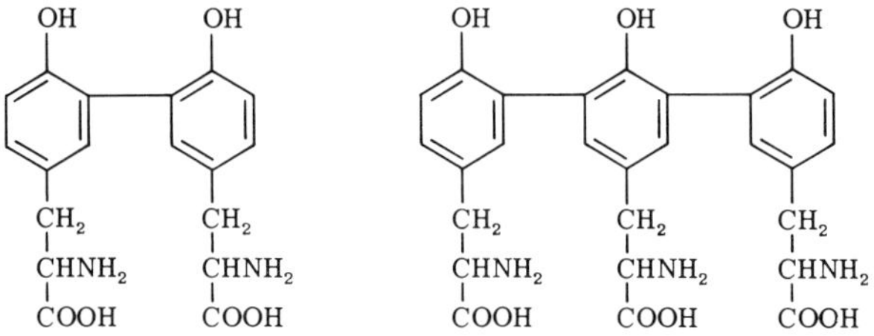

OH OH OH OH OH

CH_2 CH_2 CH_2 CH_2 CH_2

$CHNH_2$ $CHNH_2$ $CHNH_2$ $CHNH_2$ $CHNH_2$

COOH COOH COOH COOH COOH

Fig. 8. The bonding of tyrosine units for dityrosine and trityrosine.

C. Chitin/Protein Complexes

With all the discussion of chitin and protein forming a glycoprotein in cuticle it seems surprising that so little can be said about chitin/protein complexes. This reflects the difficulty of the problem and is due to our inability to extract an undegraded glycoprotein for analysis, our uncertainty about the nature of any chitin/protein bonds, and our ignorance of details about the protein chains. Some solubility data suggest that there may be both covalent and hydrogen bonding between chitin and protein. When there are so many bulky side groups on the chains of a fibrous protein the problem of fitting these together must be complicated.

While there is detailed knowledge of the configuration of chitin in purified chitin (Fig. 3), we have no corresponding information on the proteins or on the chitin/protein complexes thought to exist. However, there are considerable data on the native conformation of chitin/protein complexes primarily due to the X-ray diffraction and electron microscope work of Rudall (1962, 1967). This information is primarily recognition that something repeats at 3.1 nm or at 4.1 nm or at some other distance. An attempt is then made to rationalize the data in terms of known dimensions of units in chitin chains and the molecular weights of associated proteins. A start has been made, most significantly in identification of bonds at which allowable rotation would result in conformational changes (Fig. 7).

Important for future studies is Rudall's discovery that cleaning cuticle in water or alcohol alters the native conformation. Cleaning must be done by swabbing in the blood of the insect. While the purified chitin lattice (Fig. 3) is stable, that found in fresh, normal, soft cuticle is very unstable. It is rapidly changed to the pure chitin lattice by mild heating or by treatment with dilute acidic or alkaline solutions or even by water. However, once the normal lattice has been stabilized by sclerotization it is no longer readily altered.

D. Lipids

Cuticular lipids are extremely heterogeneous. But, then, lipids are a widely disparate group of substances lumped together for convenience because of their solubility characteristics. Electrophoretically one can separate cuticular proteins into dozens of fractions but these seem to be rather similar. The lipids show no corresponding degree of similarity. There are long-chain hydrocarbons (paraffins), branched or unbranched, saturated or unsaturated, with chain lengths of C_{25} to C_{43}; there are corresponding long-chain normal alcohols and sometimes aldehydes; there are fatty acids, saturated or unsaturated, with chain lengths of C_{13} to C_{30}; and commonly important esters of saturated acids with saturated alcohols that make up the commonest waxes. In some cases there are significant percentages of sterols and/or sterol esters. In a few cases there are significant amounts of glycerides. Phospholipids have occasionally been recorded but only in small amounts (see Table II and Fig. 9). Some of these are chemical series, the lower members of which are water soluble (e.g., the alcohols), but when the chain length exceeds C_{8-10} they cease being water soluble and become classed as lipids because they are soluble in lipid solvents!

Lipids ordinarily amount to only a few percent of the dry weight of cuticles, but in very thin cuticles the percentage may be considerably higher since such cuticles have a usual epicuticle. And, of course, specialized wax-secreting areas can give high values.

Cuticular lipids that could be obtained in considerable quantity either from exuviae or waxy secretions such as beeswax and Chermes wax were par-

TABLE II

The Readily Extractable Lipids of Certain Cuticles[a]

Type	*Lucilia* puparia free[b]	*Lucilia* puparia bound[b]	*Lucilia* adult ♀ [c]	*Periplaneta americana* larva[d]	*Pteronarcys* larva[e]	*Pteronarcys* adult[f]	*Tenebrio molitor* larva[g]
Hydrocarbons	55.0	35.7	65.7	75–77	3	12	10
Fatty acids	14.9	21.5	5.4	7–11	12	49	5
Alcohols	—	—	—	—	—	—	55
Alkyl esters	0.2	2.3	16.1	3–5	1	4	13
Aldehydes	—	0.5	—	8–9	—	—	—
Sterols	1.8	3.6	—	<1	1	18	—
Sterol esters	7.3	0.7	—	—	—	—	—
Glycerides	2.4	0.3	+	—	78	7	—
Phospholipids	—	0.8	—	—	—	—	2

[a]Values supposedly % total lipid by weight.
[b]Gilby and McKellar (1970).
[c]Goodrich (1970).
[d]Beatty and Gilby (1969).
[e]Arnold *et al.* (1969).
[f]Bursell and Clements (1967).

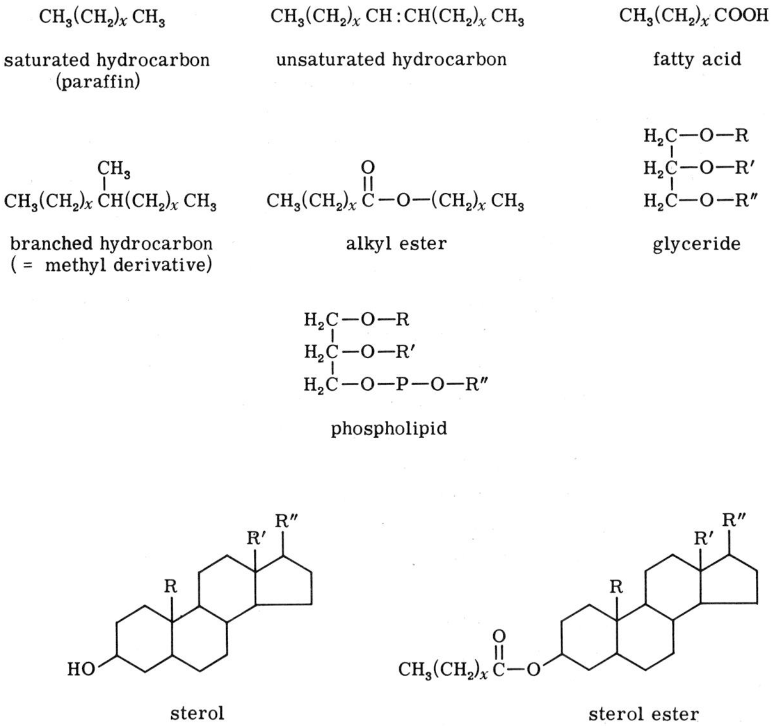

$CH_3(CH_2)_x CH_3$

saturated hydrocarbon (paraffin)

$CH_3(CH_2)_x CH{:}CH(CH_2)_x CH_3$

unsaturated hydrocarbon

$CH_3(CH_2)_x COOH$

fatty acid

$CH_3(CH_2)_x CH(CH_3)(CH_2)_x CH_3$

branched hydrocarbon (= methyl derivative)

$CH_3(CH_2)_x C(=O)—O—(CH_2)_x CH_3$

alkyl ester

$H_2C—O—R$
$H_2C—O—R'$
$H_2C—O—R''$

glyceride

$H_2C—O—R$
$H_2C—O—R'$
$H_2C—O—P—O—R''$

phospholipid

HO, R, R', R''

sterol

$CH_3(CH_2)_x C(=O)—O$, R, R', R''

sterol ester

Fig. 9. Group formulae for cuticular lipids. x, various finite numbers between 10 and 50: R, fatty acids in glycerides, hydrocarbons in sterols, and some other organic group at the phosphate link in phospholipids.

tially fractionated and some components identified as early as the 1920's. With the development of thin-layer chromatography, column chromatography, gas chromatography, mass spectroscopy, and infrared spectroscopy, the study of lipids was revolutionized, and it has become relatively easy to analyze small samples and identify minor components in mixtures.

Differences are common among species, and this can be among closely related species. For instance, Jackson (1972) reports that there is considerable difference within the hydrocarbon fraction among various species of the cockroach genus *Periplaneta*. *Periplaneta americana* has primarily *cis,cis*-6,7-heptacosadiene (71% of the hydrocarbon fraction and >50% of all the cuticular lipid); *P. japonica* lacks the above hydrocarbon and has the greatest amount of *cis*-9-nonacosene (38%) which is not found in *P. americana*; and *P. brunnea*, *P. australasiae*, and *P. fuliginosa* lack both of the above and have largely a mixture of 13-methylpentacosane plus *n*-tricosane and its 3- and 11-methyl derivatives. In contrast, cockroaches of the genera *Leucophaea* and *Blatta* have primarily *n*-heptacosane and various methyl derivatives. It does not seem credible to think that such differences are critical to the various species. Some form of hydrocarbon seems needed in the cuticle but any of many will serve—and in some cases even the hydrocarbon fraction is largely replaced by other types of lipids.

In view of the diversity of lipids from one group of insects to another (Table II), it should not be surprising that waxes secreted in bulk quantities may be considerably different from ones found as small percentage components in cuticle. The best known of such waxes is beeswax, which is secreted from certain specialized areas of the honeybee's abdomen (it does not seem to be known whether this is identical with wax in cuticle of other parts of the body). This consists mainly of alkyl esters of monocarboxylic acids (71–72%) but has also free alcohols (1–2%), free fatty acids (13–15%), free hydrocarbons (10–12%), cholesteryl esters ($<1\%$) and some impurities (2%). Slightly more than 50% of the total is myricyl palmitate (triacontanol hexadecanoate) with the formula $CH_3(CH_2)_{29}O_2C(CH_2)_{14}CH_3$.

Differences are commonly great between different developmental stages of single species, especially when larva and adult occupy different habitats (Table II).

Lipids are commonly referred to as "free" and "bound." Different degrees of bonding are involved. For the data presented in Table II, the difference between free and bound lipid is based on whether the solvent used for extraction contained a few percent of alcohol. For cuticles, as well as for tissues in general, more is obtained if the benzene or ether or other lipid solvent contains 2–5% of methanol or ethanol. The alcohol frees lightly bound lipids. But there are also more strongly bound lipids, commonly referred to as lipoproteins, which are not easily extracted and hence are not included in Table II. And it is abundantly clear from the reports of numerous workers that firmly bound lipids are present in many cuticles. These bound lipids may be unmasked by treatment with sodium hypochlorite. In extreme cases documentation depends on the release of Sudan-stainable droplets after violent chemical action such as heating in concentrated nitric acid saturated with potassium chlorate. It seems likely that these firmly bonded lipoproteins can involve several different types of lipids. Whether the proteins to which they are bonded are just "ordinary" cuticular proteins or special proteins is unknown.

The question of localization of the various lipids is an extremely difficult one, partly because localization involves the use of histochemical methods that seldom distinguish between the various lipid types. Histochemical methods do definitely show that while much lipid is in the epicuticle some is in the underlying chitin/protein cuticle. This difficult subject has been recently reviewed by Wigglesworth who has contributed much to it (see Hepburn, 1976).

Another bound lipid that is secreted by the integument of certain insects is shellac. The shellac of commerce (derived from the Indian coccid *Laccifer lacca*) is an alcohol-soluble complex lipid–carbohydrate mixture involving polyhydroxypolycarboxylic esters with laccose, but "native shellac" has in addition, some benzene-soluble wax similar to beeswax. It has been suggested that the so-called cement layer (surface layer of Fig. 1) of the epicuticle may resemble shellac.

The little information available on biosynthetic pathways for cuticular lipids suggests that both hydrocarbons and fatty acids are synthesized from acetate. However, hydrocarbons may also arise from decarboxylation of fatty acids. Dietary *n*-alkanes may be incorporated into cuticle in some species but synthesized by other species. It seems likely that biosynthetic pathways for lipids will differ from one group of insects to another, and in some cases even from larva to adult.

The skin lipids of mammals are different from those of internal tissues. Similarly, in insects, cuticular lipids differ from storage and cell-membrane lipids.

E. Pigments

Insects may be highly colored. The most brilliant colors are physical colors that depend on the microscopic construction of a reflecting part (Richards, 1951). Such physical colors may be combined with pigmentary colors or exist without pigment.

Pigments other than the blacks and browns due to sclerotization and melanization (see Section III) are seldom found in cuticles. Such pigments occur as cytoplasmic granules in the epidermal cells or even in tissues beneath the epidermis. Soluble pigments may be in the blood. Such pigments are treated in Chapter 6.

In this connection it might be well to remark that brown sclerotized cuticle may be quite transparent. In the American cockroach the abdominal tergites are uniformly brown but transparent. One readily observes the heart beating by looking through the cuticle of the living insect but the gut cannot be seen because of the interposition of opaque adipose tissue. In contrast, the orange and black pattern of the milkweed bug (*Oncopeltus fasciatus*) is due to corresponding localization of opaque black cuticle separated by transparent cuticle that allows the underlying pteridines to show through.

F. Inorganic Components

Cleaned and washed insect cuticles usually yield a very small percentage of ash on incineration. The value is commonly $<1\%$ of the cuticle dry weight but in a few cases values of several percent have been reported (calcified puparia have high values). Chitin purified in plastic vessels shows no ash but chitin purified in glass vessels may yield several percent.

Carefully cleaned cuticles of blow fly larvae (*Sarcophaga bullata*) exhaustively washed in distilled water yielded 0.365% ash (Richards, 1956). Magnesium was the commonest element, closely followed by potassium; sodium was only half as high. These cations were balanced by approximately equal amounts of phosphate and carbonate. Twenty other elements were identified in small or trace amounts by emission spectroscopy. Four points seem of potential interest: the high percent of magnesium; the lack of similarity in amounts of the various elements in cuticle and blood; the localization of most of the iron in the epicuticle as shown by parallel microincineration studies, and the loss (extrusion) of almost all ash when the larval cuticle became the sclerotized puparium.

Calcified cuticles, characteristic of decapod Crustacea, are rarely found in insects. The best known cases are the puparia of *Musca autumnalis* and *M. fergusoni* where calcification replaces sclerotization and the ash content can be 60–80% of the dry weight. For *M. autumnalis* the material was suggested to be primarily calcium carbonate (as in the calcification of higher Crustacea) but more extensive chemical determinations for *M. fergusoni* by Gilby and McKellar (1976) showed primarily a mixture of calcium and magnesium orthophosphates in a 2:1 molecular ratio. A layer of prismatic crystals of calcium oxalate, usually located just beneath the epicuticle, has been reported for larvae of a dozen lepidopterans and a single sawfly out of 147 species in 11 orders examined. In a few cases calcium deposits are reported as being on the cuticle without really being a part of the cuticle (e.g., deposits on the cuticle surface of aquatic larvae in carbonate-rich waters).

G. Enzymes

Numerous cuticular components are so large that it seems most likely they would be synthesized *in situ*. Sclerotization involves small but activated molecules of short life expectancy. It should not be surprising then that numerous enzyme activities are exhibited within cuticles. These enzymes must act on substrate molecules passed into the cuticle through the epidermis whether the substrates themselves are derived from activity of the epidermal cells, oenocytes, or other tissues whose products are transported by the blood.

On some of the known or presumed cuticular enzymes there is no significant information. This includes enzymes for protein synthesis (both fibrous proteins and enzyme proteins) and for the conjugation of lipid or chitin to proteins to form lipoproteins and glycoproteins; possibly also, enzymes for lipid synthesis. Also omitted at this point will be the chitinases, chitobiases, and proteases of the exuvial fluid, and chitin synthetase (see

Section II, A). Only the enzyme systems whose activities will be treated in the section on sclerotization will be presented.

The number of metabolic pathways entered into by tyrosine is considerable, far more than the three known for sclerotization. Tyrosinases are found throughout the plant and animal kingdoms, and the literature on them is tremendous. Even with insect material the literature goes back to about 1910. Apparently there are a number of tyrosinases in insects because that found in insect blood is inhibited by thiourea whereas that (or those) in cuticle is not.

Tyrosine, which would be better called *p*-hydroxyphenylalanine, is first changed to dihydroxyphenylalanine, commonly called dopa for short. What happens next depends on which enzyme is involved. If the enzyme dopa-decarboxylase is present, dopamine is formed (Fig. 10). This can be further changed to *N*-acetyldopamine, a change that may require a yet undescribed enzyme or perhaps just acetyl-CoA.

The *N*-acetyldopamine may be acted upon by a yet unnamed enzyme which activates the β-carbon of the side chain to attach to a protein (Fig. 11). This enzyme is always firmly bound in the cuticle framework. Before sclerotization has taken place, it can be solubilized only by a heroic procedure involving a special trypsin digestion; after sclerotization it cannot be removed, but it is not inactivated by sclerotization. This gives rise to the interesting phenomenon of a structural protein that is also an active enzyme. Andersen (1976) reports that this enzyme is thermostable to 70° C, and loses only half its activity after 5 min at 80° C (see Hepburn, 1976). It has a pH optimum of 5.0–5.5 but is stable and active over the range 4–12. It is inhibited by tetranitromethane, which nitrates tyrosine residues, by NaCN, NaF, sodium azide, and diphenols. Those diphenols that are the best substrates for polyphenol oxidases are the best inhibitors for this enzyme.

The *N*-acetyldopamine may also be acted upon by a polyphenol oxidase to

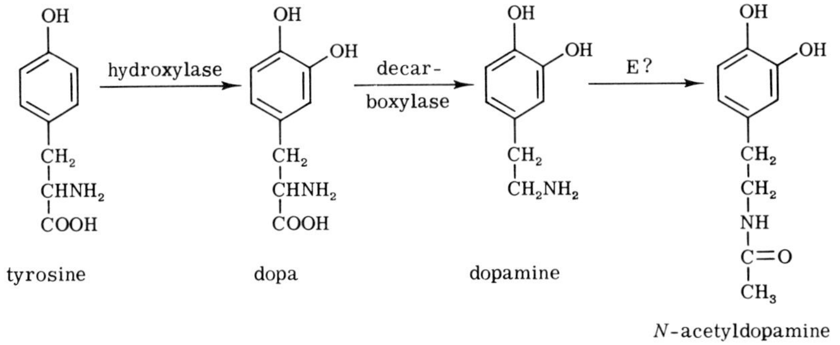

Fig. 10. Steps in the conversion of tyrosine to *N*-acetyldopamine.

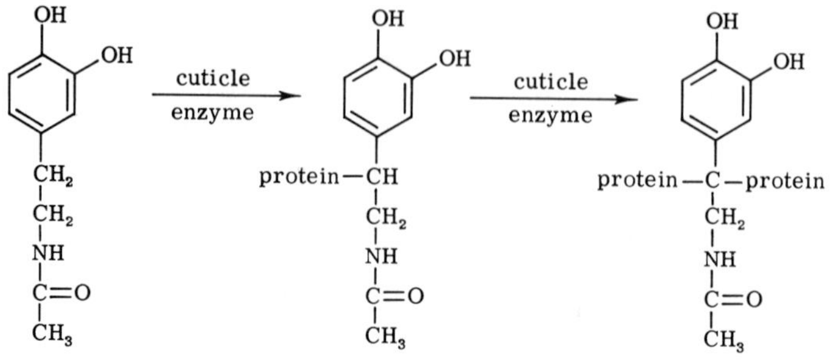

Fig. 11. Scheme for β-sclerotization.

give the quinone derivative that leads to quinone tanning (Fig. 12). The change to quinone is enzyme mediated; subsequent attachment of protein at an activated site on the ring is nonenzymatic, spontaneous,and fast. In some cases it is reported that the polyphenol oxidases are soluble, in other cases insoluble, but the soluble ones can no longer be removed after the onset of sclerotization. It follows that, either originally or during sclerotization, this enzyme becomes bound into the cuticular framework; however, it too is not inactivated by this binding.

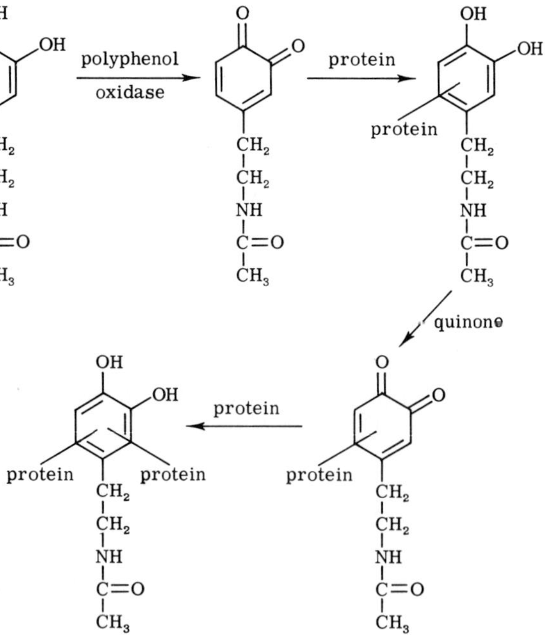

Fig. 12. Scheme for quinone sclerotization (tanning).

N-acetyldopamine (or perhaps dopa itself) may also be acted on by a polyphenol oxidase to give quinone derivatives which, following either deacetylation or decarboxylation, nonenzymatically condense to give an indole structure that can polymerize in various ways to form melanins (Fig. 13). Conceivably dopamine could be caused to condense into an indole structure. Melanin chemistry is so difficult; it is full of uncertainties.

Despite the recorded differences and variations in amount of activity in different regions, Andersen concludes it is not yet proved that the enzymes for β-sclerotization and for quinone tanning are different (for discussion, see Hepburn, 1976).

H. Other Components

Other components have been reported for some cuticles. These include small amounts of various sugars, the role of which is not known (mostly mannose but smaller amounts of arabinose, galactose, glucose, xylose, and galactosamine).

About a dozen different ortho dihydric phenols (listed in Hackman, 1974) have been identified from one cuticle or another, but it is not known how many of these are natural components. The important ones known to be involved in sclerotization will be discussed in Section III. It has been suggested that some of the diphenols may help prevent autoxidation of cuticular lipids.

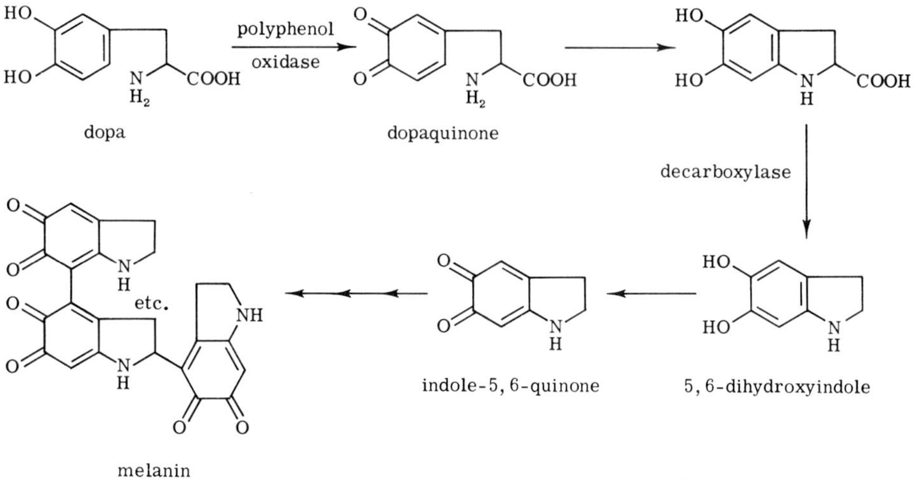

HO HO N H₂ COOH dopa — polyphenol oxidase → O O N H₂ COOH dopaquinone → HO HO N H COOH — decarboxylase → HO HO N H 5, 6-dihydroxyindole → O O N H indole-5, 6-quinone → O O N H O O N H etc. NH O O melanin

Fig. 13. Scheme for formation of indole precursors of melanin. The same results could presumably be obtained starting with *N*-acetyldopamine and having deacetylation instead of decarboxylation. Likely there are also other biosynthetic pathways leading to melanin.

III. IRREVERSIBLE CHANGES: SCLEROTIZATION AND MELANIZATION

In all older books on entomology, sclerotization was defined as a process by which the cuticle becomes hard and dark, implying either a single chemical change or perhaps two simultaneous changes. Yet it must have been apparent even then that there were many cases of hard but light-colored cuticles. In fact, if allowance is made for thinness, all wing membranes have to be classed as hard. But it was not until 1957 that Malek reported, on the basis of histological data, that hardening normally precedes darkening and therefore must be distinct from it. This led to the suggestion 10 years later that sclerotization involves a series of steps: first the procuticle which stains with aniline blue is changed to staining with acid fuchsin (becomes mesocuticle) and then to becoming unstainable (sclerotized = exocuticle), first hard but uncolored, then usually darkened, and finally sometimes blackened or melanized (see Richards, 1967). Recent chemical studies give credibility to the idea that at least three kinds of exocuticle exist but do not support the idea that they represent a developmental sequence.

Within recent years two groups of chemists, Karlson and co-workers in Germany and Andersen and co-workers in Denmark, have made great strides in determining the nature of the cross-links formed. It has been thought ever since Pryor's pioneering work in 1940 that tyrosine derivatives are involved. Indeed, numerous ones have since been identified in cuticle extracts. Karlson *et al.* (1962) reported that the important derivative is *N*-acetyldopamine derived as shown in Fig. 10. Radioactive samples of any of the above injected into insects at the correct stage become incorporated into the insoluble portion of a sclerite. This is true irrespective of whether the C^{14} label is in the ring, the side chain, or the acetyl group. Therefore the *N*-acetyldopamine molecule is used for sclerotization without any gross change. Karlson and Sekeris reported that the signal directing a shift in tyrosine catabolism to formation of *N*-acetyldopamine is supplied by the hormone β-ecdysone, which stimulates production of appropriate mRNA to induce the *de novo* synthesis of dopa-decarboxylase (see Hepburn, 1976). The hormone bursicon is also needed, perhaps, it has been suggested, to stimulate the uptake of tyrosine by the epidermal cells.

Since dopamine is a pharmacologically active substance which is also an intermediate in the synthesis of epinephrine (adrenaline), it is a surprise to find it in sclerotization sequences. But so it is.

N-Acetyldopamine seems to lead to either of two types of sclerotization. The first of these, which leads to hardening with little or no development of color, has been termed β-sclerotization by Andersen because it involves the β carbon of the side chain (Fig. 11) (see Hepburn, 1976). Attachment to the protein chains is presumed to be through the second amine of diamino acids and through terminal amines.

The acid hydrolysis of β-sclerotized proteins gives rise to various ketocatechols (Fig. 14) which are readily identified by their characteristic absorption spectrum (maxima at 280 and 310 nm).

Failure to obtain ketocatechols from hydrolysis of sclerites from representatives of other arthropod classes has led Andersen to suggest that β-sclerotization is limited to insects, but the number of noninsect arthropods examined is small (*Limulus*, scorpion, spider, and crab).

N-Acetyldopamine can also lead to sclerotization by a quite different set of steps. This is quinone sclerotization first proposed by Pryor (1940) for the tanning of cockroach oothecae (Fig. 12). Quinone sclerotization gives rise to colored products, the characteristic amber, brown, and red-brown of many insect sclerites. As for β-sclerotization, it is assumed that the attachment of protein chains to the activated carbon is through the second amine of diamino acids or terminal amines. And the product cannot be removed without hydrolysis, which in this case does not lead to the formation of characteristic and identifiable products.

Clearly, the series of steps in quinone sclerotization does not depend on preceding β-sclerotization. It is a different method for cross-linking, and the histological sequences that led to the suggestion that quinone tanning was a second step have to be reexplained in terms of sequential activity of two enzyme systems. In some cases it certainly appears that quinone sclerotization comes later, but there is no *a priori* reason why it could not come first or simultaneously or without β-sclerotization.

The fact that hydrolysis of quinone-linked protein does not lead to the formation of characteristic products (as is true for β-sclerotization) makes it difficult to determine the time of appearance of one relative to the other, though total amounts in the finished sclerite can be determined by parallel experiments involving ring ^{3}H-labeled *N*-acetyldopamine in the one case and

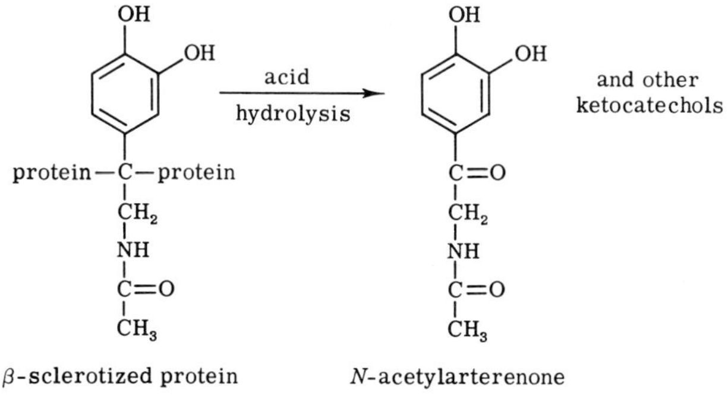

OH
OH
protein—C—protein
CH_2
NH
C=O
CH_3

acid
hydrolysis

OH
OH
C=O
CH_2
NH
C=O
CH_3

and other ketocatechols

β-sclerotized protein

N-acetylarterenone

Fig. 14. Formation of ketocatechols from β-sclerotized cuticle.

^{3}H-β-carbon labeled *N*-acetyldopamine in the other. With this method Andersen has reported ratios of ring-labeled to β-carbon-labeled ranging from 0.01 to 1.51 (see Hepburn, 1976). Since the substrate is the same, the ratios presumably represent corresponding ratios of the respective enzymes. However, ratios do not portray the whole story. Values for β-sclerotization are variable but always high; the range of ratios is primarily due to whether quinone sclerotization is low, medium, or high.

In addition to the browns from quinone sclerotization, there are also occasionally true blacks which in a few cases have been identified as true melanins (dark pigments soluble in strong alkali and giving indole derivatives after fusion with alkali). These seem likely to arise from the polymerization of excess quinones, but to get indoles requires condensation of the alanine side chain onto the benzene ring (Fig. 13). As such, even if the indole–quinone may be linked to a protein (i.e., be a melanoprotein, as mammalian melanins are thought to be), the polymer seems to be a filler without adding significantly to the number of cross-links between protein chains. Such cuticles could be called melanized; it is only important to remember that this is additional to sclerotization rather than in place of it. Melanization can sometimes be induced in normally nonmelanized cuticle (or blood) by mechanical injury which somehow induces tyrosinase activity.

While the following are not chemical tests valid for identifications, one can, at least in some cases, remove the black pigment by prolonged soaking in 2-chloroethanol, then remove the quinone tanned material with concentrated hydrogen peroxide, and finally the β-sclerotized material with warm alkali (Richards, 1967). The last step, however, removes all protein to leave a chitin residue.

The change necessary for production versus nonproduction of melanin may involve a single gene. Thus, the "black mutant" of the fly *Lucilia cuprina* produces melanin, whereas the "yellow mutant" of *Drosophila melanogaster* fails to produce melanin. And while speaking of genes, the puparium of *Musca domestica* is the usual dark brown sclerotized spheroid but those of *M. autumnalis* and *M. fergusoni* are light-colored and calcified. Obviously the genetic information permitting shifting from sclerotization to calcification is either residual in arthropods or involves a relatively slight change in the genetic code. In this connection, remember that many of the groups of Crustacea show lightly colored sclerotization with little or no calcification.

Degrees of sclerotization could be judged using any of several criteria: color, hardness, and concentration of cross-links. Hepburn (1976) thinks that physical properties such as hardness, strength, and elasticity are more biologically meaningful than structural, chemical, or staining properties. Andersen has determined that the mandibles and dorsal thoracic cuticle of locusts contain twenty times as many cross-links per unit volume as are found in abdominal sclerites (see Hepburn, 1976). Clearly there are many degrees of sclerotization both qualitatively and quantitatively.

A by-product of the formation of these cross-links is that the fibrous chains become closely packed. This results in the extrusion of water with consequent shrinking in thickness of the cuticle (exocuticle can become quite dry). Most of the small amount of ash is also extruded from the chitin/protein portions. Sclerotized cuticle will stretch very little in the plane of the surface but, unless very thick and dry, will bend considerably. It can become hard, stiff, and brittle.

Another well-known fact related to quantitative differences in sclerotization is that cuticles in larvae are commonly not sclerotized or sclerotized only in the outer portions. In contrast, adult cuticles are sometimes sclerotized throughout. This is related to a number of factors, including how long cuticle is secreted, quantity of enzymes, quantity of substrate, and the timing of events. A point of obvious biological significance is that molting fluid does not digest sclerotized cuticle. The suggestion has been made that adults can afford to stabilize their entire cuticles but that larvae cannot afford to. The suggestion seems plausible but does not explain why the adults of some species stabilize the entire thickness of sclerites whereas others do not. The fracture properties of fully sclerotized cuticle are different from those of exocuticle underlain by meso- and endocuticle (Hepburn, 1976), but if endocuticle beneath a sclerite has a significant function, how can many adult insects afford to sclerotize the entire thickness? The problem must be more complicated because larvae to a considerable extent exhibit hydrostatic skeletal mechanisms similar to those of annelids, whereas adults have a higher percentage of movements based on lever principles.

Sclerotization may occur before, during, or after ecdysis, or part before and part after. And sclerotization may occur in a sclerite that has the same dimensions after sclerotization as it had before. Or, as in wing buds, the cuticle may be in a plastic state at emergence of the adult, quickly become greatly stretched, and then stabilized by sclerotization.

The sequence of events in sclerotization can be followed in histological sections. Numerous variations have been recorded. An extensive analysis is given by Richards (1967). To some extent the histological variants can be rationalized in terms of the chemical processes presented above. By adding variations in localization of bound enzymes, temporal variations in secretion of substrate, etc., many variants seem explicable. But several points are not yet answered by the chemical data: (1) What is the nature of the change from procuticle to mesocuticle? What is mesocuticle? (2) Why is there seldom any mention of chitin in the chemical papers on sclerotization? (3) If chitin chains do contain some nonacetylated residues, could these become involved in cross-links with *N*-acetyldopamine? (4) Are the protein chains associated with chitin chains involved in sclerotization, or is sclerotization limited to protein chains not linked to chitin chains? Clearly many questions remain.

Finally, the development of patterns involves several factors, including the localization of sclerotization and degrees of sclerotization. This might be controlled by local differences in enzymes present, in substrate availability, or by the presence of inhibitors—all three have been reported in different cases. While the expression of color pattern seems as simple as cuticle chemistry, we have little idea of the chemistry of pattern determination, or of why, when the hormones β-ecdysone and bursicon initiate sclerotization, the process does not necessarily take place simultaneously over the whole body.

IV. REVERSIBLE CHANGES

Reversible changes in cuticle all concern nonsclerotized cuticles. It is possible to irreversibly extract material from sclerotized cuticle by chemical procedures, but the insect does not do this. The insect can enzymatically digest endocuticle with the molting fluid, and resorb and reuse this portion of the cuticle (see Jeuniaux, 1971). It becomes a matter of definition as to whether one can say that the organization of endocuticle is reversible when it can be digested and reused by the animal that produced it.

However, there are some cases in which changes do seem to be truly reversible, perhaps with an endocrine involvement: the extension of membranes permitting engorgement in blood-sucking bugs; changes in intersegmental membranes of male mosquitoes permitting abdominal rotation; and, Vincent's detailed analysis of the extension of abdominal intersegmental membranes for oviposition of locusts (see Hepburn, 1976). There seems no question that such phenomena involve reversal of cuticle organization, but we have no knowledge of the chemical changes involved.

V. THE EPICUTICLE AND PERMEABILITY BARRIERS

The epicuticle is the thin outermost set of layers characterized by the seeming absence of chitin (layers 1–3 of Fig. 1). Its chemistry is less known but it usually contains a wax layer which is important in water retention and several lipoprotein layers from which lipid is not readily removed by solvent extraction. The best known of these is the cuticulin layer which is thick enough to be seen in sections viewed by light microscopy. Polyphenols have been reported from this layer on the basis of a positive argentaffin reaction.

The cuticulin layer must be polymerized to a considerable extent to be so inextensible and resistant to dissolution. Whether it develops further cross-links during sclerotization of the underlying chitin/protein cuticle is not known. However, staining with the periodic acid-Schiff reaction shows that several colorations (including no coloration) can occur on different parts of a single insect. In some but not all cases, the epicuticle over a sclerite loses its stainability following sclerotization while that over nonsclerotized intersegmental membrane does not.

It is commonly stated or implied that the epicuticle is the permeability barrier between an insect and its environment, and that the chitin/protein part of the cuticle supplies a skeleton and protection. The situation is much more complicated both because several recognizable layers have been shown to be deterrents to water loss and because different groups of things may have different layers as barriers. Richards (1957) has referred to this as "the multiple barrier concept." Why an insect has a series of barriers for different substances rather than a single barrier to all chemicals is not known.

Since the subject of permeability barriers is more in the realm of physical properties rather than biochemistry, only a brief summary statement will be given. Using the numbers on Fig. 1, the surface layer of a maggot is the barrier to ions (layer 1); the wax layer is one of the important barriers to water (layer 2, sometimes absent in aquatic larvae); the cuticulin layer contains the barrier to oxygen (and presumably other gas molecules) (layer 3); the chitin/protein layers when wet are barriers to lipophilic substances (layers 4–6); the exocuticle when very thick may become dry and highly impermeable even to water (layer 4); the outer plasma membrane of the epidermal cells is being shown to be an important deterrent to water loss as well as, of course, limiting what enters the cuticle from the epidermal cells (layer 7); and the inner plasma membrane and the basement membrane limit what enters the epidermal cells from the blood (layers 8–9).

GENERAL REFERENCES

Andersen, S. O. (1971). Resilin. *In* "Comprehensive Biochemistry" (M. Florkin and E. H. Stotz, eds.), Vol. 26, Part C, pp. 633–657. Elsevier, Amsterdam.

Hackman, R. H. (1974). Chemistry of Insect Cuticle. *In* "The Physiology of Insecta" (M. Rockstein, ed.), 2nd Ed., Vol. 6, pp. 216–270. Academic Press, New York.

Hepburn, R. H., ed. (1976). "The Insect Integument," Elsevier, Amsterdam.

Jeuniaux, C. (1963). "Chitine et Chitinolyse." Masson, Paris.

Jeuniaux, C. (1971). Chitinous structures. *In* "Comprehensive Biochemistry" (M. Florkin and E. H. Stotz, eds.), Vol. 26, Part C, pp. 595–632. Elsevier, Amsterdam.

Muzzarelli, R. A. A. (1977). "Chitin." Pergamon, Oxford.

Richards, A. G. (1951). "The Integument of Arthropods." Univ. of Minnesota Press, Minneapolis, Minnesota.

Richards, A. G. (1967). Sclerotization and the localization of brown and black colors in insects. *Zool. Jahrb. Abt. Anat. Ontog. Tierre,* **84,** 25–62.

REFERENCES FOR ADVANCED STUDENTS AND RESEARCH SCIENTISTS

Andersen, S. O. (1966). *Acta Physiol. Scand.* **66** (Suppl. 263), 1–81.
Andersen, S. O. (1973). *J. Insect Physiol.* **19,** 1603–1614.
Andersen, S. O., Chase, A. M. and Willis, J. H. (1973). *Insect Biochem.* **3,** 171–180.
Arnold, M. T., Blomquist, G. J. and Jackson, L. L. (1969). *Comp. Biochem. Physiol.* **31,** 685–692.
Beatty, I. M. and Gilby, A. R. (1969). *Naturwissenschaften.* **56,** 373.
Bursell, E. and Clements, A. N. (1967). *J. Insect Physiol.* **13,** 1671–1678.
Carlstrøm, D. (1957). *J. Biophys. Biochem. Cytol.* **3,** 669–683.
Gilby, A. R. and McKellar, J. W. (1970). *J. Insect Physiol.* **16,** 1517–1529.
Gilby, A. R. and McKellar, J. W. (1976). *J. Insect Physiol.* **22,** 1465–1468.
Goodrich, B. S. (1970). *J. Lipid Res.* **11,** 1–6.
Hackman, R. H. (1958). *J. Insect Physiol.* **2,** 221–231.
Hackman, R. H. (1975). *J. Insect Physiol.* **21,** 1613–1623.
Hackman, R. H. and Goldberg, M. (1971). *J. Insect Physiol.* **17,** 335–347.
Jackson, L. L. (1972). *Comp. Biochem. Physiol.* **41B,** 331.
Johnson, L. H., Pepper, J. H., Banning, M. N. B., Hastings, E., and Clark, R. S. (1952). *Physiol. Zool.* **25,** 250–258.
Karlson, P. (1962). *Nature (London),* **195,** 183–184.
Malek, S. R. A. (1957). *Nature (London),* **180,** 237.
Pryor, M. G. M. (1940). *Proc. R. Soc. London, Ser. B,* **128,** 378–393, 393–407.
Richards, A. G. (1956). *J. Histochem. Cytochem.* **4,** 140–152.
Richards, A. G. (1957). *J. Insect Physiol.* **1,** 23–39.
Richards, A. G. (1978). *Proc. Int. Confr. Chitin/Chitosan, 1st* M.I.T. Sea Grant Program, Cambridge, Massachusetts.
Rudall, K. M. (1962). *In* "Sci. Basis Medicine." Athalone Press, London.
Rudall, K. M. (1967). *In* "Conformation of Biopolymers" (G. N. Ramachandran, ed.), Vol. 2, 751–765. Academic Press, New York.
Rudall, K. M. and Kenchington, W. (1973). *Biol. Rev. Cambridge* **49,** 597–636.
Trim, A. R. H. (1941). *Biochem. J.* **35,** 1088–1098.

6

Insect Biochromes: Their Chemistry and Role

A. E. NEEDHAM

I. INTRODUCTION

This chapter is concerned only with materials that have color because the molecule absorbs, and is excited by, one or more bands of electromagnetic radiation in the visible range. They are *chemochromes* as contrasted with *schemochromes,* or structural colors (Fox, 1976), due to physical agencies. One colorless class of materials, the purines, is included because of its close relationship, both chemically and biologically, to the pterins. In any case, "biochrome" must not be interpreted too narrowly since, for instance, the visual spectrum of insects is transposed 100–150 nm further down the wavelength scale than ours.

Insects synthesize, or acquire in their food, members of all the known chemical classes of biochrome (Goodwin, 1971): carotenoids, chromans, flavonoids, aurones, ternary quinonoids, including benzo-, naphtha-, anthra-, and polycyclic, quinones, tetrapyrroles, including porphyrins and bilins, indolic melanins, ommochromes, papiliochromes, purines, pterins, and isoalloxazines. Those acquired in the food are metabolically handled, and often chemically modified, by the insect and so become "zoochromes." As yet the biochromes of relatively few insects have been studied chemically, and no doubt other chemical classes will be identified.

Most of the solar energy reaching earth lies within or near the visible range and, probably for good evolutionary reasons, most biochemical reactions are activated by the energy quanta of radiations in that range. The importance of biochromes and their evolutionary exploitation therefore are equally understandable. Many insects are small enough for all of their biochromes to receive solar light energy but no doubt, as in other taxa, some normally are excited indirectly by chemical energy.

To be excited by the relatively small quanta of visible light, molecules need properties that lower the energy necessary for even the easiest electron transitions, from π to π^* and from n to π^* orbitals. Most effective is the possession of a long path of conjugated double bonds, a conspicuous feature of the biochromes of Figs. 1–8. The conjugated system allows delocalization of the π electrons involved, and the required excitation energy is lowered by the resonance energy of the mobilized system. This is proportional to the size of the system and also to its planarity. Most of the biochromes figured have polycyclic aromatic molecules, which closely approach this state. Carotenoids in the *all-trans* conformation (Fig. 1) have an alternative version that has the further advantage of a potentiality for a large dipole moment. The compact, condensed ring alternative has advantages for transport and other purposes.

The electron transition requiring least energy is that from n to π^* orbitals and this explains the strong auxochromic action of the O and N atoms so conspicuous in the ring and chain systems, and in the substituent groups, of the biochromes shown. They have a bathochromic effect on the molecule, shifting its absorption peaks to longer wavelengths, with smaller quanta. Since the visible color of a biochrome is complementary to that of the wavebands absorbed, the most hypsochromic color is yellow and the bathochromic sequence is yellow, orange, red, purple, violet, blue, and green. Black is regarded as the next and final stage, since the bathochromic trend affects all absorption bands of the molecule and so collectively they come to absorb over much of the visible spectrum. Various inorganic elements have colored compounds because of their nonbonding, *n*, electrons. The transitional metals also have many vacant π^* orbitals and they have been exploited as components of some insect biochromes (Fig. 5, II).

Quanta of visible light are adequate to drive reactions with activation energies between 160 and 300 kJ per mole of reagent, i.e., the redox type of reaction, involving electron transfer between molecules. Most of the classes of biochrome considered are both donors and acceptors of electrons, i.e., reversible redox agents, over some part of the biochemical redox range, −800 to +800 mV.

Free biochromes, particularly those with a condensed aromatic ring system, are fluorescent, emitting as visible light the excitation energy they have received. *In vivo,* however, the energy is usefully channelled to activate reactions, i.e., fluorescence is quenched. For entropic reasons the emission is more bathochromic than the exciting radiation and the fluorescence of biochromes is excited mainly by UV light. *In vivo* they also harness high energy quanta to biochemical reactions.

II. STRUCTURE, DISTRIBUTION, AND CHEMISTRY

A. Carotenoids

1. Structure and Distribution

This class is unique among biochromes in having a mainly open-chain molecule (Fig. 1). It is also the most fully hydrocarbon and hydrophobic. The first stage synthesized by plants (Fig. 1, I) is completely open chain (Isler *et al.*, 1971) and has conjugate double bonding only in the middle of the molecule. Extension of conjugate bonding produces lycopene (Fig. 1, II), found in some insects (Table I). Most insect carotenoids have both ends cyclized to β-ionone rings (Fig. 1, III–VI). The subclasses with oxygen substituents, mainly on the rings, are called xanthophylls (Fig. 1, IV–VI) as distinct from the purely hydrocarbon carotenes. The oxygen "functions" can be alcoholic (—OH), ketonic (=O), aldehydic (terminal ketonic), epoxy (Fig. 1, VI), alkoxy (—OR), or carboxylic (—COOH). All but the last two are known among insect carotenoids (Fig. 1, Table I).

Vitamin A and retinal 1 (Fig. 1, VII) are apocarotenoids, i.e., have the molecule secondarily shortened by the insects themselves. Extreme shortening is seen in a keto derivative (Fig. 1, VIII) found in the grasshopper *Romalea microptera;* dietary neoxanthin (Fig. 1, VI) is the probable source. A

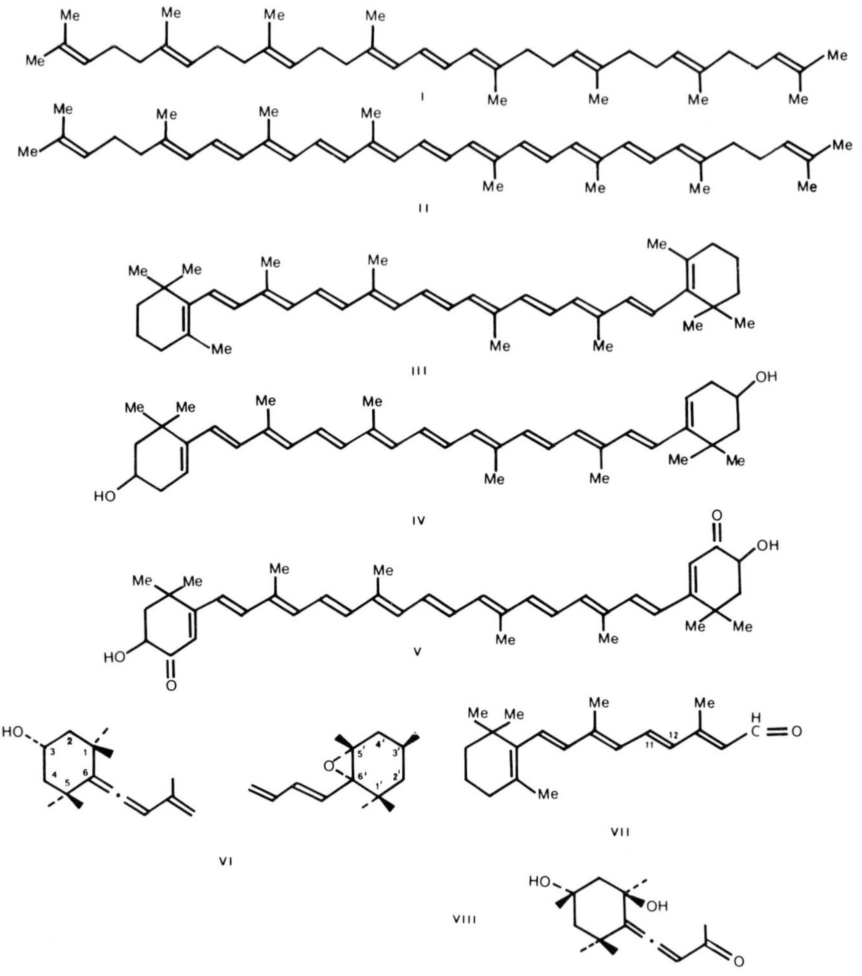

Fig. 1. Molecular formulas of carotenoids relevant to insects: I, phytoene; II, lycopene; III, β-carotene; IV, lutein; V, astaxanthin; VI, ends of neoxanthin molecule; VII, retinal 1 (R_1) = aldehyde of vitamin A; VIII, apocarotenoid of *Romalea*. In this, and in the other figures, unlabeled atoms in rings and side chains are understood to be C atoms and unlabeled substituents to be H atoms. Me, methyl.

number of carotenoprotein conjugates are formed by insects (Table I). Some have more than one carotenoid chromophore, and some have a bilin and a carotenoid.

2. Chemistry

The carotenes are insoluble in mild aqueous media and most soluble in organic solvents as apolar as themselves. Xanthophylls, similarly, are most soluble in media of their own degree of hydrophily; some, with four oxygen substituents, are quite soluble even in 70% aqueous methanol. Esterification of —OH groups, particularly of the strong —COOH acids, decreases the aquasolubility. Conjugation, even of the carotenes, with monoses or with soluble proteins makes them freely aquasoluble. *In vivo* carotenoids are mostly dissolved in glycerides and other lipid phases in intracellular vacuoles, but they are dispersed in solid form in the insect exoskeleton.

Because of their biological environment they are most accessible to mobile molecules, with the polarity of MeOH and acetone, from which they are readily partitioned according to polarities of solutes and solvents (Isler, 1971; Fox, 1976). Finer fractionation of the "isodistributive" fractions thus separated is achieved mainly by adsorption column chromatography, using metal salts of weak oxy acids or carbohydrates, as adsorbents. Purification and crystallization are particularly easy.

Free carotenoids are rarely more bathochromic than orange-red. Lycopene is a deeper red than β-carotene and some xanthophylls, since cyclization of the ends of the chain both shortens the molecular axis and resolves two double bonds. Oxy substituents have less bathochromic effect than on fully aromatic rings. By contrast, conjugation with proteins has produced hues ranging from yellow, through red, purple, violet, blue, and green to dark brown and even black, though not all of these have been found in insects. The free carotenoid is readily revealed by denaturing the protein, and its characteristic absorption peak is often evident in the spectrum of the carotenoprotein.

The visible-range absorption peak (λ_1) of free carotenoids is in the region 450–500 nm (Isler, 1971). Typically it is a triple peak but oxy substituents decrease the fine structure and canthaxanthin has a single, broader peak. In *cis*-isomers there is also a peak, λ_2, around 340 nm, in the near UV, and all have one at 260–275 nm. This indicates that even in members with the fully open chain, the ends are virtually cyclized. The λ_1 peak is very useful in identification since the peak positions are shifted systematically by substituents, positions, and numbers of double bonds, and solvents. They are most bathochromic in the most polar solvents.

Infrared spectroscopy gives information on the bonding pattern in the molecule : saturation of double bonds, two double bonds in sequence (allenic compound, Fig. 1, VI), and triple bond (acetylenic compound). It also shows —OH and unreactive =O substituents. Proton magnetic resonance spectroscopy helps to decipher the end groups, the in-chain methyl substituents and the olefinic protons. Mass spectrometry is less useful than for most biochromes because carotenoids degrade so readily on vaporization.

Free carotenoids have a weaker UV-excited fluorescence (UVF) than most classes of biochrome because they lack a condensed aromatic ring system, and the natural, *all-trans,* isomers do not absorb in the near UV.

TABLE I
Known Distribution of Carotenoids among Insects

Carotenoid and distribution	Recent and general references
1. Lycopene (ψ,ψ-carotene) *Carausius, Coccinella, Pyrrhocoris*	Fox (1976); Vuillaume (1975)
2. β-Carotene (β,β-carotene) *Carausius, Oedipoda, Schistocerca, Locusta, Leptinotarsa, Lilioceris,* most Lepidoptera, *Apis, Chironomus*	Fox (1976); Czeczuga (1971); Feltwell (1974); Mummery and Valadon (1974)
3. α-Carotene [(6*R*)-β,ϵ-carotene] *Carausius, Coccinella, Leptinotarsa, Pieris, Chironomus.*	Czeczuga (1970); Feltwell (1974)
4. Cryptoxanthin (3-OH-β,β-carotene) *Chironomus* larva, 16% of Lepidoptera	Czeczuga (1970); Feltwell (1974)
5. Isocryptoxanthin (4-OH-β,β-carotene) *Carausius, Leptinotarsa*	Willig (1969); Czeczuga (1971)
6. Echinenone (4-keto-β,β-carotene) *Leptinotarsa*	Czeczuga (1971)
7. Lutein (3,3′-diOH-β,ϵ-carotene) *Leptinotarsa, Tettigonia, Meconema Apis,* most *Lepidoptera, Chironomus*	Fox and Vevers (1960); Feltwell (1974); Czeczuga (1970, 1971); Leuenberg and Thommen (1970)
8. Zeaxanthin (3,3′-diOH-β,β-carotene) *Carausius, Leptinotarsa,* 24% of Lepidoptera	Willig (1969); Feltwell (1974)
9. Isozeaxanthin (4,4′-diOH-β,β-carotene) *Lilioceris*	Mummery and Valadon (1974)
10. Canthaxanthin (4,4′-diketo-β,β-carotene) *Lilioceris, Leptinotarsa, Papilio, Chironomus*	Ohnishi (1970); Leuenberg and Thommen (1970)
11. Astaxanthin (3,3′-diOH-β,β-carotene-4,4′-dione) *Schistocerca, Oedipoda, Leptinotarsa*	Fox (1976)
12. Astacene (β,β-carotene-3,4,3′,4′-tetrone) *Schistocerca, Chironomus*	Czeczuga (1970)
13. Taraxanthin (5,6-epoxy-5,6-diH-β,ϵ-carotene-3,3′-diol) *Pieris, Bombyx*	Fox (1976)
14. β-Carotene monepoxide (5,6-epoxy-5,6-diH-β,β-carotene) *Lilioceris,* 32% of Lepidoptera	Mummery and Valadon (1974)
15. Violaxanthin (5,6-epoxy-5,6-diH-β,ϵ-carotene-3,3′-diol) 23% of Lepidoptera	Feltwell (1974)

TABLE I (*continued*)
Known Distribution of Carotenoids among Insects

Carotenoid and distribution			Recent and general references
16. Neoxanthin (5′,6′-epoxy-6,7-dideH-5,6,5′,6′,-tetra-H-β,β-carotene-3,5,3′-triol) *Leptinotarsa,* 18% of Lepidoptera			Czeczuga (1971); Feltwell (1974)
17. Carotenoproteins			
Chromophores	Color	Site	
β,β-Carotene	Green	Locust, integument	Needham (1974)
β,β-Carotene and bilin	Blue-green	Carausius, integument and blood	Needham (1974)
β,β-Carotene, lutein, bilin	Green	Various Lepidoptera, blood	Needham (1974)
Lutein, bilin	Green	*Tettigonia, Sphinx,* integument	Needham (1974)
Astaxanthin	Blue, red, yellow	*Oedipoda,* hindwings	Fox and Vevers (1960)
Canthaxanthin	—	*Leptinotarsa*	Thommen (1971)
Lutein, bilin, taraxanthin	Green	*Plusia,* blood	Vuillaume (1975)
Ketocarotenoids	—	*Dysdercus, Tenebrio,* blood	Trautmann and Thommen (1971)

This is in contrast to their great sensitivity to visible light, particularly under aerobic conditions and at high temperatures; dark and cold laboratory conditions are even more mandatory than for most biochromes.

Alkalies form salts with carotenoic acids (—COOH substituents), and hydrolyze carotenol esters, but the skeleton of the molecule is stable. It is more sensitive to strong acids, giving a bright blue unstable product (peak absorption at 620 nm), useful for identification and estimation (Carr–Price reaction). The color varies with the number (n) of double bonds: $\lambda_{max} = (300.5 + 65.5n)$ nm. The reaction requires relatively anhydrous conditions; this and the bathochromic color imply formation of an active carbonium cation. The 5,6-epoxides and some other oxy derivatives give a similar color reaction with 20% aqueous HCl and ether, so that an active oxonium cation also may be producible. Other reactions useful in identification include reduction to the parent carotene by the active hydrogen of KBH_4 or $LiAlH_4$, and reductive saturation to the parent paraffin by PtO, PdO, or colloidal Pt in acetic acid or ethyl acetate. With special precautions the reaction can be made stepwise. The reverse process of desaturation is all-or-none in the laboratory but has stepwise control *in vivo*. Saturation by halogens, useful because of the colored products, was the method first used to resolve the isomers of carotene; it is little used now. The hydrocarbon chain can be fragmented by oxidative saturation. Iron perchloride gives a diagnostic green color reaction with carotenoids. The end groups of lycopene can be cyclized by $TiCl_4$.

B. Chromans and Flavonoids

1. Chromans

The α-tocopherols (Fig. 2, I), vitamins E, are the only members of this class relevant to insects. Their polyisopentenyl side chain, in diametric contrast to the carotenoid chain, is completely saturated, and their pale yellow color (λ_{max} = 294 nm) therefore is due to the condensed ring system and its potentially active oxygen functions.

Vitamin E is an essential dietary requirement for some insects at least. At present little is known of its acquisition and storage.

The α-tocopherols, as their molecular structure implies, are very sparingly soluble in polar media and *in vivo* therefore are usually associated with carotenoids. They are not conjugated with sugar or protein. They are stable to acids, which depress the ionization of the phenolic —OH group, but sensitive to alkalies and so form useful partners to the carotenoids in chemical protective functions. They are also much more stable to light and heat than carotenoids and stabilize the triglycerides in which both are dissolved.

They are readily oxidized, however, and the oxidation process is unusual: the pyran ring is opened (Fig. 2, IB), converting the benzene ring to the *p*-quinone form, with a pronounced bathochromic shift in color and absorption. It thus becomes structurally related to ubiquinone (Section II, C, 1), and the two may be biosynthetically related.

2. Flavonoids

a. ***Structure and Distribution.*** This class is again ternary, i.e., has only C, H, and O in the molecule but this approaches a fully condensed, aromatic, cyclic state. It is biosynthesized by a condensing cyclization of the C_3-chain connecting two benzene rings in a compound such as a chalcone (Fig. 2, II). The heterocyclic ring so added is in the unsaturated pyran state, compared with the di-H-pyran state in the α-tocopherols. Moreover the completed molecule often has a unified resonant system of conjugate bonding, and the biological members have a number of auxochrome substituents on the rings, so that most flavonoids are brightly colored.

Two main subclasses are relevant to insects, the flavones or anthoxanthins, with a keto function at position 4 (Fig. 2, III), and the anthocyanins

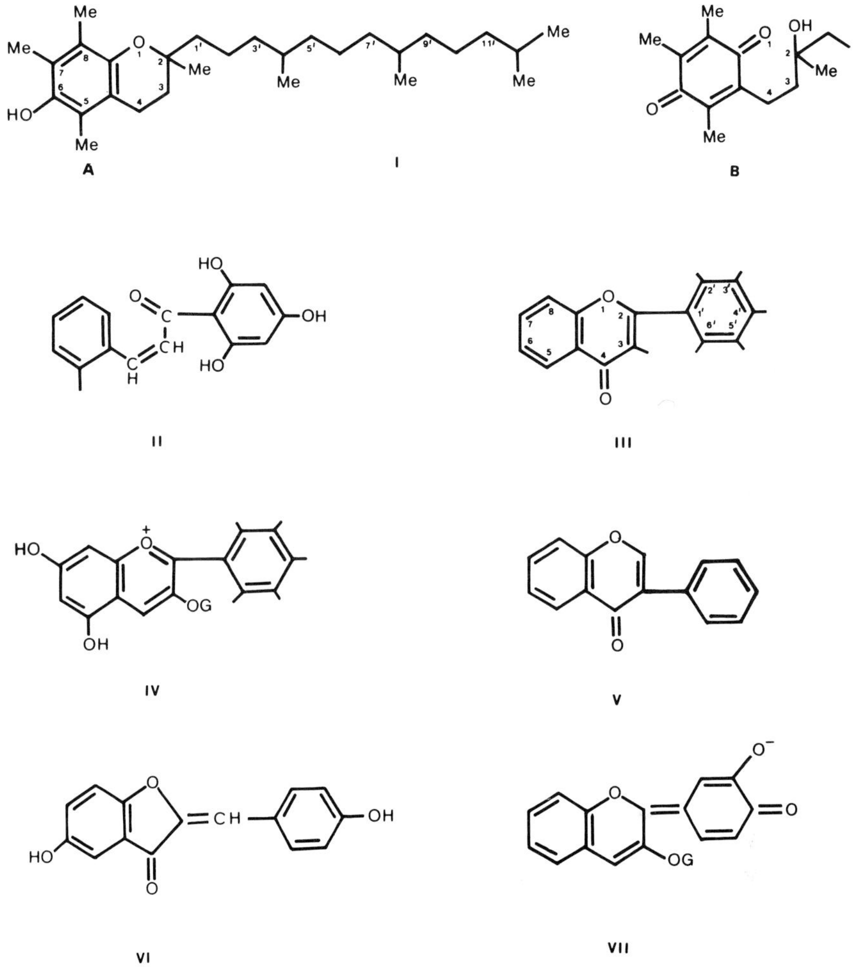

Fig. 2. Molecular formulas of chromans and flavonoids. Unsubstituted skeletons, actual members, precursors and derivatives relevant to insects: I, α-tocopherol (A, reduced form; B, ring system of oxidized form); II, chalcone; III, flavone skeleton; IV, anthocyanin skeleton; V, isoflavone skeleton; VI, marginalin, an aurone; VII, color-base of an anthocyanin. Me, methyl; G, hexose.

(Fig. 2, IV). In addition, the Common Blue butterfly, *Polyommatus icarus*, contains an isoflavonoid (Fig. 2, V) in which the lone benzene ring is bonded at position 3. The water beetle *Dytiscus marginalis* has in its pygidial gland an aurone, marginalin (Fig. 2, VI); here condensation has produced a furan ring, leaving two links in the chain. Probably all are acquired from dietary sources. In both plants and animals, flavonoids are usually conjugated, at position 3, with glucose, galactose, or rhamnose, and the free aglycones are called anthoxanthidins and anthocyanidins.

Twelve flavones are present in the wings of the Marbled White butterfly (Morris and Thomson, 1963) but contribute little to their "visual effect." They are acquired from larval food and passed on to the imago. The Common Blue has three, in addition to the isoflavonoid, free 3-OMe-kaempferol, quercetin-3,4-diglucoside, and an unidentified anthoxanthin. Again, all are found in the larval food plant. Integumental biochromes of a number of sap-feeding bugs, and of one that preys on other insects, have been identified tentatively as flavones.

The integumental color of the larva of the pugmoth, *Eupithecia oblongata,* is believed to be due to anthocyanins from the flowers of its food plants, and varies with the latter to give camouflage. The larva of the beetle, *Cionus oleus,* is similarly camouflaged by a dietary anthocyanin stored in the fat body, but showing through a translucent integument.

b. *Chemistry*. The free aglycones are insoluble in water but quite soluble in alcohols and other polar organic solvents; they are insoluble in apolar media. This is a measure of the number of polar groups in the molecule of the biological members. The melting point is correspondingly high. Conjugation with the monose makes them fully water-soluble. It also stabilizes them against light, enzyme action, and hydrolysis; boiling 2 *N* mineral acid is necessary to free the aglycone. Unlike the protein of carotenoproteins, the monose has little effect on the absorption spectrum. In plants the glycosides are usually in solution in the cell sap but sometimes in solid form, crystalline or amorphous. They are in solution in satyrid moths (Ziegler and Harmsen, 1969) but in cell granules in most other insects.

The flavones are mostly yellow to red in color and anthocyanins purple to blue. The greater bathochromicity of the latter is due mainly to the formation of a strongly resonant ionic function that completes a fully resonant system throughout the molecule (Fig. 2, IV). The large number of polar substituents also contribute—OH groups at positions 3′ and 4′ having the most effect; four or more of these give a full blue color. Metals and mordants also have bathochromic effects and they again stabilize the molecule so that natural members such as weld have long been used as dyes. Most of these properties are relevant also to other classes of biochrome.

The individual aromatic rings have an absorption peak at 250–275 nm. Most flavones and some anthocyanins have only one peak in the visible range but pelargonidin has, all told, peaks at 267, 331, 400, 450, and 505 nm. As for the pterins (Section II, H, 3, b) the most bathochromic peak tends to be at a shorter wavelength than the visible color implies, e.g., 375 nm for chrysin and 510 nm for cyanidin.

Flavonoids are amphoteric, i.e., have both cationic and anionic ionizations; electrophoretic methods therefore are useful for isolating and separating them. The ions are more bathochromic than the un-ionized acids and bases, so that there are striking color changes with pH, exploited in indicators such as litmus. The cationic function is usually the oxonium at position 1 (Fig. 2, IV), but possibly the carbonium at 4 in some anthoxanthins, while the anion is the phenolate at 3′ in the lone benzene ring. The chromatic importance of these ionizations is emphasized by the fact that the "color-base" of the intermediate pH range is colorless in some flavonoids. The color of the ionic forms is affected by their counterions, organic anions, or metal cations, respectively. The latter may partly account for the fact that the phenolates are more bathochromic than the oxonium salts in spite of the double bond added by this ionization. Phenolates of monovalent metals are soluble; the insoluble salts of some polyvalent metals, e.g., lead, are useful in separation and identification. With increase in pH the hue becomes also more fully saturated, i.e., absorption is more intense and sharply focused, and this provides a useful qualitative test. Boric plus citric acid in anhydrous acetone provide the best reagent for the purpose.

Like most biochromes, flavonoids have versatile redox properties. The anthocyanins are stronger reducing agents than the free glucose since they turn Fehling's solution red even at room temperature. Their reduction of red-brown Fe^{3+} to Fe^{2+} is another redox color test used. However, free anthocyanidins oxidize only slowly in air at room temperature. The product is colorless due to the pyran ring opening at the site of glycosidation; this is favored by alkaline pH. The large number of —OH substituents, suitably sited, make some members quinolic, readily oxidizing to quinones and polymerizing to form catechins or tannins.

Mild reduction of flavonoids with zinc dust or pyridine reversibly produces the leucobase which, unlike that of some other polyhydroxy biochromes (Section II, C, 3) is truly colorless. Reduction by the H^- of $LiAlH_4$ or of Mg plus HCl converts anthoxanthins to anthocyanins and serves as a color test. Most of them are reduced to a red oxonium form but the 6,7-diMe members produce a green-blue, and the 7,8-diMe members an orange-yellow cation (Bargellini). Sodium amalgam gives a green precipitate with polyhydroxyflavones in alcoholic solution.

C. Ternary Quinones

1. Structure and Distribution

The quinone structure is very evident in this second class of aromatic ternary biochromes. Members of most subclasses are found in insects: *o*- and *p*- benzoquinones (BQ: Fig. 3, I, II), naphthaquinones (NQ: Fig. 3, III), anthraquinones (AQ: Fig. 3, IV, V) and polycyclic quinones (PQ: Fig. 4). The BQ's, NQ's, and AQ's have maximally unsaturated cyclic systems but the PQ's are dimers of substituted NQ's and have a number of partially or fully saturated pyran rings.

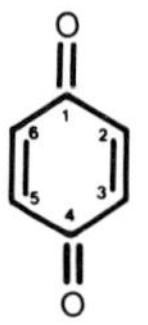

I

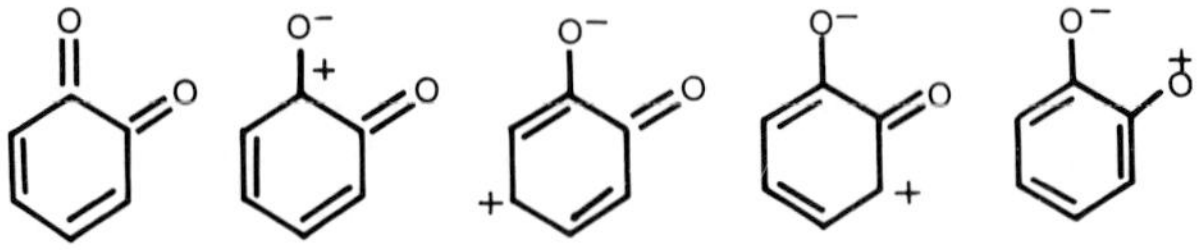

II

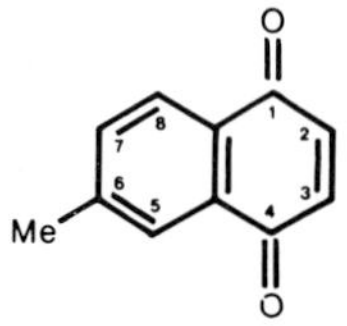

III

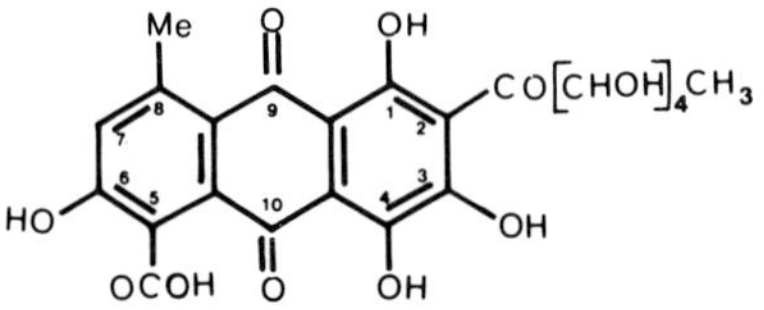

IV

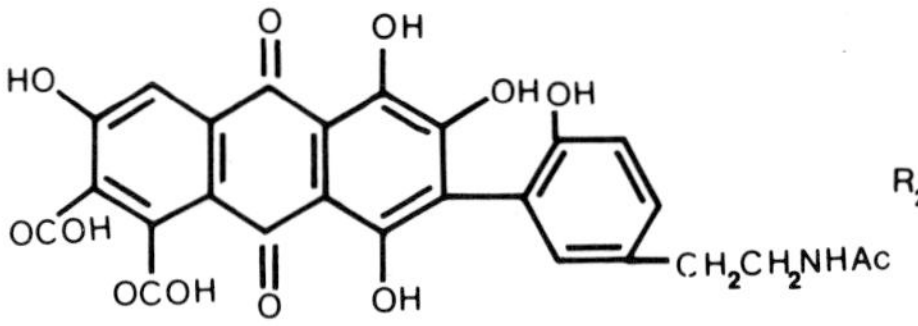

V

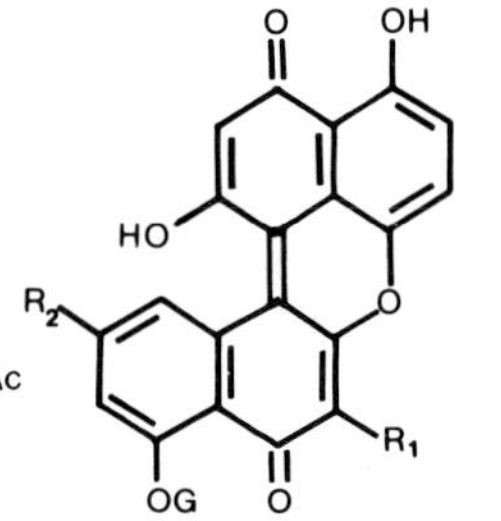

VI

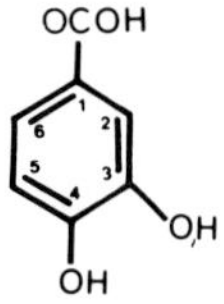

VII

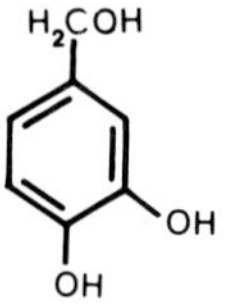

VIII

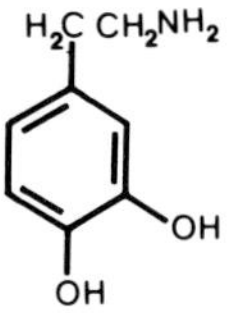

IX

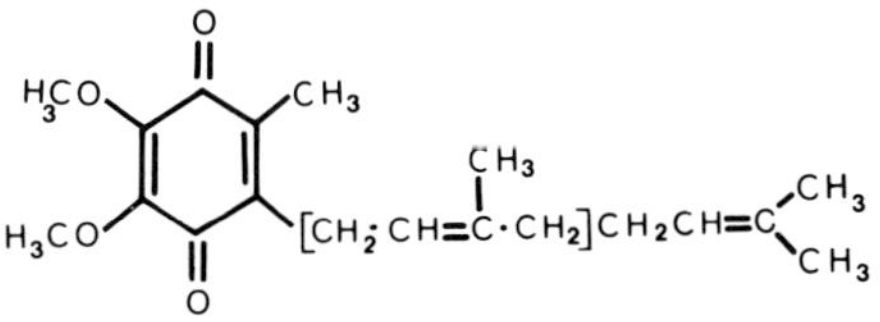

X

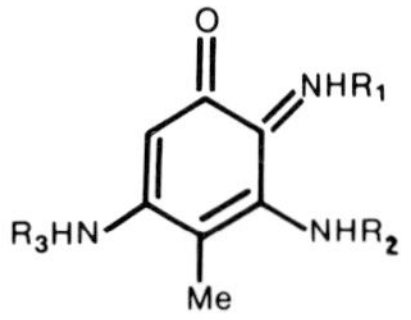

XI

Fig. 3. Ternary quinones; skeletons and actual members relevant to insects: I, *p*-benzoquinone (*p*-BQ); II, five resonant forms of *o*-benzoquinone (*o*-BQ); III, 6-methylnaphthaquinone of *Argoporis*; IV, carminic acid; V, laccaic acid A_1; VI, aphinin; VII-IX: precursors of tanning quinones: VII, protocatechuic acid (3,4-dihydroxybenzoic acid); VIII, 3,4-dihydroxymethyl · catechol (3,4-dihydroxybenzyl alcohol); IX, dopamine (3,4-dihydroxyphenyl-β-ethanolamine); X, ubiquinone (Coenzyme Q); XI, Hackman and Todd model for bonding of *o*-BQ's with amino compounds. R_1, R_2, R_3, unspecified organic substituents; Ac, acetyl; n = 9 or 10 in insects.

o-BQ's are constituents of the exoskeleton of all instars of all insects, and of oothecae, silk, and some other secretions. Their precursors are rather varied (Table IV), the most common being shown in Fig. 3, VI, VII, VIII, IX. Probably all precursors are synthesized from tyrosine by the insect (Brunet, 1967).

The *p*-BQ, ubiquinone (UQ: Fig. 3, X) also is universally distributed, as the name indicates. It has a partially desaturated polyisopentenyl side chain, intermediate between the states in carotenoids and α-tocopherols. The number of isopentenyl units is nine or ten depending on the dietary source. Much less widespread (Table V), but synthesized by the insects, are simpler *p*-BQ's secreted by various groups in chemical defense (Section III, C, 1). At least one species of beetle includes *p*-NQ's in this secretion.

AQ's are known only in one family of insects, the Coccidae (homopteran Hemiptera), and until recently were not found in any other taxon of animals. They accumulate up to 50% of the body weight of the female, in the fat body and other tissues, the eggs and the blood. In lac insects they are largely secreted in an external "scale," mixed with hard, resinous material. The substituents on the ring system (Fig. 3, III, IV) show that the three known groups of AQ's have a common origin. *In vivo* they exist as the K salts of the —COOH at position 5; the free acid of *Dactylopius coccus* is called carminic, that of *Lecanium ilicis* kermesic, and those of *Laccifer lacca* and *Tachardia lacca* laccaic A_1, A_2, and B. The substituent at 3 is the most variable and is a chain-linked aromatic ring in the laccaic acids (Burwood *et al.*, 1967). All are synthesized in the fat body, probably by symbiotic microorganisms.

PQ's also are produced only by members of one family of homopteran Hemiptera, the Aphidae (Cameron and Todd, 1967). Like the AQ's they vary in detail in the different genera but all are C_{30}-compounds and a common evolutionary origin is probable. Their naphthalene precursor is most probably 5,7-diOH-*p*-NQ or 1,3,8-triOH-naphthalene. Again the synthesis is probably by fat body microorganisms (Brown *et al.*, 1969). Concentrations reach 1% of the body weight, much lower than that of AQ's. *In vivo* PQ's are conjugated with glucose (cf. flavonoids) and are mainly in solution, in the hemolymph. Conjugation stabilizes the molecule and is particularly essential for the mechanically and chemically unstable chromophore of the protoaphins (Fig. 4, I).

Two main forms of the C_{30}/PQ-glycosides are found in aphids, (1) aphins in which the two monomers are simply and loosely bonded (Fig. 4, I) and (2) aphinin, in which the monomers are more firmly bonded by formation of a di-*H*-pyran ring (Fig. 3, IV). Aphinin is present in nearly all aphids examined, may be invariant in structure and is stable, whereas the aphins are more variable and are unstable in shed blood. The first to be investigated, from the bean aphis, *A. fabae,* and called protoaphin *fb* (PA*fb*) may be taken

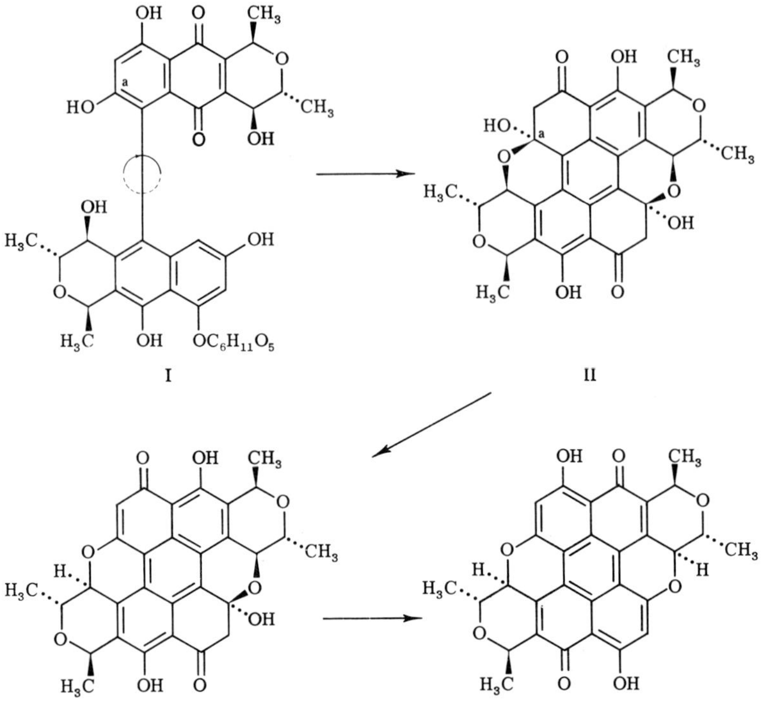

Fig. 4. Aphins of *Aphis fabae:* native glycoside (I) and series of aglycones (II–IV) formed from it in shed blood: I, protoaphin (PA*fb*); II, xanthoaphin (XA*fb*); III, chrysoaphin (CA*fb*); IV, erythroaphin (EA*fb*). (From Cameron and Todd, 1967.)

as the type (Fig. 4, I). In shed blood, two enzymes become activated, one freeing the aglycone, which by analogy with the anthocyanidins might be called protoaphidin, and the other remodeling this molecule to form a xanthoaphin (XA: Fig. 4, II). This is unstable, and in dilute acid or alkaline medium spontaneously dehydrogenates (oxidizes) to the progressively more planar, resonant and bathochromic chrysoaphin (CA: Fig. 4, III) and erythroaphin (EA: Fig. 4, IV). The aromatic core of the latter structurally has become a perylene.

Protoaphin *sl* of the willow aphis, *Tuberolachnus salignus,* is simply a stereoisomer of PA*fb* and behaves similarly in shed blood. The final EA*sl* readily epimerizes to EA*fb,* thermodynamically the more stable isomer. Dactynaphin (DA) (of the genus *Dactynaphis*) differs in having the monomers bonded via two O atoms and not directly by one C—C bond. In shed blood it is converted to a mixture of two interconvertible aglycones, xantho-DA and rhodo-DA. Heteraphin of the primitive genus *Hormaphis (Hamamelistes)* is converted to a rhodoaphin closely related structurally to a

DiOH-EA*fb*. The brilliantly orange-colored tropical species, *Aphis nerii*, has glycosides of naphthalenic monomers, and the conjugant of one is *O*-acetyl-β-D-glucose (Brown and Weiss, 1971), but even these variants are relatively minor and there is little doubt of a common origin for all.

2. Chemistry

The BQ's are sufficiently polar to be somewhat soluble in water and more so in acid and alkaline media; these provide better ionic environments for the resonant forms (Fig. 3, II) and react with them. BQ's are soluble in all organic solvents and react less with the polar members than with aqueous acids and bases. The other Q subclasses show similar solubility properties with less reactivity, and with lower solubility in nonpolar organics because of the larger number of polar substituents on the ring system. Even so they retain some of the volatility of organic materials. Their broad solubility properties facilitate extraction and separation. As for the flavonoids, mildly polar adsorbents serve in column chromatography, with ether and acetone as eluents.

The *p*-BQ's are yellow but reduce to colorless hydroquinones or quinols (Qol). This demonstrates the chromatic potency of the Q structure which has four conjugated double bonds compared with the three of the benzenoid Qol. *o*-BQ is red, the bathochromicity being due to its five resonant forms (Fig. 3, II) compared with the four of *p*-BQ. Also the proximity of the two oxygen substituents polarizes the molecule more strongly. It is less volatile, has a higher melting point, and is more reactive than *p*-BQ.

Without auxochrome substituents the *p*-NQ's are yellow, like *p*-BQ's, and similarly AQ's show no increase in bathochromicity due solely to the linear annelation of benzene rings on either side of the Q ring. This is because the Q resonance system is isolated from those of the benzene rings. Reduction of the Q unit to Qol makes that ring fully aromatic and its conjugate double bonding confluent with that of the other rings. In AQol's therefore the resonance is increased and the color shifts from yellow to red. This effect is less evident in the more stable of the PQ's since their "angular" or two-dimensional annelation provides alternative paths for electron flow; the system even of the quinone form is effectively integrated, and correspondingly bathochromic. Further contributions to bathochromicity from large molecular size and a proportionate number of auxochrome groups also affect both states in the PQ's, and even the loosely integrated protoaphins show these effects. Constraints that bring the pyran rings nearer to coplanarity with the hydrocarbon rings also act bathochromically, particularly in the compact PQ's. PQ colors are brilliant, i.e., relatively saturated, as well as bathochromic.

Unsubstituted BQ's have a single absorption peak in the visible range, but the liberally substituted insect AQ's have two or three and the EA's have six. The PA's have only two so that four are probably due to features of the large, planar condensed ring system of the compact members. The condensed ring system even of the PA's is extensive enough for the aglycones to be fluorescent but both intensity and bathochromicity of the emission increase progressively with molecular consolidation along the series: PA→XA→CA→EA (Table II). Each shows a bathochromic shift also with

TABLE II

Colors of the Fluorescence of the Series of Aphins-*fb* in Acid, Neutral, and Alkaline Media[a]

Medium	Protoaphin	Xanthoaphin	Chrysoaphin	Erythroaphin
Acid	Dull green	Blue-green	Intense yellow-green	Orange-red
Neutral	Dull green	Blue-green	Intense yellow-green	Orange-red
Alkaline	Dull violet	Brilliant yellow-green	Brilliant yellow-green	Dark ruby-orange red

[a] After Duewell *et al.* (1948).

increase in pH (Table II). In contrast to the bathochromic series for the transmission (residual) color, that of the fluorescence emission runs simply through the visual spectrum from violet to red. As usual, the fluorescence is quenched *in vivo* by the monose conjugant.

The BQ's are very sensitive to light and, particularly in the *o*-BQ's, this promotes oxidation and polymerization reactions. Annelation makes the AQ's photostable, just as it depresses most aspects of their reactivity; carmine dyes are among the most permanent. The PQ aglycones are photosensitive and are photodynamic to living organisms, the potency being related to fluorescence intensity. The AQ's are unusual in being thermally excitable, i.e., by the small quanta of heat radiation: Kermesic acid is yellow at room temperature but red in hot aqueous solution (thermochromatic response).

Quinones resemble the flavonoids in giving sharp color changes over the pH range; 2-ethyl-1,4-BQ and its methyl homologue are red in *c* H_2SO_4, yellow at pH 7.0,and green in 10% KOH. The possible cationic and anionic functions involved are indicated by the resonant forms (cf., Fig. 3, II). As among the flavonoids, the anion is more bathochromic than the cation. Among AQ's, again, kermesic acid is yellow at acid and red at alkaline pH; the anion is that of the —COOH side chain. In the aphins, as in the flavonoids, it is a phenate, and again the cation is probably an oxonium or a carbonium since *c* H_2SO_4 is necessary to ionize it. The bright red cation provides a standard test. EA remains reddish through the neutral range while the anion is green. PA is yellow-green at neutrality and has a purple anion.

The anion is particularly brilliant in a mixture of acetic acid and pyroboracetate, a test similar to that for flavonoids (Section II, B, 2, b). Again polyvalent metals form brightly colored insoluble salts with the anion, and these are technically useful.

Extremes of pH cause more extensive molecular changes, particularly in the PA's, intrinsically unstable. As for the BQ's (above) mild organic bases and NH_4OH cause less degradation than inorganic bases (Bowie and Cameron, 1967; Weiss and Altland, 1965).

The most potent property of all quinones is their reversible redox activity. The simpler Q's are the most strongly oxidizing; E_0' potentials of 1 M solutions, relative to the hydrogen electrode as zero, are as high as +950 mV. Linear annelation depresses this because conformation with the aromatic bonding of the other rings favors the quinol state. The AQ's have potentials as low as −266 mV compared with +372 mV for *o*-BQ. That of UQ is +100 mV, near the middle of the biochemical range, and so collectively the biological Q's could cover much of this range. Angular annelation increases the oxidation potential, in parallel with its increase in bathochromicity and resonance. Electron-accepting substituents on the ring system also increase it.

During the redox reactions of Q's, an intermediary with more bathochromic color and a paramagnetic signal is often detectable. This indicates an active free radical in the form of a quinhydrone complex, an association between a Q and a Qol unit, whether inter- or intramolecularly; in some of the PQ's there are intramolecular possibilities.

D. Metalloproteins

The remaining biochromes of insects contain in the chromophore system elements in addition to C, H, and O. The ligands of copper in the simple copper proteins are still unknown but theory and laboratory models indicate that they are nitrogen and oxygen functions. The iron of the simple iron proteins is ligated entirely by sulfur. Four to six ligands are usual.

The polyphenol oxidases (tyrosinases) of insects are their most important copper proteins. They contain iron, also, as cofactor. Dopamine β-hydroxylase also is a copper protein and there is reason to believe that other copper protein enzymes, uricase and amine oxidase, are present. Insect cytochrome oxidase is so similar to that of mammals that it must contain copper. Measurable amounts of copper are in the hemolymph (Wigglesworth, 1965) so that there may be an analogue of mammalian ceruloplasmin. However, unlike some of the larger terrestrial arthropods, insects have no hemocyanin.

Simple iron proteins, such as the ferredoxins, no doubt will be found in insects. Flavoproteins important in insect metabolism (Section III, A, 2), xanthine oxidase, aldehyde oxidase, and mitochondrial succinic dehydrogenase are associated, in mammals, with a group of iron proteins, consequently called flavodoxins.

The solubility of these proteins depends on site and function. *In situ* the mitochondrial and some other intracellular members are insoluble but others, including phenol oxidases, are dissolved in the cytosol, or extracellularly in the blood. Copper is usually very firmly bound to protein, the equilibrium constant being as high as 10^{18}.

The colors of these metalloproteins are mostly those of simpler compounds of the metals, blue for copper and yellow to brown for iron, implying that the protein contributes relatively little to the chromophore. However, all absorb in the region of 300–380 nm, due to charge-transfer between metal and protein, as well as to transitions between oxidation states of the metal. Also, the variations in color among the copper proteins must be due to the protein; polyphenol oxidase is colorless throughout its redox cycle of activity so that it is held by its ligands in the Cu(I) state, and all electron transitions are between *d*-orbitals of the incomplete M-shell.

E. Tetrapyrrolic Biochromes: Porphyrins and Bilins

1. Structure

This and the following classes of biochrome have N-heterocyclic aromatic rings in their molecules and some have also N-substituents on the rings. In the tetrapyrrolic chromes of insects (Fig. 5) methine bridges linking the aromatic pyrrole rings also join their conjugate double bonding so that although a free pyrrole is colorless, and a dipyrryl unit is yellow, the bilatrines (Fig. 5, I) are as bathochromic as green. Porphyrins (Fig. 5, II) have the linear tetrapyrrole chain closed to form a planar macrocycle. In biological tetrapyrroles there are also auxochrome substituents on the skeleton.

In animals most of the significant tetrapyrroles are based on one subgroup of porphyrans (metalloporphyrins), the hemes, in which an atom of iron is symmetrically ligated by the four pyrrole N-atoms of protoporphyrin IX, the key biological porphyrin, or of a near derivative. Protoheme IX itself (Fig. 5, II) is the prosthetic group of cytochrome *c* (Cyc) and hemoglobin (Hb). Most insects have no free hemes. The pyrrole side chains of insect bilins show that the latter are degradation products of the porphyrins and not by-products of their synthetic path; free porphyrins, which usually do originate in this way, are rare in insects. They synthesize most of their tetrapyrroles (Vuillaume, 1975) though some use derivatives of dietary chlorophyll.

Cobalamin, vitamin B_{12}, is a dietary requirement for insects. It is a complex analogue of the porphyrans, with cobalt the metal and the corrin

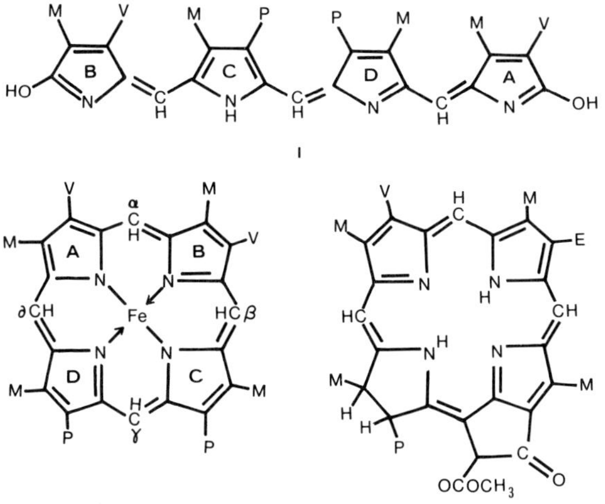

Fig. 5. Bilins and porphyrins of insects: I, biliverdin IX α; II, protoheme IX; III, phaeophorbide a. E, ethyl; M, methyl; P, propyl; V, vinyl; A,B,C,D and $\alpha,\beta,\gamma,\vartheta$, conventional labeling of pyrrole rings and methine bridges.

macrocycle biosynthetically derived from protoporphyrin by microorganisms.

The most common insect bilin is biliverdin IXα (Fig. 5, I), i.e., the macrocycle is opened at the α-bridge, as in vertebrates, but pierid butterflies produce biliverdin IXγ.

2. Distribution

Because of their direct, tracheal supply of oxygen, normal terrestrial insects do not have Hb in the blood, but a few aquatic dipteran larvae, e.g., *Chironomus* and *Tanytarsus,* have it there. It is present in the tracheal end-cells of both larvae and adults of the water boatmen, *Anisops, Buenna* and *Macrocorixa*, and of the larva of the botfly, *Gastrophilus,* parasitic in the stomach of the horse. In the young botfly larva it is present also in the fat body, muscles, and epidermis. Blood-sucking insects, such as the bug *Rhodnius* and the louse *Pediculus,* absorb varying amounts of heme from the gut but they modify and store it, and do not conjugate it to a functional Hb.

The first clear demonstration of cytochromes, by Keilin, was in insects and here the operation of the system is now fairly clear (Sacktor, 1974). Less is known of peroxidases and other heme-enzymes in insects but tryptophan oxygenase (pyrrolase), the first enzyme of the ommochrome and papiliochrome pathways, has a heme prosthetic group (Linzen, 1974).

Biliverdin IXα occurs in the integuments of various Phasmida, Mantida and Orthoptera, e.g., *Carausius, Mantis, Locusta, Schistocerca,* and *Tettigonia,* and of the larva of the privet hawkmoth, *Sphinx ligusta.* In *Tettigonia* there are also three other blue-green bilins. The larva of the tortricid moth, *Choristoneura* has biliverdin IXα in its blood (Schmidt and Young, 1971). *Chironomus* larvae store in the fat body both biliverdin IXα and bilirubin, a red-brown diene (i.e., with one methine bridge saturated). According to the local redox potential, the bug *Rhodnius* stores either biliverdin or bilirubin, formed from dietary heme. Biliverdin IXγ has been found only in the wings of pierid butterflies. In the wings of papilionids, atticids, and micro-Lepidoptera, and in the integument of the larva of the moth, *Antherea pernyi,* three new blue to blue-green bilins have been recognized and called phorcabilin, isophorcabilin and sarpedobilin (Choussy and Barbier, 1973). Since bilirubin is formed from biliverdin by intramolecular rearrangement and not by true saturation there is no independent evidence that the summer fawn color of the integuments of various orthoptera is due to partial saturation of the spring biliverdin (Vuillaume, 1975).

In the epidermis, salivary glands, and testis sheath, the squash bug, *Anasa tristis,* stores as red crystals phaeophorbide *a,* formed from chlorophyll *a* by removing the magnesium, and the phytyl side chain (Fig. 5, III). The green chrome of its fat body and related tissues also may be a chlorophyll derivative.

3. Chemistry

The tetrapyrrole skeleton is relatively insoluble in polar media but biological members retain the polar side chains of the biosynthetic precursors; uroporphyrin, the first porphyrin stage, is very soluble even in neutral aqueous media and insoluble in organic solvents even as polar as ether. Protoporphyrin has a smaller number of polar side chains but still is soluble in acid and alkaline aqueous media and in most organic solvents. The range is rather similar to that of the quinones. Bilins are less soluble both in aqueous media and in apolar lipids but in insects they either are in granules or vacuoles or are protein-conjugated. The hemoproteins are soluble or not, depending on functional site.

Comparison of the green bilatriene with porphyrin colors shows that macrocyclization has a pronounced hypsochromic effect; this is due to the shortening of the molecular axis. It is not fully compensated by ligating iron and by bonding with protein and this implies that bathochromicity is not the only desirable property. Macrocyclization produces a uniquely intense, sharply focused absorption band, the Soret band, in the region 400–420 nm (for porphyrins collectively), due to the combined action of the four symmetrically placed pyrrole nitrogens. Quanta of light in this region are adequate to drive reactions requiring 300 kJ per mole of activation energy, compared with 150 kJ for far-red wavelengths. Macrocyclization produces also a series of other strong absorption bands in the visible range, collectively in the regions 487–502, 517–537, 560–575 nm (neutral media). A band at 613–632 nm probably corresponds to that at 653 nm in biliverdin IXα and the band due to the individual pyrrole rings at 260–280 nm also is common to both. The bilin has in addition only a rounded peak at 376 nm, possibly indicating a tendency for its molecule to macrocyclize.

The planar porphyrin also permits a ligated Fe atom to complete its octahedral field by a bond in each direction perpendicular to the plane. Ligands of these bonds have effects on the absorption spectrum that are useful in identifying porphyrans; pyridine "hemochromes" are much used for this purpose.

The macrocycle is responsible also for a brilliant red fluorescence of free porphyrins, again absent from free bilins, though the Schlesinger test for bilatrienes and dienes involves the production of a fluorescence that may be due to macrocyclization, since there occurs also a hypsochromic shift in color and a number of new absorption peaks in the visible range. Fluorescence is quenched by the Fe atom, which also depresses the intensity of the Soret absorption. Proteins, which usually are directly bonded to the macrocycle as well as indirectly via the iron, also quench the fluorescence. Free porphyrins are extremely photosensitive and cause photodynamic damage in exposed tissues. In many insects all parts of the body are vulnerable and the absence of free porphyrins may be mandatory.

The porphyrin structure is exceptionally stable chemically, its biological reactions being "reversible." Porphyrins have endured in fossil material as old as the Silurian. The bonding with iron and protein also is stabilized so that by laboratory methods neither can be removed until the macrocycle has been opened. The verdoglobin so produced from Hb then is deproteinable to verdoheme, which yields biliverdin and free iron.

Like flavonoids and quinones, porphyrins are amphoteric; at their isoelectric point, pH 3.0–4.5, they are least aquasoluble and least fluorescent. Strong acids suppress the ionization of side chain —COOH and aromatic —OH acids and promote that of pyrrolic nitrogen bases as iminium cations. Acids depress the absorption peaks of the visible range and shift them bathochromically. Alkalies cause a 20–30 nm hypsochromic shift but also depress the intensity. The implication is that hemes are most potent in a hydrophobic environment.

The cyclical redox properties of the hemoproteins, including oxygen transfer, depend on the metal's ligand field and contrast with the emphasis, in quinones and flavonoids, on functions at the periphery of the chromophore. As in the simple metalloproteins, the iron may not change its oxidation level during the redox cycle: it is Fe (II) throughout the Hb cycle and Fe (III) throughout that of the cytochromes. Excitatory electronic transitions again are mainly between orbitals of the *d*-group of the M-shell.

The stability of porphyrins contrasts with the reactivity of bilins, which provides a number of diagnostic tests (Needham, 1974). Gmelin's series of reactions involve a stepwise oxidative saturation so that the color runs through a complete hypsochromic series to colorless bilanes. Color-change tests distinguishing between triene, diene, and monene, using $FeCl_3$ in methanol, involve desaturation of methylene bridges back to methine. Bingold's reaction, characterizing all but monenes, gives a red "pentdyopent" (525 nm absorption peak) product after oxidative fission by H_2O_2, followed by reduction of the two dipyrryl units with dithionite.

F. Indolic Melanins (Eumelanins) (Nicolaus, 1968)

1. Structure and Distribution

Eumelanins are high polymers of indolic quinones, i.e., of pyrrole-condensed benzoquinones, often with some heterogeneity among these monomers. They are synthesized by the insect itself in a branch (Fig. 6, I–IX) of the tyrosine oxidation pathway, which also produces the catecholic (BQ) melanins (Section III, C, 3). The pyrrole ring is formed by cyclization of the alanine side chain of tyrosine (Fig. 6, IV). In laboratory model systems, and in some *in vivo* situations, red to violet oligomer intermediaries are detected.

Because of the similarity and the inertness of the catecholic and indolic melanins, and of their common initial pathway, they are not easily distinguished; their respective distributions are still very uncertain. The general tyrosine oxidation–polymerization pathway has been found in all insects examined, including *Gryllus, Locusta, Tenebrio, Tyria, Hestina, Pieris, Drosophila, Lucila, Habrobracon, Microbracon,* and *Gilpinia*. In most it has been found locally, in the developing exoskeleton and related structures, and dispersed in the hemolymph. The former may always lead to catecholic and the latter to indolic melanins; these eumelanins may be deposited in the epidermis and other tissues (Hackman, 1974).

Both are conjugated with protein, *in vivo,* but insect eumelanin is probably always black (the pheomelanins of vertebrates not having been detected) while the catecholic melanins vary from golden to red-brown. They are diffusely distributed, whereas the eumelanins usually form microscopic granules similar to those in other animals. On this criterion the industrial melanism of *Biston betularia* and other Lepidoptera is due to eumelanin. Some insects have black granular eumelanin in the lower layers of the exoskeleton. The enzymes for this are passed up from the hemolymph in the same way as for epidermal melanin.

The catecholic and indolic branches of the common path are necessarily under discriminant controls, genetically based in *Drosophila* (Brunet, 1963);

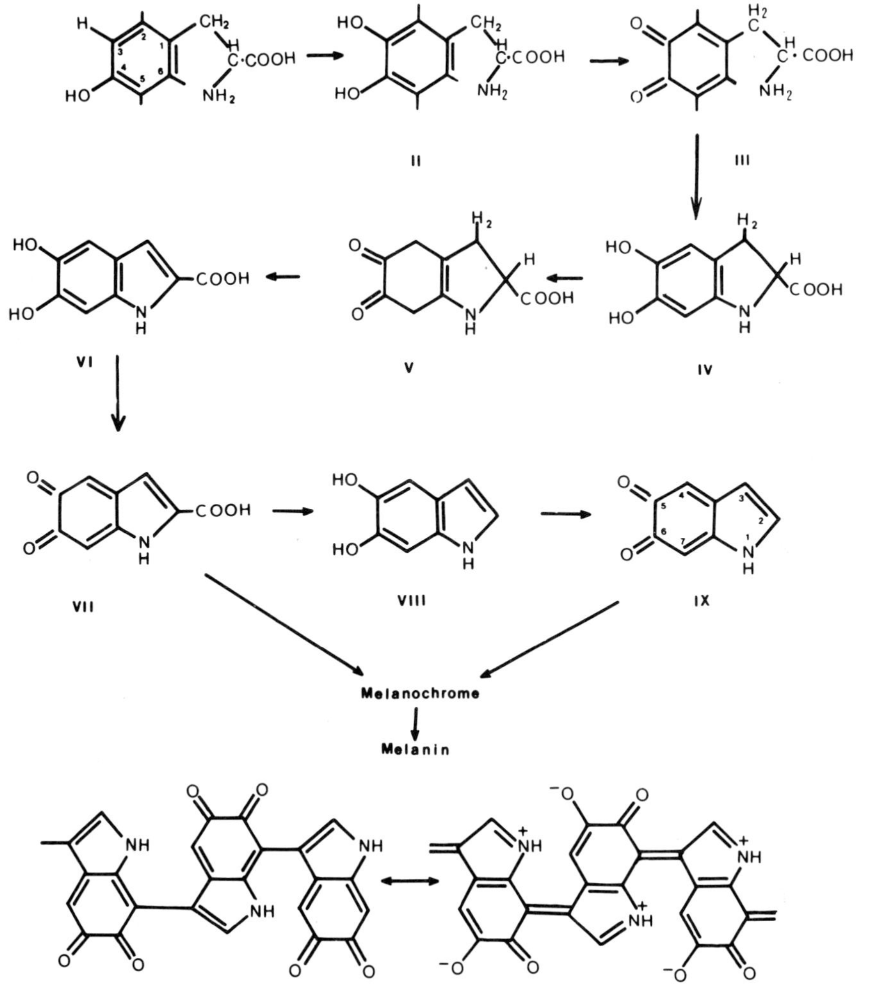

I
II
III
IV
V
VI
VII
VIII
IX
Melanochrome
Melanin
X

Fig. 6. Indolic melanin (eumelanin): I–IX: stages in biosynthesis from tyrosine (I): II, dopa (dihydroxyphenylalanine); III, dopaquinone; IV, leucodopachrome; V, dopachrome; VI, 5,6-dihydroxyindole-2-carboxylic acid; VII, indole-5,6-quinone-2-carboxylic acid; VIII, 5,6-dihydroxyindole; IX, indole-5,6-quinone. X, model of bonding of monomers in melanin polymer, and resonant forms of polymer. (From Bu'lock and Harley-Mason, 1951.)

black eumelanin can be suppressed independently of brown BQ melanin. In the BQ branch of different insects two alternative devices for preventing pyrrole ring closure have been found; one is to remove the amino group of the alanine side chain and the other to cover it by acetylation, forming acetyldopamine. Since the later steps in both paths are spontaneous, controls must be imposed before those stages (Gilmour, 1965). Experimental discriminants include the inhibition of eumelanin synthesis by phenyl thiourea (Dennell, 1958) and by ascorbic acid (Gilmour, 1965). Recently tyrosinases with different substrate specificities, relevant to the respective paths, have been recognized (Hackman, 1967), so that the two may be segregated from the outset, but in general these enzymes have a very broad substrate acceptance (Section III, F, 4).

2. Chemistry

Eumelanins are too highly polymeric to be very soluble in the common solvents, aqueous or lipid. However some macromolecular material does dissolve in hot strong acids and alkalies, in acetic and formic acids, and in some neutral and basic organic fluids such as ethylene chlorhydrin and diethylamine. The solute is amphoteric and precipitates at an isoelectric point. Solutions are yellow, brown, or black, according to concentration, absorbing throughout the visible and near UV ranges, the intensity being inversely related to wavelength. Some eumelanin solutions have a definite absorption peak around 500 nm (Hackman, 1953) and most have two between 300 and 350 nm. By contrast catechol melanins have sharp peaks below 300 nm.

Both native melanins and their solutions are chemically stable and inert, notwithstanding the number of N and O functions on and in the rings, and the extended resonance system (Fig. 6, X). Reducing agents have little detectable effect and only strong oxidizing agents such as H_2O_2 turn them through red and yellow to a colorless stage. This is irreversible and presumably involves progressive bond saturation and depolymerization. The intact resonance system is a potent charge-transfer matrix, possibly to the extent of acting as a semiconductor. The excitation energy for electron transitions is as low as 13.4 kJ per mole so that eumelanins are excited even by much of the infrared range. They also strongly bind metals, particularly Fe, Cu, Zn and Ca, and act as good cation-exchange material.

G. Ommochromes (Phenoxazines)

1. Structure

The trivial name indicates that Becker (1941) first found this class in the ommatidia of insect eyes, while the semisystematic name gives the essential

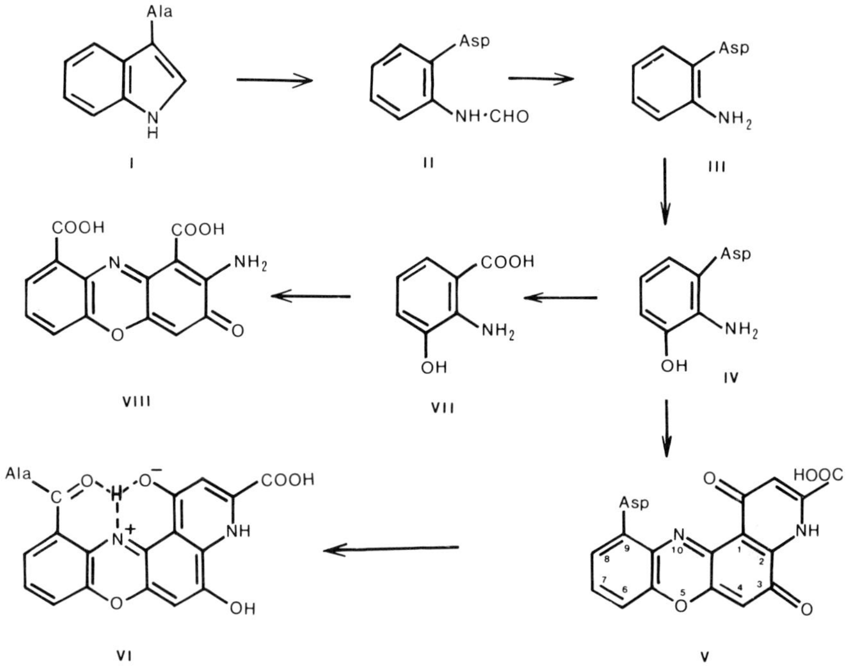

Fig. 7. Stages in pathway of biosynthesis of ommatins from tryptophan (I): II, formylkynurenine; III, kynurenine; IV, 3-hydroxykynurenine; V, xanthommatin (X); VI, dihydrogenxanthommatin (DHX); VII, 3-hydroxyanthranilic acid; VIII, cinnabarinic acid.

molecular components (Fig. 7, V) and the biosynthesis of the best-known subclass, the ommatins (phenoxazinones): an O and a N atom bridging two phenolic units, one of which bears a quinone function; this is shared between the main ring and a second formed by condensation of the aspartyl side chain (this unit is effectively 8-ketoxanthurenic acid). Ommochromes are abundant in both eyes and integument of insects (Linzen, 1974) and of other arthropods. They occur also in other phyla of the spiralian group but outside this have been found only in the medusa, *Spirocodon*. They are synthesized by these animals, and also by some fungi and bacteria. Only the pathway to the ommatins is known in any detail (Fig. 7, I–VI).

The two phenolic units are derived from 3-hydroxykynurenine (3OHK: Fig. 7, IV), a key derivative of tryptophan, along its main catabolic path. The aspartyl side chain results from opening the pyrrole ring of Trp. The initial steps in this pathway were discovered through classical studies on eye color mutants in *Drosophila* and other insects. The final dimerization step is still incompletely known but it involves the NH_2 and OH side chains of one monomer and can be catalyzed by a number of nonspecific enzymes (Section III, F, 4). The completed ring system is virtually planar.

Cinnabarinic acid (Fig. 7, VIII), structurally the simplest of the ommatins, is formed by dimerization of 3-OH-anthranilic acid (3OHA): (Fig. 7, VII). The other two subclasses, ommins and ommidines, contain sulfur derived from methionine in the molecule, but the details of their structure and synthesis are still uncertain. The ommins are macromolecular, non-dialyzable.

2. Distribution (see Linzen, 1974)

Xanthommatin (X: Fig. 7, V) is almost ubiquitous in insect eyes and integuments. In the eyes of Diptera, and in the integuments of larvae of *Vanessa (Aglais)*, *Cerura*, and *Anagasta*, it is the only member. The ommins are almost equally widespread in both sites, except in some Orthoptera and Diptera, and both subclasses occur together in most sites. Rhodommatin and ommatin D, respectively the glucoside and sulfate of reduced X (diHX) are found only in the wings and other organs of certain butterflies, which, among Lepidoptera, are unusual in not having ommins. Two other ommatins, the acridiommatins, are restricted to orthopteroid and odonatan taxa. Cinnabarinic acid is known only from the silkworm, *Bombyx*, among insects. Ommidines have been found in the eyes of orthopteroids and in the epidermis of the flightless grasshopper, *Romalea microptera*.

Ommochromes are usually bound to protein, in 0.4–0.6-μm intracellular granules. In some granules of ommatidial cells X is free and partially oxidized, but protein-bound and reduced to diHX in others. Rhodommatin and ommatin D are usually in solution, free or in vacuoles, never in granules.

3. Chemistry

Like the other aromatic polycyclic biochromes of small molecular size, the ommochromes are soluble in acid and alkaline solvents; they are most soluble in organic acids and bases. This reflects their combination of hydrophobic skeleton with hydrophilic, ionizing substituents. The most potent extractants from biological material are strong formic acid, the pyridine bases, and methanolic HCl. Mineral bases dissolve, but rapidly degrade them. Rhodommatin and ommatin D are made water-soluble by their glucose and H_2SO_4 conjugants. The insolubility of other ommochromes may be increased by a tendency to molecular aggregation (Linzen, 1974) which also may explain their dark *in situ* colors, so that they were long thought to be melanins. It may also be the reason for the apparent large number of ommins and their large particle size. Chromatographic analysis usually gives a number of bands, each perhaps representing a stage of polymerization. The degree of aggregation varies with pH and with the concentrations of other ions. Aggregation is a property common to all molecules with a number of hydrophobic interactions, as illustrated best by the melanins. These interactions also cause the strong adsorption to cellulose and other weakly polar materials. Only strongly polar solvents will elute ommochromes from these.

In solution ommochromes range from yellow through red and purple to violet, the main visible absorption peak lying between 430 and 550 nm, therefore. That of X in dilute acid is near the shortwave end but protein conjugation can shift it as far as the other end (Linzen, 1974). The ommatins, but not the ommins, have a second peak in the near UV and all have the peak of the aromatic ring at 280 nm, and one at 225–240 nm. In strong mineral acids ommochromes have a green UV fluorescence, common among Trp derivatives. It is quenched in the physiological pH range and so may be due to a resonance-enhancing cation.

This is relevant to a stabilization of ommatin D, against oxidation and fission, by its H_2SO_4 conjugant and to a very bathochromic "halochrome" formed by ommochromes in strong mineral acids. The most probable cation is the bridge iminium (Fig. 7, VI) and as usual ionic and redox properties chromatically interact. The ommins, and still more the ommidines, have greater stability than the ommatins against alkalies and oxidation, probably because of the sulfur in their molecules. They also give the halochromic response at lower acidity: indeed in *c* H_2SO_4 ommins show a marked hypsochromic shift back to yellow.

Yellow xanthommatin is readily and reversibly reduced to red diHX, the absorption peak, in neutral phosphate, shifting from 440 to 495 nm (Linzen, 1974). DiHX spontaneously reoxidizes in air, and, in the eggs of *Urechis,* X reduces to diHX simply on lowering the ambient O_2 pressure (Needham, 1974), so that this may be a functional redox system also in insects. By contrast, rhodommatin and ommatin D are held stabilized in the reduced state by their conjugants and the ommins also only slowly oxidize in air, even in neutral media. Their redox potential is as high as +196 mV and oxidants as strong as nitrite are necessary to oxidize them.

Like the AQ's (Section II, C, 3), xanthommatin reduces to a more bathochromic color and this is due again to the aromatic resonance system of the quinol-form fusing with that of the rest of the molecule. X has two alternative Q systems both involving the =O at position 2 but coupled with either the other =O of the xanthurenic unit or the imino =N at position 10. In view of the bathochromic change on reduction the second H atom should go to the xanthurenic =O since at 10 it would break the conjugation across the N-bridge. However this overlooks the possibility that 10N assumes its iminium form, further adding to the bathochromicity, and at present (Schäfer and Geyer, 1972) it is uncertain which of the three sites indicated on Fig. 7, VI actually holds the H.

4. Other Zoochromes of the Kynurenine Pathway

Unlike vertebrates, insects are unable to catabolize tryptophan completely, and ommochromes are not the only class of colored product. Papiliochromes, providing white and yellow coloring to the wings of some Papilionidae (Umebachi, 1975) have some analogy to ommochromes, since the molecule can be broken into two benzenoid moieties, one of which is a kynurenine derivative. However, the other is an *o*-diOH-benzene derivative, in fact a dopa derivative with a β-alanine side chain, as in the catechol melanin path of some insects (Section III, C, 3). Papiliochrome synthesis therefore may involve an interaction between the two pathways. The exact structure of the complete papiliochrome molecule is not yet known. The dopa derivative is not in the quinone form since dopa-Q is known to form brown or red amino-Q's with K and 3OHK.

The class is soluble in neutral aqueous media and in 70% ethanol so that it is strongly polar and not predominantly aromatic. The hypsochromic colors indicate a limited resonance system and weak bonding between the two moieties. This is confirmed by the fact that $10^{-3}M$ HCl is adequate to separate them. Papiliochromes have no detected redox properties and this again implies a limited resonance system. Their most bathochromic absorption peak is at 380 nm, not far beyond that (360 nm) of kynurenine itself.

The cocoon silk of the robin (*Cecropia*) moth, *Samia cecropia*, is light brown, and hardened by an oxidation derivative of 3-OH-anthranilic acid (Fig. 7, VII). Some relationship to cinnabarinic acid may be suspected.

H. Purines, Pteridines, and Isoalloxazines

1. Structure

The insects use uric acid (Fig. 8, I) and some other purines for visual effect; the molecule is an imidazole-condensed pyrimidine. From a purine nucleotide, guanosine 9-triphosphate (Fig. 8, II), pteridines (Fig. 8, III, IV) are biosynthesized by opening and extending the imidazole to a pyrazine ring. As the trivial name indicates, pteridines were first discovered by Hopkins in insect wings and most of the biological members belong to that particular subclass, the pterins; their 2-NH_2, 4-OH substituents indicate the guanine origin (a different numbering convention for purines and pterins tends to obscure this). Many pterins have now been found in insects (Table III) and all the unconjugated members are synthesized endogenously. The conjugated member, folic, or pteroylglutamic, acid (Fig. 8, V) by contrast is a dietary requirement.

Similarly the only common isoalloxazine, ribitylflavin (RF) (Fig. 8, VI) is demanded in the diet and is synthesized from a lumazine, i.e., a 2,4-diOH-pteridine, some of which occur in insects and are probably synthesized by them. The precursor of RF is the 6,7-diMe member (Fig. 8, VII) so that it already has part of the benzene ring that is eventually condensed on to the pyrazine. Also it is synthesized with the ribityl substituent already at position 9 and becomes the mononucleotide coenzyme, FMN (Fig. 8, VI), by phosphorylation. The flavin-adenine dinucleotide, FAD, is more versatile in insects, as in other organisms.

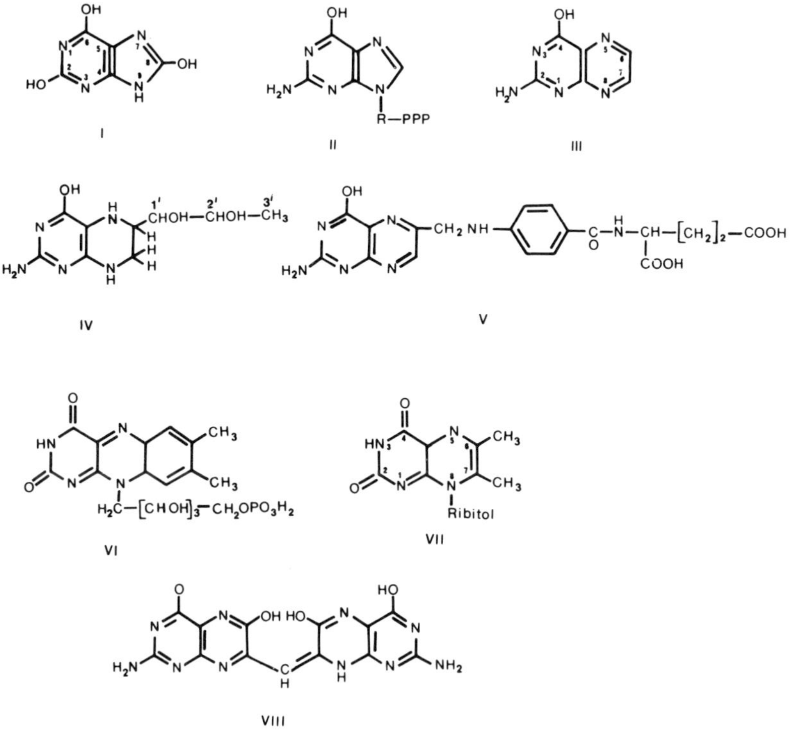

Fig. 8. Purines, pteridines and flavin of insects: I, uric acid (enol/lactim form); II, guanosine triphosphate; III, pterin (AHP, 2-amino-4-hydroxypteridine); IV, tetrahydrogenbiopterin (THP); V, pteroylglutamic acid (folic acid); VI, ribitylflavin; VII, 6,7-dimethyllumazine (keto form of 2,4-dihydroxy-6,7-dimethyl pteridine); VIII, pterorhodin (pterin dimer).

2. Distribution

In addition to pierid wings other integumental sites of insects display white purines. Colorless pterins, e.g., leucopterin, usually accompany them. Uric acid is present also, in high concentration, in the fat body, Malpighian tubules, and hindgut of many insects. Various colored pterins are present in the wings of Lepidoptera, and in the integument of larval and adult insects. They are abundant, along with the ommochromes, in the screening cells of the ommatidia and in various internal organs. Few other taxa have so extensively exploited this subclass, and the lumazines (Table III). Some are known only from insects and the interesting dimer, pterorhodin (Fig. 8, VIII) is one example. Like most of the integumental chromes, *in situ,* they are bound to a protein matrix in intracellular chromasomes ("granules" to the light microscopist). Folic acid (Fig. 8, V) being an important vitamin coenzyme (Section III, A, 2) probably is widely distributed in insect tissues. Ribitylflavin as the essential component of the prosthetic group of the

TABLE III
Known Distribution of Pteridines among Insects[a]

Pteridine	Distribution
I. Pterins (2-NH_2-4-OH-pteridines)	
1. Neopterin [6-(L-*erythro*-1′,2′,3′-triOH-propyl)-Pt]	*Apis*
2. Biopterin [6-(L-*erythro*-1′,2′-diOH-propyl)-Pt]	Phasmida, Hemiptera, Diptera, Hymenoptera, Lepidoptera
3. Sepiapterin (6-lactyl-7,8-diH-Pt)	Orthoptera, Lepidoptera, Hymenoptera, Diptera
4. Isosepiapterin (6-propionyl-7,8-diH-Pt)	Lepidoptera, *Apis, Drosophila*
5. Drosopterin, isodrosopterin, neodrosopterin (? 6-diOH-propenyl-Pt derivatives)	*Anagasta, Drosophila*
6. Ranachrome-3 (6-OH-Me-Pt)	*Drosophila*
7. 6-Carboxypterin	Phasmida, Hemiptera, Diptera, *Apis*
8. Pterin (AHP = 2-NH_2-4-OH-Ptd)	Phasmida, Orthoptera, Lepidoptera, Diptera
9. Xanthopterin (6-OH-Pt)	Orthoptera, Hemiptera, Neuroptera, Coleoptera, Lepidoptera, Hymenoptera, Diptera
10. Leucopterin (6.7-diOH-Pt)	Coleoptera, Hemiptera, Lepidoptera, Hymenoptera
11. Chrysopterin (6-OH-7-Me-Pt)	Hemiptera, Lepidoptera
12. Isoxanthopterin (7-OH-Pt)	Phasmida, Orthoptera, Hemiptera, Lepidoptera, Hymenoptera
13. Ekapterin (β-(6-OH-Pt-7-yl) α-OH-propionic acid)	*Anagasta*
14. Erythropterin (6-OH-7,8-diH-Pt-7-pyruvic acid)	Hemiptera, Lepidoptera
15. Lepidopterin (β-(6-OH-Pt-7-yl)α-NH_2-acrylic acid)	Hemiptera, Lepidoptera
II. Lumazines (2,4-diOH-Pteridines)	
16. Lumazine (2,4-diOH-Ptd)	*Formica*
17. Luciopteridine (7-OH-8-Me-lumazine)	*Luciola*
18. Violapteridine (7-OH-lumazine)	*Pyrrhocoris, Apis*
19. 6-Lactyl-7,8-diH-lumazine	Lepidoptera
20. Formicapterin (poly-OH-alkyl lumazine)	*Formica*
III. Dimeric pteridines	
21. Pterorhodin (? dichrysopterin)	Lepidoptera

TABLE III (*continued*)
Known Distribution of Pteridines among Insects[a]

Pteridine	Distribution
IV. Conjugated pteridines	
22. Pteroylglutamic acid (N-{4-[(2-NH_2-4-HO-6-pteridyl methyl)-NH_2]-benzoyl}glutamic acid)	Probably universal

[a] Me, methyl; Pt, pterin; Ptd, pteridine.

flavoprotein (FP) enzymes must be present in all insect tissues. In addition it is stored in high concentration in the Malpighian tubules and other organs of various species. In accord with its dietary origin the greatest stores are in phytophagous insects.

3. Chemistry

a. *Purines.* These are virtually insoluble in all pure, neutral solvents, polar and apolar, though guanine has some solubility in warm water. They are soluble in most dilute alkalies, forming phenates of the —OH side chains, and in acids, forming the salts of their ring N-bases; unsubstituted purine is a strong base. Guanine is distinguished from other purines by its insolubility in NH_4OH and in citric, lactic, acetic, and formic acids. The condensed aromatic ring system, with two N-atoms per ring and N and/or O functions in side chains should give a bathochromic color, but most biological members show keto-enol, coupled with lactam-lactim, tautomerism and this disrupts the resonance system. Also the molecule lacks a definite polar axis. The aromatic rings absorb in the 250–260 nm range. The structure and solubility properties show that purines are amphoteric and this is true even of purine itself, but they do not have a simple pair of ionizations in the physiological pH range. Uric acid behaves as a dibasic acid and the N-bases of its ring system ionize only at high acidities. The —OH substituents have also quinolic reducing properties used histochemically for identification and localization. Both purines and pteridines give a characteristic "murexide" color reaction, which depends on oxidative opening of the imidazole or the pyrazine ring, dimerization of the resulting pyrimidine, and formation of Na^+ or NH_4^+ salts of the dimer, purpuric acid.

b. *Pteridines.* In solubility and some other properties these are very similar to the purines. They are insoluble in cold neutral solvents, polar and apolar, but somewhat soluble in hot water and much more in dilute acids and alkalis. *In vivo* most are conjugated with insoluble proteins. They have a degree of keto-enol/lactam-lactim tautomerism, and poor molecular polarization, and in general the color is less bathochromic than the structure might imply. However, the pyrazine ring contributes one more conjugate double bond than imidazole and one more position for substituents. Even so, most of the color of the pterin subclass is due to one further extension of the conjugate bonding, by the substituent at positions 6 or 7. The most effective functions here are =O, =C (OH)— and —C(=O)—. Biopterin (Fig. 8, IV) is almost colorless while sepiapterin (Table III) is yellow and the drosopterins are red. The positions of the absorption peaks of extracted solutions are more hypsochromic than the names and the *in situ* colors imply, perhaps owing to separation from bathogenic conjugants.

In contrast to purines, free pterins have a strong UVF, the emission being maximal at a wavelength slightly shorter than that which dominates the visible transmission. The emission spectra of xanthopterin and leucopterin have as many as four peaks, i.e., more than their absorption spectra. *In vivo* fluorescence is quenched by the protein-conjugant. Strong acids quench the UVF of free pterins while alkalies enhance it, up to the pH of 5 *N* NaOH. It is quenched also by both oxidizing and reducing agents, and by visible light. Pterins are very photolabile.

The ionization constants of the acidic and basic groups have less extreme pH values than those of purines but they are weak acids and bases, forming salts only with strong monovalent metals and mineral acids respectively. Once more the phenolic ionizations have the more bathochromic effect, the main absorption peak in the visible range being 40 nm nearer the red in 0.1 *N* NaOH than in 0.1 *N* HCl. The fluorescence emission shows a much greater difference in the same sense: the color of the UVF of xanthopterin is yellow in dilute acid, blue in neutral, and blue-green in alkaline medium. Pterins give a blue product with *c* H_2SO_4 indicating a similar low-pH bathochromic cation to those of carotenoids, flavonoids, and ternary quinones.

The pterins have mild oxidizing properties and are reduced via a 7,8-diH to a 5,6,7,8-tetraH state. The 2-NH_2 and 4-OH substituents facilitate reducing activity by the TH form so that some pterins are effective cyclic redox agents *in vivo*. In its active form the folic acid coenzyme (Fig. 8, V) is in the TH state though redox changes are somewhat incidental to its main function (Section III, A, 2).

c. *Ribitylflavin (RF)*. In spite of the dimethylated benzene ring of the isoalloxazine nucleus, the 9-ribitol conjugate is much more soluble in water than free pterins. It is soluble also in neutral polar organic solvents such as acetone and ethanol. The further conjugants, to form flavoprotein (FP) enzymes, increase aquasolubility.

Addition of the dimethylbenzene unit opposite the pair of O substituents on the pyrimidine ring gives the molecule a well polarized long axis. The third ring also confers symmetry about a polar transverse axis through the two pyrazine N-atoms, analogous to that of AQ's and ommatins. Both features increase the resonance energy and RF is a most potent, if somewhat paradoxical, biochrome. Its color is less bathochromic than that of some of the pterins, due to a depression of aromaticity by the ribityl substituents and by the *m*-Q configuration on the pyrimidine ring. Indeed the functional reduced state is colorless. Nevertheless the absorption spectrum of the oxidized form has peaks at 223, 267, 373, 445 and 475 nm, three of them affecting visible color. Further, in the course of its metabolic redox cycle it assumes very bathochromic, active intermediary forms (Kamin, 1971). The 475 and 445 nm peaks are related to $\pi \rightarrow \pi^*$ and that at 373 nm to $n \rightarrow \pi^*$ transitions, conversely to what might be expected, but depending respectively on the whole system and on an N or O function.

RF has a brilliant yellow fluorescence, the emission peak being near the wavelength of maximal transmission of visible light. Both are more bathochromically situated when RF is in the solid state and both are affected in parallel by other relevant agencies. Strong acids and bases, outside the pH range 3.0–9.0, both bleach the visible color and quench the fluorescence. Reducing agents also have both actions, reversibly. Both properties are stable to oxidizing agents such as $KMnO_4$. Fluorescence is 90% quenched when FMN conjugates with the adenine nucleotide and completely quenched by further conjugation with protein. RF readily forms charge-transfer complexes with many aromatic molecules, always with quenching of fluorescence. Fluorescence is maximal in weakly acid media; in this, as in the stability toward $KMnO_4$, there is a sharp contrast in the pterins. RF is extremely photosensitive; in alkaline medium it is converted to 6,7,9-triMe-isoalloxazine (lumiflavin) and in acid media to lumichrome, actually a member of the isomeric, alloxazine class. In turn RF has a powerful photodynamic action on other biological materials and needs to be screened in small animals such as insects.

c H_2SO_4 induces the ionization of a red cation, and 50% NaOH solution a green anion, again the more bathochromic of the two. As in other classes, particularly the flavonoids, ternary quinones,and ommochromes, these ions play an important part in the essential redox properties of the biochrome. Controlled reduction, *in vitro,* reveals in sequence two bathochromic intermediaries, first a green verdoflavin, due to a flavin radical F˙AH, and then a red rhodoflavin due either to the anionic group, $F^{\cdot}A^-$, or to a cationic group $[MF^{\cdot}A]^+$, where M is one of the metals which are associated with the FP enzymes. The intermediaries are paramagnetic showing that at least one orbital, probably of the metal, temporarily has an unpaired electron (Kamin, 1971).

Free RF and its nucleotides have a redox potential in the region of −185 mV, implying that it is as strongly reducing as the purines and pterins. However the apoprotein raises the potential of the FP's to around −60 mV and the oxidized RF now accepts electrons as readily as the reduced form donates them. There is little difference in resonance energy between the two, largely due to the reduced form partially folding around the pyrazine transverse axis, decreasing the planarity. This is an outstanding illustration of functional molecular evolution. The redox change remains of the pterin type, i.e., ±2 H; the reduced form is therefore colorless because resonance across the pyrazine ring is completely broken by the bond saturated by these two H atoms.

I. Conclusions

The significant molecular properties common to insect biochromes are as follows:

1. Absorption peaks in the visible range, therefore excited by small quanta.
2. Easily excited photically or chemically; excitation transferred to other molecules, usually chemically.
3. Extensive, planar, conjugately double-bonded system, giving a continuous field of delocalized electrons, increasing resonance energy and electron excitability.
4. Excitation facilitated by electrical polarity, giving a large dipole moment.
5. Planarity usually improved by condensed ring structure, particularly two-dimensional.
6. Excitation of this type of structure readily re-emitted as fluorescence, but put to use (fluorescence-quenched) by conjugants.
7. Excitability increased by auxochrome groups based on N (mainly in ring system) and O (mainly in side-chain substituents).
8. Excitability increased also by bathochromic ionizations; these have incidental value as pH indicators.
9. Reversible redox properties; good donors and acceptors of electrons. Redox indicator value.
10. Limited aquasolubility due to hydrocarbon content, providing hydrophobic environment which facilitates electron shifts and redox activity. Adequate aquasolubility achieved by (1) N and O polar substituents, (2) salt formation, (3) conjugation with H_2SO_4, H_3PO_4, glycoses, proteins.
11. Strong polycovalent bonding affinities, mainly quinonic, producing insoluble, durable polymers and conjugates with protein.

These properties show that potentially all of them could function in energy transfer metabolism. It is well established that this is the major function of ribitylflavin, the hemes, ubiquinone, folic acid, and cobalamin, all except the hemes being obtained by insects from dietary sources. Evidence of this, for all insect biochromes, will be the first consideration of the next section.

III. ROLE OF BIOCHROMES IN INSECTS

Only biochemical functions are relevant here so that the visual effect functions—one of the most important groups to the comparative physiologist and ethologist—must be disregarded. They include camouflage (crypsis) and advertisement (semasis) effects of externally visible coloration, expounded by Poulton (1890), Thayer (1918), and Cott (1940) in particular. These concern insects as much as any taxon and an extensive literature is available, from genetic and evolutionary as well as from physiological aspects.

Some of the externally visible chromes serve other functions and therefore contravene the rules of crypsis and semasis. Some of these functions depend on the specific chemical properties of the biochrome and are relevant here, whereas for the typical visual effects these specific properties are quite incidental. Every chemical class of zoochrome has produced members that collectively include most of the spectral hues, so that, in principle, any one of the hues in a particular color scheme could be provided equally effectively by the appropriate member of any class. A matched set of yellow hues provided respectively by carotenoid, flavone, quinone, bilin, ommatin, pterin, and flavin *ipso facto* affect the visual and cerebral systems of an observer identically. Insect integumental chromes involved in visual effect include carotenoids, bilins, phenolic and indolic melanins, ommochromes, papiliochromes, purines, pterins, and sometimes flavonoids and flavin. Visible subintegumental chromes include Hb, AQ's, and aphins so that effectively all classes are involved in this function as a whole.

A. Oxidation–Reduction and Related Functions

This is the main chemical role of biochromes. A number of them cooperate as coenzymes of the central electron transport system of respiratory metabolism. Some are involved in peripheral and minor redox pathways, not all clearly defined as yet. A few insects have oxygen-transporting hemoglobin.

1. The Electron Transport System

It is now well established (Gilmour, 1965; Sacktor, 1974; Agosin and Perry, 1974) that this system (Fig. 9) is very similar to that in mammals, and shows only minor variations among insects. The coenzymes of all but the first enzyme system, i.e., NAD or NADP, are familiar biochromes considered in Section II and since the NAD coenzymes have an absorption-peak at 340–360 nm when activated, they also are biochromes, particularly for insects (Section III, D, 1).

As in mammals some substrates, e.g., succinate, are oxidized directly by the FP system but most, malate, isocitrate, diH lipoate, L-β-OH acetyl-CoA, D-β-OH-butyrate, are initially dehydrogenated by NAD or NADP enzymes.

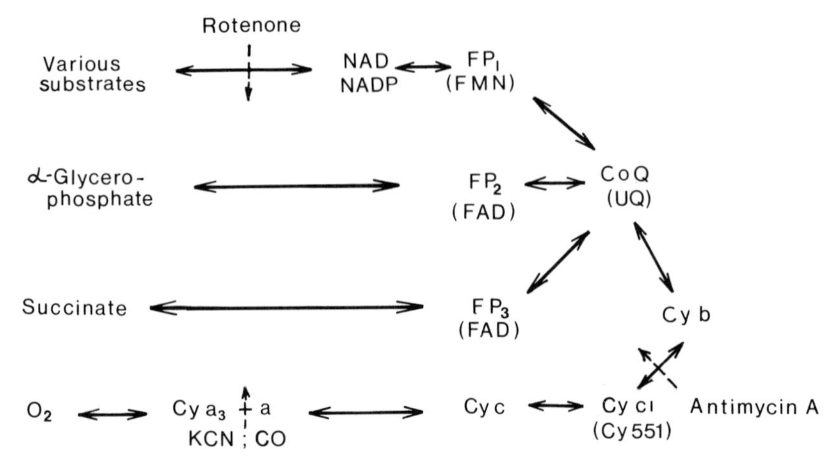

Fig. 9. Electron transport system of insects: NAD, nicotine-adenine dinucleotide; FP, flavoprotein; FMN, flavin mononucleotide; FAD, flavin-adenine dinucleotide; CoQ, coenzyme Q (ubiquinone); Cy, cytochrome; Cy *a*3+*a*, cytochrome oxidase. Broken arrows signify inhibitory action at specific steps.

Flavin mononucleotide (FMN) is coenzyme for NADH reoxidation, by "diaphorase," and the dinucleotide, FAD, for direct dehydrogenation of substrates, again as in mammals.

Ubiquinone (Section II, C, 1) is coenzyme (CoQ), at least in all the major orders of insects, for the next enzyme of the series, linking H-transfer by NAD and RF systems with electron-transfer by the cytochromes. It is rapidly reduced by the former, in the flight muscles of the blow fly, *Phormia regina*. Its concentration is about seven times that of cytochrome *c*, just as in the mammals. Similarly, all the cytochromes except Cy*c*, are in the mitochondria, in bound, water-insoluble form. The protein of Cy*c*, varies taxonomically within the Insecta, as within and between all taxa, but the properties of the holoenzyme are very similar in all.

The whole system maintains a steady state in which no member is forced to either redox extreme. The percentage of the member in the reduced state is maximal at the substrate end and decreases progressively to the oxygen end (Table IV).

TABLE IV

Percentages of Chromophores of Electron Transport Biochromes in Reduced Form at the Steady State

Insect	NAD	FP	Cy*b*	Cy*c*	Cy*a*	Cya_3	References
Musca	42	66	69	40	12	6	Chance and Sacktor (1958)
Locusta	72	60	50	30	6	—	Klingenberg and Becker (1961)

The main differences from the mammalian system are as follows: (1) as yet a nonheme iron protein has not been detected; (2) Cy*c* has its main absorption peak at 551 and not 554 nm; (3) a soluble cytochrome b_5, in the endoplasmic reticulum, is common during development and is absent from the adult; (4) substrates are more specific, either to NAD or to NADP enzymes; (5) reoxidation of NADH under anaerobic conditions is coupled with the reduction of diHO-acetone-P to α-glycerophosphate, of oxaloacetate to malate or of α-ketoglutarate to isocitrate, and not of pyruvate to lactate; (6) there is as yet no evidence that α-tocopherol (Section II, B, 1) is a member, acting between Cy*b* and Cy*c*, and promoting the oxidative phosphorylation of that stage.

Oxidative phosphorylation is coupled with the system at specific stages, as in mammals (Sacktor, 1954; Lewis and Slater, 1954); ADP therefore promotes electron flow in the oxidative direction, in proportion to its concentration (Sacktor, 1974). ATP promotes the reverse flow of electrons in the system, so as to drive reductive biosynthetic reactions (Gilmour, 1965). This is the crucial site of coupling between electronic and ionic reactions and its reversibility is noteworthy.

Cytochromes *b*, c_1, and *c* are indetectable during diapause in developing insects, except in those muscles that must function to break the diapause. Cytochrome oxidase persists, at a minimal level. The inactivity applies even when diapause is at the pupal stage, showing that aerobic respiration is not essential for the intense developmental metabolism at that stage (Shappirio, 1974).

2. Peripheral and Minor Redox Pathways

As in mammals,two special cytochromes, *P*-450 and b_5, in the microsomes of the endoplasmic reticulum are used for the oxidative destruction of drugs and other xenobiotic chemicals. They are reoxidized directly by ambient O_2, without mediation of cytochrome oxidase, and their action, like that of Cyox, is inhibited by CO. They are known as "mixed-function oxidases" (Agosin and Perry, 1974) because the O_2 has a four-electron requirement and simultaneously oxidizes the substrate(s) and NADPH:

$$S + O_2 + NADPH, H^+ \rightarrow SO + H_2O + NADP^+$$

The O_2 bonds with both substrate and *P*-450, which is reduced again by a FP.

Also as in the mammals, a number of FP's are reoxidized directly by ambient O_2: xanthine oxidase, aldehyde oxidase, pyridoxal oxidase, and amino-acid oxidase. Similarly the process is coupled with the action of two hemoproteins, peroxidase and catalase, to decompose the H_2O_2 produced. These are in particularly high concentration in insect flight muscles. Quinones are capable of peroxidase activity (Hackman, 1974) but are not known to perform that function in general insect metabolism.

Another important direct-oxidase hemoprotein, demonstrated in *Drosophila* and *Anagasta,* and probably active in all insects, is tryptophan oxidase (peroxidase, pyrrolase) which catalyzes the first step in the kynurenine pathway (Fig. 7). It opens the double bond of the pyrrole ring to form L-*N*-formylkynurenine, an oxygen atom being carbonyl-bonded to both of the C atoms exposed. This is an oxidatively catabolic step in a path which in insects subsequently becomes oxidatively biosynthetic and chromogenic.

By contrast, the first steps in the pathway from tyrosine to catecholic and indolic melanins are catalyzed by a simple copper protein, polyphenol oxidase or tyrosinase (Section II, D). The mode of action of the iron cofactor is not yet known. Other properties are considered in Section II, F. It is noteworthy that ascorbic oxidase is itself a copper protein. The quinonoid products of the later steps of this oxidative chromogenic pathway are reducible by NADH so that conceivably they could function as CoQ does in the main respiratory pathway (Fig. 9). Since a number of steps in the biosynthetic pathways of all biochromes are redox, this type of intermeshing (Section III, F, 4) has a wide feasibility (Gilmour, 1965). The chromogenic oxidation of 3-OH-anthranilic acid and gentisic acid in the synthesis of lepidopteran pupal silks involves enzymes containing copper and iron (Brunet and Coles, 1974), and presumably closely related to tyrosinase.

There is good evidence that a tetra-H-pterin (THP) acts as coenzyme in the hydroxylation of phenylalanine, tyrosine, and other substrates (Kaufman, 1967). Pterin and substrate cooperate in reducing the O_2 molecule—a four-electron requirement again. TH-biopterin and TH-sepiapterin are the most effective when tested *in vitro,* and ekapterin is another hydrogenated pterin present in insects. The diH, oxidized form of the coenzyme is reducible once more by NADPH. Biopterin is normally in the TH form *in vivo* but spontaneously oxidizes to DHB in air. Some of the other hydroxylation reactions catalyzed by THP's may not be relevant to insects but at least two are, that of kynurenine to 3OHK and that of Trp to 5-OH-Trp (Ziegler and Harmsen, 1969).

Tetrahydrofolic acid (THF, TF_4) is the coenzyme of a number of enzymes that transfer C_1-units in states between formyl (—HCO) and methyl (—CH_3). The process usually involves reduction of the C_1-unit by one step but FH_4 itself remains unchanged in redox state throughout the transfer cycle, so that there are significant differences from the hydroxylation THP's. FH_4 will hydroxylate Phe but only at 4% of the rate of TH-sepiapterin. FH_4 is synthesized via diH-folate, and the same NADPH enzyme will catalyze the regeneration of TH-sepiapterin from DHS. Through C_1-transfer, FH_4 is essential for purine synthesis and for the conversion of uridine to thymidine, i.e., for the synthesis of nucleic acids, particularly DNA.

Harano (1972) finds that 6,7-dimethyl-TH-pterin and TH-biopterin will act as coenzymes for a diaphorase present in the embryo of *Bombyx*. It reduces cytochrome *c* and is reduced in turn by an NADH enzyme. This is distinct from the NADH enzyme which reduces diH-pterins in the phenylalanine hydroxylation system. Like most pterins (Sections II, H, 3, b) the TH-pterin coenzymes are very sensitive to light which greatly increases their O_2 uptake; conceivably they mediate photooxidation reactions in the integument and/or eyes of insects or alternatively protect other materials against destruction by such reactions (Section III, C, 2). In insects the highest concentrations of pterins are in organs of highest metabolic activity (Fuseau-Braesch, 1972) so that they may have rather widespread metabolic activities.

Harano and Chino (1971) find that xanthommatin (X) also acts as a diaphorase cofactor in the embryo of *Bombyx*. It catalyzes the reoxidations of NADH and NADPH at equal rates. The phenoxazine structure is a fairly close analogue of the isoalloxazine ring system (Section II, H, 3, c), but in fact the essential coenzyme grouping is an aminophenol unit and 3-OH-kynurenine itself is an adequate substitute for X. The diHX formed is reoxidized by free O_2 in the mitochondrial system. The reactions are probably as follows:

$$\text{NADH, H}^+ + \text{X} \rightarrow \text{NAD}^+ + \text{DiHX}$$
$$\text{DiHX} + \tfrac{1}{2}\text{O}_2 \rightarrow \text{X} + \text{H}_2\text{O}$$

Evidence of further participation of the labile redox ommochromes in respiratory metabolism is extensive (Chen, 1971) but little is fully convincing. Not all insects contain significant amounts of ommatins, and various organs are devoid of them, even when the total body content is high. Albino mutants of *Locusta* are viable, and starved locusts excrete ommochromes (Brunet, 1965); certainly the class is not as ubiquitous as the enzymes of general respiration. Rhodommatin and ommatin D are held in the reduced state by their conjugants, implying that they serve only a static, visual effect function. Similarly a genetically controlled visual effect function seems the most probable explanation of the sex difference in ommochrome color in the dragonfly, *Sympetrum,* and of the systematic reduction of X to DHX at pupation in the moth *Cerura,* since it is maintained even at the high temperature of 35°C, and in pure O_2. There is no significant difference in respiratory rate between male and female *Sympetrum*. In the screen cells of the ommatidia there is oxidized ommochrome distally and the reduced form proximally (Ziegler and Harmsen, 1969) and this at least might appear to be a simple consequence of differential oxygen supply; however a tracheal O_2 supply makes such a differential doubtful. Pterin granules are sited between distal and proximal ommochrome aggregations so that the differentiation probably has some function.

Most significant is the discovery of Pryor (1962) that ommochromes are highly concentrated in the flight muscles of dragonflies and in the cells surrounding the air sacs. In the epidermal granules of the locust, acridiommatin is partially reduced and spontaneously oxidizes when extracted in air. Succinate promotes its reduction, implying a possible role for it as cofactor of the direct FP dehydrogenases. Cinnabarinic acid (Fig. 7, VIII) lacks the redox property, which depends on the xanthurenic grouping (Section II, G, 3). If ommochromes were initially selected for this property then it has been depressed in various ways in the other ommochromes, for visual effect stability. The ommins and ommidines have a restricted redox lability.

3. Oxygen Transport Biochromes

The efficient air-tracheal system of insects makes them independent of oxygen carriers in the body fluids. Only a few (Section II, E, 2) in special conditions exploit a carrier, in the blood or in special cells. It is always the familiar protohemoglobin (Hb), the iron remaining in the FeII state throughout the action cycle.

Chironomid larvae live in water often of very low O_2 content and their Hb has a very high O_2 affinity: it is fully loaded with O_2 in water only 25–40% saturated with the gas (equivalent to 40–60 mm Hg pressure), and not fully deoxygenated until in water of 5% saturation (7.5 mm Hg). Between these levels it maintains a constant supply to the tissues. Outside this range the supply depends simply on diffusion and is proportional to the ambient O_2 pressure. If the Hb is converted to CO-Hb then the simple proportionality holds also between 7.5 and 60 mm Hg pressure. Below 7.5 mm the animal suspends activity; even if the Hb is fully loaded when experimentally transferred to such low O_2 pressures, it supports activity for only about 10 minutes. Anaerobic basal metabolism continues but does not incur oxygen debt since the end products are excreted. The Hb therefore is able to mediate rapid resumption of normal motor activity as soon as the ambient pressure exceeds 7.5 mm Hg. The Hb of chironomid larvae has the highest known affinity for CO; it also has the highest known Bohr effect, i.e., its unloading is maximally stimulated by the presence of CO_2.

By contrast the larva of the botfly, *Gastrophilus,* and both larva and adult of the water boatman, *Anisops,* have typical aerobic systems; their Hb's have a low O_2 affinity and are fully loaded only near normal O_2 pressures, 150 mm Hg. Their environments in fact are well oxygenated, that of the botfly from the host's blood supply. The fly's Hb has a great advantage over that of the host in an exceptionally low affinity for such O_2 competitors as CO; this may be related to its uniquely large difference in position of the α-absorption peak between oxy-Hb and CO-Hb, viz. 95 Å, compared with an average of 44 Å. During its dive the water boatman uses the O_2 from its fully loaded Hb to replenish that diffusing into the tissues via the tracheal system, from the subelytral air bubble. In this way, it maintains its initial neutral density at the chosen depth and avoids the swimming effort that otherwise would demand much oxygen; it thus achieves a maximal duration of dive (Miller, 1964).

The Hb's of these insects are based on monomers, one heme and one globin chain, of molecular weight 15,900 daltons—about the same size as that of vertebrate Hb's and myoglobins (Mb). The intracellular Hb's of the botfly larva and the water boatman are dimers, like vertebrate Mb's, but that of the chironomid larvae also is in dimer, or even in monomer, form in contrast to the giant Hb polymers in the blood of annelids. Presumably, the simple open circulation of insects, and their exoskeleton, largely obviate the drawbacks of high viscosity and high permeation of small molecules of Hb in high concentration. *Chironomus thummi* has as many as ten different Hb molecules, monomeric or dimeric, resulting from genetic exchanges, deletions, or insertions of amino acids in the globin chain (Glossmann *et al.*, 1970). Multiple Hb's are known in other taxa, and imply functional versatility.

The globin of *Chironomus* Hb has 20% of its amino acid residues in the same primary sequence as in mammalian Mb (Hüber *et al.*, 1971). The tertiary structures, as usual, have much greater resemblance, showing that steric conformation is functionally more important than primary sequence and has evolved with considerable independence of it. The relationship between the two shows differences which students of mammalian Hb once might have considered functionally intolerable. *Chironomus* Hb has a negative Cotton effect, i.e., the direction of change of sign of the optical rotatory dispersion in the middle of the Soret absorption band is opposite to that in vertebrate Hb, indicating that the bonding of heme to protein is very different, perhaps effectively the mirror image. The main difference in tertiary structure is a break in helical region H in insect Hb.

B. Bioluminescence

As a form of chemiluminescence this involves the chemical generation of an excited state, resulting in light emission, i.e., the complete converse of the orthodox action of light on biochromes. Self-generated bioluminescence occurs in Collembola, Hemiptera, Coleoptera, Lepidoptera, and Diptera though the symbiotic use of bioluminescent fungi and bacteria is much more widespread. Among autogenous systems, only that of "fireflies" (lampyrid beetles) is well known (McElroy *et al.*, 1974). The chemical excitation requires an enzyme, luciferase, and the substrate is a pale yellow biochrome, D-(−)-luciferin (Fig. 10, I). The L-(+)-isomer is colorless and nonbioluminescent. Being energy-liberating the reaction is oxidative and the immediate products are more bathochromic than luciferin itself. In the firefly system three such products, in sequence, have been detected, the second and third

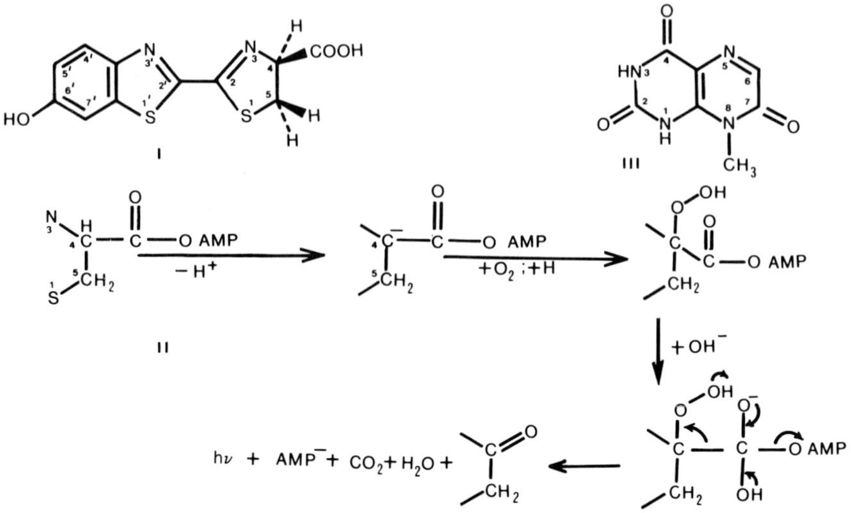

Fig. 10. Components of firefly bioluminescence system: I, substrate, luciferin; II, DeLuca and Dempsey (1971) hypothesis of the reactions at position 4 of the luciferin molecule during the enzymatic production of light ($h\nu$), involving AMP (from ATP) as coenzyme; III, 8-methyl-2,4,7-trihydroxypteridine, an associated component in the system of *Luciola*.

being stable enough for isolation and spectroscopy. Their absorption spectra, in MeOH, show λ_{max} at 278, 366, 486 nm and 304, 392, 650 nm respectively.

Luciferins in general fluoresce when illuminated, the emission being spectrally closely similar to the bioluminescence. In the present instance the 6′-OH group, ionizing as a phenate anion in the excited state, is the source of the fluorescence (Morton *et al.*, 1969). The implication is that the group donates an electron pair to the active site. Again, therefore ionization is involved in a redox process.

The essential photogenic change in the molecule involves the isolated and incompletely unsaturated thiazole ring. A study of laboratory model systems indicates that the luminescent oxidation may involve a complex series of changes at position 4 (Fig. 10, II) resulting in the replacement of the H and COOH substituents by a keto-bonded O-atom, converting the ring to the thiazolinone state. Bonding of AMP to the 4-COOH group is an essential step, which therefore demands ATP and Mg^{2+} as cofactors. Luciferase has binding sites for both luciferin and AMP. Spectral changes when the luciferin–AMP complex is subsequently hydrolyzed, *in vitro*, indicate intramolecular interactions, between the nucleotide and the ring system of the luciferin, similar to those within the ring systems of NAD and FAD. The interactions are the essential basis for the complex of electron shifts shown in the figure. The binding site for luciferase is either N-3 or the S-1 or both, so that activity is sharply focused on C-4.

Other biochromes are associated with the system. In *Luciola cruciata* there is a recently discovered pteridine, 8-Me-2,4,7-triOH-Ptd, (Fig. 10, III; Table III). In the "lantern" there is six times more ribitylflavin than in other tissues; in laboratory systems RF itself readily acts as a luciferin, with photo-induction in place of the enzyme, and H_2O_2 as the oxidant.

C. Chemical Protective Functions of Biochromes

Four types of protection may be distinguished (1) chemical deterrence, (2) screening against radiation and oxidation, (3) tanning (toughening) and (4) protection against other environmental stresses.

1. Chemical Deterrence (Eisner, 1970)

For this purpose quinones, particularly the relatively stable but volatile *p*-BQ's, have been exploited independently by various species of five different orders (Table V). They are strongly corrosive through their powerful

TABLE V
Quinones and Quinols Used by Insects as Chemical Deterrents

Deterrent	Distribution
1,4-Benzoquinone	Blattaria, Isoptera, Coleoptera
2-Methyl-1,4-BQ	Blattaria, Isoptera, Dermaptera, ?Heteroptera, Coleoptera
2-Ethyl-1,4-BQ	Blattaria, Dermaptera, Coleoptera
2-Methoxy-1,4-BQ	Coleoptera
1,4-Benzoquinol	Coleoptera
2-Methyl-1,4-BQol	Dermaptera, Coleoptera
2-Ethyl-1,4-BQol	Dermaptera
6-Methyl-1,4-NQ	*Argoporis alutacea* (Tenebrionidae)
6-Ethyl-1,4-NQ	*Argoporis alutacea* (Tenebrionidae)
6-Propyl-1,4-NQ	*Argoporis alutacea* (Tenebrionidae)
6-Butyl-1,4-NQ	*Argoporis alutacea* (Tenebrionidae)

oxidizing action; the lesions heal slowly and may become cancerous (Pavan and Dazzini, 1971). The independence of acquisition by the various taxa is shown by the variety of glands exploited for secreting them, yet nearly all use members of a small group of simple derivatives. Usually the corresponding quinols also are present (Table V), and an oxidant to convert them rapidly to the quinone. Most species have more than one Q, e.g., *Tribolium* has a mixture of 2-Me- and 2-Et-*p*-BQ's and the cockroach *Diploptera punctata* both of these and *p*-BQ itself. The tenebrionid beetle, *Argoporis alutacea,* has the usual tenebrionid *p*-BQ's and also a range of 6-substituted naphthaquinones, including the propyl and butyl homologues (Tschinkel, 1972). This is the only record of simple NQ's in an insect. They retain some volatility, essential for this purpose.

The Q's are released from storage, usually in the glands that synthesize them, either continuously in small amounts or when stimulated by potential predators. In some there is no mechanism for directing the material but in others it is ballistically aimed at the target organism. This mechanism is most sophisticated in the bombardier beetle, *Brachynus crepitans,* which secretes the Qol precursor mixed with one-third its weight of H_2O_2 and with a heat-stable catalase. The heat of its H_2O_2-decomposing reaction makes it also autocatalytic and the generation of O_2 gas is audibly explosive. The volatilized Qol is oxidized by it, under pressure, and a jet of volatile Q reaches a target as much as 45 cm away (Schildknecht, 1971).

No other classes of zoochrome are secreted in this way but some, including Q's, may be distasteful. By their redox properties they could have deleterious internal actions (Hackman, 1974). Brower (1969) suggests that this repellant action is the explanation of colorless pterins such as leucopterin and isoxanthopterin in the scales and wings of *Pieris brassicae*. Xanthomegnin, a NQ dimer of microorganisms, uncouples oxidative phosphorylation in mammalian liver (Ito *et al.*, 1970) so that protoaphins and their products in damaged aphids possibly act in this way on predators. The products are strongly photodynamic and may injure the predator also externally. Like many of these chemical defenses they may not save the individual animal but could contribute to the survival of the species. However, most aphids and even the brilliantly colored *Aphis nerii,* are prey to many animals and their chromes appear not to be aposematic, in bark or in bite.

In view of the drastic effect of the phenoxazine antibiotic, actinomycin, on the transcription of DNA to RNA it is conceivable that insect ommochromes could be used in chemical defense against predators. Actually phenothiazine, the S-analogue of phenoxazine, is used as an insecticide but this may be through acting as a competitive analogue of the ommochromes.

2. Antiradiation and Antioxidation Screening

Their absorption spectra show that all integumental melanic chromes must filter off from the deeper tissues not only visible light but also much of the shorter wave radiation. The other integumental chromes also contribute, by reflection and absorption, usually over more restricted ranges. Vuillaume (1975) stresses the value of bilins, and certainly their one visible absorption band is broad. For entropic reasons the absorbed shortwave radiation in principle should be transducible into useful energy.

On earth, shortwave radiation exacerbates a permanent menace of excessive oxidation. Most of the screening biochromes are also potential antioxidants, for more important, deeper materials. Even the inert melanins are bleached by irradiation in air. Among animals in general an antioxidant function is demonstrable for carotenoids, including vitamin A, α-tocopherol, and pterins, though as yet there is no specific evidence for this in insects.

The pigment sleeve around each ommatidium of the insect eye forms a directional screen, improving the acuity and resolution of the retinular units under photopic (bright light) conditions, by screening each from oblique light. Under scotopic (dim light) conditions many insects sacrifice the sharp apposition image for the alternative virtue of retinal sensitivity. This involves movement of the chromasomes ("pigment granules"), and in some insects also the chromatocytes themselves, so as to expose the retinulae maximally to oblique light. The precise movements vary with type and taxon of insect but most commonly the chromasomes of the proximal (retinal) chromatocytes move distad to surround the principal, or iris, cells (Autrum, 1975). Since the photopic screen is wavelength-selective (Section III, D, 1) its contribution to visual performance is significant (Corrigan, 1970; Cochran, 1974; Goldsmith and Bernard, 1974). Tapetal pterins also are waveband-selective in their reflection (Menzel, 1975). Pterins and ommochromes transmit mainly the longer wave visible rays, which are useful in regenerating the rhodopsins; they help performance in this way and give the appearance of increasing the sensitivity of the retinulae to this range.

Menzel (1975) recognizes three types of ommochromasome: B_1 contains xanthommatin, absorbing maximally at 450 nm; B_2 absorbs maximally at 560 nm and probably contains only ommins. The most abundant, A_1, contains a mixture of oxidized and reduced ommatin and ommin; it absorbs rather uniformly over the whole range, 300–700 nm, probably due to polymerization (Section II, G, 3) and perhaps to the protein conjugation. In some eyes the type of ommochrome varies between the various sectors of a single ommatidium. This may be significantly related to differences between the rhodopsins (Section III, D, 1).

Some insects have chromasomes in the retinulae themselves. As expedient they move mediad or laterad, i.e., closer to or further from the rhabdomeres. In the water boatman, *Notonecta,* however, these chromasomes move in the direction of the ommatidial axis and serve as the main photopic/scotopic commuters.

3. Tanning (Toughening) by Biochromes

The most significant example of this in insects is the sclerotization of exoskeletal and other proteins by various structural derivatives of orthobenzoquinone (*o*-BQ). Pryor's discovery of this function was helped by the knowledge that *o*-BQ's are used commercially for tanning leather. Sclerotized insect proteins, *sclerotins,* are usually colored in the process (Gilmour, 1965). Most of the *o*-BQ precursors used by insects (Table VI) are

TABLE VI
Precursors of Tanning Quinones in Insects

Precursor[b,c]	Distribution	Material tanned	References
Tyrosine-*O*-phosphate	*Drosophila*	Cuticle	Hodgetts and Konopka (1973)
N-Acetyltyramine	*Bombyx*	Pupal cuticle	Butenandt *et al.* (1959)
3,4-diOH-benzoic (protocatechuic) acid	*Blattella*, *Periplaneta*	Cuticle Ootheca	Gilmour (1965) Brunet (1967)
3,4-diOH-benzyl alcohol	*Periplaneta*	Ootheca	Pau and Acheson (1968)
3,4-diOH-acetic acid	*Locusta*	Cuticle	Gilmour (1965)
3,4-diOH-propionic acid	*Tenebrio*	Cuticle	Hackman *et al.* (1948)
3,4-diOH-phenyl-alanine (dopa)	General	Wound-scab	Hackman (1971)
3,4-diOH-phenyl-β-ethylamine (dopamine)	Cockroaches, Muscidae	Cuticle Puparium	Hackman (1974) Hackman (1974)
3,4-diOH-phenyl-β-*N*-acetylethylamine (*N*-acetyldopamine)	*Schistocerca* *Tenebrio*	Cuticle	Gilmour (1965)
	Periplaneta	Cuticle	Koeppe and Mills (1974)
	Calliphora	Puparium	Gilmour (1965)
3-*O*-P-4-OH-phenyl-β-*N*-acetyl ethylamine (*N*-acetyl dopamine-3-*O*-P)	*Periplaneta*	Cuticle	Bodnaryk *et al.* (1974)
3-*O*-S-4-OH-phenyl-β-*N*-acetylethylamine (*N*-acetyldopamine-3-*O*-S)	*Periplaneta*	Cuticle	Bodnaryk *et al.* (1974)
2-NH_2-3-OH-benzoic (3-OH-anthranilic) acid	*Samia*	Silk	Brunet and Coles (1974)
2,5-diOH-benzoic (gentisic) acid	*Samia*, *Antherea*	Silk	Brunet and Coles (1974)

[a] All but dopa produce benzoquinone tanning agents.
[b] Not all these precursors are on separate pathways and not all are immediate precursors.
[c] P, phosphate; S, sulfate.

synthesized by them from phenylalanine, via tyrosine, and the remainder from tryptophan.

In the blood of insects there are great increases in tyrosine concentration, to peaks at ecdysis, puparium formation, and other relevant events (Hackman, 1974), and a sharp decrease to a minimum when sclerotization is completed—at 24 hr after molting in *Periplaneta*. Some insects are known to use a variety of more soluble Phe and Tyr derivatives for the transport phase, e.g., β-Ala-L-Tyr (sarcophagine), Tyr-O-PO_4, and α-L-Glu-L-Phe (Chen, 1971).

The inactive protyrosinase behaves similarly. Both must be passed from the blood into the cuticle rapidly at the time of hardening. The activator for tyrosinase is first detectable in the cuticle itself. For the ootheca (egg case) of *Periplaneta* it is the substrate which prior to use is held inactive by glucosidation; activation requires a glycosidase, segregated in the right colleterial gland of the female, the other components being stored, active, in the left gland. When the contents of the two glands are extruded and mixed, the whole train of reactions follows.

The simultaneous hardening and darkening of the various scleroproteins have been repeatedly confirmed and they are also found to become chemically inert, and insoluble in all but strongly degrading solvents. When sclerotin is degraded it yields large amounts of BQ and Qol derivatives that are not detectable in the presclerotin—up to 20% of the total mass of the silk of *Samia cecropia* (Brunet and Coles, 1974). The hardness of the sclerotins is proportional to their content of BQ's (Brunet, 1963). The properties show that the bonding is covalent.

The precise nature of the bonds between Q and protein is still uncertain; it has not been possible to break down sclerotin into specific, isolable fragments consisting of Q and its immediate conjugant amino acids. The type of bond may vary between taxa and sites, and even within a particular sclerotin. Free amino groups such as the ϵ-NH_2 of lysine are considered most probable conjugants and these groups are abundant in the presclerotin of the exoskeleton, but are absent after sclerotization (Hackman, 1953). In the initial uncolored state the cockroach ootheca has 2.7 times more lysine than when it is fully darkened and hardened. Arginine and histidine also are fairly abundant in presclerotin.

Laboratory studies of the reaction of BQol derivatives with simple amino compounds such as aniline (Hackman and Todd, 1953), in the presence of polyphenol oxidase or of nonenzymatic oxidants, give products in which more than one amino groups is bonded with each Qol unit, by direct covalent bond between N atom and benzene ring. The Qol is oxidized to the Q state in the process (Fig. 3, XI). The first two amino groups are substituted at non-Q sites but the third displaces a Q =O substituent with retention of the double bond. Such multiple bonding with protein chains should give excellent toughening and insolubility, with some pliability. The amount of Q required should be very small compared with the weight of the presclerotin and it seems clear that much of the Q found in sclerotins must be simply polymerized and therefore responsible for the prevalent darkness of color of the material (Brunet and Coles, 1974). *o*-BQ's isolated in a protein matrix could give red colors, and more bathochromic shades must be due to grades of autopolymerization.

The cysteine SH group is the only other feasible conjugant. There is some sulfur in the exoskeleton of other arthropods but very little in that of insects.

The insolubility and inertness of sclerotins contribute to the waterproofing of the insect exoskeleton, a mandatory property for small terrestrial animals. Transport of water in either direction across the exoskeleton is inversely proportional to the amount of dark quinonoid material in it (Kalmus, 1941).

Eumelanins possibly contribute to both the color and the hardness of insect sclerotins. They increase the toughness and insolubility of vertebrate hair and feathers and they are a component of the hard black seal over wounds in the insect integument. Also they are present in the capsules that surround parasites and other foreign bodies in the hemolymph (Salt, 1970). It seems unlikely that grossly granular material (Section II, F, 1) can contribute to tanning but it is not ruled out that dispersed eumelanin also occurs in insect sclerotins.

Tryptophan accumulates in the exoskeleton of some insects, if tanning is inhibited (Pryor, 1962). This may indicate tanning by anthranilic acid derivatives (Table VI) but ommochromes also have been found in the exoskeleton and have the necessary quinonoid properties. There is some indication that papiliochromes also have tanning properties (Umebachi, 1975). Xanthophylls and other biochromes have been isolated from silks but at present there is no specific evidence of a tanning function. Hanser and Rembold (1968) found heavy accumulations of biopterin and neopterin in the exoskeleton of the pupa of *Apis*.

The lac resin of scale insects is obtained from crude material containing both exuviae and a secretion which includes the laccaic AQ's (Section II, C, 2). The purified shellac, amber to black, may still contain exuvial material but also has esters of polyhydroxycarboxylic acids, implying a relationship to the AQ's.

Flavonoid polymers are used as tannins both commercially and in plants, but probably not by insects.

4. Protection against Other Environmental Stresses

Apart from water exigencies, temperature extremes are probably the most critical environmental stresses for small terrestrial animals, because of their high surface/volume ratio and because air temperature fluctuates so widely and rapidly. Physical control mechanisms exist but in addition insects use integumental biochromes to assist chemically (molecularly) in controlling heat flow in and out of the body. Black and bathochromic-colored materials absorb strongly in parts of the heat-range, though not all as strongly as in the visible, while hypsochromic hues and white reflect heat most efficiently. Appropriately the nymphs of the gregarious phase of the locust, *Schistocerca,* are predominantly black at low, and yellow at high, external temperatures. A light-colored beetle, *Compus niveus,* absorbs only 26% of incident IR radiation whereas the black carrion beetle, *Silpha obscura*, absorbs 95%. Many desert insects are black; they expose themselves to solar and surface heat only in the early morning and evening.

Thermoregulatory integumental color commonly shows chromogenetic adaptation. If locusts are reared at a high temperature, 35°C, ommochrome synthesis is depressed in subsequent instars. At 50°C the adults have no dark integumental chromes at all. Similarly, if acridoid nymphs are reared at high temperature their green is replaced in later instars by a light fawn color (Frazer-Rowell, 1971; Vuillaume, 1975). The corpus allatum hormone is involved in the control of this change. In holometabolous insects such as the butterfly, *Aglais urticae*, the sensitive phase for determination of adult color is late—48 hr after the onset of pupation—probably to minimize errors due to weather vicissitudes. The butterfly, *Colias*, is multivoltine and adults are colored according to season. The chromogenetic information is stored and has effect over the several generations. By contrast, syntheses for other purposes are controlled by photoperiod, entirely within each generation (Hoffmann, 1974).

The more bathochromic of the carotenoids, pterins, etc., behave like melanins and ommochromes in this response (Ziegler, 1961; Hoffmann, 1974), whereas the paler members show the converse behavior. This explains some paradoxes within chemical classes.

Screening pigments of the insect eye show the same temperature response as the pterins and ommochromes of the integument, implying that the functioning of the eye may benefit by temperature stabilization. One gene of *Drosophila* sensitizes both pterins and ommochromes of the eye to heat while other genes affect the one class only; this may indicate that temperature is significant in more than one functional context, but see also Section III, F, 4.

Heat absorption is improved by behavioral devices. When warming up for flight, in the morning, a locust lies so as to expose maximal area to the sun while at noon it exposes only its narrow back. The spring generation of *Colias* exposes the very dark undersides of its wings to the sun.

Dark chromes also radiate more heat than light materials, from a body above ambient temperature, and in some situations this could benefit or harm insects. Dark butterflies melt snow in contact with their bodies more rapidly than white individuals (Needham, 1974). Few insects are active in conditions where this sensitive test would also be a natural event but heat loss could be useful in tropical conditions. The lacewing, *Chrysopa*, increases its integumental ommochrome content for the winter season of middle latitudes (Rudiger and Klose, 1970) and heat uptake is likely to be feasible, at times, in the highest latitudes tolerated.

Like other terrestrial animals insects have dark coloration in wet, and light colors in dry, surroundings. These hues often have camouflage value on such backgrounds, but in addition there is some correlation between damp conditions and low temperature.

D. Biochromes and Neural Functions in Insects

1. Photoperception

Since photoexcitation is the essential property of biochromes it would be surprising if photoperception were not mediated by one or more of them. Actually in all simple and compound eyes of insects examined (Goldsmith and Bernard, 1974; Menzel, 1975) photoneural transduction is by the same biochrome as in Cephalopoda, other Arthropoda, most Vertebrata: retinal 1 (R_1), the aldehyde of vitamin A (Fig. 1, VII). It is similarly conjugated with a lipoprotein, opsin, to form rhodopsin (RO). This has less lipid than in vertebrates and so is more water soluble; also there is no evidence that in insects R_1 is bonded to a phosphatide residue of the lipid during one phase of the photoperception cycle. Insects bisect the molecule of dietary β-carotene and oxidize each half via retinol to retinal. If they are made extremely deficient in carotenoids the light-sensitivity of the eye is depressed by as much as 10^4 (Goldsmith and Bernard, 1974).

Like the vertebrates, insects produce a number of RO's by variations in the opsin. In all species examined by electrophysiological and behavioral methods more than one RO has been detected; some have been isolated and characterized. In most there are three, with absorption peaks (λ_{max}) in the regions 510–570, 410–490, and 340–365 nm (Table VII). For convenience these may be called the green, blue, and UV RO's, from their positions in the human visual spectrum. The λ_{max} of free R_1 is 380 nm, confirming its identity with vertebrate R_1, and the λ_{max} therefore is shifted great distances by opsin conjugation. The explanation suggested for the prevalent bathochromic shifts (Pitt, 1971) is that the resonance system of R_1 is extended by formation of a protonated Schiff's base, —CH=N$^+$H—, between its exposed C-20 atom and an amino group of the opsin, supplemented by various induced dipoles in the hydrophobic environment of the complete chromophore. The unusual 35 nm hypsochromic shift in the UV RO of insects possibly depends on an elimination, from the chromophore's resonance system, of the aldehyde, —HC=O, group of R_1, through bonding of the C atom with both —NH_2 and —SH groups of a terminal cysteine residue of the opsin (Goldsmith, 1973).

The remarkable exploitation by insects of the near UV range for visual purposes is partly related to the small dimensions of their visual units and the finer resolution offered by short waves. Vertebrates cannot exploit this because of the restrictions of chromatic aberration in a camera-eye. In addition it permits maximal use of natural polarized light for orientation, since sky light is rich in the near UV range. The exploitation of UV colors by flowers probably is secondary to the insect's own adaptations. The UV RO of insects is usually the most sensitive of the three, compensating for the relatively low ratio of UV to visible in direct solar radiation. Aquatic arthropods make less use of this shortwave perceptor because water differentially absorbs in this range.

The screening chromes increase the apparent sensitivity to red light (Section III, C, 2) and the dragonfly, *Aeschna cyanea* (Autrum and Kolb, 1968; Menzel, 1975), may have a genuine λ_{max} in that range but in general the insect spectrum is shortened at that end by about the same amount (100–150 nm) as it is extended at the other. The narrow ommatidia of the most acute eyes act as light guides and these differentially lose long waves through the walls of the crystalline tract. Some of this red light will be transmitted by the screen, making the eye permanently scotopic for this waveband.

Not surprisingly, therefore, in insects it is the RO with the longest λ_{max} which functions in scotopic vision, in diametric contrast to the use of the one with the shortest λ_{max} by vertebrates. On increasing the intensity of illumination there is a typical Purkinje shift in the wavelength to which the eye is most sensitive, from that of the green to that of the UV RO (Fingerman and Brown, 1953). Because of the differential filtration by the screen-chromes, and perhaps of other auxillary devices, the scotopic RO shows a broader sensitivity-band *in situ* than when isolated. Like the vertebrate scotopic RO it also functions in color vision.

There are also other factors which cause differences between the absorption curves of isolated RO's and the spectral sensitivity curves, tested electrophysiologically on the perceptor units, or on the afferent nerves, or behaviorally on the whole system. The long rhabdomes cause considerable self-absorption so that proximally they receive a spectrum very different from that incident on the cornea. The differential reflection from screen and tapetum also is relevant, and in some insects there are colored reflection patterns from the corneal surface, so that it acts as a color filter. These complications are not necessarily imperfections.

Microspectroscopy has shown that the photochromes are in the membranes of the stacks of microvilli of the rhabdomeres, the functional counterpart of the stacks of discs in the outer segment of the vertebrate visual cells. Also as in the latter the molecule of R_1 is free to rotate only in the plane of the membrane; this is the tangential plane of the tubular insect microvilli, which are oriented in the transverse plane of the ommatidium. In the individual rhabdomeres all villi are parallel but the orientation varies between rhabdomeres, often being mutually perpendicular in nearest neighbors. In the tobacco-hornworm moth, *Manduca sexta,* dietary deficiency of vitamin A leads to disorganization of microvilli and the proliferation of new ones.

TABLE VII

Spectral Sensitivity Maxima of Perceptor Units of Eyes of Insects

Insect	λ_{max}(nm)[a]			Method	References
Aeschna	356	420 (356) 458	500 (356) 519	Recording from single units	Autrum and Kolb (1968); Eguchi (1971)
Libellula	380	420	500–520	Electroretinogram	Ruck (1965)
(Odonata)	—	410	450–550	Single unit records	Horridge (1969)
Periplaneta (Blattaria)	365	—	510	Single unit records	Mote and Goldsmith (1970, 1971)
Tenodera (Mantoidea)	370	—	515	Electroretinogram	Sontag (1971)
Locusta (Orthoptera)	—	430	515	Single unit records	Bennett *et al.* (1967)
Notonecta	350	420, ?464	567	Single unit records	Bruckmoser (1968)
(Hemiptera)	375	?475	520	Electroretinogram	Bennett and Ruck (1970)
Ascalaphus (Neuroptera)	350	—	530	Electroretinogram	Gogala (1967)
Manduca	350	450	500,550	Single unit records	Boëthius *et al.* (1968)
Macroglossum	348	430	500, ?620	Electroretinogram	Hasselmann (1962)
Deilephila	350	450	525	Electroretinogram	Höglund *et al.* (1973)
	345	440	520	Absorption-spectra of retinopsins	Schwemer and Paulsen (1973)

Heliconius	370–390	450–470	550–570	Electroretinogram	Struwe (1972a)
(Lepidoptera)	390–410	450–470	530–550	Single unit records	Struwe (1972b)
Drosophila	350	475	502	Behavioral	Schumperli (1973)
Calliphora	345 (470)	490 (345)	520 (345)	Single unit records	Burkhardt (1962)
(Diptera)	360	476	510	Microspectroscopy *in situ*	Langer and Theorell (1966)
Apis, ♀	345	—	535	Electroretinogram	Goldsmith (1960)
	340	430,460	530	Single unit records	Autrum (1965)
Apis, ♂	345	440	535	Electroretinogram	Goldsmith (1958, 1960)
	340	450	530	Single unit records	Autrum and Zwehl (1964)
Cataglyphus (Hymenoptera)	350	—	500,600	Behavioral	Wehner and Toggweiler (1972)
Carabus (Coleoptera)	348	430	500,?620	Electroretinogram	Hasselmann (1962)

[a]Values in parentheses are those of a second, minor retinopsin in the same unit.

The regular orientation of the retinal units gives the rhabdome dichroic properties when illuminated transversely to the axis of the ommatidium and to the axis of any one set of rhabdomeric villi. Dichroism disappears when the RO "bleaches" in the course of its action cycle and can be used as a criterion of this process. The oriented structure is the basis also for the ability of insects to analyze, and to navigate by, plane-polarized light (Autrum, 1975). For this purpose the ideal arrangement is that the microvilli of neighboring rhabdomeres are as mentioned above, mutually at right angles.

As in vertebrates and cephalopods, R_1 is in the 11-*cis* configuration in the dark-adapted (resting) eye, and the *cis*→*all-trans* isomerization is the first and essential effect of illumination by the relevant waveband. In the bee there follow steric changes in the opsin, requiring only thermal energy (Abrahamson and Wiesenfeld, 1972), and changes in absorption spectrum. The first stable product of all three RO's has a λ_{max} at 480 nm, as in vertebrate metarhodopsin I (MRO I). The *Calliphora* RO with a λ_{max} at 490 nm is exceptional in having that of its MRO I at 570 nm, near that of vertebrate prelumirhodopsin.

Free retinol 1 (λ_{max} 330 nm) is absent from the dark-adapted eye of the bee but appears immediately on illumination (Goldsmith, 1972), indicating a regeneration cycle as in vertebrates, with liberation and reduction of R_1. However, most insects have an MRO II with λ_{max} at 375 nm, indicating that R_1 regenerates RO directly. For the UV RO the reisomerization of R_1 to its 11-*cis* form is stimulated most effectively by blue light, 440–460 nm, and the other two respond similarly, in proportion to the build-up of MRO, but with an additional boost from metabolism. A complete photoreversible cycle therefore operates and regeneration is adequate under continuous illumination.

The precise mechanism of transductive coupling between the photochemical changes in the RO and the electrical generator-potential is still unknown. However, unlike the membranes of the sacs of the outer segments of vertebrate rods and cones, which have no direct contact with the cell membrane, those of the insect microvilli are part of the retinular membrane. Transduction therefore may be more direct, the *cis*→*trans* rotations of the R_1 molecule causing changes in potential or in ion flow across the membrane. The bleaching and regeneration phases of the photochrome do correspond in time respectively with the negative and positive potentials of the electroretinogram. A mutant of *Tenebrio molitor* lacking the screening chromes (Section III, C, 2) shows a larger amplitude of the negative potential and a shorter latent period than normal, thus confirming that the screen does depress sensitivity.

As in other known invertebrate responses, and in diametric contrast to that in the inverted retina of vertebrates, photoexcitation reduces the resting potential across the cell membrane, and there is increased conductance as contrasted with increased resistance. The generator potential is able to set off propagated spikes in the membrane of the retinular cell itself and not, as in the vertebrate system, only postsynaptically. Restoration of the resting membrane potential is closely correlated in time with the reconversion of the 480 nm MRO to RO. Insects with only two RO's (Table VII) appear to be color blind, and the condition is generally correlated with nocturnal habits. Those with three RO's having well spaced λ_{max} values have color vision, tested by behavioral experiments using monochromatic light of controlled intensity and saturation. Color vision has been established for members of the Blattaria, Odonata, Hemiptera, Coleoptera, Hymenoptera, Lepidoptera and Diptera, i.e., all the dominant orders. *Notonecta* has color-blind ommatidia in the anteroventral part of the eye and color vision in those of the posterodorsal region.

The careful work of Daumer (1956) and Autrum (1975) has shown a close parallel between color vision in *Apis mellifera* and that in man, apart from the 100–150 nm shift in spectral range. Bees distinguish six main hues, three due to stimulation of each of the RO's and three intermediate hues due to equal stimulation of pairs of them. As for man, this turns the visual spectrum into a color-circle (Fig. 11) closed by the hue analogously called "insect purple." The six hues therefore are green, blue-green, blue, violet, UV and insect-purple. The broken arc of the circle is to emphasize that there is no single continuous spectral range which stimulates both the UV and the green RO's in the way that the 500–480 nm range does both green and blue RO's and the 440–360 range does both blue and UV RO's. Insect-purple is simply the central recognition of a balanced stimulation of UV and green RO's by simultaneous exposure to a waveband on the long wavelength side of the green λ_{max} and one on the short wavelength side of the UV λ_{max}. In addition,

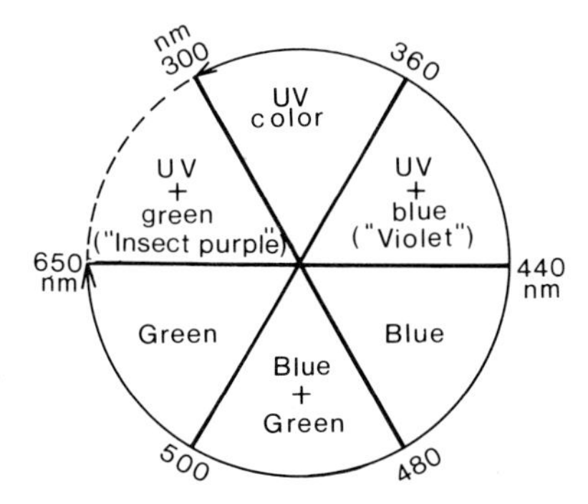

Fig. 11. Diagram of visual color circle of the honeybee, showing the six hues discriminated by the three perceptor chromes (rhodopsins) stimulated singly and in neighboring pairs. The six hues are shown as though they corresponded to equal ranges of the spectrum, but note the actual values, and see text.

bees discriminate from three to five tones between each pair of primaries and so compare in discriminability with man.

Cyclization of the subjective spectrum may account for *Calliphora* (Mote and Goldsmith, 1971) and other insects having both green and UV RO's in some of the rhabdomeres (Menzel, 1975). Moreover their sensitivity to each waveband is increased following adaptation to the other band. This will tend to keep excitation of the two RO's equally balanced whatever the ratio of incident intensities in the two bands. The balance may be further helped in some insects by the green RO itself having a second absorption peak, at 350 nm, which is not affected by the bleaching of that RO and so then should increase the differential stimulation by the UV band.

In *Calliphora* (Burkhardt, 1962) every rhabdomere has two RO's, major and minor (Table VII); one is always the UV member, and green is never the minor partner, so that three combinations, only, occur: green-UV, blue-UV, and UV-blue. This emphasizes the importance of the UV member and also may be relevant to photoregeneration, to the mutual sensitizing actions of the three, and to the virtues of the shortwave radiations. In *Musca domestica* all six outer rhabdomeres have the green and UV RO's, while the central rhabdomere has only the blue.

In addition, the relative proportions of the three in the whole ommatidium vary in a regular pattern over the eye of some insects (Menzel, 1975). Green is mainly in frontal and ventral regions of the eye, blue in the ventral region and UV in the dorsal region, receiving most of the polarized light.

The frontal, simple eye of the neuropteran, *Ascalaphus macronius,* has only a UV RO; it has λ_{max} at 345 nm and is thermostable (Autumn, 1975). The product of photostimulation is regenerated by light in the whole of the 400–600 nm range. By contrast, in the ocelli of the larva of *Aedes aegypti* there appears to be only a green RO since it has absorption peaks at ~515 and 350 nm and only the former disappears on illumination (Menzel, 1975).

The general dermal light-sensitivity of insects shows an action spectrum with a broad peak over the range 435–520 nm and therefore could be due to an RO or another carotenoprotein. However, the circadian cycle controlled by it, in *Drosophila,* is unaffected by a vitamin A deficiency severe enough to cause a 10^3 to 10^4-fold depression of sensitivity in the compound eye (Zimmerman and Goldsmith, 1971).

Although pterins and ribitylflavin are very photosensitive and both are often present in insect eyes, there is no firm evidence that either plays a direct role in photoperception. Illumination of integumental pterins increases oxygen uptake but this light reaction has a time constant as long as 5 sec, so that it could function only in slow responses. There is no positive evidence for photoperception via ommochromes (Wolken, 1975), or for the suggestion of Barbier *et al.* (1970) that bilins may function in this way in the pupa of *Pieris*.

2. Other Sense Modalities

Chemoperception is the only other insect sense modality that may involve biochromes and even so they may act only as stimuli, not as electrical transducers of the chemical signal. In some animals carotenoids and melanins may act as chemoelectrical transducers (Needham, 1974) but there is no present evidence for this in insects. The flavonoid glucoside of potato leaves, which stimulates feeding by the beetle, *Leptinotarsa decemlineata,* and the lepidopteran, *Protoparce* (Hsaio and Fraenkel, 1968), is one of the many chemical criteria of materials edible for insects. Most of them are aromatic and therefore at least subchromatic. There is no *a priori* reason why such chemical signals should not act also as chemoneural transducers.

Naphthaquinones depress both feeding and the response of the gustatory sense organs, of *Periplaneta,* and so are thought to act at the transduction stage (Norris *et al.*, 1970; Baker and Norris, 1971), since they also affect nerve conduction (Section III, D, 3). It is suggested that the Quinone (Q) acts by withdrawing electrons from a membrane protein of the sense organ, leading to increased ion flow across the membrane, as in photoneural transduction. Quinols (Qols) stimulate feeding, and the response of the taste organs, implying a Q$\rightleftharpoons$Qol cycle as the basis of a transduction cycle, and in the longer term perhaps of the feeding cycle.

3. Other Neural Functions

A number of biochromes, some unidentified, are concentrated in particular parts of the nervous systems of animals (Needham, 1974). Specific functions are implied and in some instances supported by evidence. Some information has been obtained from insects.

Ommochromes are concentrated in the connective tissue sheath of the ocelli, optic lobes, and nerve cords of the moth *Hyalophora cecropia* (Ajami and Riddiford, 1971) and of the ventral cord of the moth *Ptychopoda.* The nervous system is epidermal in origin, and in ontogenesis chromatocytes sink in with it, as they do with the vertebrate neural crest. *In vivo* ommins are usually in the reduced state but that of *Hyalophora* is covalently bonded to protein, has a redox potential somewhere between +123 and +217 mV, at pH 7.0, and at rest is oxidized. Under anaerobic conditions, in the light it reduces slowly but spontaneously, and conceivably could mediate a redox triggering of membrane conduction. The isolated ommin chelates Fe^{2+}, Cu^{2+}, and Ca^{2+}, which may increase its excitability as they do that of melanin (Needham, 1974).

In the brain of *Hyalophora* and of *Antherea pernyi,* and in aphids, pink biochromes with the properties of ommochromes are involved in photoperiodic responses. In *Antherea* and also in the moth, *Pectinophora gossypiella,* they function in controlling the rhythm of hatching of the eggs. The action spectrum for the response is similar to the absorption spectrum of ommochromes.

Dietary biopterin and neopterin accumulate in the cells of nerve ganglia of the bee (Hanser and Rembold, 1968), and L'Hélias (1957) found some evidence that pterins in the protocerebral complex may act as cofactors for neuroendocrine agents. It may be relevant, in particular, that the methyl purines, caffeine, theobromine, and theophylline have strong neurostimulatory actions and in general that a large number of neurotropic amines are related to biochromes. The outstanding examples are the tyrosine and tryptophan derivatives, tyramine, dopamine, epinephrine, tryptamine, and serotonin. Some of them, e.g., ephinephrine, spontaneously oxidize to bathochromic products, and indeed neuroamines are found in sclerotins. Neopterin and biopterin also accumulate in the bee's exoskeleton (Section III, C, 3).

Experimentally the naphthaquinones (NQ), 1,4-NQ itself, menadione (2Me-1,4-NQ) and juglone (5-OH-1,4-NQ) have been found to block nerve conduction along fibers of the ventral cord of *Periplaneta* (Baker and Norris, 1971). Their order of effectiveness parallels the order of inhibitory action on feeding (Section III, D, 2) in the living insect. The action is thought to be directly on membrane transport, by inhibiting Na-K-Mg ATPase, the chemiosmotic version of electronic–ionic coupling.

E. Biochromes Concerned in Alimentary Function, Storage, and Excretion

1. Digestion

There is a biliprotein in the digestive juice of *Bombyx mori* (Kusada and Mukai, 1971). It is formed by the action of a midgut enzyme on dietary chlorophyll and has the properties of a phycocyanin. It has antiviral action on silkworm polyhedrosis, but no detected digestive function. Some other invertebrates as well as the vertebrates have bilins in their digestive juices and various arthropods, mollusks, and annelids have hemes and porphyrins of endogenous rather than dietary origin (Needham, 1974). Hemoglobin derivatives of indirect, dietary origin are secreted from the salivary glands of *Rhodnius,* and pink chlorophyll-derivatives from those of *Anasa tristis.* Kusuda and Mukai found chromoprotein-like materials also in the gut fluids of larvae of the pine moth, *Dendrolimus spectabilis*. There is some evidence, but none specifically from insects, that these chromes have a positive function in the gut other than excretion.

2. Storage

Concentrations of biochromes are more easily detected than those of colorless materials and so have received much attention. In small animals such as insects, biochromes of internal organs and fluids can serve visual effect functions and some may be accumulated for that purpose. In various animals accumulations in internal and superficial organs have been considered "storage excretory" materials in "kidneys of accumulation." In most cases, however, accumulations in the fat body and related tissues appear to be stores for future use. This is equally probable for most, if not all, biochromes stored by the female in her eggs (Section III, F, 2). Biliverdin and bilirubin from their heme are stored in the fat body of chironomid larvae, and later used by the adults. Similarly *Rhodnius* stores in the pericardial tissue both of these bilins, from dietary heme. In general the material is stored in fluid vacuoles, ready for mobilization. In a few instances of massive accumulation, e.g., by the female coccid, *Dactylopius cacti* (Section II, C, 2), the insect does seem to have lost control of the synthesis of biochrome by the symbionts, and of its elimination, so that massive accumulation is not necessarily lethal.

Rather logically, stores of uric acid by insects have been regarded as the best evidence of storage-excretion, so that recent contrary discoveries are important. *Periplaneta* is not uricotelic, but ammonotelic, yet it stores large amounts of uric acid. The amount is proportional to the protein content of the diet (Kilby, 1963; Cochran, 1974) and when this is inadequate the stored uric acid is mobilized, mainly for the synthesis of nucleic acids. In the fat body, uric acid is protein-bound though in *Galleria mellonella* the soluble nucleotide, uric ribotide, is used for transport in the blood. Microsymbionts help in its mobilization; this fails in deflorated insects and they suffer abnormalities in color, protein metabolism, and growth. In holometabolous insects naturally there are complex changes in amount of stored uric acid through the late larval and pupal stages, and eventually a considerable amount is excreted by the moth, *Manduca sexta,* but the *Pieris* imago still has as much as 1.4 mg in its fat body (Ziegler and Harmsen, 1969).

Pterins behave rather like purines except that they are used more extensively for adult integumental and optic chromes (Ziegler and Harmsen, 1969). They are not extensively excreted except where there is a high dietary intake or when storage sites are damaged. Endogenous synthesis is geared to requirements and is decreased in the pupa if an imaginal wing-bud is excised. The imago of pierid butterflies still has 150–200 μg of pterins in its fat body (in addition to the uric acid) and only 4 μg is excreted at eclosion. The adult diet contains little nitrogen and there is a continuous turnover of these N-rich biochromes even up to old age (Linzen, 1974). Nymphalid butterflies, which do not use pterins for imaginal coloration, synthesize only one-tenth of the amount by pierids and excrete about the same amount as the latter.

Although ommochromes are used at least as extensively as pterins for integumental and optic purposes, they appear to be stored less massively. Fat body storage is recorded only in the Lepidoptera, and then only temporarily, usually in the mobile forms of rhodommatin and ommatin D. In the moth, *Cerura vinula,* they begin to accumulate in the fat body from the time the larva ceases to feed, therefore from internal sources. There is accumulation also in gut, Malpighian tubules, and hemocoel, and the total rises to 1 mg, or 0.2% of the total dry weight. This is close to the amount expected if all the tryptophan discarded when cocoon silk is synthesized from the general pool of amino acids is converted to ommochrome. At eclosion, between 75 and 470 μg of ommochrome is excreted so that much of the store is used for adult chromes. In the nymphal stages of *Locusta* some ommochrome is excreted after every molt but in *Carausius* none is detectable (Dustmann, 1964). Tryptophan is one of the least abundant of protein amino acids and a gross excess is rare.

Ommochrome synthesis is induced through starving *Cerura* larvae, presumably because body protein is then used as a source of energy. Various types of stress that also increase protein catabolism, have a similar effect. It is well established that insects cannot degrade tryptophan (Trp) completely for energy purposes, though the synthesis of ommochromes is oxidative and yields some energy. Also they are more innocuous than Trp itself (Linzen, 1974) so that, with their additional merit as biochromes, positive selection for their synthesis, and for suppressing total degradation, seems feasible. This is supported by the sophistication of structure and properties of the ommochromes, particularly as screening chromes in the eye. Some annelids, at least, can degrade Trp completely, yet some also synthesize and use ommochromes; therefore positive selection for a greater bias in insects seems more plausible than the alternative view, namely, that ommochrome exploitation was the incidental sequel to an accidental loss of the complete degradation path. Such individual accidents do not seriously affect the species.

In a number of insects there are heavy concentrations of biochromes in the wall and the lumen of the hindgut and/or Malpighian tubules. Some of these are related to excretion but others appear to be purely for storage, none of the material being passed out of the body. This is true particularly for ribitylflavin, in *Periplaneta, Bombyx,* and *Calliphora*. In *Attacus pernyi* the concentration in the Malpighian tubules reaches 1 mg/gm.

F. Biochromes Affecting Reproduction and Morphogenesis

1. Sex Differences in Content of Biochromes

Particularly among Lepidoptera there are striking sex differences in surface coloration for episematic purposes; behaviorally at least these affect reproduction, therefore, but probably none is relevant to physiological differences in reproductive activity, or in general metabolism. This has been shown (Section III, A, 2) for the red-yellow sex difference in the ommochromes of *Sympetrum*. The male of the butterfly, *Argynnis*, similarly has more red ommatin D in the wings than the female, which has a yellow flash coloration; ommatin D is not reduced merely through local lack of oxygen (Section II, G, 3), which is improbable in free-living insects. The male of *Pieris brassicae* has nearly twice, and that of *P. rapae* three times as much uric acid in the wings as the female (Tojo and Yushima, 1972). In all cases a genetically controlled visual effect difference is the only established significance.

In some instances sex differences in the biochrome content of other tissues, and of the blood, also have a visual effect. For other instances no functions are known: the male of *Drosophila* has more isoxanthopterin in eyes and gonads than the female but she has more of all other pterins (Ziegler and Harmsen, 1969). Anthraquinone concentration is much higher in the female than in the male coccid and this may be related to the generally higher rate of synthesis of materials by females. There is a more certain sex-related metabolic difference in *Musca*, the female having higher activity of cytochrome *P*-450 (Section III, A, 2) than the male (Agosin and Perry, 1974), and in *Schistocereca*, of which the female, only, has an active Trp-pyrrolase in the hemolymph.

Sex differences in some biochromes may be established very early in life. In the larvae of *Xanthia flavago* and *Bombyx mori* the male has a colorless or pale yellow hemolymph but that of the female is a strong yellow-green. The coccid *Pseudaulacaspis pentagona* shows a sex difference even from the egg stage (Seugé, 1972): female eggs are yellow and male eggs white. Dietary factors of the mother determine whether the yellow chrome is introduced into an egg, and that egg then fertilized to give a female. On a normal diet about half of the eggs are determined as female.

2. Role of Biochromes in Reproductive Processes

In holometabolous insects puberty occurs relatively quickly, in the pupa, and is confused with a major morphogenetic change not solely relevant to reproduction. Consequently there is little specific knowledge of pubertal changes in biochromes for reproductive functions. In the heterometabolous insects *Locusta* and *Schistocerca*, however, it seems clear that the replacement of astaxanthin by β-carotene in late nymphal stages has such a relevance. It is mainly β-carotene which the female packs into her eggs; the total amount increases in her ovaries, and integument, at this time.

There is a general tendency for the female to transport into her ovary and eggs the dominant, and any special, chromes of the species (Table VIII). This is expected since the embryo usually requires these, and the egg is its

TABLE VIII

Identified Biochromes in the Eggs of Insects

Biochrome	Distribution	References
β,β-Carotene	*Locusta*	Goodwin and Srisukh (1949)
β,β-Carotene and lutein	*Bombyx*	Manunta (1933)
?Flavonoid	*Bombyx*	Manunta (1936)
Carminate	*Dactylopius*	Dimroth and Kammerer (1920)
Hemoglobin	*Pediculus*, *Chironomus*	Wigglesworth (1943)
Ferrihemoprotein	*Rhodnius*	Wigglesworth (1943)
Ommochromes	*Anagasta*, *Bombyx*	Kikkawa (1953)
Pterins	*Schistocerca*, various Hemiptera, *Bombyx*	Ziegler and Harmsen (1969)
Ribitylflavin	Various insects	Trager and Subbarow (1938); Bodine and Fitzgerald (1947)

only mode of supply. Particularly instructive is the bilin of *Chironomus;* this is stored by the female since her larval stage, solely for her offspring. Transport of Hb, or of bilins, into the eggs by bloodsuckers such as *Rhodnius* and *Pediculus,* may be an essential providence until the embryo can feed. Taxonomically carotenoids are as widely distributed among the eggs as they are among the feeding stages of insects. In consequence females usually have less carotenoids than the males, in all organs but the ovary.

Since the materials transported into the egg depend on the mother's genotype they are responsible for the phenomenon of "cytoplasmic inheritance" or "predetermination" (Ziegler, 1961). It can have curious developmental effects: for instance some lethal mutants are able to develop normally until the egg store is exhausted and then die. In *Anagasta,* and possibly also in *Drosophila,* such a store of 3-OH-kynurenine induces the biosynthesis of both ommochromes and pterins in mutants lacking this power and they die once the store is exhausted. In *Bombyx mori* a similar situation depends on the ratio of isoxanthopterin to xanthopterin in the egg store. Normal embryos begin to produce their own inducers by the time the store is exhausted.

A number of biochromes have specific roles in reproductive processes. α-Tocopherol improves the reproductive output of the female of the fly, *Agria affinis* and of both sexes in the cricket, *Acheta* (House, 1965; Richet and McFarlane, 1962). *Lasioderma serricorne* and *Blattella germanica* need cobalamin (Section II, E, 1) for the production of viable eggs (Gordon,

1959): it is therefore essential also for development. Biopterin promotes the change from parthenogenic to sexual reproduction in aphid nymphs (Gilmour, 1965) and pterins are involved also in the regulation of seasonal sexual cycles in some insects (L'Hélias, 1961, 1962). Ribitylflavin is essential for yolk synthesis and egg production in *Drosophila*.

Uric acid has a striking, but enigmatic, role in reproductive activity in many species of cockroach (Corrigan, 1970; Cochran, 1974). In the form of a secretion from special "uricose" glands of the male it is poured over the spermatophores at the time of mating. Another puzzle is that the main biochrome, any special chrome, and sometimes all the biochromes, of the species are present in the testis sheath of male insects. *Papilio xuthus* has only xanthommatin but *Anagasta* has all the biochromes of the species. The special pterin-like *lampyrine* of the glowworm is abundant in the sheath, and similarly Hb in *Macrocorixa* and *Pediculus,* pterins in *Drosophila,* and pheophorbides in *Anasa tristis*. It seems improbable that these are required by the spermatozoa in their short, and often immobile, existence. As an extra screen against irradiation a single, unspecialized bathochrome should suffice.

3. Biochromes Affecting Morphogenesis

Because they are conspicuous, the morphogenetic movements of biochromes are rather well known. Some are complicated, and the color may change, indicating more than mere redistribution of material. The anthraquinone of the egg of the scale insect possibly is entirely transferred to the larval fat body, and may have no specific developmental function, but some egg-chromes (Table VIII) are metabolically important.

A severe deficiency of vitamin A in *Aedes,* maintained over three or four generations so as to deplete the egg completely, results in faulty development of the ommatidia, with subsequent degeneration (Carlson *et al.,* 1967). This effect is induced much more easily in vertebrates, which depend on retinol also for many other components of development. The precursor, β-carotene, is essential also for the integumental carotenoids of the locust, *Schistocerca gregaria,* the grasshopper, *Melanoplus bivittatus* (Dadd, 1961; Nayar, 1964) and the meal moth, *Plodia interpunctella* (Morère, 1971). In the locust, the carotenoid deficiency results also in a diminished synthesis of integumental ommochromes, indicating a joint control. This insect changes its stored β-carotene back to astaxanthin during development. In *Melanoplus* also the effect of β-carotene deficiency is on chromogenesis rather than on development in general; the effects are more drastic in *Plodia,* however, and no adults emerge from the pupa.

Plodia also requires dietary α-tocopherol for normal development. On a meridic diet containing all known growth factors, the growth rate of the cricket, *Acheta domesticus,* is accelerated threefold by adding powdered dried grass up to 30% of the weight of the food. Extraction indicates that flavonoids are responsible; known flavonoids also accelerate growth, though not as powerfully as the crude grass extracts (Neville and Luckey, 1971). Ladisch *et al.* (1967) found the benzoquinone products of tenebrionid grain and flour beetles to be carcinogenic. The total production of Q's and Qol's by some species is as much as 5% of their body weight so that the medium becomes morphogenetically dangerous to competitors and predators, as well as poisonous and deterrent. If the female of *Aedes aegypti* is injected with an inhibitor of dopa-decarboxylase, DL-3-(3,4-dihydroxyphenyl)-2-hydrazino-2-methylpropionic acid (= α-MDH), the tanning of the chorion of the eggs fails. The eggs then fail to develop, implying that sclerotization in some way is essential for development. Cobalamin is essential for development in *Lasioderma serricorne* (Pant and Fraenkel, 1954). The essential function is in C-1 transfer, as CH_3, in the biosynthesis particularly of thymidine.

For ommochromes and pterins there is much evidence of morphogenetic activity in insects. This is not entirely concerned with the development of the chromes themselves, in the integument and elsewhere, but extensively with folic acid and other possible coenzymes. Moreover the development of the integumental pterins and ommochromes is integrated, and it also affects development more widely. Wild-type eggs of *Anagasta (Ephestia)* and *Bombyx* show a higher percentage of normal development than the pale eggs of some mutants. If the eggs of the *Bombyx* mutant, *white 1,* are supplied with 3-OH-kynurenine, the embryos synthesize ommochromes, and a higher percentage hatch than among unsupplemented eggs (Kikkawa, 1953). Mutants with eggs deficient in other biochromes similarly show poor development (Needham, 1974); all may have some control of development in general, if only because this cannot be normal if any component is defective. Of course the screening chromes may function in that capacity from an early stage: in *Cerura* a copious secretion of ommatin D from the endothelium of the stalk of the ovariole is deposited over the surface of the exochorion of the eggs before laying.

Another consideration is that chromogenically deficient mutants may accumulate toxic precursors of a blocked step in the pathway. Free tryptophan itself is deleterious and experimentally added to the diet of the beetle *Oryzaephilus surina* at 0.5–1.0% of the total it depresses the rate of development and causes tumors. The intermediaries, 3-OH-kynurenine, anthranilic acid, and 3-OH-A at 1 *mM* concentration in a fluid diet also retard development; they cause tumors in *Drosophila* and *Anagasta*.

In causing aphid nymphs to mature as sexually, rather than as parthenogenically, reproducing adults, biopterin acts on a late stage of development. A relatively late action is implied also by the high concentration of biopterin in the food of larval Hymenoptera.

The development of eggs of a folate-deficient female of *Lasioderma ser-*

ricorne is subnormal and few hatch (Pant and Fraenkel, 1954). In *Aedes* the larva shows abnormalities in molting, and in biochrome synthesis, if its own diet is folate-deficient (Golberg *et al.*, 1945). When aminopterin, a potent inhibitor of folic acid, is added to the diet of larvae of *Aedes, Anopheles, Drosophila,* or *Musca* it delays pupation and inhibits imaginal emergence (Golberg *et al.,* 1944; Goldsmith *et al.,* 1948; Mitlin *et al.,* 1954). The action is essentially in C-1 transfer, to synthesize purines and thymidine for nucleic acid synthesis (Section III, A, 2), and dietary supplements of nucleic acids to developing insects make folic acid dispensable (House and Barlow, 1958; Brookes and Fraenkel, 1958).

The coenzymatically active form of folic acid, FH_4, is essential for the development of winged forms of the aphid *Myzus persicae;* di-H-folate reductase, which catalyzes its synthesis from the precursor, di-H-folate, has 42% higher activity in presumptive alates than in presumptive apterous individuals. Aminopterin has drastic effects on wing development by the nymphs, as well as on the production of nymphs by the mature female. Addition of thymidine mitigates these effects. At 10 μM concentration both folate and N^{10}-formylfolate promote cell-outgrowth from ovarian *in vitro* transplants of *Galleria melonella* and *Bombyx mori.* Most other pterins, except the 2-OH, 4-SH analogue, proved inhibitory and injurious (Zielinska and Saska, 1972).

In the embryo of the milkweed bug *Oncopeltus fasciatus* there is a protein-bound fraction of isoxanthopterin (IXP) that repressively controls the synthesis of RNA by form II of RNA polymerase. The conjugate works by binding the IXP to the relevant template-DNA. Maximal inhibition is effected by 560 ng of the conjugate per 10^7 nuclei, and extraneous IXP inhibited by 55% when it reached 0.6% of the content of dechorionated eggs (Harris and Forrest, 1967; Smith and Forrest, 1976). The amount of IXP-protein decreases to zero at the end of development while free IXP increases throughout (Ziegler and Harmsen, 1969).

Ribitylflavin is an essential vitamin for all insects tested and is abundant in the eggs of most of them. It is essential for the development of the yellow fever mosquito, and has even been found to restore to the *Drosophila* mutant, *antennaless,* the ability to develop antennae. It is depleted progressively during the development of the grasshopper, *Melanoplus differentialis* (Ziegler, 1965); the content of pterins increases progressively but probably they are already synthesized endogenously: degradation of RF would give lumazines rather than pterins (Section II, H, 1).

4. Biosynthetic Interrelationships of Biochromes

This merits further consideration since for evolutionary, selective reasons there is a genetic control of all interrelationships, so that care is needed to distinguish these indirect, systemic controls from controls by direct chemical reaction. The problem is particularly acute for visual effect systems where various chemically unrelated biochromes are closely associated topographically and behaviorally. The literature abounds in interpretations of indirect genetic interactions as direct chemical reactions. As already noted, if kynurenine is injected into *Anagasta* there is an increase not only in ommochromes but also in pterins (Ziegler, 1961), while in nymphs of *Locusta* and *Schistocerca* parental lack of carotenoid leads not only to its own absence from the integument but also to diminished ommochrome deposition there, and later to an increase in bilin (Dadd, 1970). In this last case the interrelationships must be indirect since carotenoid is dietary whereas ommochrome is endogenous and bilin adjusts only after an interval, and in the opposite direction. The synthesis of ommochromes for the imaginal eyes of *Drosophila* begins 24 hr before that of the pterins. A large number of genes are now known to control pleiotropically two or more biochromes that originally were adventitiously aggregated for visual effect, screening, and other functions. However a much larger number still control monotropically one chrome only, showing that pleiotropy is a secondary acquisition.

On the other hand, there are interactions that do appear to be direct, and no doubt in general evolution will have been concerned with materials already associated for biogenic reasons. The members of the electron transport series (Section III, A, 1) include a set that are structurally related, the cytochromes, but also others that are widely unrelated. However they are all well fitted, by redox potential, general molecular structure, etc., to react directly with their neighbors, particularly with the aid of naturally selected apoproteins. This is a feasible analogy for a similar selection in favor of the control of associated biosyntheses by direct chemical means. Moreover the number of separate chromogenic pathways is not very great and there are some direct synthetic relationships between them (Needham, 1974). Potentially, therefore, direct interactions may be rather numerous and may also include biochromes acting as coenzymes in their own pathways of biosynthesis. Further, through their redox properties and membership of the general respiratory enzyme systems (Section III, A, 1, 2) a number of them may directly control reactions in various pathways. Direct interactions are further multiplied by a broad substrate-tolerance (low specificity) of many of these enzymes.

Examples of this last property include polyphenol oxidase which converts tyrosine and other substrates to catecholic and indolic melanin precursors. It will also oxidatively dimerize 3OHK to ommatin and this may explain the reddening of shed blood of the *aka aka* mutant of *Drosophila* (Inagami, 1954). *In vitro,* at least, dopaquinone also will reversibly catalyze the 3OHK→ommatin step; effectively it becomes a (colored) coenzyme and it also illustrates how slight is the distinction between substrates and coenzymes in any predominantly redox pathway. Even more nonspecifically, the cytochromes will catalyze the 3OHK→ommatin step, and Hb the analogous step from 3-OH-anthranilic acid to cinnabarinic acid. The tetra-H-pterin enzymes that hydroxylate phenylalanine to tyrosine and dopa also convert K to 3OHK, and this may be coupled with the biosynthesis of drosopterin from a precursor pterin (Ziegler and Harmsen, 1969).

These examples mainly concern the tyrosine, tryptophan,and pterin pathways, and further aspects of their interactions are recorded. Xanthommatin may be a coenzyme for the conversion of 7,8-diH-biopterin to sepiapterin in *Drosophila* (Parisi *et al.*, 1976). By contrast,biopterin and 6-OHMe-pterin inhibit the first step in the ommochrome pathway, the pyrrolysis of Trp (Ghosh and Forrest, 1967), and isoxanthopterin also inhibits ommochrome synthesis in *Drosophila* (Forrest, 1959). It also inhibits melanogenesis in the beetle *Harmonia*. The conversion of isoxanthopterin to xanthopterin can be coupled with the conversion of dopa to a quinone that lacks the ability to polymerize to melanin, and xanthopterin itself inhibits the final stages—the oxidation and polymerization of dihydroxyindole. The subject needs further study since others have found both xanthopterin and leucopterin to promote melanin synthesis.

Perhaps the most outstanding example both of broad substrate-tolerance and of biochromes controlling their own synthetic path concerns the purine-pterin-isoalloxazine class. The flavoprotein, xanthine oxidase, catalyzes the conversion of pterin to isoxanthopterin, of xanthopterin to leucopterin, and of pterin to ribitylflavin itself (Forrest, 1962). It also converts hypoxanthine to xanthine and this to uric acid (Ziegler, 1961). It is inhibited by some pterins. Xanthommatin provides the best example of direct feedback inhibition by the product of a pathway, since it inhibits the pterin-enzyme, kynurenine hydroxylase (Linzen, 1974).

Discussion has been restricted to chromogenic paths but biochromes are at least as extensively involved in the biosynthetic paths of colorless materials, the bulk-subtrates of morphogenesis. This may prove equally true for their role in other biochemical fields, for which at present information is less abundant.

IV. CONCLUSIONS

Insects have most of the known biochromes of animals, including some, i.e., aphins, papiliochromes,and aurones, that may be peculiar to the class. They have a considerable number of carotenoids, some very widely distributed in this taxon. At present they are not known to make wide use of flavonoids, copper proteins,or ferredoxins. Hemoglobin is rare but probably all insects have the other common hemoproteins. Bilins also are common and are endogenously synthesized from the hemes. Like other arthropods, insects make much use of quinones, for tanning skeletal proteins and for other purposes. They also exploit extensively ommochromes and the purine-pterin class, mainly in eyes and integument.

Insects have an electron transport respiratory system very similar to that of vertebrates, using the same series of biochrome coenzymes. Some of the other insect biochromes also may act as redox agents of this system, or more peripherally, partly in relation to the biosynthesis of biochromes themselves. The substrate of firefly bioluminescence, luciferin, is a biochrome with a chemophotic type of energy transduction.

Insect integumental chromes primarily serve a behavioral function in crypsis or in semasis and these functions have little direct relevance to biochemical reactions. However, some of them serve in types of defense and screening where their chemical properties are involved.

As in most animals studied, one carotenoid derivative, retinal 1, has been exploited for photoperception. It behaves very similarly to that of vertebrates, and mediates color-vision in a generally similar way. Some other photosensitive chromes also may put the property to use in insects. Some biochromes may normally function in sense organs of other types, or in other components of the neural system.

A number of biochromes are stored for long periods. They remain labile and are used or excreted as expedient. Very few represent a static substitute for excretion.

Biochromes play an important part in the sexual dimorphism and mating of insects and some in other aspects of reproduction. The most important, or even all, biochromes of a species are present in the testis sheath, and in the ovary where they pass into eggs. Some of these have a significance in morphogenesis beyond their own developmental history.

There are many records of interactions between insect biochromes. Some are due to a common genetic control of the biosynthesis of those that function together in visual effect. There are also direct biosynthetic and metabolic interactions; this is because they are very much interrelated, biosynthetically and in properties, providing the coenzymes for all redox reaction chains, and because their own biosyntheses are largely oxidative, so that most of the coenzymes active in these paths are themselves biochromes.

GENERAL REFERENCES

Agosin, M., and Perry, A. S. (1974). *In* "The Physiology of Insecta" (M. Rockstein, ed.), 2nd. Ed., Vol. 5, pp. 537–596. Academic Press, New York.

Autrum, H. (1975). *In* "Traité de Zoologie" (P.-P. Grassé, ed.), Vol. 8, Part 3, pp. 742–853. Masson, Paris.

Brunet, P. C. J. (1963). *Ann. N.Y. Acad. Sci.* **100,** 1020–1034.
Brunet, P. C. J. (1965). *In* "Aspects of Insect Biochemistry" (T. W. Goodwin, ed.), pp. 49–77. Academic Press, New York.
Brunet, P. C. J. (1967). *Endeavour* **26,** 68–74.
Cameron, D. W., and Lord Todd (1967). *In* "Oxidative Coupling of Phenols" (W. I. Taylor and A. R. Battersby, eds.), pp. 203–241. Dekker, New York and Arnold, London.
Chen, P. S. (1971). "Biochemistry of Insect Development." Karger, Basel.
Eisner, T. (1970). *In* "Chemical Ecology" (E. Sondheimer and J. B. Simeone, eds.), pp. 157–217. Academic Press, New York.
Fox, D. L. (1976). "Animal Biochromes and Structural Colors," 2nd Ed. Univ. of California Press, Berkeley, California.
Fox, H. M., and Vevers, G. (1960). "The Nature of Animal Colours." Sidgwick & Jackson, London.
Fuseau-Braesch, S. (1972). *Annu. Rev. Entomol.* **17,** 403–424.
Gilmour, D. (1965). "The Metabolism of Insects." Oliver & Boyd, Edinburgh.
Goldsmith, T. H., and Bernard, G. D. (1974). *In* "The Physiology of Insecta" (M. Rockstein, ed.), 2nd Ed., Vol. 2, pp. 165–272. Academic Press, New York.
Goodwin, T. H. (1971). *In* "Chemical Zoology" (M. Florkin and B. T. Scheer, eds.), Vol. 6, pp. 279–306. Academic Press, New York.
Hackman, R. H. (1971). *In* "Chemical Zoology" (M. Florkin and B. T. Scheer, eds.), Vol. 4, pp. 1–62. Academic Press, New York.
Hackman, R. H. (1974). *In* "The Physiology of Insecta" (M. Rockstein, ed.), 2nd Ed., Vol. 6, pp. 215–270. Academic Press, New York.
Harborne, J. B., ed. (1967). "Comparative Biochemistry of Flavonoids." Academic Press, New York.
Isler, O, Gutmann, H. and Solms, U. eds. (1971). "Carotenoids." Birkhaeuser, Basel.
Kamin, H., ed. (1971). "Flavins and Flavoproteins." Univ. Park Press, Baltimore, Maryland and Butterworth, London.
Kaufman, S. (1967). *Annu. Rev. Biochem.* **36,** 171–184.
Kilby, B. A. (1963). *Adv. Insect Physiol.* **1,** 112–174.
Linzen, B. (1974). *Adv. Insect Physiol.* **10,** 117–246.
McElroy, W. E., Selliger, H. H., and DeLuca, M. (1974). *In* "The Physiology of Insecta" (M. Rockstein, ed.), 2nd Ed., Vol. 2, pp. 411–460. Academic Press, New York.
Menzel, R. (1975). *In* "The Compound Eye and Vision in Insects" (G. A. Horridge, ed.), pp. 121–153. Oxford Univ. Press (Clarendon), Oxford, London and New York.
Needham, A. E. (1974). "The Significance of Zoochromes." Springer-Verlag, Berlin and New York.
Nicolaus, R. A. (1968). "Melanins." Hermann, Paris.
Pavan, H., and Dazzini, M. V. (1971). *In* "Chemical Zoology" (M. Florkin and B. T. Scheer, eds.), Vol. 6, pp. 365–409. Academic Press, New York.
Pitt, G. A. J. (1971). *In* "Carotenoids" (O. Isler, H. Guttmann and U. Solms, eds.), pp. 717–742. Birkhaeuser, Basel.
Pryor, M. G. M. (1962). *In* "Comparative Biochemistry" (M. Florkin and H. S. Mason, eds.), Vol. 4, pp. 371–396. Academic Press, New York.
Sacktor, B. (1974). *In* "The Physiology of Insecta" (M. Rockstein, ed.), 2nd Ed., Vol. 4, pp. 272–358. Academic Press, New York.
Vuillaume, M. (1975). *In* "Traité de Zoologie" (P.-P. Grassé, ed.), Vol. 8, Part 3, pp. 77–184. Masson, Paris.
Ziegler, I. (1961). *Adv. Genet.* **10,** 349–403.
Ziegler, I., and Harmsen, R. (1969). *Adv. Insect Physiol.* **6,** 139–203.

REFERENCES FOR ADVANCED STUDENTS AND RESEARCH SCIENTISTS

Abrahamson, E. W. and Wiesenfeld, J. R. (1972). *In* "Photochemistry of Vision" (H. J. A. Dartnell, ed.). pp. 69–121. Springer-Verlag, Berlin and New York.

Ajami, A. M. and Riddiford, L. M. (1971). *Biochemistry*, **10**, 1451–1460.

Autrum, H. (1965). *In* "Colour Vision" (A. V. S. DeReuck and J. Knight, eds.), pp. 286–300. Little, Brown, Boston, Massachusetts.

Autrum, H. and Kolb, G. (1968). *Z. vergl. Physiol.* **60**, 450–477.

Autrum, H. and Zwehl, V. von, (1964). *Z. vergl. Physiol.* **48**, 357–384.

Baker, J. E. and Norris, D. M. (1971). *J. Insect Physiol.* **17**, 2383–2394.

Barbier, M., Bergerard, J., Huspin, B., and Vuillaume, M. (1970). *C. R. Acad. Sci. Paris* **271**, 342–345.

Becker, E. (1941). *Naturwissenschaften* **29**, 237–238.

Bennett, R. R., Tunstall, J., and Horridge, G. A. (1967). *Z. vergl. Physiol.* **55**, 195–206.

Bennett, R. R. and Ruck, P. R. (1970). *J. Insect Physiol.* **16**, 83–88.

Bodine, J. H. and Fitzgerald, L. (1947). *J.Exp. Zool.* **104**, 353–363.

Bodnaryk, R. P., Brunet, P. C. J. and Koeppe, J. K. (1974). *J. Insect Physiol.* **20**, 911–923.

Boëthius, J., Carlson, S. D., Höglund, G. and Struwe, G. (1968). *Acta Physiol. Scand.* **74**, 36A–37A.

Bowie, J. H. and Cameron, D. W. (1967). *J. Chem. Soc. London,* **C**, 704–723.

Brookes, V. J. and Fraenkel, G. (1958). *Physiol. Zool.* **31**, 208.

Brower, L. P. (1969). *Sci. Amer.*.**220**, 22–29.

Brown, K. S., Cameron, D. W. and Weiss, U. (1969). *Tetrahedron Letters*, Part No. 6, 471–476.

Brown, K. S. and Weiss, U. (1971). *Tetrahedron Letters,* Part No. 38, 3501–3504.

Bruckmoser, P. (1968). *Z. vergl. Physiol.* **59**, 187–204.

Brunet, P. C. J. and Coles, B. C. (1974). *Proc. Roy. Soc. London B,* **187**, 133–170.

Bu'lock, J. D. and Harley-Mason, J. (1951). *J. Chem. Soc. London,* 703–712.

Burkhardt, D. (1962). *In* "Biological Receptor Mechanisms" (J. W. L. Beament, ed.) pp. 86–109. Cambridge University Press, Cambridge.

Burwood, R., Reed, G., Schofield, K. and Wright, D. E. (1967). *J. Chem. Soc. London* **C**, 842–851.

Butenandt, A., Gröschel, U., Karlson, P. and Zillig, W. (1959). *Arch. Biochem. Biophys.* **83**, 76.

Carlson, S. D., Steeves, H. R., VandeBerg, J. S. and Robbins, W. E. (1967). *Science,* **158**, 268–270.

Chance, B. and Sacktor, B. (1958). *Arch. Biochem. Biophys.* **76**, 509–531.

Choussy, M. and Barbier, M. (1973). *Biochem. Systematics,* **1**, 199–201.

Cochran, D. G. (1974). *In* "Insect Biochemistry and Function" (D. J. Candy and B. A. Kilby, eds.), pp. 179–281. Chapman and Hall, London.

Corrigan, J. J. (1970). *In* "Comparative Biochemistry of Nitrogen Metabolism" (J. W. Campbell, ed.) Vol. I, pp. 388–488. Academic Press, London and New York.

Cott, H. B. (1940). "Adaptive Coloration in Animals." Methuen, London.

Czeczuga, B. (1970). *Hydrobiologia,* **36**, 353–360.

Czeczuga, B. (1971). *J. Insect Physiol.* **17**, 2017–2075.

Dadd, R. H. (1961). *Bull. Entomol. Res.* **52**, 63–81.

Dadd, R. H. (1970). "Chemical Zoology" (M. Florkin and B. T. Scheer, eds.). Vol. V, pp. 35–95. Academic Press, New York.

Daumer, K. (1956). *Z. vergl. Physiol.* **38**, 413–478.

DeLuca, M. and Dempsey, M. E. (1971). *Biochem. Biophys. Res. Comm.* **40**, 117–128.

Dennell, R. (1958). *Biol. Rev. Biol. Proc. Cambridge Philos. Soc.* **33**, 177–234.

Dimroth, O. and Kammerer, H. (1920). *Ber. Deutsch Chem. Ges.* **53B**, 471–480.

Duewell, H., Human, J. P. E., Johnson, A. W., Macdonald, S. F. and Todd, A. R. (1948). *Nature (London)* **162,** 759–761.
Dustmann, J. H. (1964). *Z. vergl. Physiol.* **49,** 28–57.
Eguchi, E. (1971). *Z. vergl. Physiol.* **71,** 201–218.
Feltwell, J. (1974). *J. Zool. London,* **174,** 441–465.
Fingerman, M. and Brown, F. A. (1953). *Physiol. Zool.* **26,** 59–67.
Forrest, H. S. (1959). *In* "Pigment Cell Biology" (M. Gordon, ed.), pp. 619–628. Academic Press, New York.
Forrest, H. S. (1962). *In* "Comparative Biochemistry" (M. Florkin and H. S. Mason, eds.) Vol. IV, pp. 615–641. Academic Press, New York.
Frazer-Rowell, C. H. (1971). *Adv. Insect Physiol.* **8,** 145–198.
Ghosh, D. and Forrest, H. S. (1967). *Arch. Biochem. Biophys.* **120,** 578–582.
Glossman, H., Horst, J., Plagens, U. and Braunitzer, G. (1970). *Hoppe Seyler's Z. Physiol. Chem.* **351,** 342–368.
Gogala, M. (1967). *Z. vergl. Physiol.* **57,** 232–243.
Golberg, L., de Meillon, B. and Lavoipierre, M. (1944). *Nature (London)* **154,** 608–610.
Golberg, L., deMeillon, B. and Lavoipierre, M. (1945). *J. Exp. Biol.* **21,** 90–96.
Goldsmith, E. D., Tobias, E. B. and Harnly, M. H. (1948). *Anat. Rec.* **101,** 93.
Goldsmith, T. H. (1958). *Proc. Nat. Acad. Sci. U.S.A.* **44,** 123–126.
Goldsmith, T. H. (1960). *J. Gen. Physiol.* **43,** 775–779.
Goldsmith, T. H. (1972). *In* "Photochemistry of Vision" (H. J. A. Dartnell, ed.) pp. 685–719. Springer-Verlag, Berlin and New York.
Goldsmith, T. H. (1973). *In* "Comparative Animal Physiology" (C. L. Prosser, ed.) pp. 577–632. Saunders, Philadelphia, Pennsylvania.
Goodwin, T. W. and Srisukh, S. (1949). *Biochem. J.* **45,** 263–268.
Gordon, H. T. (1959). *Ann. New York Acad. Sci.* **77,** 290–338.
Hackman, R. H. (1953). *Biochem. J.* **54,** 371–377.
Hackman, R. H. (1967). *Nature (London)* **216,** 163.
Hackman, R. H. and Todd, A. R. (1953). *Biochem. J.* **55,** 631–637.
Hackman, R. H., Pryor, M. G. M. and Todd, A. R. (1948). *Biochem. J.* **43,** 474–477.
Hanser, G. and Rembold, H. (1968). *Z. Naturf.* **23B,** 666–670.
Harano, T. (1972). *Insect Biochem.* **2,** 385–399.
Harano, T. and Chino, H. (1971). *Arch. Biochem. Biophys.* **146,** 467–476.
Harris, S. E. and Forrest, H. S. (1967). *Proc. Nat. Acad. Sci. U.S.A.* **58,** 89–94.
Hasselmann, E.-M. (1962). *Zool. Jb. Allg. Physiol.* **69,** 537.
Hodgetts, R. B. and Konopka, R. J. (1973). *J. Insect Physiol.* **19,** 1211–1220.
Höglund, G., Hamdorf, K. and Rosner, G. (1973). *J. Comp. Physiol.* **86,** 265–279.
Hoffmann, R. J. (1974). *J. Insect Physiol.* **20,** 1913–1924.
Horridge, G. A. (1969). *Z. vergl. Physiol.* **62,** 1–37.
House, H. L. (1965). *J. Insect Physiol.* **11,** 1039–1045.
House, H. L. and Barlow, J. S. (1958). *Ann. Entomol. Soc. Amer.* **51,** 299–302.
Hsaio, T. H. and Fraenkel, G. (1968). *Ann. Entomol. Soc. Amer.* **61,** 476–484.
Hüber, R., Epp, O., Steigemann, W. and Formanek, H. (1971). *Europ. J. Biochem.* **19,** 42–50.
Inagami, K. (1954). *Nature (London)* **174,** 105.
Ito, Y., Nozawa, Y. and Kawai, K. (1970). *Experientia,* **26,** 826–827.
Kalmus, H. (1941). *Proc. Roy. Soc. London B,* **130,** 185–201.
Kikkawa, H. (1953). *Adv. Genetics,* **5,** 107–140.
Klingenberg, M. and Bücher, T. (1961). *Biochem. Z.* **331,** 312–333.
Koeppe, J. K. and Mills, R. R. (1974). *J. Insect Physiol.* **20,** 1603–1609.
Kusada, J. and Mukai, J.-I. (1971). *Comp. Biochem. Physiol.* **39B,** 317–323.
Ladisch, R. K., Ladisch, S. K. and Howe, P. M. (1967). *Nature (London)* **215,** 939–940.

Langer, H. and Theorell, B. (1966). *In* "Functional Organisation of the Compound Eye" (C. G. Bernhard, ed.) pp. 145–149. Pergamon, Oxford.
Leuenberg, F. and Thommen, H. (1970). *J. Insect Physiol.* **16,** 1855–1858.
Lewis, S. E. and Slater, E. L. (1954). *Biochem. J.* **58,** 207–217.
L'Helias, C. (1957). *Bull. biol. Fr. Belg.* **91,** 241–263.
L'Helias, C. (1961). *Ann. Biol.* Series 3, **37,** 367–392.
L'Helias, C. (1962). *Bull. biol. Fr. Belg.* **96,** 187–198.
Manunta, C. (1933). *Boll. Soc. Ital. Biol. Sper.* **8,** 1278–1292.
Manunta, C. (1936). *Boll. Soc. Ital. Biol. Sper.* **11,** 50–51.
Miller, P. L. (1964). *Nature (London)* **201,** 1052.
Mitlin, N., Konecki, M. S. and Piquett, P. G. (1954). *J. Econ. Entomol.* **47,** 932–933.
Morère, J.-L. (1971). *C. R. Acad. Sci. Paris, Ser. D.,* **272,** 2229–2231.
Morris, S. J. and Thomson, R. H. (1963). *J. Insect Physiol.* **9,** 391–399.
Morton, R. A., Hopkins, T. A. and Seliger, H. H. (1969). *Biochem.* **8,** 1598–1607.
Mote, M. I. and Goldsmith, T. H. (1970). *J.Exp. Zool.* **173,** 137–145.
Mote, M. I. and Goldsmith, T. H. (1971). *Science* (U.S.A.) **171,** 1254–1255.
Mummery, R. S. and Valadon, L. R. G. (1974). *J. Insect Physiol.* **20,** 429–433.
Nayar, J. K. (1964). *Canad. J. Zool.* **42,** 11–22.
Neville. P. F. and Luckey, T. D. (1971). *J. Nutrition,* **101,** 1217–1224.
Norris, D. M., Ferkovich, S. M., Rozental, J. M., Baker, J. E. and Borg, T. K. (1970). *Science, (U.S.A.),* **170,** 754–755.
Ohnishi, E. (1970). Personal communication. See Fuseau-Braesch, 1972 (General References).
Pant, N. C. and Fraenkel, G. (1954). *J. Zool. Soc. India,* **6,** 173–177.
Parisi, G., Carfagna, M. and D'Amora, D. (1976). *J. Insect Physiol.* **22,** 415–423.
Pau, R. N. and Acheson, R. M. (1968). *Biochim. Biophys. Acta,* **158,** 206–211.
Poulton, E. B. (1890). "The Colours of Animals." Paul, Trench, Trübner, London.
Richet, C. and McFarlane, J. E. (1962). *Canad. J. Zool.* **40,** 371–374.
Ruck, P. R. (1965). *J. Gen. Physiol.* **49,** 289–307.
Rüdiger, W. and Klose, W. (1970). *Experientia,* **26,** 498.
Sacktor, B. (1954). *J. Gen. Physiol.* **37,** 343–359.
Salt, G. (1970). "The Cellular Defence Reactions of Insects." Cambridge University Press, Cambridge.
Schäfer, W. and Geyer, I. (1972). *Tetrahedron,* **28,** 5261–5279.
Schildknecht, H. (1971). *Endeavour,* **30,** 136–141.
Schmidt, F. H. and Young, C. L. (1971). *J. Insect Physiol.* **17,** 843–855.
Schümperli, R. A. (1973). *J.Comp. Physiol.* **86,** 77–94.
Schwemer, J. and Paulsen, R. (1973). *J. Comp. Physiol.* **86,** 215–229.
Seugé, J. (1972). *C. R. Seances Soc. Biol. Paris,* **166,** 530–534.
Shappirio, D. G. (1974). *J. Insect Physiol.* **20,** 291–300.
Smith, J. H. and Forrest, H. S. (1976). *Insect Biochem.* **6,** 131–134.
Sontag, C. (1971). *J. Gen. Physiol.* **57,** 93–112.
Struwe, G. (1972a). *J. Comp. Physiol.* **79,** 191–196.
Struwe, G. (1972b). *J. Comp. Physiol.* **79,** 197–201.
Thayer, G. H. (1918). "Concealing Coloration in the Animal Kingdom." Macmillan, New York.
Thommen, H. (1971). *In* "Carotenoids" (O. Isler, H. Gutmann, and U. Solms, eds.). pp. 637–668. Birkhauser, Basel.
Tojo, S. and Yushima, T. (1972). *J. Insect Physiol.* **18,** 403–422.
Trager, W. and Subbarow, Y. (1938). *Biol. Bull. Woods Hole,* **75,** 75–84.
Trautman, K. H. and Thommen, H. (1971). Unpublished results. Quoted in Thommen, 1971.
Tschinkel, W. R. (1972). *J. Insect Physiol.* **18,** 711–722.
Umebachi, Y. (1975). *Insect Biochem.* **5,** 73–92.

Wehner, R. and Toggweiler, F. (1972). *J. Comp. Physiol.* **77,** 239–253.
Weiss, U. and Altland, H. W. (1965). *Nature (London)* **207,** 1295–1297.
Wigglesworth, V. B. (1943). *Proc. Roy. Soc. London B,* **131,** 313–339.
Wigglesworth, V. B. (1965). "Principles of Insect Physiology." Methuen, London.
Willig, A. (1969). *J. Insect Physiol.* **15,** 1907–1928.
Wolken, J. J. (1975). "Photoprocesses, Photoreceptors, and Evolution." Academic Press, New York.
Ziegler, I. (1961). *Ergeb. Physiol., Biol. Chem. Exp. Pharmakol.* **56,** 1–66.
Zielinska, Z. M. and Saska, J. (1972). Quoted in J. Saska, 1972, *J. Insect Physiol.* **18,** 1733–1737.
Zimmerman, W. F. and Goldsmith, T. H. (1971). *Science (U.S.A.)* **171,** 1167–1169.

7

Biochemistry of Insect Hormones and Insect Growth Regulators

LYNN M. RIDDIFORD AND JAMES W. TRUMAN

I. INTRODUCTION

In 1917 Kopec presented the first evidence that insects use hormones to control their growth and development. Little attention was paid to this possibility until the pioneering work of Wigglesworth, Fukuda, and Williams in the 1930's and 1940's that headed the field of insect endocrinology in its present direction. Since that time, studies of the endocrine interactions controlling insect growth and metamorphosis have mushroomed to the point that it would not be possible to present a comprehensive review in the pages allotted here. Several such reviews have recently been published and appear in the General References list at the end of the chapter. This chapter presents background on the biochemistry of insect hormones as well as on some of the recent advances in the field by focusing on selected examples. The extensive use of lepidopteran examples reflects both the amount of research performed on this order of insects and the authors' research bias.

The most characteristic and obvious feature of an arthropod is its hard exoskeleton. In insects, this rigid cuticle restricts growth of the larva, and consequently at intervals a new cuticle must be made and the old one cast off. After growth is completed, metamorphosis ensues. In those animals with "incomplete" metamorphosis (the Hemimetabola), metamorphosis is a one-step process with the last stage larva transforming directly into the adult. In insects with "complete" metamorphosis (Holometabola), the changes during metamorphosis are more extensive and a pupal stage is interposed between the last larval stage and the adult. In the higher Diptera this change is extreme in that most of the larval cells die and the adult is reconstructed from small groups of undifferentiated cells (the imaginal discs and the histoblasts) that were present but nonfunctional in the larva.

Since the cuticle is such an obvious feature of insect growth and development, much attention has been given to the molting process, i.e., the production of the new cuticle and the shedding of the old one. As seen in Fig. 1, three hormones—prothoracicotropic hormone (PTTH), ecdysone, and juvenile hormone (JH)—are involved in the initiation of the molting process. PTTH is produced by the neurosecretory cells of the brain and released from the neurohemal organ for these cells, the corpora cardiaca (CC). The hormone stimulates the prothoracic glands (PTG; also called ventral glands, ecdysial glands; part of the ring gland in higher Diptera) to secrete ecdysone. Ecdysone then acts on the epidermis to initiate the steps in the production of new cuticle. The epidermis detaches from the old cuticle (apolysis) and a molting fluid is pumped into the exuvial space. Oftentimes, DNA synthesis and cell division occur around this time. After the outer layers of the new cuticle are deposited, digestive enzymes in the molting fluid are activated and begin to degrade the inner layers of the old cuticle. The type of new cuticle formed depends upon the amount of JH secreted from the corpora

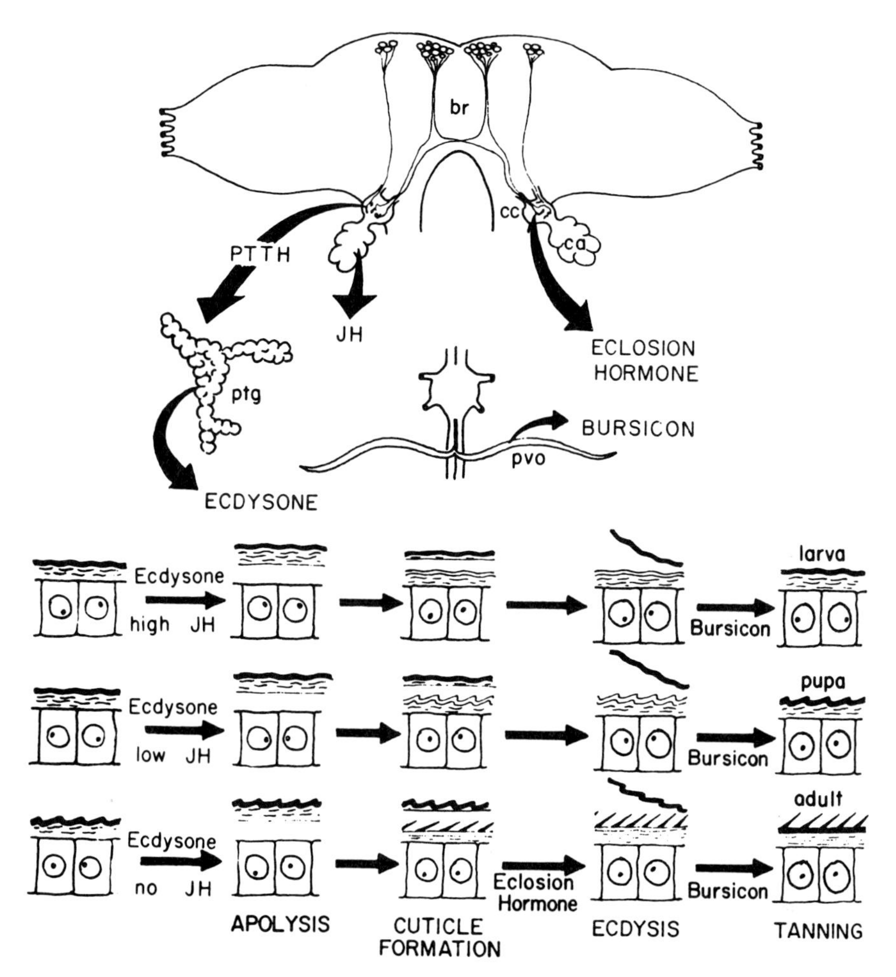

Fig. 1. The insect endocrine system (above) and the effects of its hormones on growth and development (below). br, brain; cc, corpus cardiacum; ca, corpus allatum; ptg, prothoracic gland; pvo, perivisceral organ; PTTH, prothoracicotropic hormone; JH, juvenile hormone.

allata (CA); when it is present in high concentrations, the new cuticle produced is larval. When the CA become inactive and the JH titer declines, then metamorphosis occurs.

The terminal phases of the molting process are regulated by two additional hormones, the eclosion hormone and bursicon. The ecdysis, or shedding, of the pupal skin in at least some holometabolous insects is triggered by a second hormone from the brain—the eclosion hormone. The cuticle of the newly ecdysed insect is usually soft so that it can be expanded to its proper size. Bursicon, which is released from the perivisceral organs in most insects, serves to regulate the postecdysial hardening and darkening of the new cuticle.

In most insects, after metamorphosis, the CA reactivate and again secrete JH. In feeding females this is usually necessary for some phase of oogenesis. Spermatogenesis in the male typically occurs during adult development and is substantially complete by the time the adult emerges. But, in some adults, the maturation of the male accessory glands requires JH.

First, we will consider each of these hormones as to their chemistry and their mode of action in normal growth and molting. Then hormonal control of reproduction and of diapause will be discussed. Finally, the use of these hormones and their mimics as insect growth regulators to control insect pests will be considered.

II. PROTHORACICOTROPIC HORMONE (BRAIN HORMONE, ACTIVATION HORMONE, OR ECDYSIOTROPIN)

Kopec in 1917 first demonstrated that the brain was necessary for pupation of the gypsy moth, *Lymantria dispar*. If he removed the brain before a certain time in the last larval instar, the larva would not pupate. But removal after this critical time had no effect on subsequent pupation. Relatively little attention was paid to these experiments until nearly 20 years later when Wigglesworth in 1934 reported his classic experiments on the role of the brain in the molting of the blood-sucking bug, *Rhodnius prolixus*. During larval growth this insect can go for long periods without feeding. When a proper host is found, a massive blood meal is taken and this engorgement triggers the onset of the molt to the next instar. When penultimate or earlier larval instar *Rhodnius* were decapitated within 4 days after feeding, no subsequent molting occurred. Decapitation on the fifth day after feeding was followed by a normal molt. By implantation of various cephalic organs into decapitated hosts, Wigglesworth subsequently showed that the brain was responsible for secreting a blood borne factor which initiated the molting process. At this time Wigglesworth also suggested that stretching of the abdomen by the blood meal provided the stimulus that eventually led to the release of this "brain hormone."

The brain does not act alone in triggering the molt. In the early 1940's Fukuda, working with *Bombyx mori* larvae (the commercial silkworm), performed ligation experiments that indicated that a pair of glands in the thorax, the prothoracic glands, was essential for molting. Subsequently, Williams unraveled the interaction between the brain and prothoracic glands in the stimulation of adult development in the giant wild silkmoth, *Hyalophora cecropia*. In this species, the pupal stage enters into a period of developmental arrest, diapause (Section VIII), that may last for months and that is terminated by prolonged chilling of the pupa. Using various surgical techniques, Williams demonstrated that a diapausing pupa could be stimulated to develop by implantation of an "activated" brain from a chilled pupa. Moreover, actively secreting larval brains (taken at times just before the start of a larval molt or at the onset of metamorphosis) could also initiate this development. The incisive discoveries came from studies on abdomens isolated from Cecropia pupae. These preparations would not respond to implantation of brains from diapausing individuals or to active brains from chilled pupae. Implants of prothoracic glands from diapausing pupae were similarly ineffectual, but "active" glands from developing donors stimulated the development of the abdomens. Importantly, transplantation of active brains and inactive prothoracic glands (organs which by themselves would not cause development) into isolated abdomens stimulated these fragments to undergo adult development. Thus, the brain and prothoracic glands act as an endocrine system with the brain releasing a tropic hormone that stimulates the prothoracic glands to produce a second hormone which is responsible for the initiation of development.

A. Chemistry and Source

In spite of the time since these classical experiments in the 1940's and early 1950's, little is known of the chemical nature of PTTH. Work has been carried on in a number of laboratories in Japan and Germany to purify and characterize this hormone. Kobayashi, Ishizaki, and Gersch all agree on its peptide nature, and the latest studies indicate that it may have a molecular weight of about 5000. The exact cells which synthesize this hormone are unknown. The brain contains two prominent pairs of neurosecretory cell clusters—the median and lateral neurosecretory cells. Both populations are heterogeneous and until recently attempts to ascribe PTTH activity to one group or the other have been inconclusive. A major block to both kinds of studies has been the lack of a suitable bioassay for the hormone. The main assay that has been used until recently is the "dauer" pupa of either *Bombyx* or *Samia cynthia ricini* (a pupa which has been debrained immediately after pupation to prevent the initiation of adult development). The usefulness and accuracy of this bioassay are limited because large amounts of material must be injected in order to cause development, spurious responses are sometimes triggered by solvents and other nonspecific chemicals, and the assay requires at least a week before it can be scored. A rapid, more sensitive bioassay has recently been developed by Gibbs in our laboratory using larvae of the tobacco hornworm, *Manduca sexta*. Neck ligation of penultimate stage larvae just at the beginning of PTTH release to initiate the molt to the last larval instar blocks the molt to this stage. But if these preparations are injected with active brain extracts at the time of ligation, they subsequently become last stage larvae. With this assay Gibbs demonstrated that PTTH activity in pupal brains of *Manduca* was confined to the lateral neurosecretory cell cluster.

B. Control of Release

How PTTH release is controlled is known only for a few insects. The clearest example is in *Rhodnius* in which abdominal stretch receptors send nervous signals to the brain to cause the release of PTTH. This mechanism thereby ensures that the larva will not molt unless it has sufficient nutrients to grow and produce a new cuticle. Even in this case, however, all of the steps may not be known since a number of days elapse between the initial distension of the gut and the buildup of sufficient PTTH to trigger the molt.

In larval tobacco hornworms, the release of PTTH is governed by a daily (circadian) clock which confines the possibility of release to a certain period each day. For a larval molt, this "gate" for the release of PTTH begins just before the onset of darkness and remains open for approximately 10 hr. For a given instar, larvae that have attained or exceeded a critical weight during the time of the gate then release PTTH. Those that attain this weight after the closing of the gate continue to grow and feed until the next night when PTTH release can again occur. Unlike the larval molt, the timing of the PTTH release that initiates metamorphosis in *Manduca* is governed by a number of factors. As with the other molts, PTTH release occurs only during a certain time of day and normally after the animal attains a critical weight, but this release can be blocked by a high level of JH and occurs only after the JH titer has declined. A similar delay in ecdysone release (and presumably therefore PTTH release) occurs in last instar cockroach (*Blattella*) nymphs given JH. In some insects, PTTH secretion may be blocked at a specific life stage. This induces a state of developmental arrest (diapause) which often enables the insect to survive adverse weather conditions. The brain may subsequently be reactivated by environmental factors such as photoperiod or temperature (see Section VIII).

C. Mode of Action

Even though the hormone has not been purified completely, some information is available as to its mode of action. The first step in the activation of the PTG appears to be a rise in cyclic AMP (cAMP). Indeed, Vedeckis and Gilbert showed that the stimulation of the *Manduca* PTG *in vitro* with cAMP in the presence of aminophylline (an inhibitor of the phosphodiesterase which breaks down cAMP) or with either aminophylline or 1-methyl-3-isobutylxanthine alone elicits the production and secretion of α-ecdysone. Furthermore, there is a peak of cAMP activity seen in the gland at the time that PTTH is being released to initiate metamorphosis. Thus, PTTH seems to be acting in a manner analogous to that of some vertebrate anterior pituitary hormones in its stimulation of synthesis and release of a steroid hormone. PTTH activates RNA synthesis in prothoracic glands, both *in vivo* and *in vitro,* but in no other tissue. This increase in RNA synthesis is correlated with ecdysone production *in vivo,* but a direct causative relationship has not been shown.

III. ECDYSONE

A. Chemistry

The early experiments of Wigglesworth, Fukuda, and Williams showed that the brain produced a hormone that acted on the PTG to release a second hormone that was essential for molting. At about the same time, in 1935, Fraenkel showed that blood from pupariating fly larvae contained a substance that induced pupariation of the posterior part of a ligatured mature fly larva. This finding was the basis of the *Calliphora* bioassay for the molting hormone which was used by Butenandt and Karlson for the isolation and characterization of the hormone from *Bombyx* pupae. In 1965 this hormone was chemically characterized as 2β, 3β, 14α, 22*R*, 25-pentahydroxy-5β-cholest-7-en-6-one and called ecdysone (α-ecdysone) (Fig. 2A). A second compound, 20-hydroxyecdysone (β-ecdysone or ecdysterone) (Fig. 2B), was also found in *Bombyx* pupae and has since been found in many different kinds of insects, in crustaceans, and in the parasitic nematode *Ascaris.* During the past 12 years these and other ecdysones have been extracted from various life-stages of insects including eggs. Figure 2 shows the four presently known ecdysones occurring in insects. Three additional compounds with molting hormone activity have been isolated from crustacean material. Throughout this chapter we will use the term "ecdysone" as a generic term for these compounds and whenever the particular one is not specified.

Besides the ecdysones isolated from arthropods, a host of compounds has been extracted from plants—especially the ferns and gymnosperms. Because of their plant origin and their potent molt-inducing effects, these compounds have been called phytoecdysones. Many of these phytoecdysones are more active when applied to insects than the insects' own hormones. Presumably, differences in chemical structure make them resistant to breakdown by the insect degradative enzymes. Besides producing ecdysones that are unique to themselves, some plants also produce large amounts of authentic insect hormones. For example, the rhizomes of the fern, *Polypodium vulgare,* have up to 1% of their dry weight as β-ecdysone.

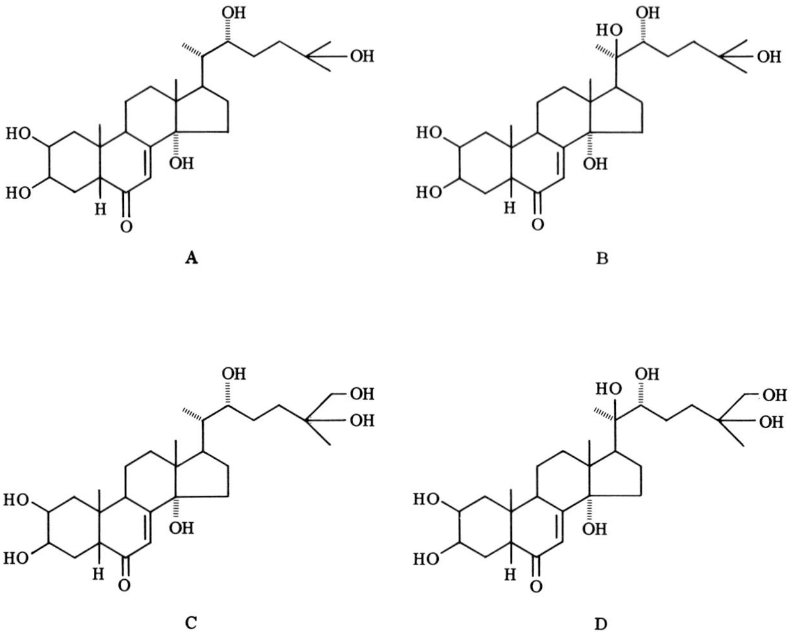

Fig. 2. The four ecdysones which have been isolated from insects. (A) α-ecdysone; (B) β-ecdysone; (C) 26-hydroxyecdysone; (D) 20,26-dihydroxyecdysone.

These presently serve as a primary source of commercially available β-ecdysone.

In the late 1960's and early 1970's there were numerous controversies over the respective roles of the two principal ecdysones (α- and β-ecdysone) and also over the glandular source of ecdysone production. Occasional reports of the molting of insect parts that lacked PTG as well as the conversion of labeled cholesterol into ecdysone in isolated abdomens challenged the generally held belief that the PTG were the only site of ecdysone synthesis. Both of these controversies have at least been partially resolved by culturing various tissues and glands *in vitro*. Cultured prothoracic glands from several Lepidoptera (*Bombyx, Manduca*), Coleoptera (*Tenebrio*), Diptera (*Sarcophaga*), and Orthoptera (*Leucophaea, Locusta*) synthesize and release α-ecdysone. The fact that the principal ecdysone extracted from insects is β-ecdysone rather than α-ecdysone is due to the subsequent hydroxylation of α-ecdysone by most of the tissues in the insect (e.g., fat body, Malpighian tubules, gut, and epidermis) with the interesting exception of the PTG themselves. Studies on cultured tissues, such as salivary glands and imaginal discs of *Drosophila* and epidermis from various Lepidoptera, have shown that β-ecdysone is effective at about $10^{-7}M$ in stimulating morphogenetic changes and is at least 100 times more active than α-ecdysone in these systems.

Ecdysones are also produced by a few other tissues. The ovaries of both mosquitoes and locusts produce α-ecdysone, and ecdysones have been found in the blood or ovaries of various other female insects (see Section VIIA). The oenocytes have also been implicated in ecdysone production in various insects, and those of *Tenebrio* have been shown to produce β- but not α-ecdysone *in vitro*. However, at this time, there is no compelling evidence that any structure besides the PTG produces sufficient ecdysone during normal larval growth to cause a molt.

B. Metabolism

The biosynthesis of α-ecdysone from cholesterol has not been completely elucidated. Since insects can not synthesize the sterol ring system, cholesterol or a plant sterol precursor is an essential part of their diet for growth and development. Injection of labeled cholesterol or 7-dehydrocholesterol into intact insects leads to the isolation of labeled α- and β-ecdysone, yet conversion of either of these compounds to ecdysone by PTG *in vitro* has not yet been demonstrated. The biosynthetic pathway depicted in Fig. 3 is for the intact animal and is based on the recent finding by Gilbert's and Horn's groups that 3β,14α-dihydroxy-5β-cholestenone (III in Fig. 3) is converted by *Manduca* PTG *in vitro* to α-ecdysone (V in Fig. 3) with 38% efficiency. Furthermore, they have postulated that it is converted to α-ecdysone via 22-deoxyecdysone (IV in Fig. 3), which is a known ecdysone precursor.

The conversion of α- to β-ecdysone (VI in Fig. 3), the active hormone, does not occur in the PTG as discussed above, but in various other tissues. The α-ecdysone produced is immediately secreted and then either is converted by a C-20-hydroxylase to β-ecdysone or is degraded by ecdysone oxidase to 3-dehydro-α-ecdysone (VIII in Fig. 4). The C-20-hydroxylase has been localized in the mitochondria, and its activity is increased dramatically in the last larval instar of many insects at the time that the ecdysone titer is increasing. In fact, in those insects in which the distinction between α- and β-ecdysone in the blood or the whole animal has been made (see below), the α-ecdysone peak is much smaller and also comes earlier than the β-ecdysone peak, suggesting that the release of α-ecdysone by the PTG may induce an increase in the C-20-hydroxylase in the other tissues.

At some times, α-ecdysone may be directly oxidized to 3-dehydro-α-ecdysone (VIII, Fig. 4), which is inactive in the induction of molting. This rate of conversion is highest in the early part of the larval instar, at least in *Locusta, Calliphora,* and *Choristoneura fumiferana*. Beta-ecdysone may likewise be oxidized to an inactive 3-dehydro form (VII, Fig. 4). Further metabolism of α- and β-ecdysone may occur via 20,26-dihydroxyecdysone

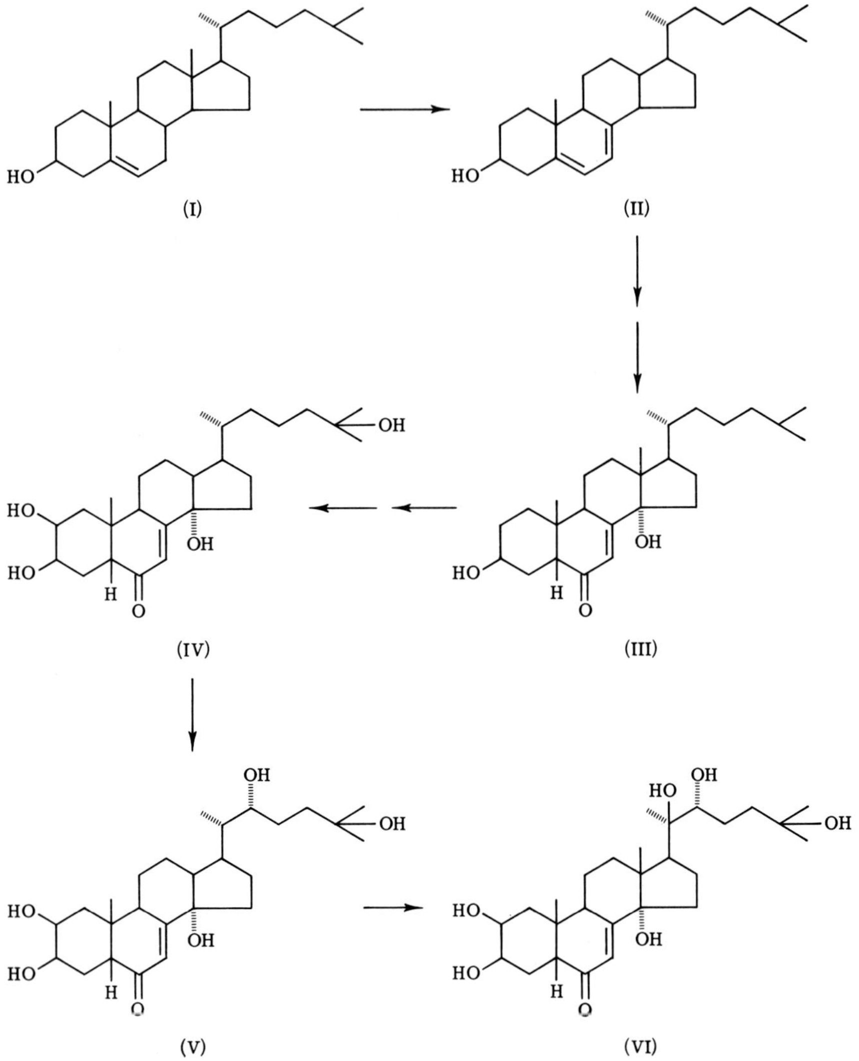

Fig. 3. Hypothetical scheme for the biosynthesis of α-ecdysone (V) and β-ecdysone (VI) from cholesterol (I). II, 7-dehydrocholesterol; III, 3β, 14α-dihydroxy-5β-cholestenone; IV, 22-deoxy-α-ecdysone. The succession of two arrows indicates two or more intermediate steps. [Based on studies by Bollenbacher *et al.* (1977) and earlier studies reviewed there.]

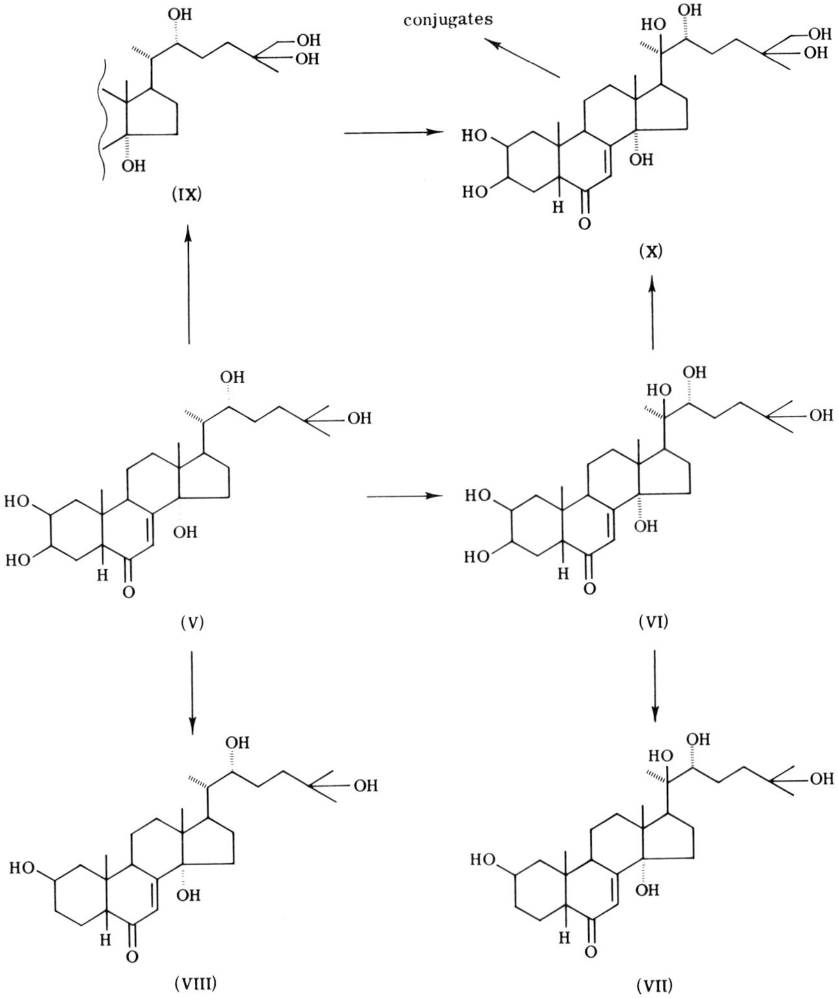

Fig. 4. Catabolism of α-ecdysone (V) and β-ecdysone (VI) in insects. VII, 3-dehydro-β-ecdysone; VIII, 3-dehydro-α-ecdysone; IX, 26-hydroxyecdysone; X, 20,26-dihydroxyecdysone.

(X, Fig. 4) to various conjugates with α-glucosides, glucuronides, and sulfate esters.

C. Titers

The blood titer of any hormone is a function of two processes: hormone release and hormone uptake and degradation. A knowledge of these titers is important because it depicts the environment to which the tissues of the animal are exposed. The first ecdysone titers were measured for *Bombyx* and *Calliphora* by Shaaya and Karlson using the *Calliphora* abdomen bioassay. This work showed that peaks of ecdysone were present at the outset of the larval molts and absent during the intermolt period. These early ecdysone determinations by bioassay were laborious, required that large amounts of blood be extracted, and did not discriminate between α- and β-ecdysone. In recent years the determination of ecdysone blood titers has become easier through the development of microchemical techniques and radioimmunoassay. Borst and O'Connor prepared antibodies to ecdysone by injecting the conjugate of the oximinoacetic acid derivative of β-ecdysone with bovine serum albumin into rabbits. In the radioimmunoassay, the unknown amount of ecdysone in an extract of blood or tissue competes with a known amount of labeled β-ecdysone for binding by the antibody. This original antibody and the one later prepared by DeReggi and co-workers did not clearly discriminate between α- and β-ecdysone or some metabolites. In order to distinguish among the naturally occurring ecdysones in the extracts, gas chromatography and mass spectroscopy of the trimethylsilyl derivatives or high-pressure liquid chromatography are used. Very recently, Horn and O'Connor have prepared a new antibody to a conjugate of bovine thyroglobulin and a hemisuccinate derivative of α-ecdysone, which has high specificity for α-ecdysone (10× greater than for β-ecdysone).

Using either chemical methods or the radioimmunoassay technique, ecdysone titers have been determined for a number of different insects. Late in the penultimate larval stage of Hemimetabola (*Locusta, Schistocerca, Blattella,* and *Panstrongylus*), there is a peak of ecdysone about 2 days before ecdysis to the final stage larva (see Fig. 5A for a representative titer for *Schistocerca*). Similarly, late in the last larval instar, another peak occurs, this time 3 days before adult ecdysis. In *Schistocerca* a small peak of α-ecdysone has been detected the day before the β-ecdysone maximum.

Figure 5B shows an example of the ecdysone titers during the last half of the life history of a representative holometabolous insect, *Manduca sexta*. During larval molts the ecdysone increases are similar to those seen during larval molts of hemimetabolous insects with one peak of ecdysone occurring to trigger apolysis and the subsequent production of larval cuticle. The onset of the larval–pupal transformation, however, shows the unique arrangement of two discrete peaks of ecdysone secretion. The first peak of ecdysone in *Manduca* has an α : β ratio of 1 : 1 and is small, but its effects are dramatic. It provokes a change in behavior of the caterpillar from feeding to seeking a proper pupation site and construction of a pupation chamber. Pigmentation changes may occur during this time; these are especially marked in the Puss moth, *Cerura vinula*. This ecdysone exposure also changes the commitment of the epidermal cells such that they are no longer sensitive to JH and can only synthesize pupal cuticle when subsequently exposed again to ecdysone.

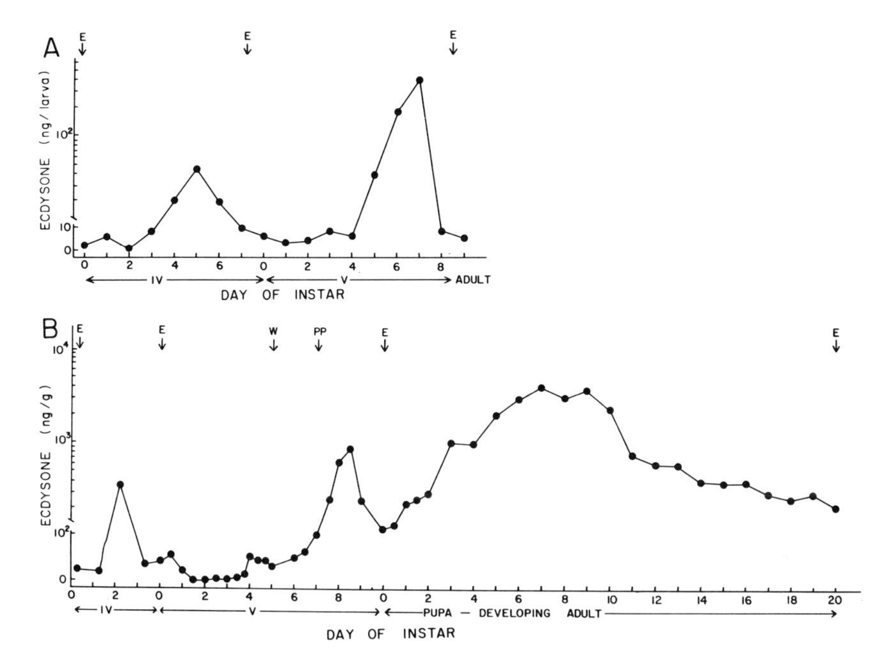

Fig. 5. Ecdysone titers through larval growth and metamorphosis of (A) a hemimetabolous insect, *Schistocerca gregaria* and (B) a holometabolous insect, *Manduca sexta*. Both curves represent ecdysone extracted from whole animals. The break and slash in the *Y* axis of (B) indicate a scale change from linear to log. E, ecdysis; W, wandering larva; PP, prepupa. [(A) based on data from Morgan and Poole (1976b); (B) based on data from Bollenbacher *et al.* (1975) and Bollenbacher and Gilbert (unpublished observations).]

The first release of ecdysone is followed by a larger, longer-lasting surge that has an $\alpha : \beta$ ratio of 1 : 5 reflecting the increased C-20-hydroxylase activity in the midgut and possibly other tissues at this time. This second release initiates epidermal apolysis, the degradation of the old larval cuticle, and the synthesis of pupal cuticle. It also causes the transformation of the internal organs into the pupal condition.

A similar picture has been reported in several other lepidopterans (*Pieris, Bombyx* and *Philosamia cynthia*). In two coleopterans (*Choleva, Tenebrio*), but not in a third (*Leptinotarsa*), two peaks of ecdysone are also seen at the onset of metamorphosis. In the last case, it may well be that the *Musca* bioassay that was used was not sensitive enough to pick up a small peak that may initiate digging behavior of the larva. Radioimmunoassay determinations of ecdysone content of *Drosophila* larvae usually find only one peak at puparium formation (which begins before the wandering phase). Yet from the ecdysone requirements for the observed puffing sequence in *Drosophila* salivary chromosomes (Section III, D), it is likely that ecdysone is required for the larva to leave the food and to initiate pupariation,and that a second release is required after puparium formation for pupation. Problems with complete synchronization of a population of larvae may account for the difficulty in detecting a small peak of short duration just before the larvae leave the food. Also, the radioimmunoassay used detects several inactive ecdysone metabolites as well as α- and β-ecdysone, so these may obscure the rise and fall in active ecdysone at this time. Most likely the appearance of two peaks of ecdysone at the onset of metamorphosis will prove to be a general feature of holometabolous insects: the first to induce premetamorphic behavior and change the commitment of the tissue and the second to provoke synthesis of pupal cuticle.

In all holometabolous insects after pupal ecdysis there is another release of ecdysone that initiates adult development (Fig. 5B). This peak of ecdysone usually lasts through at least the first third of adult development and is apparently necessary to sustain the initial stages of this process. Abdomens that are isolated during this time and therefore cut off from a continuing supply of ecdysone from the PTG will stop and not develop further. In *Pieris* the main rise in β-ecdysone follows the α-ecdysone peak that occurs 60 hr after pupal ecdysis. Either this reflects an increase in the C-20-hydroxylase, as has been seen at the initiation of metamorphosis in many insects (see above), or the α-ecdysone plays some as yet undiscovered role in the initiation of adult development.

In many female insects the ecdysone content has recently been found to increase again late in adult development (*Bombyx, Galleria*) or during adult life (*Drosophila, Aedes aegypti, Macrotermes* queens, *Leucophaea, Locusta*). At first sight this result appears paradoxical because the PTG either are degenerating or are completely absent by this time. This paradox was resolved, however, by the finding of Hagedorn and others that the ovary is capable of producing ecdysone that is responsible for certain phases of reproduction (Section VIIA).

D. Mode of Ecdysone Action

In 1965 Karlson postulated that ecdysone causes its effects by a direct action of the hormone on the genome of the epidermal cell. This conclusion was based on two lines of experimentation. In the early 1960's Clever and Karlson demonstrated that ecdysone caused "puffing" of specific regions of the giant polytene chromosomes of *Chironomus tentans*. One of these regions, band I18C, puffed within 15 min of ecdysone addition. Puff IV2B was also induced soon thereafter. Then beginning about 5 hr later another series of puffs was activated. All of these puffs are seen normally during pupal development. Induction of the first two puffs by ecdysone could be blocked by the RNA synthesis inhibitor, actinomycin D, but not by protein synthesis inhibitors; whereas the secondary set of puffs was completely prevented by both types of inhibitors. Thus, it appeared that ecdysone acted on the first puffs to produce a product that subsequently activated the next series of genes. In another series of experiments on the blowfly, *Calliphora,* Karlson and Sekeris found a correlation between the increase in ecdysone titer at pupariation and the increase in dopa-decarboxylase activity at that time. This enzyme is thought to be the key enzyme which directs the change in tyrosine metabolism from the *p*-hydroxyphenylpyruvic acid pathway to the formation of *N*-acetyldopamine, the principal molecule required for sclerotization (tanning of the puparium in this instance) (see Chapter 5 for details). From the studies using various inhibitors of RNA and protein synthesis, they suggested that ecdysone stimulated *de novo* synthesis of dopa decarboxylase.

At that time, Karlson's conclusion that ecdysone directly activated certain genes was an audacious proposal for the mechanism of hormone action, and all the details were not well documented. Vertebrate endocrinologists seized upon the idea and have shown this to be the primary mechanism for steroid hormone action. The most thoroughly studied example in vertebrates is the action of steroids on the chick oviduct. Progesterone injected into the immature estrogen-primed chicks stimulates the synthesis of ovalbumin messenger RNA (mRNA) by the oviduct. The hormone first enters the cell, combines with a cytoplasmic receptor protein and is translocated to the nucleus as a hormone–receptor complex. The receptor protein is composed of two subunits, each of which binds a molecule of progesterone. One of the subunits also binds to a specific nonhistone protein on the chromatin, allowing the other to bind to a particular region of the DNA. This binding apparently then dramatically increases the number of initiation sites for RNA synthesis. One of the main products of this increased RNA synthesis is the mRNA for ovalbumin.

Despite the fact that Karlson's theory of ecdysone action stimulated all of this work in the vertebrates, the information gained thus far about the details of ecdysone action in insects is not nearly as complete. Although it is quite clear that ecdysone enters the target cell, no cytoplasmic receptors have yet been isolated. There do, however, appear to be a small number (80–500) of high-affinity binding sites for β-ecdysone in certain *Drosophila* cell lines, *Drosophila* imaginal discs, and *Locusta* epidermal cells. These sites are thought to be in the nucleus, although clear-cut competitive binding studies have not yet been done.

The imaginal discs of flies divide and grow during larval life, then, at metamorphosis, they evaginate and differentiate into adult structures such as legs, wings, eyes, antennae, and external genitalia. Since these discs respond to β-ecdysone *in vitro,* first by evagination (in about 14 hr), then pupal cuticle formation, and finally imaginal differentiation (bristle formation is seen by about 72 hr), they have been studied extensively for a clue to the primary mode of action of ecdysone. Fristrom has developed a technique for the mass isolation of these discs that has permitted biochemical manipulations. The uptake and specific binding of β-ecdysone mentioned above were found to occur rapidly and to precede the increase in RNA synthesis which began in 30 min and reached its maximum by 2 hr. The RNA made was found to be mainly ribosomal RNA, and no specific mRNA's have been found. Yet after 2 hr exposure to ecdysone, protein synthesis increased. By using a double label technique Fristrom's group was able to show that some new cytoplasmic proteins in addition to ribosomal proteins were produced. The identity of these proteins has not yet been ascertained.

Lepidopteran wing discs also respond to ecdysone *in vitro* by evagination and the formation of pupal cuticle. Oberlander and others have shown that ecdysone induces both RNA and protein synthesis within a few hours. Yet the continued presence of ecdysone is required for at least 24 hr to obtain cuticle deposition which begins much later (about the sixth day of culture). After this 24-hr exposure to hormone, a second increase in both RNA and protein synthesis occurs and is also necessary for cuticle production. Again, none of the proteins made has been identified. Minced pupal wing epidermis of Cecropia has recently been shown by Willis to differentiate in response to ecdysone *in vitro* by forming adult cuticle with scales. Several new proteins were induced by β-ecdysone as well as a decrease in synthesis in one of the major proteins produced in the hormone's absence. Some of the new proteins do not co-migrate on SDS-urea gels with the adult cuticular proteins produced *in vivo;* since the former proteins are larger, they are thought to be precursors to the adult proteins. The initiation of the synthesis of cuticle proteins however is not the first action of ecdysone in the induction of a molt. The first events are apolysis of the epidermis and the production of molting gel, but little work has been done in these areas.

In the metamorphic molt , the change in commitment of the epidermal cell from the production of one type of cuticle to the production of a new type of cuticle occurs very soon after the ecdysone-induced detachment and may constitute one of the first actions of ecdysone. We have found that the change in commitment of *Manduca* larval epidermal cells so that they can only produce pupal cuticle requires 24 hr of exposure to β-ecdysone *in vitro*. This ecdysone-induced change is not prevented by inhibitors of DNA synthesis, but is blocked by inhibitors of either protein or RNA synthesis , including those which inhibit specifically mRNA synthesis. The effective concentrations of these latter inhibitors are not toxic to the cells and still allow larval cuticle synthesis after washout of the inhibitors. The RNA's made during this ecdysone-induced change in commitment include RNA's in the messenger size class as well as ribosomal RNA's. These new messengers could be coding for new chromosomal regulatory proteins but are unlikely to code for pupal cuticle proteins since pupal cuticle is not produced without further stimulation by ecdysone. None of these messengers has yet been identified.

The only case in which a specific protein has been shown to be induced after ecdysone stimulation has been the three-to fourfold stimulation of the messenger RNA for dopa-decarboxylase in *Calliphora* epidermal cells at the time of puparium formation. Dopa–decarboxylase does not appear until about 7 hr after the cells are exposed to ecdysone, which presents the possibility that again it is not due to a primary action of ecdysone but results after a series of events triggered by that primary action.

The events occurring during the early action of ecdysone on the insect genome may eventually be worked out using salivary glands. As described above, the first demonstration that ecdysone would induce puffing of polytene chromosomes was done with the salivary glands of *Chironomus*. More recently, Ashburner and colleagues have shown the same phenomenon in *Drosophila melanogaster*. In this instance there is a series of puffs which appear and disappear before, during, and after pupariation (the formation of the puparium just after the wandering stage; the pupa is then later formed inside the puparium). This entire series can be induced *in vitro* by a certain ecdysone regimen as seen in Fig. 6. Exposure of mid-third larval

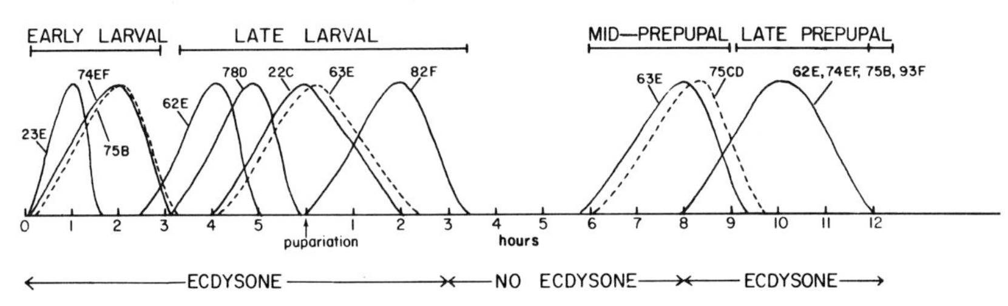

Fig. 6. The puffing pattern induced in the salivary glands of *Drosophila melanogaster* by β-ecdysone *in vitro* at 25°C. [The data are from Ashburner and Richards (1976b) and Richards (1976).]

instar glands to ecdysone *in vitro* causes the rapid induction of three early larval puffs. The size of these early puffs is dose dependent over a relatively broad ecdysone concentration with the 50% response occurring at 10^{-7} *M* β-ecdysone. In the continuing presence of ecdysone these puffs regress and are followed by a set of late larval puffs. If ecdysone is removed after the early puffs have attained their maximal size, the last three of the late larval puffs appear prematurely. Also, inhibitors of protein synthesis block the regression of the early larval puffs and the induction of the late larval puffs. Based on these observations, Ashburner has proposed a dual control model for the role of ecdysone in the sequential activation of the larval puffs. In this model, ecdysone, which is complexed to a receptor protein (E–R), acts on sites controlling the early puffs to produce a product (P) that is responsible for induction of the late puffs and for the regression of the early puffs. The E–R complex is also postulated to inhibit the induction of the late puffs. Therefore, P induces these puffs only when it has reached sufficient concentration to compete with E–R for the sites.

The puffs we have discussed thus far are the same ones that appear *in vivo* through pupariation and the formation of the early prepupa. The subsequent mid- and late prepupal puffs normally seen in metamorphosing *Drosophila* did not appear in cultured glands during continual ecdysone exposure. Richards has recently shown that appearance of these prepupal puffs requires the absence of ecdysone ($<5 \times 10^{-9}$ *M*) for at least 3 hr. Similarly, when glands are removed at pupariation and cultured with ecdysone, the mid-prepupal puffs do not appear, but they arise on schedule if the glands are maintained in hormone-free medium. By contrast, the late prepupal puffs (some of which are the same as the larval puffs) require ecdysone but may be induced by it only after the ecdysone-free period. After these puffs appear, the glands histolyse. Thus, the puffing pattern of the salivary gland chromosomes of *Drosophila* at the initiation of metamorphosis reflects a complex fluctuation in ecdysone titer at this time. The larval glands must be exposed to ecdysone to bring them to the prepupal condition; then they require a low amount or the absence of the hormone for about 5 hr, followed by reexposure to ecdysone, in order to play out their prepupal role.

IV. JUVENILE HORMONE

In 1934, Wigglesworth first presented evidence that a blood-borne factor from the head served to prevent metamorphosis of larval insects. Using penultimate instar *Rhodnius* larvae, he found that decapitation immediately after a blood meal completely prevented molting because of the lack of secretion of PTTH, whereas decapitation 5 days later resulted in a molt to the last larval instar. The crucial result, however, came with decapitation after 3 days which caused the insects to molt into precocious, miniature adults. This finding was further explored by joining fed, last-stage larvae parabiotically (so that the blood systems were common) to penultimate or earlier instar larvae. Under these conditions, the former insect did not metamorphose but molted into a larva or a larval–adult intermediate. This then confirmed that young larvae had a circulatory factor that served to prevent metamorphosis. Later, by transplanting various organs into decapitated insects, Wigglesworth demonstrated that the corpora allata were the source of this "inhibitory hormone."

In the late 1930's and early 1940's, Bounhiol and Piepho working respectively with *Bombyx* and *Galleria mellonella* obtained similar results for

Lepidoptera. Removal of the corpora allata (CA) from early instar larvae led to precocious metamorphosis and the production of miniature pupae and adults, whereas implantation of CA into last instar larvae prevented metamorphosis and resulted in giant supernumerary larvae.

In the 1950's Williams embarked on experiments directed at prolonging the lives of the nonfeeding Cecropia moths by parabiotically joining them to diapausing pupae. Supplied with the nutrients of the diapausing host, the lives of the moths were prolonged but these experiments led to a serendipitous discovery of even greater importance. In pairs in which the pupal partner had been previously chilled and subsequently began development, the pupae often developed not into moths but rather into second pupae. The factor that determined the developmental fate of the pupa was the sex of its parabiotic adult partner. When joined to a female moth, normal adult development would ensue, whereas parabiosis to a male resulted in a repeat of the pupal molt. Further experiments demonstrated that stored in the abdomens of male *H. cecropia* and *Samia cynthia* moths was a blood-borne factor that would exert a *status quo* action on the development of the partner. Moreover, this material only accumulated in males that had previously had intact corpora allata. Since implants of larval corpora allata into chilled pupae also caused development into second pupae, this *status quo* factor was named "juvenile hormone."

A. Chemistry

The abdomens of male Cecropia moths thus provided the only known extractable source of the hormone which would prevent metamorphosis. In 1956, Williams isolated a "golden oil" from these abdomens that would induce the formation of second pupae when injected into silkmoth pupae. Also, when tested on *Rhodnius*, this material prevented metamorphosis and caused the production of supernumerary giant larvae. During the following years numerous attempts were made to identify the juvenile hormone, but it was not until 1966 that Röller and co-workers finally determined its chemical structure as methyl-10,11-epoxy-7-ethyl-3,11-dimethyl-2-*trans*-6-*trans*-tridecadienoate (JH I, Fig. 7). Shortly thereafter, a second, minor (15%) constituent of the Cecropia oil was isolated by Meyer and Schneiderman and shown to be the C_{17} homologue (JH II, Fig. 7). Both hormones have identical biological activities in all of the assays thus far reported. The structure of a third juvenile hormone was obtained in 1973 by Judy and colleagues starting with material obtained from media used to culture adult female *Manduca* corpora allata *in vitro*. This hormone was the C_{16} homologue of the Cecropia hormone (JH III, Fig. 7). Since then, this last hormone has been obtained from *in vitro* cultures of the CA as well as whole

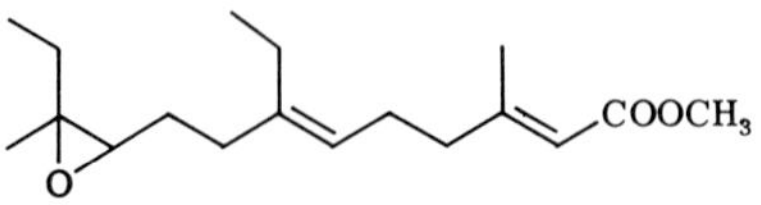

$COOCH_3$

JH-I

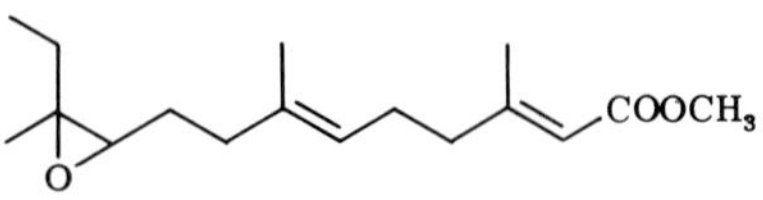

$COOCH_3$

JH-II

$COOCH_3$

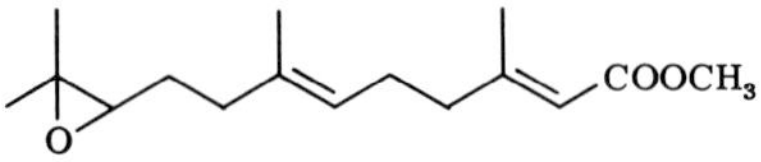

JH-III

Fig. 7. Structures of the three juvenile hormones that have been isolated from insects.

body extracts of a number of different insects (cockroaches, locusts, beetles, bees, flies, and termites) and appears to be the most common form of the hormone outside the Lepidoptera.

Most recently, the techniques of chemical derivatization followed by electron capture detection have been used by several groups to confirm the existence of all three hormones in the blood of several larval and adult insects. For example, in the hemolymph of larval *Manduca,* all three compounds are found with JH I and JH II predominating, but in the adult female mainly JH II and JH III (in roughly equal amounts) are found. In the cockroach *Nauphoeta,* JH I and JH II together comprise about 50% of the hormone titer in the larva, whereas only JH III (95% of the total hormone content) is found in the adult female. From these data, coupled with larval and gonadotropic assay data in cockroaches for the three hormones, Lanzrein has hypothesized that JH I and JH II are morphogenetic hormones while JH III serves a gonadotropic function in the adult (Section VII,A). However, this interpretation is questionable in *Manduca* where JH I and JH II are consistently 200–300 times more active than JH III in both morphogenetic *and* gonadotropic biological assays. Consequently in this lepidopteran, whenever JH I and JH II are present, they account for most of the biological activity.

B. Metabolism

The biosynthetic pathways for these hormones have not been completely worked out. *S*-Adenosylmethionine (SAM) serves as a donor of the methyl ester group for all of the JH's when added to corpora allata *in vitro* or to broken cell preparations. The biosynthesis of the carbon skeleton of JH III appears to occur through the usual isoprenoid route from acetate through mevalonate [(1) to (3) and (3) to (13), Fig. 8]. The "extra" carbons at C-11 in JH II and at C-7 and C-11 in JH I come from the C-1 group of propionate [(4), Fig. 8)] apparently through homomevalonate [(5), Fig. 8]. Since the latter

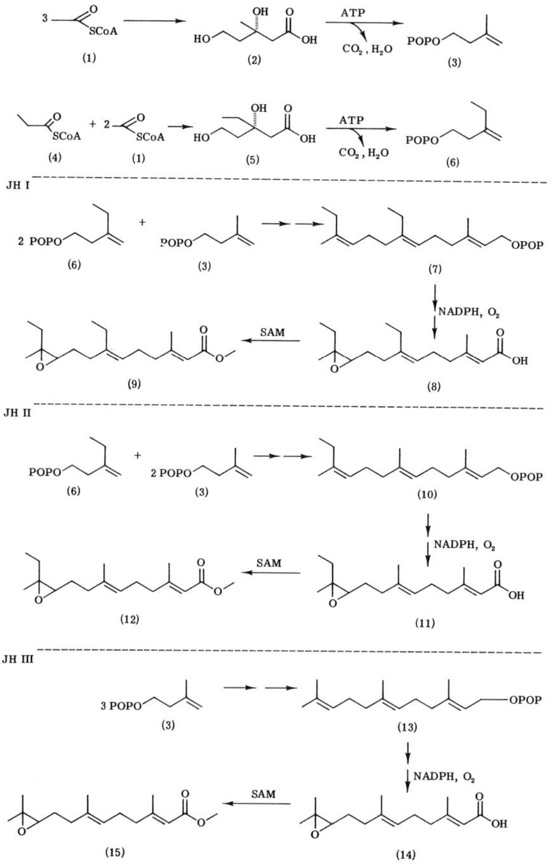

Fig. 8. Proposed scheme for the biosynthesis of the three juvenile hormones. (1) acetyl-CoA; (2) mevalonic acid; (3) isopentenyl pyrophosphate; (4) propionyl-CoA; (5) homomevalonic acid; (6) 3-methylene pentyl pyrophosphate; (8) JH I-acid; (9) JH I; (11) JH II-acid; (12) JH II; (13) farnesyl pyrophosphate; (14) JH III-acid; (15) JH III. SAM, *S*-adenosylmethionine. [After Schooley *et al.* (1976) and Reibstein *et al.* (1976).]

will serve as a precursor but is not taken up by intact CA very efficiently, these steps in the pathway are still in question. Homogenates of *Manduca* CA convert farnesyl pyrophosphate [(13) in Fig. 8] and farnesenic acid to JH III in the presence of NADPH, O_2, and SAM, but do not convert methyl farnesenate to JH III. By contrast, methyl farnesenate can be converted to JH III by microsomes from *Blaberus* CA in the presence of NADPH and O_2, and it accumulates in *Schistocerca* CA incubated with farnesenic acid. Consequently, there may be direct epoxidation of farnesenic acid to form JH III acid which is then enzymatically alkylated on the carboxyl group using SAM as a methyl donor as seen in Fig. 8. Alternatively, the methylation may occur prior to epoxidation. The major obstacle in these studies has been the impermeability of the CA to some of the labeled putative precursors. To circumvent this problem, homogenates of CA are used but these contain enzymes that rapidly degrade the newly synthesized hormone.

Of especial interest is the recent finding by Dahm and Röller that the CA of adult male Cecropia do not produce JH I *in vitro,* but rather produce only the JH I acid. Apparently, their CA lack the methyltransferase to alkylate the carboxyl group, whereas the female Cecropia CA have this enzyme and thus produce JH I. In male Cecropia the acid is taken up by the accessory glands and there converted to JH I and stored. The purpose of this sequestered JH is unknown except that it served as a serendipitous reservoir of hormone from which JH could be isolated and identified.

Once synthesized, the JH's are not stored in the corpora allata, but are immediately secreted into the hemolymph. In all insects so far studied (Lepidoptera, Coleoptera, and adult locusts), there are proteins in the hemolymph that bind JH. Some of these are high molecular weight lipoproteins with a relatively low affinity and specificity for JH (K_d ca. 10^{-5} M); but others, found particularly in larval lepidopteran hemolymph, are about 30,000 in molecular weight and show high specificity and affinity for JH (K_d ca. 10^{-7} M). The JH carrier protein (also called the JH binding protein) has been isolated from *Manduca* larvae by Law's group and shown to have one binding site for JH I ($K_d = 4.4 \times 10^{-7}$ M). Both JH II and JH III compete for this binding site but JH metabolites do not; both the ester and epoxide groups are required for binding.

In a variety of insects, the breakdown of JH has been shown to occur primarily by two mechanisms: the hydrolysis of the ester to the acid, and the hydration of the epoxide group to the diol (Fig. 9). The next step in either case is formation of the acid-diol which is then subsequently conjugated with sulfates for excretion. None of these metabolites has any JH activity. By far, the most common metabolite seen is the JH acid because the hormone can be attacked by general carboxyesterases which are ubiquitous throughout the insect. These esterases are made in the fat body and some are released into the hemolymph, but they are also found in epidermis, gut, and Malpighian tubules. In both holometabolous and hemimetabolous insects, JH-specific esterases appear in the hemolymph of the last instar larva before ecdysone is released to initiate metamorphosis. Although the JH carrier protein protects the bound JH from the general esterases, it does not protect it from the JH-specific esterases. The epoxide hydratase is membrane-bound and found primarily in the fat body, although *Manduca* imaginal discs *in vitro* also are able to metabolize JH to the acid-diol. The JH carrier protein affords some protection against this intracellular enzyme since it retards the uptake of JH by the cells. Thus, as Gilbert's and Law's groups have shown for *Manduca,* the JH produced by the larval CA is bound to the carrier

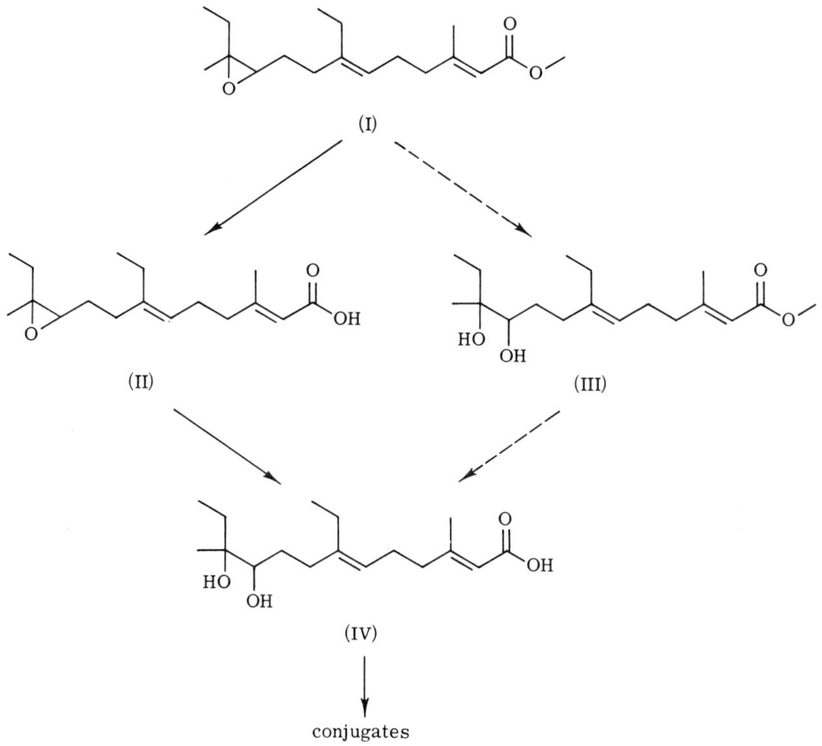

(I)

(II)

(III)

(IV)

conjugates

Fig. 9. Catabolism of juvenile hormone in the insect. I, JH I; II, JH I-acid; III, JH I-diol; IV, JH I-acid-diol. The breakdown of JH II and III occurs similarly. Dotted line represents minor pathway.

protein which protects the hormone from catabolism and retards its uptake by the tissues. Therefore, the JH titer is regulated not only by the biosynthesis and degradation of the hormone, both of which may fluctuate, but also by the presence or absence of the carrier protein.

C. Titers

The blood titers of JH during larval life have been followed in a few insects. The early work on these titers was made possible by the development of the *Galleria* wax wound test by de Wilde that provided an ultrasensitive assay for JH. Both this and the *Manduca* black larval assay are still the most widely used bioassays. Recently, various microchemical techniques involving extraction, derivatization, and analysis by gas chromatography using an electron capture detector have enabled detection of much lower concentrations than the bioassays, as well as determination of the amount of each JH present. Unlike ecdysone, which appears only at the time of the molts, juvenile hormone is present through most of larval growth. In the insects examined, there is a tendency for the JH titer to drop as the larva proceeds through successive instars with transient perturbations during the period of the molt. Late in larval life there is an abrupt disappearance of circulating JH that allows metamorphosis to occur. The pattern of JH disappearance differs in hemimetabolous and holometabolous insects. In the former, such as *Locusta* (Fig. 10A), the JH titer declines around the time of ecdysis to the last larval instar and remains undetectable throughout this

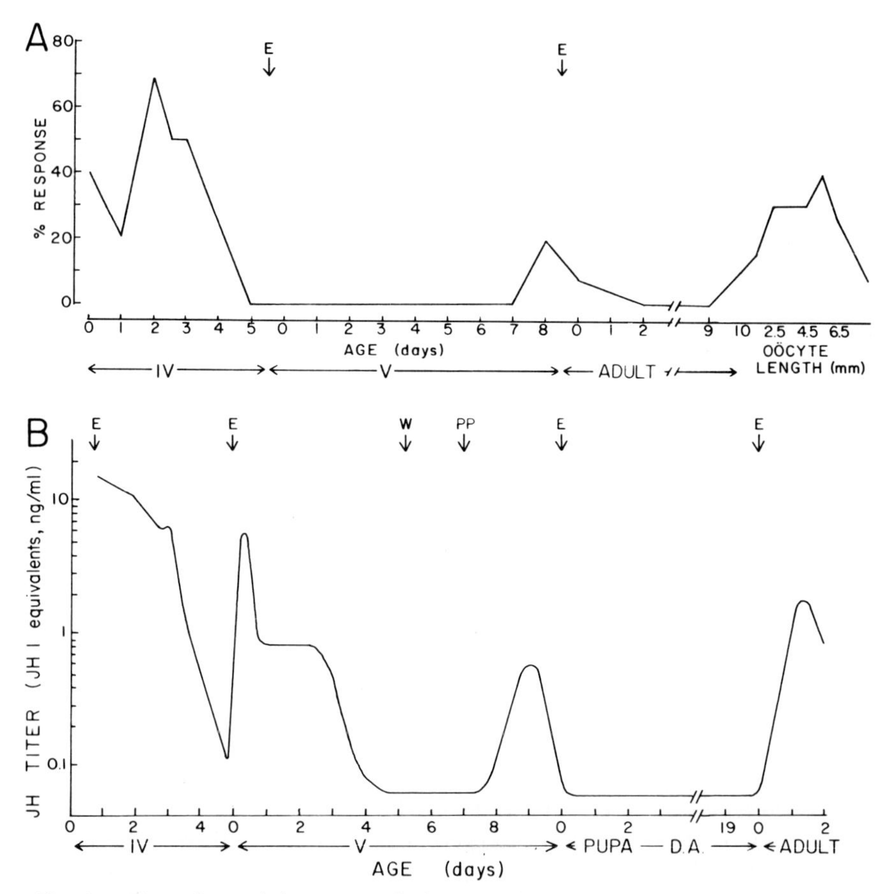

Fig. 10. Titers of juvenile hormone in the hemolymph of a representative (A) hemimetabolous (*Locusta migratoria*) and (B) holometabolous (*Manduca sexta*) insect. E, ecdysis; W, wandering larva; PP, prepupa; D.A., developing adult. In (B) the values below 0.1 ng/ml were undetectable. [(A) is based on data from Johnson and Hill (1973, 1975); % response refers to the % *Galleria* pupae which showed a positive response in the wax wound assay to hemolymph extracts. (B) is based on data from Fain and Riddiford (1975), Nijhout and Williams (1974), Dahm *et al.* (1976), and Schooley *et al.* (1976).]

instar. Just before adult ecdysis, there is a small peak, the function of which is unknown. In holometabolous species, titers have been determined for several lepidopteran species including *Manduca* (see Fig. 10B). Those insects with complete metamorphosis characteristically show an extension of high JH titers through the first half of the last larval instar, and then these titers rapidly decline. In some species the JH concentration falls below detectable levels prior to the onset of metamorphosis, then rises and falls during the late prepupal period before pupal ecdysis. This prepupal rise in JH apparently prevents the imaginal discs from prematurely initiating adult development. When either Cecropia or *Manduca* are allatectomized during the final larval instar or, in the case of the latter, even after the initiation of metamorphosis during the wandering stage, the subsequently formed pupae show some adult characters such as compound eyes and partial development of the gonads and genitalia. Thus, without some JH present during pupal development, parts of the animal may "overshoot" the pupal stage.

The JH titer measured in the hemolymph depends on several parameters: the rate of secretion by the CA, hormone catabolism in the blood, and hormone uptake and metabolism by various tissues. Undoubtedly, the most important factor is CA activity but measurement of secretion rates of the glands *in situ* have not been carried out. In the absence of direct measurements, several indirect indices of varying reliability have been used. Early studies on the activity of the CA often assumed that the volume of the gland was a good index of its activity, but this relationship is not always valid. For example, in the penultimate (fourth) and final (fifth) larval instars of *Locusta,* the volume of the CA increases in parallel with the body size of the animal, whereas the blood titer of hormone is high in the fourth instar and undetectable during the fifth. Although gland volume is not a valid measure of activity during larval stages in the adult female of the same species, there is a good correlation between CA size and blood titers of JH.

Activity of the CA has also been measured by transplanting glands into assay animals sensitive to JH and assessing the extent of juvenilization. This technique assumes that the glands will continue to function in a somewhat normal manner when removed from the neural and hormonal environment of the donor and also that their activity will not be influenced by the host. These assumptions were valid in Cecropia, in which Williams examined the CA activity throughout the life history of the insect by implanting them into chilled pupae and examining the results after development. This measurement of CA activity closely parallels titer measurements made for related moths. In some other insects the technique may be misleading. For example, in the cockroach *Leucophaea,* the severance of the nervous connections to a CA leads to its activation. Similarly, last instar larval *Locusta* glands show activity when implanted into assay animals but no hormone can be detected in the blood when they remain *in situ*. Corpora allata from adult female

Periplaneta and *Schistocerca* secrete JH into the medium during short-term (3-hr) culture at a rate reflecting their activity *in vivo* at the time they were removed, but so far no one has successfully shown this to be true of larval glands.

The other factors that are important for the regulation of JH titers are the metabolism and the cellular uptake of the hormone. The rate of metabolism becomes of special importance late in larval life. In *Manduca,* midway through the final instar, the JH titer declines quite rapidly due to the "shut off" of the CA and the appearance of a JH-specific esterase. As discussed above, this enzyme can attack the JH bound to the carrier protein, and apparently from the experiments of H. F. Nijhout is also able to hasten the decay of covert effects of the hormone in the target tissue. Its activity increases sharply before the first surge of ecdysone and thus serves to eliminate JH from the hemolymph and the tissues so that this ecdysone can cause the change in cellular commitment. Enzyme activity then falls to zero during wandering and remains so until just before pupal ecdysis when another small peak of activity is seen. Presumably, this second increase eliminates the minor JH peak which occurs just before the release.

D. Morphogenetic Action

The morphogenetic role of JH is usually considered to be its *status quo* action at the time of the initiation of a molt. In order for it to exert this effect, it need be present only during the short time of ecdysone release, but as seen in Fig. 10, JH remains at a relatively high concentration throughout the larval instar. Recent findings suggest that JH has at least two other functions in the larva that may account for its continued presence. These are to maintain larval tissues and to regulate pigmentation in some insects. The timing of these postulated roles of JH during the penultimate and last larval instar of *Manduca* is depicted in Fig. 11. Although the latter two effects have not been extensively studied, they will be briefly discussed before proceeding to a consideration of the well-known *status quo* effect.

1. Maintenance of Larval Tissues

The epidermis that forms the crochets (hooked tips of the larval prolegs) provides a good example of a larval tissue maintained by JH. During larval life the ability of this epidermis to form crochets in response to ecdysone *in vitro* is retained as long as the tissue is exposed to JH. When the JH titer declines midway through the last larval instar, this competence rapidly disappears, although cell death does not become evident until later during the larval–pupal transformation. Importantly, this competence to form crochets cannot be restored by exposure of the tissue to JH again before incubation with ecdysone. Similarly, crochet epidermis from penultimate

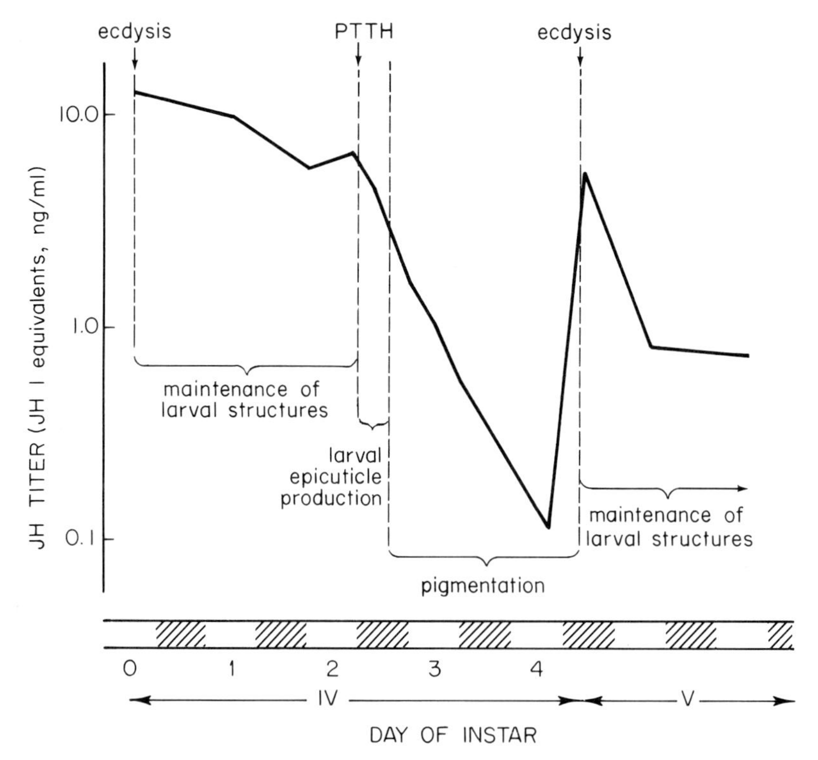

Fig. 11. The hemolymph titer and critical periods for roles of juvenile hormone during a larval instar of *Manduca sexta*.

stage larvae also loses its competence when cultured several days in the absence of hormones. But, in contrast to the last stage crochet epidermis, its competence can be restored by prolonged incubation in a young, premolt penultimate instar host (in which the JH titer is high).

A possible molecular correlate of the maintenance of the synthetic capacity of larval tissue by JH may be seen in the levels of RNA synthesis by cultured body wall epidermis. Incubation of tissue with JH for 24 hr produces an increase in the rate of RNA synthesis over that seen in tissue incubated with no hormones. Another example is the recent finding that *Drosophila* imaginal discs will grow best in a culture medium supplemented with JH (and insulin).

2. Pigmentation of Larvae

In some grasshoppers and some caterpillars, larval pigmentation is influenced by JH. For example, in *Manduca,* the color of the next larval instar is dependent on the JH titer during the time that the new cuticle is being produced (Fig. 11), a time at which JH titers can no longer influence the form of the next stage. In the presence of JH, most of the cuticle is transparent and the underlying epidermis contains blue pigment granules. With very low levels of JH at this time, the overlying cuticle becomes melanized after ecdysis and the epidermal cells lose their granules. In these animals topical application of the hormone during a short critical period will allow normal coloration. The manner by which JH acts to prevent melanization is unknown. This phenomenon serves as the basis of a rapid and sensitive bioassay for JH.

3. The *Status Quo* Effects

Although the *status quo* action of JH has been well-documented in many different types of insects, the mode of action is unknown. It is clear, however, that the hormone does not prevent ecdysone from triggering a molt but rather, it prevents a progressive change from one developmental stage to the next. As described above, when ecdysone stimulates a molt, the epidermis usually undergoes DNA synthesis followed by RNA and protein synthesis for the production of the cuticle proteins. From studies on *Galleria* and saturniid pupae, it became obvious that JH must be present before this ecdysone-induced DNA synthesis in order to have its *status quo* effects. Thus, this DNA synthesis appeared to be necessary for reprogramming of the genes just as the transformation of a vertebrate myoblast into a differentiating muscle cell may require a critical round of DNA synthesis. Therefore, Schneiderman, Sehnal, and others proposed that the presence of JH somehow prevented the activation of new genes during this special synthesis. Yet there are examples where adult transformation occurs without any apparent DNA synthesis, such as in the tergites of *Tenebrio* and the bristle cells of *Rhodnius*. Moreover, in many systems molting is rapidly triggered by ecdysone which prevents a clear distinction between the cellular events related to the commitment of the cells to a new program of synthesis and the subsequent implementation of that program. Our recent studies on the larval–pupal transformation of *Manduca* epidermis *in vitro* have indicated that RNA synthesis, rather than DNA synthesis, is the crucial event involved in the reprogramming of the epidermal cells by ecdysone (see Section III,D). This reprogramming occurs during the first release of ecdysone in the absence of JH, but the actual synthesis of pupal cuticle does not occur until the second exposure to ecdysone. Consequently, the cellular events involved in the commitment to pupal differentiation can be studied without the added complication of a simultaneous production of pupal cuticle.

When epidermis from late last–stage *Manduca* larvae is exposed to β-ecdysone either *in vitro* or *in vivo* for approximately 24 hr in the absence of JH, the cells become committed to pupal differentiation. Addition of an inhibitor of DNA synthesis does not prevent this change in commitment, but it is blocked by either JH or inhibitors of RNA synthesis. Yet JH is not preventing RNA synthesis, but rather is directing the type of RNA made. When JH is present, RNA for larval cuticle which appears at 48–60 hr of culture is made. When JH is absent, new RNA's which commit the cell to later pupal differentiation are produced. Similarly, Willis has shown that the addition of both JH and β-ecdysone to pupal wing epidermis *in vitro* causes pupal cuticle formation, whereas β-ecdysone alone stimulates adult cuticle and scale formation. In this case, RNA synthesis under the two regimes has not been studied yet, although the pupal and adult cuticular protein synthesis patterns are seen to be different.

E. Mode of Action

Not much is known about the molecular action of JH, but several theories have been proposed. In 1970, the Ilans suggested that the *status quo* action of JH resulted from its effect on RNA translation by the cell. From their studies on the pupal–adult transformation of the mealworm, *Tenebrio molitor,* it appeared that the messengers for adult cuticle were already present on the first day of adult development (a day-1 pupa in their terminology) but could not be translated by the day-1 ribosomes because of the lack of a specific leucine tRNA and its corresponding leucine tRNA synthetase. They showed that addition of tRNA's and their synthetases from developing adults (their "day-7 pupa") then allowed translation of the adult messengers in an *in vitro* protein-synthesizing system and consequently the formation of adult cuticle proteins. These proteins were identified by their high tyrosine/leucine ratio since the Ilans claimed that the tyrosine content of adult cuticle was high compared to that of pupal cuticle. Recent studies have indicated that this translational control mechanism may not be valid in *Tenebrio*. First, by amino acid analysis, Andersen and co-workers showed that in fact adult *Tenebrio* cuticle had a lower tyrosine content than pupal cuticle (adult/pupal = 0.39). Second, and more importantly for the theory, White and co-workers showed that the experimental conditions used by the Ilans did not allow for optimal aminoacylation of the leucine-tRNA. After optimization of pH, ionic, and other conditions, they found no difference between the day-1 and day-7 leucine tRNA's, although using the Ilans' conditions they could repeat their results. Furthermore, they could not demonstrate a new leucine tRNA appearing *in vivo* with the progress of adult development. Since there are no other eukaryotic systems in which translational control by tRNA has been conclusively proved, we should for the moment at least abandon this idea in insects as well.

In 1970, Williams and Kafatos presented a theoretical model for the action of JH based on the control over sporulation in *Bacillus subtilis*. They proposed that there exist three master regulatory genes which control the larval, pupal, and adult gene sets. Each master regulatory gene can be read only by a particular RNA polymerase—one for each gene set. These three RNA polymerases differ by a sigma factor. According to the theory, ecdysone serves to activate the different gene sets depending on how much JH is present. During larval life JH acts as a co-repressor of both the pupal and adult gene sets. When the JH titer drops, the pupal repressor no longer binds to the pupal master regulatory gene because its binding requires a high JH titer. With the appearance of ecdysone, the pupal sigma factors are made and the pupal gene set can be read. One of these pupal genes is thought to code for a protein that acts as a permanent repressor for the larval master regulatory gene and thus for the larval gene set. When the JH titer falls to zero, the repressor of the adult gene set no longer binds to it and ecdysone can therefore initiate adult development. At this time, little experimental evidence for or against this theory has been forthcoming. There has been a report of a peak in soluble RNA polymerase activity in *Manduca* at the time of the first release of ecdysone for metamorphosis but no similar peak has been seen at the onset of adult development. Studies using pupal wings from *A. polyphemus* show that the increase in RNA polymerase activity (mainly RNA polymerase II which is necessary for mRNA syntheses) stimulated by β-ecdysone *in vitro* is completely inhibited by the addition of JH. However, there have been no reports of differences in sigma factors or in other RNA polymerase subunits in the different metamorphic stages.

In the Diptera, puffing in the salivary gland chromosomes is a good index of the action of ecdysone (see Section III,D). As the animal approaches metamorphosis, certain other puffs disappear, suggesting that some of these puffs may be under JH control. At least one puff (I19D) in *Chironomus thummi* is thought to be caused by JH. These data support the notion that JH may direct certain types of RNA synthesis as discussed above. The question remains as to how this is done. Since JH is a lipid, it most likely acts as a steroid hormone in that it enters the cell, combines with a cytoplasmic receptor protein in the target cell, and is translocated to the nucleus where it eventually is located on a particular region of the chromatin.

Evidence is accumulating for this type of intracellular action of JH. In 1973 Schmialek, using a labeled JH mimic with high specific activity, reported that it specifically accumulated in the epidermal cells of *Tenebrio* and bound ($K_d = 10^{-10}$ *M*) to a high molecular weight macromolecular fraction (360,000) which was isolated and characterized as a ribonucleoprotein. Unfortunately, although several laboratories have tried, no one has been able to duplicate this finding. Very recently Mitsui in our laboratory has found that epidermis from young fifth instar *Manduca* larvae (who still have endogenous JH) takes up and holds the hormone for about 20 min without releasing metabolites into the medium. By contrast, epidermis from mature larvae following commitment to pupal differentiation rapidly takes up and metabolizes the hormone, releasing JH acid into the medium. Preliminary experiments using a competitive binding assay with ^{3}H-JH suggest that nuclei from the young 5th instar larvae have high-affinity binding sites for JH, whereas those from mature larval epidermis no longer do. Therefore, it appears that JH enters the cell and goes to the nucleus where it presumably acts.

V. ECLOSION HORMONE

The process of ecdysis at the end of each molting cycle may involve more than simply the shedding of the old cuticle. In holometabolous insects, the eclosion or emergence of the adult from the pupal cuticle also signals a dramatic change in the behavior of the insect and often the degeneration of muscle groups used during eclosion. In the Lepidoptera, we have shown that these changes are triggered by a hormone that is released at the end of adult development.

The eclosion hormone is produced in the median neurosecretory cell cluster of the brain, stored in the corpora cardiaca, and released under the control of a circadian clock. Reynolds has recently shown that it is a protein with a molecular weight of about 9000 daltons. The hormone has a number of actions that are related to the rapid transition to the adult way of life. Its most obvious action is the triggering of the behavior involved in adult emergence. In Cecropia the emerging adult shows a stereotyped sequence of behaviors (the preeclosion behavior) that lasts about 1 ¼ hours and involves primarily abdominal movements. This behavior can be elicited from pharate (preemergence) adults or from abdomens isolated from pharate adults by injection of extracts containing the eclosion hormone. Studies on the action of the eclosion hormone have concentrated on the preeclosion behavior and have centered around isolated abdomen and isolated abdominal nervous system preparations. In the latter preparations when the motor activity is recorded from the nerves that normally participate in the behavior, the behavior was found to begin about 15–60 min after the nervous system is exposed to the hormone and then to continue to completion in the total absence of peripheral sensory feedback. Consequently, the entire preeclosion behavior is "programmed" into the circuitry of the CNS and is released by the direct action of the hormone on the nervous system. The hormone apparently has a rapid action because its presence in the bathing medium for only 1–2 min is sufficient to trigger the subsequent onset and play-out of the preeclosion behavior.

The hormone apparently releases the behavior through an increase in cyclic AMP in the CNS. When administered to either isolated abdomen or isolated CNS preparations, dibutyryl-cAMP with theophylline (a phosphodiesterase inhibitor) will mimic the action of the hormone and release the behavior. Also, protection of endogenous cAMP by pretreatment with theophylline will greatly enhance the effectiveness of a low dose of hormone. Extracellular calcium is also required for the action of this hormone.

Besides the release of the preeclosion behavior, the eclosion hormone has other effects at the time of adult emergence. In *Antheraea pernyi* moths the pharate adult shows an impoverished repertoire of behavior; walking, the righting response, and mating behavior cannot be elicited from these animals even though they apparently have a fully developed adult CNS. The eclosion hormone causes the abrupt "turning on" of these behaviors around the time that the animal is undergoing eclosion movements. This hormone also triggers the degeneration of the abdominal intersegmental muscles that occurs after their use during emergence of adult silkmoths. Lockshin and Williams demonstrated in the 1960's that these muscles apparently degenerate because of changes that occurred in their motor neurons which now we know are triggered by the eclosion hormone. It is likely that these effects of the eclosion hormone on turning on adult behavior and on signaling muscle degeneration are mediated through cAMP and Ca^{2+} as is the release of the preeclosion behavior (Fig. 12).

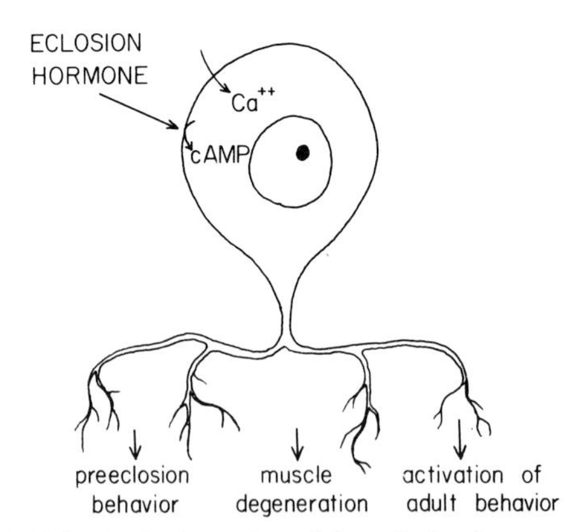

Fig. 12. Proposed initial steps in the action of the eclosion hormone on its target neurons.

Attempts to demonstrate the presence of the eclosion hormone in Lepidoptera at the time of larval or pupal ecdysis have been unsuccessful. This hormone appears also to be present in Diptera but it has not been found in any hemimetabolous insect. Thus, it may be that the eclosion hormone is restricted to the Holometabola and, moreover, used only at the time of adult ecdysis to coordinate the changes in the CNS that occur at that time.

VI. BURSICON

After ecdysis, the new cuticle of an insect is often quite soft and flexible so that it can be inflated into its proper shape (e.g., the wings of most insects). But before the animal can then begin its normal activities, the cuticle must become sclerotized—a process during which cross-linking occurs between cuticular proteins. It is presently thought that this hardening may occur by one of two mechanisms, either quinone tanning or β-sclerotization as seen in Fig. 13. Quinone tanning involves the enzymatic oxidation by a cuticular phenoloxidase of an *o*-diphenol to an *o*-quinone which in turn reacts with a free amino group of the cuticular protein. Further oxidation through reaction with other quinones allows cross-linking of several proteins. The usual *o*-diphenol made by the epidermal cell and secreted into insect cuticle is *N*-acetyldopamine, which has been shown by tracer experiments to be incorporated into the cuticle during sclerotization. Also, β-alanyltyrosine derivatives may be used, particularly by the Diptera. The process of β-sclerotization involves a ketoderivative of *N*-acetyldopamine that forms cross-links with the cuticular proteins via the β-carbon.

The triggering of this sclerotization is usually under hormonal control. In the specialized case of puparium formation in higher Diptera, this process is initiated by ecdysone which apparently activates the gene for dopa-decarboxylase (see Section III,D). But the actual tanning of the puparium has been shown by Fraenkel to be initiated after ecdysone release by a neurosecretory factor from the central nervous system. This tanning factor apparently works via cAMP as the latter will also accelerate tanning in a wandering larva, but how it initiates tanning is unknown.

In most other insects, cuticular sclerotization which occurs after ecdysis is stimulated by the neurosecretory hormone, bursicon. In the early 1960's Cottrell in England and Fraenkel and Hsiao in the United States simultaneously reported that a blood-borne factor was required for postecdysial tanning of adult blowflies, *Calliphora*. In this species, the newly emerged fly normally has to dig up out of the ground and can delay wing expansion and tanning for a number of hours until it reaches the surface. Neck ligation of newly emerged flies served to prevent sclerotization of the thorax and abdomen and provided a bioassay for material with tanning hormone activity. The tanning hormone (bursicon) was found in the blood of tanning flies and in both the median area of the brain and the fused thoracic-abdominal ganglionic mass. It is released primarily from the latter ganglia. The hormone appears to be a protein with a molecular weight of about 40,000 daltons. Bursicon has been reported to be present in many different kinds of insects and the various life stages. Its site of synthesis and/or release varies according to the insect, but in many instances it is found in the perivisceral neurohemal organs associated with the ventral chain of ganglia.

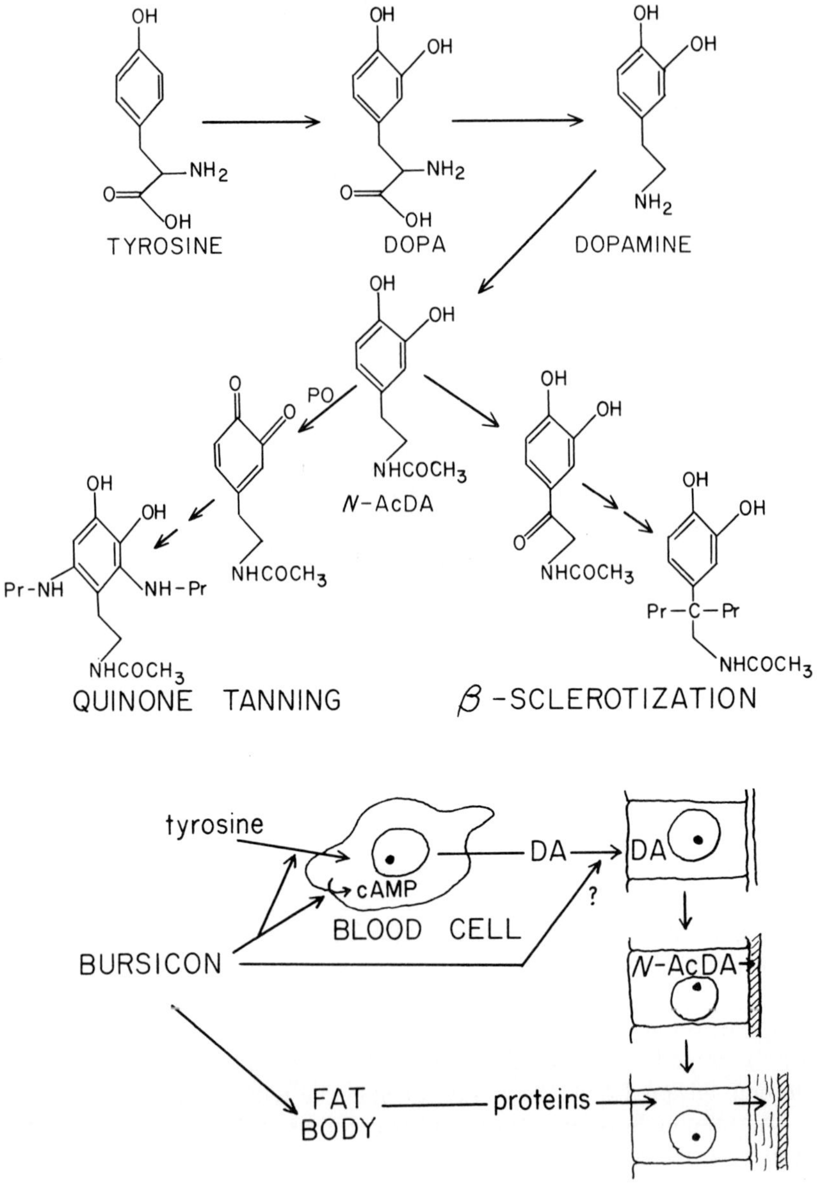

Fig. 13. Top: Proposed biochemical steps for quinone tanning and for β-sclerotization. Bottom: Possible scheme for the role of bursicon in the promotion of tanning and endocuticle deposition. DA, dopamine; *N*-AcDA, *N*-acetyldopamine; Pr, protein; PO , phenol oxidase.

The mechanism of action of bursicon has not been completely elucidated; a proposed scheme is depicted in Fig. 13. From studies on cockroaches and *Pieris,* bursicon appears to increase the permeability of the hemocytes to tyrosine; it probably also increases the permeability of the epidermis to dopamine and possibly to tyrosine and/or dopa. The epidermal cells then convert the dopamine to *N*-acetyldopamine and tanning proceeds. Cyclic AMP will cause tanning in the posterior portion of neck-ligatured *Calliphora* and *Periplaneta,* suggesting that bursicon may act via cAMP. The stimulation of cAMP in epidermal cells or hemocytes by bursicon has not yet been demonstrated. Bursicon also stimulates endocuticle deposition after ecdysis, apparently by stimulating cuticle protein synthesis by fat body and/or its uptake into the epidermis.

VII. HORMONAL CONTROL OF REPRODUCTION

A. Females

The hormonal control of insect reproduction was first shown by Wigglesworth in *Rhodnius.* After an initial batch of eggs laid soon after ecdysis, female *Rhodnius* mature a clutch of eggs after each successive blood meal. The role of the corpora allata was demonstrated by cutting through the head of newly fed females at a level that removed the brain but left the CA. These females matured their eggs normally. But when the head was removed at the neck, thereby also removing the CA, then no egg development took place. Parabiosis of a decapitated, fed female to an adult that retained its CA or implantation of active CA into such an animal similarly caused ovarian maturation. These experiments clearly indicated that the CA, which are inactivated in the last larval instar to permit metamorphosis, are reactivated in the adult to stimulate reproductive maturation. The requirement for the CA, and thus juvenile hormone, for reproduction is seen in many insects. In those females which do not feed as adults and consequently must mate and lay eggs as soon as possible (e.g., the saturniid moths), egg maturation is nearly completed during adult development and no hormonal control seems necessary. But in most feeding adult females, except for phasmids, the CA are necessary for some phase of reproductive maturation. For purposes of discussion of the various roles that JH may play, we may subdivide ovarian maturation into three phases: (1) oocyte growth, (2) vitellogenin (yolk protein) synthesis, and (3) vitellogenin uptake into the oocyte (Fig. 14).

The initial production of primary oocytes in the germarium of the ovary usually occurs during metamorphosis under the influence of ecdysone in the absence of JH. But in most insects the oocytes do not progress further until after adult emergence, when JH is again secreted. For instance, in the

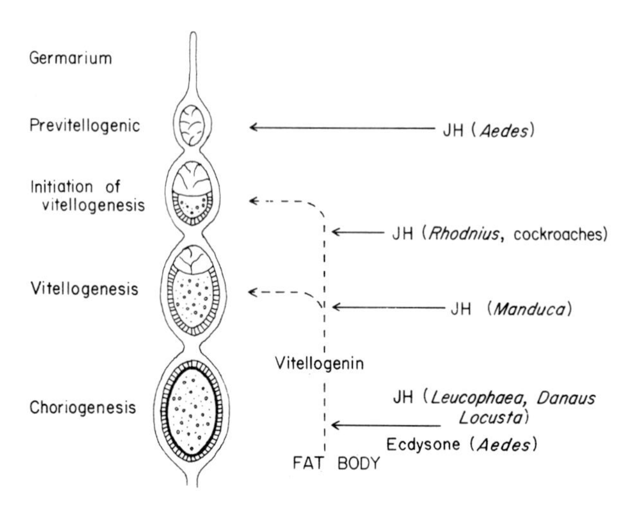

Fig. 14. Stages of oogenesis in insects at which hormones have been found to control the maturation process. (Based on drawing by M. M. Nijhout, 1975.)

mosquito *Aedes aegypti,* the JH secreted just after emergence causes the oocyte to grow from 80 μm to the 110-μm "resting stage" at which it remains until the female takes a blood meal that initiates vitellogenesis (to be discussed in more detail below). By contrast, in the moth *Manduca,* both vitellogenin synthesis and uptake are initiated during the latter part of adult development, but the oocytes cease growth at about 20% of their final volume until JH is secreted soon after emergence.

Over 20 years ago Telfer showed that the blood of Cecropia females contained a protein which was not found in males. This protein comprised a substantial percentage of the blood protein and was antigenically identical to the yolk protein found in the mature egg. Studies on many other insect species have since shown that these female-specific proteins, or vitellogenins as they have come to be called, are widespread among insects and are usually large (molecular weights of 200,000 to 500,000).

Hormonal stimulation of vitellogenin synthesis was first demonstrated by Engelmann in the cockroach, *Leucophaea maderae*. Injection of JH I, II, or III into immature females caused the synthesis of a hemolymph protein which was presumably vitellogenin. This protein could be precipitated by antisera against female blood proteins from which the antibodies against nonfemale-specific proteins had been removed by adsorption with male serum. This JH-induced synthesis could be blocked by actinomycin D and also in later experiments by α-amanitin, suggesting that the hormone was stimulating *de novo* synthesis of the protein. The direct stimulation by JH of vitellogenin mRNA synthesis in *Leucophaea* fat body *in vitro* has not yet been shown, but recently Wyatt and co-workers have shown such a stimulation with *Locusta* fat body. In the latter study, the mRNA prepared from fat body of allatectomized females injected with JH directed the synthesis of vitellogenin in a cell-free wheat germ protein–synthesizing system. RNA from fat body from unstimulated allatectomized females or from males was ineffective. Therefore, in this gonadotropic action, juvenile hormone appears to act primarily by stimulating the production of a specific mRNA for the production of yolk protein. This action is analogous to that of estrogen on the stimulation of vitellogenin synthesis in the liver of some vertebrates.

In at least one insect, the mosquito *A. aegypti*, ecdysone replaces JH in the function of stimulating vitellogenin synthesis. As seen in Fig. 15, JH is

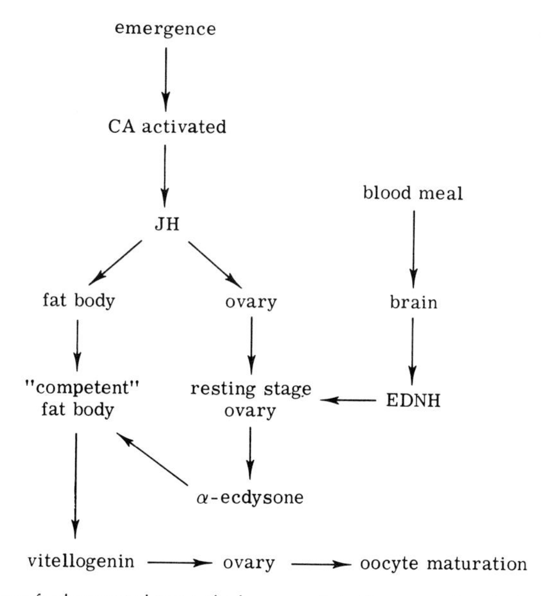

Fig. 15. Scheme for hormonal control of oogenesis in the mosquito, *Aedes aegypti*. [Based on Hagedorn (1974), and Flanagan and Hagedorn (1977).]

released at emergence to cause the growth of the oocyte to the resting stage where it then remains. After the female takes a blood meal, the brain releases a neurosecretory hormone (EDNH = egg development neurosecretory hormone) which in turn stimulates the ovary to release α-ecdysone (whether it stimulates production or secretion of stored hormone or both is unknown at this time). This α-ecdysone then acts on the fat body (presumably after being converted to β-ecdysone) to stimulate vitellogenin synthesis and release. Since fat body removed from females allatectomized at emergence will not respond to ecdysone, it appears that JH also has a role in priming the fat body for vitellogenin synthesis. Whether ecdysone turns on vitellogenin mRNA synthesis or whether it acts somewhere later in the translation of preexisting mRNA is presently in dispute.

Since Hagedorn first reported in the early 1970's that the mosquito ovary secretes ecdysone, many laboratories have looked for ecdysone in other adult insects. Ecdysones have been found in the females of such diverse species as *Leucophaea, Locusta, Drosophila, Bombyx, Galleria,* and *Macrotermes* queens. In *Locusta*, Hoffman and co-workers have shown that the follicle cells of the ovaries produce α-ecdysone toward the end of oogenesis; most of this ecdysone is not secreted into the hemolymph but rather is sequestered in the oocytes where some is metabolized to more polar compounds. A similar story is emerging in *Bombyx* and *Galleria*. Whether this ecdysone plays any role in either the final stages of egg maturation or in embryonic development remains to be seen. But it seems likely that ovarian production of ecdysones may be a general phenomenon in insects.

A third role of JH in ovarian development is to promote the uptake of the vitellogenin into the egg. In order for the large proteins to reach the oocyte surface, they must first penetrate the follicular epithelium which surrounds the oocyte. At the beginning of vitellogenin uptake the follicle cells change shape so that there are large spaces between the cells, through which the proteins can migrate to the egg surface and there be taken into the oocyte by pinocytosis. Bell and Barth showed by transplanting ovaries into male cockroaches (*Eublaberus*) followed by injection of a crude preparation of vitellogenin that no uptake of this vitellogenin into the ovaries occurred until JH was also administered. After JH treatment the oocytes accumulated yolk and grew, although none reached full maturity. The most convincing demonstration that JH has a direct action on the follicle cells has been done by Davey and co-workers. When the follicular epithelium of *Rhodnius* was incubated *in vitro*, about 3×10^{-7} *M* JH I was optimal to stimulate an increase in the intercellular spaces, thereby presumably enhancing yolk uptake. This JH-stimulated change was insensitive to inhibitors of RNA and protein synthesis such as actinomycin D and puromycin, but was inhibited by colchicine and cytochalasin B, inhibitors of microtubule and microfilament formation, respectively. Whether JH interacts directly with these subcellular elements to bring about the changes in cell shape is unknown at this time.

Juvenile hormone may also be needed for the maturation of the various female accessory glands, such as the colleterial glands in the cockroach *Periplaneta*. Also, it may be necessary for sex pheromone production as in the female cockroach, *Byrsotria fumigata*.

B. Males

In male insects the CA have no positive role in spermatogenesis. Meiosis and sperm maturation usually occur during metamorphosis in response to ecdysone in the absence of JH. In some insects meiosis may even begin during the last larval instar when the JH titer has declined.

In the early 1950's, Schmidt and Williams demonstrated that a material in the blood of silkworms contained a factor that was essential for sperm maturation. When the spermatocyte cysts from diapausing pupal testes were cultured in hanging drops in the absence of this material, they survived for long periods but would not differentiate. But addition of this factor to the culture media was rapidly followed by meiosis and by the differentiation of the spermatocytes into elongate spermatids. This material, "macromolecular factor" (MF), was nondialyzable, non-species specific and heat labile. Its appearance in the blood was also stage specific as it was virtually absent in fifth instar larvae and diapausing pupae but present in pupating animals and in developing adults. From the times of its appearance and its dramatic morphogenetic action, this factor was initially postulated to be the molting hormone from the insect prothoracic glands. Yet when ecdysone (a heat stable, dialyzable molecule) was isolated, it was found to be completely inactive in promoting the differentiation of isolated spermatocyte cysts.

The relationship between ecdysone and MF remained unresolved for almost 20 years until reexamination by Kambysellis and Williams. They confirmed the above results and went on to show that ecdysone and MF act together to promote sperm maturation in the intact animal. When whole testes were cultured in the presence of ecdysone, no development of the cysts inside was evident. But incubation of testes with both ecdysone and MF led to complete spermatogenesis. Moreover, a brief initial exposure to only ecdysone followed by incubation with only MF was effective. Thus, in this instance, the ecdysone acts on the walls of the testis such that they become permeable to MF which in turn promotes sperm development. Similar results have been subsequently reported for *Monema flavescens*.

An apparently unique hormonal control has been reported in the firefly, *Lampyris noctiluca*. The apical region of the larval male gonad releases an androgenic hormone that promotes differentiation of the testes and also of the male secondary sexual characters. Implantation of male gonads into female larvae causes masculinization.

Although JH is not required for sperm maturation, it is needed in many male insects for proper development of the accessory glands. The secretions from these glands make up the spermatophore and also have diverse roles in reproduction ranging from acting as vehicles for sperm transport, to activation of sperm motility, to causing alterations in the female's behavior. For instance, the accessory glands of allatectomized male *Melanoplus san-*

guinipes produce only about 25% of their normal protein complement. Furthermore, the glandular secretions lack two antigenically distinct proteins which are normally found, one of which appears to release oviposition behavior in the mated female. When JH is given to allatectomized animals, these proteins reappear, suggesting that the hormone has a direct effect in stimulating synthesis of specific proteins.

C. Control of the Corpora Allata

The JH titer in adult insects appears to be primarily a function of the secretory activity of the CA. These glands, however, are not autonomous and the brain acts to regulate their activity. In some instances this control has been shown to be stimulatory, in others inhibitory. A clear example of inhibitory control over the corpora allata is seen in females of the cockroach, *Leucophaea*. In this ovoviviparous roach the presence of an egg case in the brood sac results in a shutting off of the CA, thereby preventing the maturation of the next set of eggs. Scharrer and also Lüscher and Engelmann demonstrated that unilateral transection of the nervous pathways between the brain and the CA caused hypertrophy of and JH secretion by the ipsilateral CA but did not affect the contralateral gland. Thus, in the "pregnant" female the brain acts tonically to inhibit these glands. Whether this inhibition is accomplished by neurosecretory neurons is uncertain because the severed nerves apparently contained both neurosecretory and non-neurosecretory axons.

Evidence for a positive control by the brain over CA activity has accumulated for a number of different species. The most compelling evidence has been gathered for *Leptinotarsa, Schistocerca,* and *Calliphora*. In *Leptinotarsa,* reproductive activity is regulated by photoperiod. It is well established that the reproductive diapause induced by "short days" is primarily due to the lack of JH and that allatectomy of "long-day" beetles will physiologically mimic this diapause condition with one important exception. De Wilde and de Boer showed that CA implants restored egg production in allatectomized long-day beetles, but that this treatment was ineffectual in terminating diapause of short-day females. However, those allatectomized short-day females with CA implants would mature eggs when exposed to long-day conditions. Thus, the CA implants are apparently controlled by a humoral factor from the brain, and this factor appeared to be a stimulatory hormone that was continuously produced under long-day conditions and absent under short days. This conclusion was further supported by the fact that cautery of the median neurosecretory cells (prime candidates for the site of production of this allatotropic hormone) of the brain caused some of the long-day beetles to go into diapause.

Thus, the brain acts to regulate the activity of the CA through either stimulatory and/or inhibitory actions. In some cases it is clear that the brain can control implanted CA through a hormone released into the blood. Since the CA receive direct neurosecretory innervation from the brain, it is likely that under normal conditions, the allatotropic hormone is supplied directly to the CA by these axons.

VIII. DIAPAUSE

To withstand various adverse environmental conditions, many insects enter a state of dormancy or diapause. True diapause is characterized by an extremely low metabolic rate and relative inactivity as well as certain biochemical specializations, e.g., the increased glycerol content of the blood of diapausing saturniid pupae to enable them to withstand prolonged subfreezing conditions. Diapause can occur during any of the life stages of the insect—embryo, larva, pupa, or adult—but usually occurs only once in the life history of a particular insect species. Depending upon the stage at which it occurs, different hormones are involved. Therefore, each type will be discussed separately.

A. Embryonic Diapause

Many diverse kinds of insects have an egg diapause which may occur at any of several stages of embryonic development, just after germ band formation, around the time of blastokinesis (usually approximately halfway through embryonic development), or at the full-grown larval stage just before hatching. Although many such examples are known, the hormonal basis of embryonic diapause is known only from work on one species, the commercial silkworm, *Bombyx mori*. In this species the developmental fate of the egg depends on the photoperiod to which the mother was exposed during larval development. Females reared under short-day conditions (such as those grown in early spring) lay nondiapausing eggs, whereas those exposed to long-day conditions lay diapausing eggs. Under the latter conditions the subesophageal ganglion of the female secretes a peptide hormone, the "diapause hormone" which then acts on the ovary. Eggs maturing under the influence of this hormone subsequently arrest at the germ band stage early in embryonic development. The hormone also acts on the developing ovary to increase its permeability to 3-hydroxykynurenine (a precursor to the ommochrome pigment found in the serosa of diapausing eggs), to increase trehalase activity and thus enhance glycogen synthesis and accumulation in the oocyte, and to increase the synthesis and storage in an inactive form of a nonspecific esterase (esterase A). Subsequent reactivation of diapausing eggs requires prolonged chilling which somehow activates esterase A when eggs are brought again to warm temperature. The esterase then causes lysis of the yolk cells which allows continuation of embryonic development. Since the germ band will reinitiate development when explanted to a defined culture medium, it is clear that the diapause is maintained by extraembryonic factors.

B. Larval Diapause

Although larvae may diapause in any instar, the most common one is the last larval instar. Here again, most of the experimental studies have been carried out on Lepidoptera. In the early 1960's from experiments with the corn borer, *Ostrinia nubialis,* Beck suggested that a factor called proctodone from the hindgut was important for establishing diapause. This interpretation has failed to be confirmed and attention has subsequently shifted to the corpora allata as the center controlling larval diapause.

Working with the southwestern corn borer, *Diatraea grandiosella,* Chippendale and Yin have shown that juvenile hormone is responsible for the maintenance of larval diapause. In *Diatraea,* temperature serves to determine diapause or development. At 30°C, larvae progress through 6 instars and pupate after 16–18 days, whereas at 23°C, the larval stages are prolonged and the sixth stage larva then molts into a white "immaculate" form which diapauses. Larvae remain in this state for more than 20 weeks and may undergo a series of stationary molts during this period. When JH mimics were applied to nondiapausing sixth instar larvae, they subsequently molted into immaculate larvae which showed all the signs of diapause. Measurements of JH titers in nondiapausing larvae show a high titer late in the fifth instar followed by a precipitous decline with the onset of the sixth stage. By contrast, newly diapausing larvae show a somewhat intermediate level at half of that seen in the fifth instar. In diapausing larvae of the rice stem borer, *Chilo suppressalis,* a high juvenile hormone titer has also been reported.

Consequently, larval diapause is not simply due to an endocrine deficiency. The brain-prothoracic gland axis apparently can still function during diapause as manifest by the erratic stationary molts. Larval diapause is initiated and maintained by the continued activity of the CA which results in a relatively high JH titer during this period.

C. Pupal Diapause

The endocrine control of pupal diapause is probably best known because it was the basis for many of the discoveries by Williams that revealed the interaction between the brain and prothoracic glands. The Cecropia silkworm is a univoltine species that has an obligatory pupal diapause, in which stage it overwinters. Diapausing pupae are characterized by a high glycerol content to protect the animal from low temperatures and a low respiratory rate associated with a decline in the number of functional mitochondria in all tissues except muscle. The diapause condition is then broken by prolonged exposure to cold. Williams demonstrated that pupal diapause was brought about by the failure of the brain to secrete PTTH. Low temperature apparently acted directly on the brain so that the production of PTTH could be reactivated when the pupa was brought back to room temperature. Consequently, the brain from a chilled animal was sufficient to initiate adult development even when implanted into an unchilled diapausing host. The mechanism by which chilling causes this reactivation is still unknown.

In many other insects the onset of pupal diapause is facultative and dependent upon conditions experienced by the growing larva or even the embryo. The most commonly studied factor that influences diapause is photoperiod; long-day conditions typically promote continuous development while short-day conditions cause diapause. Temperature may also have an important role in promoting or averting diapause.

The termination of pupal diapause in facultative species is most commonly accomplished by chilling, presumably similar to that seen for Cecropia as detailed above. In a few species diapause can also be terminated by exposing pupae to long-day photoperiods. In one such species, the Chinese oak silkmoth, *Antheraea pernyi*, Williams and Adkisson demonstrated that the long-day photoperiod acts directly on the brain to promote the release of PTTH. It is also of interest that in this species under short-day conditions the pupal brain produces and stores large amounts of PTTH but does not release it. How the long-day photoperiods subsequently activate this release is unknown.

D. Adult Diapause

Depending on environmental conditions, the adults (both males and females) of a number of different species can enter into a reproductive diapause to avoid inhospitable times of the year. In all cases this period of developmental arrest is due to a lack of juvenile hormone. In females there is a cessation of oogenesis, whereas in males the accessory glands usually regress.

The best known example is that of the Colorado potato beetle, *Leptinotarsa decemlineata,* which was studied by de Wilde and colleagues and will be used to illustrate this phenomenon. Under conditions of warm temperature, adequate food supply, and long-day lengths, newly eclosed beetles emerge from the soil and begin to feed. This is soon followed in the female by the production of vitellogenin, the onset of oogenesis, and the development of the flight muscles. Beetles that are exposed to cool temperatures, poor food, or short days transiently emerge from the soil, but soon burrow back into it. The flight muscles do not develop, oogenesis does not proceed,and a class of specific short-day proteins appear in the blood. With the termination of diapause, the beetles reemerge from the soil, the flight muscles are reconstructed, and oogenesis resumes.

The role of the corpora allata in adult diapause was indicated by the effects of removal of the CA from long-day females. These beetles showed all of the physiological symptoms of diapause. For the reasons described above (Section VII, C), implants of the CA were relatively ineffective in breaking diapause in short-day beetles so that 11 to 12 pairs had to be implanted in order to obtain a reasonable degree of egg maturation. The implications of the gland transplantation experiments are further supported by measurements of juvenile hormone titers in long-day- and short-day-reared beetles. In the former, the JH titer begins to rise at adult emergence and has reached a high level by the onset of oogenesis. In short-day beetles, the JH titer shows the initial rise, but then declines to undetectable levels by the time the beetle reenters the soil. The termination of diapause is coincident with a rise in JH titer and can be induced by exposure to long-day conditions. Thus, in this beetle, reproductive diapause along with the other assorted physiological changes, is enforced by the absence of JH.

The neuroendocrine system is also important for the regulation of adult diapause in *Leptinotarsa:* First, as indicated above, an allatotropic hormone from the brain serves to activate the corpora allata. Second, other neurosecretory factors seem to be required for maintained vitellogenesis. Evidence for the latter comes from the fact that allatectomized beetles switched to long-day conditions show enhanced responses to applied JH, in comparison to allatectomized beetles maintained in short days. Thus, an additional factor that works in concert with JH is apparently released under long-day conditions. Cautery experiments indicate that this factor also comes from the neuroendocrine system.

As summarized in Table I, the diapause responses to potentially adverse environmental conditions are mediated through a variety of endocrine mechanisms. Embryonic and larval diapauses are each induced by the presence of a hormone, namely, the proteinaceous diapause hormone and the

TABLE I

Hormonal Mechanisms for the Control of Insect Diapause

Stage	Endocrine basis
Embryonic	Diapause hormone in mother
Larval	High juvenile hormone level
Pupal	Lack of prothoracicotropic hormone
Adult	Lack of juvenile hormone

juvenile hormone, respectively. In contrast, pupal and adult diapause result from the lack of particular hormones, PTTH and JH, respectively.

IX. INSECT GROWTH REGULATORS

In 1956, when Williams first reported the isolation of a crude extract containing JH from abdomens from male Cecropia moths, he suggested that this hormone might have potential in insect control. The material readily penetrated insect cuticle and consequently topical application could prevent metamorphosis and the subsequent multiplication of pest species. But it was not until Sláma and Williams discovered the "paper factor" in 1965 that this potentiality was seriously discussed. After Sláma brought the linden bug, *Pyrrhocoris apterus*, from Czechoslovakia to Harvard, he soon found that egg hatch was low and that supernumerary larvae, rather than new adults, appeared in the cultures. Following an examination of the differences between rearing conditions in Czechoslovakia and the United States, they found that the causative factor for this phenomenon was a material in the paper towels used in the rearing cages. This material could be found in American but not European paper products and proved to be from the Balsam fir, the main pulp tree used in the United States paper industry. The active component was subsequently identified by Bowers as the methyl ester of todomatuic acid (juvabione, Fig. 16A). Juvabione proved to be a very specific JH mimic for the family Pyrrhocoridae but not for other insects, not even the closely related lygaeid bugs. This high specificity coupled with the movement against the persistent environmental effects of DDT and other

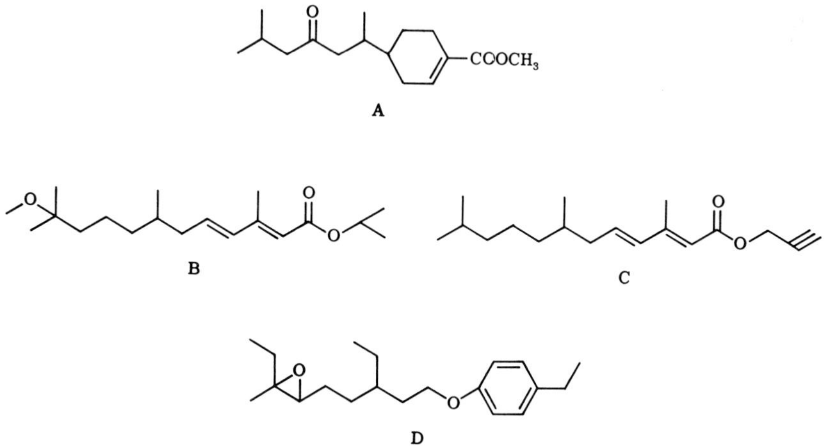

$-COOCH_3$

A

B

C

D

Fig. 16. Juvenile hormone mimics, now commonly known as insect growth regulators. (A) juvabione; (B) methoprene; (C) kinoprene; (D) epofenonane.

chlorinated hydrocarbons led to industrial interest in insect hormone mimics or "insect growth regulators" (IGR's) as they have come to be called (the IGR's presently also include the chitin synthesis inhibitors which will not be considered further). Here was an opportunity to develop quite selective compounds which potentially would act on a pest species but not affect beneficial insects.

Over the past 12 years, over 5000 of these JH mimics have been prepared. The spectrum of activity of these mimics varies widely with some acting on only one or a few families of insects and others showing broad activity. Three such compounds are shown in Fig. 16B-D. Methoprene (Fig. 16B) is now licensed by the EPA for use against floodwater mosquitoes and is effective in a microencapsulated form at less than one part per billion in the water in which they breed. At this concentration it has no effect on other aquatic insects such as dragonflies or on other aquatic invertebrates (e.g., *Daphnia*) or on vertebrates. In both acute and chronic experiments, neither this compound nor JH I has any adverse effects on the mammals tested and they seem to be metabolized rapidly. Methoprene is also active in the higher Diptera and has been used as a feed-through insecticide (usually as an additive to the mineral salt) to control horn flies and other flies that breed in cattle manure piles. Kinoprene (Fig. 16C) is a JH mimic with different specificity, acting primarily on aphids. In this case, the compound sprayed on the plant is taken up and acts systematically to control aphids through feeding as well as by direct contact. Epofenonane (Fig. 16D) has a somewhat different chemical structure, yet it is also quite effective against aphids as well as against scale insects and Lepidoptera. Of importance from a practical point of view is its lack of JH activity on honeybees.

The JH mimics serve to control insects in one of two ways: (1) prevention of metamorphosis and (2) interference with embryonic development. The prevention of metamorphosis is an extension of the normal role of JH in larval life, and the mimic acts in the last larval instar when the JH level should normally fall. This method of insect control insures that no reproductive adults are formed but problems arise when the damaging stage is the feeding larva. Often viable supernumerary larvae are formed and their continued feeding may increase crop damage.

The interference with embryonic development by JH was first discovered by Sláma with the paper factor. *Pyrrhocoris* eggs which were exposed to paper after oviposition went through embryonic development but did not hatch. Application of JH or JH mimics to the female during oogenesis or to the freshly laid eggs blocks embryonic development, primarily by interfering with blastokinesis—a movement of the embryo through the yolk that occurs midway through embryonic development before organogenesis and the formation of the first instar larval cuticle. The mechanism by which JH prevents these movements is unknown; ultrastructural studies have shown only that some cells of the serosa (the extraembryonic membrane which contracts during blastokinesis) lack their normal apical bands of microfilaments. Although the movements are blocked, many of the embryos nevertheless show further differentiation but do not subsequently hatch. When JH is applied to eggs which contain developing embryos, the abnormalities are rarely seen in embryonic development but only later during larval life or at metamorphosis. These latter effects have been shown to be due to an interference with the development of the programming of the brain–corpora allata axis such that the CA do not turn off at the end of larval growth, and metamorphosis is completely or partially prevented. Thus, the ovicidal action of JH has promise for controlling species that have destructive larval stages, but the JH mimic must be applied to the eggs early.

Several problems have arisen with the use of the JH mimics for insect control. The natural JH's and many of the mimics are chemically unstable when exposed to sunlight, high temperatures, aqueous conditions, and bacteria. Some mimics are volatile and rapidly disappear. Moreover, the insects themselves may have efficient mechanisms of ridding themselves of exogenously applied hormone mimics. Houseflies that are resistant to chemical insecticides also show cross-resistance to these JH mimics. By raising *Drosophila* on food containing a JH mimic at the end of larval life, a few JH-resistant mutants (6 per 10^5 flies) have been found. Thus, the possibility of pest species evolving resistance to JH mimics is a very real one. Because of the chemical instability of many of the JH mimics and the sensitivity of insects to these compounds only at selected times in their life history, various stabilizers and slow–release formulations have been developed to ensure a longer life of the compound under field conditions. These formulations are apparently effective in mosquito and horn fly control and will probably be put to practical use in the control of other insects.

In 1975 Bowers reported the first anti-juvenile hormone. When given to early larvae of *Oncopeltus,* this compound caused precocious metamorphosis after one intervening larval instar. When applied to adult females, it also prevented egg maturation. This compound, 6,7-dimethoxy-2,2-dimethyl chromene (precocene II, Fig. 17A), was extracted from *Ageratum* seeds along with a second compound of similar but lower activity. Precocene is effective on many hemimetabolous insects but has little effect on the Holometabola with the possible exception of Coleoptera (e.g., it induces

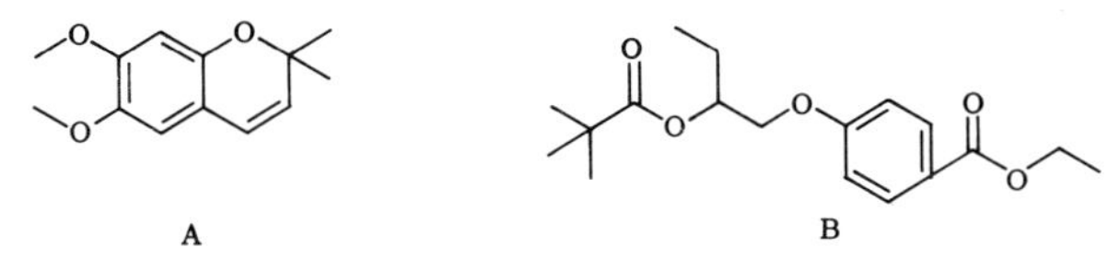

Fig. 17. Anti-juvenile hormones. (A) precocene II; (B) ethyl 4-[2-(*tert*-butylcarbonyloxy)] butoxybenzoate (ETB).

diapause in *Leptinotarsa*). The application of exogenous JH to treated bugs both prevents precocious metamorphosis and restores egg development. Precocene apparently does not act on the target cells for JH, but rather it acts somehow to shut off the CA, as recent studies by Bowers and Pratt on the production of JH by *Periplaneta* CA *in vitro* have shown. Such an anti-JH is a valuable new weapon for insect control because even if the miniature adults survive, they do not reproduce. A different compound, ethyl-4-[α-(*tert*-butylcarbonyloxy)-butoxyl]benzoate (Fig. 17B), has now been found by Staal and colleagues to have similar effects in Lepidoptera, namely, *Manduca*. Here, precocious metamorphosis is seen at the end of the fourth larval instar, following application to the third stage larva or earlier. In this instance, very few of the miniature pupae survive to adulthood.

The other insect hormone that regulates insect growth and development, ecdysone, has proved not to be a good candidate for use as an IGR. Since ecdysone and its mimics are steroids and more hydrophilic than lipophilic, they cannot penetrate the waxy layer of the cuticle and so are ineffective when applied topically. Therefore, they must be fed to the insect. In laboratory tests, the amount required to cause precocious molting (and often derangement of some of the processes involved) was quite high. Moreover, these high concentrations often deterred feeding on the treated material. Therefore, their future in insect control has not been pursued.

ACKNOWLEDGMENTS

We thank Professor L. I. Gilbert for use of unpublished data on ecdysone titers seen in Fig. 5B; Professor C. M. Williams, Drs. Andrew Chen, Stuart Reynolds, and Craig Roseland, and Ms. Nancy Beckage and Mr. Richard Vogt for their critical reading of the manuscript. The previously unpublished experimental data discussed were obtained in studies supported by the National Science Foundation (PCM74-02781A03 and PCM76-18800 to LMR; BMS 75-02272 to JWT), the National Institutes of Health (AI12459 to LMR; NS13079-01 to JWT), and the Rockefeller Foundation (RF73019 to LMR).

GENERAL REFERENCES

Gilbert, L. I. (ed.) (1976). "The Juvenile Hormones." Plenum Press, New York.

Gilbert, L., and King, D. S. (1973). In "The Physiology of Insecta" (M. Rockstein, ed.), 2nd Ed., Vol. 1, pp. 249–370. Academic Press, New York.

Gilbert, L. I., Goodman, W., and Bollenbacher, W. E. (1977). In "Int. Rev. Biochem. Biochemistry of Lipids II," (T. W. Goodman, ed.), Vol. 14, pp. 1–50. University Park Press, Baltimore, Maryland.

Goldsworthy, B. J., and Mordue, W. (1974). *J. Endocrinol.* **60,** 529–558.

Sláma, K., Romanuk, M., and Sorm, F. (1974). "Insect Hormones and Bioanalogues." Springer-Verlag, Berlin and New York.

Wigglesworth, V. B. (1970). "Insect Hormones." Freeman, San Francisco, California.
Willis, J. (1974). *Annu. Rev. Entomol.* **19,** 97–116.

REFERENCES FOR ADVANCED STUDENTS AND RESEARCH SCIENTISTS

PTTH

Brauer, R., Gersch, M., Bohm, G. A., and Baumann, E. (1977). *Zool. Jahrb.* **81,** 1–12.
Fain, M. J., and Riddiford, L. M. (1976). *Gen. Comp. Endocrinol.* **30,** 131–141.
Fukuda, S. (1944). *J. Fac. Sci., Imp. Univ. Tokyo* **6,** 477–532.
Gibbs, D., and Riddiford, L. M. (1977). *J. Exp. Biol.* **66,** 255–266.
Kopec, S. (1922). *Biol. Bull. (Woods Hole, Mass.)* **42,** 323–341.
Suzuki, A., Isogai, A., Horii, T., Ishizaki, H., and Tamura, S. (1975). *Agric. Biol. Chem.* **39,** 2157–2162.
Truman, J. W. (1972). *J. Exp. Biol.* **57,** 805–820.
Vedeckis, W. V., Bollenbacher, W. E., and Gilbert, L. I. (1976). *Mol. Cell Endocrinol.* **5,** 81–88.
Wigglesworth, V. B. (1934). *Q. J. Microsc. Sci.* **77,** 191–222.
Wigglesworth, V. B. (1940). *J. Exp. Biol.* **17,** 201–222.
Williams, C. M. (1952). *Biol. Bull. (Woods Hole, Mass.)* **103,** 120–138.

Ecdysone

Ashburner, M., and Richards, G. (1976a). *Symp. R. Entomol. Soc. London* **8,** 203–225.
Ashburner, M., and Richards, G. (1976b). *Dev. Biol.* **54,** 241–255.
Best-Belpomme, M., and Courgeon, A. (1975). *C. R. Acad. Sci., Ser. D* **280,** 1397–1400.
Bollenbacher, W. E., Vedeckis, W. V., Gilbert, L. I., and O'Connor, J. D. (1975). *Dev. Biol.* **44,** 46–53.
Bollenbacher, W. E., Galbraith, M. N., Gilbert, L. I., and Horn, D. H. S. (1977). *Steroids* **29,** 47–63.
Clever, U.,and Karlson. P. (1960). *Exptl. Cell Res.* **20,** 623–626.
Fraenkel, G. (1935). *Proc. R. Soc. Ser. B* **118,** 1–12.
Fragoulis, E. G., and Sekeris, C. E. (1975). *Eur. J. Biochem.* **51,** 305–316.
Fristrom, J. W., Logan, W. R., and Murphy, C. (1973). *Dev. Biol.* **33,** 441–456.
Fristrom, J. W., Gregg, T. L., and Siegel, J. (1974). *Dev. Biol.* **41,** 301–313.
Karlson, P., and Sekeris, C. E. (1976). *In* "The Insect Integument" (H. R. Hepburn, ed.), pp. 145–156. Elsevier, Amsterdam.
Milner, M. J., and Sang, J. H. (1974). *Cell* **3,** 141–143.
Morgan, E. D., and Poole, C. F. (1976a). *Adv. Insect Physiol.* **12,** 17–62.
Morgan, E. D., and Poole, C. F. (1976b). *J. Insect Physiol.* **22,** 885–890.
Oberlander, H. (1976). *In Vitro* **12,** 225–235.
Richards, A. G. (1976). *Dev. Biol.* **54,** 256–263, 264–275.
Yund, M. A., and Fristrom, J. W. (1975). *Dev. Biol.* **43,** 287–298.

Juvenile Hormone

Dahm, K. H., Bhaskaran, G., Peter, M. G., Shirk, P. D., Seshan, K. R., and Röller, H. (1976). *In* "The Juvenile Hormones" (L. I. Gilbert, ed.), pp. 19–47. Plenum, New York.
Fain, M. F., and Riddiford, L. M. (1975). *Biol. Bull. (Woods Hole, Mass.)* **149,** 506–521.
Gilbert, L. I., Goodman, W., and Nowock, J. (1976). *Colloq. Int. CNRS* **251,** 413–434.
Ilan, J., Ilan, J., and Patel, N. (1970). *J. Biol. Chem.* **245,** 1275–1281.

Johnson, R. A., and Hill, L. (1973). *J. Insect Physiol.* **19,** 1921–1932.
Johnson, R. A., and Hill, L. (1975). *J. Insect Physiol.* **21,** 1517–1520.
Judy, K. J., Schooley, D. A., Dunham, L. L., Hall, M. S., Bergot, B. J., and Siddall, J. B. (1973). *Proc. Natl. Acad. Sci. U.S.A.* **70,** 1509–1513.
Kramer, K. J., Dunn, P. E., Peterson, R. C., Seballos, H. L., Sanburg, L. L., and Law, J. H. (1976). *J. Biol. Chem.* **251,** 4979–4985.
Lanzrein, B., Hashimoto, M., Parmakovich, V., Nakanishi, K., Wilhelm, R., and Lüscher, M. (1975). *Life Sci.* **16,** 1271–1284.
Lassam, N. J., Lerer, H., and White, B. N. (1975). *Dev. Biol.* **49,** 268–277.
Nijhout, H. F. (1975). *Biol. Bull. (Woods Hole, Mass.)* **149,** 568–579.
Nijhout, H. F., and Williams, C. M. (1974). *J. Exp. Biol.* **61,** 493–502.
Piepho, H. (1942). *Wilh. Roux Arch.* **141,** 500–583.
Reibstein, D., Law, J. H., Bowlus, S. B., and Katzenellenbogen, J. A. (1976). *In* "The Juvenile Hormones" (L. I. Gilbert, ed.), pp. 131–146. Plenum, New York.
Riddiford, L. M. (1976). *In* "The Juvenile Hormones" (L. I. Gilbert, ed.), pp. 198–219. Plenum Press, New York.
Ruh, M. F. V., and Dwyer, K. A. (1976). *Insect Biochem.* **6,** 605–608.
Schooley, D. A., Judy, K. J., Bergot, B. J., Hall, M. S., and Jennings, R. C. (1976). *In* "The Juvenile Hormones" (L. I. Gilbert, ed.), pp. 101–117. Plenum, New York.
Sehnal, F., and Schneiderman, H. (1973). *Acta Entomol. Bohemoslov.* **70,** 289–309.
Truman, J. W., Riddiford, L. M., and Safranek, L. (1973). *J. Insect Physiol.* **19,** 195–203.
Vince, R. K., and Gilbert, L. I. (1977). *Insect Biochem.* **7,** 115–120.
Wigglesworth, V. B. (1936). *Q. J. Microsc. Sci.* **79,** 91–121.
Williams, C. M. (1956). *Nature (London)* **178,** 212–213.
Williams, C. M. (1961). *Biol. Bull. (Woods Hole, Mass.)* **121,** 572–585.
Williams, C. M., and Kafatos, F. C. (1972). *In* "Insect Juvenile Hormones: Chemistry and Action" (J. J. Menn and M. Beroza, eds.), pp. 29–42. Academic Press, New York.

Eclosion Hormone

Lockshin, R. A. and Beaulaton, J. (1974). *Life Sci.* **15,** 1549–1565.
Truman, J. W. (1973). *Amer. Sci.* **61,** 700–706.
Truman, J. W., Fallon, A. M., and Wyatt, G. R. (1976). *Science* **194,** 1432–1434.

Bursicon

Andersen, S. O. (1976). *In* "The Insect Integument" (H. R. Hepburn, ed.), pp. 121–144. Elsevier, Amsterdam.
Cottrell, C. B. (1964). *Adv. Insect Physiol.* **2,** 175–218.
Fogal, W., and Fraenkel, G. (1969). *J. Insect Physiol.* **15,** 1235–1248.
Fraenkel, G., and Hsiao, C. (1965). *J. Insect Physiol.* **11,** 513–556.
Fraenkel, G., Blechl, A., Blechl, J., Herman, P., and Seligman, M. I. (1977). *Proc. Nat. Acad. Sci. U.S.A.* **74,** 2182–2186.

Reproduction

Abu-Hakima, K., and Davey, K. G. (1977). *Gen. Comp. Endo.* **32,** 360–370.
Bell, W. J., and Barth, R. H., Jr. (1971). *Nature (London), New Biol.* **230,** 220–222.

Davey, K. G., and Huebner, E. (1974). *Can. J. Zool.* **52,** 1407–1412.
de Wilde, J., and de Loof, A. (1973). *In* "The Physiology of Insecta" (M. Rockstein, ed.), 2nd Ed., Vol. 1, pp. 97–157. Academic Press, New York.
Engelmann, F. (1974). *Am. Zool.* **14,** 1195–1206.
Flanagan, T. R. and Hagedorn, H. H. (1977). *Physiol. Ent.* **2,** 173–178.
Fong, W. F., and Fuchs, M. S. (1976). *J. Insect Physiol.* **22,** 1493–1500.
Hagedorn, H. H. (1974). *Am. Zool.* **14,** 1207–1217.
Kambysellis, M. P., and Williams, C. M. (1971). *Biol. Bull. (Woods Hole, Mass.)* **141,** 527–540, 541–552.
Lagueux, M., Hirn, M., and Hoffmann, J. A. (1977). *J. Insect Physiol.* **23,** 109–120.
Lea, A. O. (1972). *Gen. Comp. Endocrinol., Suppl.* **3,** 602–608.
McCaffery, A. R. (1976). *J. Insect Physiol.* **22,** 1081–1092.
Nijhout, M. M. (1975). Ph.D. Thesis, Harvard Univ., Cambridge, Massachusetts.
Nijhout, M. M., and Riddiford, L. M. (1974). *Biol. Bull. (Woods Hole, Mass.)* **146,** 377–392.

Diapause

Chippendale, G. M. (1977). *Annu. Rev. Entomol.* **22,** 121–138.
Chippendale, G. M., and Yin, C. M. (1975). *Biol. Bull. (Woods Hole, Mass.)* **149,** 151–164.
de Wilde, J. (1970). *Mem. Soc. Endocrinol.* **18,** 487–514.
Kai, H., and Nishi, K. (1976). *J. Insect Physiol.* **22,** 1315–1320.
Kubota, I., Isofe, M., Goto, T., and Hasegawa, K. (1976). *Z. Naturforsch. C* **31,** 132–134.
Williams, C. M. (1956). *Biol. Bull.* (*Woods Hole, Mass.*) **110,** 201–218.
Williams, C. M. (1969). *Symp. Soc. Exp. Biol.* **23,** 285–300.
Yamashita, O., and Hasegawa, K. (1976). *J. Insect Physiol.* **22,** 409–414.

Insect Growth Regulators

Arking, R., and Vlach, B. (1976). *J. Insect Physiol.* **22,** 1143–1151.
Bowers, W. S., Ohta, T., Cheere, J. S., and Marsella, P. A. (1976). *Science* **193,** 542–547.
Bowers, W. S. and Martinez-Pardo, R. (1977). *Science* **197,** 1369–1371.
Plapp, F. W. (1976). *Annu. Rev. Entomol.* **21,** 179–197.
Riddiford, L. M. (1972). *In* "Insect Juvenile Hormones: Chemistry and Action" (J. J. Menn and M. Beroza, eds.), pp. 95–111. Academic Press, New York.
Staal, G. B. (1975). *Annu. Rev. Entomol.* **20,** 417–460.

8

Chemical Control of Behavior—Intraspecific

NEVIN WEAVER

Chemicals that are released by one individual and that affect the physiology or behavior of another individual of the same species are called pheromones. They are called releasers if they cause an immediate behavioral response, and primers if they affect the physiology of the receiving animal in such a way that the behavioral response is delayed. Beyond that, though, classifications are purely matters of convenience. Pheromones are often classified according to their effect on the target animal (e.g., alarm pheromone), and such classifications usually are unambiguous. They are also classified according to their functions. For instance, the causing of alarm is a function of an alarm pheromone, but the pheromone may act to recruit nestmates to the defense of the colony and therefore can be classified as a recruitment pheromone or as one of a complex of defense pheromones.

A pheromone may be one chemical compound, but usually it is a mixture of chemicals, each of which is a component of the pheromone. Since the emitting animal may have some control over the release of components, and since there are differences in the diffusion of and sensitivity to components, it is sometimes difficult to know to which chemicals the receiving animal is responding.

Until quite recently almost all pheromones whose structures were known were relatively small organic compounds of carbon, hydrogen, usually oxygen, and occasionally nitrogen, having molecular weights between 80 and 300. Most of them were fatty acids, derivatives of fatty acids, terpenes, or terpenoid compounds. This relatively simple picture is now undergoing rapid change with the chemical characterization of both small compounds and large complex chemicals that have pheromonal activity.

I. DIFFUSION AND DETECTION

A. Sex–Recruiting Pheromones

Sex–recruiting pheromones that attract one sex, usually the male, over great distances must be fairly volatile, and yet must not diffuse so rapidly that they will quickly reach concentrations too low to elicit a response in the males. In still air the concentration gradient of pheromone within a very few centimeters of the female will be sufficiently steep that the male can detect it through exploratory movements of head and antennae, but at distances more remote, concentrations must be compared over time as the animal moves, and random movements will rarely carry the male toward higher concentrations.

Free-flying insects, though, seldom live in still air, and the wind carries the pheromone out as a plume. Again, the concentration gradient will be quite small except within a few centimeters of the emitting animal. Furthermore, air turbulence will result in regions of greater and lesser pheromone. Where the minimum concentration of pheromone necessary to elicit behavioral response (the behavioral threshold concentration) has been reached, many males automatically fly upwind and, if the concentration falls below threshold, they begin to fly acrosswind. This and other behavior responses lead the male to the vicinity of the calling female. Very high wind speed will disperse the attractant so quickly that the behavioral threshold concentration is reached only fairly close to the emitter, and the males of some species will not fly if the wind velocity is either very low or very high. If the female releases pheromone in a series of evenly spaced discreet puffs into a steady wind, the information content is increased, since at closer distance before diffusion spreads the molecules, the receiving animal perceives sharper boundaries of pheromone and shorter intervals of high concentrations of the chemicals.

On the basis of evaporation and diffusion rates and other considerations, it was predicted that most long-distance sex recruitment pheromones would have molecular weights of between 200 and 300. This has proved true of most such pheromones determined thus far.

The threshold concentration of sex recruitment pheromones can be astonishingly small. For instance, the female cabbage looper moth (*Tricholplusia ni* Hubner) produces *cis*-7-dodecen-1-ol acetate (**1**), and the

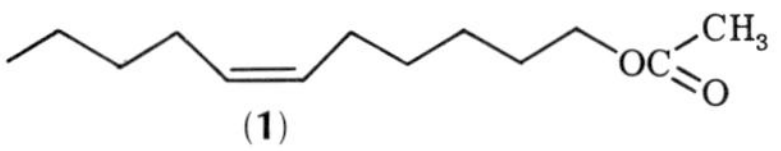

$OC(=O)CH_3$

(**1**)

male of the species will flutter its wings or take flight if he is in a concentration of an estimated 10 molecules per cubic millimeter of air. The male silkworm moth *Bombyx mori* (L.) responds to a similar concentration of the female sex attractant, bombykol, or *trans*-10, *cis*-12-hexadecadien-1-ol (**2**).

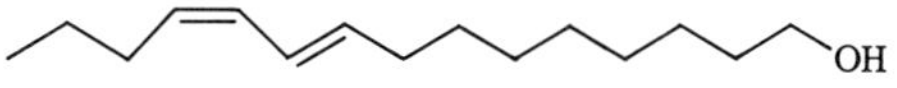

OH

(**2**)

Electrophysiological studies have shown that a sensory cell of a male will respond to the presence of one molecule on it and that an estimated 200 molecules at sensory cells will produce behavioral response of the male; this is just above the theoretical minimum signal–to–noise ratio required to transmit information. Both of these species produce rather large quantities of sex pheromone. The sex pheromone gland of the female cabbage looper contains about 1 μg of pheromone and the amount does not decrease when the gland that releases it is exposed for 30 minutes. Half of the pheromone evaporates in 8 minutes when 1 μg is placed on filter paper, and evaporation is probably equally rapid from the gland. It seems probable, therefore, that a female can release a great deal of pheromone during the 20-minute intervals that it normally exposes the gland at any one calling session.

The system is well designed to bring males to females from great distances: a pheromone that is fairly volatile but which does not diffuse rapidly to below threshold concentration and males that are very sensitive to pheromone and that have a mechanism for following a plume of pheromone to its source. There are cases known in which males found females from 4.5 km away; communication over much greater distances has been claimed and is theoretically possible. However, in the absence of observations on the behavior of individual males or statistical studies to demonstrate that the males that reached the target did not make some portion of the journey through purely random movement, we should be slow in accepting claims of long-distance communication of readiness to mate. For one thing, the long flights into the wind can only come at considerable energetic cost to the male, and the cost will be too high if another male is likely to arrive first. Very long distance recruitment might, nevertheless, occur when the population density is extremely low.

B. Trail Pheromones

A different type of recruiting pheromone is found in the imported fire ant, *Solenopsis invicta* (*saevissma*) Buren. When a worker finds a new source of food it returns to the nest, frequently touching its sting to the substrate as it goes. Pheromone from the Dufour's gland is thus deposited in closely spaced dots on the substrate and is both very attractive and very excitatory to other workers which immediately begin to follow any trail they encounter. The excitatory component(s) of the pheromone is quite volatile and a trail on glass lasts only slightly more than 100 seconds. An ant encountering a trail can, therefore, almost immediately detect a concentration change as it moves, and will quickly turn around if necessary to follow the trail back to the food. Because of the ant's zigzag manner of walking, the follower repeatedly crosses the trail, and it seems likely that one antenna or the other often leaves the threshold concentration and that the ant is therefore guided back toward the antenna that is in the higher concentration of pheromone. Evidence for this mechanism is supplied by experiments with other species that show that if one antenna is amputated, the insect repeatedly overcorrects to the other side, and if the antennae are glued into a crossed position the ant has difficulty following the trail. In order for such a system to work, there must be a steep concentration gradient at the trail boundary and such a steep gradient will only be obtained by a rapidly diffusing substance. Unfortunately the highly volatile pheromone occurs in such small amounts and is so difficult to isolate that it has not been identified.

The rapid fade-out time of the trail allows for the transfer of more information than would a longer lived trail. Most of the foragers that find the food lay a trail back to the nest, thus reinforcing it until so many workers arrive to exploit the food that some of them do not reach it or must struggle against a mass of other foragers to do so. These workers do not lay a trail. As fewer workers release pheromone during the return journey the excitatory components of the trail begin to fade and it attracts fewer workers.

The fire ants are, of course, opportunistic species that exploit transitory food resources. In contrast, higher termites which forage above ground may exploit the same food source for long periods of time. A highly stable, relatively nonvolatile trail pheromone component has been identified from several species of *Nasutitermes* and has been named neocembrene-A. This 20-carbon diterpene hydrocarbon has a 14-member ring and perhaps is derived directly from the food the termite eats, since it has been identified in at least one species of tree. Several species of termites will follow artificial trails made with neocembrene or old trails made by other species that release this material, but they will follow fresh trails of their own species only, indicating that more volatile chemicals give species specificity to the trails.

C. Alarm Pheromones

Most if not all species of social insects and some aggregating insects produce alarm pheromones. One function of most of these pheromones is to recruit other individuals to the defense of the colony. The recruitment and defense system has been studied in detail in the ant *Acanthomyops claviger,* which produces the terpenes, 2,6-dimethyl-5-hepten-1-al (**3**), 2,6-dimethyl-5-hepten-1-ol (**4**), citronellal (**5**), neral (**6**), and geranial (**7**) in the mandibular glands, and the hydrocarbons and ketones undecane (**8**), tridecane (**9**), 2-tridecanone (**10**), pentadecane (**11**), and 2-pentadecanone (**12**) in the Dufour's gland. These compounds are shown below.

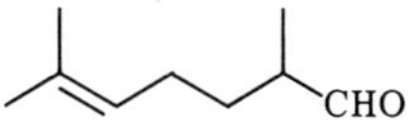

CHO
(**3**)

(**8**)

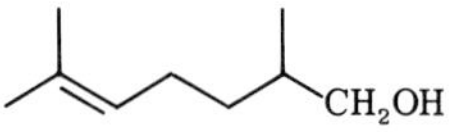

CH_2OH
(**4**)

(**9**)

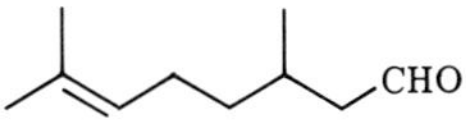

CHO
(**5**)

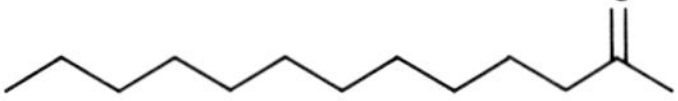

O
(**10**)

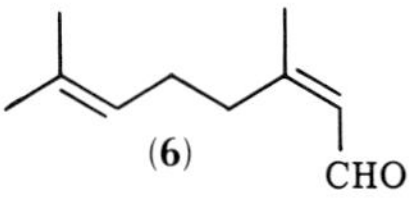

CHO
(**6**)

(**11**)

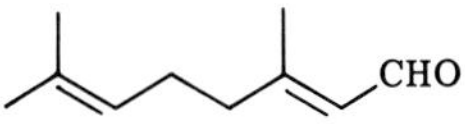

CHO
(**7**)

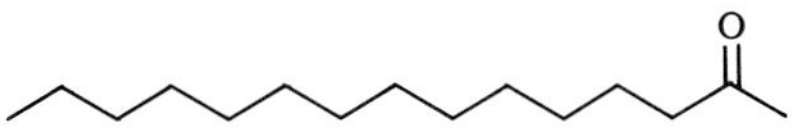

O
(**12**)

When tested singly, all except the latter two compounds are efficient alarm substances. When a worker is disturbed near the nest it releases about 1 μg of pheromone from both of these glands. Nestmates up to about 10 cm away sweep their antennae through the air in typical searching behavior and then run toward the source, becoming more excited as they grow nearer, and if they encounter the alarming stimulus, they themselves release pheromone so that an increasing amount of the chemical is discharged, and the entire colony can eventually be mobilized. If, on the other hand, the stimulus is removed, calm will soon return. The entire process can be studied quantitatively and predictions concerning the spread and disappearance of alarm can be checked against the experimental evidence.

If the air is still and if the components do not react with and are not absorbed by anything in the environment, the pheromone will disperse in accordance with the laws of diffusion. Pheromone concentration at any point remote from the source will depend on the rate of pheromone emission, the diffusion coefficient of the pheromone, the distance from the source, and the time from the initiation and/or termination of emission. By determining the evaporation rates and diffusion constants of the pheromone components and then carefully observing the onset of alarm when each component evaporates in still air from a surface of known area that is placed a measured distance from the ants, it is possible to calculate the minimum threshold concentration for alarm behavior. The ratio between the emission rate, Q, and the threshold concentration K in molecules/cm^3 can be related to the maximum radius of the active space (the space in which the pheromone will be at or above threshold concentration), the time required to reach the maximum active space, and the fade-out time of the pheromone (Wilson, 1970). In order for such calculations to be meaningful, it is necessary to know the emission rate of pheromone from the living animal; with alarm pheromones this can be done by collecting pheromone from alarmed individuals and determining its weight. The combined errors in the above techniques may be as much as two orders of magnitude, but that is sufficiently accurate to allow roughly quantitative statements on the active space of different types of pheromone.

The *Acanthomyops* alarm pheromone has a Q/K value of $10^3 - 10^5$ cm^3/sec and the effective transmission radius is about 10 cm. It takes 2 minutes for the signal to reach that distance and 8 minutes for it to fade out. The *Solenopsis* trail pheromone discussed above has a Q/K of about 1 cm^3/sec, a transmission radius of 0.6 cm, and a fade-out time of about 100 seconds. Thus we can think of a *Solenopsis* following the trail as trying to stay within a narrow tunnel of active space. The silkworm sex attractant, bombykol, on the other hand, has a Q/K ratio of $10^{10} - 10^{12}$. In still air this would give an active space several kilometers in diameter and a fade-out time of several days.

It is quite obvious that (to use a tautology) each pheromone is well adapted to its function. It should also be obvious that if an alarm pheromone has a low Q/K it will be effective only over a very narrow area, and if the Q/K is high the colony either will be in a state of continuous agitation or it will accommodate to the pheromone and be unable to respond. It is perhaps instructive that the ant, *Lasius alienus,* responds to the major pheromone component, undecane, when it is at least two orders of magnitude more dilute than *Acanthomyops* can detect, but instead of moving toward the source of pheromone to attack an enemy, the *Lasius* flee when they detect the pheromone. An early warning system is more appropriate for cowards then for those who stay to fight.

At the other extreme are some of the termites that transmit alarm only through physical contact with nestmates. Since danger to a termite colony is likely to be a highly localized event, it is best to restrict alarm to only that part of the gallery that is in danger.

II. SPECIFICITY

There are three aspects of the problem of specificity: (1) the species, population, or individual specificity of the pheromone, (2) the chemical and physical properties of the pheromone that result in its activity, and (3) the specification of the message(s) transmitted by the pheromone(s). In this section we will deal with the first two of these interlocking problems in some detail and with the third only incidentally; in the section on the complex messages of social insects we will treat the third problem in detail.

Any attempt to study specificity demands extensive, careful, and tedious chemical and biological research. All components of the pheromone should be determined and their structures confirmed through stereospecific syntheses; the activity of each component and of mixtures of components should be determined at threshold and higher concentrations, and under natural and experimentally varied conditions. Unfortunately it would be difficult to find a field in which the rush to find *the* answer has done more to deter understanding. Indeed, the procedure followed all too often has been to determine a chemical structure (usually correctly), give the chemical a "lurid" name (e.g., gyplure, muscalure), and rush into print, sometimes following a minimum of biological observation.

Many of the pheromones whose structures are known are fairly simple compounds which probably can be synthesized by most insects, and closely related species often have pheromone components that are similar or identical in structure. This raises a problem, particularly with the various sex pheromones. In some cases complete reproductive isolation of species with identical sex recruitment pheromones may be attained if the female of each species releases pheromone, and the male is responsive only at certain times of the day, while related species are sexually active only at different hours of the day. There are reports of cases in which the female of one species releases large amounts of pheromone which is attractive to males of that species, whereas a closely related species releases and is attracted to only small amounts of the same pheromone. There are also cases of multispecies attraction to the same pheromone with reproductive isolation being attained by means of various visual, behavioral, or other clues that are necessary for successful mating. Several species of moths are sometimes captured at traps baited with a single pheromone, and there are examples of males of several species being attracted to a single female. It is highly probable, however, that most phermones are multicomponent systems and that there is a great deal of species specificity in nature.

When sensitive bioassays are available it is often found that extremely minor components of the pheromone are more active than the major ones, and that the proportionate mixture of components is more active than any single component. This was discovered years ago with a honeybee pheromone. When a bee finds a rich source of food, or under some other circumstances, it exposes the Nassanoff gland, which releases geraniol (**13**), citral which is a mixture of the geometric isomers, geranial (**14a**) and neral (**14b**), geranic acid (**15**), and nerolic acid (**16**). This mixture is highly attrac-

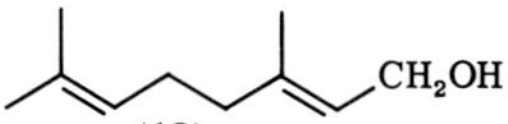

CH_2OH (**13**)

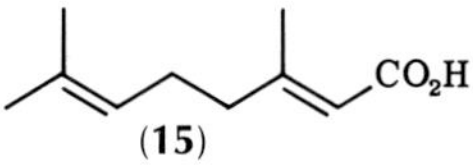

CO_2H (**15**)

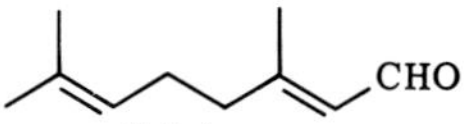

CHO (**14a**)

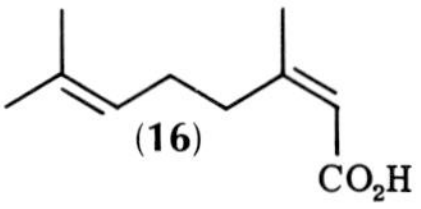

(**16**) CO_2H

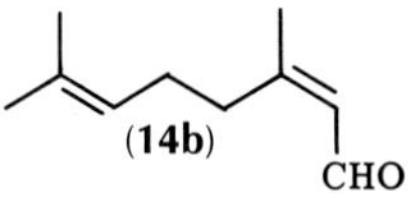

(**14b**) CHO

tive to other bees, and serves as a recruiting and marking pheromone. Citral does not comprise more than 3% of the mixture (and much of it may arise after the pheromone is secreted, presumably by the oxidation of geraniol), but it is by far the most attractive component, and the mixture of all components is slightly more attractive than citral.

The discovery of the attractiveness of citral illustrates a technical problem often encountered in biological assays of activity. When citral was presented on the ground with no visual clues to its location, bees seldom found the exact source but instead methodically explored all vegetation within several meters of evaporating chemical, and an untrained observer scored the attractiveness as zero. Although this technique demonstrated the way in which the pheromone recruits foragers in the field, it yielded little data suitable for a statistical analysis to prove attractiveness. For that purpose it was more convenient to train bees to seek food at petri dishes and then bait different dishes with different chemicals.

Recruiting pheromones are often assayed by means of automatic collecting devices. Such devices certainly collect samples free of human bias, but in the absence of careful field observations one is never sure that he is not collecting data on the reaction of the insects to traps as well as to pheromones. The same sort of objection can be raised to laboratory tests on individuals that are under unusual confinement or that are tested under unnatural conditions.

A. Aggregation Pheromones of Boring Beetles

Many of the beetles that bore into living trees have complex aggregation pheromones that attract other beetles. The pheromones are combinations of chemicals, most of which are similar to each other. Some are produced by the beetles and others by the host trees; in many cases different species of beetles produce one or several of the same components. Since a species may be repelled by a component that it does not produce, and since the host trees (and the terpenes from them) may be different, there is a considerable isolation of species brought about by the apparently similar pheromones. Some brief examples should clarify these complex interactions.

The female southern pine beetle, *Dendroctonus frontalis* Zimmerman, produces several pheromone components including frontalin (1, 5-dimethyl-6-8-dioxabicyclo[3.2.1.]octane) (**17**). The main host tree contribution to the pheromone is α-pinene (2,6,6-trimethylbicyclo[3.1.1.]hept-2-ene) (**18**). The western pine beetle (*D. brevicomis* Le Conte) female also produces several pheromone components including brevicomin (*exo*-7-ethyl-5-methyl-6,8-dioxabicyclo[3.2.1.]octane) (**19**), while a major attractant from the host ponderosa pine is myrcene (7-methyl-3-methylene-1,6-octadien) (**20**). Electrophysiological studies have been carried out on the antennae of these two species. The electroantennograms (EAG's) of the male

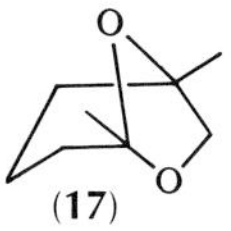

(**17**)

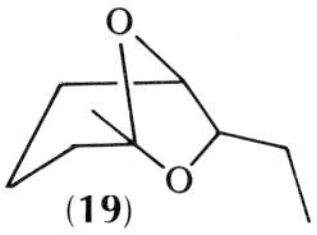

(**19**)

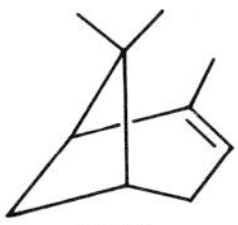

(**18**)

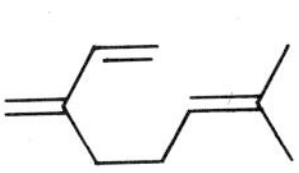

(**20**)

southern pine beetle show a response to both frontalin and brevicomin, though the animal will give a behavioral response only to the former compound. Furthermore it is possible to eliminate the response to one of these compounds by exposing the antennae to the other first; this demonstrates that the same receptors respond to the two substances. While the receptors are adapted to either of these compounds, though, the antennae will still respond to the tree terpenes, indicating that a different set of receptors is involved there.

The situation among *Dendroctonus* has further complications. Males of *D. frontalis* produce brevicomin, and males of *D. brevicomis* produce frontalin. Not only is each attractive to females of its own species, but the ternary mixture of frontalin, brevicomin, and myrcene is more attractive to both sexes of *D. brevicomis* than is a binary mixture of any two of these compounds. Furthermore, the naturally occurring optical isomers, (1*R*, 5*S*, 7*R*)-(+)-*exo*-brevicomin and (1*S*, 5*R*)-(−)-frontalin are much more attractive than are their antipodes. The presence of both enantiomers neither stimulates nor depresses the response to the active one. A similar situation exists in the leaf-cutting ant, *Atta*; only the naturally occurring enantiomer of the alarm pheromone, (+) 4-methyl-3-heptanone (**21**), is active but an

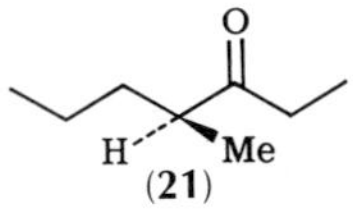

O
H Me
(**21**)

admixture of the antipode has no effect. We cannot, however, conclude that this is a general rule until more optical isomers have been tested.

As indicated above, several species of *Dendroctonus* produce some of the same pheromone components and there is about the amount of cross-attraction among species that one would expect on the basis of the pheromones each produces. The cross-attraction of species appears to be minimized by those components in the pheromone of each species that partially inhibit responses in other species (compounds which are structurally similar to pheromones (parapheromones) often are inhibitors), and by different host terpenes. Since, in this genus, there is some known intraspecific geographical variation in pheromone production and response, the role of pheromones in reproductive isolation must be studied among populations in the same geographical area. It may turn out that some of the variants are actually sibling species as seems to be the case among certain sympatric moths (see below and Chapter 10).

The situation is even more complex in the genus *Ips* and extensive taxanomic, chemical, and behavioral work has not elucidated the problem fully. Among some of the species there is considerable interspecific attraction, but most of it is between species that are not sympatric. In one case, though, there is even the interspecific mating and the production of fertile young, indicating that in those regions where the two species are sympatric they probably intergrade.

There is evidence that some of the beetles that bore into living trees are stimulated to increased pheromone production by the vapor of the terpene volatiles of the tree. This increasing invitation to competitors, even interspecific ones, makes sense since the oleoresins of the tree are able to physically expel the borers from their galleries and, unless a sufficient number of insects quickly attack a tree and weaken it, a vigorous tree will repel the attack.

There are also mechanisms to promote the dispersal of young. For instance, several species of beetles are attracted to pheromone only when flying, and *D. pseudotsugae* Hopkins is not attracted to pheromone until it has flown for about 90 minutes.

B. Some Multicomponent Systems in Moths

A simple binary sex-recruiting pheromone occurs in the smaller tea tortrix, *Adoxophyses fasciata* (Walsh). The female produces both *cis*-9 and *cis*-11-tetradecen-1-ol acetate. The male does not respond when presented with either substance individually, when the compounds are presented in sequence within a few seconds of each other, nor when the males are presented with the proper mixture of the two compounds a few seconds after being exposed to one of them. The *trans* isomers cause no response individually or in mixture, nor do they block the response to a mixture of the *cis* isomers. In the absence of electrophysiological work, the simplest explanation is that there is a set of receptors that is specific for the *cis*-9 compound and another set that is specific for the *cis*-11, and both sets must be activated for a behavioral response to occur; neither binds the *trans* compounds, but both bind to either of the *cis,* so that a single *cis* blocks activity by one of the two sets of sensillae and thus blocks a behavioral response. The blockage, incidentally, begins to disappear about 1 minute after the moths are removed from an atmosphere laden with the blocking compound.

The female redbanded leafroller *Argyrotaenia velutinana* (Walter) produces both geometrical isomers of 11-tetradecenyl acetate with a 91:9 ratio of *cis* to *trans* isomers. The mixture of the two isomers is most attractive to males at the percentage concentrations in which they occur in the female, and they are synergized by dodecyl acetate, which is also emitted by the calling female. The latter compound seems to have its greatest effect at close range and to promote landing of males near the pheromone source. Since the antennae of the males of many species rather quickly become habituated in the presence of high concentrations of pheromone, it is probable that more cases of pheromone constituents that are active only at close range will be discovered.

The males of the yellow race (or sibling species) of the gelechid moth, *Brytropha similis* (Stainton), is attracted to *trans*-9-tetradecen-1-ol acetate, while the males of the gray race are attracted to the *cis* isomer; a mixture of the two isomers attracts no males, indicating that the "wrong" isomer is repellent or that it is a masking agent for the other isomer. There are also two other well-studied cases of sibling species of moths that are reproductively isolated in exactly the same way by *cis–trans* pheromones.

The European corn borer, *Ostrinia nubilalis* (Hubner), has geographic variation in pheromone composition. The females from New York have 97% *trans*- and 3% *cis*-11-tetradecenyl acetate, while those from Iowa have 96% *cis* and 4% *trans*. The males of each type are strongly attracted to the pheromone appropriate to their place of origin. The males and females of the two strains do not mate readily even when confined to bags together, but the hybrids they produce are fertile and there is no reason to doubt that the strains are members of a single species.

C. Structural Modifications of Pheromones

Any one sensillum of the antennae of insects will only respond to a small range in configurations of odorous molecules. Different sensillae may respond to different and overlapping molecular configurations, so that a wide range of odorous compounds are detected.

Those sensillae which respond to a variety of odorants are collectively classified as generalist sensillae. In contrast, pheromone receptors are specialist sensillae. Not only do all such sensillae in any one class respond to the same compound (with slight variation by a few of them) but the specialist sensillae are connected into the nervous system in such a way that stimulation leads fairly directly to the appropriate behavioral response. It is to the advantage of the insect to save nervous tissue by pushing decisions as far to the exterior as possible, and specialist sensillae are one means of accomplishing that saving.

An extreme example is the male domesticated silkworm moth which, as we saw earlier, is remarkably sensitive to the sex recruitment pheromone, bombykol (**2**), released by the female. Only the naturally occurring geometric isomer, *trans*-10, *cis*-12, is highly attractive, the other three possible isomers being required at one hundredfold to several thousandfold greater concentrations to give either an EAG or a behavioral response. A wide variety of parapheromones, pheromones of other species, and other chemicals have been tested, and they either showed no activity or had to be present in much higher concentration than bombykol in order to cause a response. Since any compound that caused an EAG response also caused a behavioral response and no compound caused a behavioral response without causing an EAG response, the most likely interpretation is that *Bombyx* males have only specialist sense cells and that a very small percentage of them are modified sufficiently to give a response to geometrical isomers or other slight molecular modifications. Most insects, of course, cannot afford the luxury of such extreme specialization, nor does stimulation of sensillae generally lead directly to a behavioral response.

We have also seen that the cabbage looper male responds to small concentrations of its sex pheromone *cis*-7-dodecen-1-ol acetate (**1**). A large number of parapheromones were tested for EAG and behavioral response, and for the number of males that could be caught in traps baited with the parapheromone. Shortening the chain length by one carbon atom had the least effect on activity of any modification tested, but even with the 11-carbon chain about one order of magnitude more parapheromone than pheromone was required. The males were even more sensitive to changes in the position of the double bond than to chain length.

With multicomponent pheromones the situation becomes more complicated, as the following example will show. The female orange tortrix moth, *Argyrotaenia citrana* (Fernald), produces *cis*-11-tetradecenal acetate and *cis*-11-tetradecenal. Virtually none of the aldehyde can be collected from the abdominal tips, but collections of the airborne effluvia from calling females yield the aldehyde and ester in a ratio of about 15:1. When each is used singly to trap males only the aldehyde is active, but the addition of any percentage of ester to the aldehyde greatly increases the catch at the traps. Furthermore, increasing admixtures of the *trans* aldehyde with the *cis* aldehyde or with mixtures of *cis* aldehydes and esters reduces the catch of males while the admixture of *trans* ester has no effect on catch. EAG's with male antennae reveal that 14-carbon aldehydes, acetates and alcohols with double bonds all cause responses, and that *cis* double bonds give the strongest responses; the *cis*-11 compound is the most excitatory in each case, but double bonds in the 9, 10, or 12 positions also yield fairly strong responses. Of all the materials tested, the most excitatory single compound is the naturally occurring aldehyde, followed by the naturally occurring ester and the corresponding alcohol.

Precise molecular structures are important to the activity of many types of pheromones. The leaf-cutting ant, *Atta texana* (Buckley), makes trails more than 100 meters long to trees that may be exploited for months. The poison gland of this ant secretes a trail pheromone made up of rather volatile and relatively nonvolatile constituents. Of the fractions that elicit trail following, only the most volatile has been identified, and it is 4-methylpyrrole-2-carboxylate (**22**). A large number of modifications of this compound have been assayed for their activity in eliciting trail-following, and some general

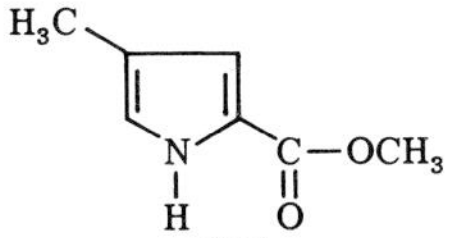

H_3C — N–H — $C(=O)$—OCH_3

(**22**)

conclusions can be drawn: The nitrogen and the two substituents must be in the 1,2,4 positions in the ring; the ring and its substituents must be essentially planar (the atoms in the ring, the bonds to substituents and the two oxygen atoms must lie in a plane); halogens can substitute for the 4-methyl with retention of full activity.

Even alarm pheromones may be quite specific. 4-Methyl-3-heptanone is the alarm pheromone of the harvester ant, *Pagonomyrmex badius* (Latr). Of ninety-nine ketones tested for pheromone activity, only 4-methyl-3-hexanone was as active as the pheromone and 3-methyl-4-heptanone, 4-methyl-3-octanone, and 2,4-dimethyl-3-heptanone were fairly active; the further compounds departed from the combination of the size and shape of the pheromone, the less active they were.

The situation with honeybees is slightly more complex. The mandibular gland produces 2-heptanone and the sting gland, isopentyl acetate. Both are alarm pheromones. As would be expected, bees react with moderate alarm to esters or ketones of about the same size, shape, and polarity as the appropriate alarm pheromone, but they do not show alarm in the presence of a fairly wide variety of aldehydes, acids, ethers, hydrocarbons, or aromatics.

The alarm pheromones of some of the stingless bees are even more complicated. Both the head and the abdomen of *Trigona pectoralis* (Dala Torre) yield alarm pheromones, but those from the head are by far the most excitatory. The head pheromone consists of 2-heptanol, 2-nonanol, 2-heptanone, 2-nonanone, 2-undecanone, 2-tridecanone, 2-pentadecanone, and benzaldehyde, with the last three components comprising about 58, 13, and 8%, respectively; there is also about 10% unidentified compounds. Each of the two alcohols alone cause some excitement and attack when presented at the nest entrance, and the ketones are more excitatory than the alcohols. Mixtures of the alcohols or of the ketones are more excitatory than are the individual components, and either mixture is much more excitatory if it also contains benzaldehyde. Mixtures of the alcohols, ketones and benzaldehyde are still more excitatory, and all of the mixtures containing benzaldehyde become more excitatory as they age. In the air, benzaldehyde readily oxidizes to benzoic acid, and of the many chemicals tested, the single compound that causes by far the most intense attack is benzoic acid. Benzaldehyde that has been kept in an atmosphere of helium causes some exploration and other mild reaction in the bees, but as the substance ages in an atmosphere of air, it causes increasing severity of attack.

We do not know the full implications of these findings, but it is obvious that an enemy attacked by a bee would be marked by pheromone from the mandibular gland and would become more excitatory as time went on. The pheromones may mediate a complex of defensive actions that have not been studied sufficiently, and there is some evidence that they serve directly as defensive chemicals.

There is also a great deal of data on the effect of each pheromone component and of a wide variety of other chemicals, but the interpretation of these data is complicated by the fact that the bees react to the chemicals of their enemies and therefore a discussion of them will be postponed to Chapter 9.

D. Mating Pheromones

Pheromones that promote mating have been called aphrodisiac pheromones, and this unfortunate terminology has become embedded in the literature. Different pheromones that promote mating may have different modes of action. In some cases it appears that the sex recruitment pheromone or the aggregation pheromone alone is sufficient to lead to mating without further chemical clues. In the cabbage looper moth, for instance, increasing concentration of the female sex recruitment pheromone is sufficient to initiate each successive step in the responses leading to male copulatory behavior. In some species copulation follows immediately upon location of the female, but in any but the most carefully studied cases, we must be circumspect in drawing conclusions. As we have seen, pheromones that act only at very close range may be so volatile that they are difficult to detect and identify, or alternatively they may be so large or polar that they are not extracted or detected by the chemical techniques usually employed.

Pomance ("fruit") flies go through elaborate stereotyped premating behavior before copulation can occur. After a male orients toward another fly he must obtain the proper pheromone in order to initiate the next step in the courtship sequence. The female must also accept the male, and at least one part of the acceptance depends upon pheromone from the male. The female *Drosophila pseudoobscura* can distinguish between males of the same genetic strain that have been raised at different temperatures. They can also recognize (by odor) different genetic strains and, if the males of two strains are in equal numbers, will mate with one strain in preference to the other. If the strains occur in unequal numbers the females show a bias toward mating with males whose proportion in the population is the lowest. Furthermore if a normally unattractive strain occurs in the proportion of 1 to 1 to 4 with two attractive strains, then the bias toward the unattractive strain is of a magnitude that would be expected if the numbers of the two attractive strains were added together.

The male of the monarch butterfly has perfected a flying tackle followed by something that strongly resembles rape, but most butterflies use a more subtle and seductive approach. Male butterflies and moths have scent-disseminating structures on the legs, thorax, or abdomen that range from simple tufts of hairlike scales to complex eversible organs.

The queen butterfly, *Danaus gilippus berenice* (Cramer), can be taken as typical. The male apparently locates the female by sight and flies over her, everting a pair of tufted stalks called hairpencils from the posterior of the abdomen. He brushes the hairpencils against the female's antennae and head, and if she is responsive she alights on herbage. The male continues to hover and hairpencil the female for a while and then he alights and mates with her. The hairpencils are covered with cuticular spherules called dust and this is transferred to the female's antennae during the hairpencilling. Two chemicals have been identified from the hairpencils. 3,7-Dimethyl-deca-*trans*-2, *trans*-6-dien-1,10-diol (**23**) serves as a glue to hold the dust to the antennae; mineral oil is sufficiently sticky to substitute for it. The dust is impregnated with 2,3-dihydro-7-methyl-1H-pyrrolizin-1-one (**24**), which induces the

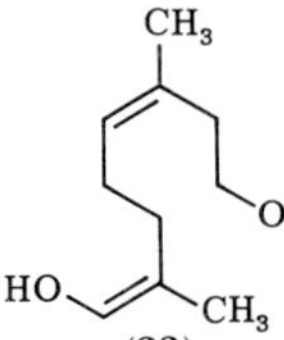

(**23**) 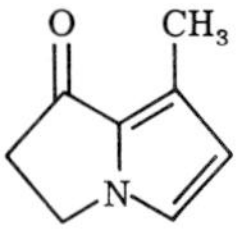 (**24**)

queen to sit quietly so mating can occur. By rearing caterpillers on selected food plants it is possible to produce males with chemically deficient hairpencils, and such males cannot successfully mate because the female will not remain still. If the deficient hairpencils are artificially treated with the ketone and with a sticky chemical, the males are then able to mate. EAG's of both males and females show a strong response to the ketone and a weak response to the diol, so it is probable that the diol has some function in addition to serving as a glue, but the bioassays are not sensitive enough to detect it.

E. Contact and Gustatory Mating Pheromones

Many cockroaches go through rather complicated mating behavior which is mediated in large part by pheromones. The male's antennae must first make physical contact with the female where, in the German cockroach *Blattella germanica* (L.), he perceives 3,11-dimethyl-2-nonacosanone and the corresponding 29-hydroxy compound. The pheromone causes the male to raise his wings, exposing an oily secretion of low volatility on his tergum. The female mounts the male to feed on the liquid and thus brings the genitalia into the proper position for copulation to occur.

There are many cases in which no chemistry is known. The female of a cricket appears to feed on the surface of the male tergum while he inserts a spermatophore; the female retrieves the spermatophore and eats it; while she is eating the male inserts a second spermatophore. A male fly produces a foam from the mouth; while the female feeds on the foam the male engages her in copulation. During courtship the males of some beetles produce gustatory pheromone from some glands over the body surface. The list could go on and on.

III. PHEROMONES OF SOCIAL INSECTS

Solitary insects need to send only limited messages to conspecifics; among aggregating species, complexity of messages increases; and among social insects the number and types of messages are very large indeed, and pheromones may mediate virtually every aspect of social life. Unfortunately some of the pheromones are so subtle and pervasive that it will prove extremely difficult to isolate and assay them.

A. Colony Odor and Food Exchange

Since natural selection in social insects operates primarily at the colony level, species in which intraspecific competition is strong must have a system for the recognition of nestmates. Recognition is mediated through distinctive odors that colonies acquire. A caged honey bee colony can be divided and if the daughter colonies are given the same food they will not acquire distinctive colony odors. If, however, one of the daughter colonies is fed on molasses or is exposed to its odor, the two daughter colonies soon begin to show antagonism to each other, and foragers from each colony will gather food from a dish visited by hivemates in preference to a dish visited by bees from the other colony. Since colonies of all social insects inhabit nests with distinctive odor profiles and each colony feeds on slightly to grossly different forage, individual colony odors are easily acquired.

The acquisition of distinctive colony odors is aided by the fact that nestmates constantly exchange food. Within a day after six bees brought 20 ml of radioactive food into the hive, 60% of the 25,000 bees in a colony became radioactive, and more than 80% of the large larvae were radioactive after a couple of days. This rapid spread of food through what has been called the common stomach of the colony not only preserves the colony odor, but it keeps all members of the society informed of the metabolic and pheromonal condition of the colony.

Food exchange also serves other functions. In both wasps and termites, mutual feeding is necessary for the complete digestion and utilization of food, and in many taxa it may play a role in the feeding of larvae. Since it would be stretching the meaning of the term pheromone to include the exchange of food as distinct from specific primers or releasers, this aspect of behavior will not be discussed further.

B. Recognition of Caste

The female hymenopteran social insects have at least a queen and a worker caste. Many ants have one or more additional worker or soldier castes; the males are always monomorphic. Among the termites both the males and females are polymorphic.

It has long been known that both workers and queen honeybees recognize the caste of queens and that workers can distinguish their own queen from a foreign one. The mandibular glands of queens contain large amounts of caste-specific fatty acids and some other lipids. By far the largest component from the mandibular glands is free 9-octodec-*trans*-2-enoic acid (**25**); 9-hydroxydec-*trans*-2-enoic acid (**28**) is the second-largest free fatty acid component; these two acids occur at concentrations of about 200 and 80 μg/head, respectively; there are closely related fatty acids and unknown fatty acids that occur in much smaller amounts. Methyl *p*-hydroxybenzoate and *p*-methoxybenzoic acid occur in fairly high concentrations but no function has as yet been determined for them. There are also unidentified chemicals from other parts of the queen's body and, in the absence of evidence that the queen can control release of different compounds or compounds from different sources, the entire odor and flavor bouquet of the queen must be regarded as queen identification pheromone.

If the queen is removed from the colony the workers will soon begin to fan their wings and show other signs of agitation. About a day later they will build queen cells. (In other words, they will enlarge and reshape cells containing young worker larvae and begin feeding the larvae on royal jelly; these larvae develop into queens.) If queenlessness persists for several weeks some of the workers will develop enlarged ovaries and will eventually begin to lay a few eggs which, being unfertilized, will normally develop into males (drones).

If one removes the queen and feeds the workers 9-ketodecenoic acid the nurse bees will start far fewer queen cells than if no queen pheromone is given; other fractions of queen extract do not depress queen cell construction if there is no 9-ketodecenoic acid in the fraction. The 9-ketodecenoic acid alone, though, is not nearly so attractive to the worker bees as is the entire queen extract, nor is it as active in preventing the bees from building queen cells. The odor of a queen or of queen extract is not sufficient to prevent the building of queen cells if 9-ketodecenoic acid is not consumed, but if the workers are continuously fed 9-ketodecenoic acid and are exposed to the odor of the remainder of the queen extract, virtually full activity is achieved. The major odorous constituent that synergizes the keto acid is the 9-hydroxydecenoic acid.

The suppression of worker ovaries has received far less attention than has the control of the building of queen cells, but the ingestion of 9-ketodecenoic acid is the major factor in suppressing ovary development, and the remainder of the queen extract is needed for full activity.

In the spring when the population of the colonies has been increasing rapidly and is approaching the maximum, the colonies often reproduce themselves by swarming. Typically the bees rear queen cells and before the new adult queens are ready to emerge, the old queen and most of the bees that are old enough to fly leave the hive and cluster on the branch of a tree or some other object until a new nest site has been selected; the swarm then moves to the new home.

If one removes the queen from the cluster, the bees almost immediately start showing unrest, and soon the cluster breaks up and the bees slowly return to the parent hive. If an airborne swarm is deprived of the queen it will be attracted to the odor of 9-ketodecenoic acid but will not settle. On the other hand bees are not attracted from a distance to the odor of 9-hydroxydecenoic acid, but the bees that are close (or are attracted from a distance by the keto acid) will cluster around the hydroxy acid and the cluster will stabilize. If both acids are present, some of the bees on the cluster will expose the Nassanoff gland which releases a recruiting and marking pheromone; the cluster will be joined by virtually all the circling bees and will hang relatively quietly. However, if one introduces a strange queen or a slight amount of the alarm pheromone, 2-heptanone, the workers will immediately quit exposing the Nassanoff gland.

C. Mating

Virgin queen honey bees are only attractive to drones if they are in the air at least 2 meters above the ground on a windy day or much higher on a still day. Quantitative assays of the factors affecting mating are exceedingly difficult to perform; accordingly for the present we can only say that the requirements for mating are the release of the odor of 9-ketodecenoic acid into moving air at sufficient height that drones can fly up-wind to it, and the presence of an object releasing the odor that is of approximately the size and shape of a queen and that has an open sting chamber or its equivalent; 9-hydroxydecenoic acid again seems to act synergistically with the keto acid and pheromone from other glands may be important, especially at close range.

Interestingly enough there are rather restricted areas where drones congregate and circle about year after year. There is an unconfirmed report of drone pheromone attractive to other drones; there is no evidence on why an area becomes a location for congregating, nor is there evidence on whether virgin queens are attracted to the areas or to the drones.

The males of some bumblebees secrete species-specific pheromone from the mandibular gland and mark territorial sites with them. The secretions are attractive to both males and females so that they tend to congregate at specific sites and thus increase the chances that mating will occur.

A male attractant is known in the carpenter ant, *Camponotus pennsylvanicus* (DeGeer). The mandibular glands produce methyl 6-methyl-salicylate, 3,4 dihydro-8-hydroxy-3-methylisocoumarin, and lesser amounts of other compounds. When males fly they release the secretion over the nest and this causes the virgin queens to swarm from the nest in mating flights. The workers show slight excitement when exposed to the secretion and the males show no response. The EAG's of workers and virgin queens are of about the same amplitude in response to the male mandibular gland secretion or either of the major components, and the male response is about one-third as great. Thus there is a general *antennal* response, but a caste-specific *behavioral* response.

D. Origin and Spread of Pheromones

In the hive the adult queen is attractive to bees that are within a few centimeters of her but not to bees further away, and she appears to be slightly repellent to some of the bees that come close to her head. Bees near the queen can often be seen to lick her, and they are presumed to obtain the contact pheromone in that way. The same pheromone constituents that occur in the queen's mandibular glands are readily extracted from the surface of the thorax and abdomen, indicating that they are dissolved in the cuticular waxes. If radioactive 9-ketodecenoic acid is placed on the thorax of a queen or worker, it quickly spreads to the head and abdomen; it reaches the tip of the abdomen in 5 minutes and attains a fairly high concentration there within an hour. It appears to be carried both over the surface of the body and in the blood or other internal tissues.

There is circumstantial evidence indicating that the mandibular glands are the main site of synthesis of the queen pheromone. However, queens that have had the mandibular glands removed are still fairly attractive to queenless bees and cause a reduction in the number of queen cells started. A few virgins, which had the mandibular glands removed soon after emergence as adults even managed to mate. Queens that have had the mandibular glands removed contain small amounts of the two most active components of queen pheromone 10 days later, but we do not know whether they continue to synthesize it. If we knew the normal precursor of 9-ketodecenoic acid these matters probably could be settled in short order. Unfortunately a precursor that readily gives rise to the pheromone is yet to be discovered.

E. Origin of Castes

Young female honeybee larvae are given a food that is derived wholly or in major part from secretions of the hypopharyngeal glands of the nurse bees. Larvae destined to develop into queens are floated in a large excess of the glandular food, called royal jelly, thoughout larval life. Worker larvae, on the other hand, receive small amounts of glandular food similar in appearance to royal jelly for the first 3–4 days of larval life; thereafter some pollen and honey are included in the food.

A highly labile compound in the dialyzable fraction of royal jelly is necessary for the development of queens. It appears to act on the endocrine system and promote the production of juvenile hormone and possibly other hormones which then control dimorphic development.

Termites have unusual plasticity of development. Older nymphs, called pseudergates, of the lower termites can undergo stationary molts or they can molt one or more times to form reproductives, supplementary reproductives, or one or more types of soldiers; individuals that have not reached the terminal molt can undergo retrogression to the pseudergate form. In the higher termites the situation is similar, but differentiation of castes begins in the early instars and there is a terminal worker caste.

Lower termites carry out extensive anal feeding. Reproductives of *Kalotermes flavicolis* (F.) pass pheromone from the anus that inhibits the development of supplementary reproductives. The inhibitory pheromone appears to arise in the mandibular glands. Female pseudergates in colonies with reproductives pass on the inhibitor of male reproductive development and male pseudergates pass on the inhibitor of females. Furthermore, each reproductive stimulates the other to produce the inhibitor and, if there is only a reproductive of one sex in a colony, both male and female supplementary reproductives will develop. There is, however, a stimulative pheromone produced by a lone male reproductive that will speed the development of a substitute female. The higher termite reproductives also release inhibitory pheromones, but the passage of the inhibitor from one individual to another is, of course, different since the higher termites do not feed anally. We know nothing concerning the nature of the inhibitor or its removal, nor do we know anything about the nature or control of the events that lead to the development of the reproductive castes.

There is evidence for weak inhibitory pheromones from the soldiers of some termites. High juvenile hormone titer favors the development of soldiers, but it is not known whether the hormone itself is passed from pseudergates to developing soldiers. A termite colony sometimes acquires an excess of one of the castes; this situation is rectified through the cannibalization of the excess caste members by the pseudergates.

We know even less about caste determination in ants than in bees and termites, but it appears that in some species at any rate, larval nutrition contributes to differentiation.

In all social insects there is recognition of developmental stages and of the caste of the developing individuals, a recognition which is necessary for the proper care of colony members. The brood of all castes of the imported fire ant *S. invicta* contains pheromone that causes the nurse ants to groom and care for them or to groom a small object treated with the pheromone. The pheromone that the nurses use to distinguish sexual from worker brood appears to be predominantly triolein, and so a piece of filter paper treated with that substance will be placed by the nurse ants among the sexual pupae.

F. Further Complex Interactions

We have seen that the message transmitted by the queen honeybee pheromone depends upon the context in which it occurs, and that different components of the pheromone may have somewhat different functions. We must now reconsider some of the other pheromones of social insects in the light of the need of societies to integrate different aspects of behavior.

It will be remembered that in the honeybee the Nassanoff secretion causes bees to explore vegetation if it is released at a source of food, and that it helps to cause bees to settle on a swarm cluster. When a swarm moves from its first resting place to a new home it is led by workers that fly with the Nassanoff gland exposed. The gland is often exposed at the hive entrance, and if one transfers a colony to a new or different hive, the marking becomes especially intense. Swarms will select as new homes empty hives treated with citral (the most attractive pheromone component) in preference to untreated hives. Bees frequently expose the gland when they gather water; a recruiting marker is probably especially needed by water carriers since a source or water usually will not have so strong and unmistakable an odor as a source of nectar or pollen.

In many species of ants both the alarm and the trail pheromones may serve more than one function. We have seen that trail pheromones are used to lead workers to a new source of food. They are also used to lay a trail to a new nest site and to an enemy. In some cases an alarm pheromone is an attractant at low concentrations and causes alarm and attack only at high concentrations; most alarm pheromones are also marking pheromones and orient the ants toward the enemy if a trail is not laid. In some cases trails are made of both alarm and trail pheromones so that excited workers are quickly recruited. Finally the more volatile components of the trail pheromone may excite the recruits and the less volatile components cause trail following but no excitement. The harvester ant *Poganomyrmex* lays a recruitment trail with pheromone from the poison gland and an orientation trail with pheromone from the Dufour's gland; the latter pheromone does not cause trail following by nestmates, but the trail is followed back to the nest by the individual that laid it. There are "footprint factors" in bees, ants, and termites, but the factor may be nothing more than animal (and colony) odor that is transferred to a surface on which the insect walks rather than being a specific exocrine gland secretion; the fact that its source is diffuse does not make an odorant any less a pheromone, of course.

The complex of interactions among many components of three glands has been extensively studied in the European common red ant *Myrmica rubra* L. The ants can lay trails with the sting using only secretions of the poison gland and other ants will rather calmly follow the trail. However, if an enemy is encountered, the ant discharges the Dufour's gland which is highly excitatory and causes other ants to run fast in a zigzag path toward the source of the pheromone and to themselves discharge Dufour's gland contents. Excited young workers will lay a trail composed of both poison gland and Dufour's gland secretions from the nest to the vicinity of the enemy, and other ants will follow the path much more rapidly than a path composed of poison gland secretion alone.

Recently it has been found that the Dufour's gland produces acetaldehyde, ethanol, acetone, and butanone in the approximate ratios of 35:3:40:25. A specific function has been assigned to three of these compounds: acetaldehyde is the attractant; acetone causes an increase in linear speed; ethanol causes the exaggerated sinuous zigzag movements (!) characteristic of the excited ants. The synergism among components is so great that the ants will react to mixtures three orders of magnitude more dilute than the threshold amount of pure chemical. In addition to the very small molecules, there are also some normal hydrocarbons 9 to 19 carbons long and some unidentified compounds of intermediate size; the hydrocarbons lead to no response if the small molecules are not present, and probably serve as solvents and extenders for them. A Dufour's gland contains about 12 ng of volatiles and 1.4 μg of hydrocarbons, and it has been estimated that only about 80 pg of volatiles are released at any one time; that is not far above the threshold for activity. The mandibular gland secretion has been classified as an alarm pheromone but in the defense of the nest it displays rather complicated roles. Both 3-octanol and 3-octanone occur in the mandibular glands. The 3-octanone is an attractant to the workers and 3-octanol is an arrestant. Aggregations form much more readily if both chemicals are presented than if either is presented alone. The 3-octanone has another function: it markedly suppresses aggressiveness. Since the combined poison gland and Dufour's gland secretion make the ants highly aggressive, it has been suggested that the mandibular gland secretion serves not only to attract ants to the site of alarm, but also to protect the emitter of alarm pheromone from being attacked herself.

IV. NEGATIVE CONTROLS

There are times when it is important for insects to be able to repel conspecifics, to stop them from acting, to cancel the effect of pheromone, or to destroy existing pheromone. Negative controls may be applied by the originator of the pheromone or by the recipient, and without careful study it is often difficult to determine how they have come about.

Sometimes the control is quite simple. For instance, the secretion of the hairpencils of the male tobacco budworm moth, *Heliothis virescens* (F.), causes the female to cease to release sex recruitment pheromone. Most negative controls, however, are more complex.

A. Repellent and Deterrent Pheromones

The cotton stainer bug, *Dysercus intermedius* Distant, lives in dense aggregations. A gland on the dorsum of the abdomen produces *n*-dodecane, *n*-tridecane, *n*-pentadecane, hexanal, 4-oxyhex-2-en-1-ol, oct-2-en-1-al, and 4-oxo-oct-2-en-1-al. The mixture is probably a defensive secretion, but it also alarms other bugs of the same species and causes them to disperse.

The parasitic wasp, *Telenomus sphingis* (Ashmead), spends about a minute and a half rubbing the ovipositor over the host egg after ovipositing in it. The same or other females are much more likely to attack an egg that has not been marked.

The braconid parasite, *Cardiochiles nigriceps* Vierech, marks larvae in which it lays eggs with a secretion of the Dufour's gland and other females will not oviposit in such marked larvae. The gland secretion can be fractionated into a hydrocarbon, a sterol ester, free fatty acid, and an unknown compound. The hydrocarbon alone is by far the most active single material, but a mixture of the four fractions is 700 times more active than the hydrocarbon alone.

The female apple maggot fly, *Rhagoletis pomonella*, and the cherry fruit fly, *R. cerasi* L., deposit a pheromone on fruit after laying an egg in it. The pheromone is a highly stable, polar, water-soluble compound and, although it does not repel other females, it does prevent them from ovipositing on the same fruit. The same pheromone promotes the aggregation of males whether it is deposited on fruit or on artificial mounds in the laboratory. The pheromone is not attractive to males from a distance, but it causes arrest so that if they alight the males tend to accumulate where the pheromone is deposited. The males then release a pheromone which is attractive to virgin females.

Heliconius erato butterfly males produce pheromones which are transferred to pouches in the female genitalia during mating. These pheromones are extremely persistent and repel other males, thus helping to insure monogamy.

In most cases the effect of the male is more indirect. The impregnated female fruit fly is not sexually receptive until the stored sperm are used up. Courtship or copulation without ejaculation has no effect on sexual receptivity or on egg laying. Male paragonia, but not testes, transplanted into the abdominal cavity of virgin females cause them to refuse to mate, so that some of their behavior in rejecting males is similar to that of fertilized females.

The paragonia produce a peptide composed of eleven amino acids that stimulates the female to lay eggs. Although it seems likely that the same polypeptide is also responsible for the rejection of males, this possibility has not yet been tested.

The presence of factors from males that stimulate egg production and cause females to refuse to copulate are fairly common in insects. They are usually polypeptides with molecular weights between 750 and 3000, or proteins with molecular weights of 30,000 to 60,000. Many are not species-specific in their activity, indicating that a common amino acid sequence may be involved, but we know very little about these male factors. In the *Cecropia* silkworm moth it appears that the presence of sperm or other material from the male in the bursa copulatrix induces that organ to produce a substance which both acts on the brain to cause the female to stop release of sex recruitment pheromone, and stimulates release of a hormone that promotes oviposition.

B. Masking Pheromones

The same gland of the female gypsy moth, *Porthetria dispar* (L.), produces both the attractant pheromone *cis*-7,8-epoxy-2-methyloctadecane and a compound, 2-methyl-*cis*-1-octadecene, which masks the presence of the pheromone. It appears that under normal circumstances the mated female does not release the inhibitor, so additional studies will be required to elucidate its exact role in nature. There is evidence for masking pheromones in several species; several compounds have been found by trial-and-error that exhibit this property even though they are not known to occur in the species which respond to them. Such compounds could act upon the same receptors as the pheromone, or on different receptors, so until detailed studies have been conducted we will not know what their mode of action might be.

Although it is advantageous to an individual wood or bark beetle to be joined by others early in attacks on a tree, it is also advantageous to have some method of turning calling signals off or of repelling rivals when the tree is saturated or during sexual rivalry. The female Douglas fir beetle, *D. pseudotsugae*, releases 3,2-MCH (3-methyl-2-cyclohexen-1-ol), which synergizes the attraction of the aggregation pheromone if it is released at low concentrations, but at high concentrations it masks the aggregating pheromone. Several other *Dendroctonus* species also produce MCH and react in the same way to it, so it is assumed to be a general anticompetitive hormone in this group. All of the factors that stimulate the production of MCH are not known, but if a female is exposed to the stridulations of a male at the entrance to her gallery, or to a silent or stridulating male within, she will release MCH, thus ceasing to call other males. The male will fight other males that it encounters at or near a gallery and will release large amounts of 2,3-MCH and an isomer, 3,3-MCH. These and other components released under the stress of intrasexual rivalry serve as anticompetitive and anticuckold pheromones.

C. Habituation

Males of *Lasioglosum zephyrum* become habituated to the odor of individual females to which they are presented repeatedly and are not attracted to them though they continue to be attracted to other females. This habituation, which amounts to recognition of individuals, probably serves to prevent males from attempting to mate with females that have already been impregnated. The same type of learned recognition may operate in species of primatively social bees so that guard bees do not react aggressively to the few nestmates that use the nest. Habituation may also account for the lack of aggression between nestmates in all social insects.

D. Metabolic Destruction of Pheromones

We have already seen that a pheromone component can oxidize to a more active compound in air; it is also likely that in air and light some components quickly undergo reactions to form inactive compounds. Furthermore pheromones may be attenuated to below the behavioral response threshold by diffusion or by adsorption on surfaces. Pheromones may also simply evaporate from the sensillae. Beyond these mechanisms, however, there are cases of active destruction of pheromone.

The sensillae of the antennae of the male cabbage looper *T. ni,* bind the sex recruitment pheromone, *cis*-7-dodecen-1-ol acetate (**1**) and degrade it to acetate and *cis*-7-dodecenol at a rate of about 34% per hour. The free alcohol causes a sensory, but not a behavioral response. The chemosensory sensillae membrane and fluid bathing the nerve endings can be isolated. The enzyme that hydrolyzes the pheromone appears to be bound to membrane and also to occur in the liquid fraction.

Early events in the absorption and metabolism of bombykol, *trans*-10, *cis*-12-hexadecadien-1-ol (**2**), by antennae of male silkworm moths *B. mori* have been extensively studied. The bombykol is first dissolved in the surface hydrocarbons of the cuticle and from there diffuses to more polar lipids; half of the bombykol diffuses from the surface in 1.6 minutes. The half-life for the migration of bombykol from the hairs to the antennal branches is only 4.0 seconds; the migration rate of hexadecane or non-*Bombyx* pheromones is considerably slower, indicating that selection of specific molecules begins even before the sensillae are reached. At the sensillae the alcohol is metabolized to the corresponding acid and several further breakdown products. Only a small percentage of the pheromone or its metabolites is covalently bound to insoluble substances.

We have seen that queen honeybee pheromone is licked from the body of the queen and is passed from one worker to another during mutual feeding of nestmates. If one removes the queen from the colony with the minimum of disturbance, it is usually about a day before the bees start to build queen cells in response to the absence of 9-ketodecenoic acid (**25**). If radioactive 9-ketodecenoic acid is fed to worker bees, more than 95% of the starting material is metabolized in 24 hours. The first steps in the inactivation of the pheromone occur in the ventriculus by either of two pathways. The double bond may be reduced to form 9-ketodecanoic acid (**26**) and then the ketone is reduced to form 9-hydroxydecanoic acid (**27**). Alternatively, the ketone may be reduced first to form 9-hydroxydecenoic acid (**28**) with the reduction of

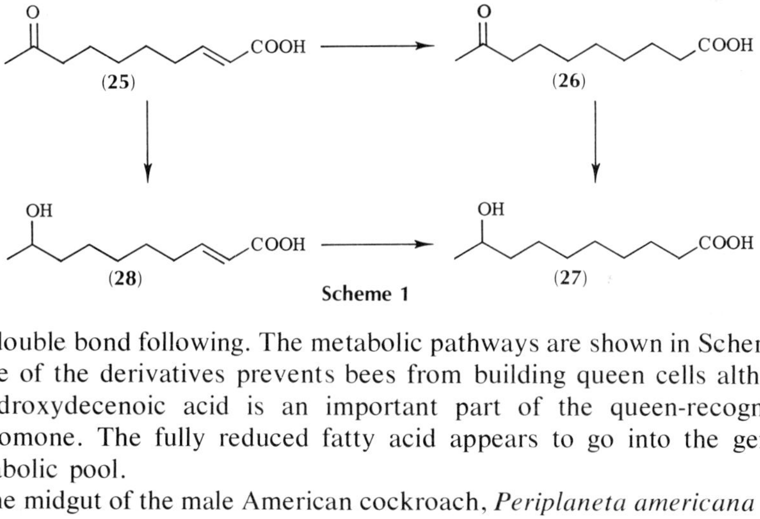

O COOH (**25**) → O COOH (**26**)

OH COOH (**28**) → OH COOH (**27**)

Scheme 1

the double bond following. The metabolic pathways are shown in Scheme 1. None of the derivatives prevents bees from building queen cells although 9-hydroxydecenoic acid is an important part of the queen-recognition pheromone. The fully reduced fatty acid appears to go into the general metabolic pool.

The midgut of the male American cockroach, *Periplaneta americana* (L.), enzymatically inactivates the sex pheromone of the virgin female. The midgut of the virgin female does not destroy the pheromone, but the mated female and last instar nymphs of both sexes do. Since cockroaches frequently clean the antennae with the mouth, the metabolism probably keeps the antennae from becoming habituated. The destruction of the pheromone by all forms except those that use it as a signal is probably a general insurance against the acquisition of the pheromone.

The pheromone is petroleum ether soluble, but when it is extracted from the virgin female a pheromone that masks it is extracted with it; nothing is known about the release of the masking pheromone.

V. BIOSYNTHESIS

Surprisingly little work has been done on the biosynthesis of pheromones. This is probably because many of the pheromones are very common compounds that occur widely in animal and plant tissues, and the most common metabolic pathways have been established in non-pheromone-producing tissue. There are some very complex and unusual pheromone components, of course, but insect tissue is hardly ideal for establishing the details of metabolic pathways.

We have already seen that in some cases chemicals from plants are used unchanged in such a way that we can properly call them insect pheromones or pheromone components. These will be discussed in Chapter 9. In other cases food materials may undergo only extremely minor modifications in the insect before they are released as pheromones. The direct utilization of food substances, or minor alterations of such substances, make a great deal of sense for the insect, since it saves the considerable amount of energy which would be required for the biosynthesis of complex molecules. Furthermore, sex recruitment pheromones from the food plant bring males and females together at the host.

Enough evidence has now accumulated on the production of bark beetle pheromone from host tree terpene to make it seem likely that all pheromone components of both *Dendroctonus* and *Ips* are derived from the host tree unaltered or with only slight modification. *Ips paraconfusus* Lanier converts the host α-pinene (2,6,6-trimethylbicyclo[3.1.1.]-hept-2-ene) supplied as a vapor to verbenol (*trans*-2,6,6-trimethylbicyclo[3.1.1.]hept-2-ene). Furthermore, the conversion is stereospecific; (−)-α-pinene is converted to (+)-*cis*-verbenol and then to (−)-myrtenol while (+)-α-pinene is converted to (+)-*trans*-verbenol and then to (+)-myrtenol according to Scheme 2.

The male waxmoth, *Galleria mellonella* L., produces a sex recruitment pheromone which attracts females. It is composed of *n*-nonanal and *n*-undecanal. Nonanal is synthesized slowly from acetate and propionate, indicating that these are not the usual precursors; much more rapid synthesis takes place when oleic acid is used as a precursor. With [$1-{}^{14}C$]oleic acid there is little incorporation of radioactivity, but with $10-{}^{14}C$, incorporation is fairly rapid. One possibility is that the oleic acid is cleaved at the 9 double bond to form nonanoic acid from the terminal end and that the acid is then reduced to aldehyde. Another possibility is that the oleic acid undergoes α- and β-oxidation to nonanoic acid which is then reduced. Incorporation of all those putative precursors into undecanal is very slow and more work is needed to establish the usual biosynthetic pathway.

The terpenes citronelal (**5**) and citral [mixtures of geranial (**6**) and its geometric isomer, neral (**7**) are called citral] can be synthesized by *Acanthomyops* from acetate or mevalonate. In the most common pathway of

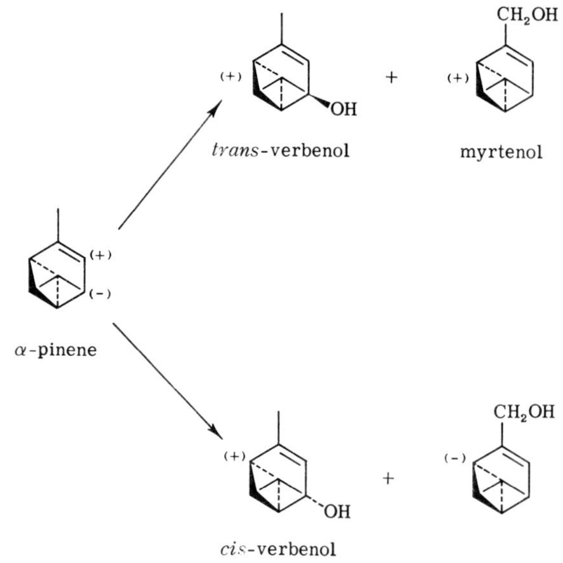

(+) *trans*-verbenol + (+) myrtenol (CH_2OH)

α-pinene (+) (−)

(+) *cis*-verbenol + (−) (CH_2OH)

Scheme 2

biosynthesis of terpenes, mevalonate is phosphorylated and the 1-carboxyl group is lost as carbon dioxide, thus forming Δ^3-isopentenyl pyrophosphate which is then used to form the terpene. Since there is no incorporation of radioactivity into citral and citronellal if [1 − ^{14}C]mevalonate is used as a precursor in the ants, it is probable that these insects follow the common biosynthetic pathway in making terpene.

No attempt has been made to determine the rate of formation of bombykol (**2**) from simple precursors, but it is readily formed from palmitic acid in late pupae of the silkworm. Furthermore an enzyme preparation from the tip of the abdomen of late pupae will stereospecifically dehydrogenate hexadecanyl pyrophosphate and form bombykol (hexadecadienol) from it.

Similarly, the complete biosynthesis of the gypsy moth sex recruitment pheromone, *cis*-7,8-epoxide-2-methyloctadecane, has not been attempted. It has been determined, though, that the moth can form the epoxide from a *cis* double bond in the 7-8 position.

The male boll weevil, *Anthonomus grandis* (Boheman), produces four terpenoid aggregation and sex attractant components, including about 40% of (+)-*cis*-2-isopropenyl-1-methylcyclobutane ethanol (**29**). The difficulty of

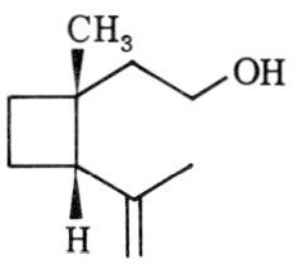

CH_3 OH H

(**29**)

synthesizing a stereospecific cyclobutane has made this compound a favorite of organic chemists who wish to test novel methods of synthesis. It has also been the subject of a study of biosynthesis. This and the other components of the pheromone are synthesized by the male boll weevil from acetate, mevalonate, or glucose, indicating that the early stages of biosynthesis, at any rate, are probably typical of terpenoids in general. The synthetic pathway leading to this unusual compound will be very difficult to determine. None of the radioactive precursors resulted in much label in the final products, and the male boll weevil produces only about 1.3 μg of total pheromone per day or about 40 μg during his lifetime.

There are many problems beside biosynthetic pathways, however, that cry out for solution. The cells of many of the glands that release pheromone have the appearance of biosynthetically active cells, but in no case do we know unequivocally the site of biosynthesis of pheromone. The sting apparatus and mandibular glands of ants and bees store rather large quantities of pheromones, but we do not know when, where, or how fast synthesis occurs. In several moths there is fairly strong circumstantial evidence that sex recruitment pheromone is synthesized as it is released, but more direct evidence is needed to distinguish simple unmasking from complete synthesis. We do not know to what extent the synthesis and release of some pheromone components are controlled independently of the synthesis and release of others. We do not know how or when synthesis and release of masking pheromones is controlled or how much pheromones affect the synthesis and release of other pheromones. Some of these problems should be soluble with the aid of radioactive precursors of pheromones.

We have a fairly long catalogue of compounds that have pheromonal activity. What is needed now is the biochemical, physiological, behavioral, and ecological work that is necessary to put chemical communication and the problem associated with it into their proper perspective.

VI. LITERATURE

A recent book edited by Birch (1974) and one by Shorey (1976) deal extensively with most subjects covered in this chapter except pheromones as information transfer systems, and that is covered by Wilson (1970) or Chapters 14, 15, and 16 of Wilson (1972). Law and Regnier (1971) have written an excellent brief review of the chemistry of pheromones and Mayer and McLaughlin (1975) have assembled a compendium of all the sex pheromones known up to the time of publication.

GENERAL REFERENCES

Birch, M. C., ed. (1974). "Pheromones," 495 pp. Am. Elsevier, New York.
Law, J. H., and Regnier, F. E. (1971). *Annu. Rev. Biochem.* **40,** 533–548.
Mayer, M. W., and McLaughlin, J. R. (1975). Fl. Agric. Exp. Stn., Monogr. **6.**
Shorey, H. H. (1976). "Animal Communication by Pheromones," 169 pp. Academic Press, New York.
Wilson, E. O. (1970). *In* "Chemical Ecology" (E. Sondheimer and J. B. Simeone, eds.), pp. 133–155. Academic Press, New York.
Wilson, E. O. (1972). "The Insect Societies," 548 pp. Harvard Univ. Press, Cambridge, Massachusetts.

REFERENCES FOR ADVANCED STUDENTS AND RESEARCH SCIENTISTS

Johnston, N. C., Law, J. H., and Weaver, N. (1965). *Biochemistry* **4,** 1615–1621.
Renwick, J. A. A., Hughes, P. R., and Krull, I. S. (1976). *Science* **191,** 199–201.

9

Chemical Control of Behavior—Interspecific

NEVIN WEAVER

Chemicals that are produced by individuals of one species and that affect the physiology or behavior of individuals of another species are called allomones, or sometimes allomonics or allelochemics. The term *kairomone* is often used for allomonics that cause reactions favorable to the receiving organism, and *allomone* is restricted to chemicals causing reactions favorable to the emitting species. The term *synomone* has recently been proposed for chemicals that are favorable to both the emitter and receiver, and apneumone for chemicals from nonliving materials that are favorable to one species but detrimental to another living in or on them. Obviously the invention of new terms can continue and it probably will. It will also continue to spread confusion in its wake. The harm or benefit from chemical messages is often ambiguous at best and may vary with circumstances. We need careful analyses of the biological functions of chemicals far more than we need tortuous attempts to fit function to a preconceived term. Those who decide to struggle along with the single term, allomone, need have no fear that they are depriving science of any blinding insights.

Chemicals from an organism that affect other species purely in a nutritional role are not considered to be allomones, but if a nutrient is used in communication then it is an allomone in that role, regardless of the other roles it may play.

I. COMMUNICATION FROM NONINSECTS

The insect must find food, and quite often is must also find a home. Since the larvae and adults usually occupy quite different niches and have different requirements for both food and living space, the search of the adults can be quite complicated. There is the further fact that it is often to the advantage of an organism to escape exploitation, so the final choice of the insect depends on a complex of information on attractants and defenses.

Many of the clues that the insect uses are chemical ones, particularly in the close-range selection of food and oviposition sites, and the selection always involves several steps. At the minimum the insect must arrest movement, it must sample the environs, and it must proceed to feed or oviposit. These steps usually are very difficult to separate from each other during observations or experiments, and most studies have only revealed something of the factors affecting the end result: feeding, oviposition, or rejection.

A. Phagostimulants of Plants

Most herbivorous insects feed on the members of only a single plant family or genus, and some of them are restricted to a single species. There is an extensive literature on host plant selection and a great deal has been learned about insect preferences, but, as we shall see, we understand little about the allomones involved.

An impressive array of nutrients or potential nutrients, alone or usually in combination with other chemicals, has been shown to be effective in promoting feeding. These include many sugars, many amino acids and related compounds, albumin, β-sitosterol, phospholipids, hydrocarbons, fatty acids, ascorbic acid, thiamine, α-tocopherol, oxaloacetic acid, citric acid, and some complex mixtures. Of all the chemicals that have been shown to be important to the promotion of feeding, sucrose is the most effective in by far the widest variety of insects. Insects also react to a large number of secondary substances—chemicals of plants that have no apparent metabolic function and that probably evolved as protective chemicals against herbivores, parasites, saprophytes, or competitors, or that evolved to promote interactions between individuals or species.

It was early observed that the number of plant species eaten by a caterpillar would be increased if the maxillae were amputated. This led to an hypothesis that monophagy is based largely on the repellency of nonfood plants. There is now, though, firm electrophysiological evidence that the selection of food is based on the interaction of many factors. Each of the two sensilla styloconica of the maxillae of caterpillars has four receptors, each of which is sensitive to one class of substances. For instance, it was reported that one sensilla styloconica of the domesticated silkworm (*Bombyx mori* L.) larvae, contains a receptor for salt, a receptor for glucose, one for sucrose, and one for inositol. The other contains two receptors for salt, one for water, and one for what were thought to be repellent substances. Probably this early electrophysiological work too narrowly specified the chemicals to which the taste receptors respond. When chemicals found in mulberry leaves were tested on caterpillars, many of them were inactive, but 2-hexenol, 2-hexenal, citral, linalyl acetate, linalool, and terpenyl acetate attracted larvae, β-sitosterol, β-sitosterol glucoside, isoquercitrin, and morin caused them to bite, and cellulose, inositol, sucrose, silicate, and phosphate induced them to swallow; the sense organs responsible for most of these responses were not determined. The caterpillar has about forty olfactory sensillae, some of which are specialist and some of which are generalist cells (see Chapter 8). It is now known that specialist sensillae fire when exposed to 2-hexenol, and that hexenol is the main attractant from mulberry leaves. It attracts caterpillars from 3 to 4 cm away.

Tobacco hornworm, *Manduca sexta* (Johan), larvae feed on plants in the family Solanaceae. First instar larvae are polyphagous and will feed on many kinds of nonhost plants, though they are unable to grow on them. They will remain polyphagous throughout larval life if they are reared on synthetic diet, but first instar larvae will become monophagous if they are allowed to feed on a host plant; this process is known as induction. The induced larvae usually reject nonhost plants, and if they are given a choice they prefer plants of the species to which they were induced over other normal host species.

The role of olfaction in the selection of food by larvae of the tobacco hornworm has been extensively studied. The earliest experiments failed to indicate any sense of smell at all. Electrophysiological experiments subsequently showed that the hornworm gives an olfactory response to a wide variety of chemicals, but behavioral observations failed to demonstrate attraction or repulsion to any chemicals. Finally it was found that when the larvae were given a choice between two closely related host plants they made the choice on the basis of olfactory clues.

The larvae also use gustatory senses in the selection of plants on which to feed. After a larva has been induced to a given host plant, the lateral gustatory chemoreceptors show decreased sensitivity to the repellent chemicals of the inducer plant. In this restricted sense the receptors, and not the central nervous system, remember and choose the inducer plant. However, some host plants elicit more spikes from the medial sensillum than do others, and the response does not change with feeding experience. Sufficient information is, therefore, available to the central nervous system for it to be involved in the choice of food.

Most species of aphids feed on plants from only a single genus, and their selection of food plants has been extensively studied. Winged forms arise in response to overcrowding. Physical contact with other aphids or other objects is a sufficient stimulus for wing production, but in some species at any rate, winged forms also develop in response to nutritional factors. For instance, if isolated *Acyrothosiphon pisum* Harris are reared on bean seedlings they will produce only apterous offspring, but if they are placed on mature bean leaves, about half of them will begin to produce alate offspring in 24 hours. If excised mature leaves are allowed to take up a mixture of amino acids and glucose, alate production is reduced but not eliminated. This indicates that alate production is probably promoted by both secondary substances and poorer nutrition from old leaves.

Winged aphids typically fly in the morning from the plants on which they were reared and are often carried to great heights by the wind. As air turbulence decreases in the evening the aphids descend onto host and nonhost plants at random. They are able to direct their flights for only the last few centimeters of the journey. The aphids usually probe the outer cells of the plant several times, apparently receiving taste stimuli, although sap from the plant that is thus sampled does not arrive at the gastrointestinal tract of the aphids. If the aphid is on a nonhost plant it usually will fly again soon and, even when it has alighted on the proper host species, it often takes flight. The preferred level of various nutrients, such as certain amino acids and sugars, will encourage aphids to stay and feed. Some of the amino acids that promote feeding are essential for aphid growth, but others are not. In addition, the secondary substances of plants are important. For instance, aphids that feed on members of the mustard family (Cruciferae) select for the mustard oil glycosides produced by the crucifers, but the level of both amino acids and glycosides are important to the selection of food.

The leaf-cutting ants carry parts of leaves into the nest, cut them into tiny pieces, partially macerate them, and raise fungus on them. Any species of these ants forages on a wide variety of plants, but there is still a considerable amount of selection; for instance, some species forage largely on grasses while others usually do not collect grasses at all. If given a choice the ants cut pieces of leaves at the site of previous cuts, indicating that a whole leaf can mask its attractiveness. The ants apparently are not attracted even to cut leaves from a distance, but have to stumble upon them. At the leaf they show preference for high levels of common constituents such as carbohydrates, amino acids, and glycerides, and they are repelled by such chemicals as the oils from citrus rinds. Although the major part of the ants' nutritional needs is supplied by the fungus and not the leaves, the choices are more complex than chemical tests have been able to demonstrate. If the ants are restricted to collecting leaves from a single species for 10 days and then are given a choice, they will overwhelmingly prefer to forage from a different species of plant.

The European elm bark beetle, *Scolytus multistriatus* (Marsham), is strongly attracted to feed on pith disks treated with *p*-hydroxybenzaldehyde which is known to be an oxidative degradation product of hardwood lignins. A number of modifications of the stimulatory compound have been tested. Molecules with 1,4 substitutions on the benzene ring are most active, with 1,3 and 1,2 substitutions decreasing in activity. Methoxy substitutions on benzaldehydes increase short-range attractiveness but reduce feeding. The substitution of a methoxyl for a hydroxyl group of 1,2 disubstituted phenols results in an unattractive compound. The substitution of a methoxy next to a 4-hydroxyl of benzaldehydes reduces attraction.

In an attempt to determine the identity of the naturally occurring phagostimulants for *S. multistriatus*, the constituents of elm bark were subjected to extensive extraction and fractionation. A benzene extract of elm bark stimulates feeding, and one fraction of the extract obtained by silicic acid chromatography is a powerful phagostimulant. This fraction contains phenolic compounds. The benzene extract of seven nonhost tree barks do not stimulate feeding and, when mixed with elm extracts, they deter feeding. Two species of nonhost bark, however, contain short-range attractants or arrestants, and beetles will feed on sucrose-soaked pith discs under discs treated with the benzene extracts, but not on the treated discs themselves. Hickory is one of the nonhosts that has been tested. Its bark contains 5-hydroxy-1, 4-naphthoquinone which deters feeding by *S. multistriatus,* but neither stimulates nor deters the closely related hickory bark beetle, *S. quadrispinosus* Say. Even insects that feed on the same tissues of the same species are not necessarily attracted by the same chemicals. The American elm bark beetle, *Hylurgopinus rufipes* (Eichoff), feeds on some of the same species as the European elm bark beetle but is not stimulated by the benzene extract of elm bark nor by *p*-hydroxybenzaldehyde. It is, instead, stimulated to feed by the 80% ethanol extract of elm bark.

The repellents of nonhost plants deserve careful study, particularly since man may be able to manipulate some of them to his own advantage. Larvae of the wheat bulb fly, *Leptohylemyia coarctata* (Fall.), will feed on most cereals, but not on oats. Wheat contains a phenolic compound that serves as an arrestant. Larvae that encounter the chemical remain in the area and thus are likely to feed. Oats contains a polyhydroxylated aliphatic substance which causes the larvae to move and will decrease the attacks of larvae on wheat grown in the presence of the antiarrestant.

Arrest alone is not, of course, sufficient to cause feeding. The vegetable weevil, *Listroderes costirostris obliquus,* is arrested by coumarin from sweet clover and feeds on the plant. *Sitona cylindricoles* (Fahraeus), on the other hand, is arrested by coumarin but does not feed on sweet clover.

The vegetable weevil is, incidentally, exceptionally polyphagous, having been recorded as feeding on 178 species or varieties in 34 families. Both larvae and adults are attracted to many of the mustard oils that occur in various host crucifers and to several of the volatile substances from umbelliferous plants. It also feeds on many composites and plants from other families whose attractive chemicals are not known.

In some cases there are complex relationships between allomones and pheromones. Females of the native silkworm moth, *Antheraea polyphemus,* will not release sex pheromone until it has been exposed to *trans*-2-hexenal. This chemical occurs in the leaves of red oak, a favored food for the larva.

The brown-rot fungus, *Gloephyllum trabeum,* produces *cis*-3, *cis*-6, *trans*-8-dodecatrien-1-ol which is highly attractive to the termite, *Reticulitermes virginicus* Kollar. The fungus grows in wood that is suitable food for the termite and thus attracts the termite to the food. The dodecenol also serves as a trail pheromone of the termite, but whether the pheromone released by the termite is derived from the fungus or is synthesized by the insect is not known.

There has been an evolutionary race between plants and phytophagous insects. Plants have evolved high concentrations of the secondary substances that are repellent or toxic to herbivorous animals and some animals have evolved to tolerate and even to be attracted by the protective chemicals. Members of the mustard family (Cruciferae) produce several mustard oil glycosides. Cellular damage causes the enzymatic hydrolysis of the glycosides to yield mustard oils. For instance, the very common glycoside, sinigrin (**1**) yields the mustard oil allyl isothiocyanate (**2**). These oils are repellent to many insects and are toxic to most insects that can be forced to feed on them, but they are attractive to both ovipositing and feeding insects of species that utilize members of the mustard family as their host plants. This can easily be shown by allowing cuttings of various noncruciferous plants to take up mustard oils from solution and watching the reactions of insects that normally feed on the plants and the reactions of insects that normally feed on crucifers.

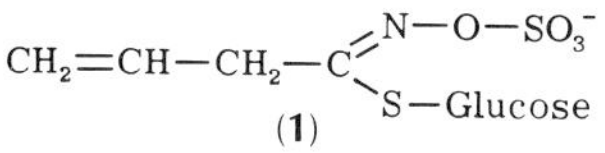

$$CH_2{=}CH{-}CH_2{-}C\begin{matrix}\nearrow N{-}O{-}SO_3^- \\ \searrow S{-}Glucose\end{matrix} \quad (\mathbf{1})$$

$$CH_2{=}CH{-}CH_2{-}NCS \quad (\mathbf{2})$$

Larvae of the cabbage butterfly, *Pieris brassica,* feed on a wide variety of crucifers, but a larva is easily inducible and prefers to feed on the species on which it has previously fed. There is even evidence that the females of some insect species prefer to oviposit on plants of the species on which they fed as larvae. As protection against being tracked too closely by the evolution of its feeders, the crucifers are quite heterozygous for the production of the various mustard oils, and in any natural population some plants will be less suitable hosts than others. It is this natural heterozygosity enforced by phytophagous insects that the plant breeder depends upon to breed insect resistance into agricultural crops.

One of the most interesting examples of apparent evolutionary tracking occurs in the danaid butterflies. The females oviposit on members of the milkweed (Asclepiadaceae) and dogbane (Apocynaceae) families. The plants produce vertebrate heart poisons (cardenolides) which are sequestered in the tissues of larvae feeding on them and render both the larvae and resulting adults toxic to vertebrate predators (see Chapter 11). The adult males seek out plants of the Boraginaceae and Asterales that produce pyrrolizidine alkaloids. The males are attracted to the plants by dissociated acids that were once fused to the ring nucleus of the alkaloids, and they are stimulated to feed by the nucleus or derivatives of it. The nuclei are used after only slight modification as hairpencil pheromones to arrest females so mating can occur (see Chapter 8), and some of the branched-chain acids are lactonized and used as both pheromones and allomones to repel other males, including males of other species. Since the four families of plants are members of the same subclass, it has been proposed that the ancestral plant produced both cardenolides and pyrrolizidenes, and that the two lines subsequently diverged with each line retaining one of the protective poisons. The butterflies feeding on the ancestral plants evolved to track one line in the adult stage and the other in the larval stage. Since the pyrrolizidines occur in several plant families and the ring nucleus is synthesized from the ubiquitous precursor, ornithine, the idea of divergent evolutionary tracking by adults and larvae must remain only an interesting hypothesis.

There are several examples of the evolutionary tracking of plants which produce powerful toxins. Plants of the genus *Hypericum* produce the crimson-colored dianthrone derivative, hypericin. This chemical causes intense photosensitivity and skin irritation in vertebrates and sometimes leads to blindness and death. It also protects the plants against most insects. Several beetles of the genus *Chrysolina*, however, not only utilize *Hypericum* as food, but search over the leaf until hypericin stimulates tarsal receptors which in turn trigger feeding.

A substance that is not required nutritionally but that causes animals to feed is a token stimulant in that it is an indicator of an acceptable host. Even a required nutrient may stimulate in a token role, and the attractive and repellent secondary substances are token stimuli. In spite of the extensive use of token stimuli, it is nevertheless true that some insects can recognize nutritionally superior foods. If larvae of the fly, *Agria affinis* Fallen, are given a choice of four synthetic foods which differ in the balance between an amino acid mixture and glucose, they show the same order of preference for the foods as the order in which the foods are ranked in promoting larval growth and development.

B. Insecticides of Plants

Although some insects have evolved to use insecticides of plants as attractants or phagostimulants,these are exceptional cases, and in the evolutionary race between plants and herbivores, most insects are either repelled or they are simply not deterred from feeding. In the latter case they must develop detoxification enzymes or other mechanisms to overcome the defense of their host plants. The insect would be expected to save the energy required for the synthesis of a wide variety of detoxification enzymes by restricting its feeding to a very few species and losing the ability to detoxify compounds not found in its host plants. There is a great deal of evidence that this has, indeed, occurred.

The microsomal epoxidases of the midgut of insects attack a variety of lipoidal chemicals and detoxify both natural and man-made insecticides. Monophagous lepidopterous larvae have significantly lower epoxidase activity than oliphagous (feeding on two to ten plant families) larvae, which in turn have lower activity than polyphagous larvae. In the southern armyworm, *Spodoptera eridania*, these mixed-function epoxidases are readily and rapidly induced by a wide variety of secondary plant substances, and following induction, a larva becomes less susceptible to dietary poisoning.

The pyrethrins are among the natural insecticides that are attacked by the mixed-function oxidases of some insects, and chrysanthemum flowers produce not only pyrethrins but also sesamin, a lignin which inhibits the appropriate oxidases. The plant thus produces both an insecticide and a specific biochemical synergist for it.

Potato and tomato plants that have been injured by the Colorado potato beetle or by mechanical means rapidly produce a powerful protein inhibitor of the intestinal proteases of animals. The factor that stimulates the production of the inhibitor is transported from the site of the injury and within a few hours adajacent leaves on the plant begin production of the antiprotease.

Sometimes defense against digestion is even simpler. For instance, old oak leaves are of little nutritive value to herbivores because tannins bind the proteins into indigestible complexes which serve as growth inhibitors for moth larvae.

Many plants produce resins, gums, or emulsions that deter insect feeding by physical means and sometimes by being toxic or repellent. Other chemicals repel invaders solely by being toxic. Some termites and other wood-boring insects feed on the wood of living trees, and many trees have developed insecticides that protect them from the boring and tunneling insects. The termiticides range from relatively simple volatile compounds that may be repellents and/or contact insecticides to large crystalline compounds which are deposited in the wood and which act as stomach poisons. The insecticides may persist as long as the wood lasts. The volatile termiticides are monoterpenoid compounds, and the crystalline stomach poisons are di- and triterpenoids. The wide variety of termiticides and the large number of tree species that produce them will doubtless continue to test the ingenuity of phytochemists for years to come.

The amino acid L-canavanine makes up 5–10% of the dry weight of leguminous seeds of species of *Canavalia* and *Dioclea* in Central America, and these seeds are remarkably free of predation by most insects and other animals. The L-canavanine is an analogue of L-arginine, and many insects will incorporate it into their proteins which are consequently nonfunctional. This unusual amino acid is thus a powerful insecticide for many insects. Larvae of the bruchid beetle, *Caryedes brasiliensis* Thunberg, however, subsist solely on the seeds on *Dioclea megacarpa* Rolfe which contain more than 8% L-canavanine. These beetles are able to metabolize some of the unusual amino acid and may have other detoxification mechanisms, but their main defense seems to be the ability of their arginyl-tRNA synthetase to distinguish between arginine and canavanine, and to reject the latter.

Seeds of the legume genus *Mucuna* of the tropics are also remarkably free of predation. They contain 6–9% free L-dopa, a phenylalanine analogue. The widely phytophagous southern armyworm, *Prodenia eridania* Cramer, has the ability to detoxify many toxic chemicals. When L-dopa was fed to armyworms in 0.25% concentration abnormal pupae resulted; in 5% concentration dopa was repellent and the armyworms refused food containing it.

Attractants and feeding stimulants are not confined to species that might serve as food. For instance, when the Colorado potato beetle was given a choice between potato, its natural host, and *Solanum nigrum* on which to oviposit, it laid more egg masses on the latter than on the former. *Solanum nigrum* was, in fact, the most attractive plant tested with gravid females, and it was also the most toxic to the larvae. Another example of a highly attractive lethal plant occurs among the *Passiflora*. *Passiflora serratifolia* is a super oviposition stimulus to *Heliconius* butterflies, but its young leaves contain a potent toxin that kills all young larvae.

The term, plant defense guild, has been proposed for the association of two or more species of plants that act together to reduce herbivore pressure or to reduce competition (Atsatt and O'Dowd, 1976). Sometimes mixed cultures of plants maintain populations of the parasites and predators throughout the growing season. Sometimes the chemicals of plants are directly involved, and four modes of action of allomones are considered in this chapter. Although a number of such guilds are known, they have not been used in agriculture nearly as extensively as they might be.

Masking odors (see Chapter 8) can serve in plant defense guilds. There are cases known in which plants grown in mixed culture do not attract as many herbivorous insects as plants grown in monoculture and, in at least some cases, the effect has been shown to be olfactory rather than due to other types of interference.

Masking odors from plants can affect parasites as well as herbivores. The percentage of larch sawflies (*Bessa harvey*) parasitized by *Pristiphora erichsonii* sometimes varies in different parts of the larch trees, and the variation depends on the position and proximity of other species of trees that produce masking chemicals that act on the sense of smell of the adult parasitoid.

C. Insect Hormones and Antihormones

The discovery of insect hormones in plants came as a surprise, although in retrospect it seems inevitable that they would be there. The best known insect hormones are ecdysone, which promotes molting and differentiation, and juvenile hormone, which suppresses metamorphosis and promotes fecundity of adult females (see Chapter 7).

Quite by accident, it was discovered that the European bug, *Pyrrhocoris apterus,* would not metamorphose when reared in America in the laboratory. The factor responsible turned out to be a compound with juvenile hormone activity for the bug. The hormone was derived from trees of the balsam fir and reached the bugs by way of paper toweling made from fir and other woods and that served as liners in the petri dishes in which the immature bugs were reared.

There are several compounds that have juvenile hormone activity for some species of insect, and many of them occur in plants where they act as defensive chemicals.

There are also several chemicals that have ecdysone activity, and chemicals with such activity have been found in a wide variety of plants, and particularly among the ferns and gymnosperms. Some insects refuse to feed on material containing large amounts of ecdysone. It may well be that the occurrence of insect hormones in plants has supplied the selective pressure necessary for the diversification of compounds that have hormone activity for insects, and that as a defense against plant defenses the insects have evolved different sensitivity to analogous compounds.

Recently a systematic search was initiated for anti-insect hormones in plants, and two such compounds were quickly discovered from *Ageratum houstonianum* (Bowers *et al.*, 1976). Precocene I, or 7-methoxy-2, 2-dimethylchromene and precocene II, or 6,7-dimethoxy-2,2-dimethyl-chromene are powerful juvenile hormone inhibitors and cause precocious metamorphosis. Further systematic search will undoubtedly reveal many additional antihormones from plants and many additional compounds with insecticidal activity.

D. Taste and Other Distortions

Man and some other mammals are unable to perceive sweetness after eating gymnemic acid which can be obtained by chewing the leaves of the asclepiad *Gymnema sylvestre*. Miraculin which is present in the berries of the African shrub, *Synsepalum dulcificum*, causes substances that would ordinarily seem sour to taste sweet instead. Gymnemic acid is a feeding deterrent to the caterpillar, *Prodenia eridania*, even on a sugar-free diet, which suggests that it does not have its effect by distorting the taste of sugar. It will be surprising if there are not taste and other sensory distorters in plants that are effective on insects.

Distortions need not be restricted to sensory modalities, of course. In this chapter we will see examples of disorienting chemicals from insects and orchids. The insect equivalents of hallucinogens and psychotomimetic agents probably occur in the plant kingdom as protective chemicals against insect attack. Although plants do not protect themselves against bees, of course, there are some suggestive observations on honeybee behavior. Bees become disoriented when exposed to sublethal doses of some of the alkaloids in toxic nectar, and they become drunk when they imbibe ethyl alcohol from fermented fruits. Drunk bees will stagger about as they try unsuccessfully to give clear directions to the source of food by means of the bee dance.

E. Attraction of Beneficial Insects

Plants produce a number of substances that seem to have evolved as attractants to insects that benefit the plants. The most common substance is nectar, a solution of sugars and slight amounts of other substances. Floral nectaries undoubtedly evolved primarily to bring pollinating insects to the flowers in such a way that pollination would be effected. The odors of flowers aid insects in locating flowers of the same species repeatedly and thus promote efficient cross-pollination. Furthermore, when a honey bee that has discovered a rich source of forage returns to the hive and dances to give the distance and direction to the discovery, the odor of the flower clings to its body and aids the other bees in finding the same source of forage.

Most of the flower odors probably are not innately attractive to honey bees and other pollinators that gather nectar from flowers, though bees can be trained to associate food rewards with flower odors but not with fetid odors. Pollen, however, contains rather large amounts of free octadec-*trans*-2, *cis*-9, *cis*-12-trienoic acid and other free fatty acids which are attractants to honey bees and cause them to gather pollen. If the free fatty acids of pollen are extracted and deposited on cellulose powder, bees will gather the powder in preference to the residue of pollen from which the lipids have been extracted. Free fatty acids (probably the same ones) are phagostimulants that cause honey bees to eat pollen in the hive. It is not known whether these fatty acids are also attractants to other species of pollinating insects, nor is it known how varied the attractant fatty acids may be among the pollens of different plant families.

Most orchid species are pollinated by only one species of solitary bee, and the flowers of many orchids produce odors that are innately attractive to the bees that pollinate them. The males of the euglossine bees, commonly called golden bees because of their brilliant colors, visit flowers at which they obtain no nectar, pollen, or other apparent food reward. They brush certain surfaces of the flowers with special pads on their front feet. The pads pick up essential oil fragrances from the flowers, and the bees squeeze the perfume into cavities in their hind legs. Each species of bee collects fragrances from only a few flower species. When a bee has collected a wide spectrum of fragrances (up to fourteen compounds can be distinguished by gas–liquid chromatography), he becomes highly attractive to other males of the same species, and a small swarm of males will buzz about the fragrant bee in some opening in the forest. The females appear to be attracted by the sight of the brightly colored bees dancing about, and they approach the swarms where mating occurs. The fragrances seem to perform no other function for the bee except to promote the formation of the swarms (or leks) and through them promote mating, but for the orchids, pollination results from the activities of the males in gathering the fragrances. Some of the orchid blossoms have rather bizarre forms that insure that a bee of the right size will effect pollination as he gathers odorous substances, but flower form alone is not a sufficient isolating mechanism, and here the fragrances are important. When fragrant compounds that occur in orchid blossoms are tested alone they attract males of many euglossine species, but as the compounds are mixed with each other, fewer species are attracted. For instance, in a 5-day test, 1,8-cineole attracted 433 males euglossine bees representing 32 species, and benzyl acetate attracted 36 bees in 6 species. When these compounds were mixed in the same proportion in which they occur in the fragrance of *Stanhopea triconis* Lidl., only 49 males of 8 species were attracted. Finally the natural odor of the orchid was approximated by adding the appropriate amount of α-pinene to the mixture, and then only the pollinator and one other species was attracted. The other bee was too small to effect pollination. Thus the fragrance spectrum of the orchid is the major isolating mechanism between species.

Another of the bizarre pollination systems of orchids depends upon the sexual deception of males. Portions of the flower are said to resemble the females of one species of bee, wasp, or fly, and when males attempt to copulate with the image of the female they effect pollination. To the human eye the resemblance of most of the flowers to females is not overwhelmingly close and some of these flowers are known to attract males by their odor. With modern techniques it should be possible to identify the attractive chemicals and to determine whether they resemble the sex pheromones of the appropriate females.

There are probably many odors that are intrinsically attractive to some pollinator. Orchids that depend on carrion flies for pollination have a fetid odor to the human nose.

In one case chemicals from an orchid appear to disorient the bee. Male *Halictus languinosus* will cling motionless for hours to the flowers of *Diuris pedunculata*. If the bee is removed from the flower he shows no sign of having been harmed in any way.

Many plants, especially tropical and subtropical ones, have extrafloral nectaries whose function was, until recently, in dispute. For instance, the trumpet creeper *Campsis radicans* L. has five distinct nectary systems. The ovarian or floral system secretes glucose and fructose and attracts hummingbirds and bumblebees which are the pollinators. The extrafloral nectaries occur on the adaxial surface of the petiole, the adaxial surfaces of the calyx and corolla, and on the fruit. They secrete sucrose, glucose, and fructose in ratios ranging from 1:1:1 to 3:2:1, and attract ants that feed on the nectar. Although some species of ants that are found on the trumpet creeper may afford little protection against the enemies of the plant, a *Formica* species of the *fusca* group defends the plants vigorously against intruders.

Anyone who has observed the aggressive behavior of ants on plants with extrafloral nectaries (or has encountered them) will have trouble believing that any other defense against herbivores is necessary. The adults of many predacious and parasitic wasps, however, feed at the nectaries, and they are also important to the protection of the plants. For instance, most *Passiflora* species have extensive systems of extrafloral nectaries, and in one study conducted in Trinidad it was found that over 90% of the eggs of *Heliconius* butterflies on *Passiflora* were killed by parasitoids.

Protection is not restricted to the plants that bear extrafloral or floral nectaries. Eighteen times as many tent caterpillar pupae may be parasitized in trees growing near nectar-producing plants as in trees lacking associated nectar producers. Tachinid and ichneumonid parasites are much more numerous in cabbage fields grown near flowering umbelliferous plants than in cabbage in monoculture. Again, such plant defense guilds should have a prominent place in agriculture.

An example of extreme mutualism between insects and plants occurs in the relationship of the bullhorn acacia of the new world tropics and ants of the genus *Pseudomyrmex*. The acacia has been said to maintain a standing army of the ants. The ants excavate the swollen thorns of the acacia and establish nests in them. They feed on the products of extrafloral nectaries and on Beltian bodies that occur at the tip of each leaflet. The Beltian bodies have no other known or suspected function except to serve as food for the ants. The ants are very aggressive and if a man or cow comes near an acacia tree the ants will rush toward the source of the odor in a threatening manner, and if a man touches the tree he receives multiple painful stings. At any time of the day or night, up to one-fourth of the ants living in a tree actively patrol its surface. If all the ants are killed or otherwise removed from a tree, it is immediately attacked by a wide variety of biting and sucking phytophagous insects. Furthermore, surrounding plants tend to overgrow the acacia and shade it out, and after only a year the unprotected plants are almost dead. The ants not only protect the plants against insect and vertebrate herbivores, but they also attack sprouts that grow within 40 cm of the acacia trunks and chew and maul them until they die. The leaves and shoots of plants whose branches touch the canopy of the acacia are also attacked.

F. Mammalian Attractants and Phagostimulants

The various blood-sucking insects that feed on man and other mammals seem to use similar clues in locating hosts and in feeding on them. One of the most extensively studied blood feeder is the yellow fever mosquito, *Aedes aegypti* (L.), which can well serve as a model species. The female mosquito is attracted by L(+)-lactic acid which reaches the surface of the human body both in sweat and by leaching through the skin. There are also other attractive substances on the skin, such as certain amino acids and human sex hormones. None of these substances is very attractive, however, unless it is offered in the presence of high concentrations of carbon dioxide. The CO_2 is not itself an attractant, but only serves as an activator to make the mosquito more sensitive to the presence of attractive compounds. There is, however, no chemical or combination of chemicals that is as attractive to mosquitoes as is an intact man, and blood-sucking insects seem to recognize man as a complex of effluvia (to paraphrase Hocking's happy phrase).

Limited work on the insects that feed on other species of mammals indicates that they find their host in much the same way that they find man. Indeed, the blood-sucking insects are, in general, polyphagous. *Stomoxys calcitrans* L. is a blood-sucking fly that normally feeds on cattle. If it is kept for a time at a constant relative humidity it will probe with the mouthparts for about 30 seconds after being exposed to higher humidity, and then it becomes habituated. It will fly upwind in the presence of an increased CO_2 concentration, and will orient upwind very precisely toward the odor of human skin, but neither of these factors will lead to probing by the fly.

Most of the blood-sucking insects will not feed on blood plasma, but they will gorge if red blood cells or platelets are added to the plasma. ATP and, to a lesser extent, other nucleotides will substitute for the formed bodies, probably because the platelets and red blood cells are especially rich in ATP. Other substances that will stimulate feeding by blood-sucking insects are sucrose, glucose, lactose, several amino acids, glutathione, and some saline solutions isotonic with blood plasma.

II. COMMUNICATION BETWEEN INSECT SPECIES

The study of chemical communication between insect species is truly in its infancy. Research on the interactions between predators or parasites and their prey usually is rather straightforward, and if enough diligence is employed some chemical message can be detected and the chemicals involved determined. Studies of other interspecific chemical communication are much more difficult, except, perhaps, among the social insects. The chemicals involved in interspecific communication have seldom been isolated and shown to be effective when separated from the insect. They have even less often been identified chemically.

A. Parasitoid–Prey Relationships

There are five steps in successful parasitization that may be mediated by chemical signals. The parasite must find a habitat that may contain the host, it must find a host, it must identify the host, it must select those individual hosts that are acceptable, and it must render the host suitable for the development of larvae. The same chemical may act in more than one of these steps.

The pine which is the host of the southern pine bark beetle, *Dendroctonus frontalis* Zimmerman, produces α-pinene which attracts the beetle and becomes a part of its aggregation pheromone (see Chapter 8). The α-pinene also attracts *Heydenia unica* Cook and Davis, the pteromalid parasitoid of the beetle. The pinene is highly effective in bringing the parasitoid to a location where the host is most likely to be found, but for finding the prey itself further clues are necessary. At least six species which are parasites or predators of bark beetles have been shown to be attracted by the aggregation pheromones of the beetles, and several others seem to use the aggregation pheromone and additional clues.

It is probable that pressure from parasites and predators that have broken the chemical code of their hosts have caused insects to reduce pheromone production to the minimum that is necessary for survival. In the case of the bark beetles, the aggregation pheromones are necessary because the first invader of the tree needs help to overcome the defenses of the host, but the beetles must also minimize the calling of its enemies. As was pointed out in Chapter 8, some of the bark beetles produce methylcyclohexenol which masks their pheromones, but we do not know whether such masking agents are active on their enemies. It would be surprising if masking chemicals were not used by insects as a protective device, and there are observations which strongly suggest that they sometimes are (see Section II, F below).

Once a parasitoid has located a host plant it will often immediately explore the area near injured tissue. Parasitoids on caterpillars have often been observed to follow odor trails left by host larvae as they moved from one feeding place to another. Some of these trails are due to proteins or other high molecular weight chemicals that are detected only on contact. Even more interesting is the fact that, in searching the ground cover for hosts, many ichneumonids that are parasitic on cocoons avoid areas where they encounter their own odor trails or those of conspecifics, congenerics, and even intergeneric individuals that parasitize the same host. The ichneumonids avoid previously visited areas for up to 4.5 hours.

It has repeatedly been observed that female parasitoids quickly learn to associate various environmental clues with success in finding a prey. One of the most interesting cases involves *Bracon mellitor* Say, which is a parasitoid on the larva of the boll weevil, *Anthonomus grandis* Boheman. Experienced adult females probed the frass from laboratory-reared boll weevil larvae with the ovipositor. They also probed toweling that had methyl *p*-hydroxybenzoate (methyl parasept) on it. The methyl parasept had been used as a fungicide in the artificial diet on which the boll weevil larvae were reared, and could be extracted from the frass of weevils. It could not be extracted from the frass of boll weevil larvae in their natural habitat, nor from the cotton buds in which the larvae lived in nature. When the female *Bracon* were allowed to parasitize laboratory-reared larvae and were then caged for variable periods of time before being tested with methyl parasept, the association remained strong for about 4 hours and then decayed so that it had virtually disappeared in 36 hours.

It is not uncommon for parasitoids to be attracted to frass from the prey. The braconid parasitoid *Orgilus lepidus* Muesebeck is attracted by the frass from its host, the potato tuberworm, *Phthorimaea operculella* (Zell.), and probes it with the ovipositor. The attractive chemical in the frass is heptanoic acid which the tuberworm ingests from the potato and concentrates during passage through the gut; the chemical that leads to ovipositor probing has not been determined.

Similarly the frass from the corn earworm, *Heliothis zea* (Boddie), contains 13-methylhentriacontane which triggers short-range host-seeking by the parasitoid, *Microplitis croceipes* (Cresson). It is reported that *M. croceipes* will not respond to 12-methylhentriacontane. If this is true, the sensillae of this insect are astonishingly specific in their structural requirements.

The eggs of the corn earworms have tricosane and other hydrocarbons associated with them and it is these materials that cause host-seeking by the egg parasitoid *Trichogramma evanescens* Westwood and other species of that genus. Since tricosane is found in high concentrations in the hydrocarbons of corn and other plant leaves, it has been proposed that the plants are the ultimate source of the attractant, though the hydrocarbons seem to reach the vicinity of the eggs via scales that are shed from the moth.

The tobacco budworm, *H. virescens* (F), is very closely related to *H. zea*. It produces several methyl-branched saturated C_{32}, C_{33}, and C_{34} hydrocarbons even when reared on a synthetic diet. The mandibular glands are the source of the hydrocarbons which ultimately show up in both salivary secretions and in the frass of the budworm. The mixture, much more than the individual components, triggers host-seeking by the braconid parasitoid *Cariochiles nigreceps* Viereck.

Some female parasitoids that attack more than one species of host prefer to oviposit in host of the species in which they themselves were reared. The identity of the chemical clues that lead to this preference would be most interesting.

If the host-finding allomone is deposited on a nonhost insect, the parasitoid may sting the insect but will not lay an egg in it. Females can sometimes be induced to lay in an artificial host by the presence of certain amino acids, sugars, and saline.

A female often marks a prey in which she has laid an egg, and other parasitoids (often of other species) are less likely to lay in marked hosts. The marking chemical usually lasts only a few hours, and a female will often sting but not oviposit in a superparasitized host after the marking chemical has lost its potency.

The blood chemistry of a parasitized host is altered in a way that makes the prey a more suitable home for the parasitoid. Although some of the changes in blood chemistry are known, we know nothing of the significance of the changes. There are two lines of evidence which suggest that materials injected by the females of certain species cause alterations in the host. If a developing larva is transferred to a host that has not been stung, it will be encapsulated and will not develop. If the poison gland is removed from the female, fewer of her eggs subsequently deposited in hosts will develop.

Many parasitoids paralyze insects or spiders which then serve as food for the developing larvae. Paralysis is brought about by neurotoxins in the sting of the female parasitoid, and some of the venoms rival the bacterial toxins in potency. For instance, one part of *Bracon hebetor* venom in 200 million parts of *Galleria mellonella* larval blood will cause permanent paralysis in the larva. Although it has been calculated that a *B. brevicomis* wasp produces enough venom to paralyze about a million and a half larvae, the record is 1698 larvae paralyzed by a single female during her 3-week adult life. The venom blocks neuromuscular transmission at a presynaptic site, and the pharmacology of the venom will doubtless reveal much concerning transmission at neuromuscular junctions of insects. There is rather tenuous evidence that the toxin may block the release of glutamate at the synapse.

B. Attractiveness of Some Predators

Some predacious insects station themselves on or near flowers and wait for the prey to come to them. Two assassin bugs (Reduviidae), however, produce their own attractive chemicals. *Apiomerus pictipes* Herrich-Schaeffer waits near the entrance tube to the nest of the stingless bee, *Trigona fulviventris* (Guerin). Bees fly slowly down to the bug, and when they get close they attack it. The bug then catches the bees and sucks the fluids from them. The bees are attracted to the bug if it is hidden, but not if it is in a sealed glass vial in plain view.

Ptilocerus ochraceus Montandon stations itself near the trails of the ant *Hypoclinea biturberculata* (Mayer). Hairlike extensions on the ventrum of the bug hold a glandular secretion that is attractive to the ants. While the ants imbibe the secretion the bug does not bother them, but the ants soon become paralyzed and then they are sucked dry. Since ants closely related to *Hypoclinea* have powerful defensive secretions which render them immune to most predators, the tranquilizing attractant is probably an effective method of avoiding the defensive sprays of the prey.

Most sucking predators that prey on other insects immobilize the prey within 5 seconds by injecting up to 10 mg of salivary venom into them. The venom of reduviids contains hyaluronidase, proteases, phospholipase, and possibly other digestive enzymes. They probably accomplish the quick paralysis of the prey by digesting the nerve sheath. The enzymes act rapidly and it is reported that the predator obtains a great deal of partially digested material from the prey. If one of the sucking insect predators bites a man, the area of the bite remains sore and extremely painful for weeks.

It is often said that some predators have broken the chemical code of insect societies. This probably is true, but in some cases it appears that the predator has managed to insert its own chemical code into that of the society that it invades.

Ants care for the larvae of a majority of the lycaenid and riodinid butterflies. The caterpillars produce substances from paired glands on the eighth abdominal segment that is highly excitatory and attractive to the ants, and they produce an exudate from a gland on the seventh segment that the ants imbibe. The relationship is obligatory for some species of lycaenids; if the exudate is not removed regularly by ants the caterpillars soon die. *Lycaena arion* feeds on wild thyme until it reaches the third instar. It then crawls to the ground and wanders about till it encounters an ant of the genus *Myrmica*. After the ant imbibes some of the glandular secretion, the caterpillar deforms its body into a peculiar humped appearance and is carried into the ant nest and placed among the brood. It immediately turns carnivore and devours ant larvae.

The larvae of several staphylinid beetles of the genus *Atemeles* are carried by *Formica* ants into the nest and placed among their brood which the grub proceeds to devour. In the brood chamber they are groomed and given regurgitated food by the nurse ants. Radioactive tracer experiments show that the ants obtain materials from the body of the grubs. If an acetone extract of the beetle grubs is deposited on filter paper, the dummy will be carried by ants and placed among its brood. If the grubs are extracted with acetone or are coated with shellac they are carried to the garage dump. It is not known how similar chemically the ant brood pheromone and the staphylinid allomone might be.

The relationships between *Atemeles* and ants can be quite complex. *Atemeles pubicollis* pupate in Formica nests, but soon after the adults emerge in the fall they migrate to nests of *Myrmica* which, unlike *Formica*, maintain brood throughout the winter. In the spring the beetles find their way back to *Formica* nests where they mate and lay eggs. They do not find the hosts' nests by following foraging trails of the ants, and they are unable to locate them in still air, but they follow windborne odors to their destinations. At the nest entrance the ant is first attracted to and licks glandular secretions from glands on the abdominal tip. The secretion of these glands seems to suppress aggression and has been called an appeasement substance. The ant next moves to paired glands along the sides of the beetle's abdomen. These have been called adoption glands because the ants will only carry the beetle into the nest and place it among the larvae if it has received the secretion of these glands. Some other staphylinid beetles have the appeasement gland but not the adoption gland and they are only tolerated in the food chambers or garbage dumps of the nests. Indeed some species can avert attack by ants by exposing their appeasement glands if they are in the garbage dump, but if they wander into some other part of the nest the appeasement is not sufficient and they are killed. It is obvious that, with allomones as with pheromones, the message transmitted by a chemical depends upon the context in which it is offered.

C. Ants and Aphids

Many insects seek and eat the honeydew excreted by a variety of homopterans, and mutualism has developed between some ants and aphids to such an extent that the aphids can be considered as domesticated animals of the ants. Aphids pierce the phloem of plants and the sap is forced into them under some pressure. As the sap passes through the intestinal tract some of the simpler sugars, amino acids, and other nutrients are removed from it, and oligosaccharides and some other substances are added back before it is expelled as honeydew from the anus. Ants sometimes build clay pens and shelters for aphids, care for the eggs over winter, move aphids to feeding places and from unattractive to more attractive feeding sites, carry them from danger, and guard them against predators. The female reproductives of some ants even carry coccids with them on their nuptial flights and care for them until a new nest is established.

Aphids that are not being milked by ants may expel the honeydew forcefully, but these same aphids, when antennated by ants, release the honeydew slowly as small droplets. Some aphids will not produce honeydew unless they are attended by ants, and others will greatly increase honeydew production when they are milked. The aphids that are most closely associated with ants have lost their jumping legs and the cornicles that produce defensive secretions; some of them have added hairs around the anus that hold the droplets of honeydew secretion in place, and some have added adaptations for riding on the backs of ants. Aphids may attain much greater populations without alates developing among them when they are herded by ants than when they are not. In a study of the reasons for this differential development, *Aphis fabae* Scop. were reared on an artificial diet with and without attendant *Formica fusca* Wheeler. There was an increase of about 35% in the number of apterae that developed when the aphids were reared with ant attendants. When unattended first instar aphid larvae were given a single topical application of the mandibular gland contents from *F. fusca* workers, about 30% more of them developed into apterae than among the controls. It is probable that the effect of the mandibular gland secretion is due to its juvenile hormone activity. Juvenile hormone is known to prevent the development of alates, and the ant, *Lasius fuliginosus* Latr., is known to have dendrolasin, a furan of farnesol with weak juvenile hormone activity in its mandibular glands. When first instar aphids were treated with topical application of dendrolasin, about 15% more of them underwent apterous development than among the control aphids.

Formica subsericea will antennate any species of aphid it encounters, but nonmyrmecophiles will attempt to escape rather than give up a droplet of honeydew. The ants will sometimes catch the fleeing aphid and carry it to the nest, apparently as prey. Myrmecophiles are not carried to the nest unless they are injured. (If presented with an excessive population of their domesticated aphids, some ant species will consume aphids until the population is reduced to the size necessary to satisfy their need for honeydew.) Upon exposure to a predator many aphid species release alarm pheromone of which *trans*-β-farnesene is a major component. Myrmecophilous species are less likely to fall or jump from the plant when exposed to their alarm pheromone than are nonmyrmecophiles. Furthermore myrmecophiles that are being guarded by ants are less likely to walk away when exposed to their alarm pheromone or to pure farnesene than those that are not attended. The aphid pheromone is an allomone for the ant, and *F. subsericea* will become alarmed and attack when exposed to aphid alarm pheromone or to β-farnesene.

D. Symbiosis among Ants

In the most extreme symbiosis in ants, every caste and developmental stage of two species share the same nest. These extreme symbiotic relationships exist between species which are very distantly related, probably because their pheromone systems have diverged so far in evolution that the two species do not jam each other's chemical communication. The types of symbiosis intergrade from this extreme form to occasional intrusions of individuals from one colony into the nest or foraging area of another. Parasitism and slave-making are common forms of symbiosis among ants, and in these cases the species involved in the symbiotic relationship are most likely to be closely related to each other. This is probably because their pheromone systems (except for the pheromones that are also used as offensive weapons by slave-making species) have not become very different during the brief evolutionary divergence of the two species. Even though the pheromone systems may be similar, though, there is behavioral evidence that habituation and/or the acquisition of the same colony odor play a major role in the invasion of the nest by members of other species.

Only in the case of slave-making species, though, do we have proof that chemicals play a role in symbiotic relationships. The chemicals involved are used in the recognition of other species, they are recruitment and alarm pheromones, or they are weapons. The slave-making ants make massed raids on the colonies of species they enslave and carry larvae and pupae back to their own nests. In several species the entire behavior that is characteristic of a raid can be brought about by making a trail of pheromone from the nest of the slave-makers to the victims' nest at an appropriate time of the year and day.

Slave-making ants of *Formica subintegra* have enlarged Dufour's glands that each contain about 700 μg of a mixture of decyl acetate, dodecyl acetate, and tetradecyl acetate. These chemicals constitute 10% of the entire body weight of the ants, or about two orders of magnitude more pheromone/allomone than is present in closely related species. The acetate mixture has many functions. It serves in defense at the nest (see Chapter 11), probably makes up part of the trail pheromone that is used to initiate and guide the raid, is an alarm pheromone that promotes excitement and attack, and is an offensive allomone that helps to subdue the colony that is raided for slaves. During slave-gathering raids on nests of *F. subsericea,* the *F. subintegra* spray resisting ants with the acetates. This causes disorientation in defenders that are sprayed directly or that are in the presence of high concentrations of the material. Since the acetates have relatively low volatility, they have a large active space (see Chapter 8) and many defenders are rendered helpless by the spray. Furthermore, potential defenders that come into low concentrations of the acetates tend to run and thus to desert the nest. The acetates have, for that reason, been called propaganda chemicals.

The use of chemicals for offensive purposes probably is fairly common among ants. *Solenopsis fugax* raids other species of ants for larvae and, when in their nests, discharge a substance from the poison gland that prevents the invaded ants from defending the brood. Similarly, when *Monomorium pharaonis* compete with other species of ants for the same food resources, they release a substance from the poison gland that repels the other ants and allows the pharaoh's ant to dominate the area.

Specific recognition of enemies has been demonstrated in the ant *Pheidole dentata* when they encounter *Solenopsis geminata* near their nest. The minor workers which make the discovery go back to the nest, laying a trail which brings soldiers to the scene. The minor workers will show the same behavior when presented with several individual ants from other genera, but a single *Solenopsis* triggers the behavior. The *Pheidole* will also react to an extract of *Solenopsis* that is deposited on a model which is agitated, but they will not react to a treated model that does not move nor to a model without the extract regardless of its movement.

E. Foraging Trails

Under laboratory conditions most ants will follow the foraging trails of closely related species, and some of them will follow trails of quite distantly related ants. However, when an ant readily follows the trail of an alien species, it will almost always use a trail of its own species if it is given a choice. Interspecific trail-following does, however, sometimes occur in nature. Large numbers of *Camponotus lateralis* sometimes follow the trails of *Chrematogaster scutellaris* to their feeding grounds and exploit the same food resources there.

Camponotus beebei regularly utilize the odor trails of *Azteca chartifex*. Since the *Azteca* forage primarily at night and *Camponotus* during the day, there is seldom conflict, but the *Azteca* treat the *Camponotus* as enemies when they encounter each other; the swift and agile *Camponotus* usually have no trouble escaping. Occasionally a *Camponotus* worker will lay a recruitment trail along an *Azteca* trail and other members of its own species will follow it closely; the *Azteca* workers, though, apparently disregard the *Camponotus* trail, indicating that the exploitation of trails is one-way only.

The inquilines that live with army ants are sometimes more sensitive to the odor trails than are the ants that laid them. A random sample of a wide variety of the guests of army ants was offered a choice of two trails. Almost all of them chose the trail of the species with which they were first encountered over that of a competing species, and some of them were actively repelled by the odor of the wrong species. A few chose the trails of nonhost species. None of the guests could distinguish between the trail of their own colony and that of another colony of the same species.

Stingless bees of some species lay trails of mandibular gland secretion from the nest to a rich source of food by alighting every 1–5 meters and placing a droplet of secretion on the ground or vegetation. Bees that have been alerted in the nest to seek food follow the trails of their own species readily, and there is very limited interspecific trail-following. For instance, *Trigona xanthotricha* Moure can follow the trail of *T. postica* Latreille, but not vice versa. The *T. postica* have one isomer of tetradecenyl acetate not produced by *T. xanthotricha* and the two species show other apparently minor differences in pheromone composition. Their pheromones are so similar, though, that their failure to accomplish reciprocal trail-following is surprising. *Trigona spinipes* (Fab.) is a member of a different subgenus from the other two bees and has only three pheromone consituents, all of which occur in the other two species. There is no interspecific trail-following between *T. spinipes* and either of the other two species.

F. Interspecific Relations of Stingless Bees

Lestrimelitta limao (F. Smith) lives by robbing honey and pollen from the nests of other bees. It uses a mandibular gland secretion of citral and a small amount of an unidentified compound as an offensive weapon. Upon exposure to high concentrations of the allomone (or to pure citral) the defending bees become disoriented and give no resistance to the invaders. Species of bees that are not disoriented by citral are reportedly able to withstand attacks of *L. limao*.

Trigona fulviventris (Guerin) is a very gentle stingless bee which apparently has no alarm pheromone. When either living or dead *L. limao* are brought near to the entrance tube of their nest, massed agitated *T. fulviventris* slowly but inexorably emerge from deep in the nest and attack the *L. limao*. If the foreign bee or an extract of it is held near to the entrance tube, the *T. fulviventris* will remain agitated. Pure citral causes the same sort of agitation, but it is less severe and slowly dies out. *Melipona beecheii* Bennett also has no alarm pheromone, and it too becomes agitated and attacks *L. limao* or citral. *Trigona pectoralis* Dalla Torre and *T. cupira* Smith have alarm pheromones which cause instant severe mass attack, but they apparently also use the offensive weapon of *L. limao* as an allomone which alerts defenders.

Judging by the reactions of other species of bees to *T. pectoralis* or to its alarm pheromone at their nest entrance, this species too must commonly rob from the nests of other bees. *Lestrimelitta limao* causes considerably more agitation in all species than does *T. pectoralis*, which in turn causes a great deal more agitation than *T. cupira* (a fierce attacker of humans), and all other species of bees virtually disregard *T. fulviventris*.

The recognition of enemies by their pheromones may be common among social insects. For instance, an unidentified ant that often lives near nests of *T. pectoralis* uses 2-heptanone as an alarm pheromone. The ants are often seen going into the entrance tubes of *T. pectoralis* nests and being gently butted out. When the bees are exposed to the odor of 2-heptanone, they mass in the mouth of the entrance tube and remain almost motionless there, so that the mouth is solidly filled with their hands and returning foragers cannot gain entrance. The ants, in the meantime, dash excitedly about.

The reactions of bees to the chemicals of their enemies makes it difficult to interpret the behavior of bees in tests of possible pheromones and allomones. For instance, the *T. pectoralis* alarm pheromone contains 2-heptanol and 2-nonanol, but the bees react about equally to those chemicals and to 1-octanol, 2-octanol, 1-nonanol, and 1-decanol. It is commonly assumed in cases of this kind that the bees cannot distinguish between molecules of approximately the same size and shape. The bees behave quite differently, however, in the presence of two other components of their alarm pheromone that are also similar in size and shape. They fill the entrance tube and remain still, as described above, when exposed to 2-heptanone but show the excitement and attack typical of alarm behavior in the presence of 2-nonanone; both of these compounds are components of their alarm pheromone.

It seems probable that, if sensillae are excited by similar molecules, the nerve impulses that are generated are processed in the same way in the central nervous system unless it is to the advantage of the insect to behave differently in the presence of the slightly different molecules. We cannot, therefore, conclude on the basis of similar behavior in the presence of similar molecules that the insect is incapable of distinguishing between them. Indeed, we must proceed with the greatest of caution in attempting to determine any aspect of the significance of chemical structure to insect behavior.

There is some evidence indicating that the reactions to the chemicals of other species are learned rather than innate. When honey bee colonies in the region where stingless bees occur were presented with *L. limao* and *T. pectoralis* at the entrance to the hive, most colonies were agitated by both species, a few were agitated by one species but not the other, and one colony was agitated by neither.

It is possible that *T. fulviventris* has a masking chemical that renders it odorless to *Eciton* army ants. It is not known that they are ever attacked by *Eciton,* and the ants have been observed to go to the lip of the entrance tube to the subterranean nest and yet make no attempt to enter.

III. CHEMICALS THAT AFFECT NONINSECTS

By far the best understood of the allomones of insects that affect organisms of other taxa are the many defensive chemicals which are the subject matter of Chapter 11. We will only consider a few examples of the other types of allomones that act on noninsects in order to illustrate some of the interesting problems that remain to be solved.

A. Microbes and Fungi

The royal jelly of honey bees is a nutritionally rich secretion of glands in the heads of young worker bees. Although there is a large excess of the food in the queen cells, it does not become contaminated with yeast, fungi, or bacteria. The lipid fraction, composed almost exclusively of polar-free fatty acids, is primarily responsible for this microbiocidal or microbiostatic activity. Over 50% of the lipid is *trans*-10-hydroxy-dec-2-enoic acid, and there are lesser amounts of 10-hydroxydecanoic acid, *trans*-dec-2-endioic acid, decan-1,10-dioic (sebacic) acid, and several other 8- and 10-carbon hydroxy acids. The structural features of the fatty acids that are responsible for the microbiocidal properties of these fatty acids are not known.

The brood food of the stingless bees apparently has some other microbiostat. The lipids of the royal jelly of *Melipona quadrifasciata* Lep. contains little free fatty acid and virtually no highly polar fatty acid.

Many ants live in habitats that would appear to be ideal for the growth of fungi and bacteria, and secretions of the metapleural gland could well account for the ability of ants to survive in such a habitat. The metapleural gland of *Atta sexdens* produces a highly acid secretion that includes phenylacetic acid. The secretion is fungistatic and bacteriostatic, and protection against microorganisms may be the function of this unique gland in all ants.

Atta and other fungus-growing ants fertilize their fungus gardens with fecal material. The feces not only directly supply the fungus with amino acids and other nitrogen sources, but they also contain proteolytic enzymes which the fungus lacks. Without the fecal enzymes of the ants the fungus is unable to digest the protein of the substrate.

B. Effects on Plants

Many insects cause galls or other deformations to develop in plants from which they suck juices or in which they lay eggs. The nature of the deformation-producing substances in the salivary secretions of hemipterans is under vigorous dispute. Some bugs produce 3-indoleacetic acid and other auxins that directly promote growth of plant tissue. The lygus bug, *Lygus disponsi* Linnavuori, seems to produce no auxin, but it produces a factor that either promotes the production of indoleacetic acid in the sugar beet leaf or that inhibits the activity of indoleacetic acid oxidase. The saliva of this bug also contains pectinases which contribute to phytotoxicity. The roles of mechanical injury, quinones produced by the plant as a result of insect attack, free amino acids in the saliva of bugs, auxin synthesis-promoting factors in the saliva of bugs, and the roles of other possible salivary constituents are all under dispute.

Soon after the boll weevil, *Anthonomus grandis* Boh., lays an egg in the flower bud of cotton, the bud is shed. *Endo*-polymethylgalacutronase, and possibly other pectinases from the larva seem to be responsible for the abscission of the bud.

More interesting are the highly regular galls that result from the oviposition of plant parasites. The insects responsible for these galls are sawflies (Hymenoptera: Tenthredinidae), gallwasps (Hymenoptera: Cynipidae), gallmidges (Diptera: Cecidomyiidae), and psyllids (Hemiptera: Psyllidae). Not only are there few families of insects involved, but the insects restrict their activities to a small number of plant genera and families. For instance, there are hundreds of known cynipid galls, but more than 90% of them occur on only two unrelated genera: the oaks and the roses. The galls are so predictable in species of plant, position on the plant, and size, shape, and morphology that they can be used as taxonomic characters to identify the insect that causes them to appear. Indeed, sibling species of insects are sometimes parasitic on closely related plant species and can only be recognized as different if cross-infestation is attempted in the greenhouse. The galls have been likened to organs of the plant that only appear as a result of insect activity. Either new genetic information is introduced into the plant or genetic information already present is expressed in an otherwise nonexistent but predictable way. Usually an egg or developing larva is necessary for a gall to develop fully, but in a few cases galls develop as a result of the insertion of an ovipositor even if no egg is laid. This is a problem that deserves attention by scientists using modern techniques.

C. Man

It would be to the advantage of a blood-sucking insect to go undetected while it extracts its meal, and it appears that none of the blood suckers has a salivary venom. Even though the saliva is very dilute, however, it does contain antigens. Man and other mammals do not react to the first bite of a flea, mosquito, or probably other blood suckers, but they become sensitized after a few bites, and the painful welt that forms is the result of an allergic reaction.

Little is known of the composition of the salivary secretions, but they should be studied from the point of view of the adaptive advantage that they bestow upon the insect.

IV. GENERAL CONSIDERATIONS

If there is one aspect of research on chemical communication that has been neglected it is the biochemistry of the processes involved. There are innumerable studies that are in an unsatisfactory state simply because the necessary biochemical work has not been done. This is particularly unfortunate since many of the techniques of the biochemist are eminently suited to the study of the type of transient phenomena that occur in chemical communication.

Most of the experiments that have been conducted have involved the identification and demonstration of some effect of one to a few chemicals. This is partly because no one scientist has the time to master and use all the techniques that should be applied, and cooperative research by scientists in different fields often is not a simple solution. Problems of timing make cooperation among specialists difficult, and even with the best of wills, attempts to work together often founder. There is, however, the problem of the lack in depth of many studies. There have been too many attempts to skim the cream from complex problems, so that we are presented with bits and pieces of information scattered about. What is needed now is a series of painstaking and exhaustive studies of single systems of communication using every technique available. We do not need a greater volume of inadequate data. When any system is studied in breadth and depth, the cream turns out to still be there.

V. LITERATURE

Whittaker and Feeny (1971) presented a brief overview of allomones of both plants and animals. Except for the pioneering effort both the original literature and the reviews are scattered. Sondheimer and Simeone (1970), Wood *et al.* (1970), and Harbone (1972) have reviewed much of the literature on chemical communication from plants to insects. They have laid particular emphasis on the role of phagostimulants. Gilbert and Raven (1975) and Wallace and Mansell (1976) are somewhat broader in their coverage of plant-insect interactions. Atsatt and O'Dowd (1976) have reviewed plant defense guilds, and Bowers *et al.* (1976) have discussed the discovery of anti-juvenile hormones.

There have been no book-length treatments of insect allomones, and extended treatment of this subject would probably be premature. Hocking (1971) and Friend and Smith (1977) have reviewed the literature on blood-sucking insects. Vinson (1976) has summarized current knowledge of host seeking and selection by parasitoids. The more extensive literature on the use of allomones among social insects has been reviewed by Wilson (1971).

GENERAL REFERENCES

Atsatt, P. R., and O'Dowd, D. J. (1976). *Science* **193,** 24–29.

Bowers, W. S., Ohta, T., Cleere, J. S., and Marsella, P. A. (1976). *Science* **193,** 542–547.

Friend, W. G., and Smith, J. J. B. (1977). *Annu. Rev. Entomol.* **22,** 309–331.

Gilbert, L. E., and Raven, P. H., eds. (1975). "Coevolution of Animals and Plants." Univ. of Texas Press, Austin, Texas.

Harbone, J. B., ed. (1972). "Phytochemical Ecology." Academic Press, New York.

Hocking, B. (1971). *Annu. Rev. Entomol.* **16,** 1–26.

Sondheimer, E., and Simeone, J. B., eds. (1970). "Chemical Ecology." Academic Press, New York.

Vinson, S. B. (1976). *Annu. Rev. Entomol.* **21,** 109–133.

Wallace, J. W., and Mansell, R. L., eds. (1976). *Recent Adv. Phytochem.* **10.**

Whittaker, R. H., and Feeny, P. P. (1971). *Science* **171,** 757–770.

Wilson, E. O. (1971). "The Insect Societies." Harvard Univ. Press, Cambridge, Massachusetts.

Woods. D. L., Silverstein, R. M., and Nakajima, M., eds. (1970). "Control of Insect Behavior by Natural Products." Academic Press, New York.

10

Chemical Control of Insects by Pheromones

WENDELL L. ROELOFS

I. INTRODUCTION

Insects ingest a rich variety of complex chemicals that can be used either by the insects directly or by their symbiotic microorganisms to produce message chemicals with amazing structural diversity. Biochemical processes in any one species can be developed to produce certain chemicals that have a unique behavioral meaning for that species. These chemicals may be ubiquitous in nature, but with different effects or messages when used in other bouquets, concentrations, or behavioral milieu. Thus, a chemical such as 11-dodecenyl acetate can function as a defensive agent with the larvae of one species (Marchesini *et al.*, 1969) and as a pheromone component with the adult moths of another species (Nesbitt *et al.*, 1973). The important point is that most species have evolved specialized biochemical processes for emitting or perceiving certain chemicals that play key communicative roles in a variety of activities, such as in locating food or oviposition sites, in the use of food supply trails and alarm signals, in the regulation of castes, marking territories, finding aggregation sites, and in locating an appropriate mate.

For several centuries naturalists have known that some insect species use a communication system with which an individual of one sex can assemble a host of individuals of the other sex from long distances for mating purposes. The French naturalist, J. Henri Fabre, conducted interesting tests with giant silkmoths to show that males appeared at night from long distances in search of caged female moths through the use of olfactory signals rather than auditory or visual signals. The attraction of hundreds of huge male moths, which were rare in the surrounding countryside, into Fabre's house in response to the female's emanations revealed the specificity and the potency of this chemical communication system. Another vivid demonstration of this mating communication system was the attraction of over 11,000 male introduced pine sawflies, *Diprion similis*, to 1 caged female within 5 days. This dependency of the vital mating process on a chemical or a set of chemicals operating over a distance has long intrigued scientists as a possible weak link that specifically could be manipulated for insect control.

Efforts in the first half of the twentieth century to decode this communication system were limited and tedious. In 1959, however, a 30-year effort by Professor Butenandt and co-workers (1959) was culminated with the revelation that the silkworm moth, *Bombyx mori*, uses (*E*,*Z*)-10,12-hexadecadien-1-ol as its sex pheromone. This domesticated insect, long exploited for making silk, was by no means a pest species, but the fact that the communication system could be broken was enough to spur interest in that area. The development of highly sensitive instruments made this a much more appealing research area than previously, and reverberations sent through federal funding agencies and the scientific community from Rachel Carson's book "Silent Spring" generated much interest in research on the mysterious codes used by pest species.

This chapter discusses various aspects of sex pheromone research as related to their use in insect control programs. A differentiation of the terms sex pheromone and sex pheromone component is made as follows:

Sex pheromone: The total mixture of chemicals that is released from one organism and that induce responses, such as orientation, precopulatory behavior, and mating, in another individual of the same species.

Sex pheromone components: The individual chemicals that are found to be part of a sex pheromone. They must be rigorously identified either from the organism or its effluvium, and it must be demonstrated that they are involved in eliciting some portion of the precopulatory behavioral activity.

II. BREAKING THE COMMUNICATION CODE

In general, insects are so highly sensitive to their sex pheromone that only minute quantities are emitted by the transmitting insect. The natural supply of pheromone is so low per individual that synthetic compounds usually are required for use in insect control programs. This implies defining the structure for each pheromone component from low quantities of material.

A. Methods for Obtaining Crude Pheromone

Pheromone has been obtained from insects for analyses by a variety of techniques, which have been developed for different types of insects and, sometimes, because of unique problems with some species.

1. Pheromone Extraction

Crude pheromone extracts can be made by extracting whole insects or large segments containing the pheromone gland. Four pheromone components were identified for the cotton boll weevil, *Anthonomus grandis,* by extracting over 4 million weevils and 54.7 kg of fecal material. The pheromone of the Indian meal moth, *Plodia interpunctella,* was identified from whole body extracts of 670,000 males and females, although simultaneously other researchers identified the pheromone by extracting filter papers exposed to 19,000 virgin female moths in glass jars.

In many cases, especially in the Lepidoptera, a crude pheromone extract can be obtained by snipping off the pheromone glandular area and extracting it with methylene chloride. The lepidopteran pheromone gland, which usually is located just anterior to the ovipositor lobes on the abdominal tip, is modified intersegmental membrane that is invaginated within the body cavity. The gland can be eversed by applying enough pressure to the abdomen to protrude the ovipositor. The eversed glandular tissue is clipped with a very small abdominal tip to provide a pheromone extract that is relatively free from the huge quantities of fats and oils found in the rest of the abdomen. In some cases the extract can be analyzed directly by gas-liquid chromatography (GLC) and the pheromone components visualized as major peaks on the GLC recordings (Fig. 1).

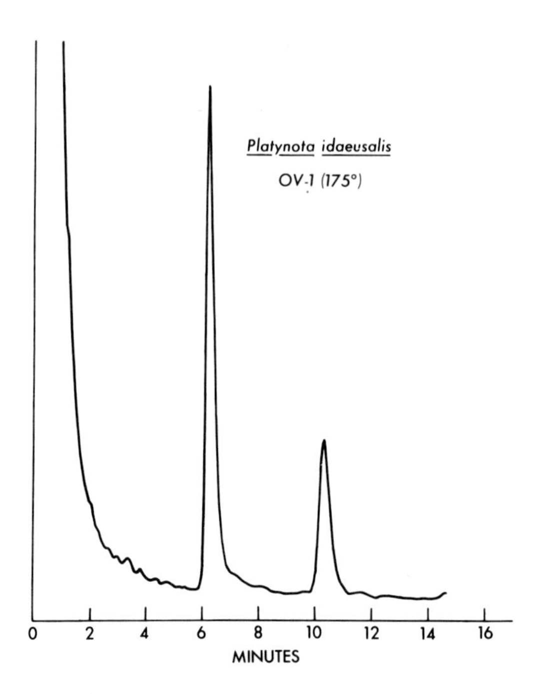

Fig. 1. Crude sex pheromone gland extract of female tufted apple bud moth, *Platynota idaeusalis,* injected directly onto a nonpolar GLC column. The GLC tracing shows the two pheromone components (*E*11-14 : OH and *E*11-14 : Ac) as the only two prominent peaks.

Many times, however, the pheromone components are present in much lower quantities than a host of other compounds in the extract and other purification steps are required. It is important to determine by appropriate bioassays which compounds are active as pheromone components and which ones are inactive for the species under study. In the Lepidoptera it is common to find saturated analogues and functional group analogues in pheromone gland extracts, even though they may not be released by the insect or used as pheromone components. For example, a certain species may use (*Z*)-11-tetradecenyl acetate (*Z*11-14:Ac) as a pheromone component, but gland extracts also could contain tetradecyl acetate and (*Z*)-11-tetradecen-1-ol (*Z*11-14:OH). It is possible that some of these other compounds function as precursors in the biosynthesis of the pheromone components. In some species, certain pheromone components, particularly aldehydes, are apparently stored in some precursor state and are not found in pheromone gland extract. In these instances the pheromone can be obtained by collecting female effluvium.

2. Collecting Emitted Pheromone

Many techniques have been used to obtain pheromone after it has been released from the insects. The pheromone of some bark beetle species has been obtained by extracting the insect's hindguts, but a number of pheromone components have been identified from extracts of huge quantities of frass produced by beetles boring into logs. Pheromone from the American cockroach, *Periplaneta americana,* has been obtained by extracting filter papers that were placed in a box of cockroaches for almost a month. Alternatively, the pheromone was obtained by passing air through a 10-gallon milk can containing 1000 virgin females and condensing the air in an alcohol-dry ice bath.

a. *Conventional Cold Traps.* Condensing pheromone-laden air in a cold trap often has been used as a means of obtaining crude pheromone. It produces a dilute aqueous condensate containing many contaminating compounds, particularly when air is passed over host material containing the insects. Condensates obtained for the red scale, *Aonidiella aurantii,* pheromone contained many terpenes and other volatiles because the virgin female scales were maintained on lemons in the air stream. Improvements were made by switching to potatoes as the host material and using a solid absorption trap instead of the dry ice-acetone trap. An aqueous emulsion containing many compounds also was collected in dry ice traps when air was drawn over ponderosa logs infested with male bark beetles, *Ips confusus.*

b. *Cryogenic Trap.* Conventional cold trapping techniques were not very efficient in condensing pheromones, and so a cryogenic trap (Fig. 2) was developed to collect quantitatively all components in an air stream. It essentially uses a liquid nitrogen trap to liquefy all the air passing through a barrel

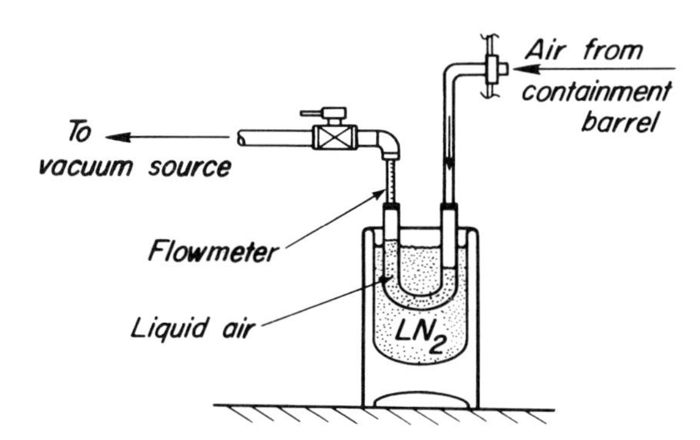

Fig. 2. A cryogenic air liquefaction trap for the continuous collection of airborne pheromones. Liquid nitrogen (LN_2) condenses air that has passed over pheromone-emitting beetles in a barrel. The amount of liquid air in the U-tube is held constant by allowing it to evaporate as fast as it condenses. (From Browne *et al.*, 1974.)

of boring beetles. An air flow is maintained by the continuous condensation of air at the inlet of a U-tube. A vacuum source on the outlet end of the U-tube lifts the liquid air in the tube above the liquid nitrogen level in the trap and allows air to evaporate from the outlet at the same rate as it condenses at the inlet. The cryogenic trap technique was used successfully with several bark beetle species to obtain crude pheromone concentrate from male-infested bolts, and was used to monitor pheromone emission from standing trees under attack by bark beetles in the forests.

c. *Absorption Traps.* Cold-trapping techniques have the disadvantage of yielding a dilute aqueous condensate that must be processed. Researchers, therefore, have investigated various types of material that absorb volatile organic compounds as they pass over or through the absorbents. An interesting, but not too useful, early example is the use of triglycerides to trap the pheromone of the greater waxmoth, *Galleria mellonella.* In this species, unlike most other lepidopterous species, the males emit a musklike odor from their wing glands to attract the females. Air from cages of males was drawn through a stack of twenty-seven fat-coated glass plates. The fat was a mixture of rendered and cleaned retroperitoneal tallow and mesenteric lard in a 2:1 mixture. Pheromone absorbed in the fat was recovered by molecular distillation and then collection from a gas chromatograph. An active com-

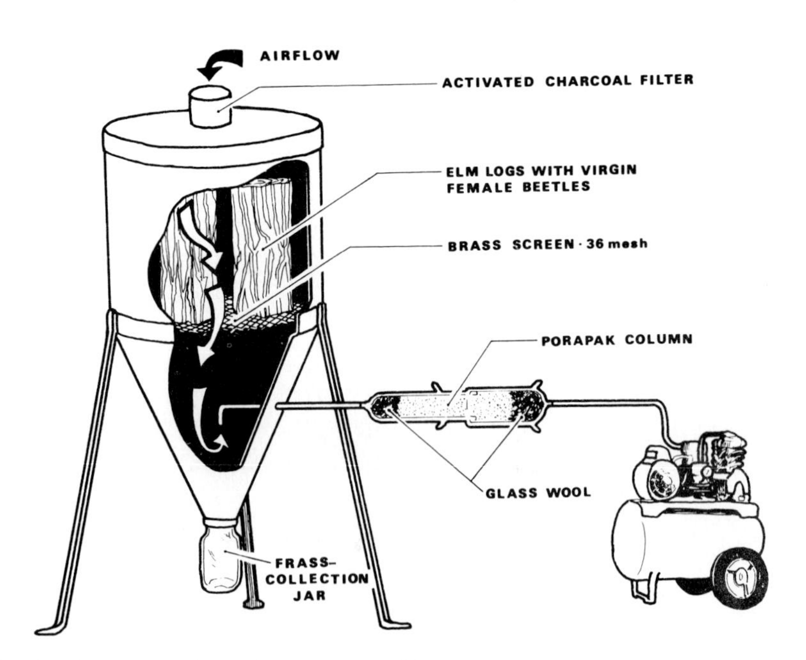

Fig. 3. Porapak trap for collecting host- and beetle-produced chemicals as female elm bark beetles tunnel in logs. (From Peacock *et al.*, 1975.)

pound was identified as *undecanal* in that study, but other investigators using the cold trap technique found that *nonanal* also is a major pheromone component of the greater waxmoth.

Many other absorbents have been used to trap pheromones, but the most complete study has been carried out with Porapak Q (ethylvinylbenzenedivinylbenzene copolymer) as a solid-phase absorbent. A short column of Porapak Q efficiently removes compounds in a wide range of volatility and stability from an air stream. Aeration of pheromone-emitting insects can be carried out over long periods and the sorbed compounds easily extracted from the absorbent. This technique was used successfully in the identification of the attractive volatiles in the air around virgin female smaller European elm bark beetles, *Scolytus multistriatus,* as they tunnel in elm logs (Fig. 3). Batches of 4000–7000 virgin females were aerated continuously for 7 days. Volatiles were extracted from the Porapak Q, fractionally distilled and finally fractionated by GLC. Bioassays with recombinations of the GLC fractions showed that there were at least three active compounds. Two were identified as beetle-produced pheromones, while the third was a host-produced synergist. The compounds are (−)-4-methyl-3-heptanol (**1**), (−)-α-multistriatin (**2**), and (−)-α-cubebene (**3**), respectively.

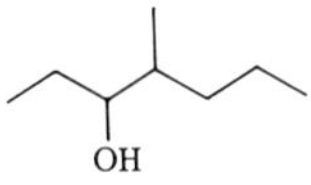

OH

(**1**)

O

(**2**)

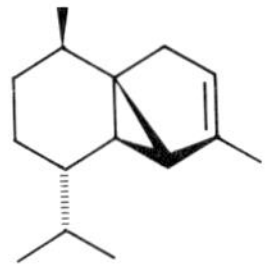

(**3**)

Pheromone collected by insect aeration should be the best method for obtaining the pheromone components in their naturally occurring blends. With many Lepidoptera, however, it is much more difficult to obtain the minute pheromone quantities by aeration than by gland extraction. Problems with aeration of these insects usually involve the vagaries of calling female moths and the large ratio of contaminating compounds to pheromone in the absorbent extract. Nevertheless, the aeration technique should be used along with the gland extraction technique to verify that all pheromone components have been identified and to check the component ratios in the female effluvium. Pheromone identification of the orange tortrix moth, *Argyrotaenia citrana,* illustrates this point. Female pheromone gland extracts were obtained from twenty-five virgin females and fractionation on a nonpolar GLC column (OV-1) yielded two biologically active compounds in a 100:1 ratio. The major component was identified as *Z*11-14:Ac, but there was too little sample to identify the second component. Field attractancy tests with the acetate component showed that this compound by itself did not lure any males into the traps. Pheromone collected on Porapak Q from air drawn over fifty females in a glass tube during their "calling" period contained the acetate as a minor component (1:15) to the second component. The second component was identified from these airborne collections as (*Z*)-11-tetradecenal, which proved to be very attractive by itself to *A. citrana* males in the field, although trap catches were more than doubled by adding the acetate component to the lure. Apparently the aldehyde was stored in the gland in a precursor state and was not extractable as the active aldehyde.

B. Identification of Pheromone Components

Structural characterization of pheromone components has been accomplished for many pest species and has revealed the use of a large variety of structural types. They range in size from ethanol, a primary attractant of several species of ambrosia beetles, to a 31-carbon pheromone, 3,11-dimethyl-2-nonacosanone, of the German cockroach, *Blattella germanica*. In 1975 a list of 129 different compounds was given in relationship to the pheromones of 232 insect species (Mayer and McLaughlin, 1975). In addition to the usual problem of limited pheromone quantities, the structural complexities of some pheromones have been a challenge to chemists. For example, at least six structures could be suggested from data on a pheromone component of the American cockroach, *Periplaneta americana*, although the most plausible one seemed to be the germacrane-type compound (**4**). Synthesis of complex pheromone structures and of pure

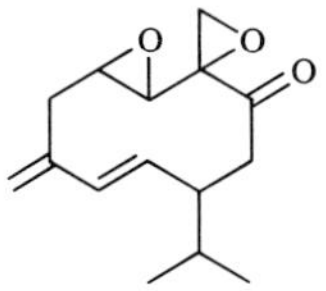

O O O

(**4**)

stereoisomers, particularly enantiomers, also has provided some challenges for synthetic chemists. In general, however, the most difficult problem is to define all the pheromone components in relationship to the natural blend and behavioral context.

A variety of behavioral assays in the laboratory and the field has been used in this research since each species can require technique modifications to fully describe the active components. Some pheromone components can be present in such small percentages that they are not detected in most analyses. Some components can elicit such subtle behavioral modifications that their presence in fractionated crude pheromone goes unnoticed in most bioassays for total activity. In mating-disruption programs it usually is not essential to know every minor component, but trap specificity and potency in other programs generally are greatly increased as the synthetic lure more closely duplicates the natural pheromone. Some considerations in detecting and defining certain blends will be discussed.

1. Electroantennogram Technique

The electroantennogram (EAG) technique has been very useful, particularly with lepidopteran species, for detecting pheromone components in fractionated extract or effluvium. The technique utilizes the response specificity of male moth's antennae to pheromone components. In most moth species the male's antennae are highly developed for detecting low concentrations of pheromone in the air. Each antenna can possess thousands of specialized olfactory hairs that efficiently filter the pheromone molecules from the air stream. These cuticular hairs are filled with sensillum liquor and are innervated with dendrites of several receptor cells. The hairs are perforated with thousands of pores connected to fine tubules, which direct molecules towards surface membranes on the dendrites. Pheromone molecules that impinge upon the sensory hairs apparently reach the acceptor sites of the receptor cells by surface diffusion to the pores and through the pore tubules.

Sensory transduction occurs when appropriate odor molecules interact with the acceptor sites on the dendrites. Interaction of the chemical stimulus with the receptor molecule can activate the receptor by changing its conformational state. This action results in a change in the membrane conductance. Increased conductance produces a depolarization of the membrane, a response that is called the receptor potential. It is the receptor potentials elicited simultaneously in many olfactory cells lying in series that is recorded in the EAG technique. These extracellular slow potentials, usually 1–10 mv, are recorded using electrodes in contact with the tip and base of an antenna (Fig. 4). The amplitude of the receptor potential is directly related to the concentration of the chemical stimulus. The receptor potential amplitude also determines the frequency of nerve impulses that are generated from the sense cells and sent to the central nervous system by unfused antennal nerves.

The process of sensory transduction has been summarized by Kaissling (1975) to involve the following six steps:

Stimulus $(S)_{gas}$	$\rightarrow S_{surface}$	Adsorption	(1)
$S_{surface}$	$\rightleftharpoons S_{at\ receptor}$	Diffusion	(2)
$A_{acceptor} + S_{at\ receptor}$	$\rightleftharpoons AS$	Binding	(3)
AS	$\rightleftharpoons A'S$	Activation	(4)
$A'S_{induces}$	$\Delta G_{membrane}$	Increase of conductance	(5)
$\left.\begin{matrix} S_{adsorbed} \\ S_{at\ receptor} \end{matrix}\right\}$	$\rightleftharpoons S_{inactivated}$	Inactivation	(6)

Receptor specificity apparently is determined in the steps following diffusion. Inactivation of the stimulus is an important, but unexplained, part of

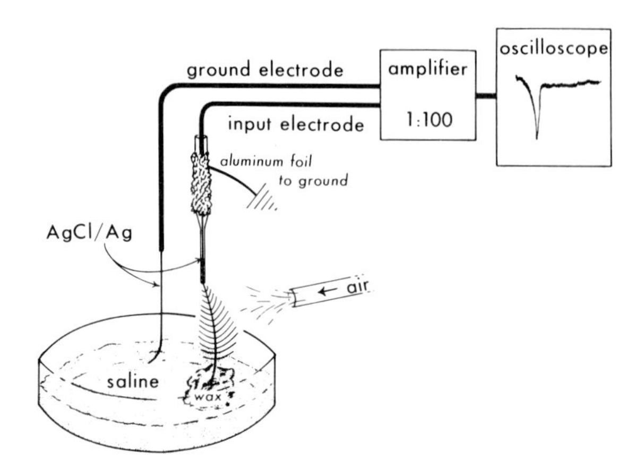

Fig. 4. Electroantennogram setup showing antenna positioned in wax in saline-filled dish. The input electrode, containing saline and a silver chloride-coated silver wire, connects to the distal antennal tip, while the ground electrode makes contact to the antennal base through the saline solution. Grounded aluminum foil around the disposable pipet used for the input electrode screens extraneous noise signals. Test chemicals are puffed into an air stream that continuously passes over the antenna. Antennal responses are amplified 100× and displayed on the oscilloscope. The depolarization amplitude gives a measure of the test chemical activity relative to a standard chemical used just prior to the test chemical.

the overall process. Postulated inactivation processes, such as diffusion, penetration, and metabolism, have been determined to be too slow to account for the apparently fast inactivation observed in EAG recordings.

In addition to EAG data on antennal responses, single sensillum recordings with microelectrodes can be made of sensory cell nerve impulses. Individual sensory cells in the olfactory hair can be monitored because of differences in nerve impulse amplitudes among the cells. One pheromone compound can produce an excitatory response (increased nerve impulse frequency) with one sense cell and an inhibitory response in another cell of the same olfactory hair. The encoding of a particular blend of pheromone components appears to result from a pattern of activity across a number of receptors with different sensitivities. There can be at least two sets of cells on an antenna with each optimally sensitive to a different pheromone component (Fig. 5), and within each set the cells can display different sensitivities to a series of related compounds.

a. *Detection of Pheromone Components.* Electroantennogram (EAG) recordings can be used successfully with fractionated crude pheromone of lepidopteran species because a majority of the male's olfactory sensilla are specifically attuned to the pheromone components. For example, the male polyphemus moth, *Antherea polyphemus,* antenna has more than 60,000 sensilla with 150,000 receptor cells. At least 60% of these cells are optimally

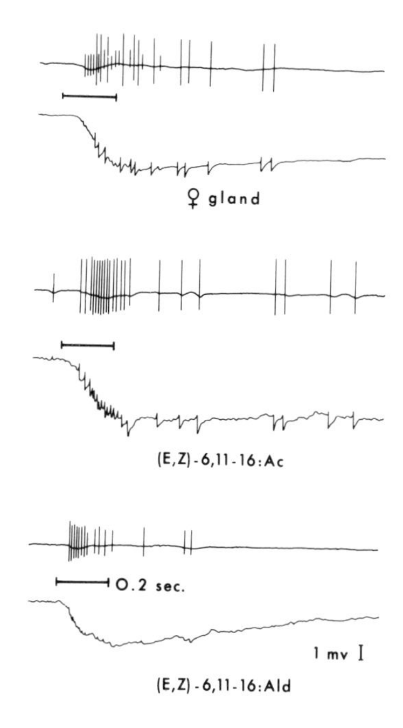

Fig. 5. Electrophysiological responses of two receptor cells from a long olfactory hair (sensillum trichodeum) of *A. polyphemus*. The stimulus was from one female gland or 10^{-3} μg of each compound on filter paper. Traces 1, 3 and 5 show the nerve impulses (a.c. recording) and traces 2, 4 and 6 show the nerve impulses superimposed on the receptor potential (d.c. recording). (From Kochansky *et al.*, 1975.)

responsive to the pheromone components. In the Lepidoptera, good EAG responses are elicited by compounds that are the pheromone components, or closely related to them. Fractions can be quickly assayed for EAG activity by puffing an aliquot across the male's antenna positioned between two electrodes. Crude pheromone can be injected onto various GLC columns and the effluent collected in glass capillary tubes in 1-minute fractions. Each tube is checked for EAG activity and the active areas correlated with standard retention times to obtain structural information on possible pheromone components. This technique can quickly locate the major pheromone components and can give indications of some less abundant components.

All compounds eliciting activity are not necessarily pheromone components. Many times corresponding hydrogenated and functional group analogues are extracted from pheromone glands and exhibit weak EAG activity. Pheromone component activity must be demonstrated in further behavioral assays, but the EAG technique provides an excellent method for locating a number of potential pheromone components. In many cases, pheromone components do not elicit behavioral responses unless mixed together in the proper proportions with other components. Thus, behavioral assays of fractions involve tedious replications of a multitude of recombinations to locate all pheromone components. Since the antennal receptors respond to the individual components, recombinations are not necessary with the EAG technique.

Pheromone components collected by GLC in capillary tubes can be recovered from the tubes and used for further analyses. The GLC effluent, however, can be monitored directly by EAG. The effluent can be vented over a positioned antenna at regular intervals to test for activity, or the EAG setup can be coupled directly to the GLC and used as a GLC detector. The latter method is particularly effective for obtaining precise retention times of EAG-active compounds from high-resolution GLC columns. Receptor specificity is demonstrated with this method in Fig. 6 as no antennal response is recorded by the electroantennographic detector (EAD) to the huge solvent peak or to the closely-related geometric isomers of the pheromone at the low

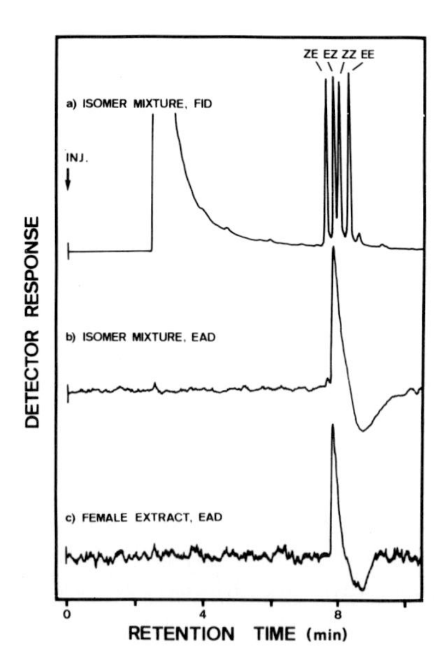

Fig. 6. A mixture of 7,9-dodecadienyl acetate isomers separated on a UCON 50 HB 5100 glass capillary GLC column and detected by (a) a flame ionization detector (FID), and (b) an electroantennographic detector (EAD) using a *Lobesia botrana* male antenna. The EAD tracing shows its sensitivity to the EZ isomer, which is the pheromone for this insect. The bottom tracing shows the presence of EZ isomer in an injection of female *L. botrana* sex pheromone gland extract (From Arn *et al.*, 1975.)

concentrations used. Responses with a signal-to-noise ratio of 3:1 can be obtained with only 0.1 pg of pheromone. This is more than a thousand times more sensitive than flame ionization GLC detectors.

b. *Prediction of Double Bond Positions and Configurations*. The specificity of pheromone receptors already was observed with the first pheromone structure. The geometrical isomers of the silkworm pheromone, (*E,Z*)-10,12-hexadecadien-1-ol, were found to be up to one thousand times less active than the pheromone at equal concentrations. Specificity most likely occurs in the binding and activation steps of the sensory transduction process listed above. Acceptor affinity would be dependent on the interaction of several active sites on the chemical stimulus and the receptor membrane. EAG analyses of hundreds of lepidopteran species have shown that maximum responses are determined by at least three molecular characteristics: the overall compound length, the functional group, and the positions and configurations of double bonds. Thus, the acceptor affinity of a sensory cell responsive to *Z*11-14:Ac as a pheromone component is the greatest for that particular arrangement of active sites. Most molecular modifications will result in a chemical with less acceptor affinity and consequently a lower EAG response. EAG recordings of a series of monounsaturated 14-carbon acetates would show a response increase as the double bond is moved closer to the 11-position and then a response decrease with the 12- and 13-tetradecenyl acetates. Likewise, functional group analogues and 12- and 16-carbon chain homologues would elicit decreased antennal responses.

The sensory cell specificity for pheromone component molecular characteristics makes it possible to use the antenna as a specialized analytical instrument. In many species, pheromone component retention times on both polar and nonpolar GLC columns can quickly be obtained and provide information concerning their carbon chain length and functional group. For example, a 16-carbon chain aldehyde could be suggested from the GLC data. EAG analyses with a complete library of monounsaturated acetates, aldehydes and alcohols then could reveal sites of unsaturation and also support the 16-carbon aldehyde prediction. The 16-carbon aldehyde standards should give higher responses in general than the other series of compounds, and a histogram of response amplitudes of that series should indicate the double bond position to which the sensory cell is optimally attuned. Maximum response to (*Z*)-11-hexadecenal (*Z*11-16:ALD), for example, would indicate that at least one set of sensory cells is sensitive to this particular arrangement of active sites and that these molecular features probably are found in the main pheromone component structure. Additional pheromone components could be involved in the pheromone but not be evident from the EAG analyses with the monounsaturated standards, although they could have been indicated in the GLC–EAG analysis.

Pheromone components possessing two double bonds also can be predicted from the antennal responses to the monounsaturated compound series. A monounsaturated standard with unsaturation in common with one of the unsaturation sites on the pheromone molecule apparently has more acceptor affinity than monounsaturated standards with double bonds in the other positions. An example is given in Fig. 7 with the European vine moth, *Lobesia botrana*. The GLC–EAG analysis of crude pheromone extract

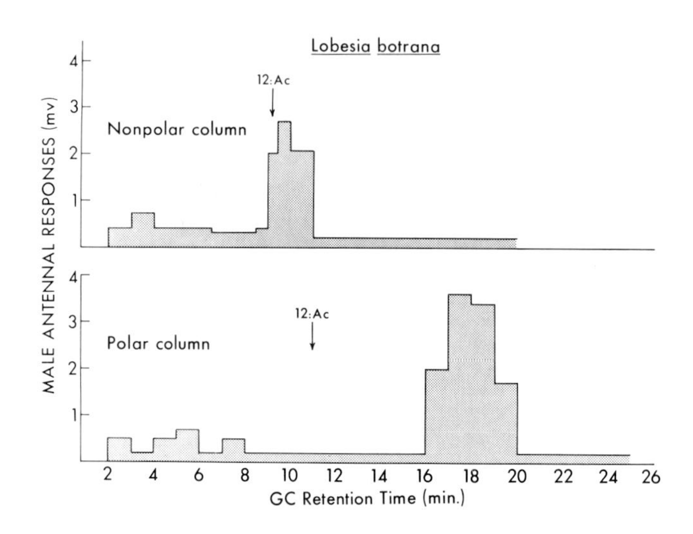

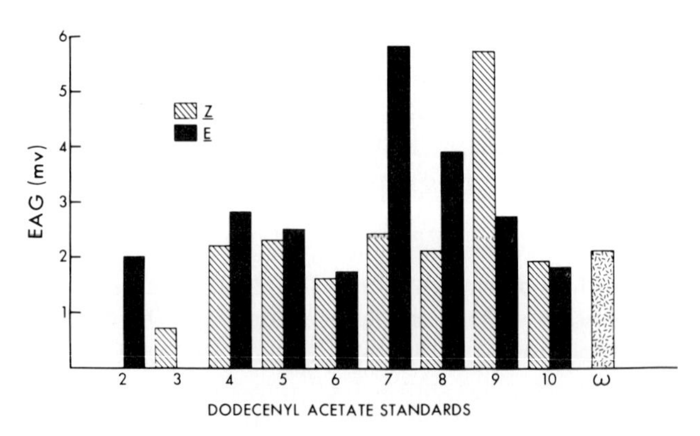

Fig. 7. The top two figures show the male *L. botrana* antennal responses (EAG) to 1-min collections of GLC effluent of female sex pheromone gland extract. The active collections are close to the retention time of dodecyl acetate (12 : Ac) on the nonpolar column, but are much later (more polar) than 12 : Ac on the polar GLC column. The EAG profile of responses to standard monounsaturated compounds shows highest activity to *E*7- and *Z*9-12 : Ac. The pheromone was found to be (*E*,*Z*)-7,9-dodecadienyl acetate.

indicated that the pheromone was a 12-carbon acetate with retention on a polar column similar to that of a conjugated double bond system. Antennal responses to a set of 12-carbon chain acetate standards revealed a sensitivity to the *E*-7- and the *Z*-9-isomers. The combined data predicted the pheromone to be (*E*,*Z*)-7,9-dodecadienyl acetate, which later was confirmed by chemical and physical analyses and behavioral tests.

2. Defining Isomer Blends

Many complex pheromone structures cannot be predicted by the electroantennogram technique, but require sophisticated chemical and physical analyses for their characterization. Regardless of structural complexity, however, it is important to know the isomeric purity of the natural pheromone components and to determine the effect of other closely related isomers. This information not only is important in defining the whole pheromone blend, but also for research on using pheromones in insect control programs.

a. *Positional Isomers.* Pheromone components that are functional group analogues or different carbon length homologues have different physical properties, and thus are easily separated by normal chromatographic procedures. Since the compounds would possess different relative vapor pressures, it would be difficult to emit a precise mixture of these chemicals at various temperatures and to maintain a specific component ratio for long distances in the field. Therefore, ratios of these components usually are not as critical in luring males as are some mixtures involving compounds of similar volatility. A precise ratio of positional isomers has been found to be essential for moth activation and upwind anemotaxis in a number of species. Specificity to a particular ratio makes it possible for several sympatric species to use the same pheromone components, but each with a ratio that is unattractive to the other species. Two tortricid moth species in the apple orchard of The Netherlands have similar mating times and also use the same two pheromone components, Z9-14:Ac and Z11-14:Ac. One species, *Adoxophyes orana,* uses a 9:1 ratio of components and the males are captured in large numbers by traps emitting that ratio, while increasing amounts of Z11-14:Ac results in a sharp decrease in trap catch (Fig. 8). The other species, *Clepsis spectrana,* utilizes the opposite ratio, 1:9, and is not attracted to the 9:1 ratio traps that capture *A. orana* males. In each case the females produce a specific ratio and the corresponding males are optimally attuned to that ratio. With *A. orana* single sensillum recordings from male antennal olfactory hairs have shown that two sense cells respond to the pheromone; one cell is more sensitive to Z9-14:Ac and the other cell is more sensitive to Z11-14:Ac. The brain evidently requires a particular activity pattern generated by the proper excitation of these two cell types in a number of olfactory hairs.

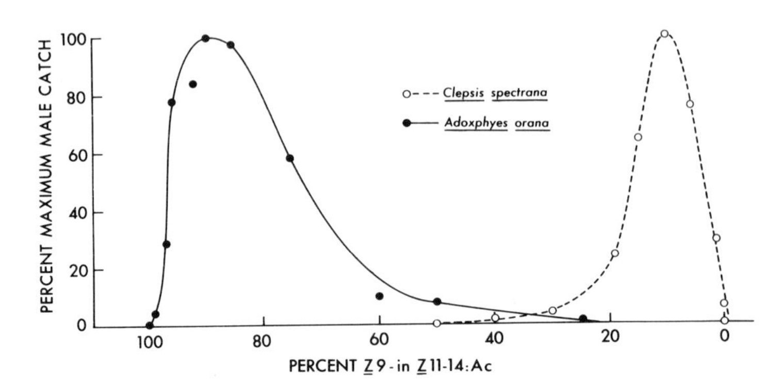

Fig. 8. Trap catches of two sympatric tortricid moth species by mixtures of Z9- and Z11-14 : Ac. Males of *A. orana* are maximally captured by a 9 : 1 ratio, whereas males of *C. spectrana* are captured best by a 1 : 9 ratio. (From Minks *et al.*, 1973.)

b. *Geometric Isomers.* Many pheromones include only one compound of a pair of geometric isomers or a specific ratio of the two isomers. In *Dendroctonus* beetles, *exo*-brevicomin is one of the main pheromone components. The beetles can differentiate between the *exo* and *endo* isomers (**5,6**), which have been isolated from the beetles, but there does not appear to

(+)-*exo*-brevicomin (**5**) *endo*-brevicomin (**6**)

be a required ratio to which they are optimally attuned. Two of the four identified cotton boll weevil pheromone components are a pair of geometric isomers (**7,8**). Again, there does not seem to be a precise ratio of isomers that is obligatory for attractancy.

(*Z*) (**7**) (*E*) (**8**)

In the Lepidoptera, however, long-distance upwind anemotaxis is often dependent on a particular mixture of geometric isomers. The leafroller pheromones listed in Table I illustrate the various blends used by a number of closely related moth species. Many of these species are sympatric and use

TABLE I

Leafroller Moth Pheromones Utilizing a Specific Blend of Geometric Isomers

Species	*Z*11-14:Ac	*E*11-14:Ac	Other components
Choristoneura rosaceana	97%	3%	Z11-14:OH
Argyrotaenia velutinana	92	8	12:Ac
Archips mortuanus	90	10	*Z*9-14:Ac, 12:Ac
Archips argyrospilus	60	40	*Z*9-14:Ac, 12:Ac
Archips podana	50	50	
Archips semiferanus	30	70	
Archips cerasivoranus	15	85	
Platynota stultana	12	88	E11-14:OH

pheromone specificity as an isolating mechanism. It is important for pheromone-trapping and other insect control programs utilizing pheromone components to know the isomeric blend used by each species.

EAG and single sensillum recordings with *Argyrotaenia velutinana* indicate that at least two types of cells are involved in the perception of the two isomers. It is not known, however, how different females of the various species regulate the production of their own unique blend. Research with *A. velutinana* has shown that the isomer ratio in several hundred moths only varies about ±4% from the mean ratio of 92:8. The moths emerged from field-collected pupae and were analyzed individually by GLC. The geometric isomer ratio also was found to be constant from its first appearance in pupal extracts and throughout the life of the female moths. The first appearance of pheromone in pupae 1 day before eclosion and the increase in pheromone production in the first 4 days following moth emergence were correlated with the appearance and increasing abundance of agranular endoplasmic reticulum as well as the appearance and accumulation of lipid droplets in the pheromone gland cells (Fig. 9). These data indicate that the pheromone is produced in the pheromone gland cells, but do not provide information on possible enzymatic processes for isomeric blend regulation. Insects reared on a diet containing oleic and linoleic as the only lipid sources still produced the normal quantity of pheromone and the same 92:8 ratio of component isomers.

c. *Optical Isomers.* The stereochemistry of biologically active compounds is important not only in terms of geometric isomers, but also in terms of optical isomers (enantiomers). Any compound that has a nonsuperimposable mirror image is said to possess chirality. Many life processes are chiral dependent, so it is not surprising to find that odor perception also utilizes chiral differences. It has been well documented that the human nose differentiates the odors of some enantiomeric pairs. For example, optically pure *R*-(−)- and *S*-(+)-carvone are detected as a strong spearmint odor and a

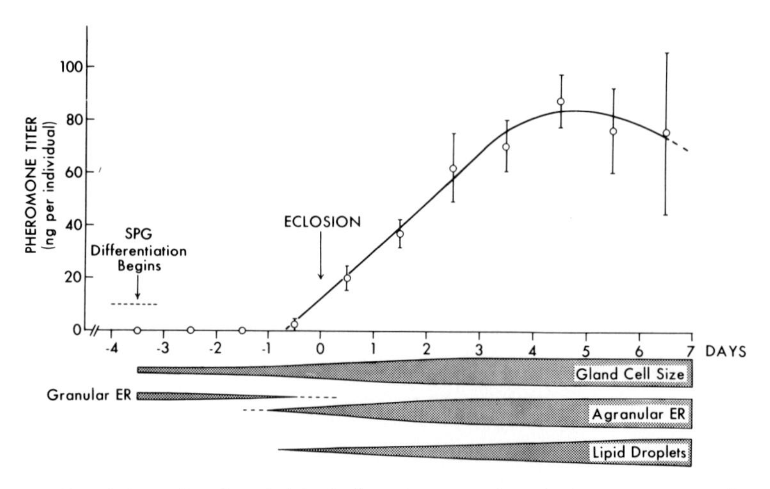

Fig. 9. Correlation of redbanded leafroller, *Argyrotaenia velutinana,* age, sex pheromone gland morphology, and titer of pheromone (*Z*11- and *E*11-14 : Ac, 92 : 8). (From Miller and Roelofs, 1977.)

caraway odor, respectively. Humans cannot distinguish between the (*R*)- and (*S*)-enantiomers of 2-octanol, but worker bees can be trained to extend their proboscis specifically to a single enantiomer of carvone and of 2-octanol. Adult female mosquitoes are much more attracted to (*S*)-lactic acid, which is the main component in mammalian sweat, than to (*R*)-lactic acid, and female spruce budworm moths, *Choristoneura fumiferana,* are stimulated to oviposit by *S*-(+)-α-pinene, but not by the *R*-(−) enantiomer. With pheromones, sensillum receptor molecules also must possess appropriate chirality to preferentially bind to one enantiomer. Specific blends of enantiomers would most likely interact with several cell types, similar to the perception processes of positional and geometric isomer mixtures.

Pheromone components possessing centers of chirality cause additional difficulties for the identification and synthesis of the possible optical isomers (enantiomers). A chiral center could be produced with an allene system, such as with (−)-methyltetradeca-(*E*)-2,4,5-trienoate (**9**), a reported sex

(**9**)

attractant of the dried bean beetle, *Acanthoscelides obtectus,* or by an asymmetric carbon atom, which is bonded to four different atoms or groups of atoms. An example is sulcatol (**10**), an aggregation pheromone of the

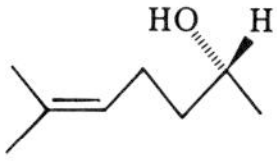

HO H H OH

(R)-(−) sulcatol (S)-(+)
(10A) (10B)

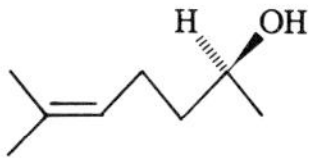

ambrosia beetle, *Gnathotrichus sulcatus*. The two enantiomers, which are nonsuperimposable mirror images, have identical physical and chemical properties in an achiral medium. However, they can be differentiated because they rotate plane-polarized light in opposite directions, but in equal amounts, and they interact differently with chiral chemical agents. Conventional synthesis of this structure yields both forms in equal numbers to give a racemic mixture (±). Natural products, however, are synthesized with optically active enzymes and can be produced as a pure enantiomer or as some particular enantiomeric mix. Sulcatol (**10**) is produced by the ambrosia beetles in a mixture of 65(+)/35(−).

If an insect utilizes an optical isomer mixture that is close to 50/50, it is quite possible that the normally synthesized racemic mixture would almost be as active in the field as the natural pheromone and would be the best to use for field programs. In addition to the 65(+)/35(−) sulcatol mixture, other optical isomer blends produced as insect pheromone components include 60(−)/40(+) *trans*-verbenol (**11**) in the southern pine beetle, *Dendroctonus frontalis*, and 50(+)/50(−) seudenol (**12**) in the Douglas fir beetle, *Dendroctonus pseudotsugae*.

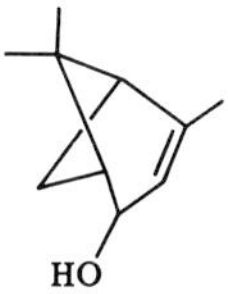

OH

HO

trans-verbenol seudenol
(**11**) (**12**)

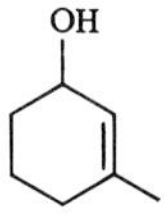

Insects that utilize only one of the enantiomers are sometimes not attracted to the racemic mixture, especially if the wrong enantiomer is effective in suppressing behavioral responses to the pheromone. Synthetic pheromone for the Japanese beetle, *Popillia japonica*, did not attract beetles in the field until the correct enantiomer [*R*-(−)] (**13**) was tested. The gypsy moth, *Porthetria dispar*, is attracted best to the pure (+)-disparlure (**14**),

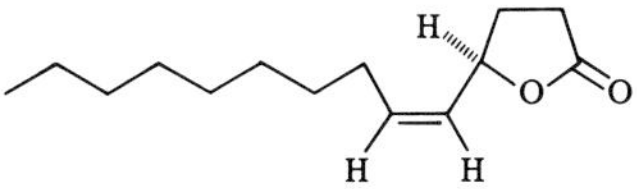

H O O
H H

(**13**)

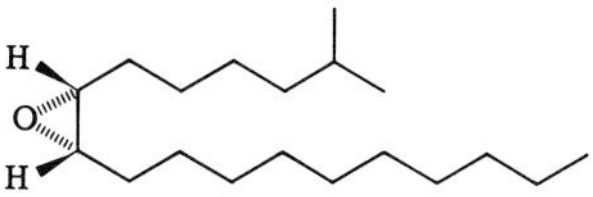

H
O
H

(7*R*, 8*S*)-(+)-disparlure
(**14**)

only poorly to the racemic mixture, and not at all to the (−)-enantiomer. In contrast, males of the nun moth, *Porthetria monacha,* a sympatric species with *P. dispar* in Europe, are responsive to mixtures of the disparlure enantiomers. The five-spined engraver beetle, *Ips grandicollis,* aggregates in response to *S*-(−)-ipsenol (**15**) and very poorly to the racemic mixture or to

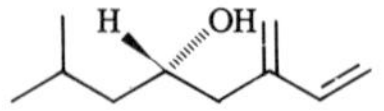

(-)-ipsenol
(**15**)

the *R*-(+)-isomer. In several species of leaf-cutting ants, *Atta texana* and *A. cephalotes,* the alarm pheromone, *S*-(+)-4-methyl-3-heptanone is 100 times more active than the (−)-enantiomer.

Pheromone components of the western pine beetle, *Dendroctonus brevicomis,* were found to be (+)-*exo*-brevicomin (**5**) and (−)-frontalin (**16**).

(-)-frontalin
(**16**)

Previously the best field-trapping lure had been found to be a mixture of myrcene and the racemic mixtures of *exo*-brevicomin and frontalis. Use of (1*R*,5*S*,7*R*)-(+)-*exo*-brevicomin or (1*S*,5*R*)-(−)-frontalin, however, significantly increased the flight response of both sexes to the lure, whereas the corresponding antipodes gave poor results. Analyses of the natural materials showed that the insects produced the single enantiomers (+)-brevicomin (**5**) and (−)-frontalin (**16**).

Reproductive isolation thus can be effected with chiral pheromone component blends as well as with the other isomeric blends. Two *Ips* species of bark beetles that produce ipsdienol (**17**) as a pheromone component actually

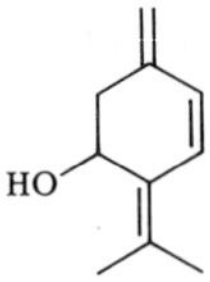

ipsdienol
(**17**)

use different chemicals since *I. paraconfusus* produces almost pure (+) and *I. pini* Idaho produces pure (−). In two genera of pine sawflies, species were found that use 3,7-dimethylpentadecan-2-ol or the corresponding acetate or propionate esters. It appears, however, that species in one genus use one optical isomer and species in the other genus use another optically active form.

The complexities involved in identifying the pheromone structures possessing chirality are illustrated in the characterizations of the two beetle-produced pheromone components of the elm bark beetle (**1**,**2**). The basic structure of 4-methyl-3-heptanol (**1**) was identified by comparison of its mass, IR and NMR spectra with those of a synthetic sample. The compound possesses two chiral centers, so the synthetic compound exists as two diastereomers, but only one was found in the natural product by GLC analysis. It was determined to be a single enantiomer by interpretation of the proton and fluorine NMR spectra of the Mosher derivatives [R(+)-α-methoxy-α-trifluoromethylphenylacetic acid derivative]. The bicyclic ketal structure of compound **2** (multistriatin) is identical to the ring structure of two other beetle aggregation pheromones, brevicomin (**5**) and bicyclic frontalis (**16**). The basic structure of **2** was assigned by mass, IR,and NMR spectral data, and the carbon skeleton was determined by the mass spectra of compounds produced by hydrogenolysis on palladium. A nonstereospecific synthesis of multistriatin (**2**) gave four diastereomers labeled α, β, γ, and σ. The α-isomer was found to be biologically active, so the pheromone component was labelled α-multistriatin. The *endo* configurations assigned to the α-isomer were based on stereospecific synthesis of the isomers, NMR signals,and isomer interconversions. The absolute configuration was determined by synthesizing the isomers from (S)-(+)-2-methyl-3-butenoic acid. Comparison of the specific rotation of the natural material with that of the pure synthetic enantiomer showed that the pheromone component was a single enantiomer, (1S,2R,4S,5R)(−)-α-multistriatin.

III. PHEROMONE-BAITED TRAPS

Chemical characterization of pheromone components must be followed by extensive behavioral testing to determine the activity of each identical component. Field evaluations can be conducted by observing the insects that respond to various test mixtures, but most investigators use trap catch as an indicator of sex attractant activity. Increases and decreases in trap catch with different mixtures and ratios provide information on blends that are the most effective in luring insects to traps, but these tests do not define the individual behavioral roles. One of the pheromone components could effect large increases in trap catch when added to an attractant mix, but the added component could be affecting the close-range processes, such as increased landing, and not long-distance upwind anemotaxis. Nevertheless, pheromone lures developed in the field–trapping tests can be important tools in pest management programs.

A. Population Monitoring

Pheromone lures have been developed for many pest species around the world. The potency and specificity of a good pheromone trap make it a valuable tool in monitoring moth flights and, in some cases, in assessing the population density. The impact on conventional insecticide usage is that sprays are applied only when the pest population is present and of sufficient density to require suppressive measures.

Precise timing of chemical sprays for each grower is particularly useful when there are only one or two major pest species that dictate the spray applications. This situation exists with the codling moth, *Laspeyresia pomonella,* on apple and pears in western United States and Canada. The pheromone (*E,E*)-8,10-dodecadien-1-ol, can be used to determine when the moth flight has been initiated. Instead of using a preventive spray schedule in which applications are made at regular intervals, decisions to spray can be made on a more rational basis. Data on the first appearance of the moths in the pheromone traps can be combined with other information to help decide when the pesticides should be used. Additional information could include previous moth-emergence patterns and weather records for the season. The first moth appearance also can be used as a biological fix point for the accumulation of thermal units. Moth catch over a certain interval of thermal units can help to predict population densities and the accumulated thermal units can be used to predict egg hatch.

Predicting egg hatch is important for some pest species because insecticide spray can be most effective on the vulnerable newly hatched larvae as they disperse to their feeding sites. An example is with the summerfruit tortrix moth, *Adoxophyes orana,* a major pest of apple in The Netherlands. The pheromone was identified as a 90:10 mixture of *Z*9-/*Z*11-14:Ac. Pheromone traps were dispensed to the apple growers and used to determine the initiation of moth flight in each orchard. A correlation between larval dispersal and moth flight was used to precisely time insecticide spray applications. Use of this system reduced the conventional five to seven preventive sprays to about three or four precisely timed and more effective sprays.

Pheromone-monitoring traps also can be useful in crop situations involving a multispecies pest complex. Apples in the eastern United States are plagued with such a complex. In addition to various diseases, mites, aphids, curculio, scale, leafhoppers, and the formidable apple maggot, there are over fifty species of leafroller moths whose larvae can be found feeding on apple in New York. Only about five or six leafroller species, however, are considered to be major pests. A New York State apple pest management project was established in 1973 to integrate all pest management techniques into a control program that would reduce pesticide use without reducing quality and quantity of fruit. The program was based upon a continuous monitoring of orchards for pertinent weather data, pests, chemicals used, and beneficial organisms. Pheromone traps were used for the five major lepidopterous pest species and sprays were applied only when necessary. Growers in this program were able to decrease insecticide and miticide usage up to 50%. In addition to conserving the beneficial organisms in the orchards, decreased pesticide usage greatly lowered the operational cost per acre. In a typical example, the cost of pesticide usage was lowered from $250/ha (hectare) to $150/ha in 1 year.

B. Pheromone Traps for Insect Suppression

A primary goal in insect pheromone research is to use the acquired data for the development of a control technique that is specific for the pest species as well as being innocuous to the environment. Pheromone traps that reduce the abundance of one or both sexes of a species would be excellent for this purpose if they could be used effectively and economically. The only killing agent would be the traps themselves, and the chemical lures not only are quite nontoxic, but are released at extremely low rates compared to insecticide applications. A typical pheromone trap emits the pheromone at a rats of ca. 10^{-5} g/hr or less. The density of traps employed, however, will depend greatly on factors that vary with each species and each situation. Important factors could include (a) the pest population density, (b) the effective range over which a pheromone trap can compete with a virgin adult, (c) the percent catch needed to effect the desired population decrease, and (d) adult emergence and mating characteristics. Pheromone traps could be more effective with species that exhibit protandry (males emergence preceding female emergence) or with species in which the responding individuals are attracted to the continuous-emitting traps over a broader activity period than that during which the organisms emit pheromone. Several examples in which pheromone traps were employed for insect suppression are given below to illustrate further the potential and some problems of this technique.

1. Redbanded Leafroller Moth, *Argyrotaenia velutinana*

The redbanded leafroller has a very broad host range, but has established itself, along with several other tortricid species, as an important pest in commercial apple orchards in the eastern United States. The pheromone was found to be a precise blending of *Z*11-/*E*11-14:Ac (92:8) (Table I) for long-distance upwind anemotaxis, and dodecyl acetate in conjunction with them to mediate close-range precopulatory behavior. Traps emitting these at a rate of approximately 1 μg/hr from polyethylene caps were calculated to be about twice as effective in luring males as were virgin females. Trap effectiveness could not be increased by faster release rates because trap catches decrease at higher rates. The male moths apparently become confused, adapted or habituated with the higher pheromone concentrations.

The pheromone traps compete with female redbanded leafroller moths for males, and so the traps must be quite efficient to prevent males from mating before eventually being captured by a trap. It has been calculated that the traps must capture at least 95% of the male population in order to effect a subsequent population decrease. If one assumes that (a) the males are polygamous but do not mate or attempt to mate more than once each 24 hours, (b) the females are monogamous, (c) the overall sex ratio is 1:1, (d) the moths have an 80-day life span, and (e) the traps are equal to 2 females, it can be calculated that the trap:female ratio needed for 95% control is 2.5 to 1. This would be feasible for low pest populations, but not for high infestations. Experience proved both points.

In 1969, a heavily infested 8-hectare commercial apple orchard was used as a test plot for pheromone trapping. Sectar traps (Fig. 10) with the three-component lure in polyethylene caps were deployed at the rate of three traps/tree. The 2400 traps captured over 17,000 redbanded leafroller males in the summer flight. This would indicate the presence of 17,000 females, or a trap:female ratio of 1:3.5 after multiplying the actual number of traps by 2. It was calculated that this low ratio would give only 48% control, which would allow 8000 females to mate and produce about 320,000 surviving larvae (20%

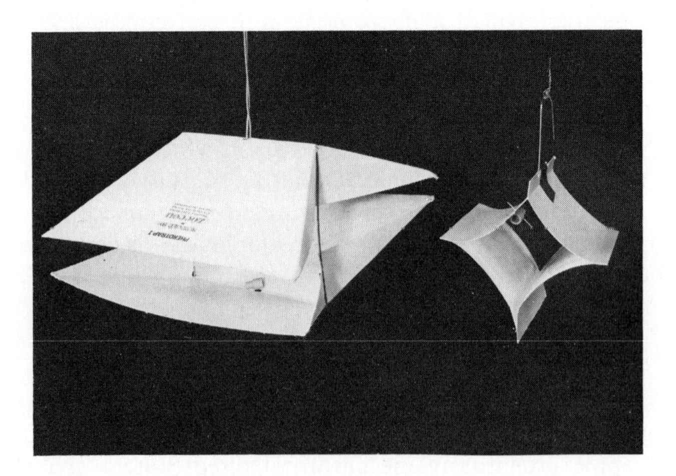

Fig. 10. Examples of commercial (Zoecon Corp.) sticky-coated pheromone traps. The Pherocon trap on the left and the Sectar trap on the right each are baited with a rubber septum charged with pheromone.

of 200 eggs/female), or about 400 larvae/tree. The theoretical percent fruit damage would be 40% and the actual fruit damage found in this test orchard was 32%. The calculated number of traps needed to provide 95% control with that population was 5000 traps/ha, or 50 traps/tree. This not only is impractical physically and economically, but would be a failure in terms of trapping males because the high concentrations of pheromone present from all the traps would cause a disruption of male orientation (see Section V,A).

Most commercial apple orchards have greater than 98% control of the insect pests, and so a second pheromone-trapping test was conducted in a 6-ha commercial orchard on the southern shore of Lake Ontario. The purpose of the test was to see if pheromone traps could maintain a commercially acceptable low population level of redbanded leafroller moths, even though there was high population pressure from surrounding woods and fields. In the first year, 2 pheromone traps were placed in each tree to give a total of 1100 traps. Only 723 males were captured in the first redbanded leafroller flight and 76 males in the second flight. There was only 0.1% fruit damage in that year. The test area was enlarged to 16 ha in 1970 and the trap density reduced to 1 trap/tree. Although thousands of male redbanded leafroller moths were captured in the surrounding areas, the pheromone traps succeeded in maintaining the orchard population at an extremely low level. The experiment continued for 4 years and the fruit damage was only 0.09% in the test orchard when the test was terminated. A nearby 6-ha check orchard was set up in 1971 and the fruit damage due to redbanded leafroller shot up to 12% in the first year.

It was shown that pheromone trapping could be an effective insect control technique, but other factors made it unusable in the case of the redbanded leafroller moth. The obliquebanded and threelined leafrollers were two salient reasons. No insecticides were used in the test orchard to control the numerous leafroller pests of apple, and some other leafrollers, particularly the obliquebanded and threelined, rapidly infested the orchards. The pheromones are known for these pests, but a separate trap must be used for each species to preserve the specific pheromone component ratios emitting from each trap. The obliquebanded leafroller moth, *Choristoneura rosaceana*, is trapped with a 97:3:10 ratio of *Z*11-/*E*11-14:Ac/*Z*11-14:OH, and the threelined leafroller moth, *Pandemis limitata,* with a 91:9 ratio of *Z*11-/*Z*9-14:Ac. The use of numerous traps in each tree to trap all the different species would be impractical from many standpoints. The mass trapping technique would be applied better in situations involving only one major lepidopteran pest species.

2. Western Pine Beetle, *Dendroctonus brevicomis*

The western pine beetle is responsible for an annual loss of an estimated billion board feet of ponderosa pine in the western United States. The beetle is dependent on these trees since they must attack and kill living trees in order to reproduce. The adult female beetle generally finds a suitable host tree and begins to bore through the outer bark into the phloem tissue. While boring in the phloem, the females release *exo*-brevicomin (**5**) and myrcene and attract predominantly males to the host tree. The males release frontalin (**16**) and the combination of pheromones attracts beetles in a 1:1 ratio throughout the mass attack. After mating, the female excavates an egg gallery and deposits eggs singly in niches. The resulting larvae feed in the phloem throughout the first instar and then move to the outer bark for the remainder of their development. If the beetles do not aggregate successfully above a certain threshold attack density, they do not reproduce.

Lures designed to release *exo*-brevicomin, frontalin, and myrcene at rates of 1, 1, and 4 mg/hr, respectively, were able to duplicate the high level of attraction of western pine beetles that occurs during the mass invasion of ponderosa pine. Research was conducted to determine if the synthetic attractants could be used to suppress mass attacks below the threshold attack density that kills the trees and allows the beetles to reproduce.

A "trap-out" strategy was tested, but was much more complex than the lepidopteran mass-trapping tests. The concentration of beetle attractants had to be regulated so that they were competitive but not too high or beetles would mass-attack and kill trees before reaching the traps. The lures and traps had to be designed so that a minimum of natural enemies of the beetles were trapped. It had been found that *exo*-brevicomin was highly attractive to one predator, *Temnochila chlorodia* (Coleoptera: Trogositidae). The placement of traps also was important because of the possible variable effects of the type, size, and physiological state of nearby trees, and of other environmental features of each trap site.

The first study was conducted in 1970 in a 65 km^2-test area at Bass Lake, California. Large vane-traps were used to capture the attracted beetles, because it had been found that the catch increased as the trapping surface was increased. The traps consisted of four sticky-coated screen panels (each 0.76×2.0 m) erected 2.3 m above the ground to a metal pipe with the panels at right angles to each other. The entire test area was surveyed with traps placed on a 0.8-km grid throughout the flight season (May through October). The survey traps released 2 mg pheromone/hr and were used to determine the beetle population density and location within the test area, and the possible correlation of this information with the infested trees before and after the beetle flight.

Two suppression plots and two check plots, each 2.56 km^2, were set up within the test area. Suppression traps releasing 20 mg pheromone/hr were dispensed at the rate of 66 traps/1.3 km^2 and operated throughout the spring flight. The suppression traps caught an estimated 405,000 beetles and the survey traps caught an estimated 189,000 beetles during the spring flight.

Tree mortality due to the beetles dropped from 227 immediately before the test to 73 trees killed by beetles during the test period.

A similar test was conducted in a larger area at McCloud Flats, California, but the beetle population was much larger than at the Bass Lake site. Beetle catches in 1971 and 1972 were 2,600,000 and 4,300,000, respectively. Tree mortality in this case was not reduced, probably due to the unfavorable ratio of attacking beetles to trapping surface. There is optimism that the trapping technique will be useful in forest pest management programs, and the data suggest that it is possible to operate a trap:beetle ratio that suppresses mass attack densities below some threshold required to kill the trees.

3. European Elm Bark Beetle, *Scolytus multistriatus*

Various bark beetle species have plagued commercial timber production in many countries, but they also are pests in shade and ornamental trees. A widely known example of the latter is the European elm bark beetle, which is the principal vector for the Dutch elm disease pathogen *Ceratocystis ulmi*. The Dutch elm disease has been primarily responsible for the disappearance of elm trees in the northeastern United States and poses a threat for the remaining American elms. The behavior of the elm beetle is similar to that of the *Dendroctonus* beetle in that there is a dispersal flight of emerged beetles that can cover several kilometers and then the virgin females find other elm trees for breeding sites.

The virgin females tunnel into the inner bark and release the aggregating pheromone (**1,2**) that attracts many other beetles to the breeding site. Both sexes are attracted in equal numbers. Females initiate new attacks in the host tree, but the males scurry over the bark to locate and mate with several tunneling females. Mated females do not attract beetles to the breeding site since they do not release the alcohol component (**1**), and release of (**2**) alone is not attractive.

Control of aggregating beetles with their sex pheromones appeared to be a specific method for reducing the spread of the Dutch elm disease. Large-scale aeration of logs infested with virgin female beetles was conducted (see Section II,A,2,c) to provide material for the identification of the three active chemicals (**1, 2,** and **3**). Following chemical syntheses of the components, research was carried out in the field on optimizing the pheromone component ratios and dosages, as well as trap design, size, placement, and height. Many different styles of traps were tried, but the best ones were flat sticky sheets of cardboard that contrasted with the background (Fig. 11). If the panels were hung with the sky as background, then dark ones were best, but if they were attached to trees, then white traps were best. Increasing the size of poster traps gave increased effectiveness in trapping beetles, but a medium size, 46 × 66 cm, was selected as the "standard" trap size. The most effective trap sites were found to be fully exposed utility poles. The

Fig. 11. An elm bark beetle poster trap positioned on a telephone pole. The sticky surface traps the pheromone-attracted beetles.

beetles readily can locate the traps fixed 3 m above the ground and are not attracted to trees that also could be attacked.

Pheromone-release studies showed that increased dosages of (**1**) and (**3**) gave increased catches, at least up to 400 μg/day, whereas the release rate of (**2**) had to be less than that of (**1**) and release rates above 300 μg of (**2**)/day strongly depressed trap catches. Regulation of these release rates over long periods (3 months) was accomplished by using controlled-release dispenses. Conrel hollow fiber dispensers and Hercon laminated plastic dispensers were found to be the most suitable for mass-trapping studies. The traps released about

400 μg 4-methyl-3-heptanol (**1**), 100 μg α-multistriatin (**2**), and 800 μg α-cubebene (**3**) per day for at least 120 days. This rate is equivalent to the release rate of 2000 females/hour.

Survey traps were found to be useful in detecting new infestations of the beetles, and to determine the density of elm bark beetles in areas under local Dutch elm disease control. The use of pheromone traps for reducing Dutch elm disease was a primary goal, however. Advantages of this situation compared to the Lepidoptera is that the traps kill both male and female beetles and they trap the destructive stage of the insect. In 1975, traps deployed at 30–50-m intervals in a 520-ha area in Detroit captured 4 million beetles, but the Dutch elm disease increased in the test plot from 4.3% in 1974 to 7.4%. In a similar test in Ft. Collins, Colorado involving 1350 ha, the traps caught 1.5 million beetles and the disease rate was reduced from 3.5% in 1974 to 2.8% in 1975. In the latter case, there was no appreciable source of immigrating beetles to be attracted into the test area, and so the trapping had a more positive effect.

In addition to the grid technique of mass trapping, an encirclement technique also was investigated. Groves of elms were surrounded with pheromone traps to prevent immigration of the beetles. At several locations this resulted in a decrease in the infection rate of the Dutch elm disease.

In 1976, the studies evaluated further the encirclement and the grid techniques. The grid tests were modified by increasing the trap intervals to 100 m, doubling the surface area of the traps, and emphasizing trap placement on fully exposed tree poles or utility poles. The encirclement technique was employed in Evanston, Illinois where two or three rows of traps completely encircled the city. The rows were separated by 50–75 m and the traps within a row were spaced 20–40 m apart. The 750 traps captured over 3.5 million beetles. The beetle population within the city was reduced, but there was no corresponding decrease in Dutch elm disease. Continued trapping, however, could eventually curb the spread of the disease.

IV. TECHNIQUES UTILIZING PHEROMONE BAIT

A. Pathogen Dissemination

Stored product insects are major pests in food-processing plants and warehouses, and have been the target for innovative pest control programs. Pheromone traps are important in detecting new infestations so that pest control personnel can minimize their spread and the product damage. At low infestation levels, the pest habitats are usually restricted, so there has been much interest in developing specific control measures that can be applied selectively where they are needed. The following case illustrates how a pheromone can be used as a lure to inoculation devices for the dissemination of a pathogen.

A number of serious stored-product pests are found within the genus *Trogoderma* (Coleoptera: Dermestidae). The dermestid beetle, *Trogoderma glabrum* will be used as an example. In this species the adult male remains concealed, except for an 8-hour active period in the middle of the photophase during which the males are most responsive to the female-released sex pheromone.

It also is during the active period that the females produce and release maximum quantities of pheromone from glands associated with the seventh abdominal sternite. Females in a typical "calling" position elevate their abdomen so that they appear to be resting on their heads. While in this inclined position, they partially expose the ovipositor segments in slow, rhythmic pulsations. Extraction of these segments during the calling periods gave extract with at least 100-fold more pheromone activity than that recovered during other times. This observation probably is due to the phenomenon discussed above that aldehyde pheromone components many times are stored in an inactive state until actively released.

The most active compound extracted from whole females was (*E*)-14-methyl-8-hexadecen-1-ol. However, this alcohol is released into the air by the beetles in very low amounts and is fairly inactive when compared to airborne collected material. Aeration of live beetles with collection of the volatiles on Porapak Q revealed that the corresponding aldehyde, (*E*)-14-methyl-8-hexadecenal, is a highly active pheromone component; at least 1000 times more active than the alcohol. The aldehyde was not detected in extracts of whole females, but is as active as the total airborne pheromone. In related species, the *Z* isomer of the aldehyde is the major component in *T. inclusum* and *T. variabile,* whereas a 92:8 mixture of *Z*:*E* is found in *T. granarium*. The importance of enantiomeric purity still must be researched.

The aldehyde pheromone component can replace a female beetle in eliciting the three stages of male mating responses during the active period: arousal/searching; preliminary recognition, and genital (copulatory). Arousal can involve elevation of the head or antennae followed by low motion toward the pheromone source. In the next stage, the male comes into contact with the female or model. Over the next 2 to 30 seconds the male will explore the female with its maxillary palps (mouthpart sensillae), as well as intermittently pushing its head against the female (head-butting). Frequent contact with the caudal end of the female abdomen will move the male to the genital phase, in which it extends the aedeagus and explores the female genitalia until it discovers the genital pore. Only 10 ng of aldehyde on a paraffin model was needed to elicit the full repertoire of mating responses in 50% of the males.

Pheromone-baited traps can be used for insect detection and surveillance in food-processing plants and warehouses, but they also can be used in conjunction with insecticide or pathogens. The stored-product insects harbor an array of pathogenic protozoans. An example is *Mattesia trogodermae*, a schizogregarine that has been found to infest almost 90% of the *Trogoderma* larvae in certain years. The mortality rate for *T. glabrum* and several related species exposed to the spores (6.25×10^7 spores/g medium) is 100%. Adult males externally contaminated with *Mattesia* transmit infective spores to females and to larvae that feed on them. Spores are transmitted further through the population by venereal transmission, by contaminating eggs during oviposition, by cannibalism of infected or dead larvae, and by elimination of spores in feces of infected larvae followed by ingestion of these spores by healthy larvae.

It has been shown that male beetles can get adequately contaminated by spores if they are lured by pheromone to *Mattesia* powder. An "inoculation device" was a 2.5×2.5 cm^2 of corrugated cardboard contaminated with *Mattesia* (2.3×10^6 spores/mg powder) baited with 10 μg of the aldehyde, (Z)-14-methyl-8-hexadecenal. In a laboratory test this device succeeded in infecting 100% of the male beetles, which then infected 92% of the females. Under simulated warehouse conditions (16 adult pairs/m^2), there was 85% mortality in treated F_1 populations and over 99% F_2 mortality. There is much promise in using this pheromone–pathogen system for insect suppression.

B. Poisoned Hosts

European foresters unknowingly utilized pheromones as a bark beetle control strategy for several centuries. They provided felled or girdled trees that aggregated beetles in preference to standing timber. The beetles were destroyed more easily in the trap trees, which were subjected to mass attacks by pheromone-responding beetles. With the availability of synthetic pheromone, it became possible to bait selected trees and concentrate the population in a predetermined area. The beetles then are harvested with the trees or killed by insecticidal treatment. A variation of that technique is to attract beetles to trees injected with cacodylic acid. The acid treatment loosens the bark and makes the trees unsuitable for maturation of beetle broods. This method is much cheaper than treating the tree with insecticide. Another type of host-baiting has been used for the control of the cotton boll weevil. This example is discussed in greater detail.

Alternative methods of insect control have been sought for the cotton boll weevil because of the huge annual usage of insecticide for this pest. It became resistant to DDT in the 1950's and now over $70 million is spent annually for organophosphate insecticides for boll weevil control. Pheromones have been researched to provide one of the tools for pest management programs on cotton.

Studies first revealed that the male boll weevil releases a pheromone that not only attracts the females, but acts as an aggregant for both sexes. Various techniques were used to collect the pheromone, including aeration with an activated charcoal trap and extraction of fecal material and insects. With the latter method, 4,500,000 weevils and 54.7 kg of fecal material were extracted with methylene chloride. Steam distillates of the extracts were extracted with methylene chloride and the extracted material fractionated on silica gel and then on silver nitrate-coated columns. Two fractions that were inactive alone, but active when combined, were fractionated further on GLC columns. Four components were identified in ratios of 13 : 10 : 1 : 1 as follows: (+)-*cis*-2-isopropenyl-1-methylcyclobutane ethanol (**18**); (*Z*)-3,3-dimethyl-$\Delta^{1,\beta}$-cyclohexane ethanol (**19**); (*Z*)-3,3-dimethyl-$\Delta^{1,\alpha}$-cyclohexane

CH_3 OH CH_3 H $\qquad$ H C CH_2OH

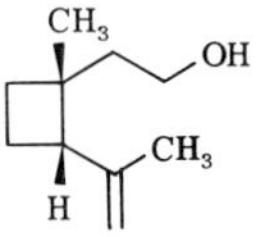

(+)-grandisol

(**18**) $\qquad$ (**19**)

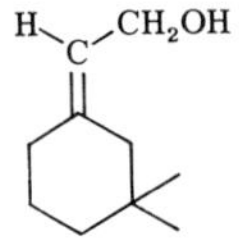

acetaldehyde (**7**); and (*E*)-3,3-dimethyl-$\Delta^{1,\alpha}$-cyclohexane acetaldehyde (**8**), respectively.

A prerequisite for the production of the pheromone appears to be male adult feeding on cotton. The insect could utilize some chemicals from the plant in the synthesis of the pheromone components, although radioactivity incorporation (0.02%) from labeled acetate, mevalonic acid, or glucose showed that at least some of the pheromone is synthesized *de novo*. It also has been suggested that a precursor of myrcene, a major constituent of the cotton bud, could be used to produce the two types of compounds (Fig. 12).

The four pheromone components have been prepared by many chemists and have been available in large quantities for use in field-trapping studies. Efforts then were directed to the development of slow-release formulations of the four components and to the design of effective boll weevil traps. Traps generally used were yellow cone-shaped traps with a collecting funnel at the top to take advantage of the boll weevils' attraction to yellow and their negatively geotaxic behavior after landing on the trap surface. The pheromone traps were immediately used for survey and detection of incipient boll weevil populations. The weevil response to pheromone, however, changed throughout the year, with minimal response during midseason, and the trap catches did not always relate directly to the population densities and movement. Nonetheless, the pheromone trap has been invaluable in evaluating the effectiveness of various boll weevil control programs, particularly in assessing overwintering spring boll weevil population densities, after fall diapause treatment programs.

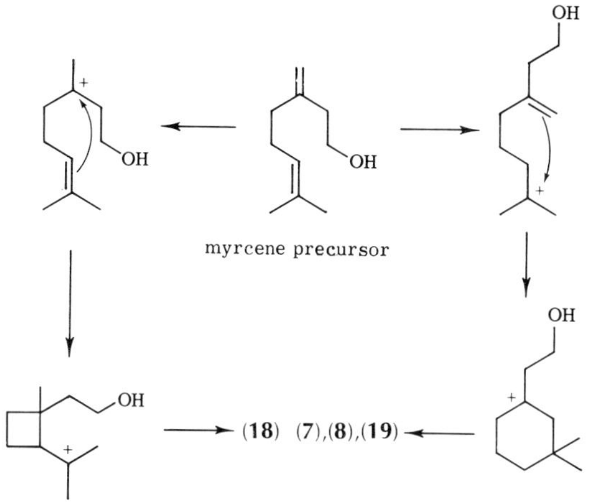

Fig. 12. Possible biosynthetic route to cotton boll weevil pheromone components from a precursor of myrcene. (See Hedin, 1976.)

Mass-trapping experiments for the boll weevil showed that this technique is useful only at low densities (10 weevils/acre or less). Programs have utilized traps around cotton fields in the spring and infield traps for subsequent adult generations. Conclusions from these programs are poorly defined for the effectiveness of pheromone traps in weevil suppression, but integration of trapping with other techniques has much potential in developing an economical management program for the boll weevil. One of the other strategies is the host-baiting technique.

It had been known for years that the greatest weevil infestations occurred in the earliest planted cotton. It was suggested that the weevils could be concentrated for control in a few rows of an early maturing cotton variety. This technique received much attention when the synthetic boll weevil could be added to the trap-crop for increased attraction of overwintered weevils. Early tests in 1971 showed that pheromone-baited traps placed throughout small restricted areas of a cotton field aggregated 71% of the weevils in the treated areas. Further studies showed that early planted cotton (3–5% of the total field area) treated with aldicarb insecticide and baited with the pheromone effectively suppressed small populations of overwintered boll weevils. Use of this technique potentially would lower the weevil population in the rest of the cotton field and maintain damage below an economic level or at least significantly delay the economic damage. This particular host-baiting technique would be used best in conjunction with a fall suppression program that reduces weevil populations to low levels. A major problem with this technique is to find a suitable cotton that is significantly more advanced in growth than the regular cotton.

C. Trail Pheromones

Insect sex pheromones have received the most attention in research directed toward insect suppression, but some efforts have been made to utilize some of the complex array of pheromones in social insects. Of particular interest have been the trail pheromones of termites and ants. Potentially they could be used to attract pest insects to poisoned baits, which could have a devastating effect if brought back to the nest.

Social insects present challenges in establishing biological activity and are characterized by complex secretions, but some interesting and active chemicals have been identified. In *Reticulitermes* termites, a trail pheromone is secreted by the sternal gland to mark the source of suitable broods for other workers. An active principal of this secretion was found to be (*Z,Z,E*)-3,6,8-dodecatrien-1-ol. Worker termites would follow a 10-cm path produced with only 1×10^{-11} g of synthetic pheromone. Although woods decayed by the fungus *Lenzites travea* also produce the same compound, it was shown that the termites are able to synthesize the compound from glucose. A much more complex compound was found to be a trail pheromone for *Nasutitermes* termites. The diterpene hydrocarbon, neocembrene-A (**20**),

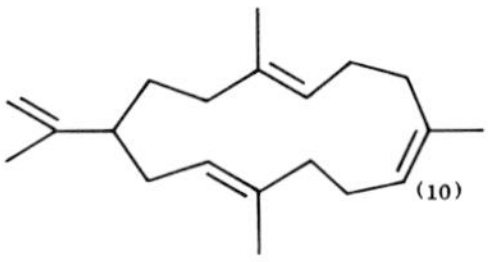

(10)

neocembrene-A

(**20**)

also was isolated from plant oils in Russia and in India. Thus, large quantities of (**20**) are available for research on its use in pest control programs.

With ants, research has been conducted on combining trail pheromones with baits. An example is with the leaf-cutting ants, *Atta texana* and *A. cephalotes*. Foraging ants of these species use trail pheromones between their colonies and plant materials, which they gather as a substrate for a fungus that they cultivate as their sole source of food. The major trail pheromone component was identified as methyl 4-methylpyrrole-2-carboxylate (**21**), a trace constituent of the proteinaceous secretion from the poison gland. The identification first was accomplished by extracting the whole bodies of 3.7 kg of *A. texana* worker ants. The extract was short-path distilled and the distillate fractionated on four GLC columns. The major active compound was identified, but at least four other fractions elicited lesser amounts of activity. However, the single compound was detected by the worker ants in a trail at 8×10^{-11} g/cm. At this rate, only 0.33 mg of the compound would be required to draw a detectable trail around the world. Slight changes in the structure of (**21**) greatly reduce or eliminate the

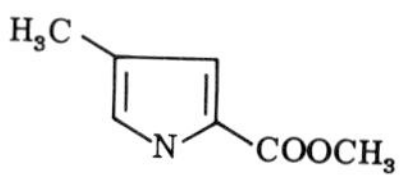

H_3C ... N ... $COOCH_3$

(21)

trail-following activity. Changes in the arrangement of substituents on the ring eliminated activity, except when the 4-methyl group was moved to the 3-position. Good activity was obtained when the 4-methyl group was replaced with a halogen atom, but all changes in the 2-position carbomethoxy moiety greatly reduced the activity.

In laboratory tests it has been shown with three species of leaf-cutting ants that the addition of trail pheromone (**21**) increases the effective size of pieces of bait and increases their chance of being found. It also could increase the foraging activity of the nest, which would increase the chance of the bait being found. In the West Indies, citrus pulp has been used as an arrestive material with baits. Preliminary experiments showed that the trail pheromone compound **21** can be used in this system to increase the effective bait size. It was calculated that 1 mg **21** in 1 kg of pulp would be the optimum ratio.

The trail pheromone of the Pharaoh ant, *Monomorium pharaonis,* also has been investigated for practical reasons. The Pharaoh ant is a typical insect, but has become a pest in heated buildings such as bakeries and hospitals, in many nontropical countries. In hospitals they can establish large colonies that are very difficult to exterminate. The workers are carriers of pathogenic bacteria and possess the ability to penetrate bandages and other minute holes. Attempts to control this pest led to investigations of the trail pheromone, which is laid by worker scouts returning from a food source and then followed by recruited workers.

Methylene chloride extracts of thousands of worker ants were fractionated by GLC to yield three alkaloids and an unidentified bicyclic unsaturated hydrocarbon that appear to be major components of the trail pheromone. The alkaloids were identified as 3-butyl-5-methyl octahydroindolizine (**22**); 2-butyl-5-pentylpyrrolidine (**23**); and 2-(5^1-hexenyl)-5-pentylpyrrolidine (**24**).

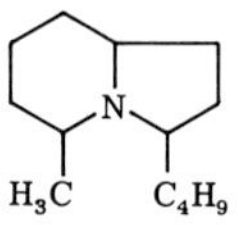

N, H_3C, C_4H_9 (22) — H_9C_4, N, C_5H_{11} (23) — $H_{11}C_5$, N, C_6H_{11} (24)

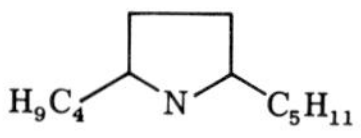

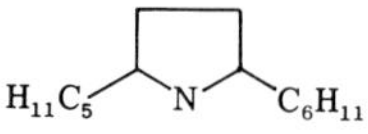

Research with various stereoisomers is being conducted to describe the best attractant system for practical applications.

V. MANIPULATING THE COMMUNICATION CODE

Once the communication code is broken, it can be duplicated and used in a variety of ways to lure the insects to traps or poison. Information on the communication system also can be used to develop air permeation techniques in which the continuous release of certain chemicals in the field can effectively interrupt the mating processes of selected pest species. The chemicals modulate pheromone perception by effecting quantitative and qualitative changes in the chemical stimuli and alter the normal behavioral responses to the natural pheromone. Research must be conducted to determine the most effective disruptant chemical and the release rate required to interrupt mating.

A. Pheromone Components

It has been observed many times in field-trapping studies that a pheromone trap becomes less attractive if the pheromone is released at rates above the optimum. The artificially high concentration of pheromone causes sensory adaptation or possibly central nervous system habituation resulting in the receiving insect becoming disoriented or nonactivated. The pheromone sensory cells are attuned optimally to the pheromone components, and so these chemicals are excellent candidates to effect changes in the sensory input.

Initial experiments with the cabbage looper moth, *Trichoplusia ni,* pheromone, (*Z*)-7-dodecenyl acetate, showed that males could not locate live female traps in areas in which the pheromone was evaporated at a rate of 100 mg/hr/ha from planchets spaced 3 m apart. Further experiments with the cabbage looper revealed that the evaporators could be spaced as far as 400 m apart as long as the total evaporation rate per hectare was at least 0.07 mg/hr. If the total pheromone release per area was below that level, the male moths had no trouble in finding the female moths.

If several pheromone components are used in a specific ratio for long-distance upwind anemotaxis, several possibilities exist for effecting communication disruption. An individual component could be used in the air permeation technique, or a certain blend of components could be used. Generally it has been found that the natural mix of components is the most effective at the lowest release rate per unit area. For example, with the redbanded leafroller moth, permeation with the natural mixture (92:8) of *Z*-11/*E*-11-tetradecenyl acetates was much more effective than with a 50:50 mix or the minor component (*E* isomer) used alone. A rate of 5 mg/hr/ha was effective with the 92:8 mixture, but a rate of 15 mg/hr/ha was required for the 50:50 blend. Although a higher rate is required with the 50:50 mixture, the fact that disruption can be effected with this unnatural blend means that a high release of this mixture could be effective in interrupting mating of all the leafrollers (Table I) that utilize these two pheromone components. An advantage of this technique is that disruptants can be effective with several species, and, also, structurally dissimilar disruptants can be combined and disseminated in the same operation for various pest species.

Developing the technology for an appropriate long-term slow release system for the disruptants has been the subject of much research. Maintaining a liquid film in a planchet or replacing wicks is useful for experimental purposes, but in practical usage, a better release system is required. One technique used with varying results is to microencapsulate the chemicals. The small plastic capsules (ca. 50 μm) can be dispensed with conventional spray equipment and provide a relatively uniform distribution of chemical over the entire area. With the gypsy moth, aerial applications of the pheromone at the rate of 20 g pheromone/ha appeared to be effective in interrupting mating for at least 5 weeks. Various methods of encapsulating are available from several commercial companies, but the drawback at present is the exponential release of chemical in the first few weeks. It also has been found that there is an inefficient release of chemical, with some formulations retaining at least 90% of the chemical. This increases the amount of chemical required and decreases the effective time period per application.

A more constant release rate is obtained by dispensing the chemicals from small hollow fibers. In Australia, mating interruption of the Oriental fruit moth, *Grapholitha molesta,* was achieved by permeating the air in apple orchards with pheromone, (*Z*)-5-dodecenyl acetate, released from polyethylene microcapillary tubes. The use of four tubes/tree (10 mg/hr/ha) resulted in no infestation in the treated orchard compared to a heavy infestation in an adjacent check orchard. Investigators in the United States showed that only 2.5 g of pheromone, (*Z*)-9-dodecenyl acetate, per ha released from hollow fibers gave excellent disruption of male grape berry moth orientation to pheromone traps in a vineyard for over 2 months. A more detailed example of this technique is given below for the pink bollworm moth, *Pectinophora gossypiella*.

The pink bollworm moth has been a prime target for alternative methods of control because the extensive use of chemical insecticides, which kill predators and parasites as well, usually leads to outbreaks of other pests, such as the cotton bollworm (also corn earworm), *Heliothis zea,* and the tobacco budworm, *Heliothis viriscens*. It is a serious pest in most of the cotton-producing countries in the world, including the desert southwest in the United States. Efforts particularly are being made to halt its northern advances in California. Pheromone traps are very useful in detecting the presence of new infestations, and the economical use of communication disruptants for control would have many advantages over the present heavy usage of organophosphate insecticides.

Pink bollworm mating–interruption research first was conducted with (Z)-7-hexadecenyl acetate, a compound that is about 100 times less attractive in traps and less effective in disruption than the natural pheromone. The pheromone then was identified as a 50:50 mixture of the *Z,E*- and *Z,Z*-isomers of 7,11-hexadecadienyl acetate and was used in further studies. In 1974, a large area test involving the 1600 hectares of cotton in the Coachella Valley of California gave encouraging results, but it was not completely satisfactory. Pheromone was dispersed from widely-spaced evaporators, which needed to be resupplied every week and which continually had to be adjusted to the height of the growing cotton.

A followup of that experiment was conducted in 1976 with excellent results. Three fields of cotton (23 ha) were involved in the test in which economically practical levels of the pheromone was rebased into the air throughout the season. The chemicals were released from 104-mm long, thermoplastic fibers (Conrel Co.), which were fashioned into hoops of 1.5 revolutions (22-mm diameter) and filled with pheromone. The hoops were distributed by hand throughout the cotton fields on a 1 × 1 m grid to stimulate dispersal of the fibers by aerial application. They were attached to the upper stems of the cotton plants every 3 weeks to provide air permeation near the tops of the growing plants. A total of 230,000 hoops were used for the five applications between mid-May and early September. Monitoring pheromone traps in the treated fields showed that the technique was almost 100% effective in reducing moth captures, and was successful in suppressing the populations of pink bollworm larvae infesting the cotton bolls. Only 33 g pheromone/ha provided pink bollworm control equal to that of neighboring fields receiving a mean number of two to six insecticide applications per hectare. At the present cost of $0.80/g for pheromone, the disruption technique would cost only $26/ha, which is comparable to the cost of the conventional program. The remaining step was to develop the technology for applying the fibers from an aircraft.

A commercial company (Conrel) developed the necessary technology for dispersal of the fibers and used it in 1976 in a large mating-interruption test for pink bollworm moths. After obtaining the required toxicological data on the pheromone chemicals, they were granted an experimental use permit from the Environmental Protection Agency to treat 1160 hectares of cotton in Arizona and the Imperial Valley of California. The pheromone was dispersed from chopped fibers (8-mil capillaries, 1.9 cm in length) that were sealed in the middle and had both ends open. Each fiber released the chemicals at the rate of 0.03 μg/hr. The fibers were dispersed throughout the fields with a ground rig, which was designed to coat the ejecting fibers with a sticky material to assure coverage at the top of the foliage.

The fibers were applied whenever pheromone-monitoring traps in the field started to catch males, indicating a need to bring the pheromone level back up to a disrupting level. Approximately 25–40 g of pheromone was used per hectare per season in this program, and the number of insecticide applications needed in these fields was drastically decreased. As an example, one farm of six fields used 12 and 11 insecticide applications in 1974 and 1975, respectively, but only three in 1976 when 8 applications of pheromone were used for mating interruption. Good results were obtained even though this area usually sustains severe pink bollworm infestations and neighboring fields followed the same trend in 1976. The success of this program has encouraged the Conrel Co. to attempt mating interruption on 8000 hectares of cotton in 1977.

Insect resistance to this method of suppression should develop very slowly, but probably depends on the amount of genetically determined natural variation in both the emitting females and the responding males. As previously discussed, there was not much variation in the 92:8 *Z*:*E* ratio of pheromone components produced by wild redbanded leafroller moth females, but continuous mating disruption by the air-permeation technique using pheromone could result in selection processes that favor other ratios. An increasing release rate of disruptant would be required if a portion of the population eventually had a higher behavioral response threshold for the 92:8 mixture and a much lower threshold level for a blend quite different from that being used to permeate the air. This population change would require changes in both pheromone production and perception. This could be a slow process in species (such as the redbanded leafroller moth) that produce a very narrow range of isomeric blends. Also, with several species no phenotypic variation was detected in the responding males. Oriental fruit moths are attracted optimally to a 93:7 mix of *Z*- and *E*-isomers of 8-dodecenyl acetate, but can be captured at ca. one-fifth the rate by 97:3 and 89:11 blends. A capture–mark–recapture technique was used to show that males coming to the unnatural mixtures did not represent disparate phenotypes, but were attracted optimally to the pheromone mixture of 93:7 if given a choice. Some species also have been trapped by a very narrow range of isomers. The oak leafroller moth, *Archips semiferanus*, was trapped only by mixtures between 60:40 and 70:30 of *E*-/*Z*-tetradecadienyl acetates, with maximum catches in traps baited with a 66:34 component ratio. Slow evolutionary changes in the pheromone system coupled with the option of raising the release rate or changing the chemicals in an air permeation program should delay resistance for many years.

B. Oviposition-Marking Pheromones

A different type of insect control system utilizing pheromones is that proposed for some economically important *Rhagoletis* flies, such as the apple maggot, *R. pomonella,* European cherry fruit fly, *R. cerasi,* walnut husk fly, *R. completa,* the black cherry fruit fly, *R. fausta,* the eastern cherry fruit fly, *R. cingulata,* the western cherry fruit fly, *R. indifferens,* and the blueberry maggot, *R. mendax.* Female flies drag their ovipositor on the fruit surface immediately after egg-laying to deposit a marking pheromone that deters subsequent attempts at egg-laying in the parasitized host. This marking behavior would disperse eggs among available fruits and provide an efficient utilization of the natural resources for larval development.

The marking pheromones are water-soluble and can be washed from the fruits. Aqueous solutions of the pheromone are effective in deterring egg-laying when reapplied to the fruit. This suggests a practical role for the pheromone in pest management programs. One technique would involve spraying the marking pheromone on six trees and then using the next tree as a trap tree. Sticky spheres or yellow ammonium-baited panels in the trap tree would capture the flies as they arrive at the only tree without oviposition deterrent. Use of this technique would depend on the chemical identification of the marking pheromone. Research with the European cherry fruit fly has shown that the marking pheromone can be partially purified from an aqueous solution by an anion exchange procedure. The molecular weight is less than 10,000 and apparently it is stable in boiling 1 *N* hydrochloric acid for at least 10 minutes and in 1 *N* sodium hydroxide for at least 1 minute. If finally characterized, the pheromone would be synthesized and potentially used in fruit fly control programs.

C. Nonpheromonal Component Disruptants

The communication system also can be disrupted by chemicals that are structurally similar to the main pheromone component and modulate that antennal input, or that are naturally occurring chemicals that function to decrease the response of the receiving insect. A few examples of these various types are given below.

1. *Heliothis* Species

A main pheromone component in a number of *Heliothis* species, including the corn ear worm moth, *H. zea,* and the tobacco budworm moth, *H. virescens,* is (Z)-11-hexadecenal. A structurally similar formate analogue, (Z)-9-tetradecenyl formate (Fig. 13), was found to affect greatly the male moth's response to the pheromone. Apparently the formate has very good affinity for the antennal receptors and is able to modulate the pheromone input signals. The formate eliminates male orientation to pheromone traps when released along with pheromone or calling females, and the formate interrupts the mating process when used in the atmospheric permeation technique. It is effective in reducing mating at the low rate of 2 mg/hr/ha.

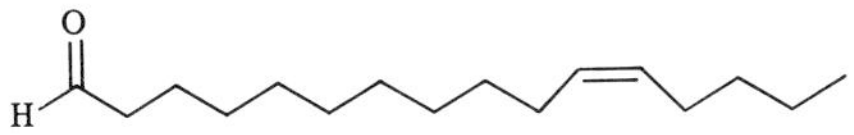

(Z)-11-hexadecenal

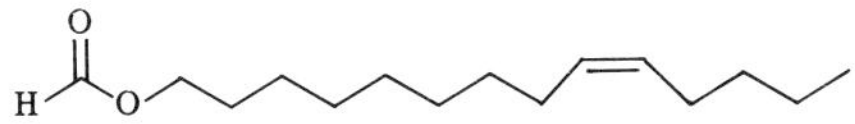

(Z)-9-tetradecenyl formate

Fig. 13. Two structurally similar compounds involved in pheromone research of *Heliothis* moth species. The aldehyde on top, (Z)-11-hexadecenal, is a primary pheromone component of several species. The formate on the bottom, (Z)-9-tetracenyl formate, is not a pheromone component, but functions as a communication disruptor at low concentrations.

Advantages of using this compound include its stability to oxidation relative to the pheromone aldehyde, and decreased male activation in the treated area by utilizing a chemical that does not elicit behavioral responses.

2. Red Bollworm Moth, *Diparopsis castanea*

The red bollworm is a major pest of cotton in Central and Southern Africa. Female moths of this species, similar to most lepidopteran species studied, release a sex pheromone to attract males for mating. The sex pheromone gland is similar to that found in other Lepidoptera in that it consists of modified intersegmental tissue located between the eighth and ninth abdominal segments. Observations of the female precopulatory behavior revealed a certain sequence of events. For over an hour after the onset of scotophase (darkness), the females are restless and exhibit sporadic locomotor activity. They then cling to a vertical surface or to the underside of a horizontal surface and extend the ovipositor to expose the translucent epithelium of the pheromone gland. When in this calling position, the female extends the wings so that they are not folded back over the abdominal tip, and the antennae are elevated. The ovipositor tip is curved upwards and moved slowly vertically and horizontally. The females will continue to call throughout the night or until mated.

Analyses of pheromone gland extract revealed the presence of a number of compounds, including (*E*)-9,11-dodecadienyl acetate, (*Z*)-9,11-dodecadienyl acetate, 11-dodecenyl acetate, (*E*)-9-dodecenyl acetate, and dodecyl acetate. The total quantity of the compounds varied roughly between 2 ng during photophase (period of light) to a maximum of 14 ng in the scotophase. Laboratory and field studies showed that the first three compounds were pheromone components and that a mixture of 15:1:4, respectively, was very potent in eliciting male activity and was a very good lure in traps. On the other hand, increasing concentrations of (*E*)-9-dodecenyl acetate progressively decreased male excitation in laboratory bioassays and trap catch in field tests. The biological significance of this naturally-occurring "inhibitor" is not known, but it possessed several advantages for use in mating-interruption programs.

Disruption of the mating communication system was effected by air permeation with the diunsaturated pheromone components. It was estimated that aerial concentrations of 10^3–10^5 times the male threshold response level (9.2 molecules/mm^3/sec for red bollworm moths) should be sufficient for mating interruption. In practice a 95–98% reduction in red bollworm oviposition was obtained when the release rate was between 106–440 mg/hr/ha, but less than 80% reduction with rates of 14 mg/hr/ha. A rate of 400 mg/hr/ha was calculated to give a release of 1.5×10^4 molecules/mm^3/sec, which is within the predicted range of aerial concentrations deemed necessary.

The "inhibitor," (*E*)-9-dodecenyl acetate, is more stable and cheaper than the pheromone components, and would have the advantage of not attracting males into the treated fields in air–permeation programs. The inhibitor appeared to be as effective as the pheromone at release rates of ca. 30 mg/hr/ha in disrupting the mating communication system in air–permeation tests, and the inhibitor effected a greater reduction in mating when evaporated into cages of moths. Further research on permeation technology should allow the inhibitor to be used in insect-suppression programs in Africa.

3. Douglas Fir Beetle, *Dendroctonus pseudotsugae*

The Douglas fir beetle is a major pest of the Douglas fir throughout western United States and British Columbia. The beetles breed in epidemic numbers in damaged trees, and the progeny mass-attack and kill selected trees. Similar to the cases of the other bark beetles discussed above, pheromones play an important role in the population aggregation of these beetles.

Females again are the pioneering beetles that make the initial attack on a new host. The beetles first go through a period of dispersal flight, during which they have a high threshold for host stimuli. After a certain duration of flight, however, the beetles are attracted to uninfested Douglas fir trees, presumably due to volatiles from the phloem tissue. It has been proposed that the change from a low threshold level for dispersal flight to a low threshold level for host volatiles is regulated by lipid metabolism during flight. Beetles with more than 20% total lipid were flight positive and host-negative, whereas beetles with less than 20% lipid were host positive. The selective oxidation of monounsaturated fatty acids could be responsible for the behavioral change by producing selective metabolic by-products.

The pioneer beetles initiate gallery construction in the host tree and release aggregating pheromones that attract both sexes. Aggregation occurs in response to female-produced frontalin (**16**) and racemic seudenol (**12**), in combination with host volatiles, such as camphene or α-pinene. Ethanol also increases the aggregation activity of (**16**) and (**12**). The attacking beetles undergo a certain period of dispersal flight, similar to the female pioneer beetles, before responding to the pheromones. The pheromones also inhibit further flight once the beetle has alighted. An important phenomenon then eventually occurs to prevent an oversaturation of the host tree. A chemical is released to decrease the response of beetles in the field to the aggregation pheromones. The chemical was found to be 3-methyl-2-cyclohexen-1-one (MCH), a ketone analogue of seudenol. It was isolated and identified from 14,500 female beetle hindguts, but further research has shown that it is much more abundant in the male beetles. Since MCH acts to suppress aggregation, it became a good candidate for use in preventing mass attack on selected Douglas fir trees.

Research was conducted to determine the optimum release rate and spacing of sources of MCH in order to prevent an increase in Douglas fir beetle populations in attractive, susceptible host trees. Three rates, 0.06, 1.12, and 91 mg/day, were used at three spacings of 1, 10, and 52 release sources per freshly felled Douglas fir tree. Data were collected throughout the summer on the treated trees.

The results were interesting in that the most effective treatments with regard to brood densities were the 0.06 and 1.12 mg/day releasers used 52 per tree. None of the single source-per tree treatments and none of the 91-mg/day treatments was significantly different from the control. It appears that there is an optimum elution rate for suppressing aggregation, and that release rates of MCH above the optimum are not effective. It is possible that sensory adaptation or habituation to MCH reduces its effectiveness.

Efforts now are being made to develop a formulation that can be applied by aircraft and that will release MCH at the optimum rate of 50–100 mg/hr/ha. A commercial formulation (Zoecon) has been developed that will release MCH at a rate of 10 μg/hr/g formulation for over 60 days. This technique has the advantages of not aggregating beetles in surrounding live trees as occurs with the pheromone frontalin, and is less disruptive of predators that are attracted to the pheromone.

A problem common to all of the cases involving communication disruption chemicals is that of registering the chemicals as insecticides. The expense and time involved in acquiring the necessary toxicological data preclude the procedure for disruptants of minor pests. It is hoped that continued research in the chemistry, biochemistry, biology, and behavior of the insects and their communication systems will show the wisdom of using naturally occurring innocuous chemicals in small quantities to manipulate specific pest species.

GENERAL REFERENCES

Beroza, M. (1976). "Pest Management with Insect Sex Attractants". Am. Chem. Soc. Symp. Ser. No. 23, Am. Chem. Soc., Washington, D.C.
Birch, M. (1974). "Pheromones." Am. Elsevier, New York.
MacConnell, J. G., and Silverstein, R. M. (1973). *Angew. Chem., Int. Ed. Engl.* **12,** 644–654.
Roelofs, W. L. (1975). *In* "Insecticides of the Future" (M. Jacobson, ed.), pp. 41–59. Dekker, New York.
Roelofs, W. L., and Cardé, R. T. (1976). *Annu. Rev. Entomol.* **22,** 377–405.
Shorey, H., and McKelvey, J. (1977). "Chemical Control of Insect Behavior: Theory and Application." Wiley, New York.

REFERENCES FOR ADVANCED STUDENTS AND RESEARCH SCIENTISTS

Arn, H., Stadler, E., and Rauscher, S. (1975). *Z. Naturforsch. B* **30,** 722–725.
Baker, T., and Roelofs, W. (1976). Electroantennogram responses of male *Argyrotaenia velutinana* to mixtures of its sex pheromone components. *J. Insect Physiol.* **22,** 1357–1364.
Bedard, W. D., and Wood, D. L. (1974). Bark beetles—the western pine beetle. *In* "Pheromones" (M. Birch, ed.), pp. 441–449. Am. Elsevier, New York.
Birch, A. J., Brown, W. V., Corrie, J. E. T., and Moore, B. P. (1972). Neocembrene-A, a termite trail pheromone. *J. Chem. Soc., Perkin Trans. 1,* pp. 2653–2658.
Borden, J. H., Chang, L., McLean, J. A., Slessor, K. N., and Mori, K. (1976). *Gnathotrichus sulcatus:* synergistic response to enantiomers of the aggregation pheromone sulcatol. *Science* **192,** 894–896.
Browne, L., Birch, M., and Wood, D. (1974). *J. Insect Physiol.* **20,** 183–193.
Burkholder, W. E. (1977). Manipulation of insect pests of stored produces. *In* "Chemical Control of Insect Behavior: Theory and Application" (H. Shorey and J. McKelvey, eds.), pp. 345–351. Wiley, New York.
Butenandt, A., Beckmann, R., Stamm, D., and Hecker, E. (1959). *Z. Naturforsch. B* **14,** 283–284.
Byrne, K., Gore, K., Pearce, G. and Silverstein, R. (1975). Porapak-Q collection of airborne organic compounds serving as models for insect pheromones. *J. Chem. Ecol.* **1,** 1-7.
Cardé, R. T., Baker, T. C., and Roelofs, W. L. (1976). Sex attractant responses of male Oriental fruit moth to a range of component ratios—pheromone polymorphism. *Experientia* **32,** 1406–1407.
Cardé, R. T., Cardé, A. M., Hill, A. S., and Roelofs, W. L. (1977). Sex attractant specificity as a reproductive isolating mechanism among the sibling species *Archips argyrospilus* and *mortuanus* and other sympatric tortricine moths (Leipdiotera: Tortricidae). *J. Chem. Ecol.* **3,** 71–84.
Furniss, M. M., Daterman, G. E., Kline, L. N., McGregor, M. D., Trostle, G. C., Pettinger, L. F., and Rudinsky, J. A. (1974). Effectiveness of the Douglas-fir beetle antiaggregative pheromone methyl cyclohexenone at three concentrations and spacings around felled trees. *Can. Entomol.* **106,** 381–392.
Gaston, L. K., Kaae, R. S., Shorey, H. H., and Sellers, D. (1977). Controlling the pink bollworm by disrupting sex pheromone communication between adult moths. *Science* **196,** 904–905.
Hardee, D. D. (1974). Cotton—the boll weevil. *In* "Pheromones" (M. Birch, ed.), pp. 427–431. Am. Elsevier, New York.
Hedin, P. (1976). *Chem. Tech.* **6,** 444–451.

Hill, A., Cardé, R. T., Kido, H., and Roelofs, W. L. (1975). Sex pheromone of the orange tortrix moth, *Argyrotaenia citrana* (Lepidoptera: tortricidae). *J. Chem. Ecol.* **1,** 215–224.

Jewett, D. M., Matsumura, F., and Coppel, H. C. (1976). Sex pheromone specificity in the pine sawflies: interchange of acid moieties in an ester. *Science* **192,** 51–53.

Kafka, W. A. and Neuwirth, J. (1975). A model of pheromone molecule-acceptor interaction. *Z. Naturforsch. B* **30,** 278–282.

Kaissling, K.-E. (1971). Insect olfaction. *In* "Handbook of Sensory Physiology" (L. M. Beidler, ed.), Vol. 4, pp. 351–431. Springer-Verlag, Berlin and New York.

Kaissling, K-E. (1975). *Verh. Dtsch. Zool. Ges.* **67,** 1-11.

Karlson, P., and Butenandt, A. (1959). Pheromones (ectohormones) in insects. *Annu. Rev. Entomol.* **4,** 39–58.

Karlson, P., and Lüscher, M. (1959). "Pheromones": a new term for a class of biologically active substances. *Nature* **183,** 55–56.

Katsoyannos, B. I. (1975). Oviposition-deterring, male arresting, fruit-marking pheromone in *Rhagoletis cerasi. Environ. Entomol.* **4,** 801–807.

Klimetzek, D., Loskant, G., and Vité, J. P. (1976). Disparlure: differences in pheromone perception between gypsy moth and nun moth. *Naturwissenschaften* **63,** 581–582.

Kochansky, J., Tette, J., Taschenberg, E. F., Cardé, R. T., Kaissling, K-E., and Roelofs, W. L. (1975). *J. Insect Physiol.* **21,** 1977–1983.

Lanier, G., Silverstein, R. M., and Peacock, J. W. (1976). Attractant pheromone of the European elm bark beetle (*Scolytus multistriatus*): isolation, identification, synthesis, and utilization studies. *In* "Perspectives in Forest Entomology" (J. F. Anderson and H. K. Haya, eds.), pp. 149–175. Academic Press, New York.

Madsen, H. F., and Vakenti, J. M. (1973). Codling moth: Use of Codlemone-baited traps and visual detection of entries to determine need of sprays. *Environ. Entomol.* **2,** 677–679.

Marchesini, A., Garanti, L., and Pavan, M. (1969). *Ric. Sci.* **39,** 874–877.

Marks, R. J. (1976). Field studies with the synthetic sex pheromone and inhibitor of the red bollworm, *Diparopsis castanea* Hmps. (Lepidoptera, Noctuidae) in Malawi. *Bull. Entomol. Res.* **66,** 243–265.

Marks, R. J. (1976). Laboratory evaluation of the sex pheromone and mating inhibitor of the red bollworm, *Diparopsis castanea* Hampson (Lepidoptera, Noctuidae). *Bull. Entomol. Res.* **66,** 427–435.

Mayer, M. S., and McLaughlin, J. R. (1975). "An Annotated Compendium of Insect Sex Pheromones." Florida Agricultural Experiment Station's Monograph Series.

Miller, J. R., and Roelofs, W. L. (1977). *Ann. Entomol. Soc. Am.* **70,** 136–139.

Minks, A. K., Roelofs, W. L., Ritter, F. J., and Persoons, C. J. (1973). *Science* **180,** 1073–1074.

Minks, A. K., and Dedong, D. J. (1975). Determination of spraying dates for *Adoxophyes orana* by sex pheromone traps and temperature recordings. *J. Econ. Entomol.* **68,** 729–732.

Mitchell, E., Jacobson, M., and Baumhover, A. (1975). *Heliothis* spp.: disruption of pheromonal communication with (*Z*)-9-tetradecen-1-ol formate. *Environ. Entomol.* **4,** 577–579.

Moorhouse, J. E., Yeadon, R., Beevor, P. J., and Nesbitt, B. F. (1969). Method for use in studies of insect chemical communication. *Nature (London)* **223,** 1174–1175.

Nesbitt, B. F., Beevor, P. S., Cole, R. A., Lester, R., and Poppi, R. G. (1973). *Nature* **244,** 208–209.

Nishida, R., Fukami, H., and Ishii, S. (1974). Sex pheromone of the German cockroach (*Blattella germanica* L.) responsible for male wing-raising: 3,11-dimethyl-2-nonacosanone. *Experientia* **30,** 978–979.

O'Connell, R. (1975). Olfactory receptor responses to sex pheromone components in the redbanded leafroller moth. *J. Gen. Physiol.* **65,** 179–205.

Peacock, J. W., Cuthbert, R. A., Gore, W. E., Lanier, G. N., Pearce, G. T., and Silverstein, R. M. (1975). *J. Chem. Ecol.* **1,** 149–160.

Persoons, C. J., Verwiel, P. E. J., Ritter, F. J., Talman, E., Nooijen, P. J. F., and Nooijen, W. J. (1976). Sex pheromones of the American cockroach, *Periplaneta americana: a tentative structure of periplanone-B. Tetrahedron Lett.* pp. 2055–2058.

Plummer, E. L., Stewart, T. E., Byrne, K., Pearce, G. T., and Silverstein, R. M. (1976). Determination of the enantiomeric composition of several insect pheromone alcohols. *J. Chem. Ecol.* **2,** 307–331.

Priesner, E., Jacobson, M., and Bestmann, H. J. (1975). Structure-response relationships in noctuid sex pheromone reception. *Z. Naturforsch. B* **30,** 283–293.

Prokopy, R. J., Reissig, W. H., and Moericke, V. (1976). Marking pheromones deterring repeated oviposition in *Rhagoletis* flies. *Entomol. Exp.* and *Appl.* **20,** 170–178.

Riley, R. G., Silverstein, R. M., and Moser, J. C. (1974). Biological responses of *Atta texana* to its alarm pheromone and the enantiomer of the pheromone. *Science* **183,** 760–762.

Ritter, F. J., Rotgans, I. E. M., Talman, E., Verwiel, P. E. J., and Stein, F. (1973). 5-Methyl-3-butyl-octahydroindolizine, a novel type of pheromone attractive to Pharaoh's ants (*Monomorium pharaonis*) (L.), *Experientia* **29,** 530–531.

Robinson, S. W., and Cherrett, J. M. (1973). Studies on the use of leaf-cutting ant scent trail pheromones as attractants in baits. *Proc. Int. Congr. Int. Union Study Soc. Insects, 7th,* London pp. 332–338.

Roelofs, W. L. (1975). Insect communication-chemical. *In* "Insects, Science and Society" (D. Pimentel, ed.) pp. 79–99. Academic Press, New York.

Roelofs, W. L. (1977). The scope and limitations of the electroantennogram technique in identifying pheromone components. *In* "Crop Protection Agents—Their Biological Evaluation" (N. R. McFarlane, ed.), pp. 147–165. Academic Press, New York.

Rothschild, G. (1975). Control of Oriental fruit moth [*Cydia molesta* (Busck)] with synthetic female pheromone. *Bull. Entomol. Res.* **65,** 473–490.

Schneider, D. (1974). The sex-attractant receptor of moths. *Sci. Am.* **231,** 28–35.

Shapas, T. J., Burkholder, W. E., and Boush, G. M. (1977). Population suppression of *Trogoderma glabrum* by using pheromone luring for protozoan pathogen dissemination. *J. Econ. Entomol.* **70,** 469–474.

Shorey, H. (1976). Interaction of insects with their chemical environment. *In* "Chemical Control of Insect Behavior: Theory and Application" (H. Shorey and J. McKelvey, eds.), pp. 1–5. Wiley, New York.

Tai, A., Matsumura, F., and Coppel, H. C. (1969). Chemical identification of the trail-following pheromone for a southern subterranean termite. *J. Org. Chem.* **34,** 2180–2182.

Taschenberg, E. F., and Roelofs, W. L. (1976). Pheromonal communication disruption of the grape berry moth with microencapsulated and hollow fiber systems. *Environ. Entomol.* **5,** 688–691.

Trammel, K. (1974). Orchard pest management. *In* "Pheromones" (M. Birch, ed.), pp. 416–421. Am. Elsevier, New York.

Vité, J. P., and Francke, W. (1976). The aggregation pheromones of bark beetles: progress and problems. *Naturwissenschaften* **63,** 550–555.

Vité, J. P., Klimetzek, D., Loskant, G., Heddin, R., and Mori, K. (1976). Chirality of insect pheromones: response interruption by inactive antipodes. *Naturwissenschaften* **63,** 582–583.

Wood, D. L., Browne, L. E., Ewing, B., Lindahl, K., Bedard, W. D., Tilden, P. ., Mori, K., Pitman, G. B., and Hughes, P. R. (1976). Western pine beetle: specificity among enantiomers of males and female components of an attractant pheromone. *Science* **192,** 896–898.

11

Biochemical Defenses of Insects

MURRAY S. BLUM

I. INTRODUCTION

Insects have evolved a multitude of chemical and behavioral defenses to blunt the attacks of an incredible variety of animals as well as microorganisms. Analyses of the chemical defense arsenals of these arthropods have shown that they are natural product chemists *par excellence*. The large number of unique compounds identified in the defensive secretions of these animals emphasizes that insect exocrinology is a highly distinctive field of biochemistry. Although the biosyntheses of most of these products have not been elucidated, the presence of compounds as diverse as alkaloids and alkyl sulfides suggests that many novel metabolic pathways occur in the defensive glandular tissues. Furthermore, the metabolic peculiarities of these invertebrates are accompanied by morphological specializations that insure that the potentially toxic compounds generated are isolated from the sensitive tissues of the producers. This combination of biosynthetic and morphological idiosyncrasies is a characteristic of the chemical defenses of insects.

In this chapter the biochemical aspects of the natural products used by insects in defensive contexts will be emphasized. Glandular adaptations that optimize the synthesis and controlled release of these compounds will be discussed when appropriate.

II. ALLOMONES AND PHEROMONES

Exocrine constituents of secretions adaptively favorable to the emitter have been divided into two classes depending on whether they are utilized in intraspecific or interspecific contexts. When these compounds are used as chemical communicative agents between conspecifics, they are referred to as pheromones. In contrast to these intraspecific chemical stimuli are allomones, compounds that play roles in interspecific interactions which can be either mutualistic or antagonistic. Allomones are exemplified by both the floral compounds that attract pollinators as well as the defensive compounds that insects unleash at their antagonists. However, this classificatory dichotomy is often inexact, especially when applied to the exocrine products of social insects. For many species of termites, bees, wasps, and ants, the same compound can function simultaneously as a pheromone for recruiting nest mates to attack an intruder, or as an allomone when applied to an assailant's body. Such parsimony means that a single compound can be used in more than one way.

Formic acid is typical of exocrine products that can function simultaneously as both a pheromone and a defensive allomone. This highly ionized acid is a stimulatory olfactant and is also a very effective predator deterrent, especially against invertebrates such as ants. Formic acid is the defensive hallmark of formicine ants, and for many of these species this poison gland product also functions as a powerful releaser of alarm behavior. When a worker ant sprays its acidic allomone on an intruder, the latter becomes a pheromonal emission source which attracts other aggressive ants to the scene of the encounter. The larvae of some species of notodontid moths and adult carabid beetles also use formic acid as a defensive compound but in these cases it has no pheromonal function.

A variety of other exocrine compounds are used by insects as either pheromones or deterrent allomones. Many species of stingless bees lay chemical trails with a mandibular gland secretion that is dominated by benzaldehyde. However, when these bees are molested, benzaldehyde becomes a key defensive element, and is one of the main constituents smeared on the body of an antagonist. Male noctuid moths also use this aromatic aldehyde as a pheromone, often liberating up to 100 μg in the presence of a female. On the other hand, some adult ants, chrysomelid beetle larvae, and millipedes utilize benzaldehyde for purely defensive purposes.

Formic acid and benzaldehyde are typical of a wide variety of exocrine compounds that have been independently evolved in many unrelated arthropod lines to function as allomones and/or pheromones. It has been suggested that defensive compounds—which are often powerful olfactants—could be exploited admirably by gregarious or social insects to function as pheromones after they had secondarily evolved a programmed responsiveness to these products. Thus, deterrent allomones may be the evolutionary precursors of many of the pheromones that are characteristic of social insects. Adapting defensive compounds, for which biosynthetic pathways are already developed, to function as pheromones would be highly adaptive since new metabolic pathways would not have to be evolved for pheromonal syntheses. Such an evolutionary development would possess especially great selective value, since the new pheromone would obviously still possess its original defensive function.

III. SOURCES OF INSECT DEFENSIVE PRODUCTS

It appears that the defensive compounds produced by insects are generally synthesized *de novo* in specific glands by endogenous enzymes. However, the possibility that some deterrent allomones may be products of microbial metabolism cannot be excluded, especially since certain insect exocrine compounds (pheromones) are believed to be synthesized by bacteria or fungi. Nevertheless, in view of the almost total lack of information on the metabolic peculiarities of arthropod symbiotes, it seems premature to offer more than superficial speculation about their role in synthesizing defensive compounds.

A. Exogenously Derived Compounds

Allomones may also arise from exogenous sources. A wide variety of lepidopterous, hemipterous, and orthopterous species sequester plant natural products in their tissues and these compounds may render their insect sequestrators emetic and/or toxic to predators. (This topic will be discussed more thoroughly in Section V,M).

Some species of insects that possess well-developed defensive glands augment the deterrent "punch" of their glandular exudates by sequestering plant natural products in them. Lygaeid bugs in the genera *Oncopeltus* and *Lygaeus,* after feeding on milkweed seeds, sequester cardenolides from this food source. Some of these compounds are concentrated to relatively high levels in the metathoracic defensive gland of adults and in the middorsal glands of nymphs. Similarly, the pyrgomorphid grasshopper *Poekilocerus bufonius* sequesters cardenolides derived from its milkweed host in its bilobed poison glands. Selective sequestration appears to occur since most of the cardenolides in the food plant of *P. bufonius* are not found in the glandular exudate. The acridid *Romalea microptera* appears to appropriate plant-derived compounds in its thoracic defensive glands; it is also likely that the main sequestered compound represents a product of carotenoid catabolism by this insect.

In at least a few cases, insects sequester plant natural products and store them in capacious reservoirs that can be discharged in much the same way as the reservoir of a true exocrine gland. Larvae of the sawfly *Neodiprion sertifer* sequester both monoterpenes and sesquiterpenes from their pine hosts, storing these compounds in large diverticular pouches of the foregut. This terpenoid concentrate, which is discharged from the mouth when a larva is tactually stimulated, serves as a highly effective defensive secretion against a variety of predators. Since none of the pine-derived terpenes are detectable in the feces of the larva, it is evident that these hymenopterans have evolved an efficient mechanism for avoiding the potentially toxic effects which might occur if these compounds reached the midgut.

B. Defensive Compounds of Enteric Origin

In some cases, enteric discharges containing intestinal products are utilized as defensive secretions. Many herbivorous insects, such as grasshoppers, will either regurgitate or defecate when disturbed. These discharges, which are fortified with plant natural products, are repellent to some invertebrate predators. In such cases it is apparently highly adaptive for these insects to feed on plants that contain compounds that constitute irritants when applied to the bodies of other animals. Furthermore, it could be of great selective value for insects not to inactivate metabolically these natural products, since to do so might nullify the repellent properties of these compounds. However, it is also possible that these plant constituents are enterically altered to compounds that are of greater repellency when regurgitated than they would be if they had not been metabolized. Ingested plant enzymes could play a major role in metabolizing these plant constituents into more active compounds during the latter's passage through the digestive tract.

While the repellent constituents in enteric discharges may frequently be of exogenous origin, these defensive compounds are sometime true products of digestive glands. In particular, a wide variety of arthropod species utilize products of the salivary (labial) glands as versatile defensive secretions. The assassin bug, *Platymeris rhadamanthus,* which normally immobilizes its prey with its enzyme-rich saliva, can accurately eject this secretion at vertebrate predators. This salivary venom, fortified with trypsin, hyaluronidase, and phospholipase, produces intense pain and edema if it strikes the mucous membranes on the head. The use of the same defensive mechanism by the spitting cobra *Naja nigricollis* emphasizes the probability that identical—and unusual—defensive systems may be evolved by species in disparate taxa.

The salivary secretions of insects may also be utilized as "glues" with which to entangle potential predators. Syrphid larvae deter aggressive ants with their viscid proteinaceous saliva, as do primitive termite species. In the latter, the proteinaceous salivary exudate is accompanied by *p*-benzoquinones that react with the proteins to produce a rubberlike material that serves admirably to entangle small invertebrates.

Characteristic of many termite species are defensive secretions which originate in the mandibular glands and other cephalic organs, and become viscous after being discharged. These exudates are frequently fortified with volatile constituents (e.g., monoterpene hydrocarbons) that function admirably as repellents. Volatilization of these low-boiling compounds results in considerably increasing the viscosity of the defensive exudate.

The presumably proteinaceous saliva of hemipterans in the genus *Velia* has been adapted to effect an escape reaction while the bugs are moving across the water surface. When disturbed, the veliid discharges its saliva posteriorly from the rostrum and the secretion effectively lowers the surface tension of the water immediately behind it. In effect, the bug rides the contracting water surface and is rapidly propelled from the site of the disturbance. This escape mechanism, termed *Entspannungsschwimmen,* has been independently evolved by beetles in the genus *Stenus,* but the pygidial gland products of these staphylinids reduce the surface tension of the water. The main glandular component, 1,8-cineole, is highly surface active and will propel a paper "beetle" through the water in the same manner as a living staphylinid.

C. Blood as a Defensive Secretion

One of the primary defensive reactions of many species of insects is to discharge blood—often only from sites immediately proximate to the region that has been tactually stimulated. This phenomenon, reflex bleeding or autohemorrhage, is well developed in adults of many species of beetles in the families Coccinellidae, Chrysomelidae, Meloidae, Lampyridae, and Lycidae. While reflex bleeding has been reported to occur in larvae of a few coccinellid and chrysomelid species, autohemorrhage appears to be much less common in immatures than adults. Significantly, autohemorrhagic species can lose substantial amounts of blood without apparent harm to themselves, indicating that this defensive mechanism is highly adaptive.

Discharged blood may not contain any demonstrable natural products, functioning purely as a physical deterrent against such small invertebrate predators as ants. In these cases the blood rapidly clots and frequently entangles the assailant to whose body it has been applied. Blood discharged reflexively by *Diabrotica* beetle larvae deters ants very effectively after it has clotted, but the palatable hemolymph does not appear to offer any protection against vertebrate predators. On the other hand, astringent compounds in the blood of many autohemorrhagic species augment its deterrent potential. (The biochemistry of these compounds will be discussed in Section V,L).

The exudates from the defensive glands of some insects are enriched with blood but the role of the bloodborne constituents is not always clear. The cervical secretions of adult arctiid moths characteristically contain, in addition to pharmacologically active choline esters, a high concentration of blood. The histamine-rich abdominal froth of nymphs of the grasshopper *Poekilocerus bufonius* is enriched with blood, but in addition, contains highly emetic cardenolides that are derived from the food plant (milkweed). Plant-derived cardenolides, along with blood proteins, also constitute the primary defensive compounds in the exudate from the dorsolateral spaces of some species of adult lygaeid bugs. Although blood may occur commonly in the exocrine secretions of insects, its significance, and that of its component blood cells and enzymes, is not readily apparent.

IV. HYMENOPTEROUS VENOMS

Insect species, particularly those in the order Hymenoptera, possess poison glands that produce venomous products which are stored in reservoirs. These venoms can be injected either with a true sting, or with a sharp seta that is driven into the body of an assailant, and qualify as true poison gland secretions in contrast to the salivary toxins of some hemipterans, dipterans, and neuropterans. The salivary "venoms" of species in the latter orders actually appear to be digestive enzymes that have been adapted to function in the immobilization and external digestion of prey. Although these oral secretions can be utilized defensively, they have been evolved primarily for offense, whereas the secretions discharged from the posterior end of insects are generally delivered as part of a defensive reaction.

The main compounds identified as constituents of hymenopterous venoms are presented in Table I. In addition to these poison gland products, it appears that the venoms of all the stinging hymenopterans contain proteins and/or peptides, some of which have been demonstrated to possess precise pharmacological properties. However, for the most part, these compounds are nomenclaturally ill-defined, both in terms of their biological effects and their structural identities. Different fractionation procedures employed by various investigators render it extremely difficult to compare the identities of isolated components whose structures have not been unambiguously established. The practice of classifying peptides by their pharmacological properties is also inexact because of species differences in sensitivity to the same peptide. Furthermore, a single peptide may possess two unrelated activities. For example, the mast cell degranulating peptide in honey bee venom was subsequently determined to be a powerful antiinflammatory agent as well. If

TABLE I

Compounds Identified in the Venoms of Hymenopterous Species

Compound	Apidae	Vespidae		Formicidae			
	Apis	*Vespa*	*Vespula*	*Formica*	*Pogonomyrmex*	*Myrmecia*	*Solenopsis*
Histamine	+	+	+	–	+	+	–
5-Hydroxytryptamine	–	+	+	–	–	–	–
Dopamine	+	+	+	–	–	–	–
Adrenaline	–	+	–	–	–	–	–
Noradrenaline	+	+	+	–	–	–	–
Acetylcholine	–	+	–	–	–	–	–
Hyaluronidase	+	+	+	–	+	+	–
Phospholipase A	+	+	+	–	+	+	+
Histidine decarboxylase	–	–	+	–	–	–	–
Esterase	+	–	–	–	+	–	–
Acid phosphatase	+	–	–	–	+	–	–
Kininlike peptide	–	+	+	–	–	–	–
Apamin	+	–	–	–	–	–	–
Melittin	+	–	–	–	–	–	–
Mast cell-degranulating peptide	+	–	–	–	–	–	–
2,6-Dialkylpiperidine	–	–	–	–	–	–	+[a]
2,5-Dialkylpyrrolidine	–	–	–	–	–	–	+[b]
Formic acid	–	–	–	+	–	–	–

[a]*Solenopsis* (*Solenopsis*) spp.
[b]*Solenopsis* (*Diplorhoptrum*) spp.

this peptide had been simultaneously classified by two independent investigators according to its defined biological effects, it would never have been recognized as being one and the same compound.

Insect venoms are a rich source of biogenic amines, five of which have already been identified in bee, wasp, and ant venoms (Table I). Histamine, a powerful algogen, is the most widespread of these compounds, whereas 5-hydroxytryptamine (serotonin) has only been identified in the venoms of a few species of wasps in the genera *Vespa* and *Vespula*. Two of the catecholamines, noradrenaline and dopamine, are characteristic components of wasp and bee venoms in contrast to adrenaline which has not been commonly encountered as a hymenopterous poison gland product (Table I).

The venoms of wasp species in the genus *Vespa* are singular in containing up to 5% of acetylcholine (ACh) and these secretions constitute the richest source of this compound in the animal kingdom. ACh and histamine, both of which occur in *Vespa* venoms (Table I), act synergistically as a potent vertebrate algogen and interestingly, these two compounds also constitute the pain-producing agents in the stinging nettles or urticaceous plants.

Among the enzymes, only hyaluronidase and phospholipase A appear widely in the poison gland secretions of hymenopterans. In contrast to the direct action of venomous constituents such as biogenic amines, both of these enzymes produce biologically active substances from the host's substrates. Phospholipase A is the major antigen in honey bee venom, constituting about 12% of the dry weight of this secretion. The indirect lytic action of this enzyme appears to be due to the production of lysophospholipids derived from the hydrolysis of structural phospholipids. Possibly histidine decarboxylase, an enzyme recently detected in wasp venoms (Table I), also acts indirectly by producing histamine from the host's histidine. This enzyme may also produce histamine in the poison gland of wasps.

A. Formic Acid

The distribution of formic acid in the Hymenoptera is limited to the poison gland secretions of species in formicine genera such as *Formica* (Table I). This compound, the most highly ionized of the fatty acids, often occurs in the poison gland reservoir as a 60% aqueous solution along with free amino acids and small peptides. While formic acid is very cytotoxic, this polar compound does not penetrate the lipophilic cuticle of arthropods very readily. However, its cuticular penetration is facilitated by nonpolar "carrier" compounds (e.g., *n*-undecane) which are applied in admixture with the formic acid from the Dufour's gland, another sting-associated structure. In addition, formic acid may be sprayed into cuticular wounds inflicted on predators with the ant's mandibles. The skin of vertebrates may be similarly broken by mandibular action before being sprayed with this concentrated acid. Formicine ants have obviously evolved devices for ensuring that their major poison gland product will reach sensitive internal tissues where its cytotoxicity will be manifested.

B. Polypeptides

Apamin, mast cell degranulating (MCD) factor, and melittin, which are restricted to the venom of the honey bee (Table I), may be representative examples of the wide scope of action of the venomous polypeptides of hymenopterans. These compounds, which possess from eighteen to twenty-six amino acid residues, each have different pharmacological activities that indicate that they have been selected primarily as vertebrate toxins. Apamin, consisting of only eighteen amino acid residues, is the smallest neurotoxic polypeptide known; it acts on synaptic pathways. MCD-peptide is a powerful cellular degranulating agent capable of releasing histamine. Melittin, a basic polypeptide of twenty-six amino acid residues, possesses strong surface activity and functions as a natural detergent or cytolytic agent. This polypeptide is sometimes referred to as the "direct" hemolysin to contrast it with phospholipase A which exerts its hemolytic action indirectly by generating cytolytic lysolecithins from hydrolyzed phospholipids.

Studies of the biosynthesis of melittin (Fig. 1) in adult bees have demonstrated that this peptide, which comprises 50% of the dried venom, is not present in the poison gland reservoir at the time of eclosion of the worker. Nor could melittin be detected in the poison gland reservoir of newly-emerged workers. However, another compound, that not only contains the entire amino acid sequence of melittin but eight additional residues, is present at the time of emergence. This polypeptide—promelittin—consists of thirty-four amino acid residues (Fig. 1). Additional peptides containing fewer amino acids than promelittin have also been detected in the venom of very young worker bees. These peptides are believed to represent intermediates in the activation of promelittin to melittin.

The N-terminus of promelittin is modified from that of melittin by a sequence containing acidic amino acids and proline (Fig. l). Significantly, the former peptide is resistant to most proteases whereas melittin is susceptible to a variety of endopeptidases. Therefore, promelittin activation is probably catalyzed only by highly specific enzymes which should not be capable of hydrolyzing the end product, melittin.

Promelittin: E P E P D P E A G I G A V L K V L T T G L P A L I S W I K R K R Q G
Melittin: G I G A V L K V L T T G L P A L I S W I K R K R Q G

Fig. 1. Comparison of the amino acid sequences of promelittin and melittin. Amino acid code: A, Ala; D, Asp; E, Glu; G, Gly; I, Ile; K, Lys; L, Leu; P, Pro; Q, Gln; R, Arg; S, Ser; T, Thr; V, Val; W, Trp.

It appears that the toxicity of melittin is related to the asymmetric distribution of basic and apolar amino acids, the former being mostly clustered at the C-terminal end (Fig. 1). The presence of acid residues at the amino end of promelittin probably results in a peptide of negligible toxicity. Modifications at the N-terminus may ensure that the ultimately toxic end product, melittin, is never present on the peptide chain during its stepwise growth. Promelittin, the first "pro-toxin" to be structurally characterized, may be typical of the precursors of toxic peptides in being rendered inactive by critically substituted residues. Such substitution would guarantee that the physiologically active entity is never present on the ribosomes during stepwise growth of the peptide from the amino to the carboxyl end.

Promelittin synthesis is detectable at the time of worker emergence and reaches its peak at about 10 days of age. Melittin cannot be detected until the bees are 2 days old, and its rate of conversion from promelittin is maximal when the bees are 10 days old. Melittin production increases until the bees are 20 days old, a highly adaptive development which ensures that their poison gland secretion is now richly fortified with a potent toxin when the workers begin foraging and are exposed to predators. On the other hand, in queen bees, the rate of conversion of promelittin to melittin is maximal at the time of emergence, a fact probably correlated with the utilization of the sting by newly emerged sister queens in colonial combats. Thus, the presence of different control systems for the rate of synthesis and conversion of venom peptides in workers and queens is of great selective value in terms of the behavioral idiosyncrasies of the members of both castes.

C. Kininlike Peptides

Naturally occurring vasoactive peptides have been identified in the venoms of wasps in several genera. These kinins all can lower mammalian blood pressure, contract smooth muscle, and release histamine from rat mast cells. Like the mammalian-derived peptide bradykinin, those present in wasp venoms are potent algogens and thus contribute markedly to the effectiveness of the secretions as vertebrate deterrents. Both polisteskinin, isolated from the venoms of *Polistes* spp., and vespulakinin 1, a product of *Vespula* spp., contain the nonapeptide bradykinin at their carboxy-terminal ends (Fig. 2). The amino terminus of polisteskinin contains pyroglutamic acid in addition to a cluster of basic amino acids. Vespulakinin 1 also contains basic amino acids at the amino-terminal end (Fig. 2), and is distinguished by the presence of carbohydrate prosthetic groups which appear to contain *N*-acetylgalactosamine. The vespulakinins appear to be the first known vasoactive glycopeptides identified in animal venoms. Both kinins are considerably more vasoactive than bradykinin.

The peptide derivatives of bradykinin produced in the poison glands of

Bradykinin:	R P P G F S P F R
Polisteskinin:	E T N K K K L R G R P P G F S P F R
Vespulakinin 1:	T A T T R R R G R P P G F S P F R (Carbohydrate 1 and Carbohydrate 2 attached at T, T)

Fig. 2. Comparison of the structure of bradykinin with polisteskinin and vespulakinin 1. Amino acid code: A, Ala; E, Glu; F, Phe; G, Gly; K, Lys; L, Leu; N, Asn; P, Pro; R, Arg; S, Ser; T, Thr.

these insects seem to occur only in some wasp venoms. As such, those kinins constitute highly distinctive biochemical characters.

D. Alkaloids

The poison gland secretions of ant species in the genus *Solenopsis* are very distinctive because these toxic exudates are fortified mostly with alkaloids and contain only traces of proteinaceous constituents. Many fire ant (*Solenopsis* spp.) venoms contain 2-methyl-6-*n*-undecylpiperidine as well as other related dialkylpiperidines. These compounds possess well-developed hemolytic and necrotic activities. Subdermal injection of these venoms causes characteristic lesions or pustules. Ant species in another subgenus of *Solenopsis* also produce venoms dominated by nitrogen heterocycles. However, these secretions are dominated by compounds such as 2-butyl-5-pentylpyrrolidine and 2-ethyl-5-pentyl-l-pyrroline rather than dialkylpiperidines. The presence of 2,5-dialkylpyrrolidines in venoms produced by species of *Monomorium* indicates that alkaloids may commonly be major constituents in the poison gland secretions of species in several ant genera.

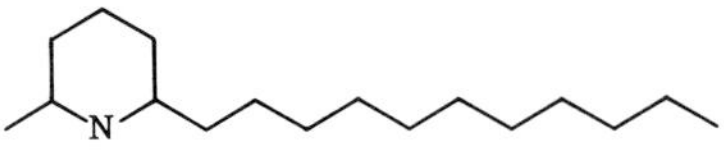

2-methyl-6-*n*-undecylpiperidine

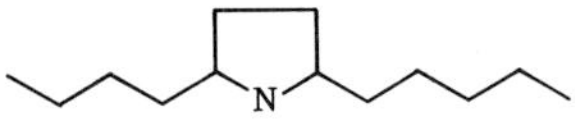

2-butyl-5-pentylpyrrolidine

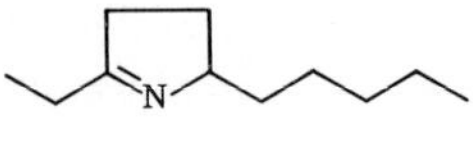

2-ethyl-5-pentyl-1-pyrroline

E. Multiple Functions of Insect Venoms

Insects' venoms, like other types of exocrine secretions, have been utilized to subserve multiple functions. Some wasp venoms contain minor constituents that release alarm behavior when perceived by nearby conspecifics. Many species of ants use trace components in their poison gland secretions as trail pheromones. In all probability, these venomous pheromones are volatile poison gland products that have been secondarily adapted to function as communicative agents. Such pheromonal parsimony, which is probably widespread among insects, especially eusocial species, is of great selective value since it enables these arthropods to exploit chemisociality with a finite number of natural products.

V. CHEMISTRY OF EXOCRINE DEFENSIVE COMPOUNDS

The defensive secretions of insects contain a large variety of compounds and each exudate usually contains a mixture of compounds belonging to several different chemical classes. The great scope of insect chemical defenses becomes evident if the exocrine compounds are grouped according to their functionalities and/or carbon skeletons.

A. Carboxylic Acids

Although carboxylic acids have been detected in the defensive exudates of insects in about five orders, they are only major constituents in those of ants, beetles, and alydid bugs. Most of the identified acids are aliphatic constituents which fall in the range of C_1–C_6. In many cases these acids are minor components that appear to be metabolically related to the compounds (e.g., alcohols, aldehydes) that dominate the exudates.

Formic acid has been identified in the exudates of notodontid larvae, adult carabids, and formicine ants. Acetic acid is characteristically a major constituent in coreid secretions and occurs as a minor pygidial gland product of some carabid species. Carabids produce nearly two-thirds of the aliphatic acids identified as insect defensive products. For example, methacrylic and tiglic acids are found in many carabid secretions, usually in admixture.

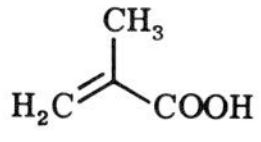

CH_3 / H_2C / COOH

methacrylic acid

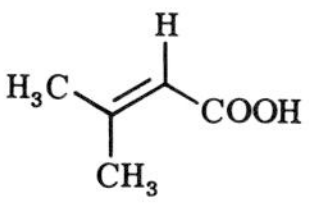

H / H_3C / COOH / CH_3

tiglic acid

Aromatic acids have not been commonly encountered as insect defensive compounds. Benzoic acid and *p*-hydroxybenzoic acid, which are synthesized in the pygidial glands of dytiscids, have been described as putative antibiotics. However, since benzoic acid is an excellent repellent for fish, these acids may be important repellents for these vertebrate predators.

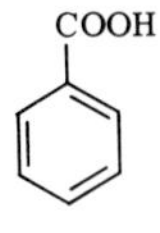

COOH

benzoic acid

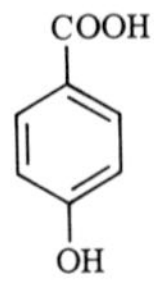

COOH

OH

p-hydroxybenzoic acid

Dihydromatricaria acid, a glandular product of adult cantharid beetles, was the first acetylenic compound to be identified as an exocrine product of animals. Since these beetles feed on plants (Compositae) that contain methyl esters of acetylenic acids, it is not certain that dihydromatricaria acid is produced *de novo* by the beetles.

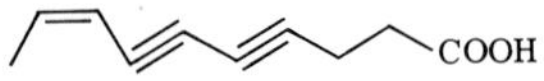

COOH

dihydromatricaria acid

B. Alcohols

Although a wide variety of alcoholic constituents have been identified as exocrine products, most of these compounds represent trace or minor products of the glandular exudates. Frequently, these alcohols are minor concomitants of aldehydic or ketonic products to which they may be biosynthetically related. This is particularly true of hymenopterous secretions which contain most of the alcohols identified as exocrine products. For example, 4-methyl-3-heptanol and 3-octanol accompany 4-methyl-3-heptanone and 3-octanone, respectively, in the defensive secretions of a wide variety of ants.

Phoracanthol [(5-ethylcyclopent-1-enyl)methanol] is an unusual alcoholic defensive product produced by adult cerambycid beetles in the genus *Phoracantha*. It is believed to be synthesized by the head-to-tail union of four acetate units.

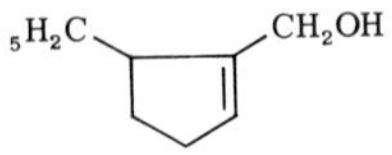

$_5H_2C$ CH_2OH

phoracanthol

Mono-, sesqui-, and diterpene alcohols are utilized as defensive compounds by a wide range of species, particularly in the order Hymenoptera. Citronellol occurs in the mandibular gland exudates of many species of formicine ants and *all-trans*-farnesol is a characteristic exocrine product of some species of andrenid and apid bees. Selin-11-en-4-ol, an osmeterial product of papilionid caterpillars (*Battus* sp.), constitutes one of the few eudesmane sesquiterpenes identified as arthropod exocrine products. Geranylgeraniol, present in the exudates of ants and bees, is one of two diterpene alcohols known to occur in defensive secretions.

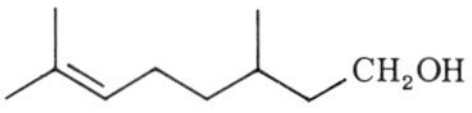

citronellol

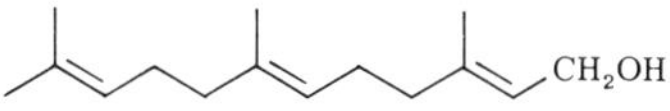

all-trans-farnesol

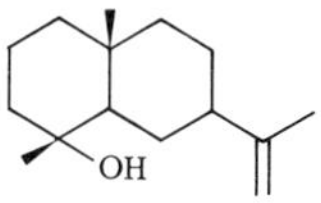

selin-11-en-4-ol

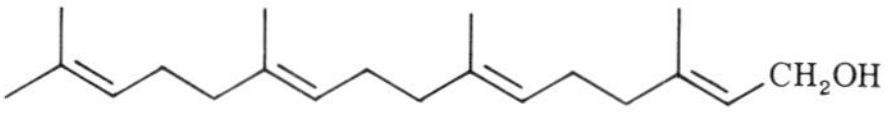

geranylgeraniol

C. Aldehydes

Hemipterous defensive secretions are dominated by aldehydes, as are those of some groups of Hymenoptera, Coleoptera, and Dictyoptera (Blattidae). The majority of these aldehydic allomones are aliphatic compounds in the range C_3–C_{10} and many of these compounds are very restricted in their distribution. As such, these compounds may constitute good taxonomic character states.

α, β-Unsaturated aldehydes characterize the secretions of true bugs and polyzosteriine cockroaches. Enals such as 2-methylenebutanal and *trans*-2-hexenal are typical of the aldehydes produced by blattids. 2-Hexenal not only is widespread in hemipterous secretions, but also is a defensive product of ants and beetles. The dicarbonyl compound, 4-oxo-*trans*-2-hexenal, is a particularly distinctive hemipterous product that is frequently produced by immature bugs. Saturated aldehydes are not commonly encountered in the secretions of true bugs, although hexanal appears to be a characteristic coreid product.

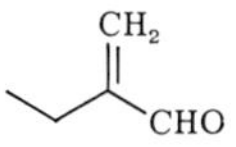

2-methylenebutanal

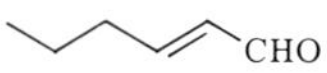

trans-2-hexenal

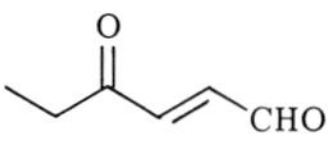

4-oxo-*trans*-2-hexenal

hexanal

Only a few aromatic aldehydes have been identified as insect defensive products. Benzaldehyde, a well known antagonistic allomone of millipedes, also fortifies the defensive exudates of some beetles, ants, and bees. On the other hand, both *p*-hydroxybenzaldehyde and salicylaldehyde are limited to the defensive secretions of larval and adult beetles.

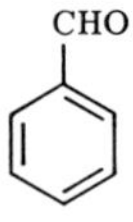

CHO CHO CHO OH

OH

benzaldehyde *p*-hydroxybenzaldehyde salicylaldehyde

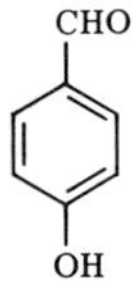

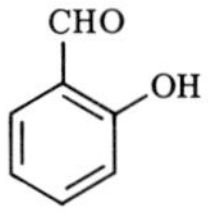

Some very distinctive terpene aldehydes are part of the chemical defense arsenals of insects. For example, the cyclopentanoid monoterpene iridodial is an exocrine product of dolichoderine ants also found in the secretions of some adult beetles. Farnesal is one of two sesquiterpenes in the mandibular gland secretions of ants, along with several mono- and diterpene aldehydes. The norsesquiterpene gyrinidal is one of several distinctive C_{14} compounds synthesized in the pygidial glands of gyrinid beetles that function as effective fish repellents.

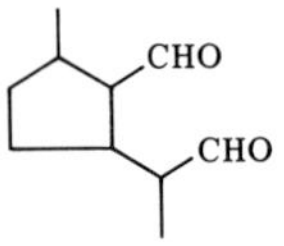

CHO CHO CHO OHC O O

iridodial 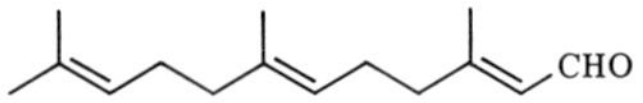 farnesal 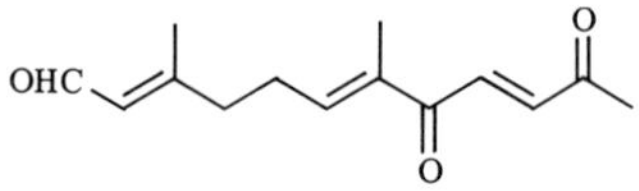 gyrinidal

D. Ketones

Although ketones are rather widespread in the glandular exudates of insects, they appear to be only characteristic ant and bee natural products. For the most part, these carbonyl compounds are only minor concomitants of other classes of compounds, or they constitute atypical major components in the exudates of Lepidoptera, Hemiptera, Coleoptera, and Dictyoptera.

Most of the aliphatic unbranched ketones in the range C_4–C_{19} identified as insect exocrine products have been encountered in the secretions of ants and bees. 2-Heptanone, also in a cockroach exudate, is commonly produced by these hymenopterans. 2-Methylcyclopentanone is representative of the few alicyclic ketones identified in these secretions. A wide variety of ethyl ketones are found in defensive exudates; 4-methyl-3-heptanone is especially typical. The only aromatic ketone that has been found in these secretions is *o*-aminoacetophenone, an ant mandibular gland product.

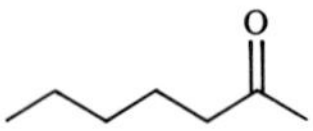

O O CH_3 O O CH_3 NH_2

2-heptanone 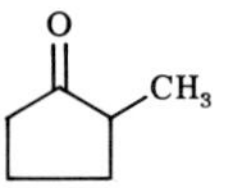 2-methylcyclopentanone 4-methyl-3-heptanone *o*-aminoacetophenone

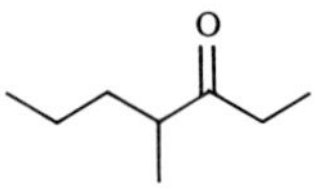

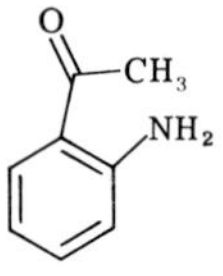

Ketonic terpenes are not especially common natural products of insects. However, 6-methyl-5-hepten-2-one has been detected in a wide range of ant species and is sometimes accompanied by 2-methyl-4-heptanone. Gyrinidone is one of four norsesquiterpenes identified as pygidial gland products of gyrinid beetles.

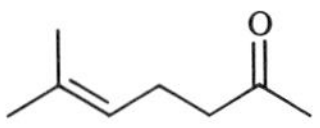

6-methyl-5-hepten-2-one

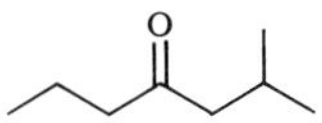

2-methyl-4-heptanone

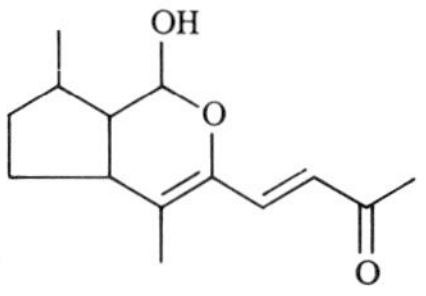

gyrinidone

E. Esters

Although more than fifty esters have been identified as insect exocrine products, they are not widespread. About half of these compounds represent ant or bee natural products with most of the remainder being derived from the defensive exudates of beetles or true bugs. Most esters are in the range C_8–C_{12} and often present as acetates. Frequently, the esters are accompanied by their alcoholic and acidic moieties to which they are probably metabolically related. Often minor constituents in defensive exudates, esters may act as wetting agents that facilitate the penetration of the main toxicants.

Ants produce acetates containing all the alcohols in the range C_9–C_{16}; however, hemipterans do not synthesize acetates with an alcoholic moiety greater than C_{10}. While *n*-hexyl acetate and *trans*-2-hexenyl acetate are typical defensive products of adult bugs, esters are generally lacking from the nymphal defensive secretions. On the other hand, some adult bugs produce distinctive nonacetate esters. *n*-Butyl butyrate and *n*-butyl hexanoate, two of the esters identified in coreid and alydid secretions, are examples of such nonacetate esters.

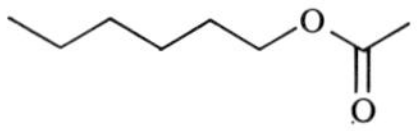

n-hexyl acetate

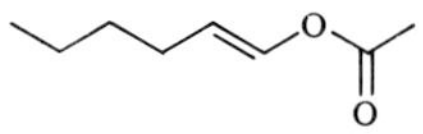

2-hexenyl acetate

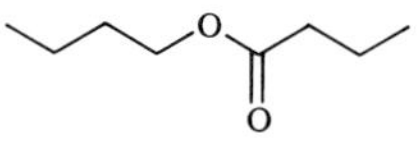

n-butyl butyrate

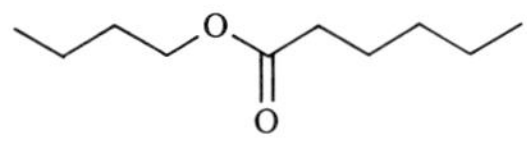

n-butyl hexanoate

Several isoprenols are esterified in the defensive exudates of ants. Farnesyl acetate and geranylgeranyl acetate occur in the Dufour's gland secretions of formicine ants as part of complex exocrine secretions.

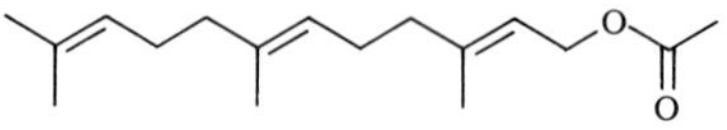

farnesyl acetate

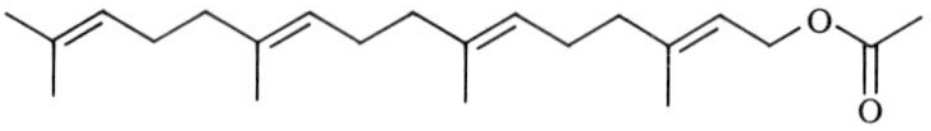

geranylgeranyl acetate

Aromatic esters appear to characterize the exudates of coleopterous species as well as the mandibular gland secretions of male ants in the genus *Camponotus*. Methyl *p*-hydroxybenzoate is produced in the pygidial glands of many dytiscids, and 2-phenylethyl isobutyrate fortifies the larval exudate of a chrysomelid species. Distinctive esters such as methyl anthranilate and methyl 6-methyl salicylate are major constituents in the mandibular gland secretions of *Camponotus* males and probably function as pheromones as well as defensive compounds.

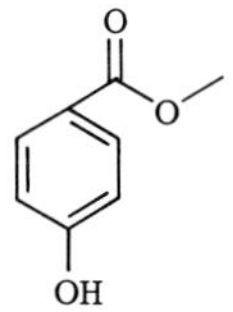

methyl *p*-hydroxybenzoate

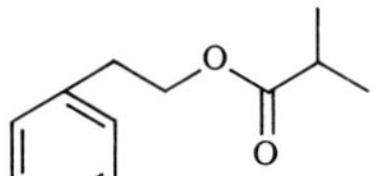

2-phenylethyl isobutyrate

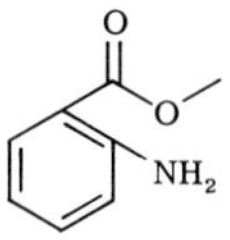

methyl anthranilate

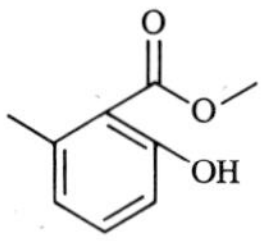

methyl 6-methyl salicylate

F. Lactones

Lactones have only been detected in the glandular exudates of ants, bees, and a few species of beetles. Iridomyrmecin is one of three cyclopentanoid monoterpene lactones identified as anal gland products of dolichoderine ants and the mandibular gland secretions of *Camponotus* species have yielded massoilactone and mellein. Macrocyclic lactones are characteristic Dufour's gland products of halictid bees.

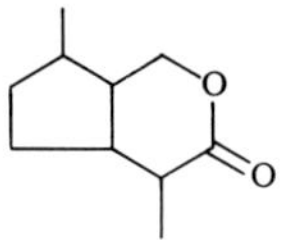

iridomyrmecin

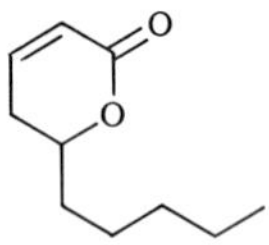

massoilactone

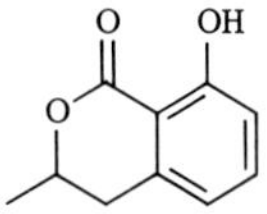

mellein

Highly distinctive lactones are produced in the pygidial glands of staphylinid and dytiscid beetles. γ-Dodecalactone and marginalin are major constituents in the exudates of staphylinids and dytiscids, respectively.

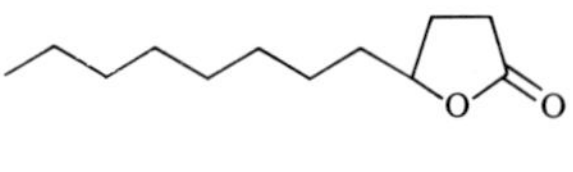

γ-dodecalactone

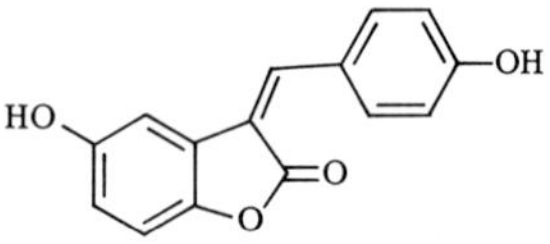

marginalin

G. Phenols

Phenolic defensive allomones have been identified in the secretions of a few cerambycid, carabid, and tenebrionid beetles, as well as one blattid species. *m*-Cresol is one three phenols produced in the prothoracic glands of a tenebrionid; 1,4-quinones are synthesized in the abdominal glands. The identification of *o*-cresol and *p*-ethylphenol in the exudates of other species demonstrates that *ortho-*, *meta-*, and *para*-substituted phenols are synthesized by these arthropods.

OH CH_3 | OH CH_3 | OH C_2H_5

m-cresol | *o*-cresol | *p*-ethylphenol

H. 1,4-Quinones

A variety of *p*-quinones have been identified in the defensive exudates of species of Dermaptera, Coleoptera, and both suborders of the Dictyoptera: Isoptera and Blattaria. These compounds are in the secretions of carabid, staphylinid, and tenebrionid beetles, species in the Tenebrionidae proving to be the richest source of 1,4-quinones in the animal kingdom. Of the eleven *p*-quinones identified as insect allomones, all are produced by some species of tenebrionids.

1,4-Benzoquinone, 2-methyl-1,4-benzoquinone (toluquinone), and 2-ethyl-1,4-benzoquinone are common quinonoid constituents of insect secretions. Some tenebrionids also produce 2-propyl-1,4-benzoquinone and 2-methoxy-3-methyl-1,4-benzoquinone, the latter a common constituent in millipede defensive secretions. 6-Methyl-1,4-naphthoquinone and three other naphthoquinones are produced by scaurine tenebrionids and are the only naphthyl derivatives identified as insect exocrine products.

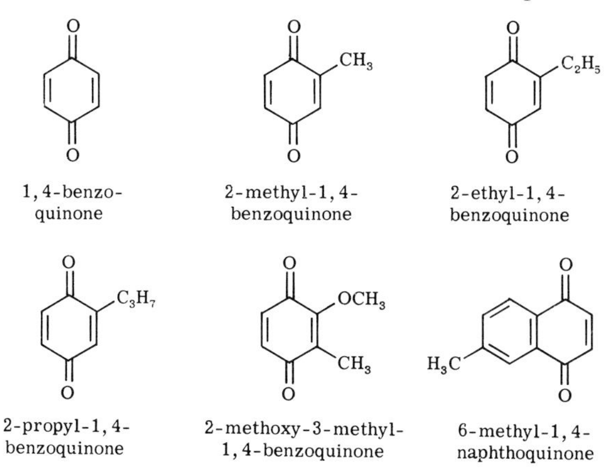

1,4-benzoquinone | 2-methyl-1,4-benzoquinone | 2-ethyl-1,4-benzoquinone

2-propyl-1,4-benzoquinone | 2-methoxy-3-methyl-1,4-benzoquinone | 6-methyl-1,4-naphthoquinone

The 1,4-quinones are frequently accompanied by their corresponding hydroquinones (quinols). Hydroquinone and 2-methylhydroquinone are in several secretions and undoubtedly serve as precursors for the 1,4-quinones that dominate these exudates.

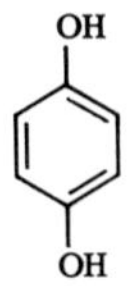

hydroquinone

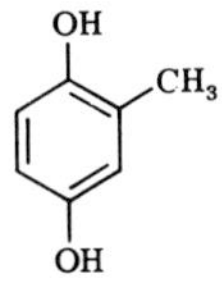

2-methylhydroquinone

I. Hydrocarbons

The more than seventy hydrocarbons found in the exocrine glands of insects demonstrate that these compounds have been emphasized as defensive agents. Because hydrocarbons are frequently present along with more polar constituents, these secretions often consist of two-phase systems. Probably a primary function of these alkanes is to act as a cuticular wetting and spreading agent which can promote the penetration of the more polar compounds in the secretions. There is also some evidence that the hydrocarbons may be important deterrents themselves, possibly by interfering with the olfactory processes of predatory arthropods.

All unbranched alkanes in the range C_9–C_{25} have been identified as insect exocrine products. Monounsaturated olefins and a number of diunsaturated hydrocarbons are also commonly produced by insects. In general, hydrocarbons are only characteristic products in the secretions of hemipterous and hymenopterous species. The Dufour's gland of ants may represent the hydrocarbon "factory" *par excellence,* since more than half the hydrocarbons identified as insect exocrine compounds are produced in this organ.

n-Undecane and *n*-tridecane constitute the most common alkanes produced in the defensive glands of insects. These compounds have been identified in the secretions of beetles, true bugs, and ants. Ants produce a variety of branched hydrocarbons including 3- and 5-methylalkanes such as 3-methylundecane and 5-methyltridecane. Alkenes such as 1-undecene and 4-tridecene, which are commonly produced by tenebrionids and ants, respectively, are frequently accompanied by other Δ^1- and Δ^4-monoolefins. Toluene, a mandibular product of a few cerambycid species, is the only aromatic hydrocarbon identified in a defensive exudate.

n-undecane 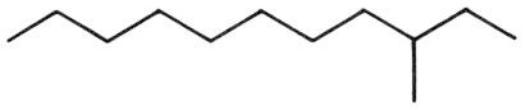 3-methylundecane 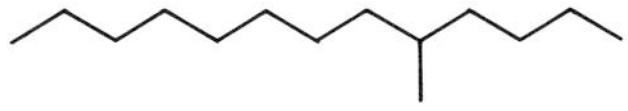 5-methyltridecane

1-undecene  4-tridecene CH_3 toluene

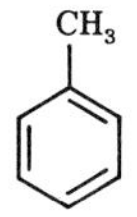

Terpene hydrocarbons do not occur widely in the glandular exudates of insects. Monoterpenes such as limonene and β-pinene are produced in the frontal glands of termitid soldiers as part of a highly effective defensive exudate. The sesquiterpene, α-farnesene, is one of four farnesenes identified as Dufour's gland constituents in both ants and bees.

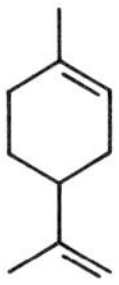

CH_2 β-pinene 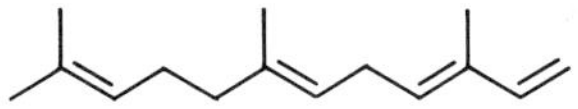 α-farnesene limonene

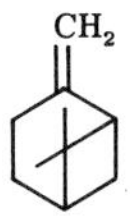

J. Steroids

The prothoracic glands of adult dytiscid beetles produce an extraordinary variety of steroids, many of which are identical to well-known vertebrate hormones. The majority of these compounds are pregnene and pregnadiene derivatives and these C_{21} steroids are reported to be outstanding defensive agents against vertebrate predators. The presence of estrone and testosterone in the glandular exudates of a few dytiscid species demonstrates that these beetles also can generate C_{18} and C_{19} steroids. In some cases, these insects produce remarkably high concentrations of these steroids; one species of *Cybister* is the richest known source of cortexone, producing up to 1 mg/beetle.

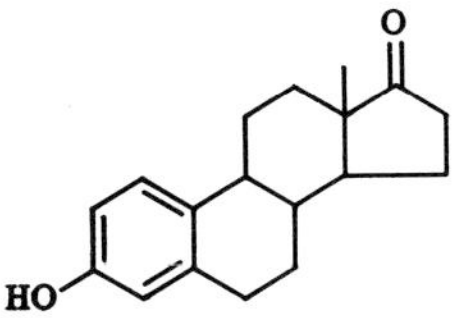

O OH CH_3 C=O HO O O estrone 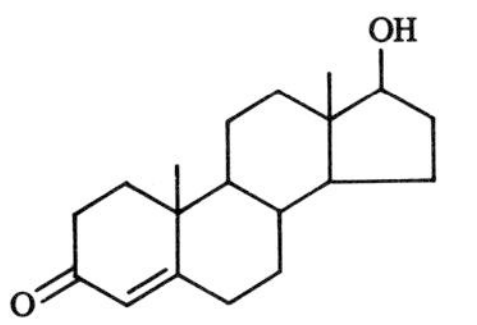 testosterone 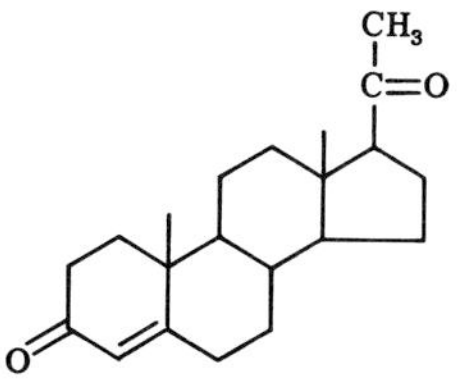 cortexone

Several of the steroidal compounds produced by dytiscids are unique animal natural products. 6,7-Dehydrocortexone, like cybisterone, is a distinctive steroid present in the secretions of a few dytiscid species. Hydroxylated pregnadienones, some of which are esterified (e.g., 4,6-pregnadien-15α,20β-diol-3-one-20-isobutyrate), are characteristic steroids of species in a few dytiscid genera.

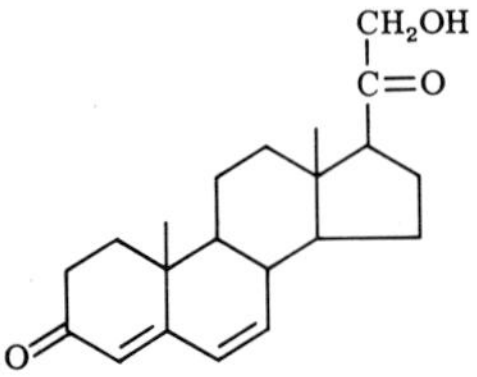

6, 7-dehydrocortexone

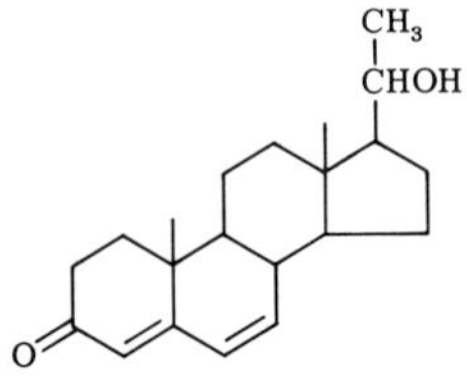

cybisterone

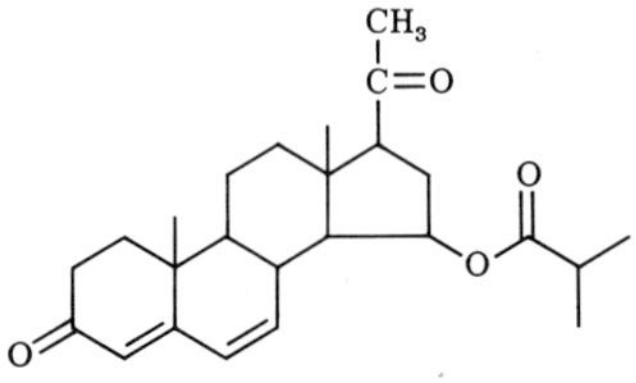

4, 6-pregnadien-15α, 20β-diol-3-one-20-isobutyrate

K. Miscellaneous Compounds

Some of the exocrine compounds produced by insects are highly distinctive natural products with a very limited insect distribution. For the sake of convenience, the defensive allomones in this chemical potpourri are grouped together as miscellaneous compounds. Although relatively few defensive secretions have been so far analyzed, the identification of compounds as diverse as alkaloids and sesquiterpenes demonstrates what a rich source of novel natural products these arthropods are.

Hydrogen cyanide occurs along with benzaldehyde in the secretions of larval chrysomelids; both compounds are probably derived from mandelonitrile, as they are in the defensive exudates of millipedes. The mandibular gland secretions of ponerine ant workers are fortified with dimethyldisulfide and dimethyltrisulfide, the only alkyl sulfides identified as animal exocrine products. Other ponerine species produce trialkylpyrazines of which 2,5-dimethyl-3-isopentyl-1,4-pyrazine has a particularly widespread formicid distribution. The mandibular gland secretion of another ant species is dominated by the sesquiterpenoid dendrolasin, one of two furans detected in this unusual exudate.

$CH_3—S—S—CH_3$

dimethyldisulfide

$CH_3—S—S—S—CH_3$

dimethyltrisulfide

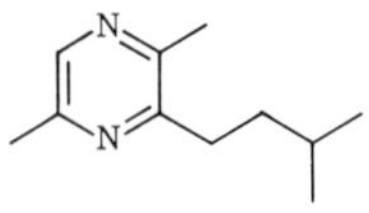

2, 5-dimethyl-3-isopentyl-1, 4-pyrazine

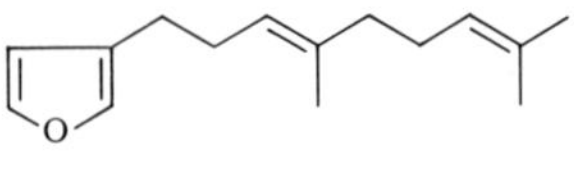

dendrolasin

Beetles have proved to be an exceptionally rich source of unusual natural products. The alkaloid actinidine has been identified as a pygidial gland product of some staphylinid beetles, whereas *N*-ethyl-3-(2-methylbutyl) piperidine has been detected in the secretions of other species. Both isomers of rose oxide, a well-known plant natural product, have been isolated from the metathoracic gland secretion of a cerambycid beetle.

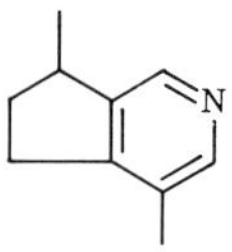

actinidine

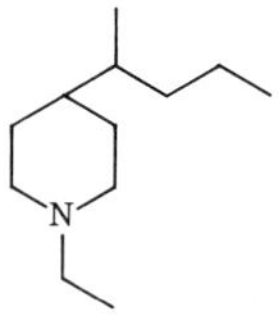

N-ethyl-3-(2-methyl-
butyl) piperidine

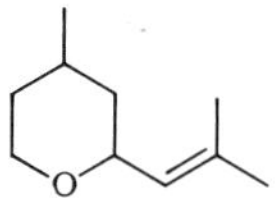

Rose oxide

The frontal gland secretions of termites are enriched with a variety of unusual exocrine products. Besides monoterpene hydrocarbons and 1,4-benzoquinones, a sesquiterpenoid ether, 4,11-epoxy-*cis*-eudesmane, has been identified as a cephalic product of termitid soldiers. Another termite natural product, 1-nitro-1-pentadecene, represents the only nitro compound identified as an insect exocrine constituent.

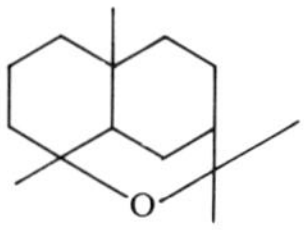

4,11-epoxy-*cis*-
eudesmane

NO_2

1-nitro-1-pentadecene

L. Nonexocrine Compounds of Insect Origin

Insects biosynthesize a wide variety of defensive compounds that are not discharged as glandular exudates. While some of these nonexocrine allomones are relatively simple compounds, others constitute the most complex defensive substances found in insects. These compounds often possess great pharmacological activity when ingested and the insects that produce them are often aposematically colored. Although not produced in exocrine glands, these compounds are usually present in hemolymph and in the case of species in some families (e.g., Coccinellidae, Meloidae), can be discharged by reflex bleeding. In general, it appears that specific nonexocrine compounds may be limited in their distribution to species in one family or a single genus. So far, all these novel compounds have been isolated from beetles.

Cantharidin, or so-called Spanish fly, is a ubiquitous natural product of meloid species. This bicyclic monoterpene, which causes severe biochemical lesions when ingested by mammals, is present in very high concentrations in the beetles (0.3–0.4%).

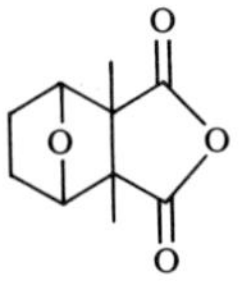

cantharidin

Coccinellid beetles synthesize a series of novel tricyclic alkaloids which are outstanding repellents of both insects and vertebrates. Precoccinelline, one of several stereoisomeric alkaloids found in the hemolymph of these beetles, is accompanied by the *N*-oxide coccinelline. The alkaloid adaline represents the only bicyclic compound identified as a natural product of these beetles.

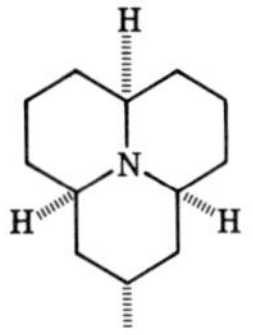

precoccinelline

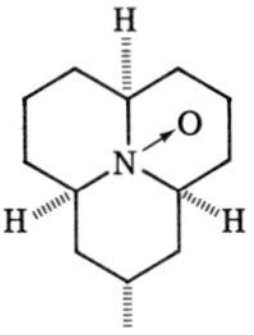

coccinelline

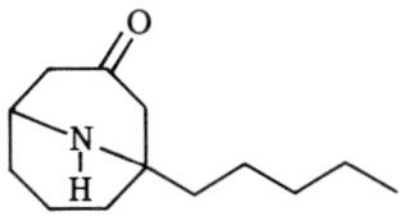

adaline

Pederin is the major vesicatory constituent in the hemolymph of staphylinid beetles in the genus *Paederus*. One of three closely related compounds present, it may constitute up to 0.25% of the wet weight of adult females.

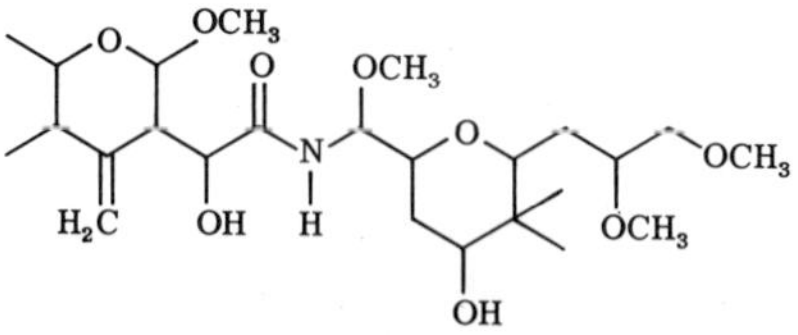

pederin

M. Nonexocrine Compounds of Plant Origin

Many insect species feed on plants containing natural products with pronounced pharmacological activities. In some cases, these arthropods not only utilize the plants as a food source, but as a source of defensive compounds. The insects then sequester the plant toxins often selectively and store these compounds in various tissues. Many endopterygote species sequester these plant allelochemics during the phytophagous larval instars, retaining them in the body through adult development. Such persistent sequestration ensures that the adult, which often does not ingest these natural products, will nevertheless be distasteful because of their presence.

VI. METABOLIC ADAPTATIONS FOR TOLERATING TOXIC NATURAL PRODUCTS

In general, the physiological bases for the resistance of insects to ingested toxic plant compounds are unknown. In some cases these natural products are detoxified and excreted; in others, the metabolically altered compounds are sequestered.

A. Microsomal Mixed-Function Oxidases and Other Enzymes

The level of microsomal mixed-function oxidases in the midgut appears to be correlated with the degree of herbivory exhibited by insects. These enzymes epoxidize, dealkylate, oxidize, and hydroxylate a wide variety of organic compounds. Polyphagous insects possess very high levels of these oxidases relative to monophagous species; oligophagous insects possess intermediate levels. Thus, mixed-function oxidases appear to be an important evolutionary correlate of polyphagy.

These enzymes may be important in the metabolism of nicotine by various species that feed on tobacco. Nicotine is both dealkylated and oxidized, and the resultant metabolite cotinine is virtually nontoxic to insects.

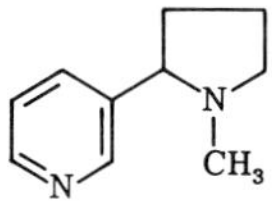

nicotine

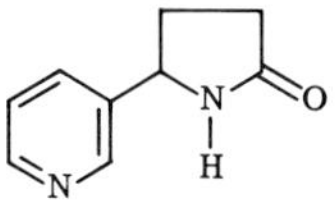

cotinine

The poisonous effects of some plant natural products are avoided by detoxifying them in the midgut prior to absorption and sequestration. Larvae of the arctiid *Seirarctia echo* convert methylazoxymethanol, a toxic compound in cycad leaves, to its β-glucoside cycasin in the midgut. Cycasin is subsequently sequestered in the Malpighian tubules and hemolymph. Significantly, whereas β-glucosidase, the enzyme responsible for the hydrolysis and synthesis of cycasin, is present in the midgut, it is absent from the blood and Malpighian tubules. The selective distribution of this enzyme guarantees that the internal tissues of the larva will never be exposed to the ingested plant toxin.

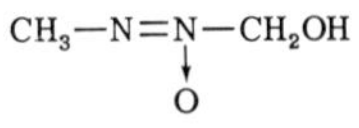

$CH_3-N=N(\rightarrow O)-CH_2OH$

methylazoxymethanol

$CH_3-N=N(\rightarrow O)-CH_2O-$(glucose)

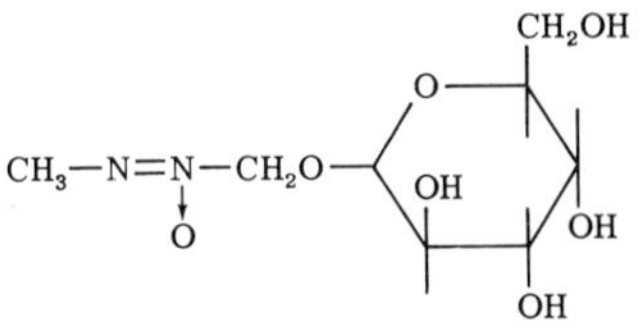

cycasin

B. Enzymatic Discrimination of Toxic Plant Natural Products

Some plant toxins bear great structural similarity to important metabolic intermediates. Serious biochemical lesions can result in animals if these analogues are utilized as surrogates for the metabolites that they mimic. However, some feeding specialists have evolved the capacity to avoid the toxic effects of these compounds by selectively discriminating against their incorporation into metabolic pathways.

L-Canavanine, a structural analogue of L-arginine, is produced in large quantities by the seeds of certain legumes. Canavanine is highly toxic to most phytophagous insects; the deleterious effects of this guanidinooxy structural analogue of arginine result from the synthesis of canavanyl proteins. These "foreign" proteins disrupt DNA and RNA metabolism as well as protein synthesis.

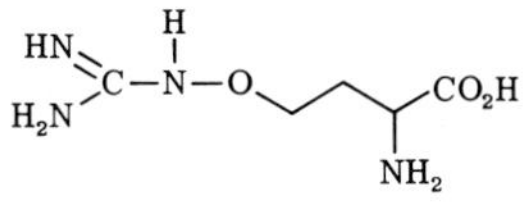

L-canavanine

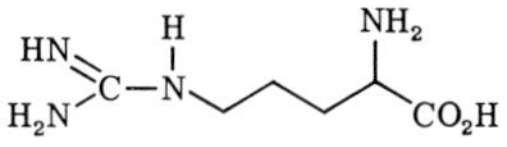

L-arginine

Certain monophagous bruchids feed exclusively on cotyledons produced from seeds that are fortified with L-canavanine. The toxic effects of this arginine analogue are avoided by the presence of an amino acid-activating enzyme that discriminates between it and L-arginine. L-Canavanine is not incorporated into proteins and bruchid larvae can tolerate large quantities of this compound without experiencing intoxication. Whereas arginyl-tRNA synthetase esterifies L-canavanine to arginyl-transfer RNA in susceptible species, these monophagous bruchids rigorously exclude this analogue from incorporation. Resistance by other feeding specialists to plant sources containing other toxic amino acid analogues may be due to the presence of highly discriminating amino acid-activating enzymes.

VII. SELECTIVE SEQUESTRATION OF PLANT NATURAL PRODUCTS

Insects in at least six orders sequester a multitude of plant allelochemics belonging to several different chemical classes. Many species in the orders Lepidoptera, Coleoptera, Homoptera, Hemiptera, Orthoptera, and Diptera are known to store plant natural products. It appears that sequestration is a selective process that often results in the retention of only a few of the compounds that are ingested. However, sequestration does not invariably result from an insect feeding on plants containing toxic allelochemics; many species do not retain these compounds in their bodies after ingesting them.

A. Cardenolides

Many insects feed on oleander, a rich source of cardiac glycosides. The main steroid present in the leaves, oleandrin, is sequestered by ctenuchid larvae, along with several other related compounds including strospeside. The last-named compound has also been detected as a sequestration product of coccids, aphids, and lygaeids.

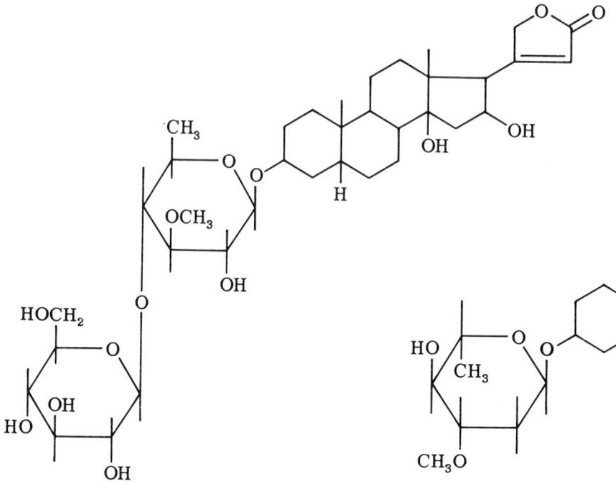

oleandrin 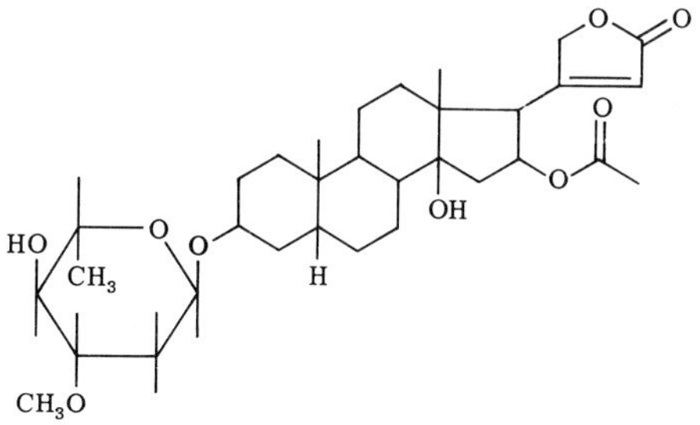 strospeside

Cardenolide-rich milkweeds (Asclepidaceae) are primary hosts for many lepidopterous, orthopterous, and hemipterous species. Many of these phytophagous insects selectively sequester cardiac glycosides in tissues and, in some cases, in secretory structures as well. The adult monarch primarily stores nonpolar steroids such as calotropin or its isomer calactin. These two cardenolides are also sequestered by some pyrgomorphid grasshoppers, the steroids being detectable in either the hemolymph or as part of an abdominal defensive secretion. Similarly, lygaeids feeding on milkweed seeds sequester cardiac glycosides in both the hemolymph or in dorsolateral defensive spaces. The lygaeids preferentially sequester polar cardenolides in their defensive spaces; nonpolar steroids may be converted to polar compounds before sequestration.

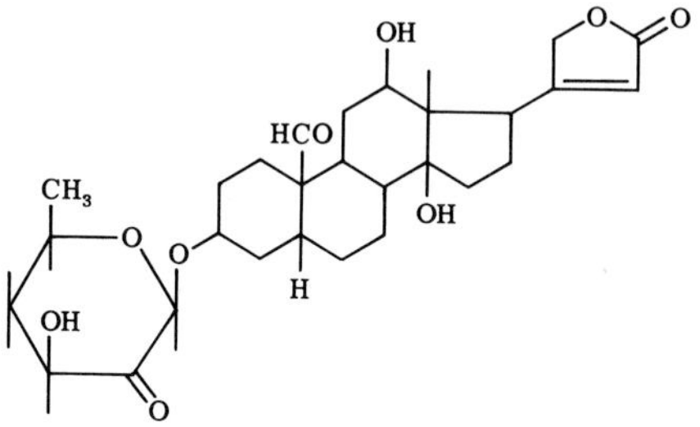

calotropin

B. Pyrrolizidine Alkaloids

Pyrrolizidine alkaloids are sequestered by the larvae of several arctiid species and are produced by plants in the genera *Senecio* and *Crotalaria*. Alkaloids such as senecionine and seneciphylline, which are readily sequestered, are retained by the adult moths. These insects often store a much greater concentration of alkaloids than are found in the leaves. Selective sequestration also seems to be the rule; the proportion of stored alkaloids differs considerably from that in the plant.

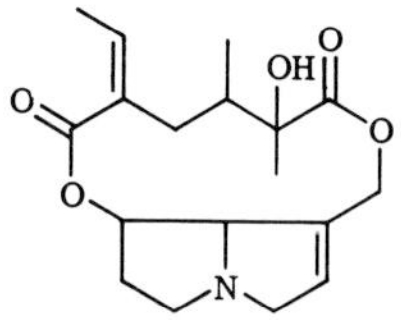

senecionine

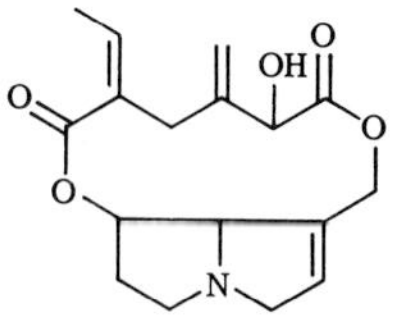

seneciphylline

C. Aristolochic Acids

Larvae of pharmacophagous swallowtails sequester a variety of aristolochic acids from their food plants, species of Aristolochiaceae. Papilionid adults may each contain up to 100 mg of aristolochic acid-1, one of several characteristic nitrophenanthrenes known for their bitter and toxic qualities.

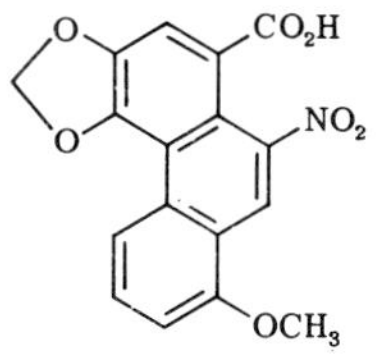

CO_2H
NO_2
OCH_3

aristolochic acid-I

D. Sequestration as an Adaptive Physiological Process

Insects vary considerably in their abilities to sequester toxic natural products that are ingested. Since sequestration is a highly selective process, only a few potential toxins may be actually absorbed from the intestine and stored in certain tissues. Nonpolar compounds, which are often more toxic to vertebrates and insects than their polar derivatives, are usually retained in the hemolymph and lipophilic tissues. Adult monarchs sequester nonpolar cardenolides in the abdomen whereas polar steroids are stored primarily in the wings. Since nonpolar cardenolides are more emetic than their polar counterparts, the selective distribution of these compounds probably reflects their adaptive value *vis-à-vis* vertebrate predators.

Lygaeids preferentially sequester polar cardenolides in their dorsolateral spaces. Nonpolar compounds are stored in the blood and probably the fat body. The secretion of the dorsolateral spaces may function as an effective gustatory and olfactory repellent to predators. The polar cardenolides in the space fluid may thus be utilized as a first line of defense. Ordinarily, a predator will only encounter the nonpolar—and more emetic—cardenolides after biting into the bug. A vertebrate can learn to avoid these insects after an unpleasant gustatory reaction and certainly after a painful emetic experience.

Polar cardenolides are also sequestered in the metathoracic glands of adult lygaeids and the dorsal abdominal glands of the nymphs. These are true exocrine glands that produce very effective defensive compounds—α,β-unsaturated aldehydes. As is the case of the dorsolateral defensive space fluid, the sequestered steroids are polar compounds. Lygaeids may be typical of many insects in augmenting the deterrent efficiency of their exocrine gland products with sequestered plant compounds.

Like arctiid moths and pyrgomorphid grasshoppers, lygaeids tend to feed on toxic plants. It appears that once it became possible for these groups of insects to exploit a plant containing one class of toxins, it was subsequently possible for them to attack plant groups with completely different classes of plant natural products. Some species can sequester completely unrelated compounds, depending on the food source. The arctiid *Arctia caja* sequesters cardenolides when reared on *Digitalis* and pyrrolizidine alkaloids from *Senecio*. When a larva is allowed to feed on *Digitalis* and then switched to *Senecio,* it sequesters both classes of compounds. The ability of an insect species to sequester such chemically disparate compounds indicates that there are some common physiological bases underlying the sequestrative process.

In at least some insects, sequestration appears to be an energy–independent process. The lygaeid *Oncopeltus fasciatus* sequesters polar cardenolides in its dorsolateral space fluid independent of steroid concentration. At high cardenolide concentrations, large amounts of these compounds are sequestered. The kinetics of uptake are consistent with a physical rather than an enzymatic mechanism. High concentrations of 2,4-dinitrophenol do not inhibit the rate of cardenolide uptake. Thus, an active transport mechanism does not appear to be involved. In addition, the sequestration of one cardenolide is independent of the concentration of other cardiac glycosides.

The dorsolateral space of *O. fasciatus* contains emulsion particles which act as a sink for cardiac glycosides. These particles are present as an emulsion phase that exists at a concentration gradient with emulsion particles in the blood. Thus, blood concentrations of cardenolides should be fixed and not rise to intoxicative levels. A physical method of sequestration guarantees rapid uptake of cardenolides regardless of their concentration in the food. Since this sequestrative mechanism is energy independent, no metabolic debt is accrued as a consequence of it. Probably this physical method of sequestration is widespread among insect species feeding on toxic plants. The ability of insects to store natural products by a physical, rather than an enzymatic process, may have been the key evolutionary development that enabled these arthropods to exploit a diversity of toxic plants.

VIII. BIOSYNTHESIS OF INSECT DEFENSIVE COMPOUNDS

The metabolic origins of most defensive compounds are unknown. It appears that most of these exocrine products are synthesized *de novo* rather than incorporated from food sources. The possible role of microorganisms in biosynthesizing these compounds has not been established, although there is some evidence that bacteria and fungi may produce some exocrine products.

In virtually all of the biosynthetic investigations undertaken, only the fact that selected compounds function as precursors for particular defensive products has been established. In general, neither the enzymes nor the metabolic intermediates involved in these biosyntheses have been elucidated. Considering the host of novel natural products produced by insects, this subject offers fertile grounds for comparative biochemical studies.

A variety of defensive compounds has been shown to be synthesized after the administration of ^{14}C-acetate. For example, hydrocarbons such as *n*-undecane and *n*-tridecane are strongly labeled after pentatomid bugs are injected with this precursor or longer-chain fatty acids. The characteristic enals (e.g., *trans*-2-hexenal) of true bugs also contain radioactivity after ^{14}C-acetate administration. The same precursor is extensively incorporated into the alkaloids coccinelline and pederin and it has been suggested that both of these compounds are produced via polyketide biosynthesis.

A. Carboxylic Acids

The biogeneses of a few carboxylic acids have been investigated by utilizing appropriate ^{14}C-amino acid precursors. In formicine ants, DL-serine[3-^{14}C] serves as an excellent precursor for formic acid, the ubiquitous poison gland product of these insects (see Scheme 1).

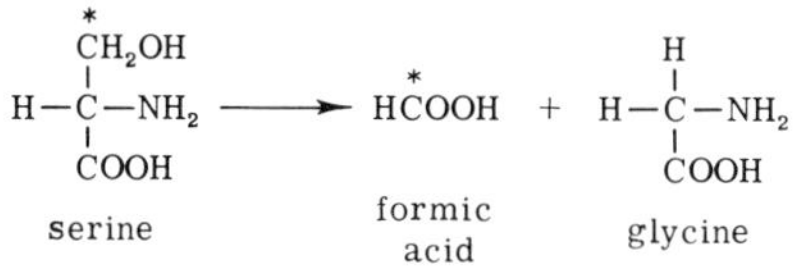

$$\underset{\text{serine}}{H{-}\overset{\overset{*}{C}H_2OH}{\underset{COOH}{C}}{-}NH_2} \longrightarrow \underset{\text{formic acid}}{H\overset{*}{C}OOH} + \underset{\text{glycine}}{H{-}\overset{H}{\underset{COOH}{C}}{-}NH_2}$$

Scheme 1

Branched fatty acids also arise from appropriate amino acid precursors. Injected DL-valine[4-^{14}C] is rapidly converted to methacrylic acid in the pygidial glands of carabid beetles. Similarly, defensive compounds produced in the osmeteria of papilionid larvae, isobutyric acid and α-methylbutyric acid, also arise from branched amino acids. [U-^{14}C] valine is converted to isobutyric acid whereas α-methylbutyric acid is synthesized from [U-^{14}C] isoleucine. These metabolic transformations presumably arise through α-keto acids which are formed by oxidative deamination. Subsequent oxidative decarboxylation would yield the acidic end products (see Scheme 2).

B. Isoprenoid Compounds

Insects produce a large variety of polyisoprenoids in their exocrine glands, many of which are produced *de novo*. Both acetate and mevalonate are utilized by insects to synthesize mono-(C_{10}), sesqui-(C_{15}), and diterpenes (C_{20}). For example, acetate is used to form mevalonic acid, the "committed" substrate in isoprenoid biosynthesis. Mevalonate is phosphorylated and eventually converted to isopentenyl pyrophosphate, the active isoprene unit. Condensation of the latter compound with an isomerized form of isopentenyl pyrophosphate yields geranyl pyrophosphate. Farnesyl pyrophosphate is produced by a repeat of this condensation and an additional condensation will yield a diterpene.

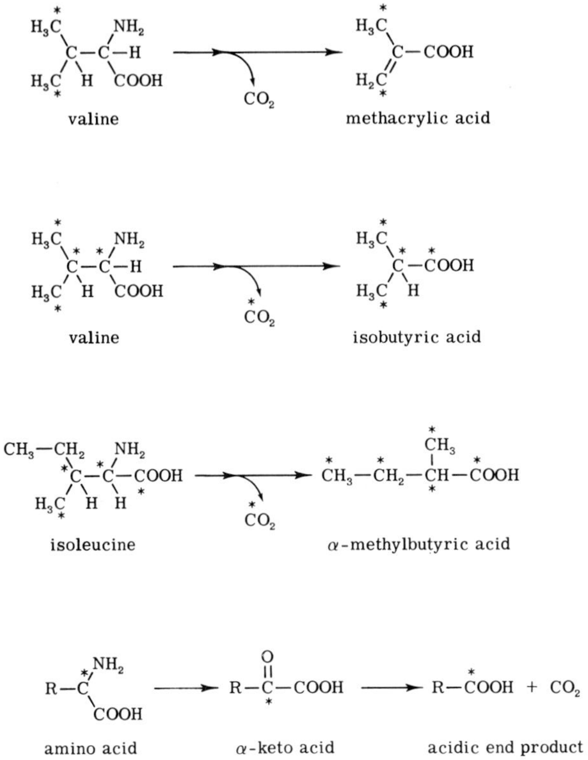

valine → methacrylic acid (CO_2)

valine → isobutyric acid (CO_2)

isoleucine → α-methylbutyric acid (CO_2)

amino acid → α-keto acid → acidic end product ($R{-}COOH + CO_2$)

Scheme 2

Formicine ants produce large quantities of the monoterpene aldehydes citral and citronellal from either acetate or mevalonate. The furanosesquiterpene dendrolasin is also a major product of mevalonate metabolism in ants. The presence of a very active system for the biosynthesis of mevalonic acid in insects is consistent with the ability of these arthropods to stress polyisoprenoid production.

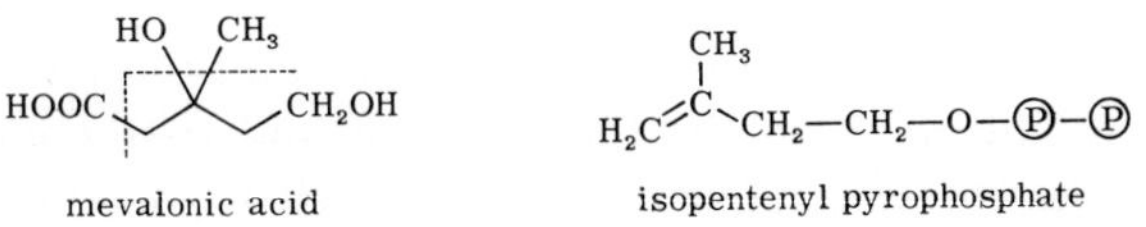

mevalonic acid isopentenyl pyrophosphate

The well-developed ability of insects to synthesize sesquiterpenes contrasts with their total lack of cholesterogenesis. Insects are incapable of synthesizing cyclic triterpenes (C_{30}) and must derive steroids from exogenous sources. They also appear incapable of producing squalene, a C_{30} isoprenoid that is the penultimate precursor of cyclic triterpenes. Squalene is formed from the head-to-head condensation of farnesyl pyrophosphate with an isomeric form of this C_{15} isoprenoid. Failure of insects to synthesize squalene from farnesyl pyrophosphate indicates that the cholesterogenic pathway is not active beyond the latter compound.

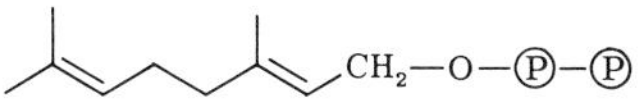

CH_2—O—Ⓟ—Ⓟ

geranyl pyrophosphate

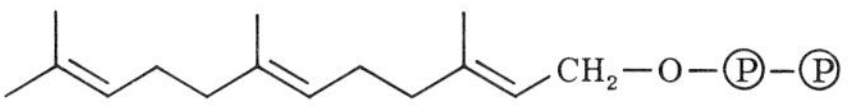

CH_2—O—Ⓟ—Ⓟ

farnesyl pyrophosphate

Nevertheless, some insects produce defensive secretions dominated by steroids. Many steroids in the prothoracic gland secretions of adult dytiscid beetles are identical to well-known vertebrate hormones. Injection of these beetles with ^{14}C-cholesterol results in all prothoracic steroids being labeled, as expected. Cholesterol, which is efficiently converted into cortexone and cybisterone, serves as an equally good precursor for both steroidal enones (cortexone) and dienones (cybisterone). On the other hand, progesterone, which is a precursor of cortexone, is poorly incorporated into dienones. These results indicate that the metabolic pathways for the enones and dienones branch before the progesterone stage.

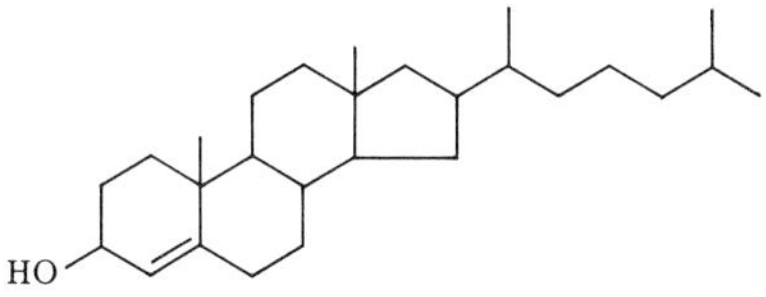

HO

cholesterol

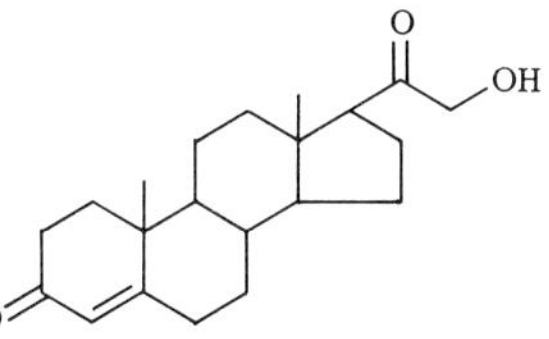

O OH O

cortexone

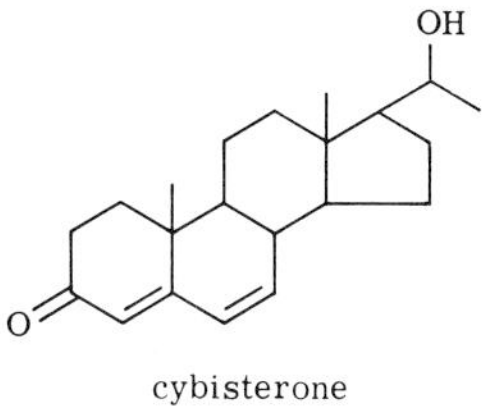

OH O

cybisterone

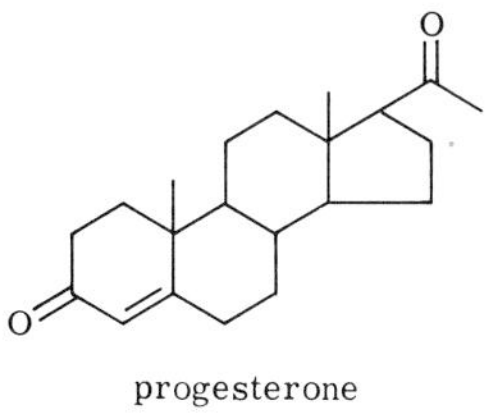

O O

progesterone

C. 1,4-Benzoquinones

Benzoquinones are particularly widespread in the defensive secretions of arthropods. These compounds have been identified as exocrine products of Dermaptera, Orthoptera, Dictyoptera, and Coleoptera, and species in other arthropod orders. Investigations on the biosynthesis of 1,4-benzoquinone in a tenebrionid beetle indicate that this compound arises primarily from the aromatic ring of amino acids; ring-labeled phenylalanine-^{14}C and uniformly labeled tyrosine-^{14}C are efficiently incorporated into 1,4-benzoquinone. The synthesis of this compound clearly involves utilization of the preformed ring of the two amino acids.

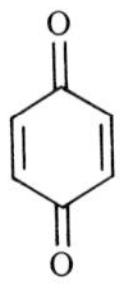

1,4-benzoquinone 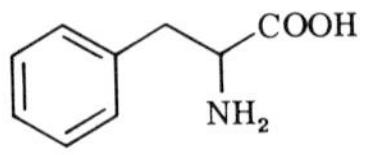 phenylalanine 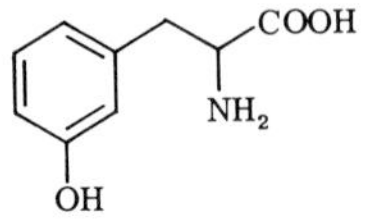 tyrosine

Methyl-1,4-benzoquinone and ethyl-1,4-benzoquinone, two other typical quinones in tenebrionid abdominal gland secretions, are not derived primarily from these aromatic amino acids. Aliphatic precursors such as sodium acetate[2-^{14}C] are extensively incorporated into these alkylated quinones; tyrosine and phenylalanine are poorly utilized for their syntheses. Since the ring carbons of both alkyl *p*-benzoquinones are extensively labeled after injection of acetate-^{14}C, these compounds clearly arise from the acetate pathway. The significance of the tenebrionid possessing different metabolic pathways for the syntheses of 1,4-benzoquinone and its alkylated homologues is unknown. It has not been established whether other insects producing these 1,4-benzoquinones utilize separate biosynthetic schemes.

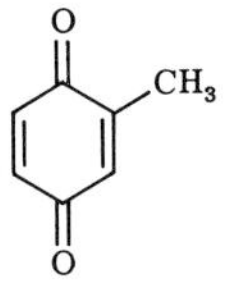

methyl-1,4-benzoquinone ethyl-1,4-benzoquinone

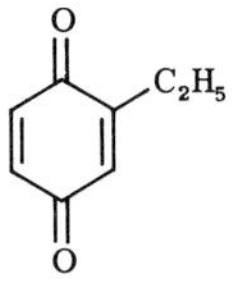

Benzoquinones in insects are produced from their corresponding hydroquinones. Bombardier beetles in the genus *Brachinus* produce these benzoquinones explosively, discharging them at a temperature of 100°C. The pygidial glands of these beetles consist of both a storage and an explosion chamber, which evacuates the hot quinones to the exterior. The storage chamber contains stabilized hydroquinones in a hydrogen peroxide solution. The peroxide concentration is frequently as high as 28%.

When this solution is discharged into the explosion chamber, the hydroquinones are rapidly enzymatically oxidized to benzoquinones and explosively discharged. Two groups of enzymes in this chamber—catalases and peroxidases—produce this double reaction. Peroxidase oxidizes the hydroquinones to quinones, and catalase decomposes the hydrogen peroxide into oxygen and water. The generated oxygen, which is under pressure, acts to discharge explosively the newly generated quinones. The catalases have the highest temperature optima recorded for this class of enzymes that ensure their stability at a reaction temperature of about 75°C. Both the peroxidases and catalases are more stable in the presence of high concentrations of hydrogen peroxide than most enzymes in their two classes.

$$\text{hydroquinone} + H_2O_2 \xrightarrow{\text{peroxidase}} \text{1,4-benzoquinone} + 2\,H_2O$$

$$2\,H_2O_2 \xrightarrow{\text{catalase}} 2\,H_2O + O_2$$

IX. ADAPTATIONS TO AVOID AUTOINTOXICATION

Insects have evolved both morphological and biochemical adaptations to avoid the autointoxicative effects resulting from the production and secretion of toxic compounds.

A. Synthesis of Toxic Products in Cuticular Organelles

The products of exocrine glands are frequently stored in cuticular reservoirs that appear to be impermeable. These cuticular membranes, which are invaginations of the body wall, insulate the sensitive tissues of the body from contact with the potential cytotoxins generated by defensive glands. The glands themselves may be lined with cuticular organelles in which the toxic end products may be synthesized in isolation from the living cells.

Tenebrionids generate quinones from phenolic glucosides that are converted to hydroquinones before being oxidized to the final quinoidal end products. The nontoxic hydroquinones are generated from phenolic β-glucosides by β-glucosidases. This reaction occurs in an inner chamber. The quinols (hydroquinones) are subsequently transferred to a reaction chamber lined with vesicular (cuticular) organelles in which a portion of the hydroquinones is oxidized to *p*-quinones by polyphenol oxidase. The quinol–quinone mixture is subsequently transferred to a more distant secretory unit in which a hemoprotein peroxidase oxidizes additional hydroquinones to *p*-quinones. The production of toxic quinones thus occurs in cuticular organelles isolated from the sensitive cytoplasm of the secretory unit. This morphological compartmentalization insures that the highly reactive 1,4-quinones will not react with the proteins present in the glandular cells. These reactions are illustrated in Fig. 3.

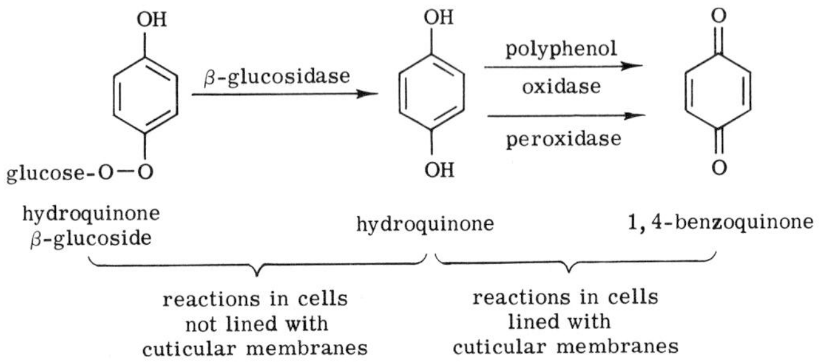

OH
β-glucosidase
glucose-O—O
hydroquinone
β-glucoside
OH
OH
hydroquinone
polyphenol
oxidase
peroxidase
O
O
1, 4-benzoquinone
reactions in cells
not lined with
cuticular membranes
reactions in cells
lined with
cuticular membranes

Fig. 3. Pathway for the synthesis of 1,4-benzoquinones in the defensive gland of a tenebrionid beetle.

The highly reactive α,β-unsaturated aldehydes produced by many adult hemipterans also appear to be generated in well-insulated cuticular chambers. High concentrations of toxic compounds such as *trans*-2-decenal are present in the scent gland reservoir, along with the less toxic acetate ester of the corresponding alcohol (i.e., *trans*-2-decenyl acetate). Significantly, the acetate is the major oxygenated compound present in the lateral glands, structures that lie free in the hemocoel and are not well isolated from the hemolymph. It appears that this less toxic ester serves as a precursor for the reactive aldehyde. Thus, synthesis of the more toxic aldehyde in the reservoir ensures that this very reactive compound will be isolated from the body cavity by an impermeable cuticular membrane (see Scheme 3).

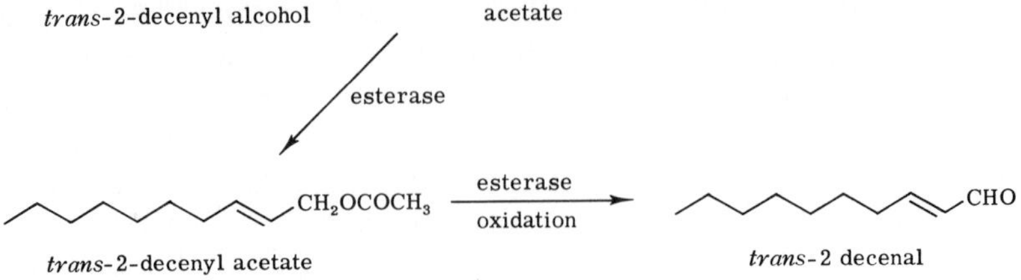

CH_2OH + CH_3COO^-
trans-2-decenyl alcohol
acetate
esterase
CH_2OCOCH_3
esterase
oxidation
CHO
trans-2-decenyl acetate
trans-2 decenal

Scheme 3

B. Detoxication of Defensive Compounds

Insects are often exposed to high concentrations of their own toxicants that can penetrate either through the spiracles or through the body wall itself after being secreted. Many of these compounds are considered to be fairly toxic and their producers obviously must possess detoxicative mechanisms for coping with them. However, little is known about the metabolic pathways utilized by insects to detoxify their own defensive compounds.

The detoxication of HCN by a few arthropod species has been studied. Zygaenid moths release HCN after tissue injury and at least part of the cyanide is detoxified by the enzyme rhodanese. This enzyme converts cyanide to relatively nontoxic thiocyanate ion in many animals as well as plants. Polydesmid millipedes also detoxify the HCN they generate with rhodanese. Additionally, they metabolize small quantities of cyanide to β-cyanoalanine and asparagine, which may be used to detoxify small quantities of HCN that leak into the millipede's body from the storage chamber which contains mandelonitrile.

$$\text{R—CN} \xrightarrow{\text{rhodanese}} \text{R—SCN}$$

thiocyanate

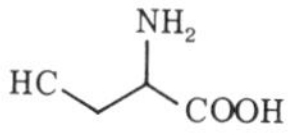

$NC\text{—}CH_2\text{—}CH(NH_2)\text{—}COOH$

β-cyanoalanine

$HOOC\text{—}CH(NH_2)\text{—}CH_2\text{—}C(=O)\text{—}NH_2$

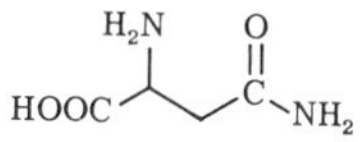

asparagine

Both insects and millipedes produce a variety of phenolic compounds that are considered to be fairly cytotoxic. Phenol, a product of cockroaches, beetles, and millipedes, is rapidly detoxified by millipedes with tyrosine phenol-lyase, the same enzyme that generates it. Since male coreid bugs detoxify phenol (a pheromone), with the same enzyme, it is probable that other phenol-producing insects utilize this enzyme similarly (see Scheme 4).

X. REGENERATION OF DEFENSIVE SECRETIONS

In general, defensive secretions, whether emitted from oozing glands (e.g., chrysomelid larvae) or spraying glands (e.g., adult carabids) are discharged very frugally. Only a fraction of the available glandular products is generally emitted during a single secretory act. Since molested insects frequently escape from their assailants after only one or a few discharges, this secretory behavior ensures that adequate quantities of secretion will be available if aggressive encounters occur soon thereafter.

Arthropods do not appear to regenerate their defensive constituents very rapidly and many have evolved mechanisms for conserving these secretions.

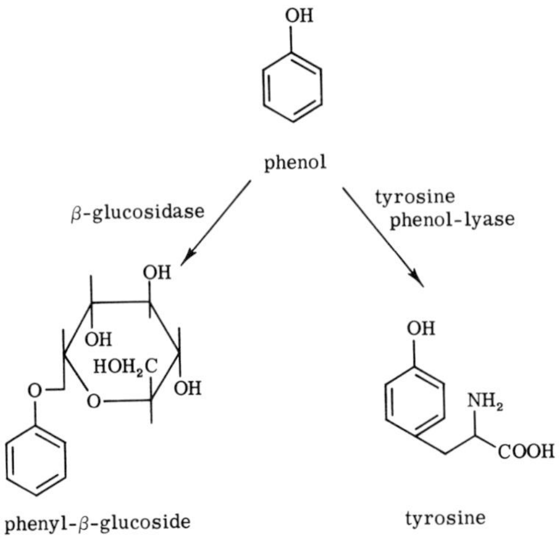

phenol

β-glucosidase

tyrosine phenol-lyase

phenyl-β-glucoside

tyrosine

Scheme 4

To illustrate, many insect species (e.g., staphylinid beetles) often only secrete their defensive products "as a last resort"—when other defensive tactics have failed to deter a predator. Much secretion is conserved by chrysomelid larvae which suck the discharged droplets back into the tubercles from which they were secreted.

Several studies indicate that defensive secretions are resynthesized very slowly. For example, pyrgomorphid grasshoppers require 1–2 weeks to regenerate their abdominal secretions. The abdominal exudate of polyzosteriine cockroaches is also slowly regenerated; blattids that initially produce large amounts of secretion yield reduced quantities of glandular fluid when "milked" subsequently. Secretory regeneration ceases completely after 4 weeks if the animals are "milked" once a week.

Polydesmid millipedes appear to regenerate their defensive products particularly slowly. Only about 10% of the original level of HCN is regenerated in a week; HCN levels are still reduced 2 months later.

Some chrysomelid larvae appear to be exceptional in being able to recharge their abdominal glands in 24 hours. A *Paropsis* species can regenerate a 0.2-mg secretion within a day of being "milked" to depletion. This exudate contains benzaldehyde, HCN, and glucose, which are presumably derived from a glycoside of mandelonitrile, as they are in millipedes.

XI. OPTIMIZING THE ADAPTIVENESS OF THE DEFENSIVE SECRETION

Insects have evolved a multitude of devices for increasing the effectiveness of their defensive exudates. While the *modi operandi* of these secretions are poorly understood, many of their adaptive features should be noted.

A. Localizing the Emission of the Deterrent Exudate

Insects with multiple exocrine glands will usually only secrete from those proximate to the source of tactile stimulation. This adaptation ensures that the exudate is delivered to the "target" from the nearest glands. Such parsimonious emission conserves the stored secretion present in glands that are spatially removed from the body of the molester.

Chrysomelid larvae possess paired thoracic and abdominal defensive glands. If probed on the thorax, secretion is generally emitted only from the stimulated segment, and primarily from the gland on the affected side. Similarly, tenthredinid sawfly larvae often only emit their aldehyde-rich secretion from the ventral gland closest to the point of stimulation.

Autohemorrhagic insects also generally limit the release of blood to the sites nearest the external contact. Adult lampyrids reflex-bleed only from the elytral pores closest to the area of stimulation. The firefly will rotate its body so as to place a bleeding site in close proximity to its aggressor. Both meloids and coccinellids reflex-bleed from femorotibial joints; usually only the stimulated joint hemorrhages in response to traumatic stimuli. Small arthropods (e.g., ants) will often be drenched with blood if the beetle pivots its leg so that the appendage makes contact with the antagonist.

A large number of blood droplets can be discharged without any obvious harm to the autohemorrhaging insect. Chrysomelid (*Diabrotica* species) larvae can lose up to 13% of their body weight via reflex bleeding and develop into normal adults.

B. Accurate Delivery of the Secretion

Insects emit their defensive exudates with remarkable accuracy. Postural adjustments ensure that the exocrine products will be delivered on target, whether the secretions ooze from the glands or are sprayed. If the rear leg of a staphylinid beetle is grabbed, it will rotate its abdomen so that the offensive object is thoroughly smeared with its pygidial gland products. Adult trichopterans (*Pychnopsyche* spp.) rotate their abdomens so that indole-rich exudate from their lateral abdominal glands is transferred to their aggressors. These caddis flies also thrust their legs into the secretion and deliver it to the bodies of their molesters. Many Hemiptera similarly employ their legs to transfer their exocrine products to the bodies of predatory animals.

Defensive secretions are almost invariably sprayed in the direction of a traumatic stimulus. Pentatomids and carabids are typical of many groups of insects in being able to place the products of their defensive glands on target. These insects position their bodies so that the glandular orifice points in the direction of the offending animal. This may involve flexing the abdominal tip (carabids) or a general postural adjustment (pentatomids). The secretions are generally sprayed with great frugality: a single discharge will often terminate an aggressive encounter.

C. Residual Effects of Discharged Secretions

In some cases, a discharged defensive exudate may exhibit prolonged deterrent activity. For example, some of the secretion emitted by either adult hemipterans or beetles may be entrapped under the elytra. The constituents in these entrapped exudates (e.g., enals, *p*-quinones) may volatilize slowly, providing the insect with a temporary odorous shield against their aggressors.

The metathoracic glandular orifices of adult hemipterans are surrounded by a finely sculptured cuticle that is believed to retard the evaporative loss of emitted defensive secretion. These specialized cuticular areas appear to be adapted to adsorb some of the discharged exudate and thus, prolong its repellent activity.

Some of the less volatile constituents (e.g., hydrocarbons) in defensive secretions may in themselves serve to reduce the rate of evaporation of the more irritating but lower-boiling compounds. *n*-Tridecane, a common major constituent in pentatomid secretions, should serve admirably to promote the residual action of prove irritants such as *trans*-2-hexenal.

D. Joint Utilization of Nonhomologous Defensive Glands

Many insects simultaneously emit the products of nonhomologous exocrine glands when they are molested. Since each of these glandular exudates usually possesses a unique chemical composition, an aggressor is sprayed or smeared with a greater variety of compounds than would have been possible if only a single gland had been utilized. The defensive benefits derived from utilizing multiple glands may be considerable.

Many formicine ant species will often typically treat a predator with constituents of three glands. Formic acid, a powerful cytotoxin, is emitted from the capacious poison gland of these formicids. In addition, the products of Dufour's gland, an accessory gland of the sting, may be secreted in admixture with formic acid. These Dufour's gland secretions, which are among the most complex exocrine secretions produced by insects, may contain more than forty compounds, including aliphatic hydrocarbons, ketones, acids, lactones, esters, and alcohols. To enhance further the effectiveness of the products of the poison and Dufour's glands, formicine ants often smear their molesters with their mandibular gland secretions. These cephalic glands produce compounds such as citral, citronellal, and 6-methyl-5-hepten-2-one. The deterrent potential of this multiple arsenal appears to be especially great.

The joint actions of these multiple glandular exudates produce a much greater deterrent "punch" than would be possible otherwise. For example, formic acid, a highly ionized carboxylic acid, will not penetrate the lipophilic epicuticle very readily. In some cases, its entry into the body of a predatory arthropod may be facilitated by the abrasion of the assailant's cuticle by the ant's mandibles which effectively removes the lipid-rich epicuticular barrier. However, abrasion is not the only means of ensuring that the cytotoxic poison gland product will gain access to sensitive internal tissues. The mandibular gland products, citronellal and citral, through their action as effective wetting and spreading agents, facilitate the penetration of formic acid through the epicuticle. The relatively nonpolar constituents emitted from the Dufour's gland may also act as cuticular wetting and spreading agents, as well as irritants per se.

The virtually simultaneous utilization of nonhomologous exocrine glands is highly adaptive for several reasons. Spatially separated glands can provide better immediate protection from predators than glands restricted to the body extremities (e.g., head, abdominal tip). Additionally, the products from different glands may be more collectively irritating than the compounds discharged from a single gland. As previously indicated, the joint actions of the products of nonhomologous glands may result in the rapid penetration of proved irritants through the cuticle. In addition, the repellent action of the products from a pair of glands is probably much greater than that of either gland alone. For example, the abdominal glands of *Zophobas* beetles produce *p*-quinones whereas the prothoracic glands synthesize a mixture of phenols; the combined deterrent action of these two classes of irritants is probably considerable. Adult dytiscids produce phenolic esters and aromatic aldehydes in their pygidial glands and steroidal compounds in their prothoracic glands. The steroids are reported to be excellent deterrents for vertebrates (e.g., fish); the aromatic products of the pygidial glands may function as defensive compounds against a wide variety of potential predators.

E. Role of Minor Secretory Constituents

Defensive exudates usually contain trace or minor constituents which may significantly increase their deterrent effectiveness. In secretions which are dominated by polar compounds, the minor accompanying products, as mentioned earlier, may be critical to facilitate the penetration of the former through the epicuticular lipid barrier. Larvae of some notodontid species discharge a cervical gland secretion that contains less than 1% of two alkanones: 2-undecanone and 2-tridecanone. The surfactant properties of these ketones ensure that formic acid rapidly penetrates the cuticle of the sprayed antagonist.

In a similar fashion, many of the Dufour's gland constituents of formicine ants may also serve as cuticular wetting and spreading agents. These ketones, esters, and alcohols originating in the Dufour's gland secretion may be sprayed with an admixture of a highly concentrated aqueous solution of formic acid (e.g., 60%).

In addition to their probable role as wetting agents, these minor constituents may possess important communicative functions. They may release alarm behavior in nearby ants and act as attractants for the excited workers. It also has been suggested that the higher-boiling minor constituents in the Dufour's gland secretions may serve as recognition markers which enable the ants to distinguish friend from foe. Conspecific ant workers marked with these trace components could be easily distinguished from foreign insects; the latter will possess, in addition, their own species-specific odors.

F. Modes of Action of Defensive Compounds

It is generally believed that the rapid deterrent action of defensive compounds often reflects their ability to stimulate the chemoreceptors of their antagonists. The receptors on cephalic sensory structures (e.g., antennae, palps) appear to be particularly susceptible to stimulation by the antagonistic allomones present in defensive secretions. Possibly the chemoreceptors identified with the general chemical sense of arthropods may also constitute effective targets for these exocrine products. An examination of some compounds emphasized by insects as defensive constituents demonstrates that highly reactive molecules occur widely in these secretions.

p-Benzoquinone, *trans*-2-hexenal, and methacrylic acid are typical defensive products of insects. All of these compounds are conjugated, very reactive, and share the ability to rapidly react with nucleophiles such as NH_2 or SH groups. Probably one of the reasons that they have been evolved to function as chemical deterrents is because of their ability to attack nucleophilic groups on proteins and thus inactivate these macromolecules. If olfactory chemoreceptors were flooded with these conjugated compounds, these sensory structures would no longer be able to function optimally. Such temporary deprivation of olfactory acuity could blunt the efficiency of insect aggressors dependent upon olfactory cues to locate their prey. In an olfactory sense, these arthropods could be blinded by these highly reactive quinones, acids, and aldehydes.

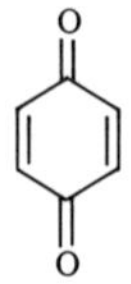

O, O, H, CHO, H, COOH

p-benzoquinone 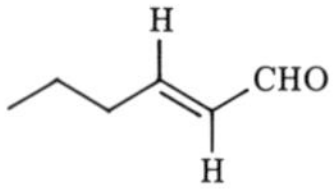 *trans*-2-hexenal 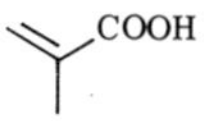 methacrylic acid

As described previously, hydrocarbons, often major and widespread constituents in defensive exudates, can function as wetting agents that promote the penetration of more reactive constituents. However, these alkanes may possess a more cryptic function as predator deterrents. It has been noted that antennal chemoreceptors are temporarily incapable of exhibiting rapid depolarization in response to their pheromonal stimuli if they are first exposed to *n*-undecane vapors. This suggests that hydrocarbons may reduce olfactory perception by altering the generator characteristics of the antennal chemoreceptors. Reduction of olfactory acuity of a predator would permit the molested insect to temporarily "hide" from its aggressor.

Defensive secretions may also figuratively "jam" the chemosensory system of potential aggressors. Overstimulation of the antennal chemoreceptors of a predator could result in the production of an uncoded array of spikes in the chemosensory neurons. This is the chemical equivalent of a "white noise," produced by bursts of uncoded information into the central nervous system. The qualitatively complex defensive secretions of many insects would appear to constitute ideal agents for generating such "negative" odors. Since the Dufour's gland secretion of formicine ants contains more than forty compounds belonging to at least five chemical classes, such an exudate should be capable of simultaneously stimulating olfactory generalists belonging to several classes. The resultant sensory input would certainly constitute a maladaptive signal for a predator. The diverse chemicals in many insect defensive exudates may actually generate cryptic odors when unleashed at an antagonist's chemosensory neurons.

XII. DEFENSIVE COMPOUNDS AND CHEMOTAXONOMY

Investigations of the chemistry of insect natural products have flourished in the last 25 years providing the basis for initial chemotaxonomic analyses. Recently, detailed comparative studies on the exocrine products synthesized by species in the families Carabidae and Tenebrionidae have demonstrated the value of these compounds as taxonomic character states for subfamily members in these two taxa. Identification of compounds from a large number of species has made it clear that while many unrelated insects produce the same compounds, many synthesize distinctive natural products. The restricted occurrence of these unusual compounds appears to have great chemotaxonomic significance.

It should be emphasized that the production of the same compound by species in widely separated phyletic groups does not imply that it was biosynthesized by the same metabolic pathway in both cases. For example, the occurrence of toluquinone in the defensive exudates of earwigs (*Forficula* species) and tenebrionids (*Tribolium* species) may simply indicate that both groups of insects have evolved the capacity to synthesize this compound by extending previously existing metabolic pathways. The independent evolution of the same compound by species in unrelated lines is comparable to the evolution of the same morphological character in unrelated insect groups. Even if the same distinctive compound has been evolved by species in unrelated taxa, the value of the natural product as a character state may still be considerable. Iridodial, a characteristic anal gland product of many dolichoderine ants, has recently been demonstrated to be a defensive product of a cerambycid beetle (*Aromia moschata*). The occurrence of this distinctive cyclopentanoid monoterpene as an exocrine product of the cerambycid would appear to constitute an excellent chemical character for this beetle.

A. Natural Products as Chemical Characters

Insects in many families characteristically produce exocrine compounds with a rather limited insect distribution. In general, sex pheromones are characteristic of lower taxonomic levels, whereas defensive compounds are characteristic of higher levels. In some cases, defensive compounds are found in all species in a family, whereas in others, they have only been detected in a single species. These biochemical idiosyncrasies appear to be useful chemical characters at different taxonomic levels for groups of insects for which good comparative data are available.

Distinctive natural product emphases clearly characterize species in a wide range of insect families. Hemipterous species in several families, for example, synthesize conjugated aldehydes (e.g., *trans*-2-octenal) in their defensive glands; these enals are especially characteristic of true bugs. Ants produce a variety of distinctive ethyl ketones (e.g., 3-octanone) in their mandibular glands which have been rarely detected as exocrine products of other insects. Characteristic compounds are also produced by termites. Soldiers of termitid species in several genera have a virtual monopoly on the production of monoterpene hydrocarbons (e.g., β-pinene) in the Insecta. Highly distinctive steroids are generated in the prothoracic glands of dytiscid beetles; these compounds have not been detected as exocrine products of species in other families. Other beetles (Carabidae) produce a large variety of conjugated acids (e.g., isocrotonic acid) in their pygidial glands and these compounds are generally absent from the secretions of noncarabids. Compounds that appear to be limited to the secretions of a single family are listed in Table II. Beetles in particular biosynthesize distinctive terpenoid natural products. Since the anhydride cantharidin has only been detected as a product of meloids, it appears to be a ubiquitous family product. Cortexone is one of a host of steroids produced by dytiscids; the prothoracic glands of these beetles are the only known insect source of these compounds. Gyrini-

TABLE II

Defensive Compounds Limited to Insect Species in a Single Family

Family	Compound	Structure
Meloidae	Cantharidin	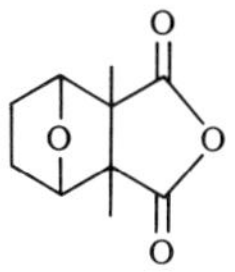
Dytiscidae	Cortexone	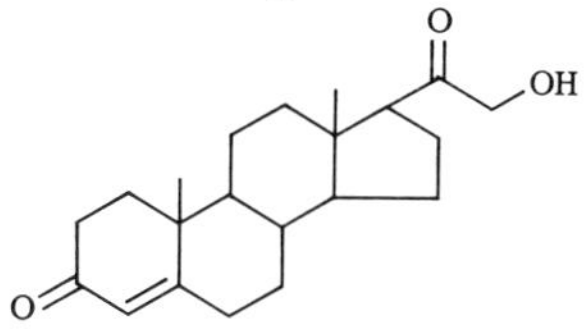
Gyrinidae	Gyrinidal	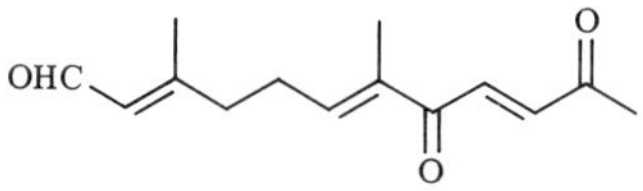
Coccinellidae	Coccinelline	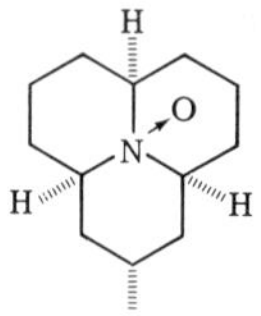
Carabidae	Tiglic acid	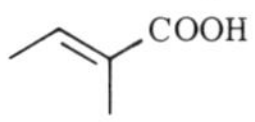
Termitidae	Trinervi-2β,3α,9α-triol 9-*O*-acetate	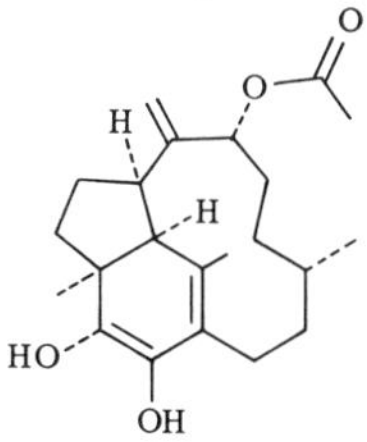

dal is one of several norsesquiterpenes that fortify the pygidial gland secretions of gyrinid beetles. These compounds are unique exocrine products.

A variety of other unusual natural products are limited to the secretions of insect species in a single family. Trinervi-2β,-3α,9α-triol 9-*O*-acetate typifies the unusual cyclic diterpenes synthesized in the frontal glands of some termite species in the genus *Trinervitermes*. Coccinelline exemplifies the equally distinctive cyclic compounds produced by adult coccinellids; these alkaloids are a hallmark of ladybugs in many genera. Carabids are distinguished by their ability to generate a diversity of α,β-unsaturated acids in their capacious pygidial glands; tiglic acid is one of four conjugated acids detected in their secretions. These conjugated acids are essentially limited to carabid species in many closely related genera.

To date, a variety of insect natural products have only been identified in the defensive exudates of species in one genus. These compounds possess obvious chemotaxonomic value as chemical characters for the species. Several are listed in Table III, along with the genera with which they are identified. The nitroalkene, 1-nitro-1-pentadecene, is a singular termite product and the only nitro compound identified in insect defensive secretions. 6-Methyl-1,4-naphthoquinone (Table III) is one of four naphthoquinones produced by *Argoporis* species; all other analyzed tenebrionids synthesize 1,4-benzoquinones. Massoilactone has been identified in two *Camponotus* species but has not been detected as an exocrine product in any of the many other species studied. Dimethyldisulfide is one of two alkyl sulfides produced in the mandibular glands of the ant *Paltothyreus tarsatus,*

TABLE III

Defensive Compounds Limited to Insect Species in a Single Genus

Genus	Compound	Structure
Prorhinotermes	1-Nitro-1-pentadecene	 NO_2
Argoporis	6-Methyl-1,4-naphthoquinone	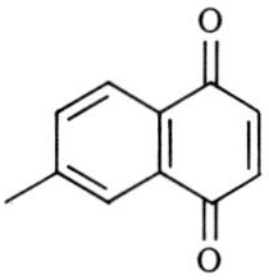 O, O
Camponotus	Massoilactone	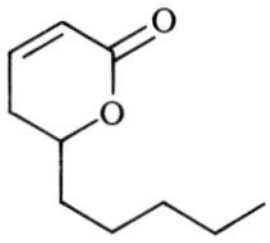 O, O
Paltothyreus	Dimethyldisulfide	$CH_3—S—S—CH_3$
Adelium	2,3-Dimethyl-1,4-benzoquinone	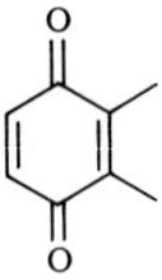 O, O
Ilybius	Testosterone	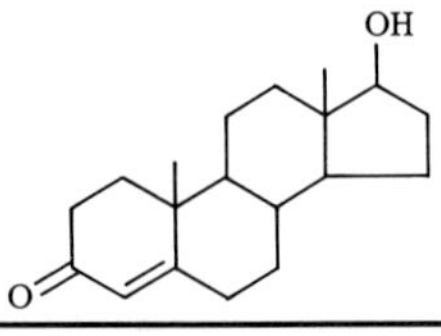 OH, O

the only known animal source of organic sulfides. 2,3-Dimethyl-1,4-benzoquinone appears to be a characteristic product of opilionids and its occurrence in species in the tenebrionid genus *Adelium* is truly exceptional; all other tenebrionids synthesize monoalkyl quinones. Testosterone is an especially good chemotaxonomic indicator for *Ilybius* species which appear to be the only dytiscids known to produce C_{19} steroids.

A limited number of insect species produce defensive compounds that have not even been detected in any other members in their genera. These natural products appear to be of outstanding chemotaxonomic value for the species that synthesize them. Several are listed in Table IV. 2-Phenylethyl isobutyrate, a natural product of *Chrysomela interrupta,* contrasts with salicylaldehyde, the defensive product identified in other species in this genus. The lactone isodihydronepetalactone is an anal gland product of

TABLE IV

Defensive Compounds Whose Known Distribution is Limited to a Single Species in a Genus

Genus	Species	Compound	Structure
Chrysomela	*interrupta*	2-Phenylethyl isobutyrate	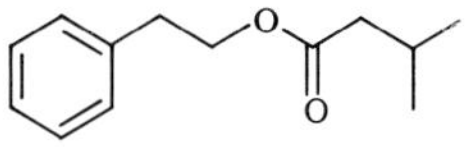
Iridomyrmex	*nitidus*	Isodihydronepetalactone	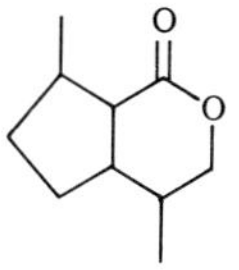
Lasius	*fuliginosus*	Dendrolasin	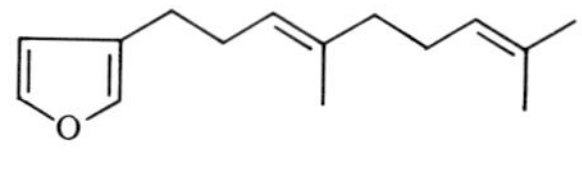
Solenopsis	*xyloni*	2-Methyl-6-*n*-undecyl-$\Delta^{1,2}$-piperideine	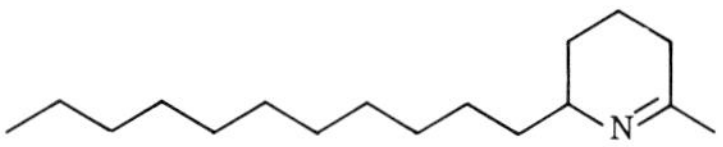
Cybister	*lateralimarginalis*	4,6-Pregnadiene-12β,20α-diol-3-one	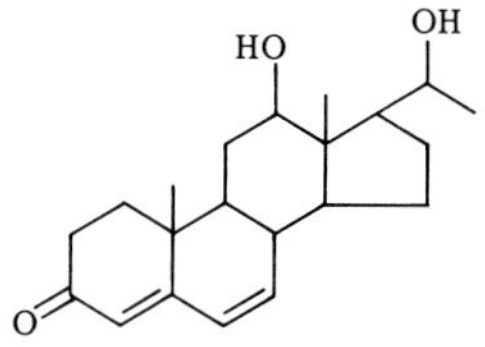
Camponotus	*clarithorax*	2,6-Dimethyl-5-heptenyl octanoate	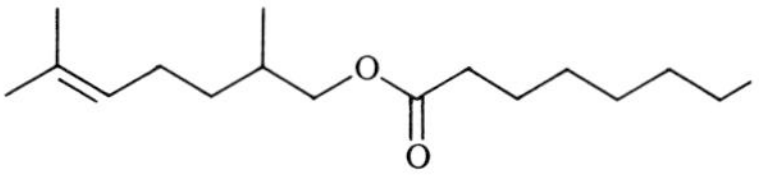

Iridomyrmex nitidus; different lactones are synthesized by other *Iridomyrmex* species. The furanoterpene dendrolasin (Table IV) is a distinctive mandibular gland product of the ant *Lasius fuliginosus;* monterpene aldehydes have been identified from other *Lasius* species. 2,6-Dialkylpiperidines are venomous constituents of many fire ant species; 2-methyl-6-*n*-undecyl-$\Delta^{1,2}$-piperideine appears to be limited to the venom of *Solenopsis xyloni.* The distinctive steroid 4,6-pregnadiene-12β,20α-diol-3-one has only been identified in the prothoracic gland secretion of one *Cybister* species. The occurrence of 2,6-dimethyl-5-heptenyl octanoate in the mandibular gland secretion of *Camponotus clarithorax* is also especially distinctive; all other species analyzed produced characteristic aromatic natural products, such as methyl anthranilate or methyl 6-methyl salicylate.

B. Defensive Compounds and Biochemical Systematics

At this juncture, there has been relatively little effort to utilize insect natural products for chemosystematic investigations. Although in themselves, these compounds cannot be employed for phylogenetic studies, they constitute potentially valuable adjuncts in such research. Exocrine compounds have been utilized effectively as systematic characters in instances in which the relationships of taxa were considered equivocal.

Apis indica has been considered to be either a subspecies of *Apis mellifera* or a separate species. The production of 2-heptanone in the mandibular glands of worker bees in all populations of *A. mellifera* and its absence in *A. indica* support the conclusion that these two forms are separate species.

In a few cases, congruity in natural products has been offered in support of the placement of the groups in a single taxon. The presence of typical coreid natural products in a *Hyocephalus* species supports placing this genus in the Coreoid complex rather than the Lygaeoid complex. Recently, species in two carabid tribes (Ozaeninae and Paussinae) were demonstrated to produce 1,4-benzoquinones and emit them by the same crepitation (explosive discharge) mechanism. These facts support the suggestion that they both be placed in the same subfamily.

The proposed phylogeny of the four honey bee species in the genus *Apis* has been fully supported by the results of studies on the amino acid sequences of melittin, the main toxin in their venoms. *Apis florea* is considered the most primitive species in this genus. *Apis mellifera* and *A. indica* are regarded as the most highly evolved of the four species, and are considered to be very closely related. *Apis dorsata* is considered to occupy an intermediate position between *A. florea* and the other two species. The amino acid sequences of the melittins produced by *A. indica* and *A. mellifera* were demonstrated to be identical. This result clearly supported their close relationship. On the other hand, the melittins synthesized by *A. florea* and by *A.*

dorsata differed considerably from that of *A. indica* and *A. mellifera*. However, that of *A. dorsata* was much closer in structure to the *indica–mellifera* melittin than that produced by *A. florea*. These biochemical systematic data clearly support the proposed phylogeny of the genus *Apis*.

GENERAL REFERENCES

Blum, M. S. (1974). *Bull. Entomol. Soc. Am.* **20,** 30–35.

Duffey, S. S. (1977). *Proc. Int. Congr. Entomol., 15th, Washington, D.C.* pp. 323–394.

Eisner, T. (1970). *In* "Chemical Ecology" (E. Sondheimer and J. B. Simeone, eds.), pp. 157–217. Academic Press, New York.

Pavan, M., and Dazzini, V. (1971). *In* "Chemical Zoology" (M. Florkin and B. T. Scheer, eds.), Vol. 6, pp. 365–409. Academic Press, New York.

Rothschild, M. (1972). *Symp. R. Entomol. Soc. London* **6,** 59–83.

Schildknecht, H. (1970). *Angew. Chem., Int. Ed. Engl.* **9,** 1–9.

Weatherston, J., and Percy, J. E. (1970). *In* "Chemicals Controlling Insect Behavior" (M. Beroza, ed.), pp. 95–144. Academic Press, New York.

12

The Biochemistry of Toxic Action of Insecticides

R. D. O'BRIEN

I. INTRODUCTION

This chapter deals exclusively with the reaction of insecticides and related compounds with their targets. It will not discuss the metabolism of any of these agents, nor their effects upon nontarget systems (which might be of importance, for instance, in the residual activity of chlorinated hydrocarbons upon the reproductive systems of vertebrates), nor with the lethal consequences which may result when one nonhuman organism attacks another chemically, as is common between insects. The emphasis will be upon the effects of insecticides upon insect targets, and for this reason will exclude such topics as the therapy of insecticide poisoning in man or effects of toxicants upon targets found only in vertebrates. Even with these restrictions, the topic is a large one, constituting a major component of several texts (listed at the end of this chapter) which deal with the overall questions of metabolism and action of insecticides.

The organization of this chapter is in terms of the targets which are attacked. Consequently, readers who come to it with such questions as "How does DDT act?" will need to come in through the doorway which the index provides.

Before embarking upon a blow-by-blow review of the targets, let us pause to get an overall view of the bullets which are directed at them. Only a few years ago, two great groups of compounds dominated all the rest: the axonic agents (of which the most important representatives were a number of chlorinated hydrocarbons) and the anticholinesterases. Because of the persistent nature of most chlorinated hydrocarbons, they have undergone a rather dramatic removal from use in recent years, so that the marketplace is now dominated by the anticholinesterases, that is to say, the organophosphates and the carbamates. Fortunately, we know a great deal about the biochemistry of the anticholinesterases, and therefore shall give a good deal of attention to them, both because of the richness of information and the importance of this class. But I would particularly like to point out to the novice that dependence upon any one class of insecticides is a danger to society. It is to be hoped that the next century will see the development of whole new categories of insect control agents, and one must certainly anticipate that these will include insecticides as well as antihormones and pheromones, and that all such agents will be used along with true biological control systems in a judicious mix.

II. ANTICHOLINESTERASES

A. The Compounds

There is a huge number of toxic anticholinesterases, but they all fall into two chemical categories, the organophosphates and the carbamates. We shall find that they both act by essentially the same mechanism, which generically could be called an acylating reaction. This concept will help in bringing order into their structures, because every such acylating agent contains an acylating group (which remains attached to the enzyme it attacks) and a leaving group which departs in the course of the reaction. It is true of both carbamates and organophosphates that a very small number of acylating groups are involved, but an enormous number of leaving groups have been designed, which bestow upon the acylating groups differing degrees of reactivity, and also of susceptibility to breakdown in the body and the environment

So far as the organophosphates are concerned, a study of Table I will show that it is common to have the acyl group contain $(C_2H_5O)_2P$ or $(CH_3O)_2P$. These two common groups are invariably associated with either a P(S)O group, in which case the compound is a phosphorothionate; or a P(S)S, in which case the compound is a phosphorodithioate; or a P(O)O, in which case the compound is a phosphate; or else a P(O)S, in which case the compound is a phosphorothiolate. Of these four classes, you can see that two can be called P(S) compounds and two are P(O) compounds, and we shall use this terminology. The P(S) compounds are invariably converted to

TABLE I

Some Common Organophosphates

Commercial compounds directly active [all P(O)]	Structure
Dichlorvos (Vapona or DDVP)	$(CH_3O)_2P(O)OCH{=}CCl_2$
Tetrachlorvinphos (Rabon)	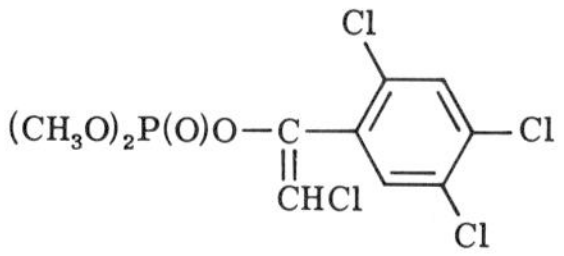 $(CH_3O)_2P(O)O{-}C({=}CHCl){-}C_6H_2Cl_3$ (2,4,5-trichlorophenyl)
Tetraethyl pyrophosphate (TEPP)	$(C_2H_5O)_2P(O)OP(O)(OC_2H_5)_2$

Latent compounds [all P(S)]

Commercial compound		Active derivative	
Parathion	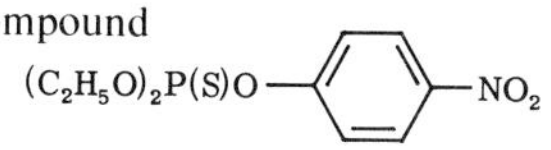 $(C_2H_5O)_2P(S)O{-}C_6H_4{-}NO_2$	Paraoxon	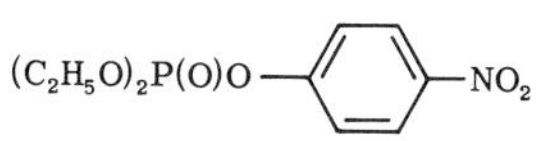 $(C_2H_5O)_2P(O)O{-}C_6H_4{-}NO_2$
Malathion	$(CH_3O)_2P(S)SCH(COOC_2H_5)CH_2COOC_2H_5$	Malaoxon	$(CH_3O)_2P(O)OCH(COOC_2H_5)CH_2COOC_2H_5$
Dimethoate	$(CH_3O)_2P(S)SCH_2C(O)NHCH_3$	Dimethoxon	$(CH_3O)_2P(O)SCH_2C(O)NHCH_3$

P(O) compounds before they have their toxic effect, and consequently although the list of practical insecticides has a high proportion of P(S) compounds in it, from a biochemist's point of view the interest is centered upon their P(O) derivatives.

The carbamates provide a much easier picture in chemical terms (Table II). The acylating group in this case is usually the *N*-methylcarbamyl group, $CH_3NHC(O)O$, although there are a few cases of older compounds which are dimethylcarbamates, having $(CH_3)_2NC(O)O$. Once again there is a lot of diversity in the leaving groups, although you should note that most of the leaving groups are aromatic.

B. Acetylcholinesterase Inhibition

It is now almost universally agreed that the carbamates and organophosphates are lethal to insects (and to vertebrates, for that matter) because they inhibit the enzyme acetylcholinesterase. The older literature contains a number of protests against this apparently simplistic view, but none of the alternate targets (which include esterases other than acetylcholinesterase) has been persuasively shown to be of importance, or to have some parallelism between their inhibition and mortality. We shall need to start with a few words about this highly important target, acetylcholinesterase, against which most of the insecticides of commerce today are directed.

Acetylcholinesterase is an enzyme which catalyzes the cleavage of acetyl-

TABLE II

Some Common Carbamates

$OC(O)NHCH_3$	$OC(O)NHCH_3$; H_3C, CH_3; $N(CH_3)_2$	$CH_3SC(CH_3)_2-CH{=}NOC(O)NHCH_3$
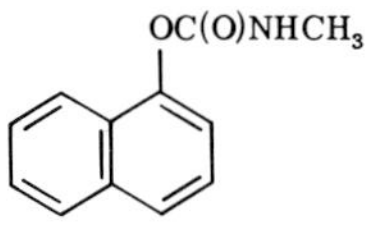 Carbaryl (sevin)	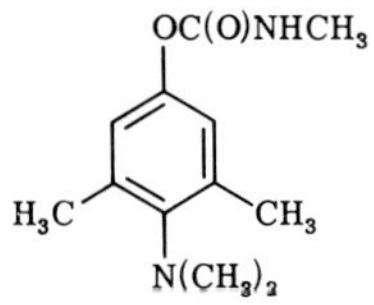 Zectran (mexacarbate)	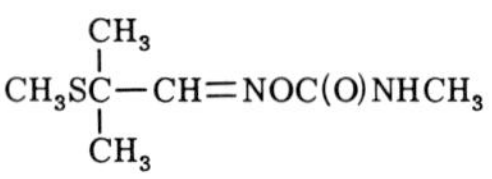 Temik (aldicarb)
$OC(O)NHCH_3$; S	$OC(O)NHCH_3$; H_3C, CH_3; SCH_3	$OC(O)NHCH_3$; H_3C, H_3C
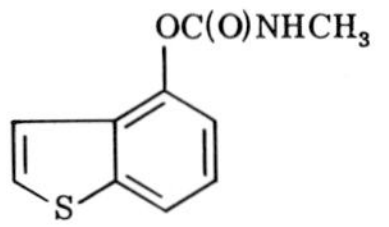 Mobam	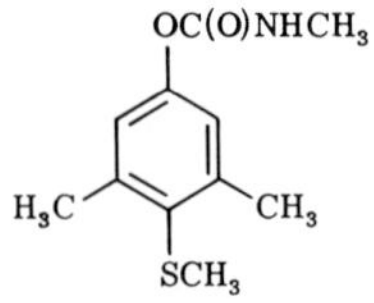 Mesurol (mercaptodimethur)	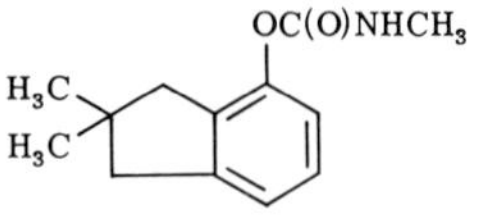 carbofuran

choline, thereby converting a physiologically potent compound to an impotent pair of compounds, acetate and choline [Eq. (1)]. Acetylcholine is the transmitter substance which enables one important class of nerve cells, called cholinergic neurons, to communicate with their neighbors. The mechanism of this communication is diagrammed in Fig. 1. When a cholinergic neuron is carrying an impulse, and when that impulse arrives at the ending of the neuron, it provokes the release of small packets of acetylcholine from the nerve ending. The acetylcholine which is released then drifts across the gap which separates the neuron from its neighboring cell. This neighbor could be another neuron (cholinergic or otherwise) or in the case of the vertebrate it could be a muscle fiber and this is an important target for the poisoning of vertebrates by anticholinesterases. In the insect nervous system, skeletal muscles are not innervated by cholinergic neurons, but probably by neurons which release glutamate. Consequently an important difference between the sensitivity of vertebrates and insects (and probably all arthropods) to poisoning is that the neuromuscular junction of vertebrates is sensitive to these poisons, whereas that of the insect is not.

$$CH_3C(O)O\ CH_2CH_2\ N^+(CH_3)_3 \xrightarrow{H_2O} CH_3COO^- + HOCH_2CH_2N^+(CH_3)_3 + H^+ \quad (1)$$

In the insect, therefore, the cholinergic neurons are located exclusively in the central nervous system, made up usually of a series of ganglia. The junction between a neuron and a cell which it affects is called a synapse, and one which is operated by acetylcholine is called a cholinergic synapse, and the neuron whose nerve ending forms a part of it is called a cholinergic neuron. The receiving cell in this synapse has, embedded in its cell membrane, a molecule called the acetylcholine receptor, which we shall discuss in detail in a subsequent section. The combination of the acetylcholine with its receptor can, if circumstances are right, stimulate the receiving neuron or

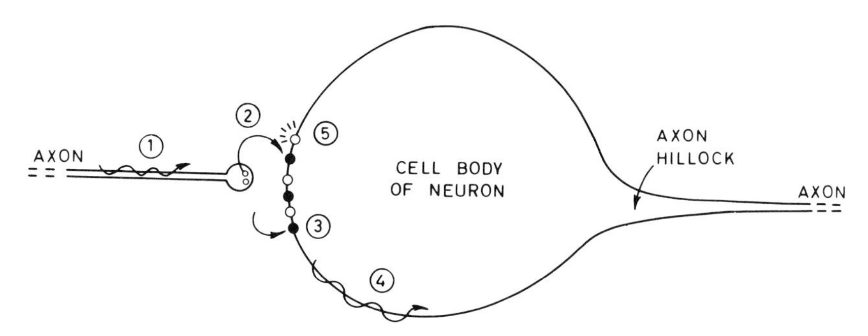

Fig. 1. Diagram of the chemical excitation of one neuron by a cholinergic neuron. The impulse (1) coming down the axon of the cholinergic neuron provokes release of acetylcholine (2) which combines with acetylcholine receptor (3) and triggers an excitation of the second neuron (4). Acetylcholinesterase (5) terminates the process by destroying acetylcholine.

gland to fire off or to secrete its appropriate material. Because a single pulse of acetylcholine must produce a single signal in the receiving cell, it is essential that the acetylcholine which has done its job should be eliminated. In cholinergic synapses, this elimination is invariably achieved by the enzyme acetylcholinesterase, which is also present in the synapse. Because acetylcholinesterase is present, the pulse of acetylcholine causes a brief appearance of acetylcholine in the synapse (or more properly, in the synaptic cleft, which is the gap between the cholinergic neuron and its receiving cell) and consequently the acetylcholine level goes from almost zero to a sharp peak and then declines rapidly. The acetylcholine receptor (which combines reversibly with the acetylcholine) passively follows this increase and decline in acetylcholine concentration, and changes its configuration in harmony with it, as we shall see later. It is not necessary to imagine that the acetylcholinesterase literally seizes the acetylcholine molecules which are attached to the receptor. Over one million molecules are released in a typical impulse, and of these about one thousand actually combine with receptor. The majority are simply destroyed. But the receptor is responding to the change in acetylcholine *concentration,* and therefore although the majority of acetylcholine molecules never see a receptor, and their destruction is physically separated from the receptor itself, yet the influence of the acetylcholinesterase upon the opening and closing of the receptor is profound.

Until very recently it was believed that the acetylcholinesterase was embedded in the same membrane as the acetylcholine receptor. There is now excellent evidence for electric fish, and preliminary evidence from the insect, that this is not the case. Instead, the synaptic cleft appears to have a noncellular basement membrane (probably made of collagen in vertebrates) and the acetylcholinesterase is anchored into it by a tail (probably collagenous in vertebrates). Isolated acetylcholinesterase in both insects and vertebrates is seen in the electron microscope as a cluster of catalytic heads attached to a relatively long tail.

In insects poisoned by inhibitors of acetylcholinesterase, the acetylcholine which is released into the cleft accumulates, and at first has the effect of causing repeated firing of the neighboring neuronal cell. But this mechanism breaks down as the persistence increases, and then the junction becomes entirely nonfunctional. In the vertebrate, these effects lead to death by blocking respiration, sometimes by paralysis of the intercostal muscles which are involved in breathing, and sometimes by direct inhibition of the respiratory center of the brain. Of course neither such mechanism can exist in the insect (where air circulates primarily by passive diffusion) and we simply do not understand the precise series of events that lead to death. But it is safe to say that the nervous system, in insects as in vertebrates, is crucial for the organized behavior which constitutes life, and the severe disruption of it caused by interference with cholinergic synapses effectively destroys the communication network, and death ensues.

Now let us take a brief look at the biochemical features of acetylcholinesterase, necessary in order to understand the mechanism of its inhibition. The overall hydrolysis of a molecule of acetylcholine by acetylcholinesterase is a three-step process. First the acetylcholine molecule is bound to the active site (Fig. 2a). This binding is believed to involve a negatively charged site, the anionic site of the enzyme, which promotes the binding by attraction to the cationic nitrogen. The affinity of this binding is not enormous, the K_d being about 10^{-5} M.*

The second step involves the chemical reaction of the acetylcholine with a part of the enzyme called the esteratic site. It is almost certain that this esteratic site is a molecule of serine which has been modified by the influence of surrounding amino acids to make it highly reactive. It is the hydroxyl group of the serine which is acetylated by the acetylcholine, and the result is to transfer an acetyl group from acetylcholine to the serine hydroxyl, giving an acetylated enzyme, and to permit the choline to drift away; choline is the leaving group (Fig. 2b). The final step involves the attack of water upon the acetylated enzyme, cleaving the acetyl–serine bond to give acetate ion and the enzyme in its original form (Fig. 2c). As in every case of a series of reactions of this type, the speed of the total reaction is governed by the slowest step. In vertebrates the deacetylation is probably the slowest, and in insects the acetylation step is probably the slowest, but in both cases the overall reaction is extremely swift, being completed in a few nanoseconds.

The phosphates and carbamates react with the enzyme in a precisely analogous way. In both cases the inhibitor sits down on the enzyme, and

*The K_d is the dissociation constant which describes the strength of binding of one molecule to another. Because we shall discuss it more below, some detail is needed for those who may be unfamiliar with the concept. If a substrate or inhibitor, X, binds to an enzyme to form a reversible complex, then the equilibrium which results is described by the equation: $K_d = [X][E]/[XE]$, where [X] is the concentration of X and [E] of enzyme and [XE] of the complex. If the binding between the two is very tight, then most of the E will be tied up as XE, and consequently K_d will be very small. For substrates, a K_d of about 10^{-4} M is a common value. For potent inhibitors, much smaller values of K_d, i.e., much tighter binding, can be found, with values of 10^{-8} M being not uncommon.

A useful property of the K_d concept is that it describes how much inhibitor is needed to half-inhibit the enzyme. At half-inhibition, the amount of free enzyme and bound enzyme is equal, so that [E] = [XE]. Substituting this into the above equation, one can see that then K_d = [X]. In other words, when the enzyme is half-inhibited, the dissociation constant is the same as the amount of the free inhibitor concentration. Because the quantity of inhibitor used in experiments is typically much greater than the quantity of enzyme, the quantity of X which is removed to make XE is almost negligible, and so it is safe to say that the concentration of inhibitor to provide half-inhibition equals the dissociation constant. Thus, if an inhibitor has a K_d of 10^{-7}, it means that 10^{-7} M of that inhibitor will serve to knock out half the enzyme.

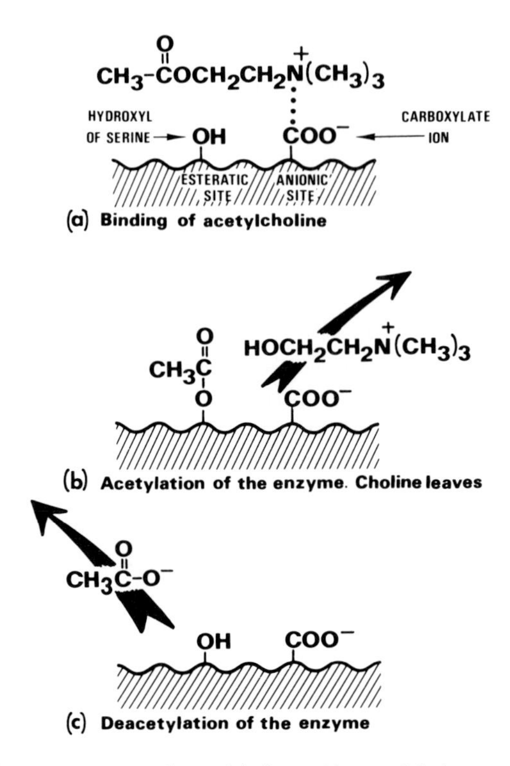

Fig. 2. Reaction of acetylcholine with acetylcholinesterase.

binds to it with a tightness which can be measured by a dissociation constant. Then in a second step, the serine hydroxyl becomes either carbamylated or phosphorylated; the efficiency of this reaction, as measured by the rate constant, is not enormously great as compared to acetylcholine. For nearly all carbamates the value is between 1 and 5 per min; for many organophosphates the rate is in the order of 50 per min, although it may go up to 126 per min in particularly effective compounds.

It is with respect to the third step in the reaction that the enormous differences between substrates and inhibitors become apparent. Whereas the deacetylation observed after acetylcholine reaction occurs within fractions of a second, the rate of dephosphorylation is typically days or even weeks. Of course the precise velocity depends upon the nature of the inhibitor (or, to be precise, the nature of the acylating group of the inhibitor) and upon the temperature and the kind of enzyme involved. In the case of carbamates, the hydrolysis is a good deal faster, being half-complete in about 20 minutes. But in both cases, these velocities are so disastrously less than with a substrate that the enzyme rapidly becomes completely phosphorylated or carbamylated, and the active sites are no longer available to do their proper job of reacting with acetylcholine.

III. ACETYLCHOLINE RECEPTOR AGENTS

As discussed in Section II,B, acetylcholine is a chemical signal which serves to "turn on" those neurons or effector cells which are part of a cholinergic function. The "turning on" occurs when the acetylcholine combines with a protein, the acetylcholine receptor, which is embedded in the cell's outer membrane, facing outwards.

The receptor occurs in many different forms, although all are stimulated by acetylcholine. Some, such as those in vertebrate skeletal muscle, can also be stimulated by nicotine and related drugs ("nicotinic receptors"); others, such as those in some vertebrate smooth muscles, are stimulated by muscarine and related drugs ("muscarinic receptors") and some are not so easily classified. The most reliable classification is based upon physiological responses observed when agents are applied directly to the cells containing the receptor. In insects, very little information exists of this kind. But biochemical evidence suggests that the acetylcholine receptor of house fly brain is of a hybrid nicotinic–muscarinic character. For instance, its binding of nicotine is blocked by atropine (a muscarinic blocker) and vice versa.

The supposed receptor from house fly heads has other unexpected properties. Much of it is water-soluble and has no phospholipid, whereas the vertebrate receptor has to be solubilized by detergent and contains phospholipid. Most disturbing of all, the supposed house fly receptor does not bind acetylcholine very tightly: its K_d is only 16 μM, compared with about 20 nM for the best vertebrate preparations as measured directly by equilibrium dialysis. But these discrepancies are not nearly enough to discredit the status of the house fly receptor. Extremely diverse claims have been made about the K_d for acetylcholine, even for the extensively studied form, from the electric skate, genus *Torpedo*. And the house fly receptor is from the central nervous system, whereas "the vertebrate receptor" which has been most studied is from mammalian muscle or (a homologous tissue) *Torpedo* electroplax. And physiological responses to acetylcholine fall in the micromolar range rather than the nanomolar range.

There are two groups of insecticides which almost certainly kill by interacting with the acetylcholine receptor. The first is a very old group: nicotine and its analogues. It has been known for over a century that, in the vertebrate, this agent acts upon the class of receptors to which it gives its name: nicotinic acetylcholine receptors, present primarily in skeletal muscles. Small doses of nicotine excite these receptors and large doses block

TABLE III
Evidence that Toxic Nicotinoids Block House Fly Head Acetylcholine Receptor[a]

Agent	% Block of 10^{-6} M muscarone binding by 10^{-4} M agent	LD_{50} House fly (μg/fly)
Nicotine	102	5
Anabasine	98	4
3-Pyridylmethyl dimethylamine	92	16
3-Pyridylmethyl diethylamine	97	11
N,*N*-Diethylnicotinamide	0	>100
N-(3-Pyridylmethyl)morphine	0	>100

[a]From Eldefrawi, M. E., Eldefrawi, A. T., and O'Brien, R. D., *J. Agric. Food Chem.* **18,** 1113 (1970).

them. In the house fly (Table III) a good correlation has been found between toxicity and the ability to bind to the receptor.

The second group is rather new. It was found in 1964 that nereistoxin, isolated from the marine worm *Lumbriconereis heteropoda* has good insecticidal activity. Subsequently a number of synthetic analogues were made, and that called Cartap is particularly effective as an insecticide. It may perhaps be converted to nereistoxin *in vivo*.

$(CH_3)_2N$–(S–S ring) $(CH_3)_2N$(–$SCONH_2$)(–$SCONH_2$)

nereistoxin   Cartap

There is direct experimental evidence that, in vertebrates, nereistoxin acts on both nicotinic and muscarinic acetylcholine receptors. A direct effect upon the insect receptor has not yet been investigated, but it is a plausible guess that this is the basis of its insecticidal action, and that of Cartap.

IV. PRESYNAPTIC AGENTS: ALDRIN AND DIELDRIN

Some toxic compounds have their effect upon the release of transmitter from the vesicles in which it is contained. Such actions have been particularly well worked out for cholinergic systems. It is likely that botulinum toxin, black widow spider venom, and the component of the banded krait snake poison called β-bungarotoxin all act by interfering with the release of acetylcholine from its storage vesicles.

The insecticides aldrin and dieldrin are believed to have a common mode of action, because aldrin is converted to its epoxide form, dieldrin, in the poisoned animal. It is very likely that the dieldrin itself requires further metabolism to its hydrated form, the diol, in the *trans* configuration.

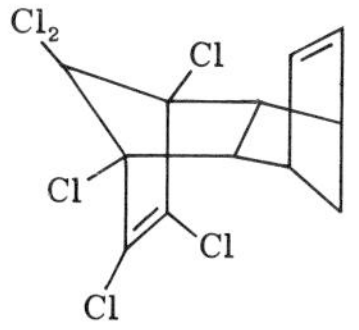

aldrin 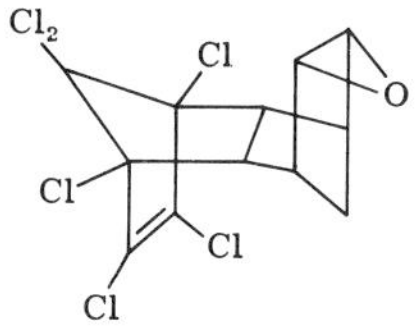 dieldrin 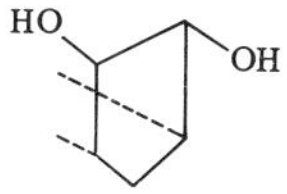 aldrin *trans*-diol

The *trans*-diol is thus probably the active toxicant derived from both aldrin and dieldrin. It was shown in the cockroach in 1973 that the physiological effects of dieldrin involve excessive release of transmitter (almost certainly acetylcholine) from its vesicular stores. Subsequently this mechanism was confirmed in the case of frog neuromuscular junction.

The physiological action is therefore fairly clear. When the molecular basis of vesicular release is discovered, the biochemical actions of aldrin and dieldrin will become clear.

V. AXONIC POISONS

Many important insecticides, including DDT and the pyrethroids, are probably toxic because they cause excitability in the axons of neurons. Neurons are the nerve calls which carry messages. Before the mechanism of this action can be described, it will be necessary to give a brief account of axonic transmission. The axon is the part of the neuron which is specialized for carrying messages (i.e., nerve impulses) rapidly, frequently for long distances, and without any change in the size or pattern of the impulse as it moves along. As in all cells, there is a difference of electric potential between the inside and the outside of the limiting membrane of the neuron. That means that if one could connect the inside and the outside by a good conductor, a current would flow. The size of this potential varies somewhat, but 70 mV is a common value, and the inside is negative with respect to the outside (see Fig. 3); we can therefore say that the neuron's cell membrane is polarized. The cause of this polarization is the difference in ion concentrations inside and outside the cell. As Fig. 3 shows, K^+ is much higher in concentration inside than out, and the reverse is true for Na^+ and Cl^-. When the axon is at rest (i.e., not conducting an impulse) its membrane is permeable to K^+, but has very little permeability to Na^+ or Cl^- ion. When a membrane is impermeable to an ion, that ion cannot influence the polarization of the membrane. Thus it is the K^+ permeability of the resting membrane that causes the polarization, by a mechanism well understood but unnecessary to recount here.

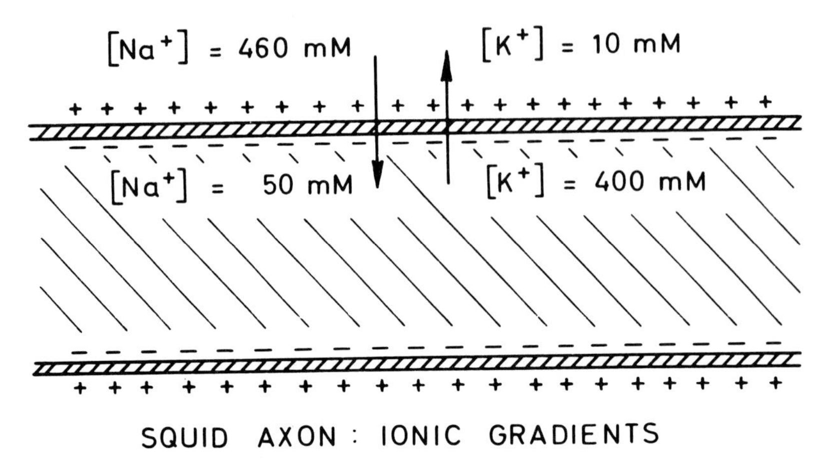

Fig. 3. Differences in ion concentrations and charge states in a nerve (the squid axon).

The impulses which the axon transmits when the neuron "speaks" to its neighbor take the form of temporary reversals of the polarity of the axon. Thus if we were to put an electrode in an axon and record the potential of the inside with respect to the outside, the resting membrane would show a picture like Fig. 4, the potential being −70 mV. But when an impulse passes by that electrode, the membrane is depolarized, which means that the potential falls to zero, and indeed a little reverse-polarization occurs, with the potential going to something like 10 mV, with the outside negative. In a very short time (about 1 msec) the potential is restored to −70 mV, and the axon has transmitted its impulse and is ready to transmit others if necessary.

The nerve impulse, also called the action potential, takes the form of a wave of depolarization that passes down the axon. Physiological experiments have provided evidence that the wave of depolarization is caused by a wave of brief Na^+ permeability. When the permeability wave arrives at any particular point, Na^+ (which is ten times more concentrated outside than in) pours into the axon, and causes the interior to become transiently positive. It is as if a gate for Na^+ were briefly flung open, and the Na^+ poured in down its concentration gradient. But the gate has the peculiar property that it can only open briefly, and after about 1 msec it slams shut again, a phenomenon called "sodium inactivation." After a brief delay, the membrane shows a substantial increase in its K^+ permeability, and therefore K^+, also following its own concentration gradient which lies in the reverse direction, rushes out and restores the membrane to its original state of polarization.

The ability to perform all these tricks depends of course upon the maintenance of ion gradients between that inside and the outside of a cell. One can

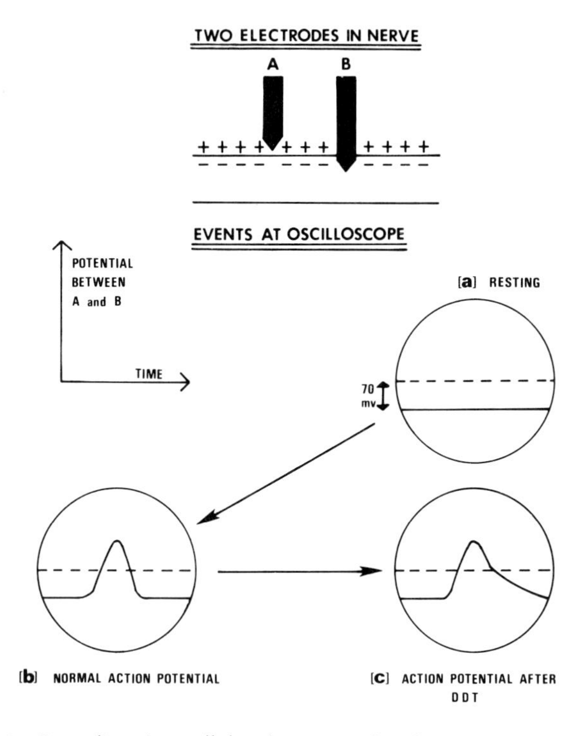

Fig. 4. Recording (intracellular) from normal and DDT-poisoned nerves.

see how the events just described tend to reduce these gradients, by letting the ions leak. Clearly there has to be some device to create the gradients in the first place, and to compensate for the leakage that occurs during the action potential. The "device" is called the Na^+–K^+ pump, and we know that it can use the energy derived from ATP to force Na^+ out of the nerve and K^+ into the nerve, both against their concentration gradients. It seems likely that the pump is identical with an enzyme which can be extracted from nerves, and is called Na^+-activated ATPase or simply Na^+-ATPase. This enzyme in the test tube splits ATP, and requires Na^+ to operate successfully, so that even in the extracted state it has some of the properties of the Na^+–K^+ pump. It should be stressed that the Na^+–K^+ pump is quite different from the Na^+ gate. The pump is present in virtually all cell membranes, but the gate is present only in excitable tissues such as nerve and muscle. The pump and gate can also be poisoned by completely different agents.

The way in which DDT affects axonic transmission is readily observed physiologically, and leads to a guess as to how it works molecularly. First it should be said that DDT is not equally effective on all axons, having lesser effects upon motor axons, i.e., axons of neurons which drive muscles. A particularly well-known example of a motor axon is the giant axon of the squid, which has been very useful to physiologists because of its large diameter (about 0.5 mm). Yet it is virtually insensitive to DDT. But the axons of sensory nerves, that is, nerves which send impulses from sense organs, show the following effect. Observed grossly, the axon is seen to become hyperexcitable, which is to say that it is too readily excited. Consequently when a single impulse is transmitted from the cell body down the axon, it gives rise to a train of impulses. It is believed that this hyperexcitability leads to the tremoring and finally the paralysis which precedes death in poisoned insects.

Looking now at the shape of an individual action potential in a DDT-poisoned nerve, one sees that the rising phase is almost normal, but the falling phase is greatly prolonged (Fig. 4). Such a prolongation could be caused by a delay of closing of the Na^+ gate, or by an inadequate opening of the K^+ gate. This question has been studied by a technique known as the voltage clamp, from which one can determine separately the effects upon Na^+ and K^+ fluxes. These experiments permit one to examine separately the three ionic factors which together account for the action potential, (1) the increase in sodium permeability, causing a "transient sodium current," whose molecular basis is the opening of the sodium gate and which leads to the rising phase of the action potential, (2) the subsequent increase in K^+ permeability which coupled with (3) "sodium inactivation" (the cutoff of Na^+ permeability; the molecular basis of this effect is the closing of the sodium gate) causes the decline in the action potential.

The effect of DDT is primarily to delay sodium inactivation (an effect which it shares with the drug veratridine) and to a lesser extent to suppress the increase in K^+ permeability. These two effects both have the same consequence: the falling phase of the action potential is delayed (Fig. 4) and because the axon does not readily recover to its resting state, it can be too readily excited again.

The pyrethroids also cause excessive excitability of axons, followed by block. The precise ionic mechanism was described for the synthetic pyrethroid, allethrin, using the voltage clamp technique. Similar to DDT, allethrin delayed sodium inactivation and suppressed the K^+ permeability. But in addition, it suppressed the extent of sodium permeability increase, resulting in a decreased height of the action potential. Thus the excitability phase of allethrin action is probably caused as in the case of DDT, but the suppressed increase of sodium permeability is a likely factor in the ultimate blockade.

In attempting to determine if a particular physiological action is responsible for a lethal effect, the use of analogues may be useful. For instance, numerous DDT analogues exist, and (as a first approximation) one would expect that the nontoxic analogues would not show the physiological effect of the lethal parent. The problem is confused by the fact that nontoxicity may be caused by failures of penetration or distribution, or by excessive metabolic degradation. But, used with care, this approach can be valuable: Thus one would be concerned if a highly toxic analogue showed none of the supposedly important physiological effect (following direct application to nerve) of the parent.

In the studies on physiological actions of DDT analogues, a complication occurs. It appears that, amongst the whole set, one can discern three distinguishable actions: excitatory, blocking, and "dualist," as Wu, van den Bercken, and Narahashi classified them in 1975. The 14 analogues are shown in Table IV.

TABLE IV

Effect of DDT Analogues ($10^{-5}M$) on Crayfish Giant Axons[a]

General formula: R—C₆H₄—X—C₆H₄—R

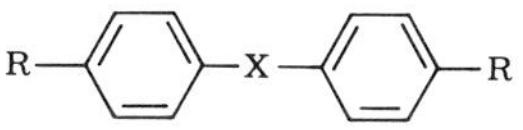

R	X	
Excitatory		
Cl	$>CHCCl_3$	(DDT)
NO_2	$CHCCl_3$	
CH_3O	$CHCCl_3$	
C_2H_5O	$CHCCl_3$	
C_3H_7O	$CHCCl_3$	
o-Cl, *p*-Cl	$CHCCl_3$	
Cl	$>CHCHCl_2$	
C_2H_5	$CHCHCl_2$	
Blocking		
NH_2	$>CHCl_3$	
OH	$CHCl_3$	
Dualist		
CHO	$>CHCl_3$	
Cl	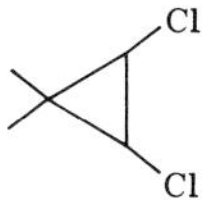 cyclopropane ring with Cl, Cl	
C_2H_5O	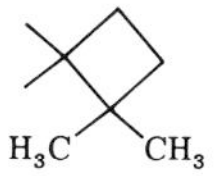 cyclobutane ring with H_3C, CH_3	

[a] Data of Wu, C. H., van den Bercken, J., and Narahashi, T. *Pestic. Biochem. Physiol.* **5,** 142 (1975).

Even if it is accepted that DDT itself kills by delaying Na^+ inactivation, this explanation only carries us to the physiological level. What is the molecular explanation? Because the Na^+ gate has never been isolated, and only indirect biochemical proof exists for its very existence, the question cannot yet be answered. It is believed that the axonic membrane is penetrated by channels which can permit Na^+ movement through the membrane, and that the ability for Na^+ to pass through is controlled by the Na^+ gate. It is plausible that this gate, probably a protein, exists in two conformations: an "on" state in which Na^+ can pass, and an "off" state in which Na^+ cannot pass.

The "off" state must be the normal resting condition, and the depolarization caused by invasion of the moving action potential must convert it to the "on" state, which is only transient, its return to "off" being the basis for sodium inactivation. Presumably DDT and pyrethroids stabilize the "on" configuration. There are other potent and specific agents, such as the toxins tetrodotoxin and saxitoxin, which have the opposite physiological effect, i.e., they block the rising phase of the action potential. We may presume that they stabilize the gate in its "off" configuration. The molecular basis for these stabilizations will doubtless be revealed in the next 10 years.

VI. BIOCHEMICAL STUDIES ON CHLORINATED HYDROCARBONS

The physiological work described above under Section V suggests that DDT and pyrethroids, and no doubt other compounds, act by interfering with the Na^+ gate. But experimental problems have prevented clear biochemical experiments upon the Na^+ gate itself, although these should be expected to become available within the next 10 years.

Meanwhile, there have been a number of studies on the effect of chlorinated hydrocarbons upon ATPases. There are at least two different ATPases. The one described above is the Na^+-K^+-ATPase, which as stated is probably identical with the Na^+ pump. But there is a different enzyme, called Mg^{2+}-ATPase. Instead of being located in the outer membranes of cells, this Mg^{2+}-ATPase is located in the mitochondria and is probably related to the production of ATP from the mitochondria; it requires Mg^{2+} for maximal activity.

Several workers have pointed to the extreme sensitivity of these ATPases to inhibition by chlorinated hydrocarbons. In some cases there have been quite good correlations between the insecticidal potency of the hydrocarbons and their ability to block such ATPases. However, it is hard to understand how the axonic effects of DDT could be caused by blockade of either of these ATPases. If the Na^+-ATPase were affected, the result would be that the Na^+-K^+ pump would fail to operate effectively, leading to a slow

TABLE V

Inhibition of Oligomycin-Sensitive Mg^{2+}-ATPase from Blue Gill Fish[a]

General formula: $R-C_6H_4-X-C_6H_4-R$

Compound	*R*	*X*	μM Concentration for 50% inhibition
DDT	Cl	$>CHCCl_3$	1.3
Dicofol	Cl	$>C(OH)CCl_3$	0.8
TDE or DDD	Cl	$>CHCHCl_2$	2.7
Perthane	C_2H_5	$>CHCCl_3$	3.8
Methoxychlor	CH_3O	$>CHCCl_3$	4.9
DDE	Cl	$>C{=}CCl_2$	8.8

[a]Data of Cutkomp, L. K., Yap, H. H., Vea, E. V., and Koch, R. B. (1971). *Life Sci.* **10,** Part II, 1201.

reduction in the ion gradients, and a *loss* of excitability in the nerve. Just such an effect can be seen with the poison called ouabain, which is a specific inhibitor of the pump. This effect is of course physiologically the opposite of what has been described above. Similarly, blockade of the production of ATP by an effect upon the Mg^{2+}-ATPase would reduce the ability of the nerve to create and sustain its ion gradients, in this case the lesion being at the level of ATP production rather than of the pump itself. Such an effect is produced by the reagent 2,4-dinitrophenol which prevents ATP production by another mechanism (see Section X,C), and again the effect is to greatly reduce the excitability of the nerve, in opposition to the physiological effect we are accustomed to in DDT.

It would therefore seem that the axonic effects of DDT cannot be caused by effects upon either ATPase. But there is little doubt that numerous chlorinated hydrocarbons are effective inhibitors of the Mg^{2+}-ATPase, as Table V shows. Furthermore, the inhibitory effect shows a negative temperature coefficient (that is, it is more effective at lower than at higher temperatures), which is observed also in the gross effect upon insects. Thus one can select doses of DDT which will be ineffective at room temperature, but which produce clear poisoning if the temperature is dropped 10°C. Perhaps one should not make too much of this evidence: Any inhibition caused by complex formation between agent and target should show such an effect; at elevated temperatures such complexes are destabilized by thermal agitation, so affinity decreases and K_d increases.

VII. ARE PYRETHROIDS AXONIC POISONS?

Naturally occurring pyrethroids have been available for over 100 years, but have become of even greater interest with the recent preparation of

TABLE VI

Some Pyrethroids

Naturally occurring

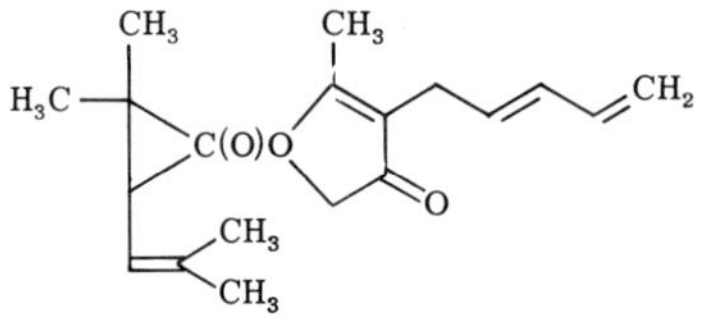

pyrethrin I

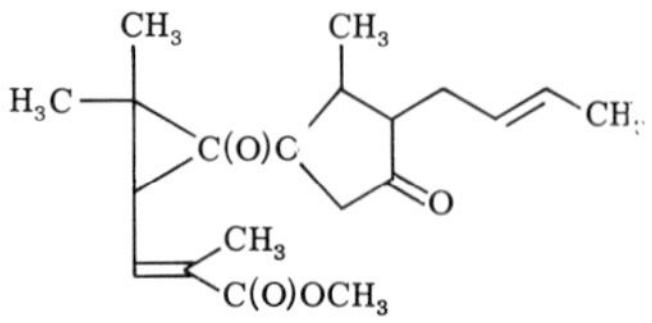

cinerin II

Synthetic

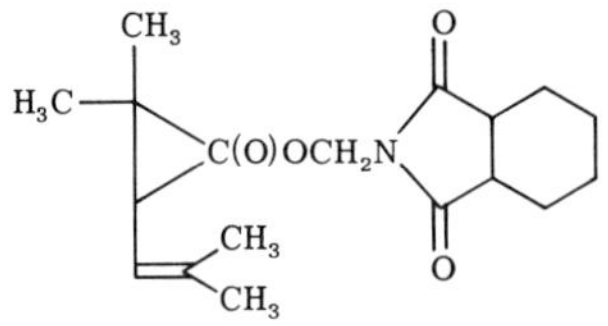

tetramethrin

extraordinarily potent synthetic analogues (Table VI). The most widely accepted view today is that pyrethroids are axonic agents, acting in a DDTlike way, with the modifications described above. Very recently, evidence has suggested that although they can undoubtedly be axonic poisons, especially when applied to isolated nerve, yet the cause of death in poisoned insects may be quite different. For instance, cockroach ganglia were more sensitive to pyrethrin I than was cockroach nerve, and the giant nerve fibers of cockroaches severely poisoned by pyrethrin I were substantially normal.

The most recent work suggests that pyrethroids vary in the relative importance of peripheral (i.e., axonic) effects and central (i.e., ganglionic) effects. Compounds of lesser toxicity such as barthrin may be primarily axonic agents, whereas fast knock-down compounds such as tetramethrin may have both ganglionic and axonic effects in poisoned insects.

VIII. A MUSCLE POISON

The insecticide called ryanodine, an alkaloid extracted from plants of the *Ryania* species, acts directly upon insect muscle. Axonic transmission in poisoned cockroaches was unaffected, but that stimulation of the muscle directly or via the nerve elicited no contraction.

More recently it was found that ryanodine reacts with isolated contractile proteins of rabbit muscle. The effect is a complicated one and, because high ryanodine concentrations are needed *in vitro,* its relationship to poisoning is uncertain.

In normal muscle relaxation, the actomyosin complex which was formed during contraction dissociates, and the actin component which was in the G (or globular) form is converted to the F (or filamentous) form, a conversion which is accompanied by an increase in ATPase activity. The conversion is slowed by a protein called B-actinin, which also causes dispersion of the F-actin, an essential feature of restoration of the relaxed state. Ryanodine at 20 μM blocked the ability of B-actinin to disperse the F-actin.

These effects have been elegantly shown in rabbit muscle. Before concluding that they account for ryanodine's insecticidal action, one must consider that (a) muscles vary in their physiological response to ryanodine: sea urchins are insensitive, as are the uterine and bladder muscles of mammals. (b) A feature of ryanodine poisoning of insects is a massive increase in oxygen consumption, reacting ten times normal in 25 min, a dramatic effect seemingly unrelated to muscle action. (c) Ryanodine causes flaccid paralysis in insects, but (after initial flaccid paralysis) final rigor in frogs and mammals.

IX. CHLORDIMEFORM: A MONOAMINE OXIDASE INHIBITOR?

Chlordimeform is a relatively new miticide and insecticide, first made in 1963. It is also known by its trade names Galecron and Fundal. Numerous analogues have been developed.

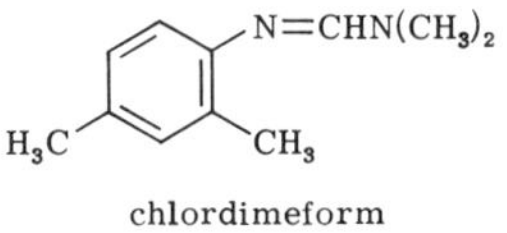

$N{=}CHN(CH_3)_2$

H_3C CH_3

chlordimeform

Its gross effect involves a loss of coordination of the insect, suggesting that it acts upon the nervous system. It is an inhibitor of monoamine oxidase (MAO), and it has been argued that this accounts for its toxicity. MAO is an enzyme that inactivates several neural transmitters of the amine type, including serotonin, dopamine and norepinephrine.

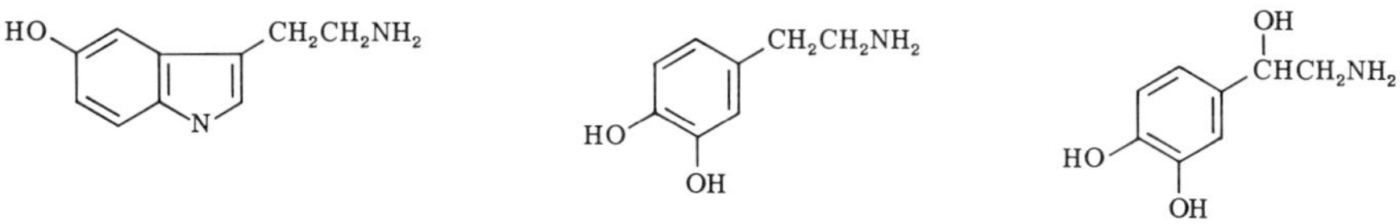

HO $CH_2CH_2NH_2$ N

serotonin

$CH_2CH_2NH_2$ HO OH

dopamine

OH $CHCH_2NH_2$ HO OH

norepinephrine

In every case, MAO inactivates the transmitter by deaminating the $—CH_2CHO$, which is subsequently easily oxidized to the acid$—CH_2COOH$, which is neurally inactive. This process in some ways resembles the inactivation of acetylcholine by acetylcholinesterase. But in fact MAO is *not* involved in the termination of the nervous impulse; termination of amine transmitter action involves reuptake rather than degradation. Nevertheless, MAO is a factor in controlling levels of the amines, and its inhibition leads to elevated neuronal levels of the amines.

Chlordimeform half-inhibits MAO of rat liver (which is used as a readily available form of the enzyme, hopefully indicative of the sensitivity of brain MAO) at 14 μM, which compares favorably with a value of 6.3 μM for the classical MAO inhibitor, iproniazid.

However, even in mammals, the course of symptom development (i.e., rapid and soon reversed) is quite unlike that of classic MAO inhibitors, suggesting that chlordimeform's primary effect may be directly on monoamine receptors, the effect being reinforced by elevation of monoamine levels caused by MAO inhibition. But it should be recognized that chlordimeform has a low toxicity to mammals, with intraperitoneal LD_{50} to rats, mice and rabbits of about 200 mg/kg, so these phenomena may not relate to the insect response.

In insects the course of death is even less clear. Chlordimeform uncouples oxidative phosphorylation, much as described below for nitrophenols. But the symptomatology of poisoning is quite unlike that of nitrophenols. Undoubted effects on amines in cockroach tissues have been described, including blockade of metabolism of tryptamine (a precursor of serotonin), and an elevation of norepinephrine after applying its precursor (dopa or dihydroxyphenylalanine) externally. But the findings are confused by the fact that the cockroach is rather insensitive to chlordimeform (it is Lepidoptera that are the sensitive insects), and substrate-disappearance effects in cockroaches once attributed to MAO have been found to be caused by *N*-acetylation. Consequently, an involvement of monoamines or their receptors is strongly suspected, but the evidence does not yet permit a clear conclusion. The problem is worsened by the limited knowledge of the existence, localization, metabolism, and function of monoamines in insects.

Experiments on frog sciatic nerves have shown that chlordimeform acts on them as a local anesthetic, with a potency and pharmacology much like the well-known local anesthetic procaine. If its primary action is indeed

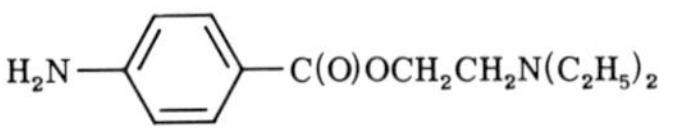

$H_2N-C_6H_4-C(O)OCH_2CH_2N(C_2H_5)_2$

procaine

procainelike, then it would block axonic transmission by a nonspecific interference with Na^+ and K^+ transport.

Quite apart from its direct actions, there are fairly potent synergistic effects of chlordimeform in insects, for instance a seventeenfold enhancement of the toxicity of mixed pyrethrins for the tobacco budworm.

X. METABOLIC POISONS

A. Rotenoids

These compounds, of which the commonest is rotenone, are derived from "tuba-root," the root of leguminous plants of the genera *Derris* or *Lonchocarpus* or *Tephrosia*. The crude material is called "cubé root" or derris dust. Rotenone acts by blocking the oxidation of $NADH_2$, and consequently

rotenone

prevents the utilization of substrates whose oxidation proceeds through $NADH_2$ as an intermediate; examples are glutamate and pyruvate and α-ketoglutarate (Fig. 5). By contrast, succinate is oxidized by a route which bypasses $NADH_2$, and so is not affected.

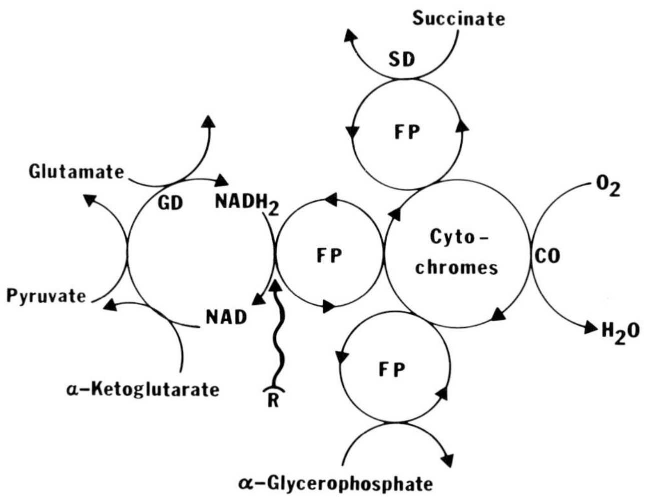

Fig. 5. Pathways for the oxidation of the intermediates of carbohydrate metabolism. Rotenone blocks at "R." Note that succinate metabolism escapes rotenone blockade. SD, succinic dehydrogenase; FP, flavoprotein; GD, glutamic dehydrogenase; CO, cytochrome oxidase.

B. Fluoroacetate and Fluoroacetamide

Sodium fluoroacetate is rarely used as an insecticide, although it is a common rodenticide, under the name 1080. But its amide, fluoroacetamide, has found practical insecticidal use in Japan and the United Kingdom. It is sometimes called Tritox. There is little doubt that fluoroacetamide is hydrolyzed *in vivo* to fluoroacetate, which is the actual toxicant.

In insects as in mammals, fluoroacetate is believed to kill only after it has been metabolized to fluorocitrate, a step requiring acetate thiokinase (to give fluoroacetyl-CoA) followed by condensation with oxaloacetate, catalyzed by condensing enzyme (citrate synthetase). In these two steps the fluoroacetate masquerades as acetate, a trick which is of double importance: it occurs in spite of the (otherwise) stringent specificities of the two enzymes; and it is essential if the organism is to be affected, and hence is a case of biosuicide or "lethal synthesis." [See Eq. (2).] But the enzyme aconitase, which utilizes

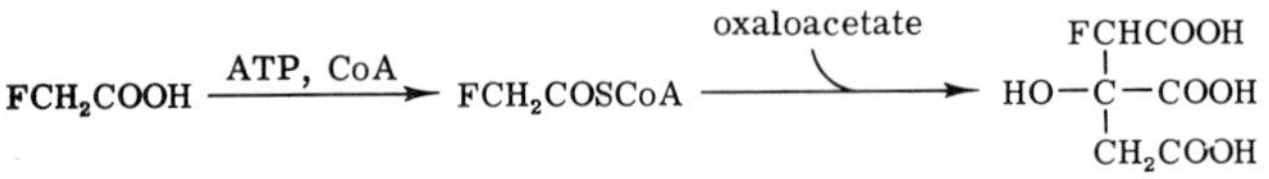

$$\underset{\text{fluoroacetate}}{FCH_2COOH} \xrightarrow{\text{ATP, CoA}} \underset{\text{fluoroacetyl-CoA}}{FCH_2COSCoA} \xrightarrow{\text{oxaloacetate}} \underset{\text{fluorocitrate}}{\begin{array}{c} FCHCOOH \\ | \\ HO-C-COOH \\ | \\ CH_2COOH \end{array}} \qquad (2)$$

citrate, refuses to be "bamboozled" by the phony citrate. Not only will it not convert it to fluoro-α-ketoglutarate, but it is blocked by it. The aconitase block leads to a pile-up of citrate in the organism. Because aconitase is a vital step in the citric acid cycle, which is of major importance in carbohydrate oxidation, the blockade may well be the cause of the lethality of fluoroacetate. But the accumulations of citrate are so great that one cannot exclude the possibility that they cause death, for instance by complexing free calcium.

C. Dinitrophenols

The commonest nitrophenols for insecticidal and miticidal use are the cresols DNOC (4,6-dinitro-*o*-cresol) and DNCHP (3,4-dinitro-6-cyclohexylphenol).

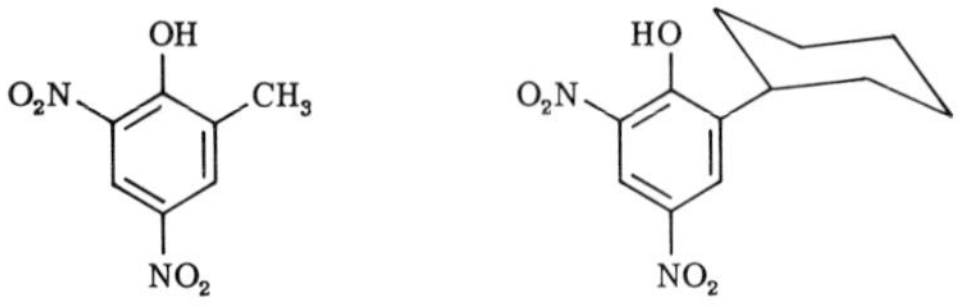

DNOC   DNCHP

DNOC was once commonly used as an ovicide and for locust control. Now these two agents are primarily used as miticides. Their biochemical action is the same as that of their parent, 2,4-dinitrophenol or DNP, which has been widely used by biochemists because of its ability to interrupt a vital pathway of energy metabolism.

The principal final stepladder in the oxidation of carbohydrates to energy (trapped in the form of ATP) is the respiratory chain. As Fig. 5 shows, the oxidation of various intermediates released by carbohydrate and protein degradation (such as glutamate, pyruvate, α-glycerophosphate, succinate) is linked to the reduction of a variety of alternative flavoproteins, and then to a series of cytochromes. These in turn are coupled to the reduction of O_2 to H_2O, with simultaneous conversion of ADP to ATP. Living cells or good *in vitro* preparations are said to be "tightly coupled," so that when all the available ADP has been converted to ATP, the system comes to a halt; it automatically starts up again when an energy usage depletes the ATP and replenishes the ADP.

Dinitrophenols are "uncouplers." They break the tight coupling and oxidation proceeds briskly, with the disappearance of substrate, but ADP is not needed, and ATP is not produced. In the intact animal, poisoned by DNOC, one sees increases in oxygen use (up to tenfold in *Tribolium* beetles) and fever in mammals, both caused by "useless" oxidation.

The precise uncoupling mechanism is unclear. The ADP→ATP reaction is tied to the oxidation of cytochromes via a cycle of unknown intermediates X and Y. Possibly X binds Y to form X~Y ("X squiggle Y"), the squiggle symbolizing a need for considerable energy to make it, and considerable energy production in breaking it. The X~Y may react with PO_4^{3-} to yield Y~P, which can then convert ADP to ATP.

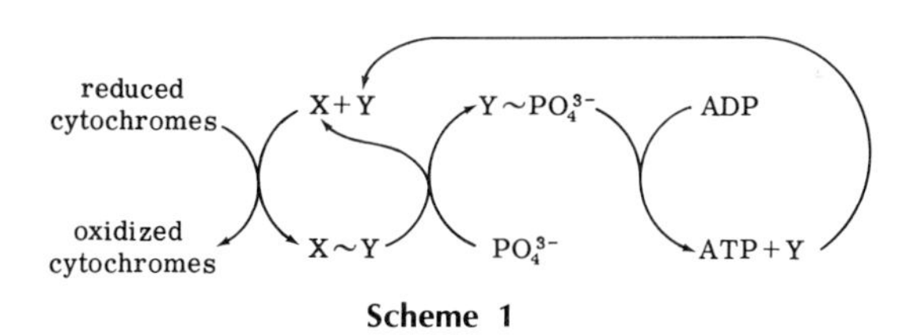

Scheme 1

It is believed that dinitrophenols catalyze the useless hydrolysis of X~Y or Y~PO_4^{3-}, and so break the above chain.

XI. INHIBITORS OF CHITIN SYNTHESIS

For many years, an attractive potential target for insecticidal action has been chitin metabolism. Chitin is a polymer of *N*-acetylglucosamine, and is absent from mammals, but crucial to insects and fungi. Consequently, agents which affect chitin metabolism might be highly selectively toxic to these organisms.

Recently an insecticide DU 19111 has been introduced which appears to block chitin synthesis; it emerged as a result of a program by the Dutch firm, Philips-Duphar, designed to produce new herbicides.

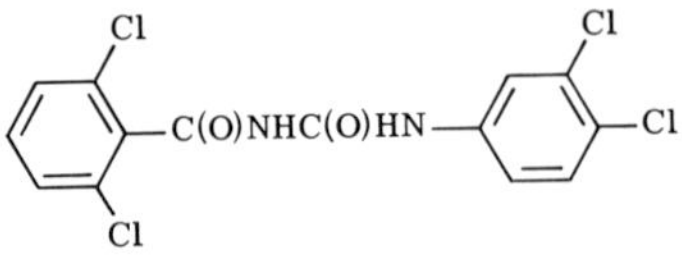

Cl Cl

—C(O)NHC(O)HN— —Cl

Cl

DU 19111

This agent was shown in 1972 to prevent proper molting of various insects. As a result of further syntheses, the compound Dimilin or diflubenzuron was developed.

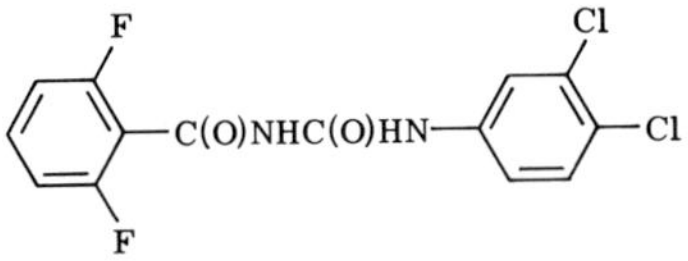

F Cl

—C(O)NHC(O)HN— —Cl

F

diflubenzuron

It has been established that these larvicides do not affect the hormonal action of bursicon, the molting hormone. But the incorporation of glucose into chitin was found to be blocked by DU 19111. Furthermore, there was a considerable accumulation of *N*-acetylglucosamine in the treated larvae; this compound does not itself occur in the synthetic pathway from glucose to chitin, although its derivative UDP-*N*-acetylglucosamine is indeed the substrate for chitin synthetase, the chitin-forming enzyme (Scheme 2).

glucose ⟶ glucose 6-P ⟶ fructose 6-P ⟶ glucosamine 6-P

↓

N-acetylglucosamine 1-P ⟵ *N*-acetylglucosamine 6-P

↓

UDP- *N*-acetylglucosamine ⟶ chitin

Scheme 2

It is therefore likely that DU 19111 uncouples the polymerization of UDP-*N*-acetylglucosamine, and leads to the useless accumulation of *N*-acetylglucosamine. Surprisingly, diflubenzuron seems to act rather differently: UDP-*N*-acetylglucosamine accumulates, but *N*-acetylglucosamine does not. A direct blockade of chitin synthetase is therefore suspected.

It should be stressed that those two biochemical actions (uncoupling and blockade) have not yet been confirmed directly in cell-free systems. The proposed actions are based on an observed failure of chitin synthesis, and on the accumulations (reported above) in poisoned larvae. Thus indirect effects upon chitin synthetase, via hormonal mediation, have not yet been ruled out.

ACKNOWLEDGMENT

The author gratefully acknowledges support from NIH grant ES 00901 for research in this laboratory, which is a part of the results described in this chapter.

GENERAL REFERENCES

Corbett, J. R. (1974). "The Biochemical Mode of Action of Pesticides." Academic Press, London and New York.
Elison, C. *Biochem. Pharmacol.* **22,** 113–120.
Kohn, G. K., ed. (1974). "Mechanism of Pesticide Action," Am. Chem. Soc. Symp. Ser. No. 2, Am. Chem. Soc., Washington, D.C.
Matsumura, F. (1975). "Toxicology of Insecticides." Plenum, New York.
Wilkinson, C. F. (1976). "Insecticide Biochemistry and Physiology." Plenum, New York.

13

Detoxication Mechanisms in Insects

W. C. DAUTERMAN AND ERNEST HODGSON

I. GENERAL INTRODUCTION

Detoxication enzymes or enzyme systems, as a general rule, increase the water solubility of foreign compounds (xenobiotics) and thus render them more easily eliminated by the excretory mechanisms of the organism. These substances are adapted for the elimination of the water-soluble end products of metabolic processes. Although most of the scientific investigations of detoxication mechanisms in insects have involved either insecticides or compounds related to them, it should be recalled that almost all of these compounds are synthetic organic chemicals of recent origin and the mechanisms themselves must have evolved in response to selection pressure by other, naturally occurring, toxic substances. These would include primarily the secondary plant substances such as terpenes, alkaloids, and methylenedioxyphenyl compounds, many of which are lipophilic and of low but appreciable toxicity. Since the number of potential toxicants is large, the enzymes which have evolved are nonspecific. Their not-infrequent role as intoxifying enzymes would appear to be the result of this lack of specificity, the many variations of chemical structure metabolized, including some in which the metabolites are more toxic than the parent compound. It would be interesting to know whether intoxication reactions are more common among naturally occurring toxicants, in response to which the enzymes were evolved, or among the synthetic organic toxicants, to which they have been exposed more recently.

With the exception of those few foreign compounds which resemble endogenous metabolites closely enough to be taken up by active transport mechanisms in the gut, the ability of xenobiotics to enter the animal body is a function of their lipophilicity. This property enables them to partition into the lipid membranes of the cell and also, in the case of insects, to pass into the waxy epicuticle of the integument of either the external surface or of the tracheal system.

The processes by which such compounds are eliminated are divided by toxicologists into two general groups, frequently referred to as Phase 1 and Phase 2 reactions. Oxidations, reductions, and hydrolyses are typical of Phase 1 reactions which introduce a hydrophilic functional group into the molecule. Although the products of Phase 1 reactions are more water soluble and may be excreted to some extent, they usually undergo one of the conjugating reactions typical of Phase 2. In these reactions the functional group introduced in Phase 1 is combined with a highly water-soluble endogenous metabolite such as glucose, glutathione, and various amino acids, to give rise to conjugation products which are readily excreted.

Although the mechanism of xenobiotic excretion in the insect has not been widely studied they are, presumably, excreted via the Malpighian tubes and the hindgut. Exceptions to this include certain volatile degradation products such as isopropanol which is excreted via the tracheal system.

II. PHASE 1 REACTIONS

A. Mixed-Function Oxidases

1. Definition

This class includes all of those enzymes in which one atom of a molecule of oxygen is reduced to water while the other is used to oxidize the substrate.

The most important such oxidase for xenobiotic metabolism is cytochrome *P*-450, a component of the microsomal mixed-function oxidase system. In this case the electrons involved are derived from NADPH via a flavoprotein reductase and the overall reaction may be written as shown in Eq. (1).

$$NADPH + H^{+} + SH_2\ O{:}O \rightarrow NADP^{+} + SHOH + H_2O \qquad (1)$$

2. Organization of Microsomal Mixed-Function Oxidase Systems

Although cytochrome *P*-450 may also occur in mitochondria, it is the microsomal enzyme system which appears to be involved in xenobiotic metabolism. The relationship between the microsomal cytochromes, their flavoprotein reductases and cofactors is shown in Fig. 1. As indicated, the relationship of cytochrome b_5 to cytochrome *P*-450 is not entirely clear, although it may be the source of the second electron in the oxidation/reduction cycle of cytochrome *P*-450.

3. Distribution of Microsomal Mixed-Function Oxidase Systems

Microsomal mixed-function oxidase systems involving cytochrome *P*-450 are extremely widespread, being found in animals, plants, and microorganisms. In the mammal, where they have been studied most intensively, many different tissues have been shown to contain cytochrome *P*-450, the function of which, as well as the location within the cell, varies from one tissue to

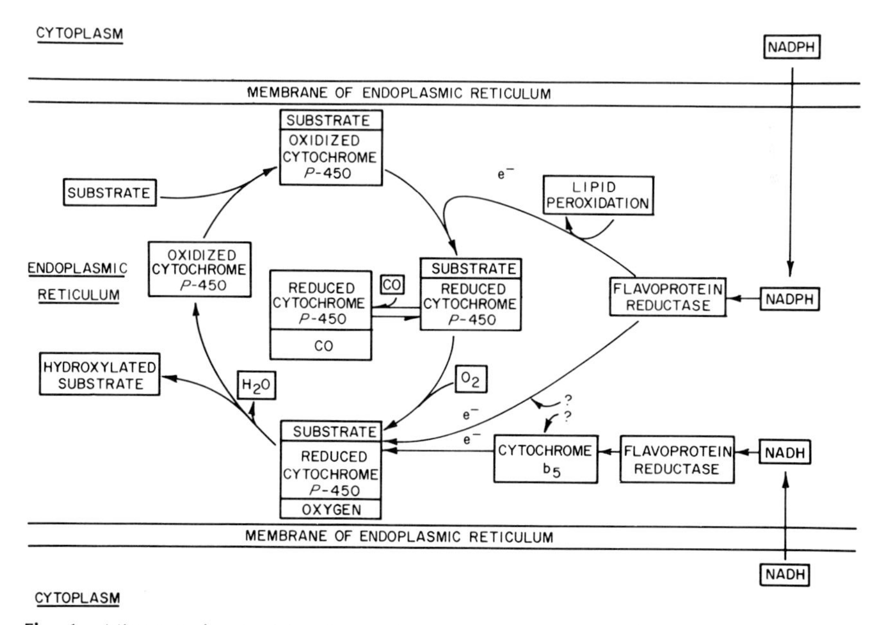

Fig. 1. Microsomal mixed-function oxidase system. Reprinted from Hodgson, E. *Essays Toxicol.* 7, 73–97 (1976).

another. For example, the cytochrome *P*-450 from liver is microsomal and is concerned primarily with xenobiotic metabolism, while in the placenta both mitochondrial and microsomal cytochromes *P*-450 can occur. In the latter case, however, both are concerned with the metabolism of steroid hormones.

Cytochrome *P*-450 and the microsomal mixed-function oxidase system are less well studied in insects but are known to occur in over twenty species, principally in the Diptera and Lepidoptera. Since it has been found in every species examined, it is, presumably widespread. Only activity toward xenobiotics has been studied to any significant extent and it is not known whether insects also have specialized mixed-function oxidase systems concerned with hormone metabolism. Mitochondrial cytochrome *P*-450 has not been reported from insects. By far the most intensively studied mixed-function oxidase system is that of the housefly, *Musca domestica,* studied primarily by Agosin at the University of Georgia, Hodgson at North Carolina State University, Terriere at Oregon State University and their numerous co-workers. This system, as it occurs in the Lepidoptera, has been studied extensively by Wilkinson and his co-workers at Cornell University. The general references should be consulted for comprehensive listings of these authors contributions.

Although much of this work has been done with whole insects or isolated abdomens, the mixed-function oxidase system has been shown to occur in fat body, Malpighian tubes and the midgut, with the last-mentioned being most important in the Lepidoptera.

Oxidative activity toward xenobiotics is always associated with the microsomal fraction of whole insect or organ homogenates. The microsomal fraction is the particulate fraction obtained by ultracentrifugation (100,000–200,000 *g*) of a postmitochondrial supernatant for an hour or more. Comparison of electron micrographs of whole cells and microsomal preparations (Fig. 2) has led to the conclusion that microsomes are derived primarily from the endoplasmic reticulum and may be rough (with attached microsomes) or smooth (without attached ribosomes). Such fractions are also heavily contaminated with free ribosomes, glycogen granules, and fragments of other cellular organelles.

4. Methodology

As mentioned above, microsomes are prepared by homogenization and differential centrifugation. It is important that homogenization media and centrifugation protocols be optimized for each particular species and/or

---

Fig. 2. (A) Midgut cells of the house fly, *Musca domestica* L., to show endoplasmic reticulum. Magnification: ×19,000. (B) Section of a microsomal pellet prepared from the isolated abdomen of the house fly. Magnification: ×40,000. Electron micrographs courtesy of Ling-yi Lucille Chang.

A

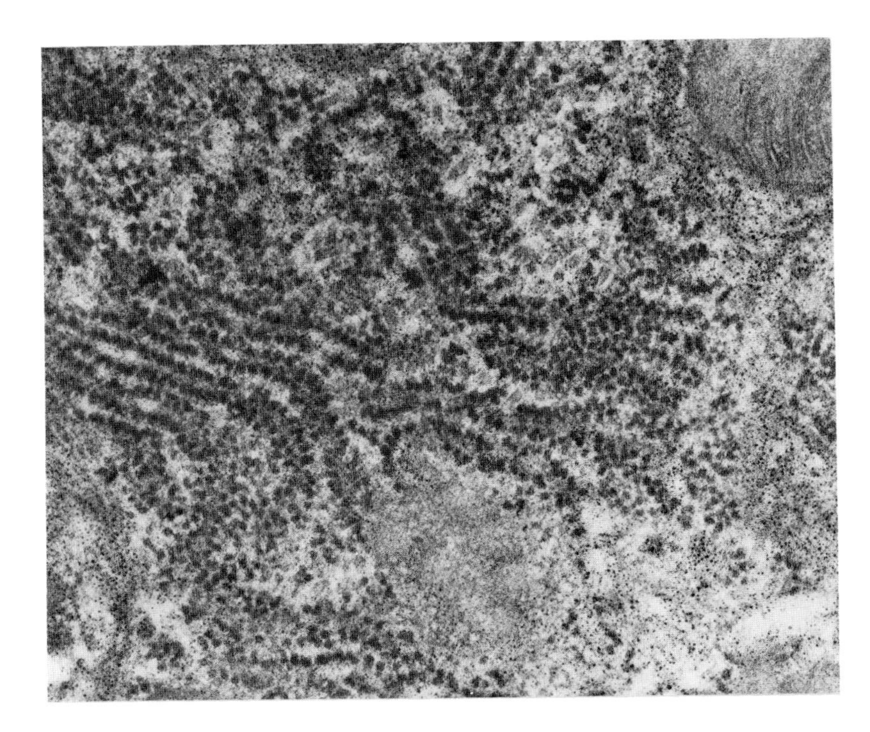

B

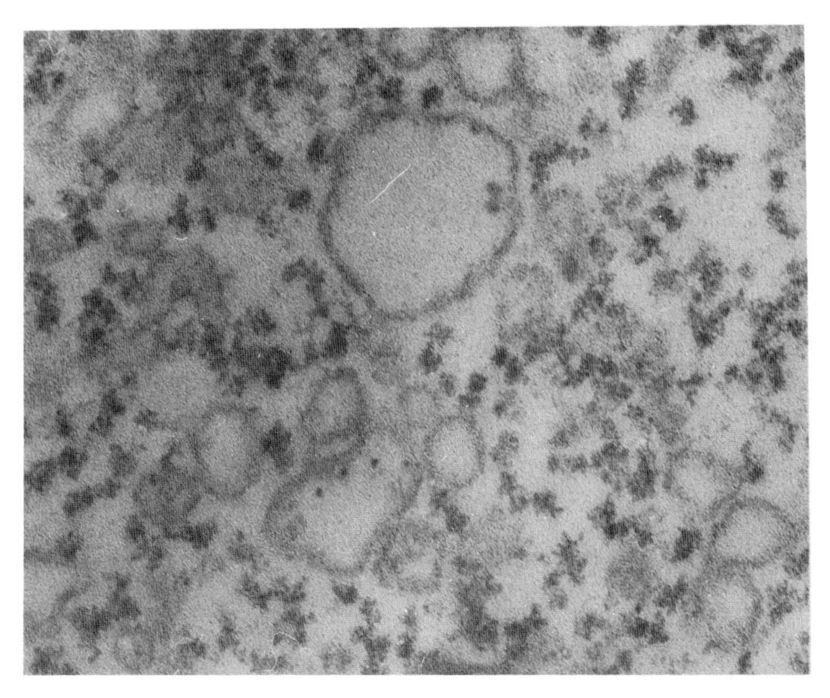

tissue. Factors of importance have been shown to be such things as ionic strength, pH, chemical nature of the buffer, type of homogenizer, homogenization time, temperature, and the presence or absence of glycerol.

The measurement of oxidative activity has been carried out by many of the classic techniques of biochemistry, all of which in one way or another estimate the appearance of product or the disappearance of substrate. For example, the epoxidation of aldrin to dieldrin is measured either by gas-liquid chromatography of the dieldrin produced or by counting ^{14}C-dieldrin produced from ^{14}C-aldrin and separated by thin-layer chromatography. O-Demethylation of *p*-nitroanisole can be estimated by direct colorimetric measurement of the *p*-nitrophenol produced. Substrates which are metabolized at several sites on the molecule give rise to complex mixtures of products. In these cases, examples of which are such insecticides as rotenone, carbaryl and diazinon, the use of radiolabeled substrates followed by chromatographic separation and counting of the separated products is the only useful method.

The most important tool in the measurement and characterization of cytochrome *P*-450 itself, rather than the products of its activity, is optical difference spectroscopy. Since the heme group of cytochrome *P*-450 is a chromophore, it, like other cytochromes, absorbs light in the visible region. However, absolute spectroscopy of turbid microsomal suspensions is complicated by light scattering and, since this is a function of wavelength, spectra are seen against a sloping baseline, the characteristics of which are dictated more by the physical characteristics of the microsomal particles than by the cytochrome. Optical difference spectroscopy avoids the problems due to light scattering and other nonspecific absorption by recording only the change in light absorption caused by the addition of a ligand and not the absolute spectra themselves. This is accomplished by placing the microsomal suspension in both reference and sample cuvettes of a split-beam spectrophotometer, thus balancing both these effects and the absolute spectra, and permitting the recording of a flat baseline. The ligand under investigation is then placed in the sample cuvette and a difference spectrum can be recorded in which only perturbations in the absolute spectrum are seen and not the absolute spectrum itself (Fig. 3A).

The difference spectrum of greatest general importance is the carbon monoxide spectrum of the reduced cytochrome (Fig. 3B), the spectrum which forms the basis for the estimation of cytochrome *P*-450 and its degradation product, cytochrome *P*-420. This shows a peak at or around 450 nm with cytochrome *P*-420 appearing, if present, as a shoulder at 420 nm. The method for the demonstration of this spectrum includes saturation of both cuvettes with carbon monoxide (CO) followed by reduction of the sample cuvette with dithionite.

Type I and Type II spectra (Fig. 3C) are formed by the addition of various

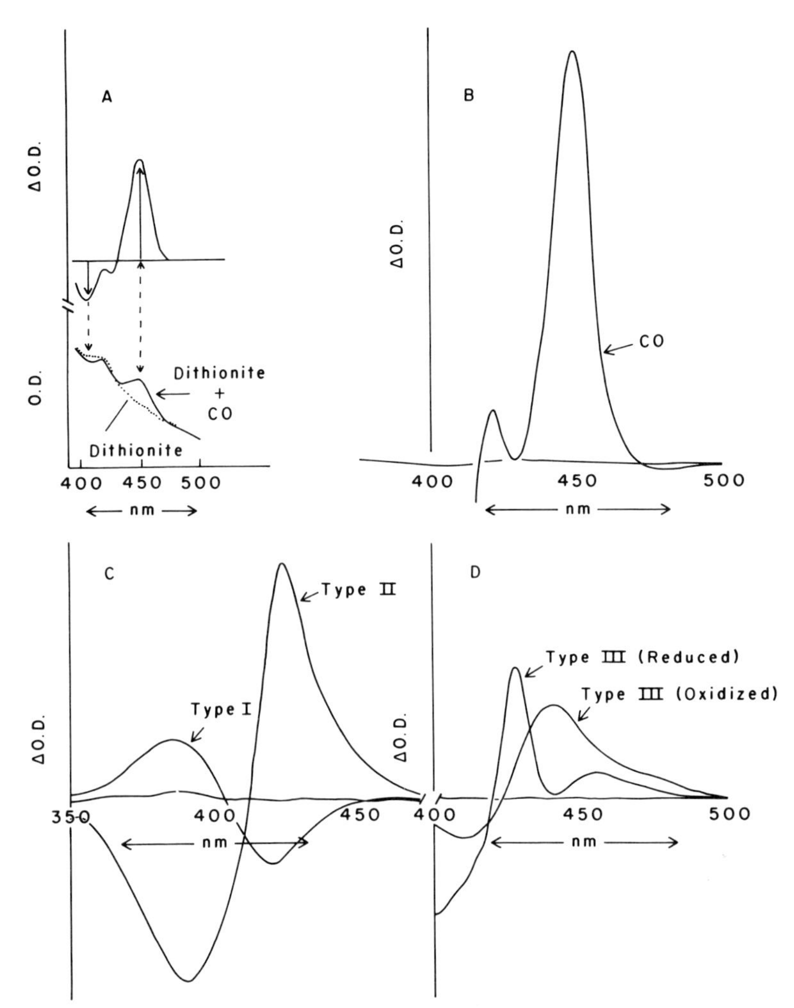

Fig. 3. Difference spectra obtained by the addition of various ligands to microsomes prepared from isolated abdomens of the insecticide-resistant Fc strain of the housefly, *Musca domestica* L. (A) Principle of optical difference spectroscopy. (B) Carbon monoxide spectrum with dithionite-reduced microsomes. (C) Type I (benzphetamine) and Type II (pyridine) spectra with oxidized microsomes. (D) Oxidized and reduced (NADPH) Type III spectrum of piperonyl butoxide. Spectra courtesy of Dr. Arun P. Kulkarni.

ligands to the oxidized form of cytochrome *P*-450. Type I has a peak at 385 nm and a trough at 420 nm while Type II has a peak at about 430 nm and a trough at about 393 nm. Type I is formed by a large variety of different ligands and is believed to represent binding to a lipophilic site somewhat removed from the heme iron, while Type II, formed primarily by organic nitrogen compounds, is believed to represent binding to the heme iron. The modified Type II (415–420-nm peak, 390-nm trough), caused primarily by alcohols, has also been called the reverse Type I, since one hypothesis holds that it results from the displacement of an endogenous Type I ligand in the sample cuvette. Other workers believe that the modified Type II spectrum may represent the binding of a nucleophilic oxygen atom to the heme iron. The Type II spectrum of *n*-octylamine is of particular interest since it occurs in two forms, one with a double trough at 410 and 394 nm and the other with a single trough at 390 nm. These forms have been used in the characterization of qualitatively different cytochromes in both mammals and insects.

Ethyl isocyanide and Type III spectra are the result of interactions with the reduced form of the cytochrome and characteristically have two peaks in the Soret region at, or close to, 455 and 430 nm (Fig. 3D). The size of the peaks is pH dependent and a pH equilibrium point can be calculated at which the two peaks are of equal magnitude. Differences in the pH equilibrium point have been used, in studies of the induction of both mammalian and insect cytochrome *P*-450, to determine whether the induced cytochrome is qualitatively different to control cytochrome. Following the demonstration that similar spectra are formed by methylenedioxyphenyl compounds it was proposed that the term Type III be used as a general designation for all pH-dependent double Soret spectra.

Unusual spectral perturbations which complicate interpretation may be caused by nonlinear baseline changes due to turbidity differences between sample and reference cuvettes, by mixed Type I and Type II interactions, by denaturation of cytochrome *P*-450, and by native absorbance of added ligands. Although studied most intensively in mammalian preparations, all of these spectra have been demonstrated using insect microsomes.

Electron paramagnetic resonance spectroscopy (EPR) has come into increasing use in recent years as a tool for the characterization of cytochrome *P*-450 and for the investigation of its reaction mechanism, particularly since the observation that substrate binding causes a change from a low-spin to a high-spin form. As yet this technique has not been used in the study of insect cytochrome *P*-450.

Although solubilization and complete purification of cytochrome *P*-450 have proved elusive, a number of more or less satisfactory methods for the preparation of purified mammalian cytochrome *P*-450 have been described. They involve solubilization with detergents, with or without sonication, followed by such well-known procedures as ammonium sulfate precipitation, absorption on calcium phosphate gel, and DEAE-cellulose chromatography. Glycerol is usually added as it appears to stabilize cytochrome *P*-450 to some extent. Even though these procedures have permitted some reconstitution, both yield and degree of purification are low and it may be that neither the best solubilization nor purification techniques have yet been found. Studies in insects involve similar techniques (see below) but are still fragmentary.

5. Biochemical Characterization of Insect Microsomes

Quantitative characterization of oxidative activities of insect microsomes is difficult since, as in mammalian preparations, reciprocal plots such as Lineweaver–Burk and Hofstee are almost invariably curvilinear except for restrictively narrow ranges of substrate and/or inhibitor concentration. Thus, such commonly used constants as K_m and K_i have only a restricted meaning and are best used for comparative purposes only. Similarly, in studies of spectral binding, K_s (ligand concentration at half-maximum spectral size) is also of restricted utility. These problems are due, presumably, to the simultaneous difficulties brought about by particulate enzymes, by substrates and inhibitors of low solubility in water, by nonspecific lipophilic binding, and by multiple forms of the enzyme. In spite of these difficulties, however, certain general conclusions can be drawn, namely, that microsomal oxidations all require NADPH and molecular oxygen and are inhibited by CO.

Although the cytochrome *P*-450 concentration of insect microsomes tends to be lower than that of mammalian liver, insect preparations of high oxidative activity, such as those from the abdomen of insecticide-resistant high oxidase strains of the housefly, show comparable values. The NADPH-cytochrome *P*-450 reductase and cytochrome b_5 values may generally be lower, relative to cytochrome *P*-450, than in most mammals but in some insect preparations (e.g., midgut microsomes from the tobacco hornworm, *Manduca sexta*) the b_5 to cytochrome *P*-450 ratio is much higher than that of mammalian liver microsomes.

Many inhibition studies have been carried out on intact microsomes from insects and it is possible, by indirect methods, to determine at which part of the electron transport chain the inhibition occurs. For example, sulfhydryl inhibitors, such as *p*-chloromercuribenzoate and *p*-chlormercuriphenylsulfonate, act primarily at the reductase level. Compounds such as cytochrome *c* or the eye pigment, xanthommatin, act as electron sinks, diverting electrons from cytochrome *P*-450. Enzymatic inhibitors, such as proteolytic or phospholipase enzymes, attack the structure of the microsomes and/or their constituent enymes. Another important class of inhibitor is the insecticide synergists, which mainly interact with cytochrome *P*-450 to form stable inhibitory complexes.

Williams and Millburn (1975) have classified oxidative Phase 1 reactions as follows:

Aromatic hydroxylation
Aliphatic hydroxylation
Dealkylation

O-dealkylation
S-dealkylation
N-dealkylation
N-oxidation
Formation of N-oxides
Formation of hydroxylamine
Formation of oximes
S-oxidation
P-oxidation
Replacement of S by O

Most of the oxidation reactions listed are due to the microsomal mixed-function oxidase system. While not all of these reaction classes have been demonstrated *in vitro* in insects, many of them have and some examples are listed below.

a. *Aromatic Hydroxylation.* The principal mechanism appears to be via epoxidation followed by subsequent hydration of the epoxide ring to yield a dihydrodiol or by rearrangement to yield a monohydroxy compound. Aromatic hydroxylation has not been studied extensively in insects but is known to occur in the metabolism of naphthalene to 1-naphthol [Eq. (2)]. The epoxide may also be conjugated with glutathione by the glutathione *S*-transferase. A similar reaction sequence appears to be one of the routes for the metabolism of carbaryl (Fig. 4).

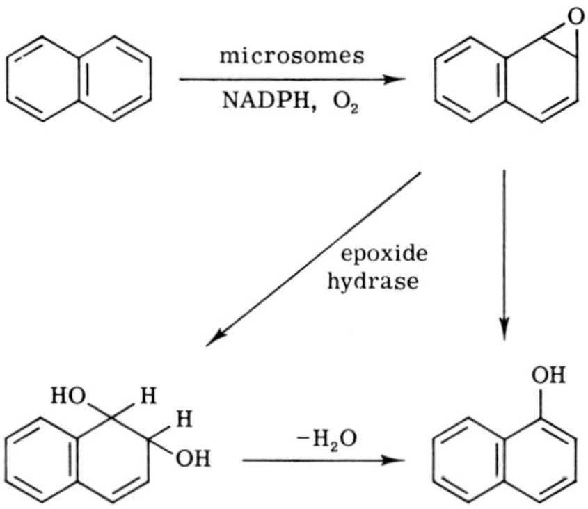

microsomes
NADPH, O_2
epoxide hydrase
HO H H OH
$-H_2O$
OH

(2)

b. *Aliphatic Hydroxylation.* This has been demonstrated *in vitro* using housefly microsomes in the presence of NADPH and O_2 and characteristically occurs at an aliphatic side chain. The oxidation of DDT to kelthane [Eq. (3)] is probably the best known example but others include the side chain oxidation of *n*-propylbenzene and isopropylbenzene as well as the oxidation of side chains on several carbamate insecticides.

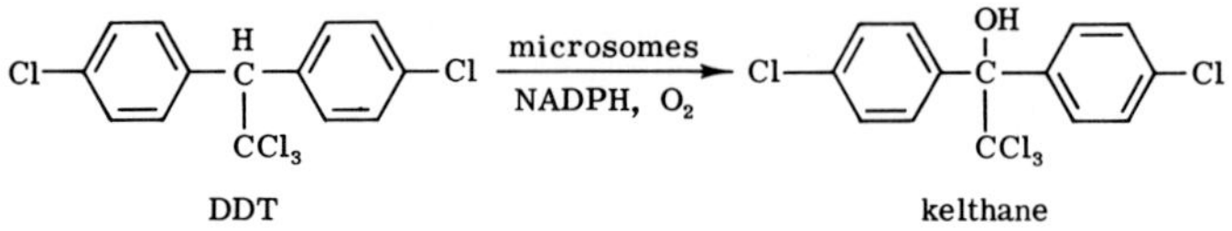

Cl H C CCl$_3$ Cl → microsomes, NADPH, O_2 → Cl OH C CCl$_3$ Cl

DDT kelthane

(3)

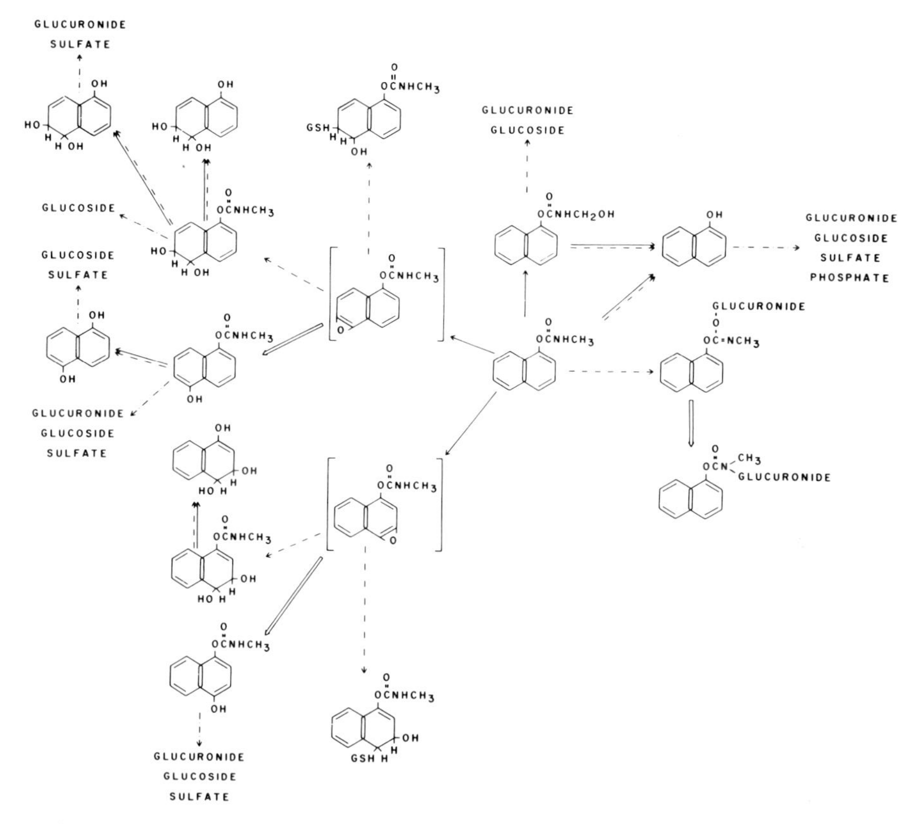

Fig. 4. General scheme for the metabolism of carbaryl in insects, mammals and plants. ⟶, Oxidation pathway; ⟶, hydration or hydrolysis pathway; –.–.→, conjugation pathway; ⇛, intramolecular rearrangement; [ ], hypothetical intermediate. Those reactions in which both oxidation and hydrolysis are indicated have been, in the past, generally supposed to be hydrolytic. However, in view of recent observations that the apparent hydrolysis of carbaryl is, in at least some organisms, a microsomal oxidation requiring NADPH and O_2, it is entirely possible that they may be oxidative, hydrolytic, or both.

c. *Dealkylation.* O-dealkylation, N-dealkylation and S-dealkylation have all been reported *in vitro* using insect microsomal preparations. O-Dealkylation of *p*-nitroanisole [Eq. (4)] is commonly used as the basis of a rapid assay method for mixed-function oxidase activity since the product *p*-nitrophenol can be measured colorimetrically. The methylol derivative is believed to be an intermediate in the reaction.

$$O_2N-C_6H_4-OCH_3 \xrightarrow[\text{NADPH, } O_2]{\text{microsomes}} [O_2N-C_6H_4-OCH_2OH] \xrightarrow{\text{nonenzymatic}} O_2N-C_6H_4-OH \quad (4)$$

A less common type of O-dealkylation involves splitting of an ester bond rather than an ether, e.g., the deethylation of paraoxon [Eq. (5)]. This reaction probably proceeds via hydroxylation of the α-carbon.

$$(C_2H_5O)_2P(=O)O{-}C_6H_4{-}NO_2 \xrightarrow[\text{NADPH, } O_2]{\text{microsomes}} (C_2H_5O)(HO)P(=O)O{-}C_6H_4{-}NO_2 \qquad (5)$$

N-Dealkylation has been demonstrated *in vitro* using insect microsomal preparations, particularly in the case of *N*-methyl- and *N,N*-dimethylcarbamate insecticides. In these cases the methylol derivative is frequently stable enough to be identified. A useful substrate for the demonstration of this reaction is *N,N*-dimethyl-*p*-nitrophenylcarbamate [Eq. (6)].

$$O_2N{-}C_6H_4{-}OC(=O)N(CH_3)_2 \xrightarrow[\text{NADPH, } O_2]{\text{microsomes}} O_2N{-}C_6H_4{-}OC(=O)N(CH_3)CH_2OH \qquad (6)$$

S-Dealkylation has not been demonstrated *in vitro* with insect microsomes but *in vivo* investigations of aldicarb metabolism by houseflies suggest that this reaction occurs.

d. *N-Oxidation.* The various nitrogen oxidations known from mammals, N-oxide formation, hydroxylamine formation, and oxime formation, are catalyzed in mammalian liver either by flavoprotein enzymes or by cytochrome *P*-450. These reactions have not been investigated in insects.

e. *S-Oxidation.* This reaction is not widely studied in insects but it does occur when Mesurol (4-methylthio-3,5-xylyl methylcarbamate) is incubated with housefly microsomes and NADPH in the presence of oxygen, resulting in the formation of a sulfoxide.

f. *P-Oxidation.* This reaction (7) has only recently been described from studies using mammalian liver microsomes and it results in the formation of a phosphate from a trisubstituted phosphine. It has not yet been looked for in insects.

$$(C_6H_5)_2P{-}CH_3 \xrightarrow[\text{NADPH, } O_2]{\text{microsomes}} (C_6H_5)_2P(=O){-}CH_3 \qquad (7)$$

g. *Replacement of S by O.* This reaction is extremely important in insect toxicology since it results in the formation of highly potent phosphate cholinesterase inhibitors from the corresponding phosphorothioates. Although numerous examples are known, the one most studied is the formation of paraoxon from parathion [Eq. (8)].

$$\underset{\text{parathion}}{(C_2H_5O)_2P(=S)O-C_6H_4-NO_2} \xrightarrow[\text{NADPH, } O_2]{\text{microsomes}} \underset{\text{paraoxon}}{(C_2H_5O)_2P(=O)O-C_6H_4-NO_2} \quad (8)$$

Recent studies have shown that this is probably associated, via a common intermediate, with the oxidative dearylation of parathion and possibly other phosphorothioates [Eq. (9)].

$$(C_2H_5O)_2P(=S)O-C_6H_4-NO_2 \longrightarrow \left[(C_2H_5O)_2P(SO)O-C_6H_4-NO_2\right]$$

$$\swarrow \qquad \searrow \quad (9)$$

$$\text{"S"} + (C_2H_5O)_2P(=O)O-C_6H_4-NO_2 \qquad (C_2H_5O)_2P(=S)OH + HO-C_6H_4-NO_2$$

The sulfur released is covalently bound, at least in mammals, to some component of the microsomes. The evidence for the proposed intermediate is so far indirect and it may yet be shown that there are two reactions catalyzed by two different forms of cytochrome *P*-450.

7. Characterization of Individual Enzymes of the Microsomal Mixed-Function Oxidase Pathway

Individual enzymes of the microsomal mixed-function oxidase pathway have not yet been studied extensively in insects. In mammals, solubilization has been effected by a variety of detergents and separation and purification by such techniques as ammonium sulfate fractionation and fractionation on columns of DEAE-cellulose, CM-cellulose, hydroxylapatite, and ω-amino-*n*-octyl Sepharose 4B. Since cytochrome *P*-450 is extremely labile after solubilization, various protectants such as glycerol, EDTA, and dithiothreitol are routinely added. The results of many such studies have shown that several different forms of cytochrome *P*-450 as well as an NADPH-cytochrome *P*-450 reductase can be purified. Xenobiotic oxidizing activity can be reconstituted by the addition of phosphatidylcholine to a solution containing these two components, NADPH and oxygen.

Much of the work on the purification of cytochrome *P*-450 and its reductase from insects has been carried out by Agosin and his co-workers at the University of Georgia. House fly cytochrome *P*-450 has been solubilized with

Triton X-100 followed by ammonium sulfate fractionation and column chromatography on either Sephadex G-25 or Sephadex G-200 and then DEAE-Sephadex A-50. Two cytochromes, designated *P*-450, with a λ_{max} in the CO spectrum of 450 nm, and *P*-450_1, with a λ_{max} in the CO spectrum of 448.5, are separated. It is reported that cytochrome *P*-450 is a tetramer of 45,000 dalton units while cytochrome *P*-450_1 is a decamer. Since cytochrome *P*-450 is an extremely hydrophobic protein with a strong tendency to reaggregate, it is likely that these are unorganized aggregates rather than organized oligomers which preexisted in the membrane. Other studies on housefly microsomes using cholate or deoxycholate have yielded two different forms of the cytochrome.

The initial attempts to purify NADPH-cytochrome *P*-450 reductase from the housefly involved the use of cytochrome *c* as an artificial electron acceptor and solubilization by isobutanol. These studies yielded an essentially pure reductase which was a flavoprotein with a molecular weight around 57,000 daltons which would reduce cytochrome *c*, 2,6-dichlorophenol indophenol, and ferricyanide, and accept electrons from NADPH but not NADH. This enzyme would not, however, reconstitute with purified cytochrome *P*-450 to yield an active mixed-function oxidase system. Subsequent studies have shown that detergent-solubilized reductases have a somewhat higher molecular weight and are capable of reconstitution. This situation is analogous to that in mammals, namely, that two forms of the reductase can be isolated, both of which show the same behavior with regard to electron donors and to artificial electron acceptors. The smaller form, however, lacks a hydrophobic polypeptide chain which is essential to the reconstitution process.

There is no evidence to date that phosphatidylcholine is necessary for reconstitution of mixed-function oxidase systems in insects as it is known to be in mammals.

Although some progress has been made it is clear that solubilization and purification of the cytochrome *P*-450 of insects and its subsequent use in reconstitution studies are more difficult than similar studies using mammalian liver.

8. Multiplicity and Specificity of Cytochrome *P*-450

This is currently an area of great interest. It is clear from investigations on mammalian liver microsomes that several different forms of cytochrome *P*-450 exist and that individual forms can be preferentially increased by the use of different inducers. Reconstitution experiments indicate that each form is relatively nonspecific with regard to xenobiotics but that the relative rates at which different substrates are oxidized vary from one cytochrome to another.

More preliminary and, therefore, less comprehensive studies have indicated that the same situation exists in insects or, at any rate, in the house fly. The evidence may be summarized as follows:

First, purification studies, as indicated above, yield two forms of cytochrome *P*-450.

Second, genetic studies of high oxidase, insecticide-resistant strains show that the cytochrome *P*-450's of these strains are qualitatively and quantitatively different.

Third, controlled tryptic digestion of housefly microsomes indicates that different forms of the cytochrome are digested at different rates. The characteristic form of the cytochrome, which shows Type I binding, disappears first while a significant part of the form also found in susceptible houseflies, which shows very little Type I binding, still remains.

Fourth, separation of microsomal fractions on density gradients demonstrate that different forms of the cytochrome are characteristic of microsomes of different densities, a finding which may mean that each form has its own characteristic distribution within the cell.

As indicated below, the genetic studies were unable to resolve the problem of whether the differences seen were due to single different cytochromes in each strain or to be different proportions in mixtures of the same cytochromes. It is now apparent that all strains contain more than one cytochrome *P*-450 but it cannot yet be determined whether all forms are present in all house fly strains with the individual concentrations as the only variant.

Since purification and reconstitution of the different forms have not yet proved possible, their specificity remains to be investigated.

B. Reduction

Williams and Millburn (1975) classified the reduction reactions of xenobiotics into nine classes, namely: reduction of azo compounds; reduction of nitro compounds; reduction of ketones; reduction of aldehydes; reduction of double bonds; reduction of pentavalent arsenic to trivalent arsenic; reduction of disulfide bonds to sulfhydryl groups; reduction of sulfoxides; reduction of N-oxides. Of these reactions only two, nitro reduction and azo reduction, are known to occur in insects.

An NADPH-dependent nitroreductase is found in the postmicrosomal supernatant fraction of adult house flies that reduces parathion to aminoparathion [Eq. (10)]. The enzyme was unaffected by the presence of oxygen while the effect of flavins was not investigated.

$$(C_2H_5O)_2P(=S)O-C_6H_4-NO_2 \xrightarrow{\text{NADPH, } O_2} (C_2H_5O)_2P(=O)O-C_6H_4-NH_2 \quad (10)$$

parathion — aminoparathion

Other studies have shown nitrobenzene reduction in both microsomes and supernatant fractions from several tissues of the Madagascar roach, *Gromphadorhina portentosa*. The activity in the postmicrosomal supernatant is NADH dependent while that in the microsomes can utilize either NADH or NADPH. The reductase activity was dramatically stimulated by flavins such as FAD, FMN, and riboflavin and it is probable that these flavins are in fact the substrates, the reduction of nitrobenzene being a nonenzymatic reaction with reduced flavins. Similar results have been obtained for azo reductase in this insect.

C. Hydrolysis

A large number of xenobiotics to which insects are exposed are esters such as the phosphoric acid esters and carbamate esters while certain other foreign compounds also contain ester or amide groups. In general, the titer of the enzymes metabolizing these xenobiotics is low except among some resistant strains.

Esterases are hydrolases which split ester compounds by the addition of water to yield alcohol and acids [Eq. (11)].

$$\underset{\text{malathion}}{(CH_3O)_2\overset{\overset{S}{\|}}{P}-S\underset{\underset{CH_2COOC_2H_5}{|}}{C}HCOOC_2H_5} \longrightarrow \underset{\text{malathion-}\alpha\text{-monoacid}}{(CH_3O)_2\overset{\overset{S}{\|}}{P}-S\underset{\underset{CH_2COOC_2H_5}{|}}{C}HCOOH} \qquad (11)$$

The presence of carboxyl ester groups in malathion and acethion make these compounds vulnerable to hydrolysis by carboxylesterases. Hydrolysis results in the cleavage of one carbethoxy group to form the nontoxic monoacid.

It is generally acknowledged that the selective toxicity of this compound, favoring mammals over insects and resistant insects over susceptible ones, is the result of carboxylesterase activity.

Increased degradation of malathion by carboxylesterase has been established in malathion-resistant house flies, blow flies (*Chrysoma putoria*), mosquitoes (*Culex tarsalis*), bedbugs (*Cimex lectularius*), rust-red flour beetles (*Tribolium castaneum*), and spider mites (*Tetranychus urticae*).

The carboxylesterases from malathion-susceptible and -resistant strains of mosquito (*Culex tarsalis*) have been purified by DEAE column chromatography and found to have the same molecular weight (16,000); properties such as K_m, energy of activation, and temperature coefficient were also similar in both strains and *n*-propyl paraoxon and the oxygen analogues of EPN inhibited the formation of malathion monoacid. It was concluded that in this case the interstrain difference was of a quantitative nature.

On the other hand, a twentyfold difference in the K_m relative to malathion between partially purified carboxylesterases from resistant and susceptible strains of the two-spotted spider mite indicated a qualitative difference between the two enzymes.

Pyrethroid-hydrolyzing enzymes have also been found in insect preparations. Those obtained from the milkweed bug (*Oncopeltus fasciatus*) and cabbage looper (*Trichoplusia ni*) cleaved the (+)-*cis* compounds, although the specificity of the isomers was less pronounced with enzyme preparations from the house fly and German cockroach (*Blattella germanica*). The relative rate of hydrolysis of (+)-*trans*-resmethrin in various preparations was of the order of mouse liver >>> milkweed bug >> cockroach > cabbage looper > house fly. This clearly favors the mammal as far as selective toxicity is concerned. It would not be surprising to find that the carboxylesterase which hydrolyzes (+)-*trans*-resmethrin is the same carboxylesterase which hydrolyzed malathion.

The enzymes that catalyze the hydrolytic attack on the phosphorus ester or anhydride bond should be referred to as phosphoric triester hydrolases. Triester hydrolysis results in the formation of an anionic metabolite which is a poor cholinesterase inhibitor; the overall result is a detoxication of the parent compound. At present it appears that the phosphotriester hydrolases have a higher activity in certain strains of resistant insects. It also seems unlikely that the hydrolases attack the phosphorothioates to any extent, even though early reports claim to have demonstrated direct esteratic cleavage.

$$(C_2H_5O)_2\overset{O}{\overset{\|}{P}}O-C_6H_4-NO_2 \longrightarrow (C_2H_5O)_2\overset{O}{\overset{\|}{P}}OH + HO-C_6H_4-HO_2 \quad (12)$$

paraoxon — diethyl phosphoric acid — *p*-nitrophenol

TEPP, DFP, and DDVP were readily hydrolyzed by house fly homogenates. The reaction was activated by 1 m*M* Mn^{2+} and Co^{2+}, but Ca^{2+}, Ba^{2+} and Sr^{2+} do not have any effect. Other compounds shown to be hydrolyzed by house fly preparations are diazoxon, paraoxon [Eq. (12)] and malaoxon. At the present time no phosphoric triester hydrolase has been purified from insects and compared to a purified mammalian hydrolase.

D. Epoxide Hydrases

Epoxide rings of certain alkene and arene compounds are hydrated enzymatically to form the corresponding *trans*-dihydrodiols [Eq. (13)].

$$-\underset{|}{C}-C- \text{(epoxide, O bridge)} \xrightarrow[H_2O]{\text{epoxide hydrase}} \;\rangle C(OH)-C(OH)\langle \quad (13)$$

The mechanism of hydration is still obscure, but activity is optimal at alkaline pH and probably involves a nucleophilic attack of the OH group on the oxirane carbon.

Cleavage of the epoxide ring of certain cyclodiene insecticides and their analogues has been demonstrated in the house fly, blow fly (*Calliphora erythrocephala*), mealworm (*Tenebrio molitor*), southern armyworm (*Prodenia eridania*), and Madagascar cockroach. The inactivation of juvenile hormone and its analogues also involves the hydration of the terminal epoxide rings. The action of many juvenile hormone analogues may be synergistic rather than intrinsically hormonal, and their activity may be due to the stabilization of endogenous juvenile hormone.

A number of inhibitors of insect HEOM epoxide hydrase includes SKF 525-A, piperonyl butoxide, sesoxane, *Cecropia* hormone, and a number of organophosphorus compounds. This is in contrast to the findings with mammalian liver styrene oxide hydrase, where piperonyl butoxide and SKF 525-A do not inhibit this enzyme.

Recent studies on some of the properties of insect HEOM epoxide hydrase have established a pH optimum of 9.0 for the enzyme from blow fly and southern armyworm and 8.1 for the hydrase from midgut microsomes of the Madagascar roach. The pH optimum for the conversion of juvenile hormone acid to the corresponding acid diol in the southern armyworm was 7.9, indicating a difference in the insect hydrases responsible for hydration of HEOM and juvenile hormone. Hydrase activity from the blow fly was not affected by Mn^{2+}, Fe^{3+}, Co^{2+}, Mg^{2+}, and Ca^{2+} ions at 10^{-3} *M* but was 80% inhibited by 10^{-3} *M* Cu^{2+} ions. Metyrapone and 1-(2-isopropylphenyl)imidazole had no effect on the insect epoxide hydrases, whereas both compounds stimulated the mammalian styrene oxide hydrase.

E. DDT-Dehydrochlorinase

In the early 1950's it was demonstrated that DDT-resistant house flies detoxified DDT mainly to its noninsecticidal metabolite DDE. The rate of dehydrohalogenation of DDT to DDE was found to vary between various insect strains as well as between individuals.

DDT-Dehydrochlorinase, a glutathione-dependent enzyme, has been isolated from the high-speed supernatant of resistant house flies. Although the enzyme-mediated reaction requires glutathione, the glutathione levels are not altered at the end of the reaction [Eq. (14)].

$$\underset{\text{DDT}}{Cl-C_6H_4-\underset{CCl_3}{\overset{H}{C}}-C_6H_4-Cl} \xrightarrow[\text{GSH}]{\text{enzyme}} \underset{\text{DDE}}{Cl-C_6H_4-\underset{Cl-C-Cl}{\overset{}{C}}\!\!=\!\!-C_6H_4-Cl} + HCl \tag{14}$$

The lipoprotein enzyme has a molecular weight of 36,000 as a monomer and 120,000 as the tetramer. The K_m for DDT is 5×10^{-7} M with optimum activity at pH 7.4.

The enzyme system catalyzes the degradation of *p,p*-DDT to *p,p,*-DDE or the degradation of *p,p*-DDD to the corresponding ethylene TDEE. The *o,p*-DDT is not degraded by DDT-dehydrochlorinase suggesting a *p,p*-orientation requirement for dehalogenations. In general, DDT-resistance of housefly strains is correlated with the activity of DDT-dehydrochlorinase, although other resistance mechanisms are known in certain strains.

At one time, it was believed that DDT-dehydrochlorinase, GSH *S*-aryltransferase, and γ-BHC-metabolizing enzymes were one enzyme since experimental results suggested the existence of a nonspecific enzyme which catalyzed all these reactions. However, it has been demonstrated that DDT-dehydrochlorinase is different from glutathione *S*-aryltransferase based on the following: difference in electrophoretic mobility; difference in stability and response to inhibitors; genetic studies; by purification of housefly glutathione *S*-aryltransferase which lacked DDT-dehydrochlorinase activity. The glutathione *S*-aryltransferase enzyme preparation, however, still had the ability to metabolize γ-BHC.

III. PHASE 2 REACTIONS

Phase 2 reactions involve the biosynthesis of natural or foreign compounds or their metabolites with endogenous compounds to form conjugates. Conjugation reactions may be grouped into three types. Examples of these reactions are glutathione conjugations (15), glucoside formation (16) and amino acid conjugation (17). In (15) the foreign compound or its metabolite has the reactivity to react enzymatically with glutathione; in (16) the foreign substrate reacts enzymatically with an endogenous activated donor; and in (17) the foreign substrate becomes activated to form a donor and then reacts enzymatically with an endogenous acceptor. Types (16) and (17) require formation of a "high energy" or "active" intermediate involving ATP whereas the reaction of xenobiotics with GSH does not. The activation of Type (15) reactions is derived from the intrinsic reactivity of the xenobio-

$$\text{Benzyl chloride} \xrightarrow[+\text{ enzyme}]{\text{GSH}} \text{benzyl glutathione} \tag{15}$$

$$\alpha\text{-Naphthol} \xrightarrow[+\text{ enzyme}]{\text{UDPG (active donor)}} \alpha\text{-naphthyl glucoside} \tag{16}$$

$$p\text{-Nitrobenzoic acid} \xrightarrow[\text{CoA}]{\text{activation}} \underset{\text{(active donor)}}{p\text{-nitrobenzoyl-CoA}} \underset{\text{enzyme}}{\overset{\text{glycine}}{\rightleftarrows}} \underset{\text{glycine}}{p\text{-nitrobenzoyl}} \tag{17}$$

tic or its metabolites. However, it should be stressed that ATP is utilized in the synthesis of reduced glutathione as well as in the subsequent formation of mercapturic acids.

For conjugation to occur, the Phase 1 metabolite or other xenobiotic must possess certain chemical groups through which the union or coupling of two substances can occur. The chemical groups required for Type (15) reactions include substitution of H, F, Br, I, Cl, NO_2, epoxides and ethers in certain aliphatic and aromatic compounds. The chemical groups necessary for Type (16) reactions are OH, NH_2, COOH and SH, and for Type (17) reactions the COOH group which is involved in amino acid conjugation.

A. Glutathione Transferases

A group of enzymes which catalyze the conjugation of electrophilic substances with endogenous reduced gluthathione (GSH) have been referred to as glutathione *S*-transferases. The function of these enzymes is to conjugate potentially harmful electrophiles with endogenous GSH and thus protect other nucleophilic centers such as proteins and nucleic acids, as well as a means of excretion after conjugation with GSH. Some of the general transferase reactions have been classified by Boyland and Chasseaud (1969) as shown in Eqs. (18)–(22).

$$\underset{\text{methyl iodide}}{CH_3I} + GSH \xrightarrow[\text{S-alkyl-transferase}]{\text{glutathione}} GS{-}CH_3 \tag{18}$$

$$\underset{\text{3,4-dichloronitrobenzene}}{\text{3,4-}Cl_2C_6H_3NO_2} + GSH \xrightarrow[\text{S-aryl-transferase}]{\text{glutathione}} \text{3-Cl-4-(SG)}C_6H_3NO_2 \tag{19}$$

$$\underset{\text{benzyl chloride}}{C_6H_5CH_2Cl} + GSH \xrightarrow[\text{S-aralkyl-transferase}]{\text{glutathione}} C_6H_5CH_2SG \tag{20}$$

$$\underset{\text{diethyl maleate}}{CHCOOC_2H_5{=}CHCOOC_2H_5} + GSH \xrightarrow[\text{S-alkene-transferase}]{\text{glutathione}} GS{-}CH(COOC_2H_5)CH_2COOC_2H_5 \tag{21}$$

$$\text{1,2-epoxyethylbenzene } (C_6H_5\text{–}HC(O)CH_2) + GSH \xrightarrow[S\text{-epoxide-transferase}]{\text{glutathione}} C_6H_5\text{–}CH(OH)CH_2\text{—SG} \tag{22}$$

Initially the enzymes were classified according to the type of substrate conjugated, i.e., glutathione *S*-alkyltransferase for methyl iodide conjugation, glutathione *S*-aryltransferase for 3,4-dichloronitrobenzene (DCNB). With the early work it was assumed that these reactions were associated with group transfer and the enzymes could be classified on this basis. Recent findings by Jakoby and co-workers (see Arias and Jakoby 1976) with purified rat liver glutathione *S*-transferases demonstrated a broad overlapping substrate specificity indicating that classification based on the group transferred (e.g., alkyl, aryl) was not appropriate. However, the use of group transfer is well entrenched and therefore will be used in this chapter.

Glutathione *S*-transferases mediate the initial reaction in the formation of mercapturic acids (Fig. 5). The synthesis of mercapturic acids occurs in several steps and involves the formation of a GSH conjugate which is degraded by the loss of glutamic acid and glycine to cysteine conjugate which is subsequently acetylated before excretion in the urine. The glutathione *S*-transferases are present in the soluble fraction of mammalian liver and/or kidney, while γ-glutamyltransferase is more active in the kidney than in the liver. Cysteinylglycinase is found in kidney, liver, and pancreas

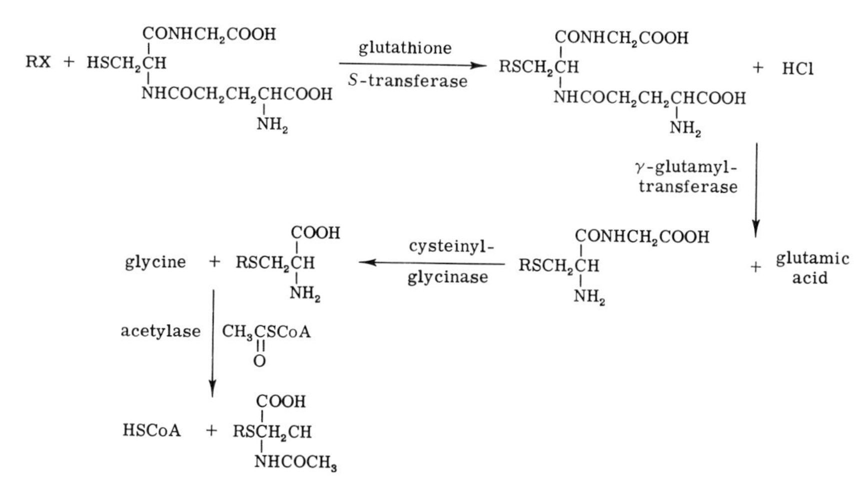

Fig. 5. Mercapturic acid formation.

while the acetylating enzyme, acetyl-CoA acetyltransferase, is present in the liver. In the locust (*Schistocerca gregaria*) partial hydrolysis of premercapturic acids to the cysteine derivative occurred in the Malpighian tubules and the gut and both the mercapturic acid and the cysteine derivative were found in the excreta.

Glutathione *S*-transferase activity has been demonstrated in all insects investigated. With certain insect strains the level of activity to a variety of substrates is higher than with other strains. In a comparison of three insecticide-resistant house fly strains and two susceptible strains, both alkyl- and *S*-aryltransferase activity is much higher in the resistant strains than in the susceptible strains (Table I). With some organophosphate-resistant strains of house flies and mites, glutathione *S*-transferases have been reported to be one of the mechanisms of resistance, especially with certain dimethyl organophosphates. All glutathione *S*-transferase activity has been found in the 100,000 *g* supernatant of insect homogenates. Alkyltransferase activity has been demonstrated in fat body and midgut tissue of horn beetle and silkworm larvae (*Bombyx mori*) and in house flies. Aryltransferase occurs in grass grubs (*Costelytra zealandica*) and in the head and thorax–abdomen portions of the house fly. This would indicate some *S*-transferase activity is associated with nervous tissue, possibly a barrier to nerve poisons. Aralkyltransferase occurs in locust, house fly, German cockroach, flour beetle, and turnip beetle (*Phaedon*). Both epoxidetransferase and alkenetransferase have been demonstrated in houseflies.

Transferase activity appears to be absent in house fly eggs but present in the 1-day-old larva. A sex difference has also been found in the housefly with female having about twice the alkyl- and aryltransferase activity of the male. Transferase activity has been shown to be inducible in the rat kidney and one would assume that this enzyme(s) is probably also inducible under certain conditions in insects.

Insect and vertebrate aryltransferases differ in their reactivity to sulfo-

TABLE I
Interstrain Comparison of Glutathione S-Transferases in House Flies

Strain	Methyl iodide (nmoles GS-methyl/♀/min)	DCNB (nmoles 2-chloro-4-nitrophenyl-GS/♀/min)
Resistant		
Rutgers	15.0 ± 1.8	36.2 ± 0.72
Fc	10.9 ± 1.2	16.5 ± 0.6
Baygon	10.0 ± 0.6	20.9 ± 0.3
Susceptible		
NAIDM	3.9 ± 0.3	8.8 ± 0.2
sbo	5.2 ± 1.0	7.6 ± 0.4

bromophthalein, which is the basis of the sulfobromophthalein liver function test. This compound is a weak competitive inhibitor with the vertebrate enzyme, but an excellent noncompetitive inhibitor of the locust enzyme with 1,2-dichloronitrobenzene as the substrate. Kinetic studies with a variety of sulfonaphthaleins showed that they compete with glutathione for its binding site on the insect enzyme. Kinetics studies also showed that the insect enzyme has two groups in the active site region with a pK of 9 while the sheep enzyme had only one group with the same pK.

Glutathione transferases are involved in the metabolism of organophosphorus insecticides. The reactions involving organophosphorus insecticides are mediated by these enzymes.

$$(C_2H_5O)_2P(=S)O-C_6H_4-NO_2 + GSH \longrightarrow (C_2H_5O)(HO)P(=S)O-C_6H_4-NO_2 + C_2H_5SG \quad (23)$$

$$(C_2H_5O)_2P(=S)O-C_6H_4-NO_2 + GSH \longrightarrow (C_2H_5O)_2P(=S)OH + GS-C_6H_4-NO_2 \quad (24)$$

In reaction (23) the compound is deethylated to form desethyl parathion and ethyl glutathione. In reaction (24) the compound is dearylated to form diethyl phosphorothioate and *S*-(*p*-nitrophenyl)glutathione. Both reactions result in the organophosphate insecticide being detoxified. The enzyme appears to be most active toward organophosphates of the dimethyl series. Enzymatic dealkylation of diethyl and higher dialkyl organophosphates has been reported but the reaction rates are generally lower than with the dimethyl compounds. Compounds which have been reported to be metabolized by insect glutathione *S*-transferases are methyl parathion, methyl paraoxon, Sumithion, sumioxon, dimethoate, Azinphos-methyl, Dipterex, parathion, diazinon, diazoxon, isopropyl diazinon, *N*-propyl diazinon and, ethyl chlorthion.

In certain resistant strains of insects, a higher rate of dealkylation via the glutathione *S*-transferase has been found in comparison to susceptible strains. Higher glutathione *S*-transferase activity parallels higher mixed-function oxidase activity in the resistant strains. At present, no strain of housefly has been found which has only high transferase activity. The levels of enzyme activity for both the mixed-function oxidases and the glutathione transferases are controlled by gene(s) on chromosome II.

A glutathione transferase purified from a resistant and a susceptible housefly strain had a molecular weight of 50,000 and consisted of two equal subunits. Kinetic analysis indicates that the housefly enzyme, in contrast to the mammalian enzyme, binds initially with DCNB and then forms an enzyme–DCNB–GS complex. A difference was observed in the kinetic constants between the resistant and susceptible strains, but did not explain the higher rate of overall reactions in the resistant strain. Therefore, only a quantitative difference was found between the housefly transferases. These enzymes were active in both O-dealkylation and O-dearylation of organophosphorus insecticides, depending upon the structure of the compound. The purified housefly enzyme also degrades γ-BHC to a water-soluble metabolite without prior metabolism to γ-PCCH as is required by mammalian glutathione *S*-transferase.

B. Glycoside Formation

Glucose conjugation is a detoxication mechanism generally associated with plants, insects, and other invertebrates. Only recently has glucosylation also been shown to occur in mammals. Although previous studies have suggested that insects can also utilize glucuronic acid, normally used in mammals, for certain conjugation reactions, subsequent studies by other investigators failed to demonstrate a glucuronyltransferase in insects. Therefore, based on present evidence, one must assume that glucosides are the only glycosides formed in insects.

Although O-glucosides, N-glucosides and S-glucosides may all be biosynthesized, it has been suggested that O-glucosides are the major class in both plants and insects. To date, only O- and S-glucosides have been demonstrated in insects.

The primary mechanism of O-glucosylation or S-glucosylation in insects appears to involve UDPG as the glucosyl donor, a UDP-glucosyltransferase, and various hydroxy or mercapto acceptor groups. [Eqs. (25) and (26)].

$$\text{D-glucose 1-PO}_4 + \text{UTP} \xrightarrow[\text{pyro-phosphorylase}]{\text{UDPG}} \text{UDP-}\alpha\text{-D-glucose} + \text{P}_2\text{O}_7^{4-} \quad (25)$$

$$\text{UDP-}\alpha\text{-D-glucose} + \text{ROH} \xrightarrow[\text{glucosyl-transferase}]{\text{UDP}} \text{RO-}\beta\text{-D-glucoside} \quad (26)$$

UDP-α-D-glucose RO-β-D-glucoside

The apparent lack of N-glucoside formation may be a reflection of the lack of studies with amino compounds or it may be that N-acetylation reactions predominate over N-glucoside formation in insects.

Examples of exogenous aglycones which have been shown by Smith and co-workers (see General References for citation) to form β-glucosides in insects are *o*-aminophenol, *p*-nitrophenol, 1-naphthol, 4-methyl umbelliferone, thiophenol, and *S*-mercaptouracil.

In insects only a limited number of studies have been conducted on glucosylation. In contrast to the mammalian glucosyltransferases, which are localized in the microsomes, the distribution of the insect enzyme is quite varied. With the house fly, the American roach (*Periplaneta americana*), and the locust, the enzymatic activity was reported to be in the 15,000–20,000 pellet while with the tobacco hornworm most of the activity was localized in the 100,000 *g* supernatant. Whether this variable distribution occurs in other insects needs to be studied further. With regard to tissue distribution of the insect glucosyltransferase, the enzyme is associated with the fat body preparations of the American roach, locust, and giant roach (*Blaberus craniifer*). Appreciable enzyme activity was also found in the tobacco hornworm midgut.

The formation of S-glucosides in insects has been reported. Thiphenol and 5-mercaptouracil are metabolized to S-glucosides which can be isolated from the excreta and treated tissues. *In vitro* studies using locust fat body preparations established that S-glucoside biosynthesis is mediated by a UDP-glucosyltransferase.

It has generally been assumed that O-glucosides represent a major class of insecticide metabolites in insects. A number of insecticides are substituted phenols and many are metabolized by oxidation or hydrolysis to phenols or alcohols. It is generally assumed that these phenols and alcohols are conjugated as O-glucosides in insects. However, in only a few studies have the metabolites actually been isolated and identified as glucosides. Generally the identification has been based on enzymatic hydrolysis and/or analysis of the aglycones after hydrolysis. In only a few examples has the glycone or carbohydrate moiety been identified. In Table II there is a list of insecticides from which an O-glucoside metabolite has been identified.

The formation of naphthyl glucoside from 1-naphthol in resistant and susceptible house flies revealed no difference in the ability of the two strains to form glucosides. In another study, the β-glucoside conjugate of *p*-methylthiophenol, a metabolite of dimethyl *p*-(methylthiophenyl) phosphate, was formed at a higher rate in the resistant strain of tobacco hornworm than in the susceptible. It is not clear whether an increase in glucosyltransferase activity would lead to increased resistance, especially since this is a Phase 2 reaction.

C. Amino Acid Conjugation

Conjugation of aromatic acids with amino acids is a well-known mechanism for detoxication. The detoxication of these xenobiotic acids via glycine conjugation has been demonstrated for several species of insects. The excre-

TABLE II
Conjugates Formed by Insects in the Metabolism of Insecticides or Synergists

Insecticide	Insect	Conjugate
Baygon	House fly	Phosphate, glucoside, and sulfate of 5-hydroxy-*N*-hydroxymethyl and *O*-depropyl Baygon
Carbaryl	Cattle tick (*Boophilus microplus*)	Sulfate and/or glucosides of 1-naphthol and 1,5-dihydroxy-naphthalene
Dimethyl-*p*(methylthio)-phenyl phosphate	Tobacco hornworm	Glucoside of substituted phenols
Furadan	Saltmarsh caterpillar (*Estigmene acrea*)	Glucosides of 3-hydroxy-Furadan
Landrin	House fly	Glucoside and/or sulfate of various hydroxymethyl dimethyl phenyl methylcarbamates
p,p-DDT	Grain weevil (*Sitophilus granarius*)	Glucosides of 3-hydroxy-4-chloro-benzoic acid, 4-hydroxybenzoic acid, and others
Piperonyl butoxide	House fly	Glucoside of 6-propylpiperonylic acid
2-Propynyl 1-naphthyl ether (synergist)	House fly	Glucoside of 1-hydroxy-2,3-methylene-dioxynaphthalene
Allethrin	House fly	Glucosides of hydroxylated allethrin

tion of glycine conjugates upon exposure to various substituted benzoic acids has been reported for houseflies, silkworms, mosquitoes (*Aedes*), and locusts. Some of the aromatic acids studied were *p*-aminobenzoic acid, *p*-nitrobenzoic acid, benzoic acid, salicylic acid, and anthranilic acid.

Amino acid conjugation occurs in two stages. The first requires the activation of the xenobiotic acid by an enzyme system requiring ATP and coenzyme A (27). This is followed by condensation of the activated xenobiotic acid with the endogenous amino acid (28).

$$\text{RCOOH} + \text{ATP} + \text{HSCoA} \longrightarrow \text{RCOSCoA} + \text{AMP} + \text{P}_2\text{O}_7^{4-} \quad (27)$$

$$\text{RCOSCoA} + \underset{\text{Amino acid}}{\text{H}_2\text{NCHRCOOH}} \longrightarrow \text{RCONHCHRCOOH} + \text{HSCoA} \quad (28)$$

The mechanism for reaction (28) has not been studied in detail in insects, but the necessity for the coenzyme A derivative and amino acid *N*-acyltransferase is common to amino acid conjugation. The enzyme involved in the conjugation of benzoic acid with glycine has been shown to occur primarily in the gut and to a lesser extent in the fat body of silkworm larvae.

With a number of organisms, it is known that more than one amino acid may be simultaneously involved in the detoxication of a xenobiotic acid. For example, the tick excretes both arginine and glutamine conjugates upon injection of aromatic acids while some arachnids form arginine, glutamine and glutamate conjugates of the same aromatic acids.

Studies on the metabolism of tropital, a methylenedioxyphenyl synergist, in the housefly showed seven products in the ether-soluble fraction of the excreta. Five amino acid conjugates of piperonylic acid, an oxidation product, were identified. They were alanine, glutamate, glycine, serine, and glutamine. It was also shown that some interconversion of amino acids occurred. It is interesting that these five amino acids are among the nonessential amino acids for many insects. The finding that a single species utilized five amino acids for conjugation is unique and may be associated with the high levels of free amino acids in insects when compared to other animals. The formation of a number of amino acid conjugates in one species has also been reported to occur in plants. The conjugating enzymes have not been studied in insects. It is not known whether the reactions with the amino acids were mediated by a number of amino acid acyltransferases or one enzyme with a low specificity for the conjugating amino acids.

D. Sulfate Conjugation

Sulfoester conjugates have been detected *in vivo* in various insect species examined after treatment with a number of phenolic compounds. Studies using *m*-aminophenol, 8-hydroxyquinoline, and 7-hydroxycoumarin and a large number of species of insects showed that both sulfate and glucoside conjugation of phenolic compounds was a common detoxication mechanism in insects. Sulfate conjugation appears to be a primitive form of detoxication based on the finding that *Peripatus,* a primitive arthropod, detoxified phenols by phosphate and sulfate conjugation and that glycoside conjugation was absent.

In insects the sulfate-activating and sulfotransferase enzymes are associated with the 100,000 *g* soluble fraction, require ATP and inorganic sulfate. Therefore, the enzyme system in insects appears to be similar to that in mammals and follows the 3-step reaction sequence given in Eqs. (29)–(31). However, utilization of (adenosine-*S*′-phosphosulfate) APS and (3′-phosphoadenosine-*S*′-phosphosulfate) PAPS in the mechanism of sulfoconjugation in insects has not been definitely established.

$$\mathrm{ATP} + \mathrm{SO_4}^{2-} \underset{\text{adenylyltransferase}}{\overset{\text{ATP-sulfate}}{\rightleftharpoons}} \mathrm{APS} + \mathrm{P_2O_7}^{4-} \tag{29}$$

$$\mathrm{APS} + \mathrm{ATP} \underset{\text{3-phosphotransferase}}{\overset{\text{ATP-adenylyl sulfate}}{\rightleftharpoons}} \mathrm{PAPS} + \mathrm{ADP} \tag{30}$$

$$\mathrm{PAPS} + \mathrm{ROH} \overset{\text{sulfotransferase}}{\rightleftharpoons} \mathrm{ROSO_3H} + \mathrm{ADP} \tag{31}$$

Sulfoconjugates are formed by the transfer of sulfate from PAPS to the phenolic hydroxyl group, the aromatic amino group or the aliphatic alcoholic hydroxyl group. The insect sulfotransferases were found to have similar properties to the mammalian enzymes with regard to cofactor requirements, pH optima, and subcellular localization.

The sulfotransferases are found in the gut tissues of southern armyworm larvae and also in preparations of the gut and the Malpighian tubules from the Madagascar cockroach. Here, as with the mammalian liver, the enzyme activity is associated with the soluble fraction of the cell. The enzyme system from the southern armyworm was active toward *p*-nitrophenol as well as various steroids from plant, mammalian, and insect sources such as cholesterol, α-ecdysone, and β-sitosterol.

The presence of the sulfotransferase system was demonstrated in high-speed supernatant fractions from eight species of insects representing Diptera, Lepidoptera, Hymenoptera, and Orthoptera. The sulfotransferase system required ATP, Mg^{2+} and inorganic sulfate for the sulfuration of *p*-nitrophenol, dehydroepiandrosterone and 22,25-bisdeoxyecdysone.

The sulfate conjugation system appears to be well distributed throughout the insect kingdom and therefore an important mechanism for detoxication.

E. Other Conjugations

The biosynthesis of phosphate esters is a general phenomenon found in all intermediary metabolism. However, phosphate conjugates of xenobiotics are rarely encountered except in insects. At present it would appear that insects are the only major group of animals which utilize this mechanism as a Phase 2 reaction.

Phosphate conjugates have been isolated by Smith (1968) from house flies, blowflies, and New Zealand grass grubs when dosed with 1-naphthol, 2-naphthol or *p*-nitrophenol. In the excreta the β-glucosides and ethereal sulfates of the phenols were also identified. An active phosphotransferase preparation was obtained from the 100,000 *g* supernatant of a gut tissue homogenate from the Madagascar cockroach. The enzyme required ATP and Mg^{2+} for phosphorylation [Eq. (32)].

$$ROH + ATP \xrightarrow[Mg^{2+}]{\text{phosphotransferase}} ROPO_3^{2-} + ADP \tag{32}$$

Similar but slightly less active preparations were also obtained from the tobacco hornworm and the housefly. The Madagascar cockroach phosphotransferase was shown to be induced by phenobarbital treatment.

Acetylation is a means by which certain xenobiotics containing amino groups may be detoxified. Both aromatic amines and aliphatic amines can be acetylated. Blow fly larvae have been shown to acetylate tyramine and dopamine by transfer of acetyl from acetyl-CoA by *N*-acetyltransferase. Evidence of acetylation of aromatic amines by three insect species (locust, silkworm, and wax moth) has also been demonstrated. The acetylation reactions appear to be common for endogenous substrates in insects but have not been encountered commonly in insecticide metabolism.

IV. MULTIPLE PATHWAYS

It is clear from the above that there are many possible pathways for the metabolism of foreign compounds. It is worth emphasizing that as a result of this many xenobiotics are metabolized by several diferent pathways and their metabolites are subject to further metabolism. This can result in a large and, from an analytical point of view, bewildering array of metabolites. An excellent example of this can be seen in the case of carbaryl (Fig. 4). Also, the same metabolite can be formed by different enzymatic routes; therefore identification of a metabolite does not necessarily indicate the enzyme involved.

V. INDUCTION OF DETOXICATION ENZYMES

Many diverse organic compounds have been shown to be inducers of mixed-function oxidases in mammals. The characteristics and effects of the two general types of inducers were reviewed extensively by Conney in 1967. Polycyclic aromatic hydrocarbons induce specific alterations in enzyme activity with simultaneous qualitative changes in the spectral characteristics of cytochrome *P*-450. The predominant type of induction is caused by compounds such as phenobarbital and is nonspecific and characterized by overall quantitative increases in all mixed-function oxidase components, including cytochrome *P*-450.

The induction of xenobiotic metabolizing enzymes has been studied intensively in mammals during the last two decades, most of the studies being concerned with the hepatic microsomal mixed-function oxidase system.

Although induction of microsomal mixed-function oxidase activity by insecticides and other organic compounds has been demonstrated for several insect species, detailed studies of cytochrome *P*-450 have been largely confined to the housefly. A broad spectrum of insecticides, including DDT, cyclodienes, phosphoric acid esters and juvenile hormone analogues are cytochrome *P*-450–inducers in insects. Interpretation of inductive effects such as changes in the quantitative and qualitative characteristics of cytochrome *P*-450 may be complicated by innate sex, age, or strain differences. For example, uninduced resistant and susceptible house fly strains exhibit striking differences when characteristics of their cytochrome *P*-450's are compared.

Increased levels of cytochrome *P*-450 have been reported after insecticide induction in several house fly strains. Dieldrin increases both cytochrome *P*-450 content and oxidase activity in the Orlando-R strain while DDT induces cytochrome *P*-450 in the Orlando-R and Diazinon-R strains. Triorthocresyl phosphate, triphenyl phosphate and tributyl phosphorotrithioate increased cytochrome *P*-448 (*P*-450) levels in the Diazinon-R strain but had no effect on levels in the Malathion-R strain.

Recently, Terriere and Yu demonstrated induction of cytochrome *P*-450 in house flies by juvenile hormone analogues. *Cecropia* juvenile hormone and hydroprene (ethyl-3,7,11-triethyldodeca-2,4-dienoate) stimulated microsomal oxidation and increased cytochrome *P*-450 levels by 31% in the Isolan-B strain. Several other compounds are also inducers of cytochrome *P*-450 in insects. Phenobarbital and butylated hydroxytoluene are inducers of *P*-450 in several house fly strains while in the Fc strain, cytochrome *P*-450 is also induced by naphthalene and phenobarbital. *In vivo* treatment with alkylbenzenes resulted in increases in mixed-function oxidase activity in gut, Malpighian tubules, and fat body of the southern armyworm with concomitant increases in cytochrome *P*-450.

Ascertaining qualitative alterations in chemically-induced insect cytochrome *P*-450 is confused by methodological differences among various laboratories. Difference spectroscopy is generally used to measure variations in insect cytochrome *P*-450's by comparisons of Type I, Type II, and Type III binding spectra and CO spectral maxima. In uninduced mammals, cytochrome *P*-450 characteristics are fairly uniform among strains; this makes detection of qualitatively different cytochromes easier. Some uninduced resistant house fly strains possess cytochrome *P*-450 which resembles in many respects that seen in 3-methylcholanthrene-induced mammals. This differs from the susceptible type of cytochrome *P*-450 in several respects: (1) the CO maximum is shifted several nm to the blue; (2) Type I binding is present; (3) Type II binding is increased in relation to the CO spectrum; (4) the ethyl isocyanide spectrum is changed in magnitude; (5) the *n*-octylamine spectrum has a single trough at 390 nm. As indicated above, these cytochromes are almost certainly mixtures of two or more cytochromes in which one ("restraint" or "susceptible") predominates.

In the Diazinon-R strain of house fly, *in vivo* administration of malathion, tropital, MGK-264 and three substituted phenyl 2-propynyl ether synergists produce a shift in the carbon monoxide spectrum from 448 (uninduced) to 450 nm, although the same compounds did not affect microsomal cytochrome *P*-450 (CO maximum 452) in the Malathion-R strain. In the Fc strain, a housefly strain previously reported to contain the "resistant" cytochrome *P*-449, the induction of a qualitatively different cytochrome with naphthalene and phenobarbital has been reported. After induction, increased titers of cytochrome *P*-450 were measured that exhibited lower CO maxima, increased Type I and Type II ratios to the CO peak, and a different ethyl isocyanide equilibrium point, than those in control microsomes. Apparently phenobarbital and naphthalene are specific inducers in certain insect strains, much like 3-methylcholanthrene in mammals.

Treatment of southern armyworm with alkylbenzenes had no effect on the CO difference spectral maxima of cytochrome *P*-450.

Little is known of the induction in insects of detoxifying enzymes other than cytochrome *P*-450. Recent reports have indicated that phenobarbital methylenedioxyphenyl compounds and juvenile hormone analogues all induce a higher level of DDT-dehydrochlorinase in the housefly and that phenobarbital induces a higher level of glutathione transferase in the same insect. In all of these cases of DDT-dehydrochlorinase induction, this enzyme and microsomal mixed-function oxidase were induced together; DDT-dehydrochlorinase has not been induced alone even though it is structurally and biochemically unrelated to the microsomal mixed-function oxidase system.

The mechanism of induction in insects is not well understood. It appears to involve derepression of the genes controlling the synthesis of detoxifying enzymes and a resultant increase in the synthesis of new enzyme. Suggestions have also been made that regulatory genes are involved. In at least one species an increase in DNA-dependent RNA polymerase has been noted.

VI. DETOXICATION ENZYMES IN INSECTICIDE RESISTANCE

In the last several decades, many species of insects have acquired resistance to insecticides. This resistance is inherited and has proved to be the biggest single barrier to successful chemical control of insects. Several enzymes have been shown to be involved, including target enzymes such as cholinesterase and detoxifying enzymes such as the mixed-function oxidase and glutathione transferase. In addition to biochemical mechanisms, penetration mechanisms and behavioral avoidance patterns may also contribute to resistance. Resistance is usually a complex phenomenon with several mechanisms operating simultaneously in the same resistant strain.

Resistance in many instances is due at least in part to increased levels of mixed-function oxidase activity, and this aspect has been investigated by several workers. Since the discovery of cytochrome *P*-450 in insects by Ray in 1967, there has been considerable interest in its role in detoxication of xenobiotics by resistant insects but all of the comparative studies on cytochrome *P*-450 in susceptible and resistant strains have been confined to the housefly.

Increased cytochrome *P*-450 levels have been reported for a number of resistant house fly strains possessing high oxidase levels. However, there is not necessarily a correlation between increased cytochrome *P*-450 and mixed-function oxidase activity. For example, Plapp and Casida showed that the house fly strains Fc, *bwb, stw,* and R-Baygon, *bwb, ocra,* have high NADPH-dependent oxidase activity yet their cytochrome *P*-450 does not exceed the levels seen in susceptible strains. In no resistant strain assayed has an increase in cytochrome *P*-450 greater than twofold been measured while detoxication rates are frequently several times those observed in susceptible strains. This fact has led some investigators to suggest cytochrome *P*-450 may not be the limiting factor in the resistant mixed-function oxidase activity.

Several recent comparative studies have indicated the existence of qualitatively different cytochrome *P*-450's in several resistant housefly strains. The "resistant" form differs from the "susceptible" in the following manner: (1) presence of Type I binding; (2) the CO maximum is shifted to a lower wavelength; (3) the Type II spectral magnitude is increased relative to the CO spectrum; (4) the ethyl isocyanide 455-nm peak of the Type III spectrum is lowered with respect to the CO magnitude; (5) the *n*-octylamine Type II spectrum has a single trough at 390 nm in contrast to the susceptible *P*-450 which binds with *n*-octylamine to form a double trough at 410 and 394 nm, respectively.

This qualitatively different cytochrome *P*-450 was first described in the multiresistant Diazinon-R housefly strain and subsequently in such strains as Fc and Dimethoate-R, while the susceptible cytochrome *P*-450 was detected in another Fc strain as well as R-Baygon and Orlando-R.

The significance of this qualitatively different cytochrome *P*-450 in some housefly strains is unclear. No specific enzymatic reaction or resistant characteristic has been correlated with this cytochrome. NADPH-dependent oxidase levels in strains containing this cytochrome *P*-450 are not substantially higher than those seen in resistant strains containing the susceptible type of cytochrome. However, high oxidase activity in strains having the resistant type of cytochrome can be correlated with the presence of Type I binding, whereas other resistant cytochrome characteristics are not necessarily related to high oxidase activity. Evidence from genetic studies suggests that in strains exhibiting Type I binding, increased oxidase activity segregates on the same chromosome which confers Type I binding regardless of the cytochrome *P*-450 titer.

While the genetics of resistance in insects has been studied extensively, until recently no data were available on cytochrome *P*-450 genetics in insects. The discovery of a qualitatively different form has made possible genetic analyses of cytochrome *P*-450 and its relation to the genetics of resistance. High oxidase activity segregates with chromosome II and/or chromosome V in resistant houseflies.

Tate and co-workers studied the genetics of cytochrome *P*-450 in Diazinon-R, a strain of previously unknown genetics, and Fc, one containing high-oxidase genes on chromosome V. Resistant strains were crossed with a susceptible strain carrying visible, recessive markers on chromosomes II, III and V, the chromosomes generally responsible for increased mixed-function oxidase activity in houseflies. F_1 males were backcrossed to females of the marked susceptible strain. Each of the eight phenotype progenies resulting from the backcross contained specific combinations of resistant and susceptible chromosomes. The substrains were analyzed for cytochrome *P*-450 variations. In the Diazinon-R strain, high oxidase activity and the cytochrome *P*-450 qualitative characteristics seen in the resistant strain were inherited as semidominants on chromosome II. On the other hand, in Fc, only Type I binding segregated with chromosome V, the chromosome which was previously shown to confer high oxidase activity in that strain. Chromosome II in Fc contained gene(s) which were quantitative in their effects on the expression of cytochrome *P*-450. These results indicate that in one strain, Fc, and possibly others, at least two or more loci are involved in phenotype expression of the resistant cytochrome *P*-450.

The biochemistry and genetics of other detoxifying enzymes involved in the resistance have also been studied. Differences in the target enzyme, cholinesterase, between susceptible and resistant strains have been demonstrated. These differences render the acetylcholinesterase of the resistant strain less susceptible to inhibition by cholinesterase inhibitors such as the organophosphate and carbamate insecticides. Such resistance was first shown in ticks and mites but has since been shown in such insects as *Nephotettix cereticeps,* the green rice leafhopper, and the house fly. Since this is closely related to mode of action, the subject of another chapter, it will not be discussed further at this point.

Carboxylesterase, discussed above, can also be involved in resistance. Unusually high levels of this enzyme are seen in several organophosphate-resistant strains of a number of insect species as compared to susceptible strains of the same insect. These include *Culex tarsalis, Heliothis virescens* and representatives of several other orders. The carboxyl-esterase of resistant strains may be qualitatively as well as quantitatively different from that of susceptible strains.

Glutathione transferases, also described above, have been shown to be involved in resistance to organophosphate insecticides. This has been studied most intensively in the house fly and in this species it is clear that a high level of this enzyme is controlled by a gene on chromosome II. More recent studies have indicated that there are qualitative as well as quantitative differences in the enzyme from susceptible and resistant strains, particularly a difference in the equilibrium constants which is reflected in a faster catalysis of the overall conjugation reaction by the enzyme from the resistant strain.

DDT-dehydrochlorinase, long known to be one of the mechanisms of DDT resistance, is controlled, in the house fly, by a gene on chromosome II.

VII. GENERAL CONCLUSIONS

Although the detoxication systems of insects have not received the attention given those of mammals, it is clear that the marvelously complex yet flexible systems that both possess are basically similar. The Phase 1 reaction oxidations are the most obvious and probably the most important. The microsomal mixed-function oxidase system of insects resembles that of mammals. The Phase 2 reactions, are, in both cases, mainly conjugations and it is here that we see one of the most obvious differences between the two groups, insects synthesizing glucosides while in analogous reactions mammals synthesize glucuronides. The importance of glutathione and glutathione transferases is currently being realized by investigators of both groups of organisms.

The great impetus to the study of detoxication reactions in insects has been the investigation of insecticide metabolism and it is to this that we owe the fact that in two areas of investigation knowledge derived from insects is in advance of that derived from mammals. The first of these areas is resistance to toxic compounds and the second the biochemical genetics of detoxication enzymes. Due to the widespread use of insecticides and the short generation time of insects, numerous cases of resistance have arisen. This in turn has led to studies of the genetics of resistance and of the biochemical mechanisms involved. More recently these fields have come together in studies of the biochemical genetics of the detoxication enzymes.

The emphasis on detoxication of synthetic organic chemicals should not blind us to the fact that the detoxication systems which evolved over many millennia enabled the insect to survive naturally occurring toxicants. These include the secondary plant substances such as alkaloids, terpenoids, methylenedioxy compounds, of low but still appreciable toxicity. Since their lipophilicity would cause eventual accumulation to toxic levels, the survival value of enzyme systems which are relatively nonspecific and which render such compounds less lipophilic is obvious. One indication of the truth of this hypothesis is that polyphagous caterpillars have, in general, a higher level of mixed-function oxidase activity than monophagous caterpillars. Thought of in these evolutionary terms, lack of specificity, which often troubles the biochemist, is a natural and useful adaptation to the wide array of potentially toxic compounds in nature. The low turnover numbers also characteristic of this system are probably a natural consequence of the lack of specificity.

This same lack of specificity has led to one biologically unfavorable feature, that some exogenous compounds of low toxicity are metabolically activated to highly toxic metabolites. This involuntary self-destruction is best known from studies of the conversion of phosphorothioates to potent cholinesterase inhibitors in insects and the conversion of aromatic hydrocarbons into carcinogenic epoxide dihydrodiols in mammals. In view of the evolutionary history of the enzymes involved it might be instructive to inquire how many naturally occurring toxicants are metabolically activated in either insects or mammals.

In conclusion we can say that enough is known of detoxication processes in insects to see the broad outlines but much remains to be done and many surprises await those who are interested in looking further.

VIII. GLOSSARY OF CHEMICAL NAMES

Common name	Chemical name
Acethion	*O,O*-Diethyl-*S*-carboethoxymethyl phosphorodithioate
Aldrin	1,2,3,4,10,10-Hexachloro-1,4,4a,5,8,8a-hexahydro-1,4-*endo,exo*-5,8-dimethanonaphthalene

Common name	Chemical name
Azinphos-Methyl	*O,O*-Dimethyl *S*-(4-oxo-1,2,3-benzotriazin-3(4*H*)-ylmethyl) phosphorodithioate
Baygon (Propoxur)	*o*-Isopropoxyphenyl *N*-methylcarbamate
γ-BHC	γ Isomer of 1,2,3,4,5,6-hexachlorocyclohexane
Carbaryl	1-Naphthyl *N*-methylcarbamate
DDD	2,2-Bis(*p*-chlorophenyl)-1,1-dichloroethane
DDE	2,2-Bis(*p*-chlorophenyl)-1,1-dichloroethylene
DDT	2,2-Bis(*p*-chlorophenyl)-1,1,1-trichloroethane
Diazinon	*O,O*-Diethyl *O*-(2-isopropyl-4-methyl-6-pyrimidyl)phosphorothioate
Diazoxon	Oxygen analogue of diazinon
Dieldrin	1,2,3,4,10,10-Hexachloro-*exo*-6,7-epoxy-1,4,4a,5,6,7,8a-octahydro-1,4:5,8-*endo, exo*-dimethanonaphthalene
Dimethoate	*O,O*-Dimethyl *S*-*N*-methylcarbamoylmethyl phosphorodithioate
Dipterex (Trichlorfon)	*O,O*-Dimethyl-(1-hydroxy-2,2,2-trichloroethyl) phosphonate
EPN	*O*-Ethyl *O*-*p*-nitrophenyl phenylphosphonothioate
Ethyl chlorthion	*O,O*-Diethyl *O*-(3-chloro-4-nitrophenyl phosphorothioate
Furadan (Carbofuran)	2,3-Dihydro-2,2-dimethyl-7-benzofuranyl methylcarbamate
HEOM	1,2,3,4,9,9-Hexachloro-6,7-epoxy-1,4,4a,5,6,7,8,8a-octahydro-1,4-methanonaphthalene
Kelthane (dicofol)	1,1-Bis(*p*-chlorophenyl)-2,2,2-trichloroethanol
Landrin	75% 3,4,5-Trimethyl phenyl methylcarbamate and 18% 2,3,5-trimethylphenyl methylcarbamate
Malaoxon	Oxygen analogue of malathion
Malathion	*O,O*-Dimethyl *S*-(1,2-dicarboethoxy) ethyl phosphorodithioate
Methyl paraoxon	Oxygen analogue of methyl parathion
Methyl parathion	*O,O*-Dimethyl *O*-(4-nitrophenyl) phosphorothioate
Paraoxon	Oxygen analogue of parathion
Parathion	*O,O*-Diethyl *O*-4-nitrophenyl phosphorothioate
γ-PCCH	γ-2,3,4,5,6-Pentachlorocyclo-l-ene
Piperonyl butoxide	3,4-Methylenedioxy-6-propylbenzyl-*n*-butyl diethyleneglycol ether
Resmethrin	(5-Benzyl-3-furyl)methyl 2,2-dimethyl-3-(2-methylpropenyl) cyclopropane carboxylate

Common name	Chemical name
Rotenone	1,2,12,12a-Tetrahydro-2-isopropenyl-8,9-dimethoxy[1]benzopyrano[3,4-*b*]furo[2,3-*b*][1]benzopyran-6-(6a*H*)-one
Sesoxane (sesamex)	2-(3,4-Methylenedioxyphenoxy)-3,6,9-trioxaundecane
SKF-525A	2(Diethylamino)ethyl-2,2-diphenyl-pentanoate
Sumioxon	Oxygen analogue of Sumithion
Sumithion (fenitrothion)	*O*,*O*-Dimethyl*O*-(3-methyl-4-nitrophenyl) phosphorothioate
TDEE	2,2-Bis(*p*-chlorophenyl)-1-chloroethylene
Tropital	Piperonal bis[2-(2-butoxyethoxy)ethyl]acetal

GENERAL REFERENCES

Ahmad, S., and Forgash, A. J. (1976). *Drug Metab. Rev.* **5,** 141–164.

Arias, I. M., and Jakoby, W. B. (1976). "Glutathione: Metabolism and Function." Raven, New York.

Boyland, E., and Chasseaud, L. F. (1969). *Advan. Enzymol.* **32,** 173–219.

Conney, A. H. (1967). *Pharmacol. Rev.*, **19,** 317–362.

Smith, J. N. (1968). *Adv. Comp. Physiol. Biochem.* **3,** 173–232.

Wilkinson, C. F., ed. (1976). "Insecticide Biochemistry and Physiology." Plenum, New York.

Wilkinson, C. F., and Brattsen, L. B. (1972). *Drug Metab. Rev.* **1,** 153–227.

Williams, R. T., and Millburn, P. (1975). Detoxication mechanisms—the biochemistry of foreign compounds. *In* "Physiological and Pharmacological Biochemistry," (H. K. F. Blaschko, ed.) Series One, Vol. 12, pp. 211–226. University Park Press, Baltimore, Maryland.

14

Chemical Genetics and Evolution

FRANCISCO J. AYALA

I. INTRODUCTION

Evolution, consisting of changes in the genetic constitution of populations, can only occur if there is genetic variation, so that alternative genetic variants increase or decrease in frequency over the generations. Thus a fundamental question in evolutionary genetics is how much genetic variation exists in natural populations. The amount of genetic variation in a population measures its evolutionary potential. Section II of this chapter deals with the methods and the results of studies aimed at ascertaining the amount of variation in natural populations, especially of insects. Early studies failed to provide quantitatively precise estimates of genetic variation; however, these have become possible in recent years through the application of biochemical concepts and biochemical methods, particularly gel electrophoresis.

Evolutionary genetics is also concerned with how much genetic change takes place in evolution. One fundamental step in the evolutionary process is the formation of new species, because these are independent evolutionary units. The amount of change taking place in the process of speciation is the subject of Section III of this chapter; we shall see that here again quantitative measures were only possible through the application of biochemical methods. Section IV, the final one in this chapter, considers the biochemical methods that have been developed in recent years to ascertain the evolutionary history of organisms and the amount of genetic change that takes place in evolution.

II. GENETIC VARIATION IN NATURAL POPULATIONS

A. Genetic Variation and Rate of Evolution

In "The Origin of Species," Darwin summarized his theory of evolution by natural selection as follows:

> As more individuals are produced than can possibly survive, there must in every case be a struggle for existence, either one individual with another of the same species, or with the individuals of distinct species, or with the physical conditions of life .... Can it, then, be thought improbable, seeing that variations useful to man have undoubtedly occurred, that other variations useful in some way to each being in the great and complex battle of life, should sometimes occur in the course of thousands of generations? If such do occur, can we doubt (remembering that more individuals are born than can possibly survive) that individuals having any advantage, however slight, over others, would have the best chance of surviving and of procreating their kind? On the other hand, we may feel sure that any variation in the least degree injurious would be rigidly destroyed. This preservation of favorable variations and the rejection of injurious variations, I call Natural Selection.

Darwin's explanation of the evolution of organisms by natural selection is, like many other great achievements of the human mind, magnificently simple and yet powerful. The starting point is the occurrence of natural variation; this is for Darwin an incontrovertible fact; although he did not know the mechanisms of genetic mutation and recombination by which hereditary variation arises. Darwin argued that some natural variations must be more advantageous than others for the survival and reproduction of their possessors, i.e., organisms having advantageous variations are more likely to survive and reproduce than organisms lacking them. This process—called *natural selection* by Darwin—leads to the spread of useful variations and to the elimination of harmful or less useful ones.

Evolutionists have, nowadays, a more complete, and more profound, understanding of the processes of organic evolution. Natural selection remains the fundamental process directing evolutionary change. However, natural selection can only occur if there is hereditary variation, and the more genetic variation there is in a population, the greater the opportunity for the operation of natural selection. This relationship is intuitively obvious. Assume first that at a certain gene locus there is no variation at all in a certain population, i.e., all individuals are genetically identical. Then evolutionary change cannot occur at that locus—all individuals will remain identical from one generation to the next. Assume now that in another population there are two allelic variants at the same gene locus as above, and that these alleles affect differently the chances of survival and/or reproduction of their carriers. Then evolutionary change will occur—one allele may increase in frequency, and the other decrease, from generation to generation. The direct relationship between the amount of genetic variation at a gene locus and the rate of evolutionary change was demonstrated mathematically by R. A. Fisher in what is known as The Fundamental Theorem of Natural Selection, which says: "The rate of increase in fitness of a population at any time is equal to its genetic variance in fitness at that time."

It follows that the first datum that evolutionists need to know, the most basic question they must ask is: How much variation is there in natural populations of organisms? It may come as a surprise to find out that for many years evolutionists were unable to find precise, quantitative answers to that question. They had evidence showing that hereditary variation was a pervasive phenomenon, but they could not tell *how much* variation there was.

Fortunately, the situation has changed during the last decade. The advances of molecular biology and the development of certain research technologies have made it possible to obtain quantitative estimates of hereditary variation in populations. Moreover, these estimates have startled evolutionists; it turns out that natural populations of organisms possess considerably more genetic variation than anybody had suspected! This is a pleasant surprise but it has created new problems.

B. Early Evidence of Genetic Variation

Biologists have known for a long time that individual variation is a pervasive phenomenon of natural populations. Human populations, for example, exhibit conspicuous variation in facial features, height and weight, body configuration, blood groups, and so on. Discontinuous variants (sometimes called "morphs" or "sports") are often so strikingly different in plants as well as in animals as to have been mistakenly described as separate species. Examples of well-studied morphological variation include color and pattern polymorphisms in snails, butterflies, grasshoppers, ladybird beetles, mice, and birds; dextral versus sinistral winding of the shell in snails; flower and seed color and pattern, as well as growth habit in many plants. This variation has been shown, in numerous cases, to be due to genetic differences.

Natural populations contain much more genetic variation than is observed by inspection of organisms. This can be shown by mating close relatives (inbreeding) which increases the probability of homozygosis, i.e., the probability that an individual will carry two identical copies of the same gene. Recessive genes become, then, expressed. For example, inbreeding wild-caught *Drosophila* has shown that each fly carries, on the average, between two and five allelic variants having conspicuous effects on the phenotype, such as abnormal eye color, extremely short wings, and abnormal bristles. About 21.3% of corn plants carry gene variants that in homozygous condition result in abnormal chlorophyll.

The pervasiveness of genetic variation in natural populations is also evident from the success of artificial selection for a great variety of traits in many different organisms. In artificial selection those individuals are chosen to breed the next generation that exhibit the greatest expression of the desired characteristic. Heritable changes over the generations in the phenotypic distribution of a population indicate that the population has genetic variation for the selected trait.

Artificial selection has been successful for an innumerable variety of commercially desirable traits in many domesticated species, including cattle, swine, sheep, poultry, corn, rice, and wheat. The changes obtained by artificial selection are often staggering. For example, the egg production of White Leghorn chickens increased from 125.6 eggs per hen per year in 1933 to 249.6 eggs per hen per year in 1965. Selection for high protein content in a variety of corn increased the protein content from 10.9 to 19.4%, while selection for low protein content reduced it from 10.9 to 4.9%. In *Drosophila* flies, artificial selection has succeeded for more than fifty different traits.

Sophisticated genetic techniques have shown in several species of *Drosophila* that genetic variation is also pervasive with respect to traits such as survival, rate of development, and fecundity. These techniques need not be described here, but the general conclusion can be simply stated: every chromosome carries several allelic variants that in homozygous condition reduce the fitness of the individuals carrying them. Gene variants that reduce the fitness of their carriers are also common in man. About 4% of all humans suffer from physical or mental disabilities caused by deleterious genes. The incidence is much higher among the children whose parents are relatives (e.g., first cousins) because such children are more likely to be homozygous for deleterious genes, than among children of unrelated parents.

C. The Problem of How to Measure Genetic Variation

The evidence reviewed in Section II,B indicates that large stores of genetic variation exist in natural populations, and hence that there is ample opportunity for evolutionary change. Yet, we would like to know how many gene loci, or what proportion of all gene loci, are variable; the evidence reviewed so far does not make this possible. In fact, evolutionists were until recently unable to obtain quantitative estimates of the proportion of gene loci that are polymorphic. There is a methodological handicap in the traditional approaches to study genetic variation. In order to measure the amount of genetic variation in a population, one must be able to detect allelic variants in gene loci that represent a random sample of all genes. However, with the traditional methods of Mendelian genetics the presence of a gene is ascertained by studying segregation in the progenies of matings between individuals differing in a given trait. Therefore, only variable genes can be studied; invariant genes cannot even be shown to exist. Thus, as long as the traditional methods of genetics were used, there was no way to obtain a sample of genes that would be random with respect to variation, i.e., that would include variable as well as invariant genes hopefully in the same proportions in which they exist.

The solution to this problem was provided by two breakthroughs, one conceptual, the other technological. The conceptual breakthrough came from molecular genetics and biochemistry; it is now known that variation in the amino acid sequence of a protein reflects variation in the gene coding for it. Variable as well as invariant proteins can of course be studied, and therefore variant and invariant genes can be sampled in natural populations. The technological breakthrough also came from biochemistry: the development of gel electrophoresis, a technique that permits the analysis of protein variation in many individuals without prohibitive requirements of time and expense. We shall now briefly look at these two breakthroughs.

In the early 1950's it became known that genes are molecules of deoxyribonucleic acid (DNA), and that the genetic information is contained in the sequence of the units (nucleotides) that make up DNA (Fig. 3). There are four kinds of nucleotides, each containing a different kind of nitrogen base: adenine (A), cytosine (C), guanine (G), and thymine (T). The nucleotides may be considered as the letters of the genetic alphabet. The genetic information contained in the sequence of nucleotides in the DNA controls the development and metabolism of the organism by determining the amino acid sequence in enzymes and other proteins. This determination takes place through two processes, called *transcription* and *translation*. The nature of these processes was ascertained in the 1960's.

Transcription is the process by which the nucleotide sequence of one strand of DNA is transcribed into a complementary sequence of ribonucleic acid (RNA). RNA differs from DNA in three ways: the nucleotides contain the sugar ribose (rather than deoxyribose); it is single-stranded; and there is no nucleotide containing the nitrogen base T but there is a nucleotide containing the base uracil (U). An RNA molecule that carries the information that directs the synthesis of a protein is called a "messenger" RNA (mRNA).

The rules of complementarity that govern the transcription of a DNA sequence into an RNA sequence of nucleotides are as follows:

DNA:	A	C	G	T
RNA:	U	G	C	A

That is, whenever A is present in the DNA, the complementary RNA will have T; whenever DNA has C, RNA will have G; and so on.

Translation is the process by which the nucleotide sequence of an mRNA molecule becomes converted into a sequence of amino acids making up a protein. Proteins are polymer molecules usually consisting of between about fifty and several hundred amino acids of twenty different kinds. The information that becomes translated into the amino acid sequence of the protein is contained in the mRNA in discrete sequences of three nucleotides, called *codons*. Since there are four different letters in the RNA (or DNA) alphabet, there are 4^3 or 64 different combinations of three nucleotides. Of these sixty-four codons, three are termination signals, i.e., they do not code for any amino acid, but tell where the process of translation should stop. The other sixty-one nucleotides code for the twenty amino acids, so that from one to six different codons code for one given amino acid. The rules that govern the translation of a nucleotide sequence of mRNA into the amino acid sequence of a protein are known as the *genetic code*.

The genetic code is redundant or, as it is sometimes said, "degenerate," because several different codons may code for a single amino acid. Hence, if we know the nucleotide sequence of mRNA (or of DNA), we can specify precisely the amino acid sequence of the protein coded by it. However, the reverse process involves some ambiguity; from knowing the amino acid sequence of a protein we cannot determine precisely the nucleotide sequence of mRNA. For all but one amino acid, ambiguity exists only with respect to the third nucleotide of each codon. In the case of the amino acid *leucine*, ambiguity exists with respect to the third nucleotide, and also with respect to the first nucleotide. The important point, however, is that genetic information flows DNA → mRNA → protein; this relationship has come to be known as the "Central Dogma" of molecular biology.

D. Protein Variation as Genetic Variation

The Central Dogma of molecular biology frees us from the constraints of Mendelian genetics that allows only the study of variable genes. It is now possible to study genes whether or not they are variant simply by obtaining the amino acid sequence of proteins. A gene with no allelic variation codes for a protein that is identical in all individuals of the population; but if we find out that proteins coded by a given gene locus differ in amino acid sequence, we know that they are coded by different alleles (i.e., different DNA sequences).

Since the early 1950's biochemists have known how to obtain the amino acid sequence of proteins. Therefore, one conceivable way to measure genetic variation in a natural population would be to pick up a fair number of proteins, say thirty, chosen without knowing whether they are variable in a given population so that they would represent an unbiased sample with respect to variation. Then, each of the thirty proteins could be sequenced in a number of individuals, say one hundred, to ascertain how much variation, if any, exists for each one of the proteins. The average amount of variation found in the one hundred individuals for the thirty proteins would be an estimate of the amount of variation found in the genome of the population.

Unhappily, obtaining the amino acid sequence of a protein is a very demanding task so that several months, or even years, are usually required to sequence each protein. Thus it is not practically feasible to sequence 3000 proteins (thirty in each of one hundred individuals) for estimating genetic variation in each population we want to study. Fortunately, there is a technique, gel electrophoresis, that makes possible the study of protein variation with only a moderate investment of time and money. Since the late 1960's, estimates of genetic variation have been obtained for many natural populations of insects and other organisms using gel electrophoresis.

Figure 1 shows the apparatus and procedures used in gel electrophoresis for studying genetic variation in natural populations. Tissue samples from organisms are individually homogenized to release the enzymes and other proteins. The homogenate supernatants are placed in a gel made of starch, polyacrylamide, or some other jellylike substance. The gel with the tissue samples is then subjected for a given length of time to an electric current. Each protein in the gel migrates in a direction and at a rate that depend on net electric charge and molecular size. After removing the gel from the electric field, it is treated with a solution that contains a specific substrate for the enzyme to be assayed, and a coloring salt that reacts with the product of

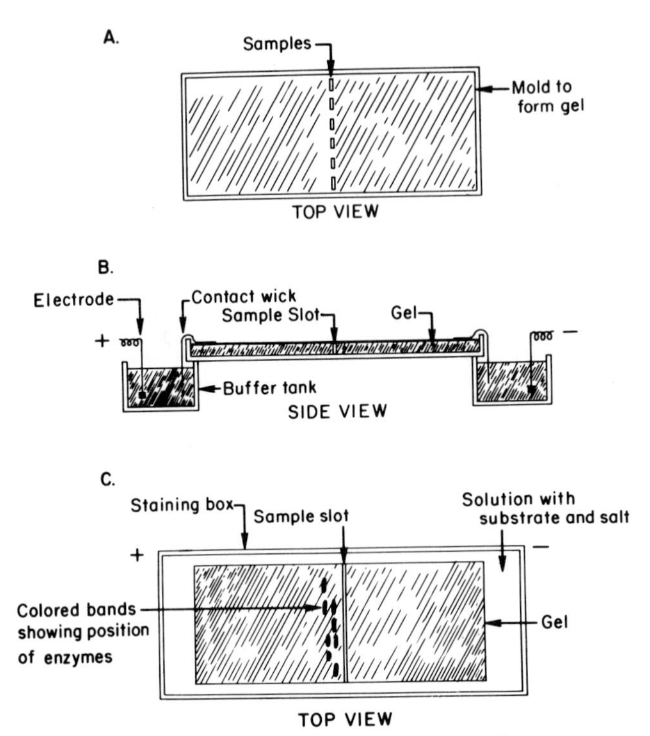

Fig. 1. Techniques of gel electrophoresis and enzyme assay used to measure genetic variation in natural populations. (A) A tissue sample from each of the organisms to be surveyed is homogenized to release the proteins in the tissue. The homogenate supernatants are placed in a gel made of starch, agar, polyacrylamide, or some other jellylike substance. (B) The gel with the tissue samples is then subjected, usually for a few hours, to an electric current. Each protein in the samples thus will migrate in a direction at a rate which depends on the protein's net electrical charge and molecular size. (C) After removing the gel from the electrical field, it is treated with an appropriate chemical solution containing a substrate specific for the enzyme to be assayed, and a salt. The enzyme catalyzes the reaction from the substrate to its product and this product then couples with the salt giving colored bands at the positions where the enzymes had migrated. Enzymes with different mobilities in the electric field have different structures and therefore are coded by different genes. The genotype at the gene locus coding for a given enzyme can thus be established for each individual from the number and position of the bands in the gels, as shown in Figs. 2 and 3.

the reaction catalyzed by the enzyme. At the place in the gel to which the specific enzyme had migrated, a reaction will take place which can be symbolized as follows

$$\text{Substrate} \xrightarrow{\text{Enzyme}} \text{Product} + \text{Salt} \rightarrow \text{Colored spot}$$

The usefulness of the method is due to the fact that the genotypes of the individuals in a sample are simply given by the patterns observed in the gels.

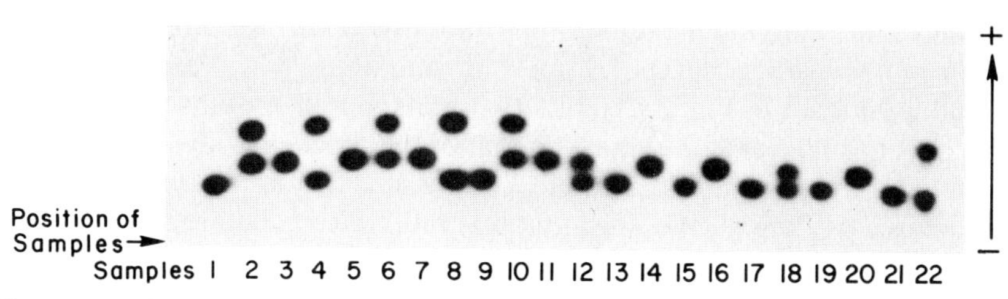

Fig. 2. An electrophoretic gel stained for the enzyme phosphoglucomutase. The gel contains tissue samples from each of twenty-two flies of *Drosophila pseudoobscura*. Flies with only one colored band are inferred to be homozygotes; flies with two bands are inferred to be heterozygotes. Enzymes with different migration are different in amino acid sequence and thus are coded by different alleles. There are three different bands in the gel; which may be represented as 100 (the band closest to the origin), 104 (the intermediate band, which migrates about 4 mm farther than 100), and 112 (the farthest moving band, which migrates about 12 mm farther than 100). These numbers may also be used to represent the alleles coding for the three enzymes. If we represent the gene locus coding for phosphoglucomutase as *Pgm*, the genotypes of the first five flies, starting from the left, are as follows: $Pgm^{100/100}$, $Pgm^{104/112}$, $Pgm^{104/104}$, $Pgm^{100/112}$ and $Pgm^{100/104}$. The genotypes of the other seventeen flies can be inferred from the gel patterns in the same fashion.

Two illustrative gels are shown in Fig. 2 and 3. The gel in Fig. 2 contains the homogenates of twenty-two *Drosophila pseudoobscura* females assayed for the enzyme phosphoglucomutase; the gene locus coding for this enzyme may be represented as *Pgm*. The first and third individuals in the gel, starting from the left, have enzymes with different electrophoretic mobilities, and thus with different amino acid sequences. Let us represent the alleles coding for the enzymes in the first and third individuals as Pgm^{100} and Pgm^{104}, where the superscripts indicate that the enzyme coded by allele Pgm^{104} migrates 4 mm farther in the gel than the enzyme coded by Pgm^{100}. Since the

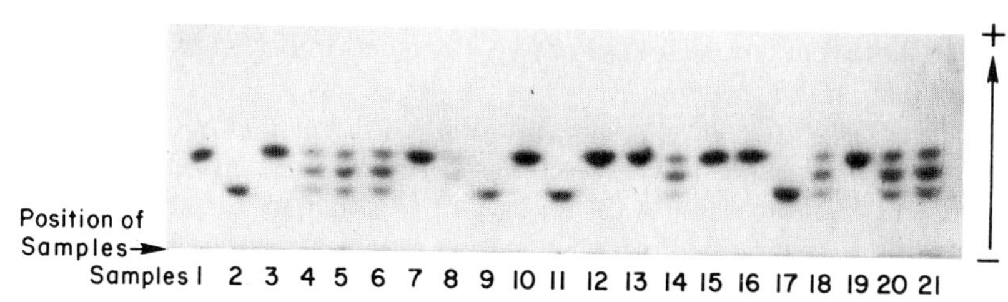

Fig. 3. An electrophoretic gel stained for the enzyme malate dehydrogenase. The gel contains tissue samples from each of twenty-one flies of *Drosophila equinoxialis*. Flies with only one colored band are, as in Fig. 2, inferred to be homozygotes; but the heterozygotes exhibit three bands because malate dehydrogenase is a dimer enzyme. If the gene locus coding for this enzyme is represented as *Mdh*, the genotype of the first and third flies is inferred to be $Mdh^{104/104}$, the genotype of the second fly is inferred to be $Mdh^{94/94}$, flies fourth, fifth and sixth all have the heterozygous genotype $Mdh^{94/104}$, and so on. As in Fig. 2, the numbers representing alleles refer to different amounts of migration of the enzymes coding for them.

first and third individual each exhibit only one band, we infer that they are homozygotes with genotypes $Pgm^{100/100}$ and $Pgm^{104/104}$, respectively. The second individual in Fig. 2 exhibits two phosphoglucomutase bands, one of the bands has the same migration as that of the third individual, but the other migrates 8 mm farther; the allele coding for the latter enzyme may be represented as Pgm^{112}. We infer that this individual is heterozygous with the genotype $Pgm^{104/112}$. Proceeding similarly, the genotypes of the twenty-two *Drosophila pseudoobscura* flies in Fig. 2 going from left to right are as follows (for simplicity, I give only the superscripts identifying the alleles): 100/100, 104/112, 104/104, 100/112, 104/104, 104/112, 104/104, 100/112, 100/100, 104/112, 104/104, 100/104, 100/100, 104/104, 100/100, 104/104, 100/100, 100/104, 100/100, 104/104, 100/100, 100/112.

Figure 3 shows a gel with samples from twenty-one *Drosophila equinoxialis* flies assayed for the enzyme malate dehydrogenase. The first two individuals exhibit only one band each. If we represent the gene locus coding for malate dehydrogenase as *Mdh,* their genotypes can be written as $Mdh^{104/104}$ and $Mdh^{94/94}$. The active form of malate dehydrogenase is, however, a dimer consisting of two units, and as a consequence heterozygous individuals exhibit three bands. Why this is so can be simply understood: if an individual has two types of units, say *a* and *b*, of a dimer enzyme, there are three possible associations of two units, namely, *aa, ab,* and *bb*. In Fig. 3, individuals 4, 5, 6, 8, 14, 18, 20, and 21 are heterozygotes, with genotype $Mdh^{94/104}$, individuals 2, 9, 11, and 17 have genotype $Mdh^{94/94}$, and individuals 1, 3, 7, 10, 12, 13, 15, 16, and 19 have genotype $Mdh^{104/104}$.

Tetramer and even higher order enzymes are known. The genetic interpretation of electrophoretic phenotypes is based on the same principles just illustrated. In general, the number of bands in a heterozygous individual is one more than the number of subunits making up the enzyme (assuming that the subunits are coded by only one gene locus); for example, a heterozygote for a tetramer enzyme will exhibit five electrophoretic bands, namely, *aaaa, aaab, aabb, abbb, bbbb*.

Electrophoretic techniques permit detection of allelic variants in individual genes. Variant as well as invariant gene loci can be identified, and therefore a random sample of genes with respect to variation is possible. Proteins and enzymes for which the appropriate assay techniques exist can be chosen for study without knowing *a priori* whether they are variable or how variable. A moderate number of proteins studied in a moderately large number of individuals is sufficient to estimate the amount of variation over the whole genome in a population. However, not all allelic variants are detectable by gel electrophoresis. Generally, only those amino acid substitutions which alter the net charge of the protein will change its mobility in an electrophoretic gel. Considerations of the genetic code and of electrical properties of amino acids suggest that only about one-third of all amino acid replacements are detectable by gel electrophoresis. Moreover, as pointed out above, a given amino acid may be coded by two or more codons, and therefore not all changes in the nucleotide sequence of the DNA result in different amino acid sequences in the corresponding proteins. Consequently, the amount of allelic variation is underestimated.

E. Statistics for Measuring Genetic Variation

The basic information obtained from electrophoretic surveys of protein variation consists of the genotypic or the allelic frequencies at each locus studied in a given population. For simplicity, only the allelic frequencies are usually given, because the number of different genotypes is usually greater than the number of different alleles; if there are n different alleles at a locus, the number of possible different genotypes is $n(n + 1)/2$.

A variety of measures can be used to express in a single statistic the amount of genetic variation in a population. For a random mating population, the most informative measure is the overall incidence of heterozygosity, which can be expressed in a variety of ways. The proportion of polymorphic loci in a population is another commonly used measure, but one which is less precise and informative.

In random mating populations, the *expected* frequency of heterozygotes, H, at a locus can be directly calculated from the allelic frequencies. If there are n alleles with frequencies $f_1, f_2, f_3, \ldots, f_n$, the expected frequency of homozygotes is simply, $f_1^2 + f_2^2 + f_3^2 \ldots + f_n^2$. Therefore, the expected frequency of heterozygotes is $H = 1 - (f_1^2 + f_2^2 + f_3^2 + \ldots + f_n^2)$. Alternatively, one may measure heterozygosity using the *observed* frequency of heterozygotes, rather than the expected frequency. The observed frequency of heterozygotes is calculated by counting the number of heterozygous individuals, and dividing it by the total number of individuals in a sample. For the sample of flies in Fig. 2, the observed frequency of heterozygotes is $8/22 = 0.364$; for the sample of flies in Fig. 3, it is $8/21 = 0.381$. In large samples of individuals from random mating populations, the observed and the expected frequency of heterozygotes usually agree quite well, although differences may exist because of natural selection or other factors.

Once the heterozygosity, H, has been calculated for each of the gene loci under study, the overall amount of variation in a population is estimated by the average frequency of heterozygotes per locus, $\bar{H}$. This is simply obtained by averaging H over all loci sampled. $\bar{H}$ may be expressed with its standard error, which reflects the amount of heterogeneity among the loci sampled.

The heterozygosity of a population can also be expressed as the average frequency of heterozygous loci per individual, $\bar{H}_i$. This is estimated by averaging (over all individuals) the proportion of heterozygous loci observed in each individual. The values of $\bar{H}$ and $\bar{H}_i$ are the same but their variance and standard error will generally be different. The variance of $\overline{H}_i$ reflects the heterogeneity among individuals for the set of loci studied. The variance of $\overline{H}$ measures heterogeneity among loci. In a random mating population $\overline{H}_i$ will be normally distributed, and its variance will be, relatively speaking, small. There is no *a priori* reason, however, why $\overline{H}$ should be normally distributed, and generally it is not; the variance of $\overline{H}$ is generally larger than the variance of $\overline{H}_i$. In order to estimate the amount of genetic variation in a population *over the whole genome,* $\overline{H}$ with its standard error is preferable since this statistic reflects the great amount of heterogeneity among loci, and its large standard error indicates that $\overline{H}$ may vary in value substantially when different sets of loci are sampled.

Genetic variation can also be measured by the proportion of polymorphic loci, *P,* in a population. This statistic is to a certain extent arbitrary and imprecise. It is arbitrary because it must first be decided when a locus will be considered polymorphic. Two criteria commonly used are (1) when the frequency of the most common allele in a population is no greater than 0.950 and (2) when it is not greater than 0.990. Criterion (1) is more restrictive than criterion (2); every locus polymorphic by criterion (1) is also polymorphic according to (2), but not vice versa.

P is also imprecise, because for each locus it establishes whether it is polymorphic, but not how polymorphic it is. A locus with two alleles with frequencies 0.95 band 0.05, and a second locus with ten alleles each with a frequency of 0.10, contribute equally to *P,* although the latter locus has more genetic variation. If several populations of a species are studied, the average proportion of polymorphic loci per population, $\overline{P}_p$, can be calculated as the average of *P* over all populations. Alternatively, the proportion of populations in which a locus is polymorphic may be calculated first, and the average, $\overline{P}_l$, estimates the average proportion of populations in which a locus is polymorphic. Generally, $\overline{P}_p$ has a smaller variance than $\overline{P}_l$.

F. Estimates of Genetic Variation

The electrophoretic techniques were first applied to estimate genetic variation in natural populations in 1966, when three studies were published, one dealing with man, the other two with *Drosophila*. Numerous populations of insects and other organisms have been surveyed since that time, and many more are studied every year.

As an example, we shall review the results of an extensive study of five neotropical species of *Drosophila* (Ayala *et al.*, 1974a). The kind of information obtained for each gene locus is illustrated in Table I, which shows the allelic variation at four loci in each of three populations of *D. tropicalis*. The four loci have been chosen to illustrate the varying degrees of genetic polymorphism that are usually observed at different loci: *Est-7* is a very

TABLE I

Variation at Four Gene Loci Coding for Enzymes in Three Populations of *Drosphila tropicalis* from Venezuela

		Localities		
Gene locus[a]	Alleles	Catatumbo	Barinitas	Caripito
Est-7	95	0.00	0.00	0.02
	96	0.04	0.02	0.00
	98	0.08	0.14	0.11
	100	0.48	0.64	0.47
	102	0.32	0.18	0.35
	105	0.04	0.02	0.05
	107	0.04	0.00	0.00
	Heterozygosity	0.656	0.538	0.642
Fum	97	0.06	0.07	0.17
	100	0.94	0.93	0.80
	102	0.00	0.00	0.01
	106	0.00	0.00	0.02
	Heterozygosity	0.117	0.130	0.331
Odh	96	0.00	0.02	0.01
	100	0.98	0.97	0.98
	104	0.02	0.01	0.01
	Heterozygosity	0.039	0.059	0.039
Adh	100	1.00	1.00	1.00
	Heterozygosity	0.000	0.000	0.000

[a]*Est-7* codes for an esterase enzyme, *Fum* codes for fumarase, *Odh* codes for octanol dehydrogenase, and *Adh* for alcohol dehydrogenase.

polymorphic gene locus, *Fum* has an intermediate degree of genetic variation, *Odh* has little variation, and *Adh* has no variation in these populations. The heterozygosities are the expected frequencies of heterozygous individuals, H, calculated according to the formula given above. In these populations of *D. tropicalis* the observed and expected frequencies of heterozygotes agree very well.

Table II gives a summary of genetic variation at thirty gene loci coding for enzymes in four populations of *D. tropicalis* from Venezuela. There is great heterogeneity in the amount of genetic polymorphism per locus. At two loci, *Adk-1* and *Est-7*, the heterozygosity is greater than 0.50, while no allelic variation exists at the *Adh* and *To* gene loci. Figure 4 shows the distribution of heterozygosity, H, among all the gene loci studied in five species of *Drosophila* collected from Venezuela. The distribution is unimodal, with the mode at the lower end; about 27% of the loci exhibit little or no allelic variation, while there are several loci with more than 60% heterozygotes. It is apparent from Fig. 4 and Table II why a fairly large number of gene loci must be studied in order to obtain an acceptable estimate of genetic variation

TABLE II

Variation at Thirty Gene Loci Coding for Enzymes in Venezuelan Populations of *Drosophila tropicalis*[a]

Gene locus	Enzyme coded	Frequency of polymorphic populations 1	2	Frequency of heterozygotes
Acph	Acid phosphatase	0.00	0.25	0.018
Adh	Alcohol dehydrogenase	0.00	0.00	0.000
Adk-1	Adenylate kinase	1.00	1.00	0.574
Adk-2	Adenylate kinase	0.00	0.50	0.015
Ald-1	Aldolase	1.00	1.00	0.327
Ald-2	Aldolase	0.75	0.75	0.124
Ao-1	Aldehyde oxidase	1.00	1.00	0.464
Ao-2	Aldehyde oxidase	1.00	1.00	0.313
Est-2	Esterase	1.00	1.00	0.366
Est-3	Esterase	0.25	0.75	0.073
Est-4	Esterase	0.00	0.25	0.005
Est-7	Esterase	1.00	1.00	0.555
Fum	Fumarase	1.00	1.00	0.248
Got	Glutamate-oxaloacetate transaminase	0.00	0.50	0.015
αGpd	α-Glycerophosphate dehydrogenase	0.00	0.25	0.005
G6pd	Glucose-6-phosphate dehydrogenase	0.75	1.00	0.134
Hbdh	Hydroxybutyrate dehydrogenase	0.00	0.50	0.027
Hk-1	Hexokinase	0.00	0.50	0.031
Hk-2	Hexokinase	0.25	0.25	0.019
Hk-3	Hexokinase	0.00	0.25	0.015
Idh	Isocitrate dehydrogenase	0.00	0.25	0.005
Lap-5	Leucine aminopeptidase	1.00	1.00	0.287
Mdh-2	Malate dehydrogenase	0.00	0.00	0.000
Me-1	Malic enzyme	0.25	0.75	0.058
Me-2	Malic enzyme	1.00	1.00	0.176
Odh	Octanol dehydrogenase	0.00	0.75	0.034
Pgm-1	Phosphoglucomutase	0.25	0.75	0.115
To	Tetrazolium oxidase	0.00	0.00	0.000
Tpi-2	Triosephosphate isomerase	0.00	0.50	0.020
Xdh	Xanthine dehydrogenase	1.00	1.00	0.278
Average frequency of polymorphic populations		0.625	0.433	
Average frequency of heterozygotes per locus				0.143

[a]The criteria used to decide whether a population is polymorphic are as follows: (1) when the frequency, p, of the most common allele is no greater than 0.950, and (2) when p is no greater than 0.990.

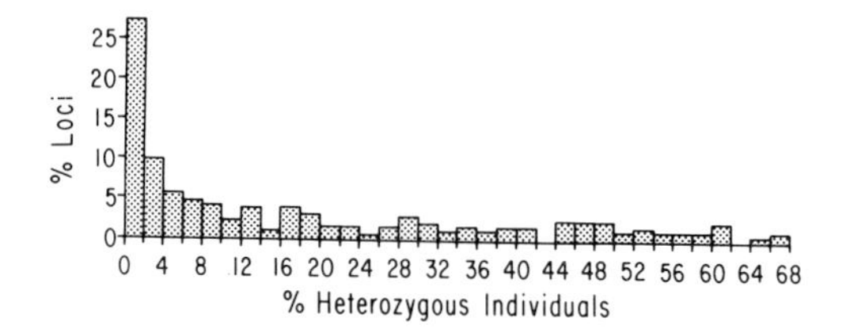

Fig. 4. Distribution of heterozygosity, *H*, among 186 gene loci coding for enzymes in five closely related species: *Drosophila willistoni, D. tropicalis, D. equinoxialis, D. paulistorum*, and *D. nebulosa*. The abscissa represents the expected frequency of heterozygous individuals; the ordinate shows the number of loci within each interval of heterozygosity. The distribution of heterozygosity is unimodal with the mode at the lowest level of heterozygosity. Because the distribution of *H* is so widespread, it becomes necessary to sample a fair number of gene loci in order to obtain unbiased estimates of genetic variation in a given population.

in a given population or a species. If only a few loci are surveyed, a very distorted estimate might be obtained, because of the great heterogeneity among loci. Experience with electrophoresis indicates that a sample of about twenty gene loci is required and is usually sufficient. Estimates of genetic variation generally change little when the number of loci studied increases beyond twenty.

Table II gives the average frequency of polymorphic populations per locus, namely, 0.625 or 0.433, depending on whether criteria (1) or (2) are used, and the average frequency of heterozygous individuals per locus, namely,0.143. There is a great deal of genetic polymorphism in *D. tropicalis*. High levels of genetic polymorphism are also observed in four other neotropical species of *Drosophila* surveyed in the study under review (Ayala *et al.*, 1974a). A summary of the genetic variation found in all five species is given in Table III. On the average, the frequency of heterozygous individuals per locus is 0.177; this average is also an estimate of the average frequency of heterozygous loci in an individual, i.e., at nearly one out of every five gene loci a fly will have two different alleles.

The amount of genetic variation observed by electrophoretic studies in other insect species as well as in other organisms is generally quite large, although it varies from species to species. Twenty or more gene loci have been studied in each of more than thirty species of *Drosophila;* the average heterozygosity for all these species in 0.150. The average heterozygosity in many other insects is of about the same magnitude as in *Drosophila*. Among species in which twenty or more loci have been studied, the greatest degree of genetic polymorphism has been observed in sexually reproducing populations of the coleopteron, *Otiorrhynchus scaber*, for which $\bar{H} = 0.309$. Other invertebrates have heterozygosities comparable to those observed in insects. Vertebrates, however, are genetically less polymorphic than invertebrates: the average heterozygosity among sixty-eight well-studied vertebrate

TABLE III

Summary of Genetic Variation in Venezuelan Populations of Five Species of *Drosophila*[a]

Species	Loci studied	Populations sampled	Frequency of polymorphic populations 1	Frequency of polymorphic populations 2	Frequency of heterozygotes
D. willistoni	31	6	0.699 ± 0.066	0.462 ± 0.079	0.177 ± 0.039
D. tropicalis	30	4	0.625 ± 0.069	0.433 ± 0.084	0.143 ± 0.032
D. equinoxialis	31	5	0.783 ± 0.059	0.540 ± 0.075	0.165 ± 0.031
D. paulistorum	32	4	0.688 ± 0.064	0.503 ± 0.077	0.203 ± 0.026
D. nebulosa	30	2	0.667 ± 0.073	0.550 ± 0.073	0.195 ± 0.037
Average	30.8 ± 0.4	4.2 ± 0.7	0.692 ± 0.026	0.498 ± 0.022	0.177 ± 0.004

[a]The criteria to decide whether a population is polymorphic are as in Table II. The standard error is given after each mean value.

species is 0.060, less than half the average heterozygosity of invertebrates. In man, seventy-one gene loci have been studied with an average heterozygosity of 0.067, which is not very different from the vertebrate average.

The staggering amount of genetic variation found in natural populations has some important implications. One implication is that there is ample opportunity for evolution to occur. Thus, it is not surprising that whenever a new environmental challenge appears—whether due to a climatic change, to man-made pollution, to the introduction of a predator, parasite, or competitor, or to any other cause—populations usually are able to adapt to it. The evolution of resistance of insect species to pesticides in recent decades provides a spectacular example. The story is always the same: when a new insecticide is introduced a certain relatively small amount is sufficient to achieve a satisfactory control of the insect pest; the necessary concentration, however, gradually increases, until it becomes totally inefficient or economically impractical. Insect resistance to a pesticide was first reported in 1947 for the housefly, *Musca domestica,* with respect to DDT. Since then, resistance to one or more pesticides has been recorded in at least 225 species of insects and other arthropods. The genetic variants required for resistance to the most diverse kinds of pesticides were apparently present in every one of the populations subject to the pesticides.

Another significant implication of the high levels of genetic polymorphism is that in outcrossing organisms no two individuals (with the trivial exception of twins developed from a single fertilized egg) are genetically identical. Consider a typical insect species, say of *Drosophila,* with an average heterozygosity of 0.15. If we assume that the genome of *Drosophila* consists of 20,000 gene loci coding for proteins, it follows that a fly would be heterozygous at about 3000 structural genes. Such an individual can potentially produce $2^{3000} > 10^{900}$ different kinds of gametes. Even it we assume that the number of structural gene loci in *Drosophila* is only 5000 (almost certainly an underestimate), the number of different gametes that can potentially be produced by a *Drosophila* fly would be $2^{750} > 10^{225}$, a number immensely large. (The number of atoms in the known universe is estimated to be "only" 10^{70}.)

The implications of electrophoretic measures of genetic variation become even more dramatic when we realize that electrophoresis underestimates the amount of genetic variation. It was pointed out above that the genetic code is redundant: not all nucleotide differences (particularly in the third codon position) result in amino acid differences in proteins. Moreover, electrophoresis distinguishes proteins with different amino acid composition by their differential migration in an electric field. But not all amino acid substitutions in a protein result in different electrophoretic mobilities, and, therefore, proteins may be thought identical when they are not. Although we are certain that electrophoretic measures of variation are underestimates, at present we do not yet know by how much.

III. GENETIC DIFFERENTIATION DURING SPECIATION

A. Anagenesis and Cladogenesis

The process of evolution has two dimensions: *anagenesis* or phyletic evolution, and *cladogenesis* or splitting. Changes occurring in a population, or group of populations, as time proceeds are anagenetic or phyletic evolution. As a rule they result in increased adaptation to the environment, and often reflect changes in the physical and biotic conditions of the environment. Cladogenesis occurs when a lineage of descent splits into two or more independently evolving lineages. The great diversity of the living world is the result of cladogenetic evolution.

It might seem that there is no way to measure in any precise way the amount of genetic change that has taken place in phyletic evolution. The ancestral organisms of a living species died in the past; they cannot be intercrossed with their living descendants in order to ascertain how many genetic changes have taken place. The DNA, RNA, and proteins of ancestral organisms are not usually preserved in their original constitution.

There is, however, a way to ascertain the amount of anagenetic change, namely, by comparing two or more species that share a common ancestor. Assume that it is found that two species, S_1 and S_2, differ at a certain proportion, x, of their genes. If we ignore (or have a way to correct for) parallel and convergent changes, as well as the fact that a gene may have changed more than once, x is an estimate of the amount of anagenetic change that has taken place in the phyletic lineage going from the last common ancestral population to S_1, plus the anagenetic change in the phyletic lineage going from the last common ancestor up to S_2. To a first approximation one may, then, assume that $x/2$ genetic changes took place in each of the two phyletic lineages.

It is in fact possible to obtain more precise estimates of the fraction of the x genetic changes that took place in each lineage by comparing S_1 and S_2 with a third species, S_3, which shares an older common ancestor with the other two species. This situation is illustrated in Fig. 5. Assume that a number of gene loci are compared between S_1, S_2, and S_3, and that the number of differences found are 8 between S_1 and S_2, 13 between S_1 and S_3, and 11 between S_2 and S_3. The amount of genetic change from common ancestor A to S_1 plus the amount of genetic change from A to S_2 is 8. However, more change must have taken place between A and S_1 than between A and S_2 because S_1 is more different from S_3 than S_1. (Note that whatever amount of change has taken place from B to A, and from B to S_3, will affect identically the comparisons between S_1 and S_3, and between S_2 and S_3.) The most reasonable interpretation of the data is to assume, as shown in Fig. 5, that 5 genetic changes took place between A and S_1, only 3 between A and S_2, and a total of 8 between B and A plus between B and S_3.

Anagenetic evolutionary change can, therefore, be determined simply by studying cladogenetic change, i.e., the genetic differences between extant species. But how to determine the amount of genetic differentiation and similarity between different species? This could not be done using the traditional methods of genetic analysis. As pointed out in Section II,C above, Mendelian genetics ascertains the presence of genes by studying segregation in the progenies of crosses between individuals carrying different alleles. But individuals of different species do not usually produce viable progeny. Fortunately, the advances of molecular genetics and biochemistry during the last 25 years have made possible genetic comparisons between

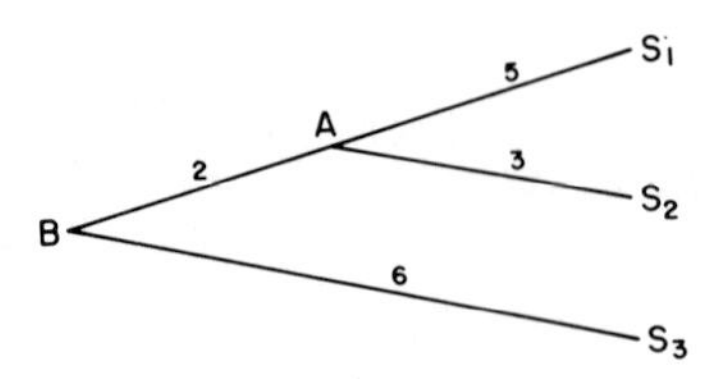

Fig. 5. A hypothetical phylogeny of three living species, S_1, S_2, and S_3. A is the last common ancestral species to S_1 and S_2; B is the last common ancestral species to S_1, S_2, and S_3. The phylogeny shown is the most likely one if the only information available is that the number of genetic differences are: 8 between S_1 and S_2, 13 between S_1 and S_3, and 11 between S_2 and S_3. The figure also gives the inferred number of genetic changes in each branch.

different species. As discussed earlier, genes with identical nucleotide sequences code for identical proteins, while proteins with different amino acid sequences are coded for by different genes. In recent years, gel electrophoresis, protein sequencing, and other biochemical techniques have provided estimates of genetic changes in anagenesis as well as in cladogenesis.

B. Geographic Speciation

Among cladogenetic processes the most decisive is speciation, the process by which one species splits into two or more species. In outbreeding sexual organisms, a species is an array or Mendelian populations among which gene exchange can occur without impediments other than geographical separation. A mutation or other genetic change originating in a single individual of a species may spread to all members of the species by natural selection or genetic drift. However, owing to reproductive isolation, such a change cannot be passed on to different species. Speciation is, then, a highly significant stage of evolutionary differentiation: species are discrete and independent evolutionary units. The question of how much genetic differentiation occurs during the process of speciation is, therefore, a most fundamental one in evolutionary genetics.

There are a number of ways in which new species may come about including, for example, polyploidy. In insects (and in outcrossing sexual organisms in general) the most common mode of speciation requires, for its inception, that populations become geographically isolated. If there is little or no gene flow between them, populations may gradually become genetically differentiated particularly as a result of their adaptation to different environments. Different races arise by this process which is a necessary albeit not sufficient condition for species differentiation. It is not sufficient because the process of geographic or race differentiation is reversible. Not every race is a future species; without reproductive isolation, differentiated populations may converge or fuse by gene exchange. This is precisely what is happening in the human species.

What processes bring about the development of reproductive isolating mechanisms? Two, not mutually exclusive, answers have been suggested. One answer is that reproductive isolation is a by-product of the accumulation of genetic differences between geographically separated diverging populations. There is little doubt that if the process of genetic divergence continues long enough, geographically separated populations may become so different as to be unable to interbreed if the opportunity would eventually arise.

The second answer considers the conditions under which the development of reproductive isolation might be accelerated. If two populations geographically isolated for some time come again into geographic contact, two outcomes are possible: (1) a single gene pool comes about, because the populations hybridize readily without significant loss of fitness (or because one of the populations is eliminated by the other through ecological competition); and (2) two species ultimately arise because natural selection favors the development of reproductive isolation between the populations. If matings between individuals of different populations leave progenies with reduced fertility or viability, natural selection would favor genetic variants promoting matings between individuals of the same population. Reproductive isolation, and therefore speciation, may ensue.

Which one of the two possible outcomes just mentioned will, in fact, occur in a particular instance depends, of course, on the degree of genetic differentiation achieved by the populations previous to regaining sympatry. There is ample evidence showing that natural selection often favors the development of reproductive isolation among genetically differentiated populations when these exchange genes by hybridization.

The question ''How much genetic differentiation is concomitant to the process of speciation?'' needs, therefore, to be partitioned into two separate parts. The first concerns the amount of genetic differentiation during the first stage of speciation, when allopatric populations of the same species become differentiated to the point that they are likely to evolve into different species if they come into geographic contact. The second refers to the amount of genetic differentiation during the second stage of speciation, when reproductive isolation is completed by natural selection between genetically differentiated populations that have come into geographic contact.

C. A Paradigm of the Process of Speciation

The most extensive and complete study of genetic differentiation during the process of geographic speciation has been carried out in the *Drosophila willistoni* group of species (summary in Ayala *et al.*, 1974b). The *willistoni* group of *Drosophila* consists of at least fifteen closely related species endemic to the tropics of the New World. Six species are siblings, morphologically nearly indistinguishable, although the species of individual males can be identified by slight but diagnostically reliable differences in their genitalia. Two sibling species, *D. insularis* and *D. pavlovskiana,* are narrow endemics; the former in some islands of the Lesser Antilles, and the latter in Guyana. Four other siblings, namely, *D. willistoni, D. equinoxialis, D. tropicalis,* and *D. paulistorum* have wide, and largely overlapping geographic distributions through Central America, the Caribbean, and much of continental South America (Fig. 6).

Some sibling species consist of at least two subspecies. Populations of *D. willistoni* west of the Andes near Lima, Peru, belong to the subspecies *D. w.*

Fig. 6. Geographic distribution of four sibling species of *Drosophila* living in the American tropics. All four species are morphologically virtually indistinguishable, although they are completely reproductively isolated. Two or more species are sympatric throughout much of the common territory in which these species are found.

quechua, while east of the Andes and elsewhere the subspecies is *D. w. willistoni*. Incipient reproductive isolation in the form of partial hybrid sterility exists between these two subspecies. Laboratory crosses of *D. w. willistoni* females with *D. w. quechua* males yield fertile males and females. However, crosses between *D. w. quechua* females and *D. w. willistoni* males from continental South America east of the Andes produce fertile females but sterile males. Laboratory tests show no evidence of ethological (sexual) isolation between the subspecies.

Two subspecies are also known in *D. equinoxialis: D. e. caribbensis* in Central America, north of Panama, and in the Caribbean islands; and *D. e. equinoxialis* in eastern Panama and continental South America. Crosses between the two subspecies yield fertile females but sterile males, independently of the subspecies of the female parent. As in *D. willistoni,* there is no evidence of sexual isolation between the subspecies of *D. equinoxialis*.

Evolutionary divergence beyond the taxonomic category of subspecies, but without complete achievement of speciation exists in a third sibling species, *D. paulistorum*. This "species" consists of at least six semispecies, or incipient species, named Centroamerican, Transitional, Andean–Brazilian, Amazonian, Orinocan, and Interior. Laboratory crosses between the semispecies generally yield fertile females but sterile males. Reproductive isolation between some semispecies is essentially complete, so that two and even three semispecies coexist sympatrically in many localities (Fig. 7). Sexual isolation is essentially complete between some semispecies, particularly when sympatric populations are tested. Gene flow among the semispecies is nevertheless possible, particularly through populations of the Transitional semispecies.

In summary, five increasingly divergent levels of evolutionary divergence or cladogenesis can be recognized in the *D. willistoni* group:

1. Between geographic populations of the same taxon.

2. Between subspecies. These are allopatric populations that exhibit incipient reproductive isolation in the form of partial hybrid sterility. If populations of two subspecies were to come into geographic contact, intersubspecific matings would leave fewer fertile descendants than intrasubspecific matings. Therefore, natural selection would favor the development of reproductive isolating mechanisms between the subspecies. Whether two species would ultimately result, or one subspecies be absorbed (by introgression) or eliminated (by competition), we of course do not know. In any case, the subspecies are allopatric populations in the first stage of the speciation process.

3. Between the semispecies of *D. paulistorum*. The process of speciation is being completed between the semispecies. Sexual isolation is being superimposed over the preexisting hybrid sterility, and is nearly complete in many cases; some semispecies are sympatric in many localities without having any substantial gene exchange. The semispecies are populations in the second stage of the speciation process.

4. Between sibling species. In spite of their morphological similarity, the sibling species are completely reproductively isolated. Study of genetic differentiation between them will show how much genetic differentiation may occur after speciation without noticeable morphological diversification.

5. Between morphologically distinguishable species of the same group. *D. nebulosa* is a close relative of *D. willistoni* and its siblings, but can be

Fig. 7. Geographic distribution of the six semispecies of *Drosophila paulistorum*. The semispecies represent groups of populations in the second stage of speciation, when reproductive isolation is being completed. Male hybrids between different semispecies are sterile. Sexual isolation between semispecies is generally quite pronounced, and is complete in places where two or three semispecies are sympatric.

easily distinguished from them by external morphology. Comparison of *D. nebulosa* with the siblings will show how much genetic differentiation occurs at this level of evolutionary divergence.

D. Biochemical Differentiation in the *Drosophila willistoni* Group

Using electrophoretic techniques, thirty-six gene loci coding for enzymes have been studied in each of the sibling species of the *D. willistoni* group and in the morphologically differentiated *D. nebulosa*. The genotypes of large numbers of individuals (from several hundred to several thousand) have been ascertained at each locus in each species, except for the two narrow endemics, *D. insularis* and *D. pavlovskiana,* of which only a few genomes were sampled. Genetic similarity and differentiation between populations are measured using two statistics: *I (genetic identity)* which estimates the proportion of genes that remain identical in two populations, and *D (genetic distance)* which estimates the proportion of gene substitutions that have taken place in the separate evolution of two populations. *I* may range in value from zero to one; *D* may range in value from zero to infinity, because it is possible that the complete replacement of one allele (or sets of alleles) by another may have happened more than once at any gene locus (Ayala *et al.*, 1974b).

The results obtained are summarized in Table IV. The first level of comparison is between local populations of the same taxon. These are populations that give no evidence of having developed any degree of reproductive isolation: when intercrossed they produce completely fertile and viable offspring. There is very little genetic differentiation among local populations of the *D. willistoni* group, $I = 0.970$; $D = 0.031$.

The second level of comparison is between subspecies. The amount of genetic differentiation is substantial. On the average, $D = 0.230$—about 23 complete gene substitutions have taken place for every 100 loci in the separate evolution of two subspecies, i.e., during the first stage of speciation. The genetic distance is $D = 0.246$ between the two subspecies of *D. equinoxialis,* and $D = 0.214$ between the two subspecies of *D. willistoni.* This agrees well with the degree of evolutionary divergence within each pair of subspecies, because as pointed out in Section III, C crosses between the *D. equinoxialis* subspecies always produce sterile male hybrids, but crosses between the *D. willistoni* subspecies produce sterile male hybrids only when the mothers are from the *quechua* subspecies.

The third level of comparison is between the semispecies of *D. paulistorum,* which are populations in the second stage of speciation. The average genetic distance between the semispecies is $D = 0.226$, not significantly

TABLE IV

Average Genetic Identity, *I*, and Genetic Distance, *D*, between Taxa of Various Levels of Evolutionary Divergence in the *Drosophila willistoni* Group[a]

Taxonomic level	*I*	*D*
Local populations	0.970 ± 0.006	0.031 ± 0.007
Subspecies	0.795 ± 0.013	0.230 ± 0.016
Semispecies	0.798 ± 0.026	0.226 ± 0.033
Sibling species	0.563 ± 0.023	0.581 ± 0.039
Nonsibling species	0.352 ± 0.023	1.056 ± 0.068

[a]The standard error is given after each mean value.

different from the value observed between subspecies. It appears that the second stage of speciation, when sexual isolation develops, does not require changing a large fraction of the genes in addition to those differentiating subspecies.

The fourth level of comparison is between the sibling species. The mean genetic distance between them is $D = 0.581$, i.e., on the average about 58.1 electrophoretically detectable allelic substitutions for every 100 loci have occurred in each pair of siblings since their divergence from a common ancestral population. In spite of their morphological and ecological similarity, these species are genetically quite different.

The fifth and final level of comparison is between morphologically different (but closely related) species, i.e., between *D. nebulosa* and each one of the sibling species. The average genetic distance for these comparisons is $D = 1.056$; thus, on the average, about one electrophoretically detectable allelic substitution per locus has taken place in the separate evolution of *D. nebulosa* and each of the sibling species since they diverged from the last common ancestors.

We may summarize the results of the study just reviewed as follows. Apparently, considerable genetic differentiation takes place during the first stage of speciation, when incipient reproductive isolation in the form of partial hybrid stability has developed. However, little additional genetic differentiation seems to take place during the second stage of geographic speciation, when reproductive isolation is being completed. On the average the semispecies of *D. paulistorum* are not genetically more different from each other than subspecies are, although reproductive isolation has in fact been completed among semispecies in the places where they are sympatric. Sexual (ethological) isolation is the main mechanism maintaining reproductive isolation between *Drosophila* species. Perhaps sexual isolation may come about by changing only a few genes that affect courtship and mating behavior.

An alternative explanation may, however, be advanced for the observation of little additional genetic differentiation during the second stage of speciation, namely, that a substantial fraction of genes may evolve at this stage, but they are not the kinds of genes that are studied by electrophoretic methods. The genes studied by electrophoresis code for enzymes involved, for the most part, in basic cell metabolism. Conceivably, such genes might not be the genes that affect courtship and mating behavior, which might be modified, for example, primarily by regulatory genes.

After reproductive isolation is completed, species evolve independently. As expected, they continue to diverge genetically. The average genetic distance between sibling species is more than twice as large as that between subspecies and semispecies. Morphologically distinct species are genetically much more different than sibling species.

How general are the results obtained with the *D. willistoni* group? Is it the case for other insects as well as other organisms that the first stage of geographic speciation involves substantial genetic change while little genetic differentiation takes place in the second stage? Many studies have, in recent years, investigated the amount of genetic differentiation between populations at various stages of speciation or between closely related species. We cannot examine these works in detail here, but a recent review can be found in Ayala (1975).

The organisms surveyed include mammals, reptiles, amphibians, and fishes among vertebrates. Among invertebrates most of the work has involved *Drosophila,* although comparisons have been also made between closely related species of butterflies and cicadas. Not surprisingly, the amount of genetic differentiation found at a given level of taxonomic divergence (local populations, subspecies, semispecies, species) varies from group to group of organisms. Yet the results obtained in the *D. willistoni* group appear to have considerable generality. On the basis of electrophoretically detectable differences, the genetic distance, *D,* between subspecies or between semispecies usually ranges between 0.10 and 0.25, while between closely related species *D* is around 0.40 or greater in most cases.

The results summarized in the previous paragraph, as well as those obtained in the *D. willistoni* group, refer to the geographical mode of speciation. The situation is expected to be different whenever polyploidy or other modes of "saltational" speciation are involved. The origin of new species by polyploidy is a common phenomenon in plants but rare in animals. Among insects polyploid species are known in cases where the females are parthenogenetic, including some beetles, moths, and sow bugs. Adequate genetic comparisons between diploid species and their polyploid derivatives are lacking, but there is little doubt that few or no allelic differences need to be involved; a polyploid organism, whether it is an autopolyploid or an allopolyploid, will initially possess all the genes present in its ancestor(s) and no new ones. New species may also arise through the fixation in one or a few organisms of translocations or other chromosomal rearrangements. In these cases also, speciation may come about with little or no genetic change in the form of allelic differences at individual gene loci.

IV. RECONSTRUCTION OF EVOLUTIONARY HISTORY

A. Conceptual Basis of Comparative Studies

The study of rates of evolution as well as the reconstruction of cladogenetic events (i.e., phylogenetic history) depends on several complementary sources of evidence. On the one hand, we have remnants or organisms living in the past, the "fossil record," which sometimes provides definitive evidence of evolutionary events, but is on the whole far from complete and often seriously deficient.

Comparative studies of living forms also provide information about anagenesis and cladogenesis. Comparative anatomy is the branch of science that in the past contributed most information about evolutionary events on the basis of the study of living organisms, but additional knowledge has been obtained from embryology, cytology, ethology, biogeography, and other biological disciplines.

In recent years, the comparative study of informational macromolecules (proteins and nucleic acids) has become a powerful tool for the study of evolution. Indeed, comparative biochemistry is rapidly becoming the single most informative branch of biology for studying evolutionary history.

The rationale of the comparative studies of living organisms in order to ascertain evolutionary history is simple. Evolution is a gradual process; consequently, organisms sharing a recent common ancestor are likely to be more similar than organisms with a common ancestor only in a more remote past. Relative degrees of similarity can, therefore, be used to infer the recency of common ancestry. However, the ascertainment of phylogenetic history is a task far from simple: rates of evolutionary change may be different at different times, or in different groups of organisms, or with respect to different features of the same organisms. Moreover, resemblances due to common descent must be set apart from resemblances due to similar ways of life, to life in the same or similar habitats, or to accidental convergence. It is not possible to discuss here in detail all these potential sources of error, and how to correct for them, but we shall briefly consider some of the biochemical techniques used to infer the evolutionary history of insects as well as of other organisms.

B. Electrophoretic Phylogenies

In Section III it was shown how the electrophoretic study of a number of proteins can be used to measure the amount of genetic differentiation at various stages of evolutionary divergence. The same information can be used to construct a dendrogram (i.e., a treelike diagram) that reflects genetic affinities among taxa. If it is assumed that the greater the degree of genetic differentiation between any two taxa the longer the time since their divergence from a common ancestor, then the dendrogram may be interpreted as a diagram of phylogenetic history.

Consider the study described in Section III in which thirty-six gene loci coding for enzymes in the *Drosophila willistoni* group of species were studied. Such a study provides us with a matrix of distances, D, for all pairwise comparisons between the taxa of the group. Various mathematical methods may be employed to construct dendrograms based on measures of genetic distance. Different methods do not always yield identical outcomes, but the resulting dendrograms are at least fairly similar. The idea behind these methods is to cluster the taxa, so that the most similar any two taxa are the more closely connected they are in the cluster.

Figure 8 shows a dendrogram of the *D. willistoni* group based on the genetic distances obtained by the electrophoretic studies described in Section III. The dendrogram also shows the amount of genetic change estimated between any two branching points or between a branching point and a living taxon. The figure is constructed so that the vertical distances between

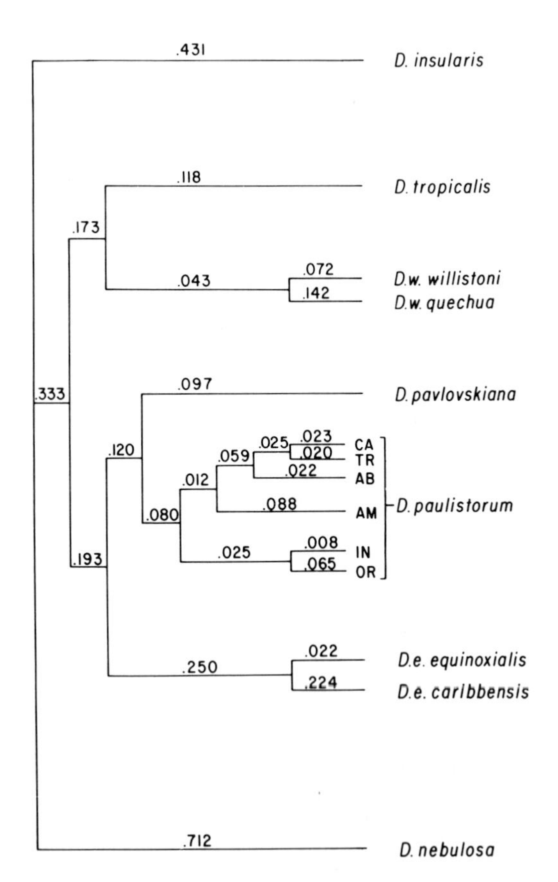

Fig. 8. Phylogeny of the species related to *Drosophila willistoni* based on electrophoretic differences at thirty-six gene loci coding for enzymes. The number on the branches are the estimated allelic substitutions (detectable by electrophoresis) per locus that have taken place in evolution. There are seven species in the phylogenetic tree. Two species, *D. willistoni* and *D. equinoxialis* consists of two subspecies each. *D. paulistorum* represents a complex of six semispecies or incipient species.

neighboring taxa are approximately proportional to their genetic distance. We see that subspecies are closely clustered indicating their recent evolutionary divergence; the incipient species of the *D. paulistorum* complex are also closely clustered. The most divergent species is *D. nebulosa* which is also morphologically clearly different from the other species.

It is worth pointing out that prior to the development of electrophoretic techniques, the taxa of the *D. willistoni* group were studied for several decades with respect to their reproductive affinities, geographic distribution, ecology, chromosomal arrangements, sexual behavior, and morphological traits. A phylogeny constructed on the basis of these nonbiochemical studies had the same configuration as the phylogeny later obtained using electrophoretic data. This coincidence reinforces the validity of the two independent results. Needless to say, biochemical evidence of phylogeny should always be incorporated with other available evidence.

C. Immunology and Protein Sequencing

The techniques of gel electrophoresis are useful in the study of phylogeny only when the taxa compared are genetically not very different, i.e., when species of the same genus or closely related genera are compared. In electrophoretic gels, two forms of a protein are judged to be the same if they have identical migrations. Whenever two proteins migrate at different rates, we know that they are structurally different but we do not know whether they differ in one, two, or several amino acid replacements. The average number of amino acid replacements per protein is estimated from the proportion of proteins that are identical in two species, by assuming that the number of replacements follows a Poisson distribution. That is, estimates of genetic distance between taxa are based on the zero class (the proportion of unchanged loci) of the Poisson distribution. If two taxa are electrophoretically different in every protein, the zero class is empty and no estimates can be made of the average number of amino acid substitutions per protein (= number of allelic replacements per locus). If the proportion of proteins with identical electrophoretic mobility (i.e., the zero class in the Poisson distributions) is very small, estimates of genetic distance are unreliable because they are potentially subject to large errors. Moreover, when genetically very different taxa are compared, it is likely that proteins with ostensibly identical electrophoretic mobilities may in fact differ in amino acid composition, since not all amino acid replacements are detectable by electrophoretic techniques.

There are biochemical techniques that can be used to estimate the degree of genetic differentiation between species even when these are not closely related. One such technique is DNA hybridization which will be discussed in the next section. Here, we shall consider two techniques that, like electrophoresis, are based on the study of proteins: immunological methods and amino acid sequencing.

Estimates of the degree of similarity between proteins can be obtained by immunological techniques such as immunoelectrophoresis, immunodiffusion, quantitative precipitation, complement fixation, and turbidimetry. In outline, the immunological comparison of proteins is performed as follows. A protein is purified from an animal, say the housefly *Musca domestica*. The purified protein is injected into a mammal such as a rabbit. The rabbit develops an immunological reaction and produces antibodies against the foreign protein (antigen). The antibodies produced by the immunized rabbit will therafter react not only against the specific antigen used (the housefly protein in the example), but also against other related proteins (such as similar proteins from other diptera). The greater the similarity between the protein used to immunize the rabbit and the protein tested, the greater the extent of the immunological reaction. The degrees of dissimilarity between the protein used in the original immunization and similar proteins from different species are expressed as "immunological distances." Immunological distances can be used to construct dendrograms in a similar way as genetic distances obtained by electrophoresis.

Immunological studies of albumin, lysozyme, and other proteins have been used to infer phylogenetic relationships among mammals and other vertebrates, but to date immunological techniques have been little applied to phylogenetic studies of insects.

The amino acid sequence of proteins provides more precise information than immunological tests about genetic differences between species. The primary structure of a polypeptide is determined by the nucleotide sequence in the DNA coding for it. The number of differences between two homologous polypeptides or proteins reflects the number of differences in the corresponding genes (although because of the redundancy of the genetic code not all nucleotide changes in the DNA result in amino acid substitutions).

The common procedure to establish the amino acid sequence of a polypeptide requires, first, breaking the protein into small fragments or peptides, and then determining the amino acid sequence in each peptide. The polypeptide is broken into fragments with enzymes that hydrolyze the bonds between contiguous amino acids at specific sites. The resulting peptides are separated from each other using procedures such as column chromatography and two-dimensional paper chromatography. The amino acid sequence in each peptide is ascertained usually with the help of an apparatus known as a "protein sequencer" or "sequenator."

The procedure just described is carried out at least twice for each polypeptide to be sequenced, using two enzymes that hydrolyze the protein at different points. Trypsin and chymotrypsin are the most commonly used enzymes. Trypsin hydrolyzes the bonds between the carboxyl group of either lysine or arginine and the amino group of the contiguous amino acid. Chymotrypsin hydrolyzes polypeptides at the carboxyl ends of either tryptophan, phenylalanine, or asparagine. Two different sets of peptides are obtained with trypsin and chymotrypsin. The complete sequence of the protein is determined from the overlaps between the amino acid sequences of the two sets of peptides.

The first protein to be sequenced was insulin, that consists of fifty-one amino acids. During the early 1950's it was shown that the amino acid sequences of the insulins of cattle, pig, sheep, horse, and sperm whale are identical except for replacements in three consecutive amino acids. The procedures to establish the primary sequence of proteins are extremely laborious. Yet, more than 500 sequences or partial sequences are presently known through the efforts of scores of investigators working in many laboratories. Many additional protein sequences are determined every year.

Cytochrome *c* is a protein involved in cell respiration that in higher animals and plants is found in the mitochondria. The amino acid sequences of the cytochromes *c* of several dozen organisms, from yeast to man and including such insects as screw worm flies and some moths, are known and have been used to reconstruct the phylogeny of these organisms. One way to do this reconstruction is to calculate the number of amino acid differences between any two species, and to obtain a dendrogram based on these differences.

A second procedure, which yields more information than the first, makes use of the genetic code. The replacement of one amino acid by another may required as a minimum either one, or two, or three nucleotide substitutions in the corresponding DNA triplet. The minimum number of nucleotide differences is determined for each amino acid replacement between two proteins, and all such differences are added. This yields the minimum number of nucleotide changes that have occurred in the separate evolutions of the genes coding for the two proteins being compared.

Table V gives the minimum number of nucleotide differences between the genes coding for cytochrome *c* in twenty different species. Dendrograms may be constructed using matrices of nucleotide (or amino acid) differences. Figure 9 is a phylogeny inferred from the data given in Table V (Fitch and Margoliash, 1967). Overall, the relationships shown in Fig. 9 agree fairly well with the phylogeny of the organisms involved as known from the fossil record and other sources. There are, however, disagreements: chickens appear more closely related to penguins than to ducks and pigeons; the turtle, a reptile, appears more closely related to birds than to the rattlesnake; men and monkeys diverge from the other mammals before the marsupial kangaroo separates from placental mammals. In spite of these and other erroneous relationships, it is remarkable that the study of a single protein yields a fairly accurate representation of the phylogeny of twenty organisms

TABLE V

Minimum Numbers of Nucleotide Differences (Inferred from Amino Acid Sequences) among the Genes Coding for Cytochromes c of Twenty Organisms

	1	2	3	4	5	6	7	8	9	10	11	12	13	14	15	16	17	18	19
1. Man																			
2. Monkey	1																		
3. Dog	13	12																	
4. Horse	17	16	10																
5. Donkey	16	15	8	1															
6. Pig	13	12	4	5	4														
7. Rabbit	12	11	6	11	10	6													
8. Kangaroo	12	13	7	11	12	7	7												
9. Duck	17	16	12	16	15	13	10	14											
10. Pigeon	16	15	12	16	15	13	8	14	3										
11. Chicken	18	17	14	16	15	13	11	15	3	4									
12. Penguin	18	17	14	17	16	14	11	13	3	4	2								
13. Turtle	19	18	13	16	15	13	11	14	7	8	8	8							
14. Rattlesnake	20	21	30	32	31	30	25	30	24	24	28	28	30						
15. Tuna	31	32	29	27	26	25	26	27	26	27	26	27	27	38					
16. Screw worm fly	33	32	24	24	25	26	23	26	25	26	26	28	30	40	34				
17. Moth	36	35	28	33	32	31	29	31	29	30	31	30	33	41	41	16			
18. *Neurospora*	63	62	64	64	64	64	62	66	61	59	61	62	65	61	72	58	59		
19. *Saccharomyces*	56	57	61	60	59	59	59	58	62	62	62	61	64	61	66	63	60	57	
20. *Candida*	66	65	66	68	67	67	67	68	66	66	66	65	67	69	69	65	61	61	41

as diverse as those shown in Fig. 9. No presently available method, other than protein sequencing, can produce a phylogeny so nearly correct on the basis of a single trait of any kind.

The study of the primary structure of proteins is a very powerful method indeed for ascertaining phylogenetic relationships. Unfortunately, protein sequencing is a very laborious procedure, and the number of known sequences is as yet limited. As more and more proteins are sequenced in more and more organisms our knowledge of evolutionary history will improve considerably. In particular, many moot questions in the phylogeny of insects are likely to be resolved whenever at least a few proteins become sequenced in the groups of insects involved.

D. DNA Hybridization and Sequencing

Estimates of the proportion of nucleotide differences over the whole genome can be obtained by making "hybrid" molecules between single strands of DNA from different organisms. Two different measures of DNA differentiation can be obtained by hybridization: (1) the extent of the hybridization reaction, i.e., the fraction of the DNA of two species that forms hybrid molecules, and, (2) the proportion of nucleotide pairs that are differ-

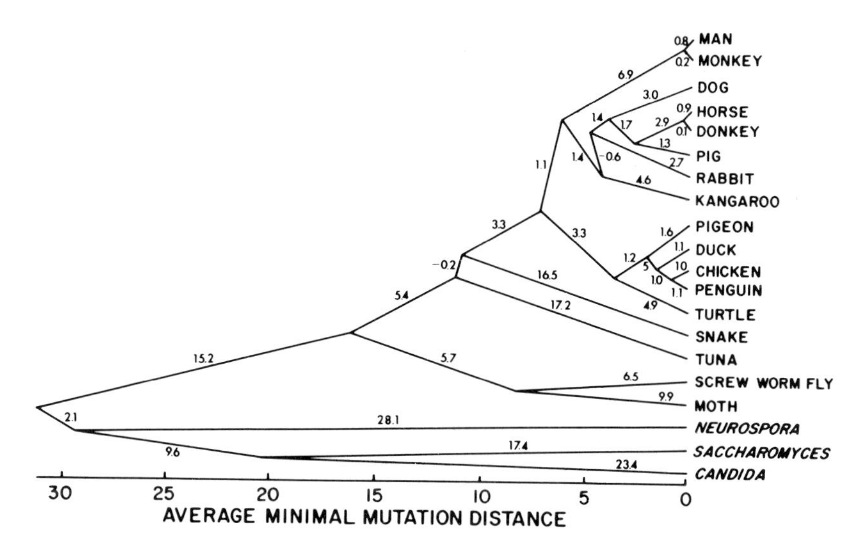

Fig. 9. Phylogeny of twenty organisms based on the amino acid sequence of their cytochromes *c*. The good agreement with fossil records and other sources is remarkable since it is based on the study of a single protein and encompasses organisms ranging from yeast, through insects, fish, reptiles, amphibians, birds, and mammals, to man. The numbers on the branches are the estimated number of nucleotide substitutions that have taken place in evolution.

ent in the hybrid molecules. A brief description follows of the methods used to make these measurements.

Double-stranded DNA can be denatured or "melted" into single strands by heating it to about 100°C. The heat breaks the hydrogen bonds between the two complementary strands. If the solution is rapidly cooled, the strands remain separated. One common technique to measure the proportion of DNA sequences that are homologous in two species starts by trapping the single-stranded polynucleotide segments of one species, A, in a homogeneous matrix such as agar or nitrocellulose membrane filters. The agar or filter is sheared into small pieces containing single-stranded DNA fragments.

The filter- or agar-bound DNA of species A is then incubated at temperatures around 60°C in a solution containing single-stranded DNA segments of the same species, A, and a second species, B. The unbound DNA of species A and B in solution has been previously denatured and sheared into small segments about 500 nucleotides in length. The free single-stranded DNA of species A is labeled with a suitable radioactive isotope, such as tritium, ^{3}H, or phosphorus, ^{32}P. The solution contains a small constant amount of the radioactive DNA of species A, but the amount of DNA of species B is varied in separate experiments. The solution is incubated for several hours to permit association between the free and the bound DNA. The remaining free DNA is then washed out. The amounts of free DNA of species A and B that have formed duplexes with the bound DNA can be determined since the DNA of species A is radioactive.

The results of a typical experiment are shown in Fig. 10. Denatured DNA from *Drosophila melanogaster* was immobilized in a nitrocellulose membrane filter, later cut into circular pieces 7 mm in diameter. The filters holding DNA were incubated in a solution containing 1 μg (1 μg = 10^{-6} grams) of ^{3}H-labeled DNA from *Drosophila melanogaster,* and variable amounts of DNA from some other species. In the control experiments, the two DNA's in solution are from *Drosophila melanogaster.* As the amount of unlabeled DNA increases, the amount of radioactive DNA that pairs with the filter-bound DNA decreases and gradually approaches zero.

When the solution contains 1 μg of labeled DNA from *Drosophila melanogaster* and variable amounts of DNA from *Drosophila simulans,* the results are different. As the amount of *Drosophila simulans* DNA increases up to about 100 μg, the amount of labeled DNA from *Drosophila melanogaster* pairing with the filter-bound DNA gradually decreases to about 20% of the total paired DNA. But the proportion of labeled *Drosophila melanogaster* pairing with the bound DNA is not reduced below 20% when the amount of *Drosophila simulans* DNA in solution is increased beyond 100 μg. This implies that about 20% of the filter-bound DNA of *Drosophila melanogaster* has no complementary sequences in *Drosophila simulans.* When the species tested is *Drosophila funebris,* the proportion of labeled *Drosophila melanogaster* DNA forming duplexes never decreases below 80%. Therefore about 80% of *Drosophila melanogaster* DNA has no homologous sequences in *Drosophila funebris* (Laird and McCarthy, 1968).

The techniques used to determine the proportion of noncomplementary

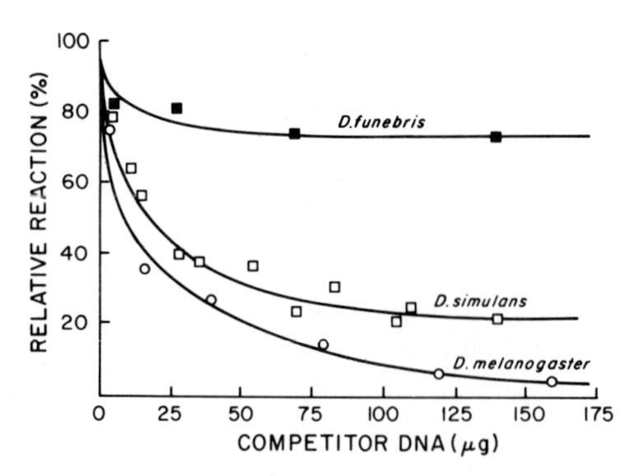

Fig. 10. Homology between the DNA sequences of *Drosophila melanogaster* and those of two other species. The experiment is carried out by hybridizing 1 μg of filter-bound DNA from *D. melanogaster* with 1 μg of radioactively labeled DNA from *D. melanogaster* and various amounts of DNA from some other species. All DNA's are first separated into single strands by heat denaturation.

nucleotides in hybrid DNA duplexes are conceptually simple. DNA duplexes are formed using bound DNA as described above, or simply placing two types of single-stranded DNA in free solution for several hours to allow hybrid double strands to form. The hybrid DNA is then heated by increasing the temperature at a rate of 1°C every few minutes. The duplex DNA gradually dissociates into single strands that are collected at regular intervals. The proportion of duplex DNA dissociated at each temperature is plotted as shown in Fig. 11. The proportion of noncomplementary nucleotides is determined by comparing the dissociation curves of hybrid DNA and of control DNA duplexes, the latter formed by reassociation of DNA strands from a single organism. The critical parameter, called thermal stability (*TS*), is the temperature at which 50% of the duplex DNA has dissociated. The differences, ΔTS, between the *TS* of hybrid DNA and of the control is known to be approximately directly proportional to the proportion of unpaired nucleotides in the hybrid DNA, so that approximately 1°C ΔTS = 1% mismatched nucleotides.

The thermal stability profiles shown in Fig. 11 were obtained using duplexes between ^{3}H-labeled DNA from *Drosophila melanogaster* and DNA of other species bound to nitrocellulose membrane filters as described above (see Fig. 10). The *TS* for homologous duplexes of *D. melanogaster* DNA is 78°C. The *TS* for the *D. melanogaster–D. simulans* hybrid DNA is 75°C,

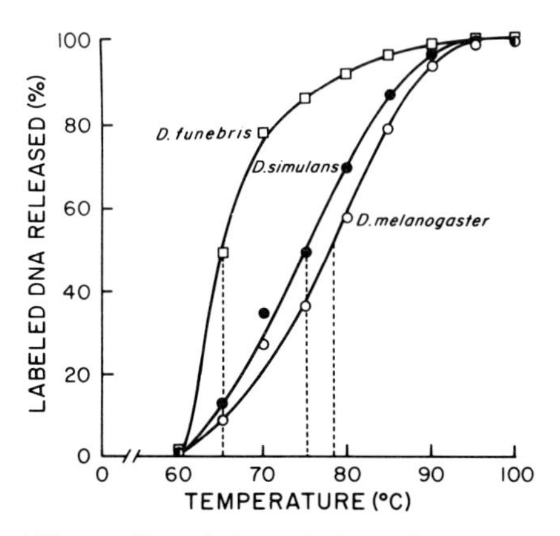

Fig. 11. Thermal stability profiles of DNA duplexes having one strand from *Drosophila melanogaster* and the other from the species indicated. The critical parameter called thermal stability (*TS*) is the temperature at which 50% of the duplex DNA has dissociated. The difference (ΔTS) between the *TS* of hybrid DNA and that of nonhybrid DNA (*D. melanogaster* with *D. melanogaster*) in degrees centigrade corresponds approximately to the percent mismatched nucleotide pairs in the hybrid DNA duplex. The *TS* for the nonhybrid duplex DNA is 78°C, for the *D. melanogaster–D. simulans* duplex DNA is 75°C and for the *D. melanogaster–D. funebris* duplex DNA is 65°C. Thus, the proportion of nucleotide pairs different from *D. melanogaster* is 3% for *D. simulans* and 13% for *D. funebris*.

therefore $\Delta TS = 3°C$. For *D. melanogaster–D. funebris* hybrid DNA, $TS = 65°C$, and $\Delta TS = 13°C$ (Laird and McCarthy, 1968). Therefore, the hybrid DNA's of *D. melanogaster* and *D. simulans* are noncomplementary in about 3% of their nucleotides, while those of *D. melanogaster* and *D. funebris* have about 13% mismatched nucleotides.

Thermal stability profiles, as well as measurements of the proportion of single-copy DNA that forms hybrid molecules, have been used to infer phylogenetic relationships among various vertebrates, particularly the primates. Although little systematic application has been made of these techniques to the study of insect phylogeny, the method is powerful and is likely to be used in the future for that purpose.

Most recently, techniques have become available to obtain directly the nucleotide sequence of specific genes. The amount of work involved in sequencing molecules of DNA (or of messenger RNA) is apparently less than the work required to obtain the amino acid sequence of proteins. As these techniques develop and become applied to the study of phylogeny, they will doubtless be another powerful biochemical tool for ascertaining evolutionary history.

E. The Molecular Clock of Evolution

A basic assumption made in order to use protein or nucleic differences in the reconstruction of phylogenetic events is that the amount of differentiation in these informational macromolecules is for any group of organisms greater the longer the time since their evolutionary divergence. If it were the case that the amount of differentiation in any one of these molecules, or a set of them, were approximately *constant* over time, then the amount of change could be used as an evolutionary clock. If the actual geological time of any of the events in a phylogenetic tree were known from some outside source (such as the paleontological record), the times of all other events could be determined by a simple proportion. That is, once it were "calibrated" by reference to one event, the molecular clock would be used to measure the time of occurrence of all other events in a phvlogeny.

Whether such a molecular clock exists is a question subject to considerable debate at present. The molecular clock is not, of course, a "metronomic clock," such as the timepieces used in ordinary life that measure exact time with theoretically perfect precision. Rather, the question is whether it is a "stochastic clock," such as radioactive decay; that is, whether the *expected* rate or probability of change is constant. Over long periods of time, a stochastic clock is quite accurate even if some variation occurs.

Early studies of protein (and DNA) evolution seemed to suggest that amino acid (or nucleotide) replacements occurred with a constant probability. More detailed and sophisticated studies have recently shown that

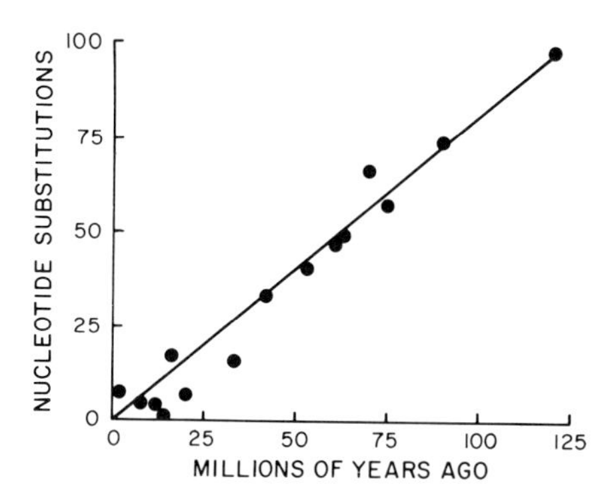

Fig. 12. Nucleotide substitutions versus paleontological time. The total nucleotide substitutions for seven proteins (cytochrome *c*, fibrinopeptides A and B, hemoglobins alpha and beta, myoglobin and, insulin C-peptide) are calculated for comparisons between pairs of species whose ancestors diverged at the time indicated in the abscissa. The solid line has been drawn from the origin to the outermost point, and corresponds to a total rate of 0.41 nucleotide substitutions per million years (or 98.2 nucleotide substitutions per 2 × 120 million years of evolution) for the genes coding for all seven proteins. The fit between the observed number of nucleotide substitutions and the expected number (as determined by the solid line) is fairly good in general.

molecular evolution is not stochastically constant. This, however, does not imply that protein (and DNA) evolution cannot be used as an evolutionary clock. Because molecular evolution is not stochastically constant, the use of a single protein or gene as an evolutionary clock is subject to considerable error. But as shown in Fig. 12 the *average* amount of molecular change over many proteins in many organisms appears to be sufficiently constant to be used as an approximate clock of evolutionary events (Fitch, 1976).

The application of biochemical studies of informational macromolecules to the study of phylogeny is a fairly young scientific field. However, there is little reason to doubt that in the future the study of proteins and nucleic acids will make increasingly significant contributions to the knowledge of past evolutionary events, by giving us both the sequence of divergence events and estimates of the time when such events took place.

GENERAL REFERENCES

Ayala, F. J., ed. (1976). "Molecular Evolution." Sinauer, Sunderland, Massachusetts. A review by thirteen experts of the contributions of biochemical studies to evolution.

Dobzhansky, T., Ayala, F. J., Stebbins, G. L., and Valentine, J. W. (1977). "Evolution." Freeman, San Francisco, California. A fairly comprehensive account of the current theory of evolution including the contributions of biochemical studies.

Lewontin, R. C. (1974). "The Genetic Basis of Evolutionary Change." Columbia Univ. Press, New York. This is an excellent review of the early attempts to measure genetic variation in natural populations, and of the accomplishments of electrophoretic studies.

REFERENCES FOR ADVANCED STUDENTS AND RESEARCH SCIENTISTS

Ayala, F. J. (1975). *Evol. Biol.* **8,** 1–78.

Ayala, F. J., Tracey, M. L., Barr, L. G., McDonald, J. F., and Pérez-Salas, S. (1974a). *Genetics* **77,** 343–384.

Ayala, F. J., Tracey, M. L., Hedgecock, D., and Richmond, R. R. (1974b). *Evolution* **28,** 576–592.

Fitch, W. M. (1976). *In* "Molecular Evolution" (F. J. Ayala, ed.), pp. 160–178. Sinauer, Sunderland, Massachusetts.

Fitch, W. M., and Margoliash, E. (1967). *Science* **155,** 279–284.

Laird, C. D., and McCarthy, B. J. (1968). *Genetics* **60,** 303–322.

Index

D

H

I

Q

R

S

T

A
B 8
C 9
D 0
E 1
F 2
G 3
H 4
I 5
J 6